Introduction to Bioengineering

Introduction to Bioengineering

Edited by

S. A. Berger
Professor of Engineering Science
University of California, Berkeley

W. Goldsmith
Professor Emeritus of Mechanical Engineering
University of California, Berkeley

E. R. Lewis
Professor of Electrical Engineering and Computer Science
University of California, Berkeley

OXFORD
UNIVERSITY PRESS

This book has been printed digitally and produced in a standard specification in order to ensure its continuing availability

OXFORD
UNIVERSITY PRESS

Great Clarendon Street, Oxford OX2 6DP

Oxford University Press is a department of the University of Oxford.
It furthers the University's objective of excellence in research, scholarship,
and education by publishing worldwide in

Oxford New York

Auckland Cape Town Dar es Salaam Hong Kong Karachi
Kuala Lumpur Madrid Melbourne Mexico City Nairobi
New Delhi Shanghai Taipei Toronto
With offices in
Argentina Austria Brazil Chile Czech Republic France Greece
Guatemala Hungary Italy Japan South Korea Poland Portugal
Singapore Switzerland Thailand Turkey Ukraine Vietnam

Oxford is a registered trade mark of Oxford University Press
in the UK and in certain other countries

Published in the United States
by Oxford University Press Inc., New York

Reprinted 2004

ISBN 0-19-856515-1

Antony Rowe Ltd., Eastbourne

PREFACE

There are many definitions of bioengineering. Other labels, such as medical engineering, biomechanics, bioelectronics, etc., each clearly refer only to part of the subject. In the use of the term bioengineering in this book we exclude genetic engineering; that is, the systematic design of phenotypes by manipulation of genotypes. Bioengineering here is taken to be the application of the concepts and methods of the physical sciences and mathematics in an engineering approach to problems in the life sciences. Most often, but not exclusively, bioengineering is concerned with human problems. Thus, while traditional engineering is the application of these sciences and mathematics to the design and analysis of inanimate, manufactured objects and structures, bioengineering may be viewed as the application of these disciplines to the study of living structures and organisms. The intent of such studies is to understand the physical processes and engineering aspects of performance under both normal and abnormal conditions, and to design, develop and use diagnostic or artificial devices meant to measure, improve, safeguard or replace life functions.

The material in this book originated in a course first introduced at the University of California at Berkeley over twenty years ago. From the beginning, it was recognized that it would be impossible to cover the entire field of bioengineering in a single course or book. Even when the text was scaled down to what was considered to be of broadest interest, it was felt that coverage of all the material was beyond the expertise of any one individual and several instructors were enlisted to do the teaching. This had the advantage of exposing the students to a wide range of bioengineering topics. However, this teaching arrangement may not be followed by other institutions. In fact, a principal reason for writing this book was to provide sufficient material to make it possible for a single instructor to teach such a course successfully.

Students in this course are expected to have had basic university-level training in physics, chemistry and mathematics. In writing this book, we had the same expectation of other users. Prior training in basic biology (including organ-systems physiology and anatomy) will enhance the value of this book to the reader. No separate chapters on anatomy and physiology are presented. Where appropriate, sections of the book include relevant physiological and anatomical background material.

The text contains contributions from instructors in mechanical, electrical, chemical, and nuclear engineering as well as from orthopedics and human biodynamics. The topics mirror the fundamental engineering science courses taught in the several engineering areas, usually at the intermediate university level, but as applied to problems in the biological world. Nevertheless, the basic principles of engineering science are presented, albeit in abbreviated form, so that the students specializing in fields not closely related to any of the subjects will be able to grasp the essence of that particular topic and be able to integrate it with one or more additional subjects necessary for the comprehension of a particular bioengineering problem. The order of presentation consists of basic engineering science subjects, followed by discussion of applications. Linear transforms, a topic applicable to a number of chapters, have been included as an appendix.

The basic mechanics of solids is presented first as the cornerstone upon whose principles many subsequent topics are built. This is followed by chapters on fluid mechanics, mass transfer, and heat transfer, which are components of mechanical and chemical engineering programs. The focus then switches, with emphasis on modeling, to systems and circuits—using theories derived largely from electrical engineering science but which apply many of the principles of mechanics. Next come two chapters on biomaterials, which lean heavily on the mechanics section. The text then covers a group of topics whose content might appear to be more directly related to the common perception of bioengineering, but which could not be reasonably presented without the underpinning of the earlier, more basic science subjects. These include human locomotion, electrophoretic separations, medical imaging, and radiation applied to diagnostics and treatment. These topics are included to give the student a flavor of some of the more important and currently very active areas of research. Brief treatments of bioinstrumentation, a broad and wide-ranging subject in its own right, are integrated in some of the chapters where appropriate.

Since the course attracts students from outside engineering, particularly from the life sciences, the chapters are presented at the simplest possible level consistent with rigor. It has been our experience that the student who has not taken the first-level engineering mechanics and electrical circuit courses usually must exert extra effort entailing outside reading; a substantial bibliography for this purpose and for providing exposure to more advanced subjects within each area is included.

For a single-semester course, there will have to be much culling of material, but we see this as an asset, giving the instructor leeway in the material he or she chooses to cover or emphasize. Those bioengineering programs that offer multiple specialized courses may find that some of the chapters are extensive enough to serve alone, or perhaps with some supplementation, as texts for those courses. That may be the case for the chapters on biomechanics, fluid mechanics, mass transfer, systems, and networks. Alternatively, the book may be used as a self-study text and reference for readers who have had formal training in the physical sciences or engineering and who wish to learn something about or review the application of these subjects to the life sciences.

Experience has shown that successful mastery of the material demands the solution of a significant number of homework problems, requiring a substantial effort on the part of both student and instructor (or assistant). This often takes the form of a one-on-one discussion or alternative presentations involving the basic approaches for this problem-solving procedure, particularly for those students that have not been previously exposed to the segment of engineering science that applies to the topic currently being covered.

Some of the chapters include sample problems to assist the student in developing the necessary skills. Instructors may wish to supplement the problems presented in the text with their own, or else choose questions adapted from the very wide literature that exists for this subject. Audiovisual aids or brief experimental demonstrations, particularly if they involve student participation, have also assisted in stimulating the interest of the audience. Each chapter also includes reference lists of books and papers relevant to the text material and for supplementary reading. These can also serve as a basis for further study in each area.

The authors would like to acknowledge and express their appreciation to Drs J. M. Kabo, Jack Winters, R. Peterson, D. Lee, Kathy Cortopassi, Steve Moore and, in particular, Kurt Eto for their invaluable contributions to this volume during the course of their participation as graduate instructors. The course has been substantially enriched by their suggestions, comments, and scrutiny of both the text and the problems.

Berkeley, California
October, 1995

S. A. B.
W. G.
E. R. L.

CONTENTS

CONTRIBUTORS

Professor Stanley A. Berger
Department of Mechanical Engineering, University of California, Berkeley, CA 94720-1740, USA.

Professor Thomas F. Budinger
Center for Functional Imaging, Lawrence Berkeley Laboratory, Berkeley, CA 94720, USA.

Dr Takeshi K. Eto
Department of Mechanical Engineering, University of California, Berkeley, CA 94720-1740, USA.

Professor Claire T. Farley
Department of Human Biodynamics, University of California, Berkeley, CA 94720-4480, USA.

Professor Werner Goldsmith
Department of Mechanical Engineering, University of California, Berkeley, CA 94720-1740, USA.

Dr Paul D. Grossman
Perkin Elmer, Applied Biosystems Division, Foster City, CA 94404, USA.

Professor Selig N. Kaplan
Department of Nuclear Engineering, University of California, Berkeley, CA 94720-1730, USA.

Professor E. L. Keller
Department of Electrical Engineering & Computer Sciences, University of California, Berkeley, CA 94720-1770, USA.

Professor Rodger Kram

Department of Human Biodynamics, University of California, Berkeley, CA 94720-4480, USA.

Professor Steven L. Lehman

Department of Human Biodynamics, University of California, Berkeley, CA 94720-4480, USA.

Professor Edwin R. Lewis

Department of Electrical Engineering & Computer Sciences, University of California, Berkeley, CA 94720-1770, USA.

Dr Howard Maccabee

Department of Radiation Oncology, University of California, San Francisco, CA 94143 USA.

Professor R. B. Martin

Orthopaedic Research Laboratories, Orthopaedic Surgery, University of California, Davis, Medical Center, 4815 Second Ave., Sacramento, CA 95817, USA.

Professor Boris Rubinsky

Department of Mechanical Engineering, University of California, Berkeley, CA 94720-1740, USA.

Professor Harry B. Skinner

Department of Orthopaedic Surgery, University of California at Irvine Medical Center, Orange, CA 92668-5382, USA.

Professor David S. Soane

Department of Chemical Engineering, University of California, Berkeley, CA 94720-1462, USA.

Professor Charles Süsskind

Department of Electrical Engineering & Computer Sciences, University of California, Berkeley, CA 94720-1770, USA.

Professor Lawrence Talbot

Department of Mechanical Engineering, University of California, Berkeley, CA 94720-1740, USA.

Professor Michael C. Williams

Department of Chemical Engineering, University of Alberta, Edmonton, AB, Canada T6G 2G6.

1

BIOMECHANICS OF SOLIDS*

Werner Goldsmith

Contents

Symbols

a, $\boldsymbol{a}$	acceleration
A	area
A'	projected area
b	damping parameter, $c/2m$
c	damping coefficient; clearance
C_{D}	drag coefficient
C_{ijkl}	elastic constants
d	distance
D	drag
e	coefficient of restitution
$\hat{e}$	unit vector
e_i	principal strain
E	energy; Young's modulus
f	friction coefficient
$\boldsymbol{f}_{ij}$	internal force
$\boldsymbol{F}$	force
g	acceleration of gravity
G	shear modulus
h	height
$\boldsymbol{H}$	angular momentum
$H\langle\ \rangle$	Heaviside step function
$\boldsymbol{i}, \boldsymbol{j}, \boldsymbol{k}$	unit vectors in directions x, y, and z
I_{ii}	moment of inertia about axis ii
I_{ij}	product of inertia about axes i and j
I'_k	area moment of inertia
J'	polar area moment of inertia
k	spring constant
K	universal constant of gravitation; bulk modulus
L	length
L^*	current length
$\boldsymbol{L}$	linear momentum
m	mass
M	magnification factor
$\boldsymbol{M}_0$	moment about 0
N	normal force
q	damped frequency
r, R	radius; radius distance

* The author is indebted to Dr Kurt Eto for his invaluable assistance in the production of the text and figures from computer diskettes.

$\mathbf{r}$	displacement		axes		μ	absolute viscosity
Re	real part	α	angular acceleration		ν	Poisson's ratio; frequency ratio
s	distance	α, β	shear angles		ξ	damping parameter; eigenvalue
s_{ij}	stress deviator	α, β, γ	direction cosines		ρ	density
S	mean stress	γ_{ij}	total shear angle		$\boldsymbol{\rho}$	displacement
t	time; thickness	δ	extension; logarithmic decrement; delta function		σ_i	principal stress
T	kinetic energy; torque				τ	stress; period
TR	transmissibility	δ_{ij}	Kronecker delta ($\delta_{ij} = 1$, $i = j$; $\delta_{ij} = 0$, $i \neq j$)		τ_{ij}	stress on plane perpendicular to i in direction j
v, $\mathbf{v}$	velocity					
V	potential energy; volume; shear force	Δ	difference		ϕ	spherical coordinate; phase angle; angle of twist
		ϵ_{ij}	strain on plane perpendicular to i in direction j			
w	load per unit length				ϕ_{ij}	relaxation function
W	weight; work	θ	angle between vectors; angular displacement		ψ_{ij}	creep function
x, y, z	rectangular coordinates; positions along coordinate axes				$\boldsymbol{\omega}$	angular velocity
		λ	Lamé constant		ω	natural frequency; angular speed
		λ_i	principal moment of inertia		$\boldsymbol{\Omega}$	angular velocity
X, Y, Z	body forces along the x, y, z axes	Λ	extension ratio		Ω	forcing frequency

Subscripts

a	axial	i	inside	R, ϕ, θ	along spherical coordinate axes
c	approach; critical	max	maximum		
d	dynamic	n	normal	rot	rotational
e	earth	o	outside	t	tangential, transverse
f	frictional; final	p	particular	tr	translational
G	mass center	r	radial; along r; restitution	x, y, z	along the x, y, z directions
h	homogeneous	r, θ, z	along cylindrical coordinate axes	0	initial

Prologue

Biomechanics describes that discipline which applies the science of mechanics to biological systems, i.e. those encompassing both the animal and the vegetable kingdom. Mechanics, in turn, is the field that describes the response of bodies—discrete or continuous, rigid or deformable, and solid, liquid, or gaseous—to the action of forces and/or couples. When attention is focused on the fluid phase, this domain is also frequently designated as physiological fluid mechanics. However, a parallel term, 'solid biomechanics', which might describe the use for biological systems of the field of 'applied mechanics' (which, by custom, is primarily concerned with solid bodies), is not usually employed. In practice, the greatest utility of biomechanics is in relation to humans, who represent the most important and interesting member of the animal kingdom (Gans 1974; Aerts 1992; Alexander 1992). Nevertheless, similar studies for other species are also highly relevant and are beginning to draw increased attention (Webb and Weihs 1983). The agricultural field, which is just now emerging as a significant component of biomechanics, presents an array of diverse and challenging questions where mechanics can play a critical role (Phipps 1983; Mohsenin 1986; Reznicek 1988; Niklas 1992).

Even when restricted to human activities, such an enormous variety of problems have come within the purview of biomechanics that it is impossible to completely catalog all subjects that comprise its sphere. However, several broad sub-specialties of the field have emerged that can be separately categorized. These include, but are not limited to: kinesiology, the science of human motion and locomotion (Hinson 1981; Hay and Reid 1988; Vaughan *et al.* 1987; Adrian and Cooper 1989; Rasch 1989; Winter 1990, 1991; LeVeau 1992); the mechanical, structural and geometric properties of the human body and its individual components (Yamada 1970; Evans 1973; Easterby *et al.* 1982; Gray 1984; Silver 1987; Hirokawa 1993); human achievement; sports mechanics (Plagenhoef 1971; Miller and Nelson 1973; Vaughan 1989; Hay 1993); exercise and leisure time activities (A. T. Johnson 1991); orthopedics (Frost 1973; Mow and Hayes 1991; Niwa *et al.* 1992), applications of which will be further discussed in subsequent chapters; clinical, diagnostic and surgical procedures, including component replacement (Ghista 1981; Dowson and Wright 1981; Marcotte 1990); and the effect of environment, such as vibrations (Lippert 1963). For example, specific achievements in the area of orthopedics have included an increased understanding of the function of bones, muscles, ligaments, and tendons (Mow and Hayes 1991; Schafer 1987); this has embraced stress growth and the interaction of stress and piezo-electric behavior in bones. These efforts have led to an improvement or even elimination of traumatic conditions such as scoliosis or bone healing. They have also resulted in the production of substantially more effective prosthetic and orthotic devices, including joint and whole hip replacement, as well as in the restoration of normal joint lubrication and other corrective processes (Journal of Bone and Joint Surgery, Inc.). A special challenge has been the development of biocompatible implantable materials and the satisfaction of the necessary interface conditions. A major triumph is represented by the success of organ replacement, where, however, much more needs to be done. Advanced surgical procedures such as arthroscopy, microsurgery, and novel methods of disk repair using suction have evolved, including nerve control replacement, that permit handicapped persons to at least partly regain normal functioning. In other instances, commands for the operation of assist devices such as wheelchairs have been communicated by breathing processes, representing a highly successful example of rehabilitation methods.

Substantial progress has been made in an understanding of traumatic conditions, their diagnosis and correction. This has included the domain of head and neck injury (Gurdjian 1970) as well as damage to other areas of the body (Pergamon Press 1969–; Aldman and Chapon 1983; Gozna and Harrington 1982; Nahum and Melvin 1993), such as the abdominal region, organs, and extremities. Studies have been conducted to attempt to delineate tolerance levels and, very importantly, to develop devices providing maximum protection for hazardous environments. Controls have been provided to ensure occupational health and safety in the workplace and elsewhere (Chaffin and Andersson 1991). However, one of the most critical areas where further study is needed is in the safe operation of vehicles, whether travelling on highways, in the sky, or on water.

The discipline of biomechanics incorporates a broad array of experimental techniques, frequently of a novel character. It also represents biological systems by mathematical models of their anatomy, physiology, and function (Morecki 1987), and examines their behavior and assesses their response to load (King and Mertz 1973; Fung 1993); this may require solution of numerical programs executed on high-speed computers. A satisfactory correlation of the experimental and theoretical results provides a certain degree of confidence in the measuring technique and the analytical representation, but is no assurance of precise predictability in view of the enormous variations in the biological properties of the human race and other species. The study of the field of biomechanics can provide a better understanding of the manner, requirements and limitations of all forms of human physical endeavors. It will also permit, in many cases, a reasonable quantitative estimate of the consequence of specified mechanical actions. In addition, its pursuit presents an enormous intellectual challenge for an expansion of our knowledge of human, animal and plant properties, functions and growth that will lead to an improved quality of life.

Biomechanics of solids rests upon an engineering science foundation encompassing well-recognized disciplines portrayed by standard courses in an engineering curriculum. A major subdivision involves mechanics of particles and rigid bodies that include courses in statics, dynamics and vibrations. When objects can no longer be idealized as being rigid—muscles being an outstanding example—the study concerns the mechanics of deformable or continuous media. Here, courses have long been developed in the strength of materials (a term that really signifies the reaction of deformable materials to loading), continuum mechanics, elasticity, viscoelasticity, and rheology. The first category is analytically represented by ordinary differential equations, while response of continuous media to loads must be described by partial differential equations. As in all disciplines, the degree of difficulty in obtaining an engineering solution in biomechanics depends upon the degree of approximation embedded in a description of the material properties, geometry, deformations, and processes incorporated in a model of the system. An excellent example of this evolutionary process of modeling may be found in an analysis of the pole vault, which is described, in successive stages of complexity, by (a) Hay (1968), (b) W. Johnson *et al.* (1975), (c) Hubbell (1980), and (d) Walker and Kirmser (1982).

The material in this section is divided into the basic science text, which presents the fundamental analytical formulations, applications involving example problems using these relations drawn from the field of biomechanics, and problems to be solved. In general, the topical treatment is not exhaustive, being limited to concepts and the simplest forms of their analytical representation; some of the material on continuum mechanics can be omitted without loss of continuity. More comprehensive or advanced descriptions of the various subjects, as well as supplementary information concerning geometric and mechanical properties of humans and animals can be found in the bibliography cited at the end of the text. The level of the subject is designed for college students in their senior year or at the graduate level, in the biological or physical sciences, in mathematics or in any engineering discipline with a solid foundation in basic physics courses and two years of college mathematics.

1.1 Statics of particles and rigid bodies

1.1.1 Introduction

The objective of this discipline is the study of the behavior of solid objects under the application of loads. The phenomena occurring in such processes are ordinarily first observed experimentally (although they may also be hypothesized) and then described by a set of principles (laws and axioms) that predict the behavior of a system that is subjected to such loads. The system is represented by a model (with different degrees of idealization) that is amenable to quantitative analysis. The mathematical tools for this process utilize vector (and tensor) operations, ordinary and partial differential equations, matrix algebra, integral representations, statistical and probabilistic concepts, operator theory, special functions, and many other well-known analytical methods. The present scope will include only a limited number of these fields, primarily covering the first two categories. A very important engineering function is the judgment as to whether the analytical predictions for a given model fit the observed behavior of its counterpart in the real world with a sufficient degree of approximation for the purposes of the user, taking full account of possible experimental errors incurred in the testing and data collection process.

1.1.2 Fundamental concepts

A rigid body is defined as a solid object all of whose infinitesimal constituent components (particles) remain at the same distance from any other specified particle regardless of what external loads may be applied. A deformable body, on the other hand, experiences changes in the distance separating any two given particles under the action of such loads. Solid objects are those whose configuration and relative particle position remain unchanged when moved to a different location or orientation in the absence of applied external loads except gravity. The response of rigid bodies to loads is described by ordinary differential equations, whereas that of deformable bodies is expressed by partial differential equations.

The events with which mechanics is concerned take place in a four-dimensional manifold involving the three-dimensional region of space and time. Dimensions in this space are measured by comparison to a standard reference length, while time, which measures a succession of occurrences, is recorded by a standard clock. Spatial locations are oriented with respect to a coordinate system; the axes of such a system are generally orthogonal. The most common is the rectangular Cartesian system xyz. Matter occupies space and reacts to the application of loads. Matter may be idealized as a particle which occupies no space but has mass; by a system of particles enveloped by a closed surface defining the volume in which the particles move; or represented by a rigid or, alternatively, by a deformable body of finite (or, in the limit, of infinite) extent. Mass is the property associated with a particle or a rigid body defining its resistance to changes in its motion when the body does not rotate. The mass of an object can be described by considering the mutual attractions between all combinations of three particles, one of which has been arbitrarily assigned a unit magnitude. The other masses are determined from Newton's law of gravitation by observing the relative accelerations of all three mass pairs. For a uniform body, mass is the product of density and volume V for both solids and fluids.

1.1.3 Scalars and vectors

A scalar is a quantity characterized solely by a magnitude, or number, as exemplified by the mechanical concepts of work or energy. A vector, represented here by a boldface symbol or else by a term with a single subscript, such as r_i, is a quantity portrayed both by a magnitude and a direction. Further, vectors must obey the parallelogram law of addition, which states that the effect of the sum of two vectors (or resultant) has a magnitude and direction given by the diagonal of the parallelogram formed by the extension of these vectors meeting at a common origin O, as shown in Fig. 1.1. An example of a vector is velocity (speed is its scalar or numerical value). Vectors are further divided into three classes: (a) a free vector, that can be drawn along any properly oriented line in space, such as the displacement of a rigid body without rotation; (b) a sliding vector, that can be located along a given line of action at any arbitrary point, such as a force acting on a body considered to be completely rigid; and (c) a bound vector, for which there exists a unique point of application along a particular line of action, as exemplified by a force acting on a deformable body. These three types of vectors are illustrated in Fig. 1.2. The negative of vector $\boldsymbol{A}$ is $-\boldsymbol{A}$, Fig. 1.3(a), directed from the same origin, but in the opposite sense as shown in the diagram. Unit vectors along Cartesian axes xyz are represented by the symbols i, j, and k, respectively (without superscript ^); for other coordinates, they will be denoted by the symbol $\hat{e}_n$ where n defines the desired direction. A vector quantity whose direction is immaterial may be represented by its scalar equivalent.

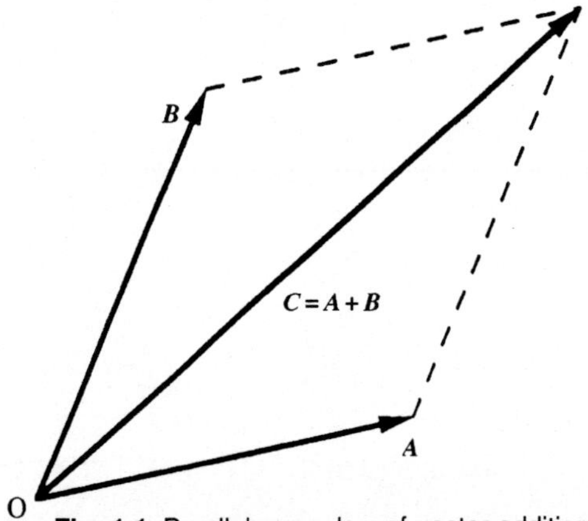

Fig. 1.1 Parallelogram law of vector addition.

1.1.4 Addition of vectors

The addition of several vectors can be performed by adding them two at a time as in Fig. 1.1, the vectors of each successive sum $\boldsymbol{R}_i$ being combined with the next vector, as shown in Fig. 1.3(b); the final addition yields the resultant $\boldsymbol{R}$. If all vectors are free, they can be added tip-to-tail as shown in Fig. 1.3(c). Vector subtraction is the addition of a negative vector, $\boldsymbol{C} = \boldsymbol{A} - \boldsymbol{B} = \boldsymbol{A} + (-\boldsymbol{B})$; this operation is shown in Fig. 1.3(d). Multiplication of a vector by a scalar n represents an extension of contraction of the original vector by the numerical value of the multiplier, depicted in Fig. 1.3(e).

Vectors may be decomposed into constituents; these need not be rectangular components (as, for example, indicated in Fig. 1.3(d): $\boldsymbol{C}$ may be resolved into components $\boldsymbol{A}$ and $-\boldsymbol{B}$). However, rectangular components along an orthogonal set of space axes xyz (a Cartesian coordinate system) are most frequently employed; in this case, a vector $\boldsymbol{A}$ is resolved into components A_x, A_y, and A_z by means of

$$\boldsymbol{A} = iA_x + jA_y + kA_z \tag{1.1}$$

as pictured in Fig. 1.4. Furthermore, the components can be expressed in terms of the magnitude of $\boldsymbol{A}$, $|\boldsymbol{A}| = [(A_x)^2 + (A_y)^2 + (A_y)^2]^{1/2}$ by

$$A_x = \alpha|\boldsymbol{A}|; \quad A_y = \beta|\boldsymbol{A}|; \quad A_z = \gamma|\boldsymbol{A}|, \tag{1.2}$$

where α, β, and γ are the direction cosines of $\boldsymbol{A}$ with respect to the Cartesian coordinate system. From the definition of $|\boldsymbol{A}|$ and eqn (1.2), the direction cosines obey the relation $\alpha^2 + \beta^2 + \gamma^2 = 1$.

Vectors may also be added by combining their components:

$$\boldsymbol{A}+\boldsymbol{B} = i(A_x + B_x) + j(A_y + B_y) + k(A_z + B_z);$$
$$|\boldsymbol{A}+\boldsymbol{B}| = [(A_x + B_x)^2 + (A_y + B_y)^2 + (A_z + B_z)^2]^{1/2}. \tag{1.3a}$$

Vector addition is commutative,

$$\boldsymbol{A}+\boldsymbol{B} = \boldsymbol{B}+\boldsymbol{A}; \tag{1.3b}$$

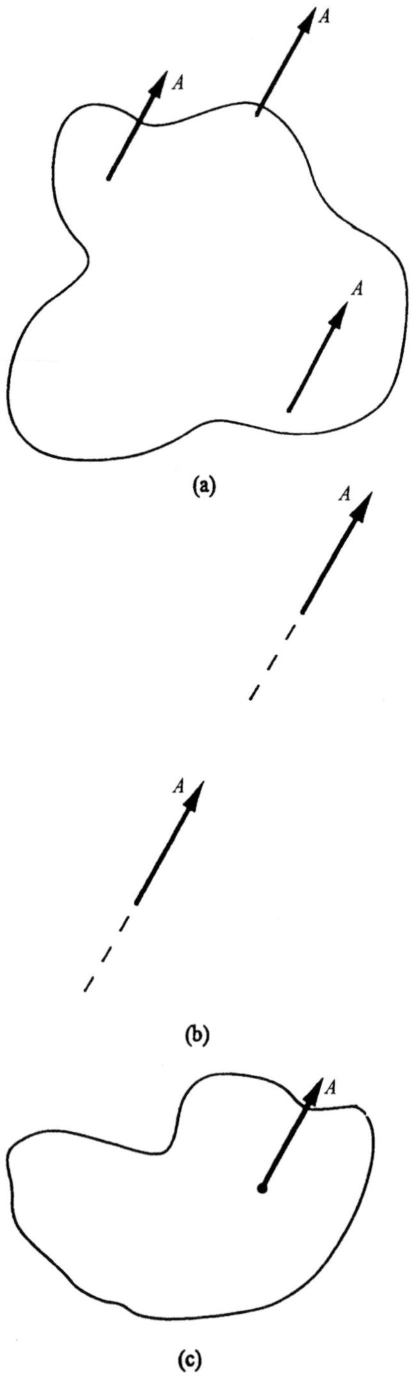

Fig. 1.2 Classification of vectors: (a) free; (b) sliding; (c) bound.

it is also associative,

$$\boldsymbol{A}+(\boldsymbol{B}+\boldsymbol{C})=(\boldsymbol{A}+\boldsymbol{B})+\boldsymbol{C}. \tag{1.3c}$$

1.1.5 Vector multiplication

Two types of vector multiplications exist: (a) The scalar or dot product $\boldsymbol{A}\cdot\boldsymbol{B}=AB\cos\theta$, where θ is the angle between the two vectors, and the result $\boldsymbol{A}\cdot\boldsymbol{B}=C$ is a scalar quantity. This may also be viewed as $\boldsymbol{A}$ multiplied by the projection of $\boldsymbol{B}$ on the line of action of $\boldsymbol{A}$, or vice versa, as indicated in Fig. 1.5(a). Furthermore, $\boldsymbol{A}\cdot\boldsymbol{B}=A_xB_x+A_yB_y+A_zB_z$ and $\boldsymbol{A}\cdot(\boldsymbol{B}+\boldsymbol{C})=\boldsymbol{A}\cdot\boldsymbol{B}+\boldsymbol{A}\cdot\boldsymbol{C}$. (b) The vector or cross product $\boldsymbol{A}\times\boldsymbol{B}=\boldsymbol{C}=\hat{e}|\boldsymbol{A}|\,|\boldsymbol{B}|\sin\theta$, where $\hat{e}$ is the unit vector (formed by using the right-hand rule) normal to the plane of $\boldsymbol{A}$ and $\boldsymbol{B}$ in the sense of a right-handed screw, and θ is the angle between vectors $\boldsymbol{A}$ and $\boldsymbol{B}$, as sketched in Fig. 1.5b. Here $\boldsymbol{A}\times\boldsymbol{B}=-\boldsymbol{B}\times\boldsymbol{A}$, so that the commutative law does not hold. However, the distributive law does hold: $\boldsymbol{A}\times(\boldsymbol{B}+\boldsymbol{C})=\boldsymbol{A}\times\boldsymbol{B}+\boldsymbol{A}\times\boldsymbol{C}$; furthermore, in terms of components A_x, A_y, A_z and B_x, B_y, B_z,

$$\boldsymbol{A}\times\boldsymbol{B}=i(A_yB_z-A_zB_y)+j(A_zB_x-A_xB_z)+k(A_xB_y-A_yB_x)=\begin{vmatrix} i & j & k \\ A_x & A_y & A_z \\ B_x & B_y & B_z \end{vmatrix}. \tag{1.4}$$

Clearly, the following relations hold among unit vectors:

$$\begin{aligned} &i\cdot i=j\cdot j=k\cdot k=1;\quad i\cdot j=i\cdot k=j\cdot k=0\\ &i\times i=j\times j=k\times k=0;\quad i\times j=-j\times i=k;\\ &j\times k=-k\times j=i;\quad k\times i=-i\times k=j. \end{aligned} \tag{1.5}$$

The following relations hold for the products of three vectors:

(a) the scalar triple product

$$(\boldsymbol{A}\times\boldsymbol{B})\cdot\boldsymbol{C}=\boldsymbol{C}\cdot(\boldsymbol{A}\times\boldsymbol{B})=-\boldsymbol{C}\cdot(\boldsymbol{B}\times\boldsymbol{A})=\begin{vmatrix} A_x & A_y & A_z \\ B_x & B_y & B_z \\ C_x & C_y & C_z \end{vmatrix}; \tag{1.6}$$

(b) the vector triple product

$$(\boldsymbol{A}\times\boldsymbol{B})\times\boldsymbol{C}=-\boldsymbol{C}\times(\boldsymbol{A}\times\boldsymbol{B})=\boldsymbol{C}\times(\boldsymbol{B}\times\boldsymbol{A})=(\boldsymbol{C}\cdot\boldsymbol{A})\boldsymbol{B}-(\boldsymbol{C}\cdot\boldsymbol{B})\boldsymbol{A}. \tag{1.7}$$

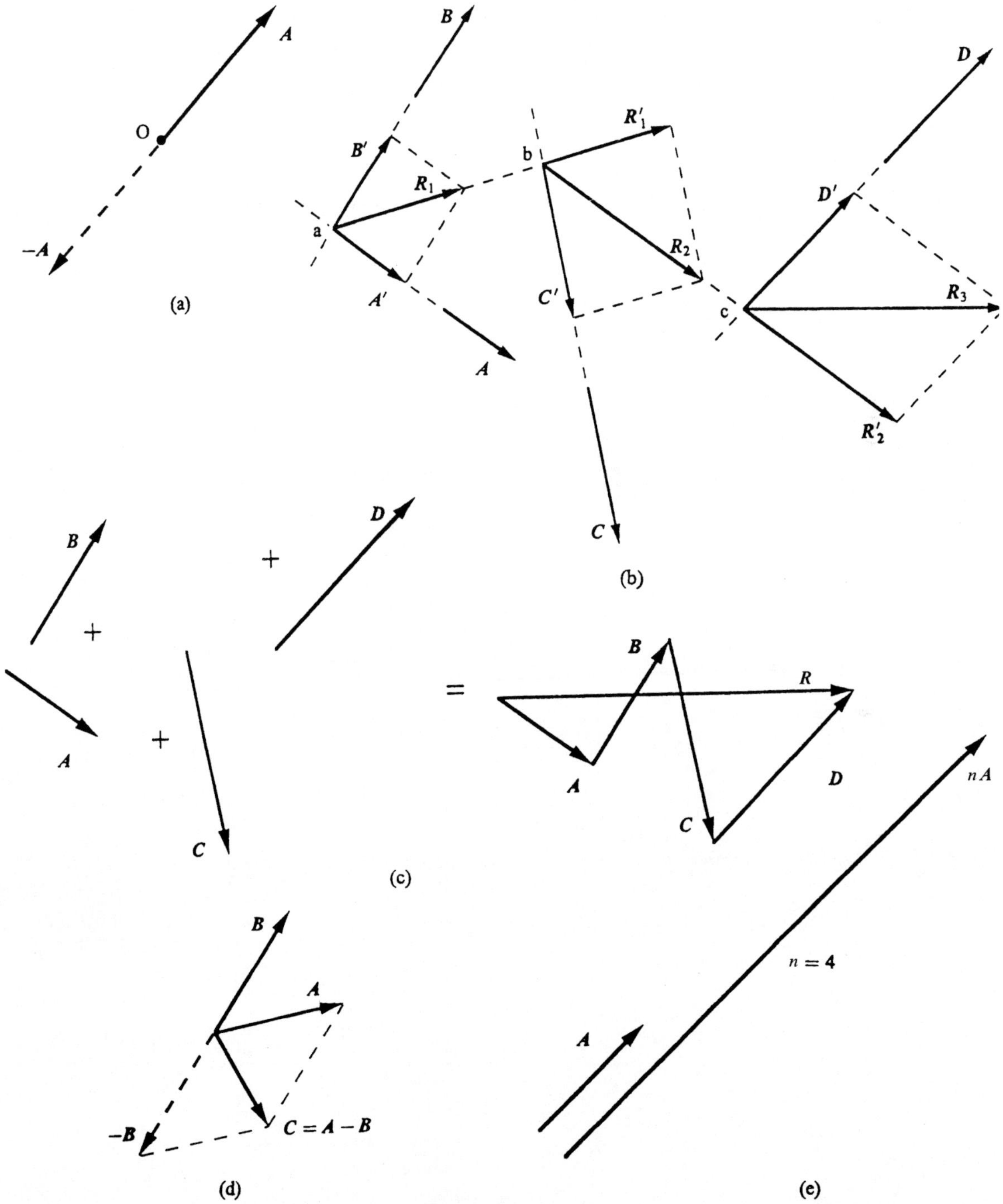

Fig. 1.3 Vector operations: (a) negative of a vector; (b) sum of sliding vectors; (c) sum of free vectors; (d) vector subtraction; (e) multiplication of a vector by a scalar *n*.

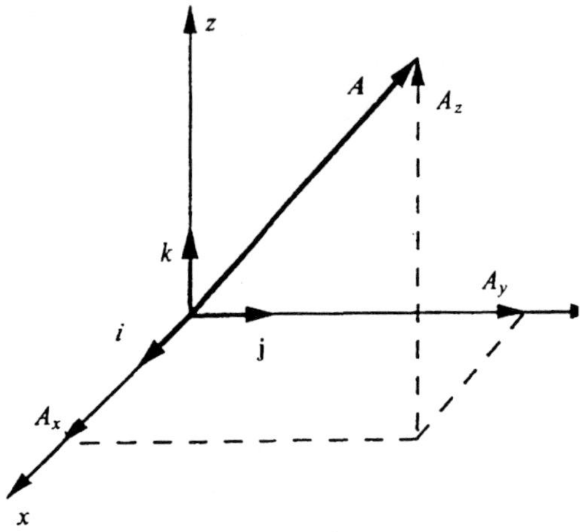

Fig. 1.4 Components of a vector.

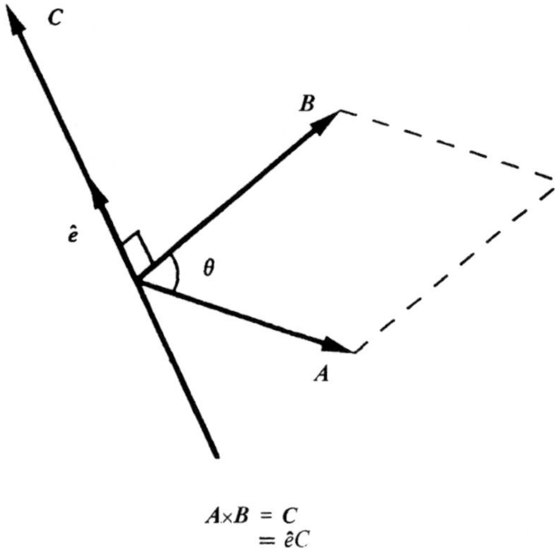

Fig. 1.5 (b) Vector multiplication: vector product.

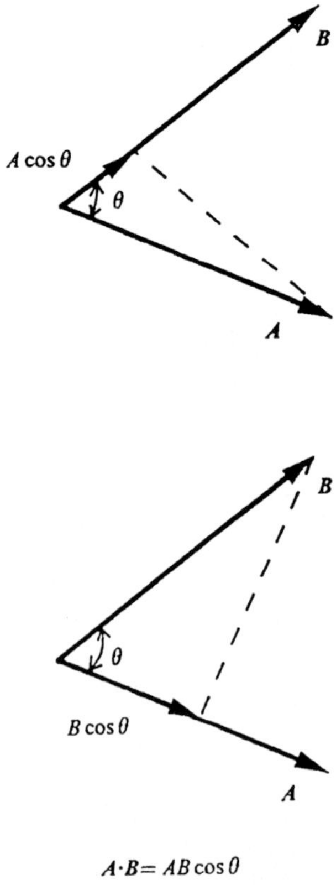

Fig. 1.5 (a) Vector multiplication: scalar products.

1.1.6 Vector differentiation and integration

The derivative of a vector A with respect to a quantity t (frequently time),

$$\frac{dA}{dt} = \lim_{\Delta t \to 0} \frac{\Delta A}{\Delta t}. \tag{1.8a}$$

Then

$$\frac{dA}{dt} = i\frac{dA_x}{dt} + j\frac{dA_y}{dt} + k\frac{dA_z}{dt} \tag{1.8b}$$

and the same holds for higher derivatives. Furthermore, with a as a scalar,

$$\frac{d(aA)}{dt} = \frac{da}{dt}A + a\frac{dA}{dt}; \quad \frac{d(A+B)}{dt} = \frac{dA}{dt} + \frac{dB}{dt};$$

$$\frac{d(A \cdot B)}{dt} = \frac{dA}{dt} \cdot B + A \cdot \frac{dB}{dt} \tag{1.9a}$$

$$\frac{d(A \times B)}{dt} = \frac{dA}{dt} \times B + A \times \frac{dB}{dt} = -B \times \frac{dA}{dt} + A \times \frac{dB}{dt}. \tag{1.9b}$$

The integration of a vector function $A\langle x, y, z\rangle$ over the volume V of a region is expressed as

$$\int_V A\langle x, y, z\rangle \, dV = i\int_V A_x \, dV + j\int_V A_y \, dV + k\int_V A_z \, dV. \tag{1.10}$$

The operation

$$i\frac{\partial}{\partial x}+j\frac{\partial}{\partial y}+k\frac{\partial}{\partial z}$$

is called the gradient ∇; it transforms a scalar quantity ϕ into a vector; thus,

$$\boldsymbol{A}=i\frac{\partial\phi}{\partial x}+j\frac{\partial\phi}{\partial y}+k\frac{\partial\phi}{\partial z}\equiv\nabla\phi.$$

The dot product of the gradient and a vector $\boldsymbol{A}$ is called the divergence of $\boldsymbol{A}$:

$$\boldsymbol{\nabla}\cdot\boldsymbol{A}\equiv\operatorname{div}\boldsymbol{A}=\frac{\partial A_x}{\partial x}+\frac{\partial A_y}{\partial y}+\frac{\partial A_z}{\partial z}. \tag{1.11}$$

The dot product of two gradients is the Laplacian

$$\boldsymbol{\nabla}\cdot\boldsymbol{\nabla}=\frac{\partial^2}{\partial x^2}+\frac{\partial^2}{\partial y^2}+\frac{\partial^2}{\partial z^2}. \tag{1.12}$$

The cross product of the gradient operator and a vector $\boldsymbol{A}$ is called the curl:

$$\begin{aligned}\boldsymbol{\nabla}\times\boldsymbol{A}\equiv\operatorname{curl}\boldsymbol{A}&=i\left(\frac{\partial A_z}{\partial y}-\frac{\partial A_y}{\partial z}\right)\\&\quad+j\left(\frac{\partial A_x}{\partial z}-\frac{\partial A_z}{\partial x}\right)+k\left(\frac{\partial A_y}{\partial x}-\frac{\partial A_x}{\partial y}\right)\\&=\begin{vmatrix} i & j & k\\ \frac{\partial}{\partial x} & \frac{\partial}{\partial y} & \frac{\partial}{\partial z}\\ A_x & A_y & A_z\end{vmatrix}.\end{aligned} \tag{1.13}$$

1.1.7 Basis of the mechanics of rigid bodies (including particles)

The subject of the mechanics of rigid bodies is divided into statics, where the bodies either do not move or move with a constant translational velocity, and dynamics, where the bodies either accelerate, rotate, or both. In dynamics, we distinguish between kinematics, which is a study only of the motion and not how this is caused, and kinetics, where both the cause and the character of the motion are examined. The standard international system of units is normally employed; the fundamental quantities of mass, length, and time are expressed in kilograms (kg), meters (m), and seconds (s).

The cornerstone of the subject is represented by Newton's three laws of motion, which were initially developed for the behavior of a single particle and then extended to a system of particles and a rigid body, and by his law of gravitation. These are:

First law of motion: A body at rest remains at rest and a body in motion remains in motion in the same straight line and with undiminished speed unless acted upon by an external force.

Second law of motion: The motion of a particle subjected to a resultant external force $\boldsymbol{F}$ is given by

$$\boldsymbol{F}=m\boldsymbol{a} \tag{1.14}$$

where m is the mass and $\boldsymbol{a}$ is the acceleration. Note that (a) this requires the force and the acceleration to be in the same direction and along the same line of action, and (b) the second law contains the first law as a special case.

Third law of motion: For every action, there is an equal and opposite reaction.

Law of gravitation: The force of attraction $\boldsymbol{F}$ between two bodies is proportional to the product of their masses and inversely proportional to the square of the distance between them. Expressed quantitatively, this is

$$\boldsymbol{F}=-Km_1m_2\hat{e}_r/r^2 \tag{1.15}$$

where K is the constant of gravitation, whose value is 6.67×10^{-11} N m^2 kg^{-2}, m_1 and m_2 are the masses of the bodies, r is their separation distance and $\hat{e}_r$ is the unit vector along r. Since the mass of the earth $m_e=5.98\times10^{24}$ kg and its radius $R_e=6.37\times10^3$ km, the Earth's gravitational constant $g=Km_e/R_e^2=9.81$ m s^{-2} (or 32.17 ft s^{-2}). Elsewhere in the universe, when the attractive mass and its radius are given by M and R, respectively, the gravitational constant

$$g'=g\left(\frac{M}{m_e}\right)\left(\frac{R_e}{R}\right)^2;$$

the ratio of moon to earth mass is 0.012 and the radius of the moon is 1740 km, so that $g_{moon}=0.16g$. The much lower effective weight of the astronauts on the moon required considerable prior training.

Mass has been quantitatively defined by using the law of gravitation and comparing the accelerations due to mutual attraction between any two of a set of three masses. Accelerations can be determined with the aid of standard measures of distance and time. Once mass is so defined, force is found from Newton's second law of motion; a force of 1 newton (N) $=1$ kg $\cdot$ 1 m s^{-2} in the mks system. The weight of an object, W, is the force produced by the local gravitational attraction on its mass: $W=mg$ on the earth.

In the field of solid mechanics, it is customary to depict the state of the system by means of a free-body diagram which isolates the system under consideration and shows all the forces acting *on* the system so isolated.

1.1.8 Force and force systems

Force has been experimentally determined to be a vector quantity and thus can be vectorially added or resolved into components. Forces are variously classified as: (a) produced by contact, or else by remote action, such as gravitation; (b) external (crossing the boundary of the system), or else internal (remaining entirely within the boundary). External forces are further classified as active or reactive. For a rigid body, the applied force is a sliding vector; for a deformable body, it is a bound vector. This leads to the principle of transmissibility for a single rigid body: an external force can be applied anywhere along its line of action and produce the same effect. This principle is neither valid for a system of joined objects nor for any deformable body.

The action of forces may be idealized as being concentrated at a single point and directed along a given line, a good approximation when the actual loading region is small compared to body dimensions. If this is not true, they are regarded as being distributed in some manner over the entire or a portion of the surface area or volume of the body. In the last case, the effect produced on the rigid body is the same as that generated by a single force that acts at an appropriate position and in the correct direction. The weight of a body, due to the gravitational attraction of the earth, is such a distributed force; its net effect is given by a single force vector acting at the center of mass (or mass center) of the object.

Application of a force acting in a given direction will also produce a tendency to rotate a rigid body (unless this line passes through the mass center). This tendency is measured by a moment

$$\boldsymbol{M} = \boldsymbol{r} \times \boldsymbol{F} \tag{1.16}$$

about an axis, where $\boldsymbol{r}$ is the vector distance from the line of action of the force to an axis of rotation, as shown in Fig. 1.6; the direction of $\boldsymbol{M}$ is that of a right-hand screw and perpendicular to the plane of $\boldsymbol{r}$ and $\boldsymbol{F}$. The magnitude of this moment is sometimes (improperly) called a torque; the vector form of $\boldsymbol{M}$ is given by

$$\boldsymbol{M} = \begin{vmatrix} i & j & k \\ x & y & z \\ F_x & F_y & F_z \end{vmatrix} \tag{1.17}$$

where a Cartesian coordinate system has been emplaced at the origin O of the intersection of the axis of rotation and the plane formed by $\boldsymbol{r} = ix + jy + kz$ and force $\boldsymbol{F}$. Similarly, the moment $\boldsymbol{M}_n$ along any axis defined by unit

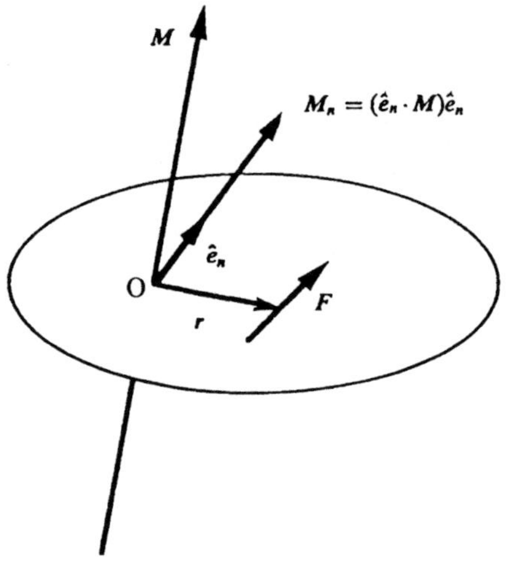

Fig. 1.6 Moment of a force: $\boldsymbol{M} = \boldsymbol{r} \times \boldsymbol{F}$.

vector $\hat{e}_n$ passing through O is given by

$$\boldsymbol{M}_n = [(\boldsymbol{r} \times \boldsymbol{F}) \cdot \hat{e}_n]\hat{e}_n = \begin{vmatrix} x & y & z \\ F_x & F_y & F_z \\ \alpha & \beta & \gamma \end{vmatrix} \hat{e}_n \tag{1.18}$$

where α, β, and γ are the direction cosines of line $\hat{e}_n$ with respect to the coordinate system employed. Moments are also vectors and may hence be added vectorially; two concurrent forces at any point produce a moment about an axis at a distance $\boldsymbol{r}$ from this point of $\boldsymbol{M} = \boldsymbol{r} \times (\boldsymbol{F}_1 + \boldsymbol{F}_2)$.

A couple $\tilde{\boldsymbol{C}}$ consists of two equal, opposite, non-colinear forces $\boldsymbol{F}$ and $-\boldsymbol{F}$ producing a moment of magnitude $|\boldsymbol{M}| = Fd$ directed anywhere normal to the plane of $\boldsymbol{F}$ and d, where d is the perpendicular distance between the forces. Couples are *free* vectors and thus can be combined vectorially, whether acting in parallel or non-parallel planes.

For a system of n concurrent forces at point A, the resultant force $\boldsymbol{R}$ and the consequent moment $\boldsymbol{M}_O$ about some other point O are given by

$$\boldsymbol{R} = \boldsymbol{F}_1 + \boldsymbol{F}_2 + \boldsymbol{F}_3 + \ldots \boldsymbol{F}_n = \sum_{i=1}^{n} \boldsymbol{F}_i, \tag{1.19a}$$

$$\boldsymbol{M}_O = \boldsymbol{M}_1 + \boldsymbol{M}_2 + \ldots \boldsymbol{M}_n = \sum_{i=1}^{n} \boldsymbol{M}_i = \sum_{i=1}^{n} r_{i_{O/A}} \times \boldsymbol{F}_i, \tag{1.19b}$$

which may be decomposed into constituents along the x, y and z axes. Any couples present may be added vectorially to the resultant moment.

For a coplanar (x–y) system, the force components along z and the moments (or couples) about the x and y axes are absent. A force resultant $\boldsymbol{R}$ is obtained by sequential sliding and vector addition of all forces, as shown in Fig. 1.3(b). For a parallel force system due only to forces acting in, say, the y-direction, $\boldsymbol{R} \equiv R_y = \sum_{i=1}^{n} F_{y_i}$ and its location x^* is obtained from the moment balance: $x^* = \left(\sum_{i=1}^{n} F_{y_i} x_i\right)/\boldsymbol{R}$. For a system consisting only of concurrent forces, the moment about the point of concurrency vanishes.

1.1.9 Equilibrium

The conditions of equilibrium for a body are:

$$\begin{aligned} &\sum_{i=1}^{n} \boldsymbol{F}_i = 0 \quad \text{and} \quad \sum_{j=1}^{m} \boldsymbol{M}_j = 0, \\ \text{or} \quad &\sum_{i=1}^{n} F_{i_x} = \sum_{i=1}^{n} F_{i_y} = \sum_{i=1}^{n} F_{i_z} = 0; \qquad (1.20) \\ &\sum_{j=1}^{m} M_{j_x} = \sum_{j=1}^{m} M_{j_y} = \sum_{j=1}^{m} M_{j_z} = 0. \end{aligned}$$

For the proper application of these equations, it is necessary to provide appropriate idealizations for system components; for example, cables or muscles can not carry compressive loads, smooth contact between bodies produces no forces tangent to their common surface, rollers or rockers produce only normal contact forces, etc. A system may either be inadequately constrained, in which case it may assume an infinity of equilibrium configurations, or it may be overconstrained, in which case removal of one or more constituent members will not produce collapse of the system. A properly constrained system has a single equilibrium configuration and will exhibit as many unknown reactive forces as equations of equilibrium. Three forces in equilibrium must be concurrent.

1.1.10 Structures

Many engineering and biomechanical systems can be represented by structures which consist of a series of rigid and deformable bodies; the latter are here modeled by springs. The objective of a structural analysis is the evaluation of support safety and hence the determination of the load carried internally by the system components or transmitted at connections. For simple structures, by definition, the rigid elements are connected by pins that can transmit only forces, but not moments; further, the components consist of rodlike members that can carry only axial forces N, but not transverse or shear forces V or bending moments M. The loads borne here are either tensile or compressive, T or C, that act along the axis and, in reality, produce either stretching or shortening of the member, respectively, as shown in Fig. 1.7(a). However, when the bodies are considered to be rigid, such deformations do not exist, by virtue of the required infinite stiffness of the member. From the equations of equilibrium, eqns (1.20), these forces produce an internal tensile or compressive stress, $\tau_{xx} = N/A$, acting on any cross-section of area A perpendicular to the axis of this two-force member that resists the external load. The first subscript in the stress term defines the axis perpendicular to the section on which the stress acts, while the second indicates the direction of loading; identical subscripts indicate a normal stress.

In a simple structure composed of two-force members, the forces in the system are solved either by isolating each pin joint or else by cutting through the members of the structures exposing internal cross sections in such fashion that the number of unknown internal forces does not exceed the number of equilibrium equations available. If this is not possible, the structure is statically indeterminate, and additional conditions beyond the realm of statics must be invoked to solve the loading pattern within the structure. However, compound structures are not limited in their composition to two-force members. They may be isolated as individual multi-force elements that can sustain bending moments M and a shear force V which produces a shear stress $\tau_{xy} = V/A$ on an internal section as well as normal force N, corresponding to stress τ_{xx}, as shown in Fig. 1.7(b). Each of these multi-force members can be analyzed by the equations of equilibrium. Again, for a determinate system, there should be an equal number of unknowns and available equations. An example of a set of pinned two-dimensional multi-force members is shown in Fig. 1.8, modeling an arm action. The equations of static equilibrium are very useful in determining the forces and moments on load-bearing members of humans and animals, and if the areas are known, also the corresponding stresses. This applies to a host of activities or tasks where dynamic effects can be ignored; here, either

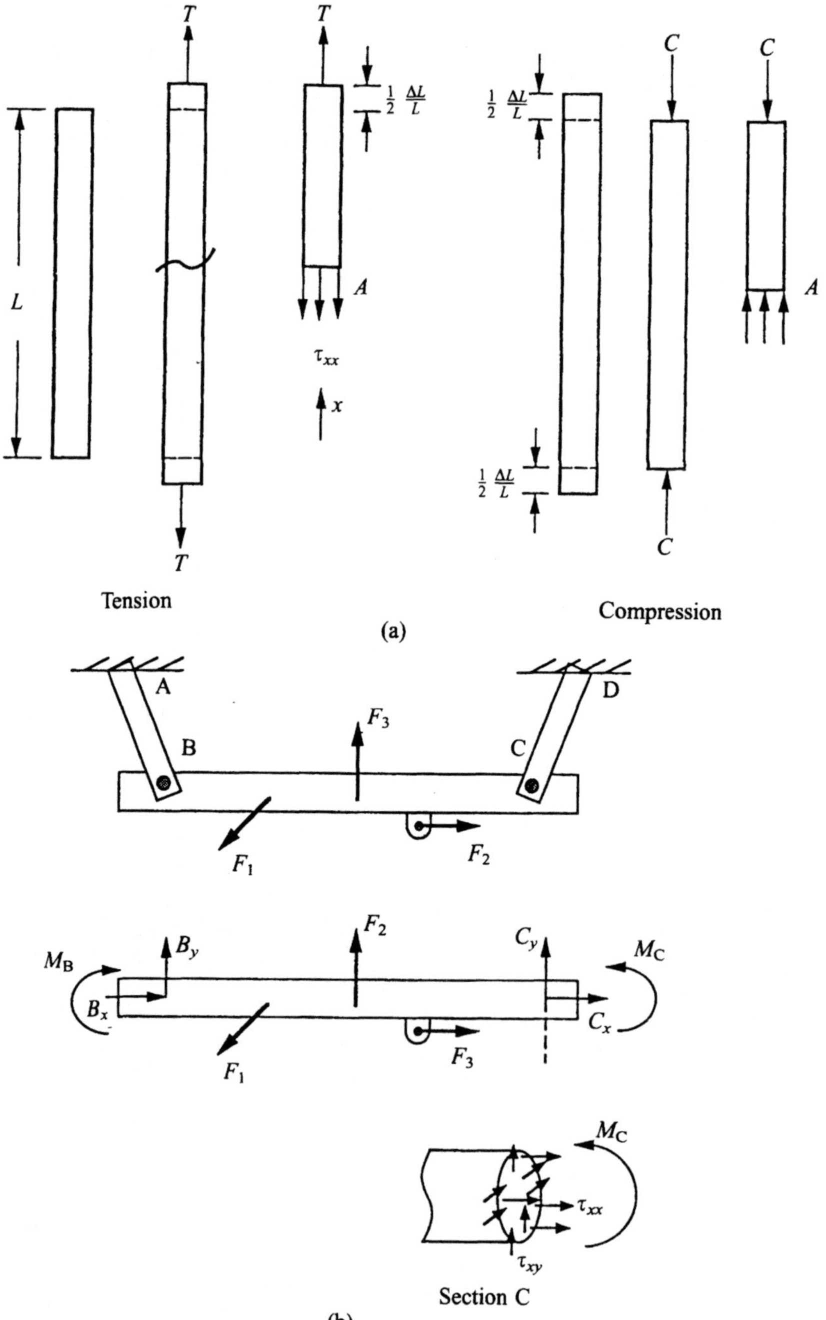

Fig. 1.7 Structural members: (a) two-force members sustaining only tension (*T*) or compression (*C*) with corresponding stresses and changes in length; (b) structural member capable of resisting axial loads (*T* or *C*), shear (*V*), and bending moments (*M*)

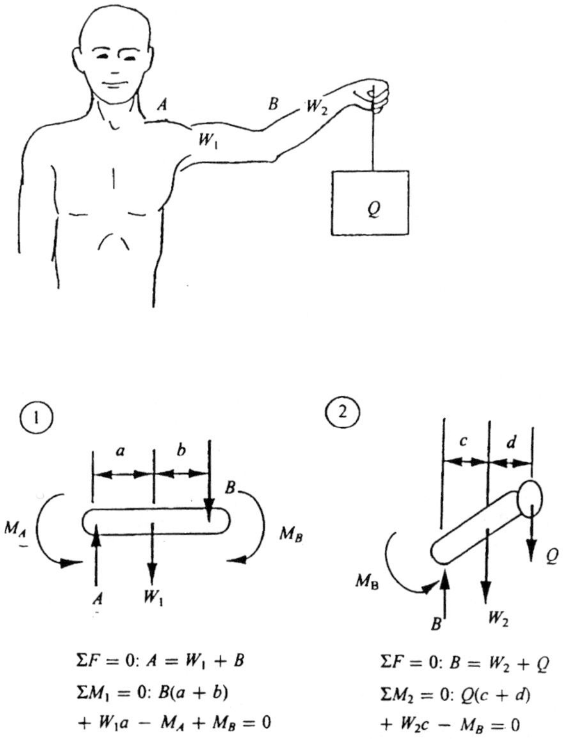

Fig. 1.8 Human arm modeled as a compound structural member with forces and moments transmitted at the joints.

no motion exists or the inertial terms are very small compared to other forces or moments present.

1.1.11 Distributed forces

Distributed forces act over a region or the entire surface or volume of an object and are measured by local intensity in terms of stress or pressure. The most common distributed force is the weight of a body which results from the gravitational attraction by the earth (or by a planet, sun, or star, if measured there) of all of its constituent particles; it is assumed that this attraction constitutes a parallel force system. The location of the action of the equivalent force representing the total weight of the object, W, is called the center of mass (or gravity) and is given by the set of relations:

$$x_G = \int_V (x\,dW)/W; \quad y_G = \int_V (y\,dW)/W;$$
$$z_G = \int_V (z\,dW)/W \tag{1.21}$$

or $$\mathbf{r}_G = \int_V (\mathbf{r}\,dW)/W = \int_V (\mathbf{r}\,dm)/m$$

Here, $W = mg = \sum dW$ is the total weight (m is mass and g is the local acceleration of gravity, approximately 9.81 m s^{-2} on the earth). This can be simplified to yield the location of the centroid of objects of very small, uniform thickness, t, by replacing dW by dA and W by A, the area of the object, or, in the case of lineal shapes, by replacing dW and W by dL and L, the line length, respectively. In the case of composite bodies, the location of the overall center of gravity is given by the coordinates

$$X_G = \sum_{i=1}^{n} (W_i x_{i_G})/W; \quad Y_G = \sum_{i=1}^{n} (W_i y_{i_G})/W;$$
$$Z_G = \sum_{i=1}^{n} (W_i z_{i_G})/W; \quad W = \sum_{i=1}^{n} W_i \tag{1.22}$$

with W_i, x_{i_G}, y_{i_G}, and z_{i_G} as the weights and centroidal positions relative to the same fixed set of Cartesian axes of the constituents of the composite body.

1.1.12 Friction

A variety of expressions have been developed to account for resistance to motion due to frictional effects. When two solids slide on each other in the absence of significant lubrication, a dry friction (or Coulomb friction) law is considered to apply where the friction force assumes the value required for static equilibrium, provided that this force, F_f, is no larger than the maximum which the materials can develop. This maximum value is $f_s N$, where f_s is the static coefficient of friction and N the normal reaction between the two surfaces. The situation is illustrated in Fig. 1.9, which shows sliding of a flat object on a horizontal surface, where the normal force $N = W$, the weight of the object. The body remains stationary until the applied force F attains the value $f_s N$, when the object begins to slip and the coefficient of friction acquires its dynamic value f_d. This coefficient is generally smaller than f_s and, further, depends on the instantaneous relative speed between the two contact surfaces. The friction phenomenon is a

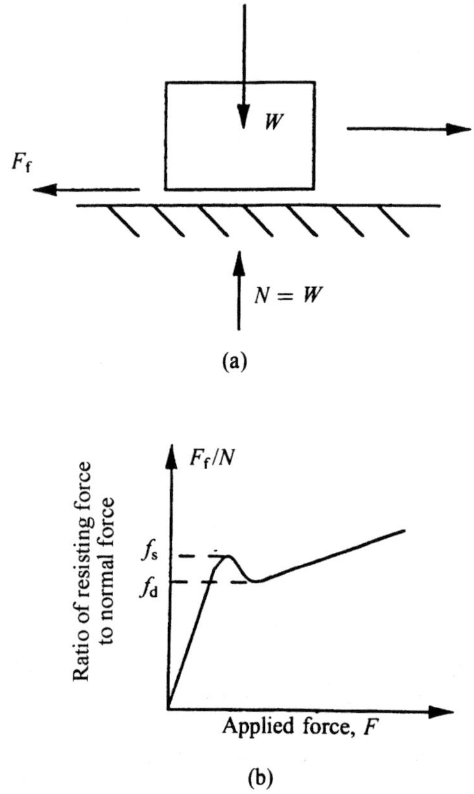

Fig. 1.9 Solid (dry) friction: (a) equilibrium of applied and friction forces; (b) variation of friction coefficient.

critical aspect of all human and animal locomotion. No comprehensive theory for this effect has thus far been developed, but for most engineering purposes, this relatively qualitative and empirical representation yields adequate results. In many instances, it is assumed with good approximation that $f_s = f_d$ and that f_d is independent of the relative speed or acceleration of the surfaces.

Another type of friction law governs when a solid moves within a fluid, as occurs in most primate joints. The frictional resistance or drag force, which opposes motion, is expressed in terms of the relative velocity v by the equation.

$$F_d = cv^n \tag{1.23}$$

where c is a constant. Exponent n is a parameter which depends on the properties of the solid and fluid and may change during a given event, such as during free fall or a parachute jump in the atmosphere, where n varies continuously with height above sea level. One of the most common hypotheses is the choice of $n = 1$ for the motion of a solid in a very viscous fluid. In a fully lubricated bearing of radius r, a shaft rotates under the actions of a load Q and a bearing reaction R, while subjected to a moment M, Fig. 1.10. The resistance to motion under non-turbulent conditions is governed by a sliding or shear stress, τ_{xy}, which, for an ideal or Newtonian fluid, is linearly related to the velocity gradient dv/dy through a constant of proportionality μ, the absolute viscosity (expressed, for example, in poises); hence $\tau_{xy} = \mu \, dv/dy$, as discussed in Chapter 2. If τ_{xy} varies with local area, the force is given by $F = \int_A \tau \, dA$. In the present case, $\frac{dv}{dr} = \frac{r\omega}{c}$, where c is the clearance of the bearing and ω is the angular velocity of the shaft. The frictional moment for a bearing of length L and surface area $A = 2\pi rL$ is

$$M_f = \tau Ar = 2\pi\mu r^3 L\omega/c. \tag{1.24}$$

A variation of this concept is represented by the fluid drag

$$D = \tfrac{1}{2} C_D A' \rho v^2 \tag{1.25}$$

where C_D is the drag coefficient, A' the presented area and ρ the mass density; this expression is also often used for the motion of a human in a fluid medium.

1.1.13 Moments of inertia

A crucial property of rigid bodies representing their resistance to change in rotational motion is their mass moment of inertia about a specified axis LM, as indicated in Fig. 1.11(a); this is defined as $I_{LM} = \int_V r^2 \, dm$, where r is the distance of mass element dm from this axis. Alternatively, if the continuous mass element dm is replaced by a finite set of n mass units m_i, this definition becomes $I_{LM} = \sum_{i=1}^{n} r_i^2 m_i$, where the summation has been

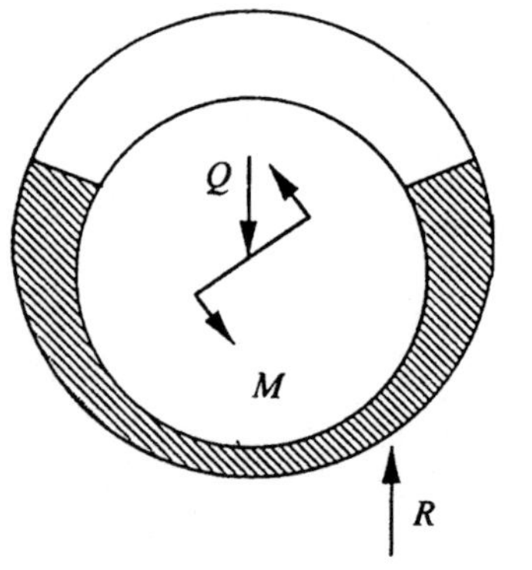

Fig. 1.10 Fluid friction: shaft rotating in a bearing.

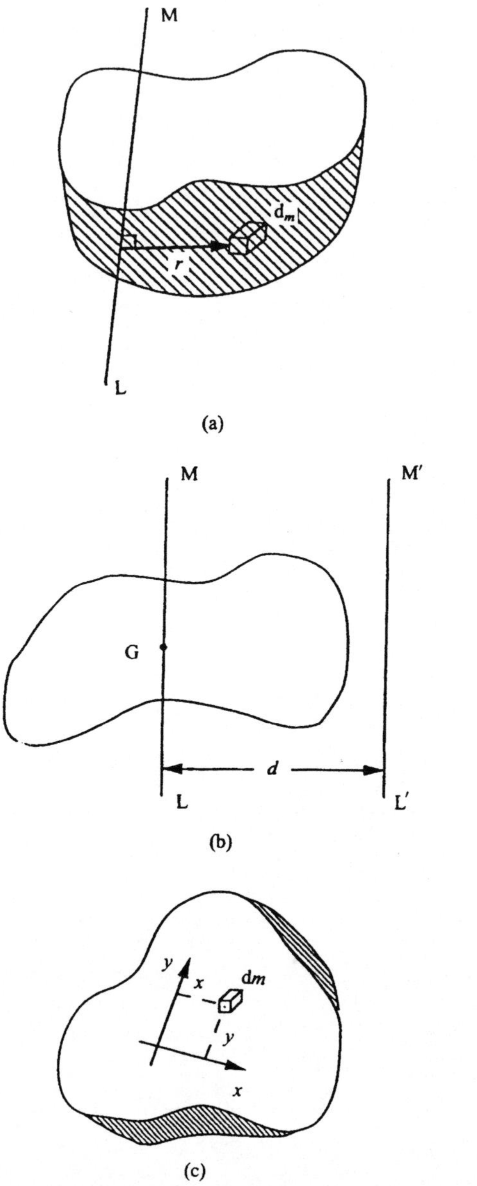

Fig. 1.11 Inertial properties: (a) definition of moment of inertia; (b) parallel axis theorem; (c) definition of product of inertia.

carried out over the entire region of the body or system of masses. If the mass density of the object, ρ, is constant over its volume V, then $I_{\mathrm{LM}} = \rho \int_V r^2 \, \mathrm{d}V$. The radius of gyration about this axis, k, is defined by $k \equiv (I_{\mathrm{LM}}/m)^{1/2}$.

It is possible to transfer the initially calculated moment of inertia to an axis parallel to the original one by using the simple relation $I_{\mathrm{L'M'}} = \mathrm{I}_{\mathrm{LM}} + md^2$, where d is the perpendicular distance between the axes, *but only if LM passes through the mass center G*, Fig. 1.11(b), or, alternatively, by use of $I_{\mathrm{LM}} = I_{\mathrm{L'M'}} - md^2$, if LM passes through G. If neither line passes through the mass center, the transfer must be performed in two steps using an auxiliary parallel line passing through the mass center. If the object is thin and of constant thickness t, then a moment about an arbitrary x-axis can be expressed by

$$I_{xx} = \rho t \int_A r^2 \, \mathrm{d}A = \rho t I'_x \qquad (1.26)$$

where I'_x is the areal moment of inertia. For composite bodies, the moment of inertia is calculated by first computing the moment of inertia of each constituent about an axis parallel to that about which the composite moment is desired, passing through the center of mass of each element. The results are then calculated for the common axis using the transfer formula described. For a void or hole inside a larger body, the mass of that component is taken as negative. An example of such a calculation for the areal moment I'_x of a homogeneous plate of constant thickness t is shown in Fig. 1.12. The mass moment is obtained by multiplying I'_x by density ρ and thickness t.

Another resistive property of rigid bodies comparable to moments of inertia, or

$$I_{ii} = \int_V x_i^2 \, \mathrm{d}m,$$

with $i = 1, 2, 3$, is the product of inertia, i.e. the integral over the body volume of the product of the elemental mass and the perpendicular distances to a pair of orthogonal axes, as shown in Fig. 1.11(c),

$$I_{ij} = \int_V x_i x_j \, \mathrm{d}m, \quad \text{with } i \neq j. \qquad (1.27)$$

In a Cartesian system, the moments and products of inertia are given by

$$I_{xx} = \int_V (y^2 + z^2) \, \mathrm{d}m; \quad I_{yy} = \int_V (z^2 + x^2) \, \mathrm{d}m;$$

$$I_{zz} = \int_V (x^2 + y^2) \, \mathrm{d}m; \quad I_{xy} = I_{yx} = \int_V xy \, \mathrm{d}m; \qquad (1.28)$$

$$I_{xz} = I_{zx} = \int_V xz \, \mathrm{d}m; \quad I_{yz} = I_{zy} = \int_V yz \, \mathrm{d}m.$$

Products of inertia can also be transferred using the parallel axis theorem in a manner analogous to that indicated for moments (that is, $I_{x'y'} = I_{xy} + md_1d_2$);

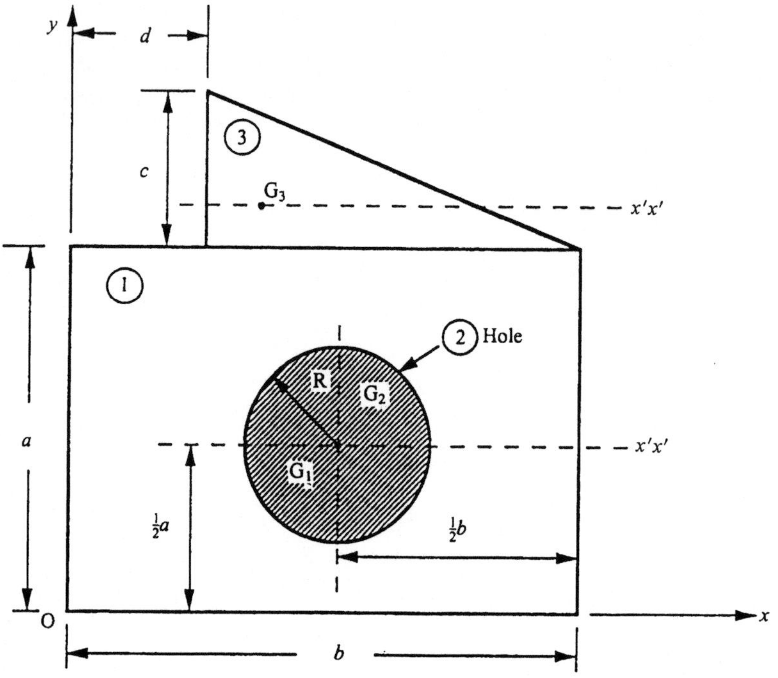

(a)

Part i	x_{G_i}	y_{G_i}	A_i	$y_{G_i}^2 A_i$	$I_{x'x'_{G_i}}$	$I_{xx} = I_{x'x'_{G_i}} + y_{G_i}^2 A_i$
1	$\frac{1}{2}b$	$\frac{1}{2}a$	ab	$\frac{a^2 b}{4}$	$\frac{1}{12}ba^3$	$\frac{1}{3}a^3 b$
2	$\frac{1}{2}b$	$\frac{1}{2}a$	$-\pi R^2$	$-\frac{\pi a^2 R^2}{4}$	$-\frac{1}{4}\pi R^4$	$-\frac{\pi}{4}R^2(a^2+R^2)$
3	$\frac{b+2d}{3}$	$a+\frac{c}{3}$	$\frac{c}{2}(b-d)$	$\left(a+\frac{c}{3}\right)^2\frac{c}{2}(b-d)$	$\frac{h^3}{24}(b-d)$	$\frac{h^3}{24}(b-d)+\left(a+\frac{c}{2}\right)\left[\frac{c}{2}(b-a)^2\right]$
Σ	—	—	—	—	—	$\frac{1}{3}a^3 b-\frac{\pi R^2}{4}(a^2+R^2)+\frac{h^3}{24}(b-d)+\left(a+\frac{c}{2}\right)\left[\frac{c}{2}(b-a)^2\right]$

(b)

Fig. 1.12 Calculation of the moment of inertia for a composite body about axis Ox (a) Geometry (b) Calculation procedure.

however, here, the reference value must also pass through the mass center and the algebraic signs of d_1 and d_2 must be considered.

Moments of inertia enter the analysis of two-dimensional rotational motion of rigid bodies, and products of inertia play a further role in three-dimensional analysis of this motion, as occurs in diving or jumping. Moments of inertia for many uniform bodies are found in many handbooks. For example, the transverse moment of inertia of a thin, uniform rod of mass m and length L through the mass center and through an end are given by $\frac{1}{12}mL^2$ and $\frac{1}{3}mL^2$, respectively.

Elimination of the products of inertia from the equations of motion greatly simplifies the analytical procedure. To do this, it is necessary to determine at the chosen coordinate origin the direction of a set of axes, known as principal axes, along which the products of inertia vanish and along which the moments of inertia are known as principal moments. Such a set of axes exists for every point of the three-dimensional space with respect to any body contained within that space. In most cases, such a set of axes at a given point is unique. However, if the body exhibits an axis of symmetry, this axis constitutes one of the principal axes for every origin located along that axis; the other two principal moments are of equal magnitude, and their directions are arbitrary in the plane perpendicular to the symmetry axis. If a body exhibits a point of symmetry (such as the center of a uniform sphere or spherical shell), then the principal moments are all equal and their directions through this point are arbitrary, as long as they are all orthogonal. Determination of principal axes is most easily achieved by considerations of symmetry, when present; this is often the case.

When symmetry considerations are inapplicable, the values of the principal moments λ_i are obtained from the solutions of the determinantal equation

$$\begin{vmatrix} I_{xx}-\lambda_i & -I_{xy} & -I_{xz} \\ -I_{xy} & I_{yy}-\lambda_i & -I_{yz} \\ -I_{xz} & -I_{yz} & I_{zz}-\lambda_i \end{vmatrix} = 0. \qquad (1.29)$$

The moments and products of inertia I_{ij} with respect to any orthogonal set of axes x, y, and z, i.e. the inertia tensor must have been computed initially at the point of interest. Equation (1.29) is a cubic, and its three roots λ_1, λ_2, and λ_3, all real and positive, represent the magnitudes of the principal moments at this position. The directions of the corresponding principal axes are obtained by substitution of each of the principal moments *separately* into the relations.

$$\begin{aligned} (I_{xx}-\lambda_i)x_1 - I_{xy}x_2 - I_{xz}x_3 &= 0 \\ -I_{xy}x_1 + (I_{yy}-\lambda_i)x_2 - I_{yz}x_3 &= 0 \\ -I_{xz}x_1 - I_{yz}x_2 + (I_{zz}-\lambda_i)x_3 &= 0 \end{aligned} \qquad (1.30)$$

and finding the ratios x_2/x_1 and x_3/x_1 for the directions corresponding to *each* of the principal moments. This scheme is straightforward, albeit tedious.

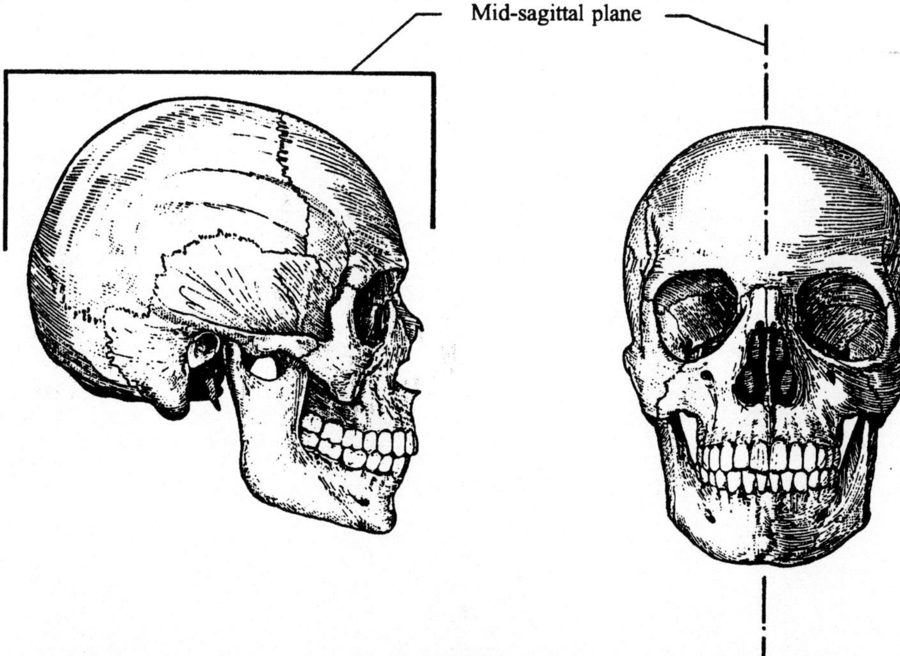

Fig. 1.12 Calculation of the moment of inertia for a composite body about axis Ox (a) Geometry (b) Calculation procedure.

The mid-sagittal plane of the human head (and, approximately, that of the body) is a plane of symmetry, Fig. 1.13. Thus one principal axis is perpendicular to that plane; the other two must be determined by the mass distribution. At the mass center, one is nearly vertical, but inclined slightly forward.

1.2 Dynamics of particles and rigid bodies

1.2.1 Kinematics of particles

The motion of a particle can be described either with respect to a fixed frame of reference (generally, an inertial frame which is defined as a coordinate system in which Newton's laws hold), where the absolute motion is analyzed, or in terms of a moving coordinate frame where relative motion is examined. A particle can move either along a rectilinear or along a curvilinear path, but, by virtue of its point mass property, is incapable of rotation. As shown in Fig. 1.14(a), the displacement, $\boldsymbol{r}$, of a particle is the vector connecting its position at two instances of time; the distance travelled, s, is a scalar measuring the total length traversed during this time. The instantaneous velocity of the particle, $\boldsymbol{v}$, is the derivative of the displacement with respect to time; its magnitude

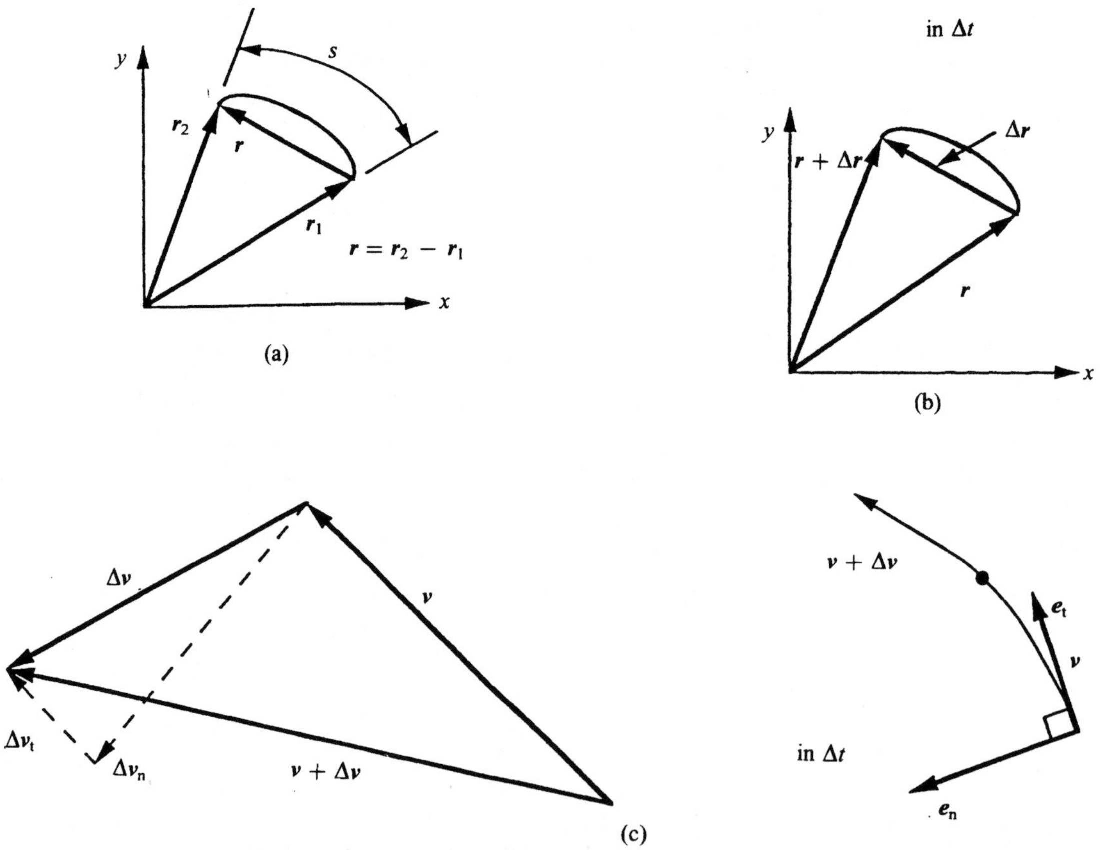

Fig. 1.14 Curvilinear motion: (a) displacement; (b) displacement change; (c) velocity change.

(or scalar value) is called the speed. The average velocity v_{avg} is the ratio of the increment of displacement and the corresponding time, as indicated in Fig. 1.14(b); hence

$$\boldsymbol{v} = \frac{d\boldsymbol{r}}{dt} = \dot{\boldsymbol{r}} \quad \text{and} \quad v_{avg} = \frac{\Delta \boldsymbol{r}}{\Delta t}. \tag{1.31}$$

Since, in the limit as $\Delta t \to 0$, $\Delta s \to \Delta \boldsymbol{r}$, $\boldsymbol{v}$ will always be tangent to the curve. The acceleration of the particle (for whose scalar equivalent there is no common name) is defined as the time rate of change of velocity, Fig. 1.14(c), so that

$$\boldsymbol{a} = \lim_{\Delta t \to 0} \frac{\Delta \boldsymbol{v}}{\Delta t} = \frac{d\boldsymbol{v}}{dt} = \dot{\boldsymbol{v}} = \ddot{\boldsymbol{r}} \tag{1.32}$$

In rectilinear motion, Fig. 1.15, the particle path is a straight line, chosen here arbitrarily as the x-direction. In this case, the definitions of the velocity and acceleration, which are now treated as scalars since their directions are assigned, eqns (1.31) and (1.32), lead to the kinematical relations

$$\dot{x} = v = \frac{dx}{dt}; \quad a = \ddot{x} = \dot{v} = \frac{d^2x}{dt^2}; \quad \ddot{x}\, dx = v\, dv. \tag{1.33}$$

Equations (1.33) must be used when the acceleration a is variable; if it is constant, $a = C$, then

$$\begin{aligned} v &= v_0 + Ct; \\ s &= s_0 + v_0 t + \tfrac{1}{2}Ct^2; \\ v^2 - v_0^2 &= 2C(s - s_0), \end{aligned} \tag{1.34}$$

where s_0 and v_0 are the initial position and velocity at time $t = 0$. If the dependence of the acceleration on time (or, alternatively, on position or velocity) is specified, the velocity and displacement can be determined as functions of time by integration, provided the initial conditions s_0 and v_0 are known.

In general two-dimensional (plane) motion, the position of a line may be described by the angle formed with respect to a fixed reference; the angular motion is described by the displacement $\theta = \theta_2 - \theta_1$, as shown in Fig. 1.16. Equations like the kinematical relations for a

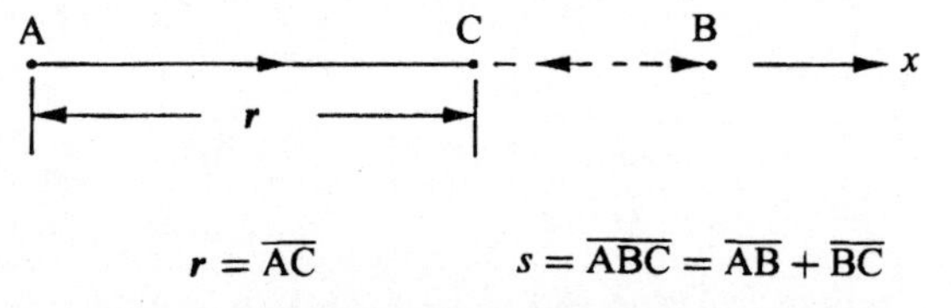

Fig. 1.15 Rectilinear motion of a particle.

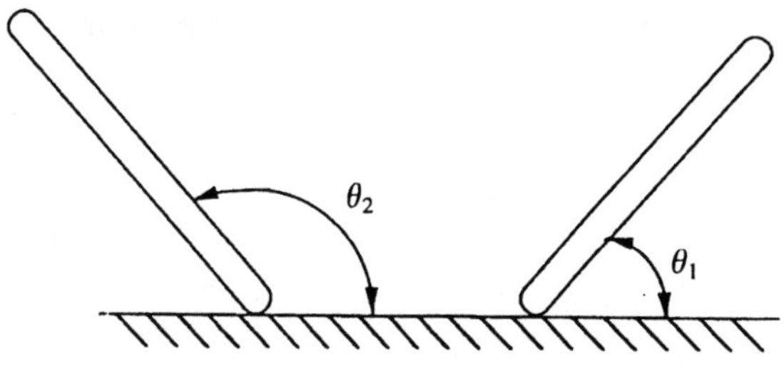

Fig. 1.16 Angular motion in a plane.

particle, eqns (1.31) and (1.32), hold for the angular speed ω and the angular acceleration, i.e. $\omega = d\theta/dt$ and $\alpha = d\omega/dt = d^2\theta/dt^2$; from these, one obtains the third relation $\alpha\, d\theta = \omega\, d\omega$. If α = constant = D, then, in analogy to eqns (1.34).

$$\begin{aligned} &\omega = \omega_0 + Dt; \quad \theta = \theta_0 + \omega_0 t + \tfrac{1}{2}Dt^2; \\ &\omega^2 - \omega_0^2 = 2D(\theta - \theta_0). \end{aligned} \tag{1.35}$$

Note that this type of two-dimensional motion occurs only when the axis of rotation remains either fixed or parallel to itself (perpendicular to the plane of movement, defined here by an x–y coordinate system); hence, the angular displacement, velocity, and acceleration directions will occur in the direction of the unit vector k perpendicular to x and y. However, these concepts do *not*

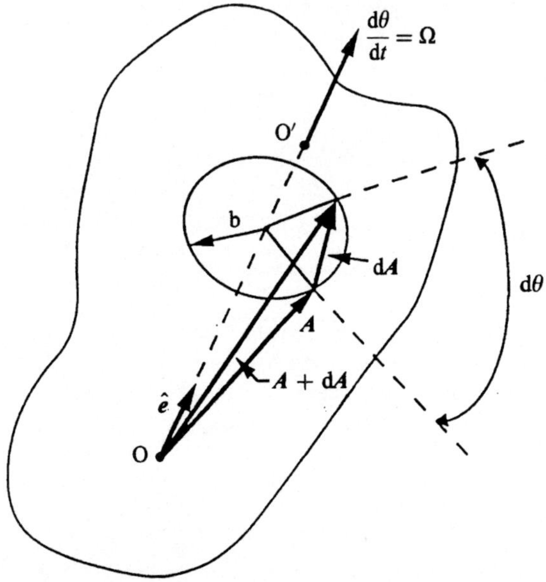

Fig. 1.17 Rate of change of a vector of fixed magnitude in a rotating coordinate system.

hold for general three-dimensional motion of a rigid body.

It is now possible to describe the rate of change of a vector $\boldsymbol{A}$ fixed either in a rigid body or in a frame of reference which has an instantaneous angular velocity Ω about an instantaneous axis of rotation OO′, as shown in Fig. 1.17; both the magnitude and direction of Ω may change with time. Here, the tip of the vector $\boldsymbol{A}$ tends to rotate in a circle about axis OO′ with radius b. In time $\mathrm{d}t$, the displacement $|\mathrm{d}\boldsymbol{A}| = b\,\mathrm{d}\theta$; but $b = |\hat{e} \times \boldsymbol{A}|$, where $\hat{e}$ is the unit vector along the axis of rotation. Thus, $\mathrm{d}\boldsymbol{A}/\mathrm{d}t = \Omega \times \boldsymbol{A}$ in a rotating system when vector $\boldsymbol{A}$ is of fixed magnitude. If, further, the magnitude of $\boldsymbol{A}$ depends on time, with its rate in a fixed (inertial) coordinate system, xyz, given by $\partial\boldsymbol{A}/\partial t$, then the total rate of change of the vector in a system rotating with instantaneous angular velocity Ω is given by

$$\frac{\mathrm{d}\boldsymbol{A}}{\mathrm{d}t} = \frac{\partial \boldsymbol{A}}{\partial t}\Big|_{\mathrm{O}xyz} + \Omega \times \boldsymbol{A}. \qquad (1.36)$$

The change indicated by eqn (1.36) also applies to unit vectors where the magnitude remains fixed and hence $\partial\boldsymbol{A}/\partial t = 0$.

In three-dimensional motion of a particle, it is customary to use orthogonal reference frames. The most common types include the rectangular Cartesian, the cylindrical, the spherical, and an intrinsic coordinate system, as shown in Figs 1.18(a)–(d). To obtain the velocity and acceleration of the particle in these coordinate systems, it should be noted that the axes of the Cartesian frame, used as the reference base for the other representations, remain fixed, so that the derivatives of the unit vectors associated with it vanish. The cylindrical coordinate system may be regarded as having an angular velocity $\dot{\theta}k$ relative to the Cartesian system. The spherical coordinate system can be regarded as being subjected to this angular velocity $\dot{\theta}k$ as well as an additional angular velocity of $\dot{\phi}\hat{e}_\theta$ relative to the cylindrical system, so that the total angular velocity of the unit vectors resolved along their axes in the spherical system in the directions $\hat{e}_R$, $\hat{e}_\theta$, and $\hat{e}_\phi$ in the directions of increasing R, θ, and ϕ, respectively, is

$$\begin{aligned}\Omega &= \dot{\theta}k - \dot{\phi}\hat{e}_\theta \\ &= \dot{\theta}(\sin\phi\ \hat{e}_R + \cos\phi\ \hat{e}_\phi) - \dot{\phi}\hat{e}_\theta. \qquad (1.37)\end{aligned}$$

The position vector for a typical point P and the corresponding velocity and acceleration relations for these systems, the latter obtained by applying successive differentiation of position vector $\boldsymbol{r}$ with respect to time, are

$$\begin{aligned}&\boldsymbol{r} = ix + jy + kz && \boldsymbol{r} = R\hat{e}_R + kz \\ &\dot{\boldsymbol{r}} = \boldsymbol{v} = i\dot{x} + j\dot{y} + k\dot{z}; && \dot{\boldsymbol{r}} = \boldsymbol{v} = \dot{R}\hat{e}_R + R\dot{\theta}\hat{e}_\theta + \dot{z}k; \\ &\ddot{\boldsymbol{r}} = \dot{\boldsymbol{v}} = \boldsymbol{a} = i\ddot{x} + j\ddot{y} + k\ddot{z} && \ddot{\boldsymbol{r}} = \boldsymbol{a} = (\ddot{R} - R\dot{\theta}^2)\hat{e}_\theta \\ & && \quad +(R\ddot{\theta} + 2\dot{R}\dot{\theta})\hat{e}_\theta + \ddot{z}k\end{aligned} \qquad (1.38)$$

$$\begin{aligned}&\boldsymbol{r} = R\hat{e}_R \\ &\dot{\boldsymbol{r}} = \boldsymbol{v} = \dot{R}\hat{e}_R + R\dot{\theta}\cos\phi\ \hat{e}_\theta + R\dot{\phi}\hat{e}_\phi \\ &\ddot{\boldsymbol{r}} = \boldsymbol{a} = (\ddot{R} - R\dot{\phi}^2 - R\dot{\theta}^2\cos^2\phi)\hat{e}_R \\ &\quad + (2\dot{R}\dot{\theta}\cos\phi + R\ddot{\theta}\cos\phi - 2R\dot{\theta}\dot{\phi}\sin\phi)\hat{e}_\phi \\ &\quad + (2\dot{R}\dot{\phi} + R\ddot{\phi} + R\dot{\theta}^2\sin\phi\cos\phi)\hat{e}_\phi.\end{aligned}$$

For the intrinsic coordinate system, Fig. 1.18(d), it can be shown (Synge and Griffith 1959) that the rate of change of the unit tangent vector, $\hat{e}_\mathrm{t}$, is given by $\mathrm{d}\hat{e}_\mathrm{t}/\mathrm{d}s = \hat{e}_\mathrm{n}/R$ with $\hat{e}_\mathrm{n}$ as the direction of the principal normal and R the local radius of curvature. Also, the velocity is always tangent to the curve and the acceleration has two components:

$$\boldsymbol{a} = a_\mathrm{t} + a_\mathrm{n} = \ddot{s}\hat{e}_\mathrm{t} + \frac{\dot{s}^2}{R}\ \hat{e}_n, \qquad (1.39)$$

where s is the distance measured along the curve. Equations (1.38) can be readily reduced to the two-dimensional case by eliminating z for the rectangular and the cylindrical cases, and eliminating either θ or else ϕ for the spherical coordinate system. Thus, for fixed-axis rotation about an endpoint O of a rod of length L, the distance travelled s, the linear velocity $\boldsymbol{v} = v\hat{e}_\mathrm{t}$, and the tangential and normal acceleration of the tip, a_t and a_n, are given by

$$s = L\theta; \quad v = L\dot{\theta}; \quad a_\mathrm{t} = L\ddot{\theta}; \quad a_\mathrm{n} = L\dot{\theta}^2. \qquad (1.40)$$

When the motion of a particle B is described with reference to the motion of another particle A, both located in a frame that is either at rest or translates with constant velocity, the position, velocity, and acceleration of B may be expressed relative to that at A by

$$\begin{aligned}&\boldsymbol{r}_\mathrm{B} = \boldsymbol{r}_\mathrm{A} + \boldsymbol{r}_{\mathrm{B/A}}; \quad \boldsymbol{v}_\mathrm{B} = \dot{\boldsymbol{r}}_\mathrm{B} = \boldsymbol{v}_\mathrm{A} + \boldsymbol{v}_{\mathrm{B/A}}; \\ &\boldsymbol{a}_\mathrm{B} = \ddot{\boldsymbol{r}}_\mathrm{B} = \boldsymbol{a}_A + \boldsymbol{a}_{B/A},\end{aligned} \qquad (1.41)$$

where the subscript B/A has the meaning of B relative to A. Figure 1.19 illustrates the meaning of these relations.

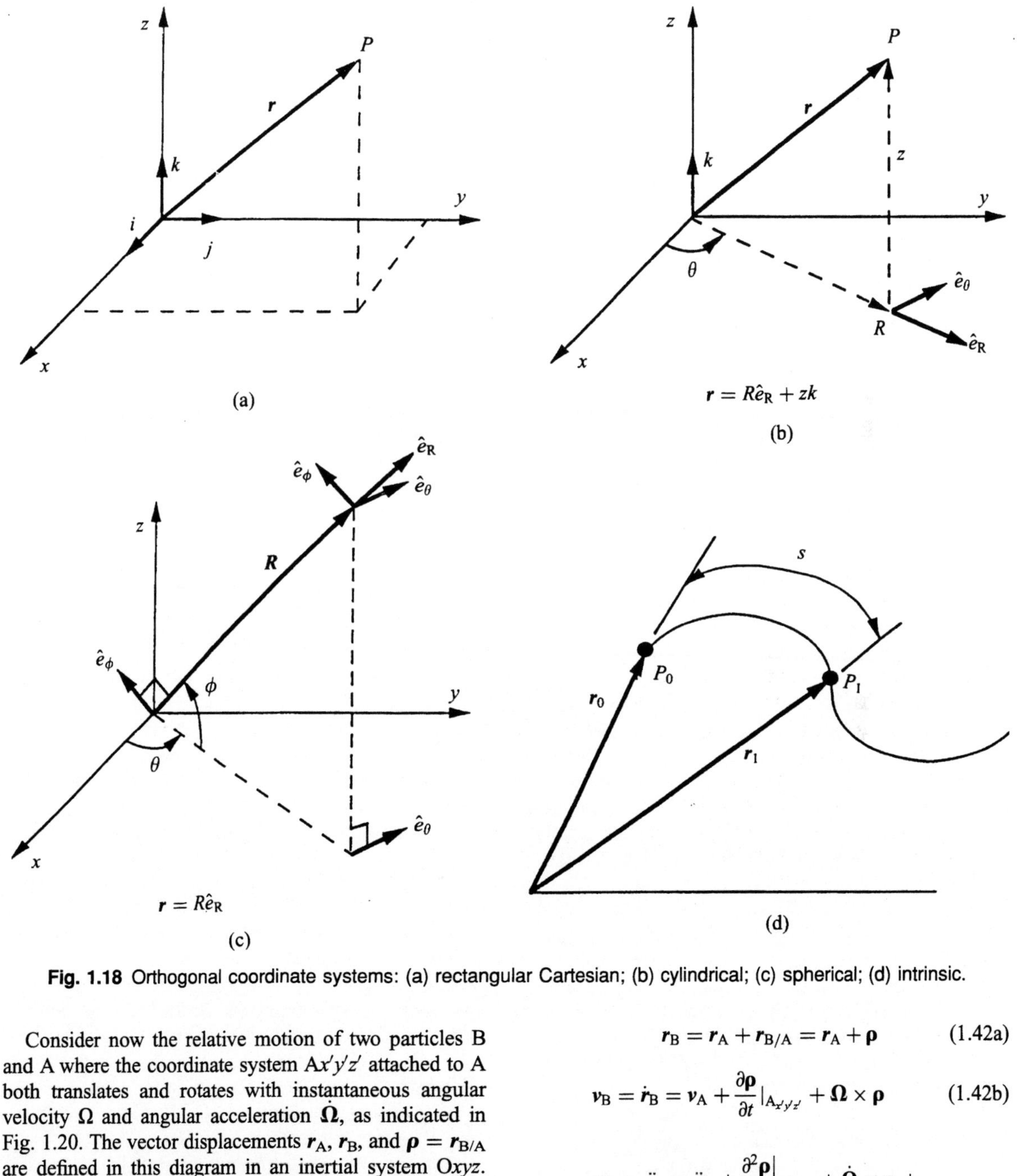

Fig. 1.18 Orthogonal coordinate systems: (a) rectangular Cartesian; (b) cylindrical; (c) spherical; (d) intrinsic.

Consider now the relative motion of two particles B and A where the coordinate system $Ax'y'z'$ attached to A both translates and rotates with instantaneous angular velocity Ω and angular acceleration $\dot{\boldsymbol{\Omega}}$, as indicated in Fig. 1.20. The vector displacements $\boldsymbol{r}_{\mathrm{A}}$, $\boldsymbol{r}_{\mathrm{B}}$, and $\boldsymbol{\rho} = \boldsymbol{r}_{\mathrm{B/A}}$ are defined in this diagram in an inertial system $Oxyz$. The absolute velocity and acceleration of B in terms of those of A are obtained by two successive differentiations with respect to time using eqn (1.36). Since point A does not rotate, these operations yield

$$\boldsymbol{r}_{\mathrm{B}} = \boldsymbol{r}_{\mathrm{A}} + \boldsymbol{r}_{\mathrm{B/A}} = \boldsymbol{r}_{\mathrm{A}} + \boldsymbol{\rho} \tag{1.42a}$$

$$\boldsymbol{v}_{\mathrm{B}} = \dot{\boldsymbol{r}}_{\mathrm{B}} = \boldsymbol{v}_{\mathrm{A}} + \frac{\partial \boldsymbol{\rho}}{\partial t}\Big|_{Ax'y'z'} + \boldsymbol{\Omega} \times \boldsymbol{\rho} \tag{1.42b}$$

$$\boldsymbol{a}_{\mathrm{B}} = \ddot{\boldsymbol{r}}_{\mathrm{B}} = \ddot{\boldsymbol{r}}_{\mathrm{A}} + \frac{\partial^2 \boldsymbol{\rho}}{\partial t^2}\Big|_{Ax'y'z'} + \dot{\boldsymbol{\Omega}} \times \boldsymbol{\rho} + \boldsymbol{\Omega} \times [\boldsymbol{\Omega} \times \boldsymbol{\rho}] + 2\boldsymbol{\Omega} \times \frac{\partial \boldsymbol{\rho}}{\partial t}\Big|_{Ax'y'z'}. \tag{1.42c}$$

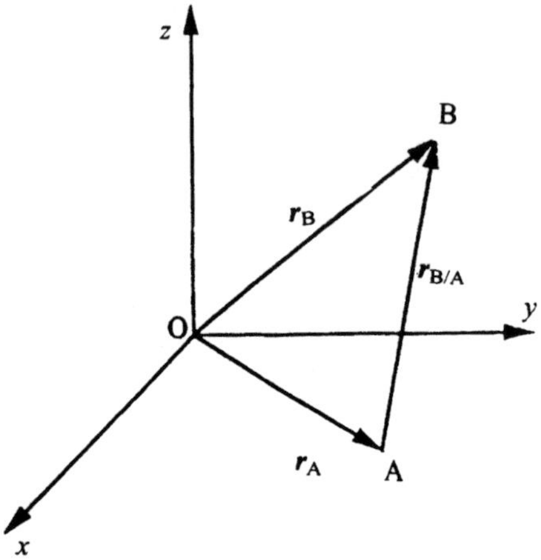

Fig. 1.19 Relative displacement of point B with respect to point A.

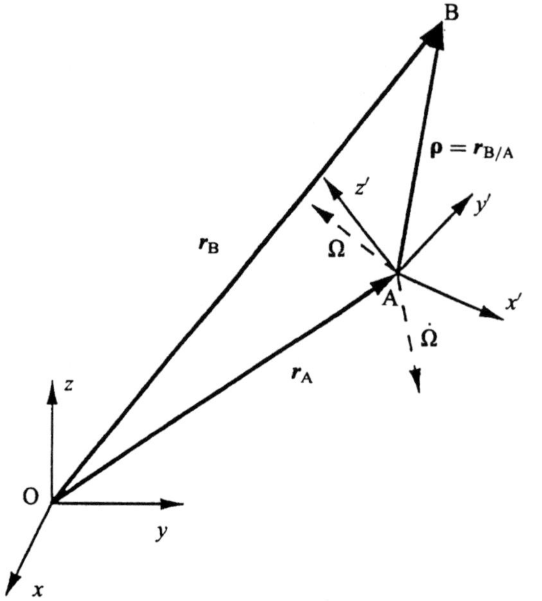

Fig. 1.20 Displacements in a rotating system $Ax'y'z'$ moving instantaneously with angular velocity $\boldsymbol{\Omega}$ and angular acceleration $\dot{\boldsymbol{\Omega}}$. $Oxyz$ is an inertial coordinate system.

The first term on the right-hand side of eqn (1.42c) is the acceleration of point A, measured in the inertial coordinate system $Oxyz$ in the usual way using eqns (1.38) or (1.39); the second term is the acceleration of B as seen by an observer located at A and moving with the coordinate system $Ax'y'z'$. The third and fourth terms are the acceleration of B relative to A due to rotation of the system $Ax'y'z'$ representing tangential and normal components. The last term is the so-called Coriolis acceleration, produced by the change of location of the unit vectors due to rotation, and represents the difference in accelerations viewed from a non-rotating and rotating coordinate system, respectively. This last term often dominates the magnitude of the total (or absolute) acceleration; since $\boldsymbol{\Omega}$ changes continuously when an angular acceleration is present, the value of $\boldsymbol{a}$ also changes accordingly.

Applications of the kinematics of relative motion are found in the movement of human or animal limbs when measurement of their motion is referred to a coordinate system fixed in the torso or an upper limb which is rotating itself about some axis. If the hand or foot can rotate independently relative to the lower portion of the extremity, the relative motion equations (1.42) may have to be applied repeatedly. Of course, such an analysis can be carried out using absolute accelerations.

Figure 1.21 presents a schematic of an arm executing a throw or pitch and the corresponding mass center accelerations. This motion can be considered to occur in a plane, and the hand can be considered to be integral with the lower arm segment. The arm is then modeled as a two-component system with the upper arm rotating about a fixed axis perpendicular to the plane of the paper through the shoulder joint O, the origin of an inertial system. The forearm rotates about elbow A, and its motion is defined relative to that of the upper arm in which a rotating coordinate system is embedded. The angular positions, velocities, and accelerations can be found using data from a framing camera or by records from transducers or light sources attached to critical locations of the limb and torso; when these are known, the joint forces and moments can be determined from the appropriate equations of motion developed in the sequel. Numerous examples for a variety of sports and typical calculations for a basketball throw can be found in Plagenhoef (1971), as well as appropriate geometric and mass distribution data for humans that is also cited in Dempster (1955), Drillis and Contini (1966), Contini (1972), Miller and Morrison (1975), and Kreighbaum and Barthels (1990).

1.2.2 Kinetics of particles

A particle of mass m moves in accordance with Newton's second law, eqn (1.14), where the acceleration used is

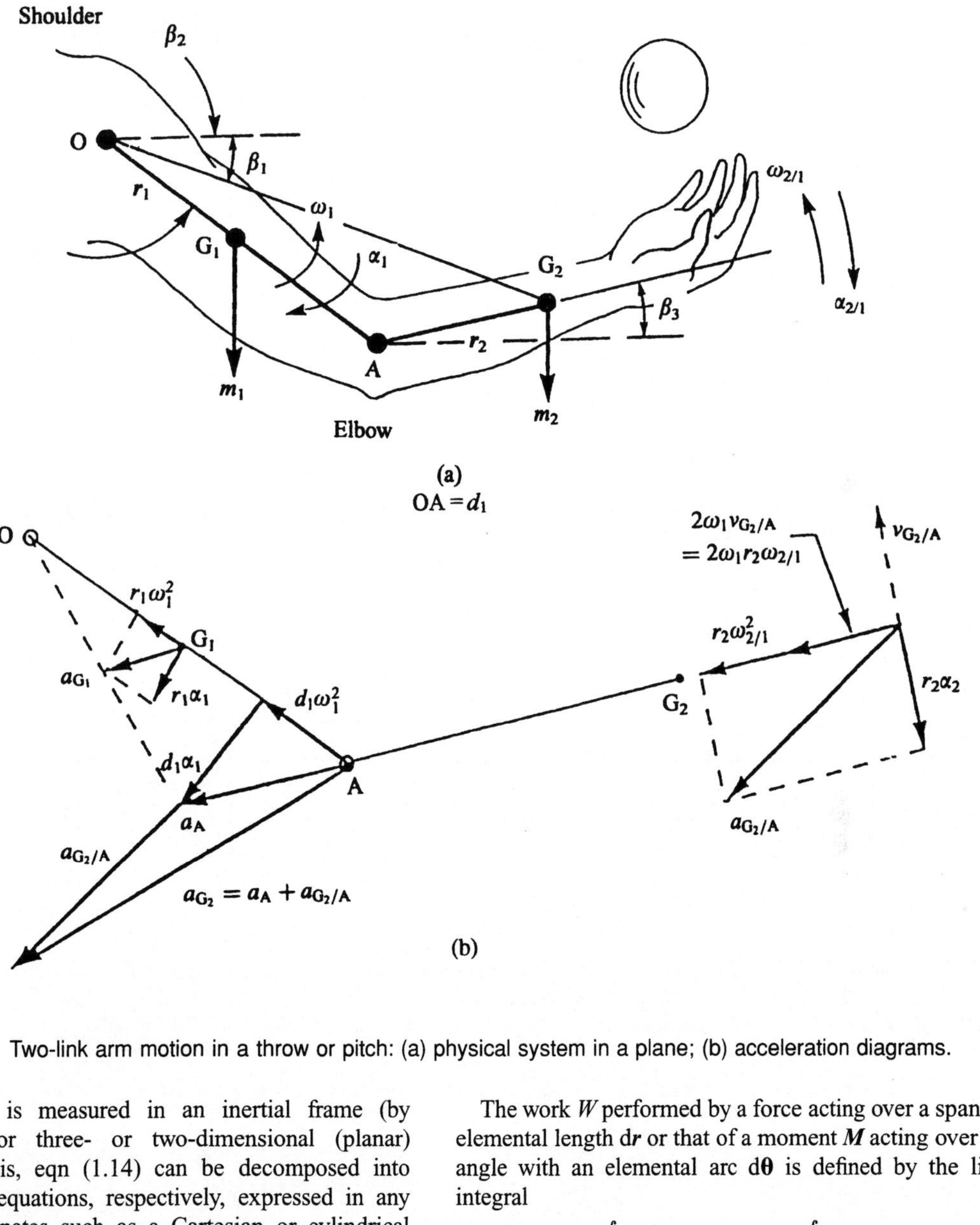

Fig. 1.21 Two-link arm motion in a throw or pitch: (a) physical system in a plane; (b) acceleration diagrams.

absolute and is measured in an inertial frame (by definition). For three- or two-dimensional (planar) motion analysis, eqn (1.14) can be decomposed into three or two equations, respectively, expressed in any desired coordinates such as a Cartesian or cylindrical system:

$$F_x = ma_x = m\ddot{x}; \quad F_y = ma_y = m\ddot{y};$$
$$F_z = ma_z = m\ddot{z};$$
$$F_r = m(\ddot{r} - r\dot{\theta}^2); \quad F_\theta = m(r\ddot{\theta} + 2\dot{r}\dot{\theta}); \quad F_z = m\ddot{z}. \tag{1.43}$$

The work W performed by a force acting over a span of elemental length d$\boldsymbol{r}$ or that of a moment $\boldsymbol{M}$ acting over an angle with an elemental arc d$\boldsymbol{\theta}$ is defined by the line integral

$$W = \oint \boldsymbol{F} \cdot \mathrm{d}\boldsymbol{r} \quad \text{or} \quad W = \oint \boldsymbol{M} \cdot \mathrm{d}\boldsymbol{\theta}, \tag{1.44}$$

where $\boldsymbol{F}$ may also be interpreted as the resultant of several concurrent forces. For the motion of a particle,

$$W = \oint \boldsymbol{F} \cdot \mathrm{d}\boldsymbol{r} = \oint m\boldsymbol{a} \cdot \mathrm{d}\boldsymbol{r} = \oint m\boldsymbol{v} \cdot \mathrm{d}\boldsymbol{v}$$
$$= \tfrac{1}{2}mv_2^2 - \tfrac{1}{2}mv_1^2 \equiv \Delta T = T_2 - T_1, \tag{1.45}$$

where T is the kinetic energy of the particle; thus work done equals the change in kinetic energy. Mechanical work or energy is expressed in joules (J): $1\ \text{J} = 1\ \text{N m}$; and electrical energy is expressed in watt-seconds (W s): $1\ \text{W s} = 1\ \text{J}$.

Forces may or may not have associated with them a potential field; if they do, they can be derived from a potential V, expressed in units of joules, by

$$\boldsymbol{F} = -\nabla V = -\left(i\frac{\partial V}{\partial x} + j\frac{\partial V}{\partial y} + k\frac{\partial V}{\partial z}\right) = iF_x + jF_y + kF_z. \tag{1.46}$$

The portion of the work done by conservative forces, $W_{\text{c.f.}}$, then equals the negative change in potential energy: $W_{\text{c.f.}} = -\Delta V$; and the work–energy equation that involves non-conservative forces whose work is $W_{\text{n.c.f.}}$ can be written as

$$\begin{aligned} &W_{\text{n.c.f.}} = \Delta T + \Delta V, \quad \text{or,} \\ &\text{iff } W_{\text{n.c.f.}} = 0, \quad T + V = E = \text{constant}, \end{aligned} \tag{1.47}$$

where E is the total energy. A very common type of conservative force field results from potential energy stored in springs, which for a linear spring is given by $V_s = \frac{1}{2}k\delta^2$, where spring force $F = k\delta$, with k as the spring constant and δ as the extension or, alternatively, the compression from the equilibrium position. The other frequently encountered potential energy is due to gravitational attraction and is given by $V_g = mgh$, where h is the height above an arbitrary reference plane.

The work–energy equations are obtained by integration of the scalar product of $\boldsymbol{F}$ and $\text{d}\boldsymbol{r}$, leading to the scalar results of eqns (1.47). Integration of Newton's laws with respect to time yields

$$\int_0^t \boldsymbol{F}\,\text{d}t = \int_0^t \frac{\text{d}}{\text{d}t}(m\boldsymbol{v})\,\text{d}t = m(\boldsymbol{v}_2 - \boldsymbol{v}_1) = \boldsymbol{L}_2 - \boldsymbol{L}_1 = \Delta\boldsymbol{L}, \tag{1.48}$$

where $\boldsymbol{L}$ is the linear momentum of the particle; the left-hand side of eqn (1.48) is the linear impulse. This is a vector relation and may be resolved into components as required. The effect of the moment of the force resultant $\boldsymbol{F} = \sum_{i=1}^{n} \boldsymbol{F}_i$ about a fixed point O produces a change in the angular momentum $\boldsymbol{H}_\text{O}$, i.e. the moment of the linear momentum $\boldsymbol{L} = m\boldsymbol{v}$ of the particle, given by

$$\begin{aligned} \boldsymbol{M}_\text{O} &= \boldsymbol{r}\times\boldsymbol{F} = \boldsymbol{r}\times m\dot{\boldsymbol{v}} = \frac{\text{d}}{\text{d}t}(\boldsymbol{r}\times m\boldsymbol{v}) = \frac{\text{d}}{\text{d}t}(\boldsymbol{H}_\text{O}) \\ &= \dot{\boldsymbol{H}}_\text{O} \quad \left[\text{since } \frac{\text{d}}{\text{d}t}(\boldsymbol{r}\times m\boldsymbol{v}) = \dot{\boldsymbol{r}}\times m\boldsymbol{v} + \boldsymbol{r}\times m\dot{\boldsymbol{v}}\right. \\ &\left. = \boldsymbol{v}\times m\boldsymbol{v} + \boldsymbol{r}\times m\dot{\boldsymbol{v}} = \boldsymbol{r}\times m\dot{\boldsymbol{v}}, \text{ since } \boldsymbol{v}\times\boldsymbol{v} = 0\right], \end{aligned} \tag{1.49}$$

as indicated in Fig. 1.22. The angular momentum may be written in terms of a Cartesian coordinate system centered at O as

$$\boldsymbol{H}_\text{O} = m\begin{vmatrix} i & j & k \\ x & y & z \\ \dot{x} & \dot{y} & \dot{z} \end{vmatrix}. \tag{1.50}$$

Integration of eqn (1.49) yields the relation equating the angular impulse and the change in angular momentum

$$\int_{t_1}^{t_2} \boldsymbol{M}_\text{O}\,\text{d}t = \Delta\boldsymbol{H}_\text{O}\Big|_1^2 = \boldsymbol{H}_{\text{O}_2} - \boldsymbol{H}_{\text{O}_1}. \tag{1.51}$$

For a single particle, eqn (1.51) is seldom employed.

Like energy, linear momentum is conserved iff the sum of the forces on the particle vanishes, and angular momentum is conserved iff the sum of the moments vanishes during the period of interest; this yields for the two situations

$$L = \text{constant} \quad \text{and} \quad \boldsymbol{H}_G = \text{constant}. \tag{1.52}$$

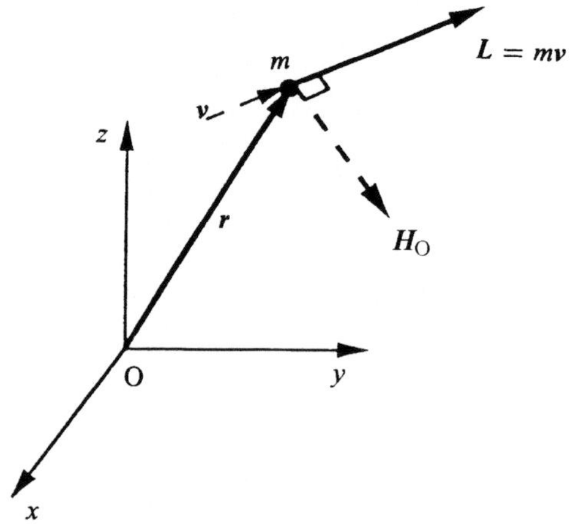

Fig. 1.22 Moment of linear momentum about point O.

1.2.3 Kinetics of a system of particles

When a surface S encloses a system of n particles whose location is identified by position vectors $\boldsymbol{r}_i$ in an inertial frame and by vector $\boldsymbol{\rho}_i$, relative to mass center G, as shown in Fig. 1.23, this aggregate can be subjected to external forces $\boldsymbol{F}_i$ which act on mass m_i across surface S, and by a set of internal forces $\boldsymbol{f}_{ji}$ exerted by particle i on particle j. Summation over all the particles of the system for the external and internal forces, and noting that $\sum_{i=1}^{n}\sum_{j=1}^{n} \boldsymbol{f}_{ij} = 0$ by Newton's third law, yields

$$\sum_{i=1}^{n}\left[\boldsymbol{F}_i + \sum_{j=1}^{n}\boldsymbol{f}_{ij}\right] = \sum_{i=1}^{n}\boldsymbol{F}_i + \sum_{i=1}^{n}\sum_{j=1}^{n}\boldsymbol{f}_{ij} = \sum_{i=1}^{n} m_i\ddot{\boldsymbol{r}}_i = m\ddot{\boldsymbol{r}}_{\mathrm{G}} \equiv \boldsymbol{F}, \tag{1.53}$$

where $\boldsymbol{F}$ is the resultant external force and $\ddot{\boldsymbol{r}}_{\mathrm{G}}$ is the acceleration of the mass center G; the total mass of the system is $m = \sum_{i=1}^{n} m_i$. The term $\boldsymbol{f}_{ii} = 0$, as a particle can not exert a force in itself. The resultant force imparts to the mass center of the system—which may not coincide with a particle—an acceleration in the direction of the

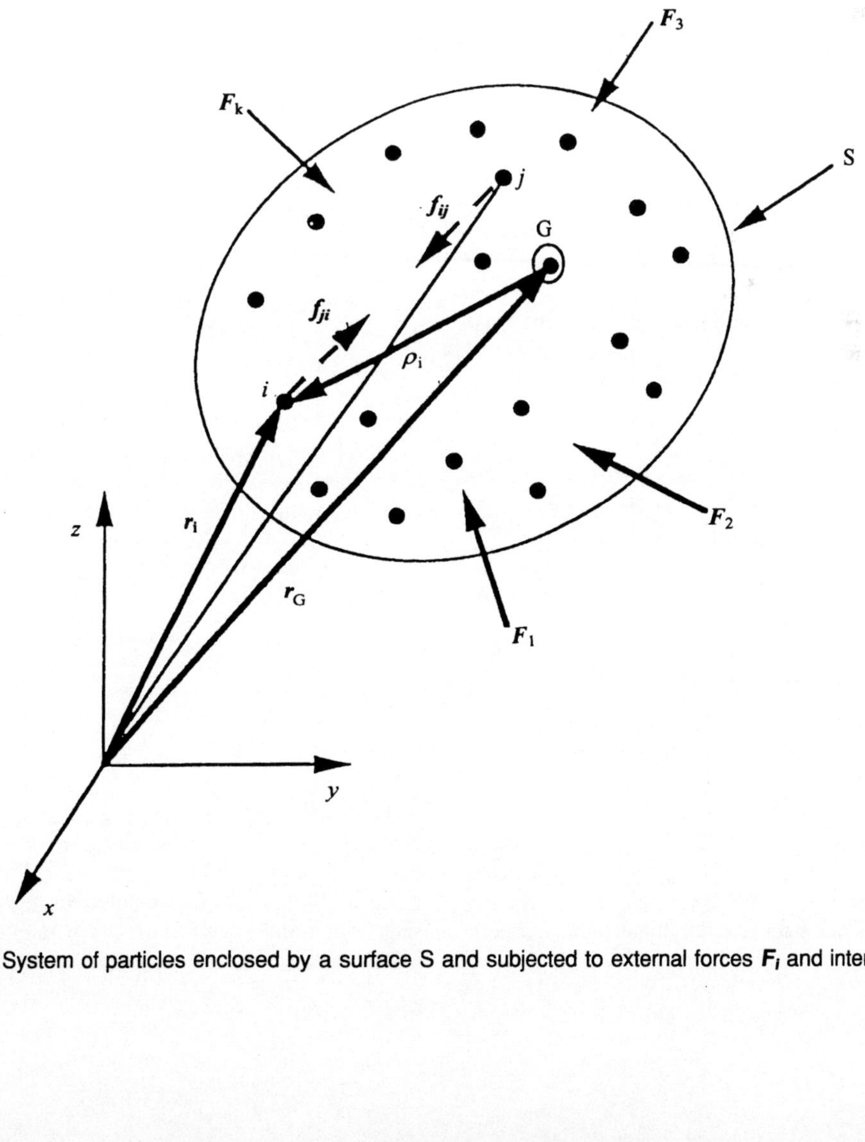

Fig. 1.23 System of particles enclosed by a surface S and subjected to external forces $\boldsymbol{F}_i$ and internal forces $\boldsymbol{f}_{ij}$.

force, equal to its magnitude divided by the system mass; however, the resultant force need not actually pass through G.

The work–energy relations developed for the single particle also apply here; however, it is frequently more convenient to analyze the system with respect to the behavior of the mass center. Thus, if the position vector of any particle is expressed as $\boldsymbol{r}_i = \boldsymbol{r}_G + \boldsymbol{\rho}_i$, the total work done is given by

$$\begin{aligned} W &= \oint \sum_{i=1}^{n} \left(\boldsymbol{F}_i + \sum_{j=1}^{n} \boldsymbol{f}_{ij} \right) \cdot \mathrm{d}\boldsymbol{r}_i \\ &= \oint \sum_{i=1}^{n} \left(\boldsymbol{F}_i + \sum_{j=1}^{n} \boldsymbol{f}_{ij} \right) \cdot (\mathrm{d}\boldsymbol{r}_G + \mathrm{d}\boldsymbol{\rho}_i) \\ &= \oint \boldsymbol{F} \cdot \mathrm{d}\boldsymbol{r}_G + \oint \sum_{i=1}^{n} \left(\boldsymbol{F}_i + \sum_{j=1}^{n} \boldsymbol{f}_{ij} \right) \cdot \mathrm{d}\boldsymbol{\rho}_i \end{aligned} \qquad (1.54)$$

(since $\boldsymbol{F} = \sum_{i=1}^{n} \boldsymbol{F}_i$, and $\mathrm{d}\boldsymbol{r}_G \cdot \sum_{i=1}^{n}\sum_{j=1}^{n} \boldsymbol{f}_{ij} = 0$). Furthermore, the kinetic energy T given by

$$\begin{aligned} T &= \sum_{i=1}^{n} T_i = \sum_{i=1}^{n} \frac{1}{2} m_i (\dot{\boldsymbol{r}}_G + \dot{\boldsymbol{\rho}}_i)(\dot{\boldsymbol{r}}_G + \dot{\boldsymbol{\rho}}_i) \\ &= \frac{1}{2} m v_G^2 + \frac{1}{2} \sum_{i=1}^{n} (m_i \dot{\boldsymbol{\rho}}_i^2) = T_{tr} + T_{rot}. \end{aligned} \qquad (1.55)$$

It comprises the translational kinetic energy of the mass center, T_{tr}, and the kinetic energy of rotation about the mass center, T_{rot}. Between states 1 and 2,

$$\int_1^2 \boldsymbol{F} \cdot \boldsymbol{r}_G = W_{tr}\Big|_1^2 = \left(\frac{1}{2} m v_G^2 \right)\Big|_1^2$$

$$\text{and} \quad \int_1^2 \sum_{i=1}^{n} \left(\mathbf{F} + \sum_{j=1}^{n} \boldsymbol{f}_{ij} \right) \cdot \mathrm{d}\boldsymbol{\rho}_i = \sum_{i=1}^{n} \left(\frac{1}{2} m_i \dot{\boldsymbol{\rho}}_i^2 \right)\Big|_1^2. \qquad (1.56)$$

In mechanical systems, the term $\sum_{i=1}^{n}\sum_{j=1}^{n} \boldsymbol{f}_{ij} \cdot \mathrm{d}\boldsymbol{\rho}_i$ often, but not always, vanishes.

By summing over the system of particles with respect to linear and angular momentum, there results:

$$\boldsymbol{F} = \sum_{i=1}^{n} \boldsymbol{F}_i = \dot{\boldsymbol{L}} \quad \text{and} \quad \boldsymbol{M}_O = \frac{\mathrm{d}}{\mathrm{d}t} \sum_{i=1}^{n} (\boldsymbol{r}_i \times m \boldsymbol{v}_i) = \dot{\boldsymbol{H}}_O. \qquad (1.57)$$

The second of eqns (1.57) can also be applied to the mass center: $\boldsymbol{M}_G = \dot{\boldsymbol{H}}_G$. However, for any generally moving point A that is chosen as the origin of a moving coordinate system, the angular momentum relation becomes:

$$\boldsymbol{M}_A = \dot{\boldsymbol{H}}_G + \boldsymbol{r}_{G/A} \times m \dot{\boldsymbol{v}}_G \quad \text{with} \quad \boldsymbol{H}_G = \sum_{i=1}^{n} \boldsymbol{r}_{i/G} \times m_i \boldsymbol{v}_i. \qquad (1.58)$$

Such expressions can also be developed for the relative angular momentum with respect to translating axes, where

$$\boldsymbol{H}_{A_{rel}} = \sum_{i=1}^{n} \boldsymbol{r}_{i/A} \times m_i \dot{\boldsymbol{r}}_{i/A} \qquad (1.59a)$$

so that

$$\boldsymbol{M}_A = \dot{\boldsymbol{H}}_{A_{rel}} + \boldsymbol{r}_{G/A} \times m \boldsymbol{a}_A \qquad (1.59b)$$

and this relation reduces to $\boldsymbol{M}_A = \dot{\boldsymbol{H}}_{A_{rel}}$ only if (i) $\mathbf{a}_A \equiv 0$, (ii) $\boldsymbol{r}_{G/A} = 0$ (so that A is the mass center G), and (iii) $\boldsymbol{r}_{G/A}$ is parallel to $\boldsymbol{a}_A$. As in the case of the work-energy equation, linear or angular momentum is conserved if the sum of the forces or the sum of the moments vanishes during the interval of interest. From eqns (1.57),

$$\boldsymbol{L} = \text{constant if } \boldsymbol{F} = 0 \quad \text{and}$$

$$\boldsymbol{H}_G \text{ (or } \boldsymbol{H}_O) = \text{constant if } \boldsymbol{M}_G = 0 \text{ (or } M_O = 0). \qquad (1.60)$$

1.2.4 Impact of two bodies

The impact of two bodies idealized as two particles moving in a plane, as shown in Fig. 1.24, involves large forces of short duration acting on each particle. Due to Newton's third law, these forces are equal and opposite and thus their sum vanishes, leading to conservation of linear momentum. The approximate treatment of this situation described here is standard procedure in elementary dynamics. The velocity components are resolved along directions normal and tangent to the contact surfaces, denoted by subscripts n and t, and momentum conservation along these directions then yields the equations

$$m_1 v_{1,n_0} + m_2 v_{2,n_0} = (m_1 + m_2) u_n = m_1 v_{1,n_f} + m_2 v_{2,n_f}$$

$$m_1 v_{1,t_0} + m_2 v_{2,t_0} = m_1 v_{1,t_f} + m_2 v_{2,t_f}. \qquad (1.61)$$

Subscripts 1 and 2 refer to the two bodies, 0 and f are initial and final values, and u_n is the common velocity reached during one instant of the process.

The linear impulse due to F_n for the compression phase of the impact, when the centers of mass of the bodies 'approach' each other, is L_c. This phase ends when u_n is reached; the impulse during the restitution stage, when the bodies tend to separate, is labelled L_r. (These impulses are not designated as vectors, since their

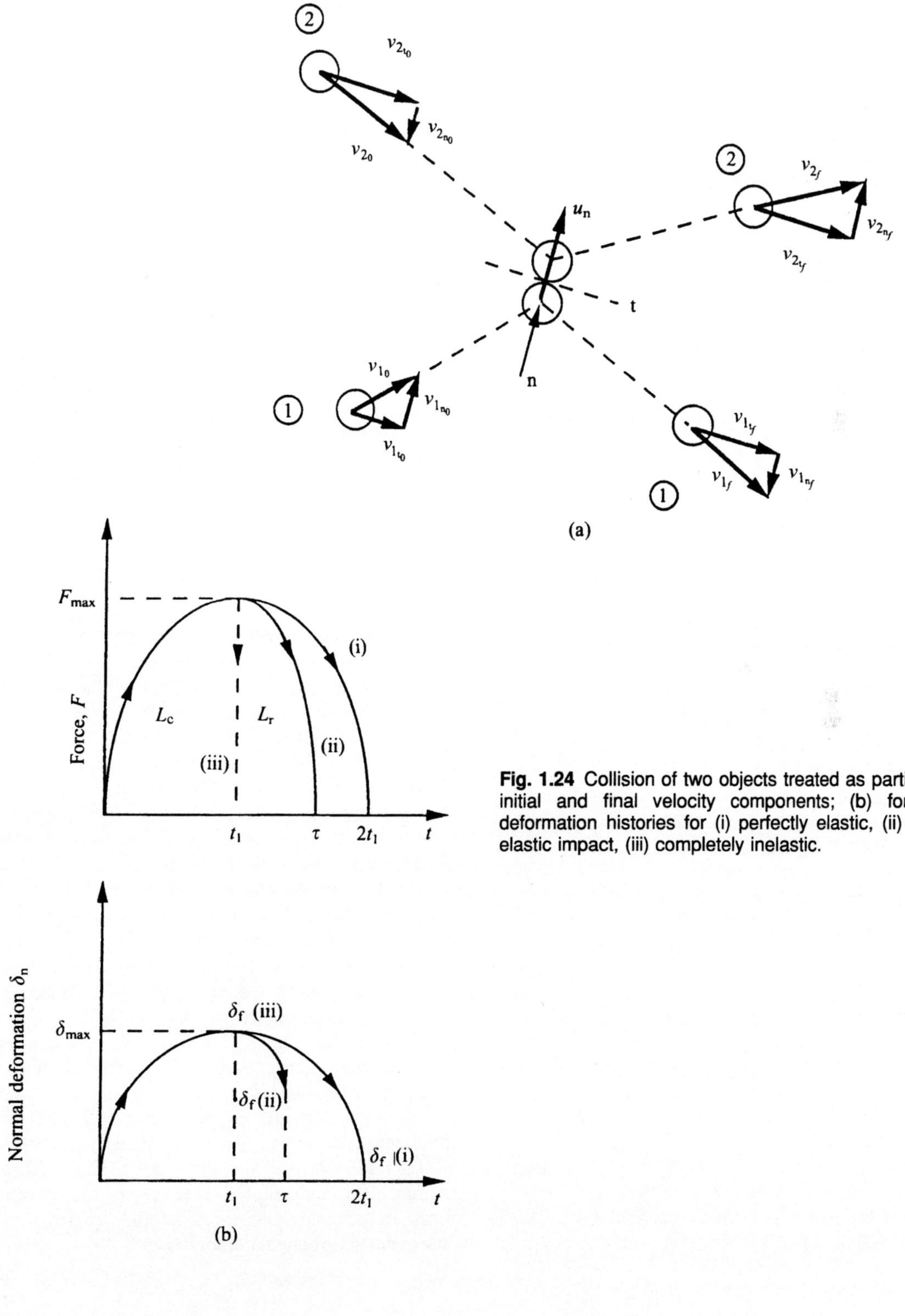

Fig. 1.24 Collision of two objects treated as particles: (a) initial and final velocity components; (b) force and deformation histories for (i) perfectly elastic, (ii) partially elastic impact, (iii) completely inelastic.

directions are specified.) The ratio L_r/L_c defines a quantity e called the coefficient of restitution which, in the absence of a much more difficult (and perhaps intractable) analysis, is employed to calculate the loss of energy during the impact. The actual value of e depends not only on the material characteristics of the objects and geometry of the striking surfaces, but also on the impact speed. The dependence on speed, and even the shape of the strikers, are usually ignored, and a constant value of e is ordinarily specified based solely on the empirically determined material behavior of the colliding bodies. As examples, typical values range from 0.6 to 0.75 for a golf ball when struck by various clubs, and from 0.45 to 0.5 for a baseball hit by a bat.

The elementary definition of e and its immediate consequences (see Stronge, 1990, for a more rigorous treatment) are

$$e = L_r/L_c = (v_{2,n_f} - v_{1,n_f})/(v_{1,n_0} - v_{2,n_0}). \quad (1.62)$$

Equation (1.62) may be interpreted in terms of the normal force history shown in Fig. 1.24(b); L_c is the area under the curve up to F_{max}, the peak value of the normal force occurring at time t_1 and corresponding to the maximum normal deformation δ_n. The quantity L_r is the area under the curve from t_1 to the end of impact at $t = \tau$. Mathematically, the model requires that the applied impulse is a delta function, while, actually, a finite contact time and actual deformations exist; this paradox illustrates shortcomings of the model.

For perfectly elastic bodies, $e = 1$; for perfectly plastic bodies, where the terminal normal velocity of both is the common velocity u_n attained at time t_1, $e = 0$; and for bodies with intermediate characteristics (in practice, all cases), $0 < e < 1$. The kinetic energy loss in such an impact is

$$-\Delta T = T_1 - T_2 = \frac{1}{2}\frac{m_1 m_2}{m_1 + m_2}(1 - e^2)(v_{1,n_0} - v_{2,n_0})^2. \quad (1.63)$$

As an example of the determination of e, this coefficient for the case of an object dropped onto a very large (infinite) mass which exhibits no initial or final velocity is $e = -v_{2_f}/v_{2_o} = [2mgh_f/2mgh_o]^{1/2} = [h_f/h_o]^{1/2}$, where h_o and h_f are the drop and rebound height, respectively.

The determination of the final tangential components also requires the specification of an additional parameter to describe the effects of friction. For the idealized case of completely smooth surfaces, the initial tangential velocity components are unchanged; for completely rough surfaces, the final tangential velocity for both bodies, u_t, is the same. Actual values about which very little is known, lie somewhere between these extremes.

Biomechanical impacts occur naturally in all kinds of endeavors: in sports, in many accidents, and in industrial operations involving human performance.

1.2.5 Kinematics of rigid bodies

Rigid bodies are by definition undeformable; they represent a special case of a system of particles where the distance between *any* two particles remains invariant. Such bodies can move in (i) translation, where all particles of the object execute identical, parallel paths and the velocity and acceleration of every particle are the same at a given instant of time, although generally time-dependent, (ii) rotation about some axis at some instant of time where all the particles move in circular arcs about this axis, along which no motion exists (in general, this axis changes its position with time relative to an inertial frame), and (iii) general three-dimensional motion combining (i) and (ii).

Euler's theorem states that the motion of a rigid body with a fixed point is the same as rotation about an axis passing through this point (illustrated by the rotation of a stiff arm in the shoulder joint). Chasles's theorem states that the most general finite or infinitesimal motion of a rigid body consists of a translation of a reference (base) point simultaneous or subsequent to a rotation about an axis through this point, as illustrated in Fig. 1.25. For identical motions, a change of base point results in a different displacement of this point, but does not change the direction of the axis nor magnitude of the rotation.

Although infinitesimal rotations are vectors, obeying the commutative law of vector addition, so that $\mathbf{d\theta}_1 + \mathbf{d\theta}_2 = \mathbf{d\theta}_2 + \mathbf{d\theta}_1$, finite rotations are *not* vectors, so that $\hat{\theta}_1 + \hat{\theta}_2 \neq \hat{\theta}_2 + \hat{\theta}_1$. This fact causes major complications in the analysis of three-dimensional rigid-body motion, since the location of the body is generally identified by the position of its mass center $\boldsymbol{r}_G$ in an inertial system, say $OXYZ$, but its orientation is defined by an embedded coordinate system $Gxyz$, whose axes were parallel to (or coincident with) $OXYZ$ at some earlier reference time t_0. Thus, the situation is that shown in Fig. 1.26, which is similar to that of Fig. 1.20, and requires the specification of three time-dependent angles which are generally chosen as the Euler angles, a non-orthogonal set of angular coordinates.

Euler angles are generated as follows: To attain a terminal position Oxyz of a set of orthogonal axes from an initial fixed orientation OXYZ with the same origin, three successive rotations are performed as shown in Fig. 1.27. The first rotation involving a precesssion angle ψ occurs with angular velocity $\dot{\boldsymbol{\psi}}$ about axis OZ to the new position O$x'y'z'$. The second rotation, defined by nutation angle θ with corresponding angular velocity $\dot{\boldsymbol{\theta}}$ along Ox', moves the coordinate triad to the second position O$x''y''z''$. The last rotation from this set of axes to the final position Oxyz is accomplished by a spin motion of angle ϕ corresponding to a spin velocity $\dot{\boldsymbol{\phi}}$ directed along the Oz'' (or z) axis. The total angular velocity of the body resulting from these rotations relative to body axes Oxyz is

$$\begin{aligned}\boldsymbol{\omega} &= \dot{\psi}\hat{e}_Z + \dot{\theta}\hat{e}_{x'} + \dot{\phi}\hat{e}_z \\ &= (\dot{\theta}\ \cos\phi + \dot{\psi}\sin\theta\sin\phi)i \\ &\quad + (-\dot{\theta}\sin\phi + \dot{\psi}\sin\theta\cos\phi)j + (\dot{\phi} + \dot{\psi}\cos\theta)k,\end{aligned} \tag{1.64}$$

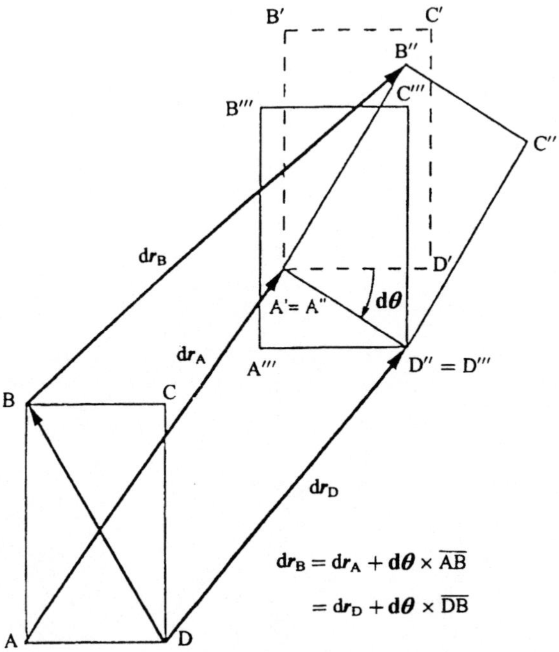

Fig. 1.25 General motion of a rigid body illustrating Chasles's theorem.

which may be derived by resolution of unit vectors $\hat{e}_Z$ and $\hat{e}_{x'}$ onto Oxyz.

The absolute motion of a point B of a rigid body rotating about point O is given by

$$\boldsymbol{v} = \boldsymbol{\Omega} \times \boldsymbol{r} \quad \text{and} \quad \boldsymbol{a} = \dot{\boldsymbol{\Omega}} \times \boldsymbol{r} + \boldsymbol{\Omega} \times (\boldsymbol{\Omega} \times \boldsymbol{r}), \tag{1.65}$$

where $\boldsymbol{\Omega}$ and $\dot{\boldsymbol{\Omega}}$ are the instantaneous angular velocity and angular acceleration of the body, and $\boldsymbol{r}$ is the position of the point measured from O, the origin of inertial system OXYZ. For relative motion, it should be noted that the distance from reference point A and that of the point in question B, $\boldsymbol{\rho} = \overline{AB}$, is fixed, as indicated in Fig. 1.28; hence $\partial\boldsymbol{\rho}/\partial t$ and $\partial^2\boldsymbol{\rho}/\partial t^2$ in eqn (1.42) vanish; thus

$$\boldsymbol{v}_B = \boldsymbol{v}_A + \boldsymbol{\Omega} \times \boldsymbol{\rho};\ \boldsymbol{a}_B = \boldsymbol{a}_A + \dot{\boldsymbol{\Omega}} \times \boldsymbol{\rho} + \boldsymbol{\Omega} \times (\boldsymbol{\Omega} \times \boldsymbol{\rho}). \tag{1.66}$$

These are the equations for the motion of B relative to A in a nonrotating frame attached to A. If the frame has an angular velocity, the Coriolis acceleration term must be added.

In biological systems, motion is almost always three-dimensional, such as the movement of a cat, parachutist, diver, high jumper, pole vaulter, or skilled or manual laborer, as well as the gyrations of leaves on trees or bushes. In many bioengineering cases, the kinematics can be approximated by plane motion, where the paths of all particles of the body remain either in or else parallel to a given reference plane. The absolute and relative motion equations developed are simplified by removing dependence on z, the axis perpendicular to the plane of motion, in the force relations, and involving rotational motion only about this axis, so that $\boldsymbol{\omega} = \omega k$ and $\dot{\boldsymbol{\omega}} = \dot{\omega}k$. For

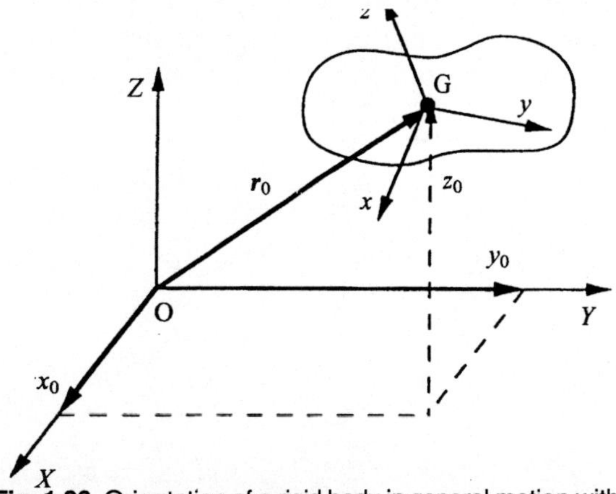

Fig. 1.26 Orientation of a rigid body in general motion with respect to an inertial system OXYZ.

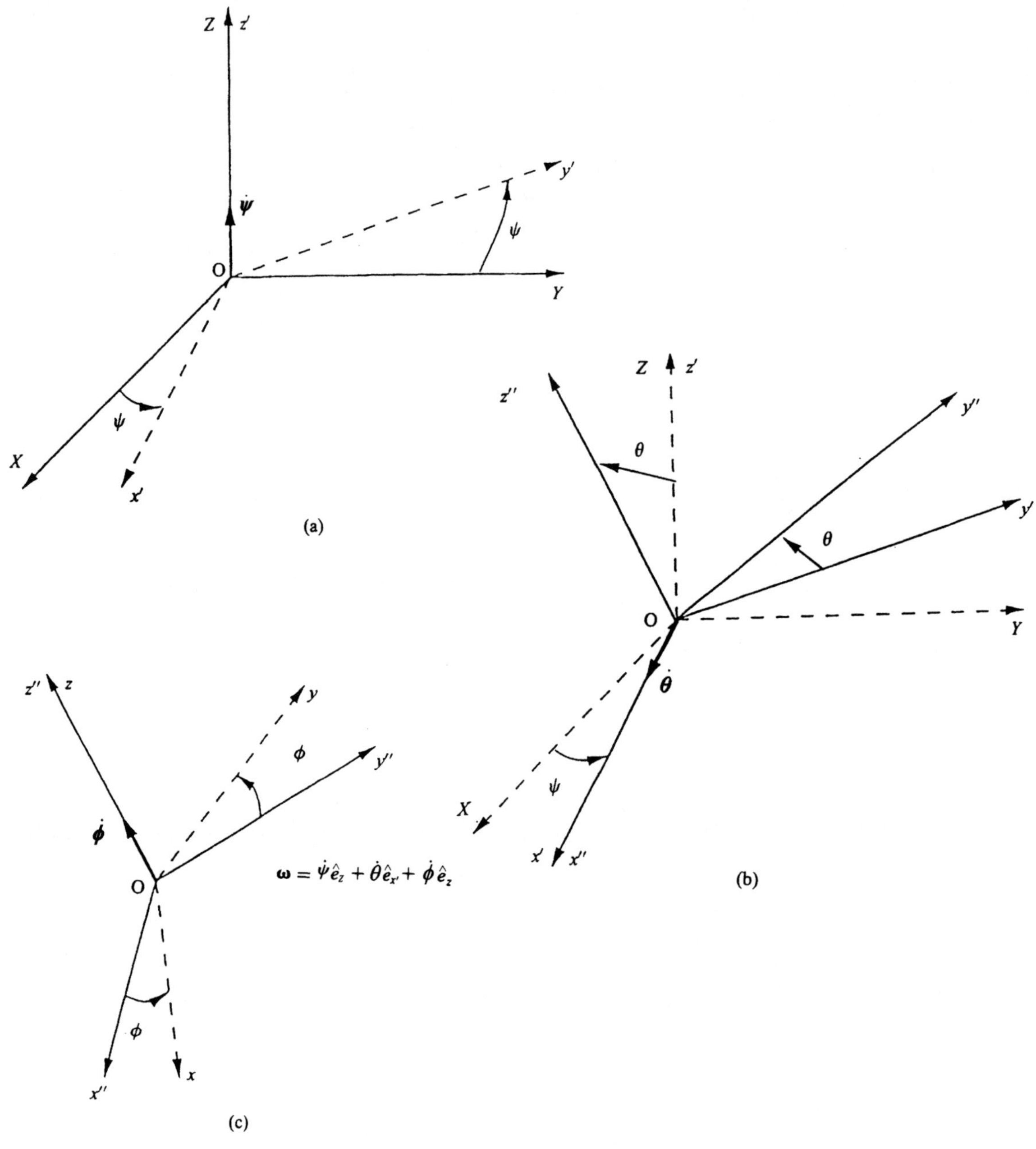

Fig. 1.27 Definition of Euler angles ψ, θ, and ϕ and corresponding angular velocities. Successive rotations are: (a) about OZ with angular velocity $\dot{\boldsymbol{\psi}}$, precession; (b) about Ox' with angular velocity $\dot{\boldsymbol{\theta}}$ nutation; and (c) about Oz'' with angular velocity $\dot{\boldsymbol{\phi}}$, spin.

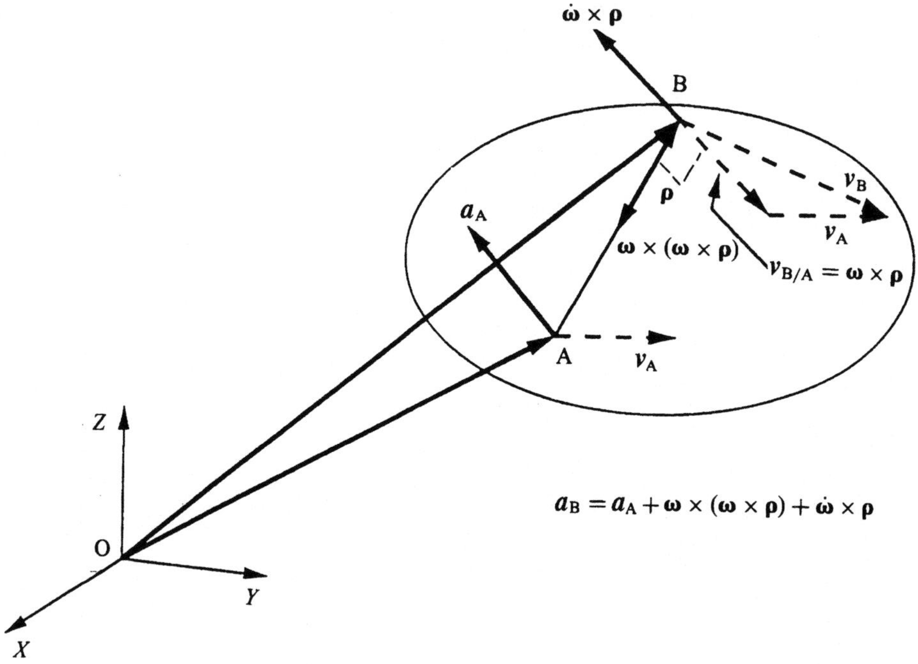

Fig. 1.28 Relative motion of two points on a rigid body.

this situation, there will exist an instant center of rotation C, a point of zero velocity (but not necessarily zero acceleration) which moves with time. As a further specialization, rotation may be absent in the case of pure translation, when all moments on the system must vanish; also, many cases of fixed-axis rotation occur where C remains permanently fixed.

1.2.6 Kinetics of rigid bodies

The equations governing the kinetics of rigid bodies are readily obtained from those for a system of particles for the special case when the distance between any two arbitrary particles remains permanently fixed. Equation (1.53) or the first of eqns (1.57) also applies directly for the motion of the mass center of a rigid body. However, the second of eqns (1.57) or, equivalently, $\boldsymbol{M}_G = \dot{\boldsymbol{H}}_G$ (only moment equations referred to the mass center or a fixed point are used) can be simplified due to the constraints in a rigid body. The absolute velocity of any particle P of the body is $\boldsymbol{v}_P = \boldsymbol{\omega} \times \boldsymbol{r}$, where $\boldsymbol{r} = \overline{OP}$ is the vector to P from fixed point O and $\boldsymbol{\omega}$ is the instantaneous angular velocity; relative velocity $\boldsymbol{v}_{P/G} = \boldsymbol{\omega} \times \boldsymbol{r}_{P/G}$. The angular momentum about either of these points can then be expressed by

$$\begin{aligned} \boldsymbol{H} = {} & (I_{xx}\omega_x - I_{xy}\omega_y - I_{xz}\omega_z)i \\ & + (-I_{yx}\omega_x + I_{yy}\omega_y - I_{yz}\omega_z)j \\ & + (-I_{zx}\omega_x - I_{zy}\omega_y + I_{zz}\omega_z)k, \end{aligned} \tag{1.67}$$

where the inertial quantities have been calculated with respect to a Cartesian system with an origin at the reference point. If principal axes are permanently embedded in the body either at a fixed point O or the mass center G, the angular momentum simplifies to the form

$$\boldsymbol{H} = I_{xx}\omega_x i + I_{yy}\omega_y j + I_{zz}\omega_z k, \tag{1.68}$$

where the directions i, j, k (or x, y, z) are now principal axes. However, this also involves the use of a rotating reference frame that permits the retention of fixed magnitudes for the principal moments instead of making them time-dependent. The angular relation must therefore be expressed as

$$\boldsymbol{M} = \left.\frac{\partial \boldsymbol{H}}{\partial t}\right|_{XYZ} + \boldsymbol{\Omega} \times \boldsymbol{H}, \tag{1.69}$$

where the first term on the right-hand side is the change

in $\boldsymbol{H}$ in an inertial system and the second term is the contribution due to the rotation of the reference axes. With the aid of eqn (1.69), there result the so-called Euler equations, valid for a coordinate system fixed at O or G with permanent principal axes xyz embedded in the body:

$$M_x = I_{xx}\dot{\omega}_x + (I_{zz} - I_{yy})\omega_y\omega_z;$$
$$M_y = I_{yy}\dot{\omega}_y + (I_{xx} - I_{zz})\omega_z\omega_x; \quad (1.70)$$
$$M_z = I_{zz}\dot{\omega}_z + (I_{yy} - I_{xx})\omega_y\omega_x.$$

If the body has an axis of symmetry, z, then the last of eqns (1.70) reduces to $M_{zz} = I_{zz}\dot{\omega}_z$. The energy relation for this case can be expressed in the form

$$W\Big|_1^2 = \int_1^2 \boldsymbol{F}\cdot\boldsymbol{v}\,\mathrm{d}t + \int_1^2 \boldsymbol{M}\cdot\boldsymbol{\omega}\,\mathrm{d}t = [W_{\mathrm{tr}} + W_{\mathrm{rot}}]\Big|_1^2$$
$$= [\tfrac{1}{2}\boldsymbol{v}_{\mathrm{G}}\cdot\boldsymbol{L} + \tfrac{1}{2}\boldsymbol{\omega}\cdot\boldsymbol{H}_{\mathrm{G}}]\Big|_1^2$$
$$= [T_{\mathrm{tr}} + T_{\mathrm{rot}}]\Big|_1^2 = \Delta T\Big|_1^2. \quad (1.71)$$

Power, the time rate of doing work, is $P \equiv \mathrm{d}W/\mathrm{d}t = Fv$ (if $\boldsymbol{M} = 0$ and $\boldsymbol{F}$ is parallel to $\boldsymbol{v}$), or similarly, $P = M\boldsymbol{\omega}$ (if $\boldsymbol{F} = 0$ and the direction of $\boldsymbol{M}$ is parallel to $\boldsymbol{\omega}$) expressed in kilowatts (kW) or horsepower (hp) with 1 hp = 0.746 kW (kJ s^{-1}).

For plane motion relative to a surface x–y, eqns (1.53) and (1.70) become

$$F_x = ma_x; \quad F_y = ma_y; \quad F_z = 0$$
$$\text{and} \quad M_x = M_y = 0; \quad M_z = I_{zz}\dot{\omega}_z \equiv I\alpha \quad (1.72)$$

provided that the axis of rotation is a principal axis so that eqns (1.70) (Euler's equations) apply. If the axis is not a principal axis, the moment equations read:

$$M_x = -I_{xz}\dot{\omega}_z + I_{yz}\omega_z^2; \quad M_y = -I_{yz}\dot{\omega}_z - I_{xz}\omega_z^2;$$
$$M_z = I_{zz}\dot{\omega}_z = I_{zz}\alpha. \quad (1.73)$$

Equations (1.73) may be used to analyze the joint forces and moments required to produce observed angular velocities and accelerations as exemplified by the case shown in Fig. 1.21 when the lengths and masses of the limbs and the mass distribution are known (Drillis and Contini 1966). Values of such forces and moments exerted by humans in sports activities are given by Plagenhoef (1971).

Several simplifications of plane motion can be considered. For pure translation, $\boldsymbol{M}_{\mathrm{G}} = \boldsymbol{M}_z = 0$; for centroidal rotation about a fixed axis passing through O,

$$\sum_{i=1}^{n} F_{\mathrm{o}_x} = \sum_{i=1}^{n} F_{\mathrm{o}_y} = 0$$

and

$$M_{\mathrm{o}} = I\alpha \text{ whose time integral is } \int_0^t M_{\mathrm{o}}\,\mathrm{d}t = \Delta I\omega_z. \quad (1.74)$$

For non-centroidal rotation, such as waving a stiff arm with the shoulder considered as a fixed point (Fig. 1.29).

$$F_{\mathrm{n}} = ma_{\mathrm{n}} = mr_{\mathrm{G}}\omega^2; \quad F_{\mathrm{t}} = ma_{\mathrm{t}} = mr_{\mathrm{G}}\alpha;$$
$$M_{\mathrm{G}} = I_{\mathrm{G}}\alpha \quad \text{or} \quad M_{\mathrm{o}} = I_{\mathrm{o}}\alpha. \quad (1.75)$$

For general plane motion, $M_{\mathrm{G}} = I_{\mathrm{G}}\alpha$ may be replaced by $M_{\mathrm{C}} = I_{\mathrm{C}}\alpha$, where C is the instant center of the body, such as the contact point of a rolling cylinder.

The work–energy relations for plane motion simplifies to

$$W|_1^2 = [W_{\mathrm{tr}} + W_{\mathrm{rot}}]|_1^2 = [T_{\mathrm{tr}} + T_{\mathrm{rot}}]|_1^2 = [\tfrac{1}{2}mv_{\mathrm{G}}^2 + \tfrac{1}{2}I_{\mathrm{G}}\omega^2]|_1^2. \quad (1.76)$$

The impulse–momentum equations remain essentially the same as for the three-dimensional case. However, in the impact of bodies of finite size, if the normal to the surfaces does not pass through both centers of mass, then both linear and angular momentum conservation equations must be written relative to the motion of the mass center G. For example, as shown in Fig. 1.30, if a baseball bat whose initial angular velocity is ω_0 strikes a ball whose initial velocity is v_{1_0} at point Q, the correct kinematic relation, with the ball treated as a particle, is

$$v_{2_{\mathrm{Q}}} = v_{2_{\mathrm{G}}} + d\omega_0 \quad (\text{where } d \text{ is distance GQ}). \quad (1.77)$$

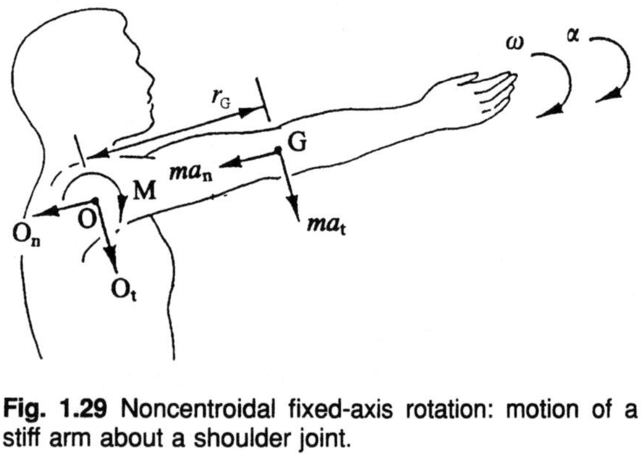

Fig. 1.29 Noncentroidal fixed-axis rotation: motion of a stiff arm about a shoulder joint.

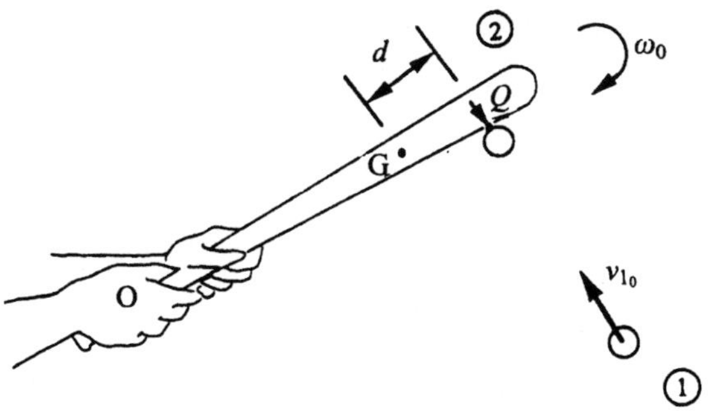

Fig. 1.30 Noncentral impact: baseball struck by a bat.

The kinetic relations applied to the compression phase up to time t_1 are

$$m_1 v_{1_0} - m_1 v_{1_Q} = m_2 v_{2_G} - m_2 v_{2_{G_0}} = \int_1^2 F_n \, dt;$$
$$I_{2_G}(\omega - \omega_0) = d \int_1^2 F_n \, dt, \tag{1.78}$$

and a similar set of equations applies for the restitution phase. This type of analysis is required to determine the moments at the shoulder joint needed to produce a given drive for the baseball.

1.2.7 Vibrations of a single-degree-of-freedom system

The term single-degree-of-freedom means that only one coordinate, either a distance or an angle, is needed to uniquely specify the position of a system; multi-degree-of-freedom systems require more coordinates. The model discussed here, shown schematically in Fig. 1.31(a), is an idealization of a distributed mass configuration, such as the biomechanical example of Fig. 1.31(b). It is analyzed as a lumped-parameter system consisting of (point) mass m subject to displacement x, a spring, regarded (here) as linear with constant k, and a dashpot exhibiting viscous damping with constant c, resulting in damping force $F_d = c\dot{x}$. When acted upon by an external time-dependent force $F\langle t \rangle$, the equation of motion is

$$F\langle t \rangle - c\dot{x} - kx = m\ddot{x} \quad \text{or} \quad m\ddot{x} + c\dot{x} + kx = F\langle t \rangle. \tag{1.79}$$

If there is no force acting, the system executes free vibrations due to an initially imposed displacement or velocity (or both); if there is no dashpot, the system is said to be undamped. The free vibration of an undamped system is obtained from eqn (1.79) by setting $F\langle t \rangle$ and c equal to zero to yield

$m\ddot{x} + kx = 0$ whose solution is

$$x = x_0 \cos \omega t + \frac{\dot{x}_0}{\omega} \sin \omega t, \tag{1.80}$$

where $\omega = [k/m]^{1/2}$ is the natural frequency of undamped vibrations of the system and x_0 and $\dot{x}_0$ are the initial displacement and velocity, respectively. The solution can also be written as

$$x = C \sin(\omega t - \phi) \quad \text{with} \quad C = \left[x_0^2 + \left(\frac{\dot{x}_0}{\omega}\right)^2\right]^{1/2}$$
$$\text{and} \quad \phi = \tan^{-1}(x_o\omega/\dot{x}_o) \tag{1.81}$$

or, in complex form, as

$$x = C\,\mathrm{Re}[e^{i(\omega t - \phi)}], \tag{1.82}$$

where C is the amplitude and ϕ is the phase angle. This simple harmonic motion is shown in Fig. 1.32 as a function of time; the period τ represents the time between corresponding displacements and the circular frequency (in Hz) is given by $f = (1/\tau) = (\omega/2\pi)$, with ω in rad s^{-1}. This motion can also be described by the energy relation $T + V = \text{constant} = E$. From eqn (1.82), $\dot{x}_{max} = \omega x_{max}$; thus $T_{max} = V_{max}$ and $T_{max} = \frac{1}{2}m(x_{max}\omega)^2$ when $V = 0$ and $V_{max} = \frac{1}{2}kx_{max}^2$ when $T = 0$.

The equation of motion for the free damped case is given by eqn (1.79) with $F\langle t \rangle = 0$. The solution can be obtained by assuming its form to be $x = Be^{st}$ with B as the amplitude and exponent s to be so determined as to satisfy the equation of motion. Substitution of this value of x in eqn (1.79) with $F\langle t \rangle = 0$ satisfies the equation provided there are two solutions for $x(t)$, i.e.

$$x\langle t \rangle = B_1 e^{s_1 t} + B_2 e^{s_2 t} \quad \text{and}$$
$$s_{1,2} = [-c \pm (c^2 - 4mk)^{1/2}]/2m \quad \text{(in general).} \tag{1.83}$$

If (i) $c^2 - 4mk > 0$, s_1 and s_2 are real and negative. If the initial conditions are $x\langle 0 \rangle = x_0$ and $\dot{x}\langle 0 \rangle = \dot{x}_0$ at time $t = 0$, then $B_1 = (-s_2 x_0 + \dot{x}_0)/(s_1 - s_2)$ and $B_2 = (-s_1 x_0 + \dot{x}_0)/(s_1 - s_2)$. This motion, shown in Fig. 1.33 for different initial conditions, is aperiodic and represents overdamping, where damping predominates over inertial effects and the displacement asymptotically approaches zero as time increases indefinitely.

If (ii) $c \equiv c_c = 2(mk)^{1/2}$ then $s_1 = s_2 = c/2m = -\omega_{\hat{c}}$, the general solution, eqn (1.83), does not hold, because

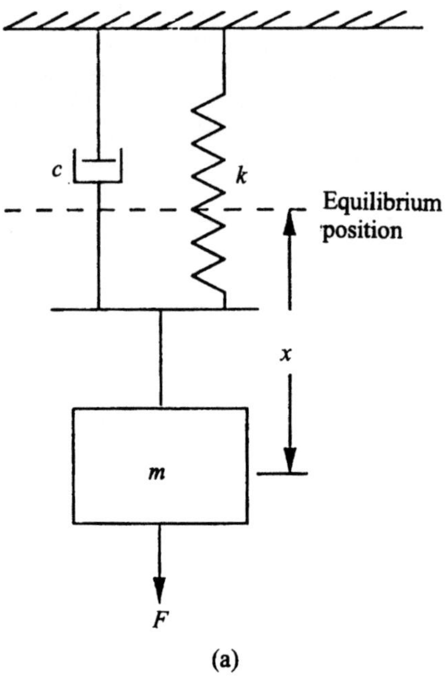

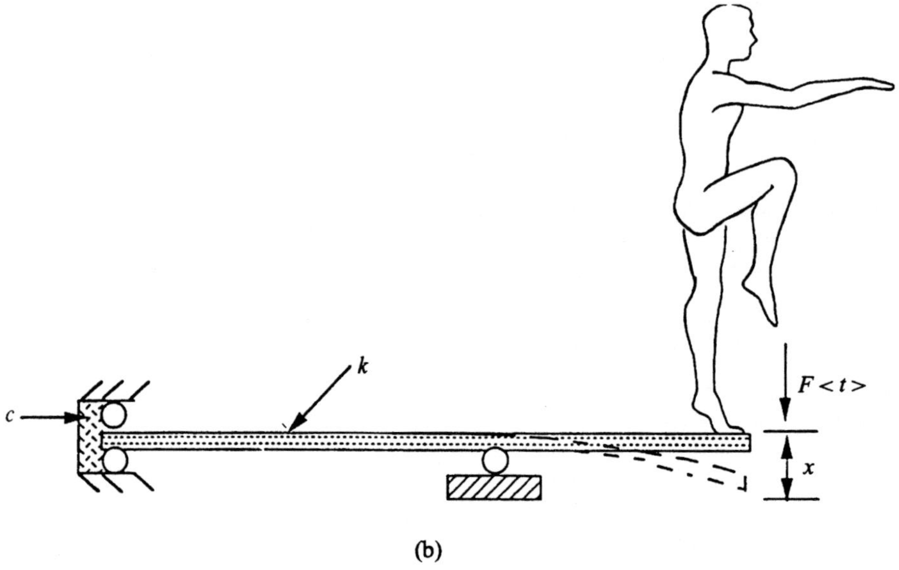

Fig. 1.31 Single-degree-of-freedom lumped parameter system: (a) schematic; (b) example; diver on a high board.

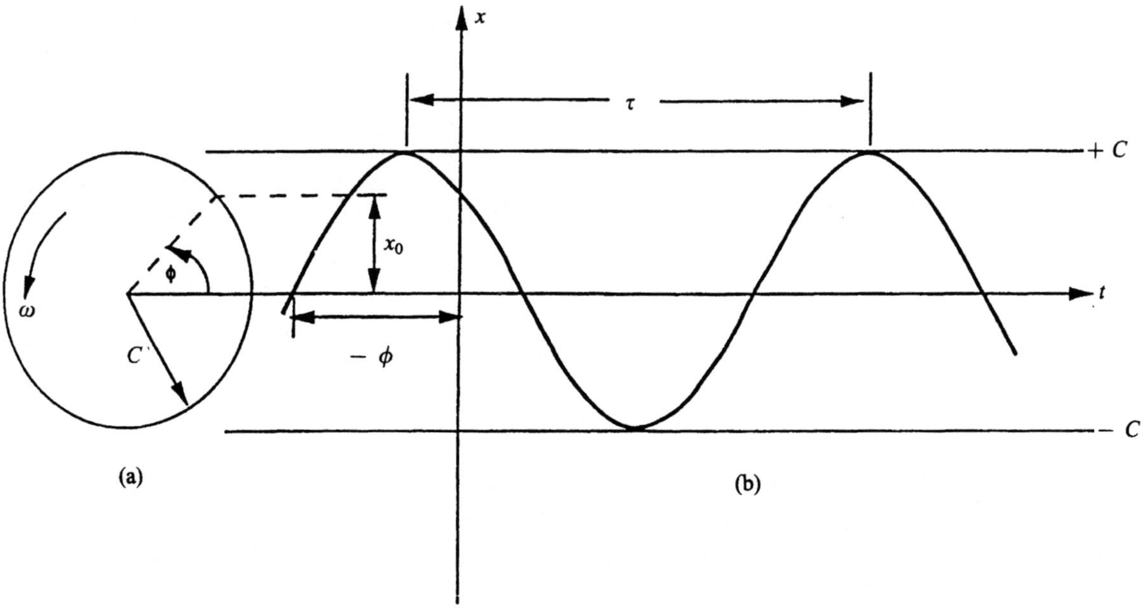

Fig. 1.32 Displacement history of a free, undamped single-degree-of-freedom system.

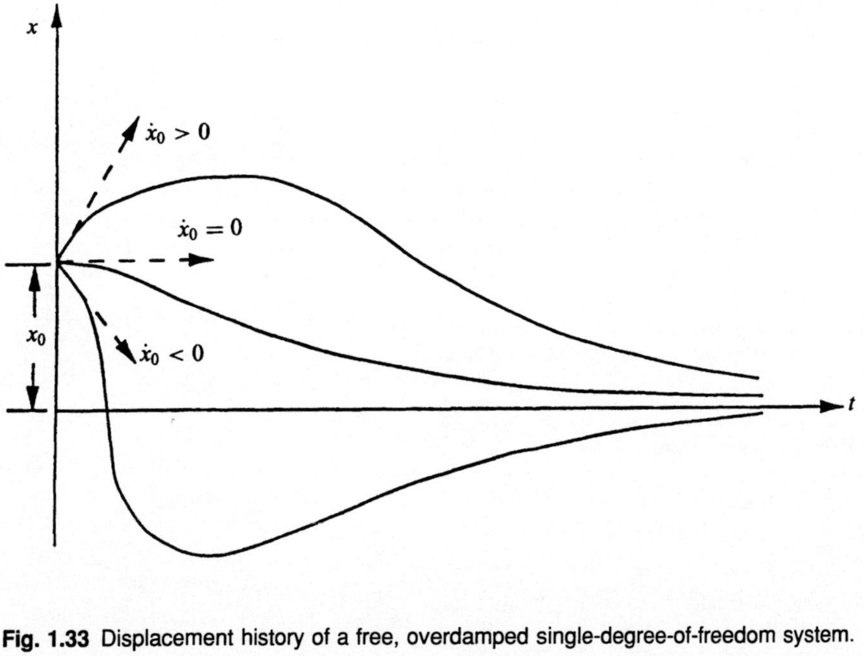

Fig. 1.33 Displacement history of a free, overdamped single-degree-of-freedom system.

the two roots of s are not independent. Then the displacement takes on the singular form

$$x\langle t\rangle = B_1 \mathrm{e}^{-\omega t} + B_2 t\mathrm{e}^{-\omega t} \quad \text{with} \quad B_1 = x_0$$
$$\text{and } B_2 = \dot{x}_o + \omega x_0. \tag{1.84}$$

This aperiodic motion is the limiting case of (i) and may also be considered to be represented by Fig. 1.33.

If (iii) $c^2 - 4mk < 0$, the roots s_1 and s_2 are complex conjugates. It is convenient to change the constants to read: $b = c/2m$ and $q = (\omega^2 - [c/2m]^2)^{1/2} = (\omega^2 - c_c^2)^{1/2}$. The solution for this case may then be written as

$$x\langle t\rangle = \mathrm{e}^{-bt}(B_1 \sin qt + B_2 \cos qt) = C\mathrm{e}^{-bt}\sin(qt + \phi), \tag{1.85}$$

where $B_1 = (\dot{x}_0 + bx_0)/q$, $B_2 = x_0$, $C = (B_1^2 + B_2^2)^{1/2}$, and phase angle $\phi = \tan^{-1}(B_2/B_1)$. As shown in Fig. 1.34, the displacement oscillates with decreasing amplitude between the limit curves $\pm C\mathrm{e}^{-bt}$; inertia predominates, although damping returns the displacement towards equilibrium with increasing time. The damping frequency, q, whose magnitude is less then ω, can be measured from a record of the decay of such a system, as can the ratio of two successive peaks, $x_1/x_2 = \mathrm{e}^{b\tau}$, where τ is the measured period. Term $\delta \equiv \ln(x_1/x_2) = b\tau$ is called the logarithmic decrement, and $\delta = 2\pi\xi/(1-\xi^2)^{1/2}$, where $\xi = c/c_c$ is the damping parameter, with c_c defined above; then $\xi = \delta/([2\pi]^2 + \delta^2)^{1/2}$. The treatment presented above can be used to approximately describe the vibrations of reeds produced by a gust of wind, the motion of a tree branch displaced from equilibrium, or the oscillations of an acrobat swinging from a support high above the ground.

Forced vibrations are analyzed by separating the displacement x into two components, $x = x_h + x_p$. Here, x_h is the solution of the homogeneous differential equation (with $F\langle t\rangle = 0$) for the free system motion, and x_p, the particular solution, is *any* solution of the complete differential equation; the specific form of x_p depends on $F\langle t\rangle$. The following cases will be considered here:

(i) $F\langle t\rangle$ = constant = A. The solution for the undamped and damped cases are, respectively,

$$\begin{aligned} x\langle t\rangle &= B_1 \sin \omega t + B_2 \cos \omega t + A/k \\ x\langle t\rangle &= B_1 \mathrm{e}^{s_1 t} + B_2 \mathrm{e}^{s_2 t} + A/k, \end{aligned} \tag{1.86}$$

where B_1 and B_2 are determined from initial conditions x_0 and $\dot{x}_0$ at $t = 0$ as previously indicated.

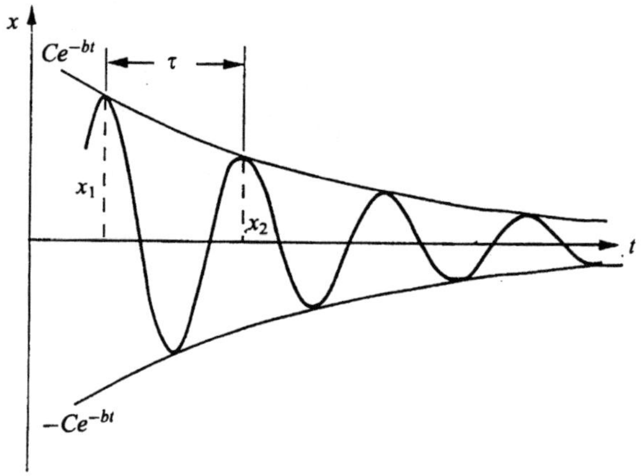

Fig. 1.34 Displacement history of a free, underdamped single-degree-of-freedom system (damped oscillation).

(ii) $F\langle t\rangle = F_0 \sin \Omega t$, varying harmonically with forcing frequency Ω. For x_p, the solution will be assumed in the form

$$x_p = A \sin \Omega t + B \cos \Omega t, \tag{1.87}$$

where A and B must be determined to satisfy eqn (1.79); this is done by equating the coefficients of terms $\sin \Omega t$ and $\cos \Omega t$ separately to zero, yielding

$$A = \frac{\left(\frac{F_0}{m}\right)\left(\frac{k}{m} - \Omega^2\right)}{\left[\left(\frac{k}{m} - \Omega^2\right)^2 + \left(\frac{c\Omega}{m}\right)^2\right]^{1/2}}$$

and

$$B = \frac{-\left(\frac{F_0}{m}\right)\left(\frac{c\Omega}{m}\right)}{\left[\left(\frac{k}{m} - \Omega^2\right)^2 + \left(\frac{c\Omega}{m}\right)^2\right]^{1/2}}; \tag{1.88}$$

$$\begin{aligned} x &= x_h + (A^2 + B^2)^{1/2} \sin(\Omega t - \phi) \\ &= \mathrm{e}^{-bt}(B_1 \sin qt + B_2 \cos qt) \\ &\quad + \frac{(F_0/m)\sin(\Omega t - \phi)}{[(\omega^2 - \Omega^2)^2 + (c\Omega/m)^2]^{1/2}}. \end{aligned} \tag{1.89}$$

Terms B_1 and B_2 are evaluated from the initial conditions.

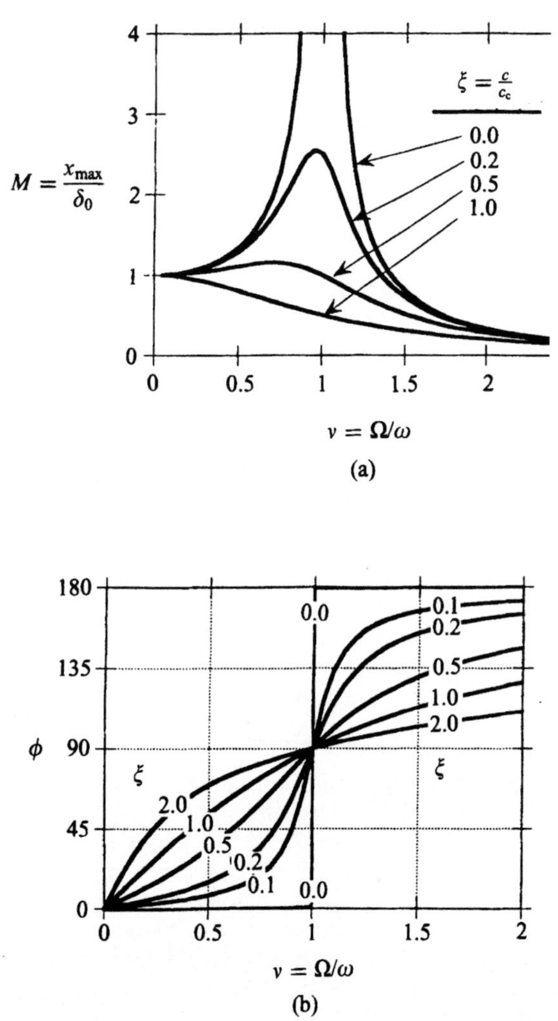

Fig. 1.35 Response of a damped single-degree-of-freedom system to forced harmonic excitation: (a) magnification factor; (b) phase angle.

It is customary to consider only the steady-state motion represented by the particular solution, since in practice the transient, from the homogeneous part of the solution of eqn (1.79), will vanish with time due to damping, even if the system is modeled as 'undamped'. If the static deflection due to the applied force, $\delta_0 = F_0/k$, is introduced into the second term of eqn (1.89), then the magnification factor M defining the ratio of maximum deflection x_{max} to δ_0 is given by

$$M \equiv x_{max}/\delta_0 = [(1 - \nu^2)^2 + (2\xi\nu)^2]^{-1/2}$$
$$\text{and } \phi = \tan^{-1}\left(\frac{2\xi\nu}{1-\nu^2}\right) \qquad (1.90)$$

where $\nu = \Omega/\omega$ is the frequency ratio and $\xi = c/c_c$ is the damping ratio.

The magnification factor M and phase angle ϕ are presented in Fig. 1.35(a) and (b), respectively, for various values of ξ. It should be noted that $\xi = 0$ corresponds to the undamped case; for this condition, $\nu = 1$ represents 'resonance', where the amplitude of the system increases without limit as time progresses, according to the relation $x = \frac{1}{2}\delta_0 t$. Such a situation, even with some damping present, is considered to be undesirable. For the undamped case, force and deflection are in phase up to resonance and 180° out of phase beyond.

(iii) $F = m_1 e\Omega^2 \sin \Omega t$, a force that arises from rotation of a motor with an unbalanced mass m_1 at distance e from the axis. The solution is obtained by replacing F_o by $m_1 e\Omega^2$ in (ii), yielding the same relation for phase angle ϕ, but a new magnification factor given by

$$M = (\nu^2)/[(1 - \nu^2)^2 + (2\nu\xi)^2]^{1/2} \qquad (1.91)$$

as shown in Fig. 1.36.

Another parameter of importance is the transmissibility TR, defined as the ratio of the maximum

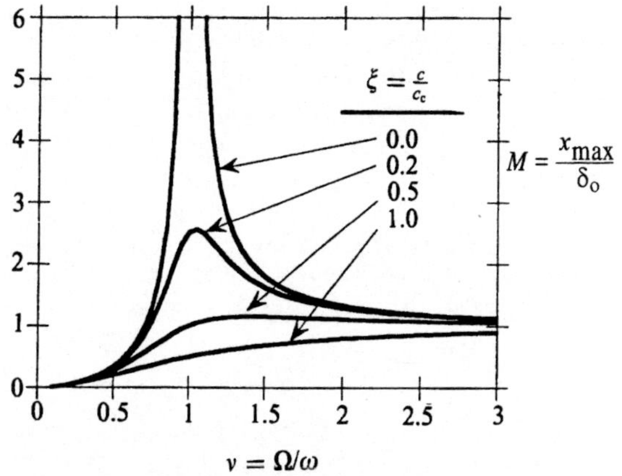

Fig. 1.36 Magnification factor for a damped single-degree-of-freedom system subjected to a harmonic force with frequency-squared amplitude dependence.

transmitted to the applied force, given for case (ii) by

$$TR = \frac{[(kx)^2 + (cx\Omega)^2]^{1/2}}{kx\left[\left(1-\left[\frac{\Omega}{\omega}\right]^2\right)^2 + \left(\frac{c\Omega}{m}\right)^2\right]^{1/2}} = \frac{[1+(2\nu\xi)^2]^{1/2}}{[(1-\nu^2)^2 + (2\nu\xi)^2]^{1/2}}. \quad (1.92)$$

Forced oscillations with bioengineering applications include floor vibrations, human–machine interactions such as a jackhammer operation, as well as ship or airplane motions which, in certain frequency ranges, disable personnel by upsetting their stomachs. These situations can be modeled at an elementary level as systems with a single degree-of-freedom.

The material presented above is the simplest representation of periodic phenomena involving discrete objects. In general, physical and biomechanical systems must be described by more than a single degree-of-freedom, and also involve different damping mechanisms and a wider class of applied forces. Treatments ranging from elementary to advanced can be found in a multiplicity of references (Steidel 1979; Minorski 1983; Dimarogonas and Haddad 1992).

1.3 Mechanics of deformable bodies—continuum mechanics

1.3.1 Fundamental concepts

When the deformation of a body under load is the principal concern of the analysis or significantly affects the overall response of the object, the idealization of infinitely stiff members can no longer be employed. As in the case of rigid bodies, such deformations result either from surface loading (contact forces) or from actions at a distance, such as gravitational attraction (body forces) whose magnitude is proportional to the mass of the body. It was shown earlier that the concept of stress arises from the need for applying an appropriate set of internal forces to a member upon sectioning in order to maintain equilibrium. The forces applied to the cross-sectional area A are either normal (N) [tension (T) or compression (C)], or else tangential (V), creating shear; they produce either normal or shear stresses denoting internal load intensity.

The stresses are usually referred to the original section A; they are then also called engineering stresses. When the load distribution over area A is uniform, the tensile, compressive and shear stresses are given by

$$\begin{aligned} \tau_{xx} &= T/A \text{ tension;} \\ \tau_{xx} &= -C/A \text{ compression;} \\ \tau_{xy} &= V/A \text{ shear.} \end{aligned} \quad (1.93)$$

The sign convention for the normal stresses is arbitrary, but is suggested by the corresponding changes in length, ΔL for (T) and $-\Delta L$ for (C). It is also possible to define the load intensity in terms of the current cross-sectional area A^*, i.e.

$$\tau^*_{xx} = T/A^*; \quad \tau^*_{xx} = -C/A^*; \quad \tau^*_{xy} = V/A^*, \quad (1.94)$$

which are called true stresses. For small strains and hence small area changes, where $A^* \approx A$, as is often true, eqns (1.93) and (1.94) give identical results.

For deformable bodies, these stresses are accompanied by length changes ΔL (elongation) or $-\Delta L$ (contraction) for tension or compression, respectively; this is also shown in Fig. 1.37. In this simple case of uniaxial loading of a rod, corresponding normal engineering strains are defined as

$$\begin{aligned} \epsilon_{xx} &= (L^* - L)/L = \Delta L/L \text{ for tension;} \\ \epsilon_{xx} &= -\Delta L/L \text{ for compression,} \end{aligned} \quad (1.95)$$

where L^* is the current and L the initial length; shear strain will be specified later. When the reference is the current rod length L^*, the corresponding changes per unit length, called true strains, are then represented by

$$\begin{aligned} \epsilon^*_{xx} &= \Delta L/L^*; \text{ for tension;} \\ \epsilon^*_{xx} &= -\Delta L/L^* \text{ for compression.} \end{aligned} \quad (1.96)$$

The definitions in eqn (1.96) are primarily employed for large strains, where several other measures of strain have also been proposed and utilized.

For the most frequent occurrence of small strain

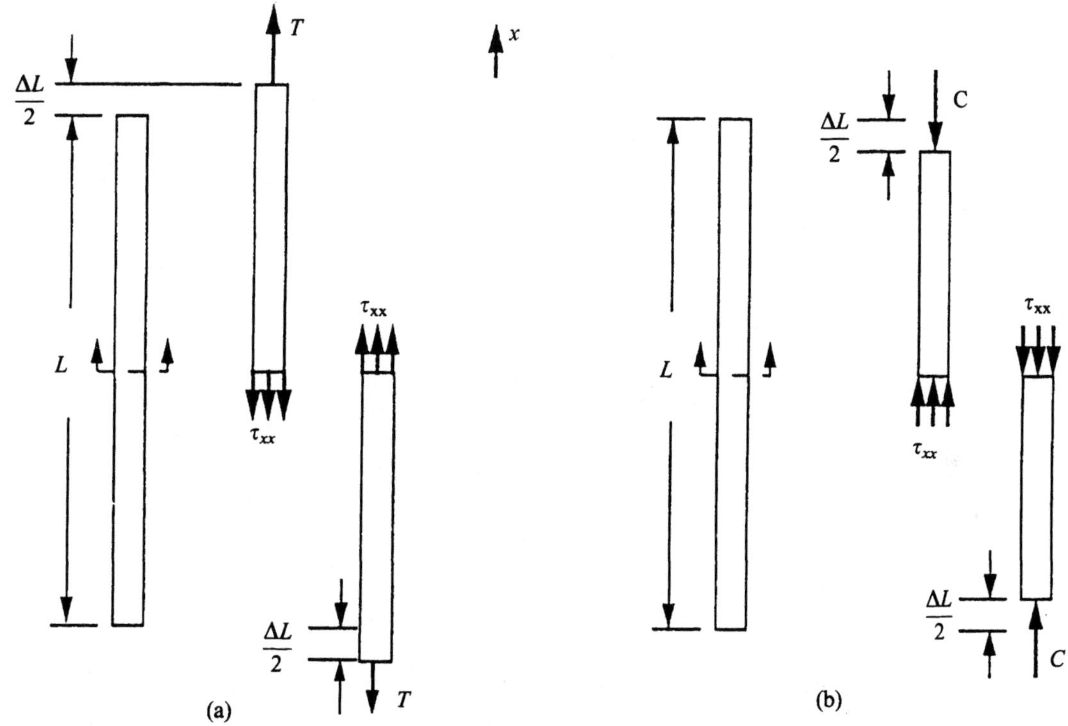

Fig. 1.37 Equilibrium and deformation of a two-force member at a section normal to the axis: (a) tensile loads (*T*) and stresses; (b) compressive loads (*C*) and stresses.

values, $\epsilon_{xx} \ll 1$, the strains described by eqns (1.95) and (1.96) are essentially identical, and the engineering designations are then customarily employed. Load intensities and the dimensionless strains are used to minimize the effect of sample size on material response; a relationship between these quantities (and their time derivatives), called a constitutive equation, defines the mechanical behavior of any material. This relation is required for analyzing the reaction of any deformable object to loading, but must be determined experimentally.

1.3.2 Stress

Figure 1.37 portrays the internal forces when the section is normal to the direction of the exterior loads. When the cut is oblique, at angle θ, as shown in Fig. 1.38 (for a compressive external load), equilibrium on area $A' = A/\cos\theta$ can only be maintained by the combination of normal forces and shear forces, N and V. If uniformly distributed, these loads produce engineering normal and shear stresses $\tau_{x'x'} = N/A'$ and $\tau_{x'y'} = V/A'$, given by

$$\begin{aligned} \tau_{x'x'} &= -(C\cos\theta)/A' = -(C/A)\cos^2\theta; \\ \tau_{x'y'} &= (C\sin\theta)/A' = (C/A)\sin\theta\cos\theta. \end{aligned} \qquad (1.97)$$

An extension of this concept leads to the representation of *all* components of the internal forces acting on a section of an object cut normal to a reference axis, as indicated in Fig. 1.39. The formal definitions for a normal stress τ_{xx} or a shear stress τ_{xy} refer to stresses acting on an infinitesimal area ΔA perpendicular to direction x, as shown in this diagram; these are given by

$$\tau_{xx} = \lim_{\Delta A \to 0} \frac{\Delta N}{\Delta A} \quad (a); \quad \tau_{xy} = \lim_{\Delta A \to 0} \frac{\Delta V}{\Delta A} \quad (b), \quad (1.98)$$

where the first subscript indicates the direction normal to the plane of the cut and the second subscript refers to the direction of the incremental load (as previously indicated in Section 1.1.10). Additional stresses may be defined by

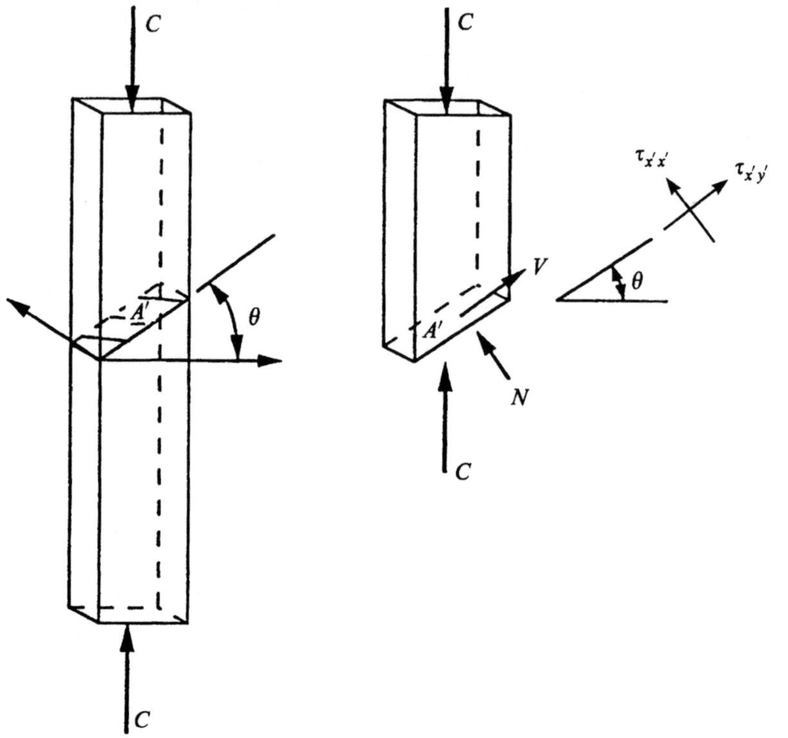

Fig. 1.38 Equilibrium of an axially loaded slender rod at a section cut obliquely to the axis involving normal and shear loads and stresses.

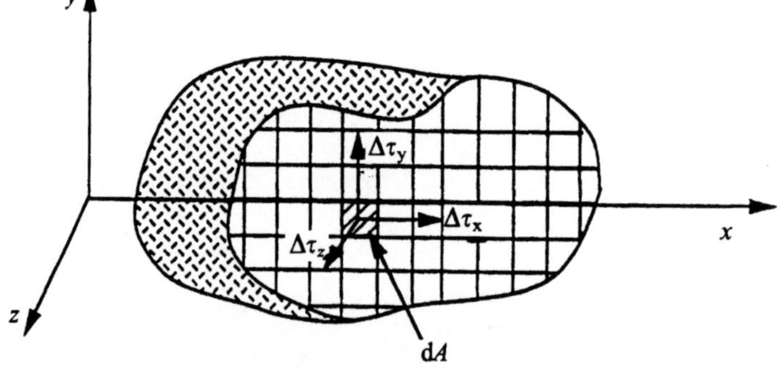

Fig. 1.39 Stress components in a deformable body on an internal section parallel to a coordinate plane.

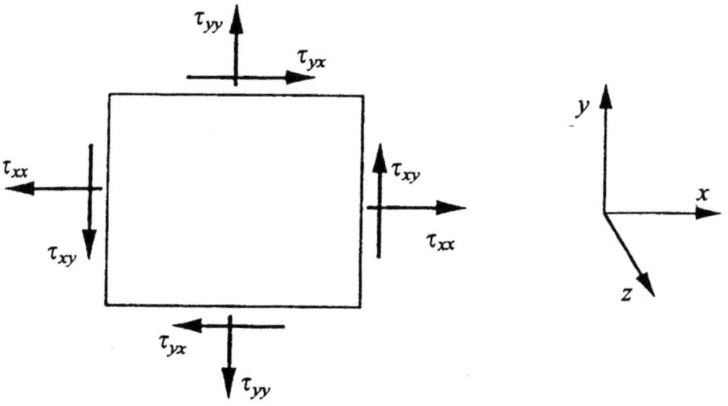

Fig. 1.40 Planar element in equilibrium under the action of normal and shear forces.

choosing the section perpendicular to the axes y and x, respectively.

A total of nine possible components of stress can be defined in this manner, involving three normal stress components τ_{xx}, τ_{yy}, and τ_{zz} (frequently represented elsewhere by a different symbol using a single subscript σ_x, σ_y, and σ_z), as well as the shear stresses τ_{ij} with subscripts i and j that take on successively the values x, y, and z, provided $i \neq j$. By considering a planar element in equilibrium, as shown in Fig. 1.40, it is evident that $\tau_{ij} = \tau_{ji}$, as otherwise rotation would occur; hence, $\tau_{xy} = \tau_{yx}$, $\tau_{xz} = \tau_{zx}$, and $\tau_{yz} = \tau_{zy}$. The stress components can be incorporated in an array known as the stress tensor in a manner that parallels the similar grouping for the moments and products of inertia. This tensor can be diagonalized to produce principal stresses, defined here by symbols σ_1, σ_2, and σ_3 acting on planes on which there is no shear stress

$$\begin{vmatrix} \tau_{xx} & \tau_{xy} & \tau_{xz} \\ \tau_{yx} & \tau_{yy} & \tau_{yz} \\ \tau_{zx} & \tau_{zy} & \tau_{zz} \end{vmatrix} \rightarrow \begin{vmatrix} \sigma_1 & 0 & 0 \\ 0 & \sigma_2 & 0 \\ 0 & 0 & \sigma_3 \end{vmatrix}, \qquad (1.99)$$

using the relation

$$(\tau_{ji} - \sigma_k \delta_{ji})\hat{e}_i = 0 \quad (i, j = x, y, z \text{ and } k = 1, 2, 3), \qquad (1.100)$$

where σ_k are the principal stresses and δ_{ji} is the Kronecker delta. It should be noted that the stress components have been defined in terms of a rectangular Cartesian coordinate system. However, the definitions apply to any orthogonal set of coordinates, such as the cylindrical or spherical system, and the notation is thus not dependent on any particular set of axes. This can be emphasized by use of the symbology τ_{ii}; τ_{ij} $(i \neq j)$ for stresses, and similarly for strains.

The stresses acting on an elemental volume $dx_1dx_2dx_3$ in a state of static equilibrium are shown in Fig. 1.41; when this stress field is unbalanced, motion will result. While the equations of motion for rigid bodies were represented by ordinary differential equations, they are now expressed by partial differential equations resulting from a force balance on a volume element such as shown in Fig. 1.41. In this case, for rectangular coordinates with stress gradients present in the three directions, the equations of motion are

$$\begin{aligned} \frac{\partial \tau_{xx}}{\partial x} + \frac{\partial \tau_{yx}}{\partial y} + \frac{\partial \tau_{zx}}{\partial z} + X &= \rho \ddot{u} \\ \frac{\partial \tau_{xy}}{\partial x} + \frac{\partial \tau_{yy}}{\partial y} + \frac{\partial \tau_{zy}}{\partial z} + Y &= \rho \ddot{v} \\ \frac{\partial \tau_{xz}}{\partial x} + \frac{\partial \tau_{yz}}{\partial y} + \frac{\partial \tau_{zz}}{\partial z} + Z &= \rho \ddot{w} \end{aligned} \qquad (1.101)$$

$$\frac{\partial \tau_{ji}}{\partial x_j} + X_i = \rho \ddot{x}_i \quad (i = 1, 2, 3), \qquad (1.102)$$

where u, v, w (or x_i) are the displacements, and X, Y, Z the body forces (force per unit volume, such as weight) along axes x, y, and z, respectively; ρ is the density of the material. Equation (1.102) is the more general form of the stress equations of motion valid for any orthogonal coordinate system and encompasses eqns (1.101). The right-hand sides of eqns (1.101) and (1.102) vanish for static loading and then represent static equilibrium. Furthermore, in two dimensions, say x and y, all stress components and dependence on z (such as a derivative) in eqns (1.101) vanish. The force balance for the x-direction of the three-dimensional case is shown in Fig. 1.42.

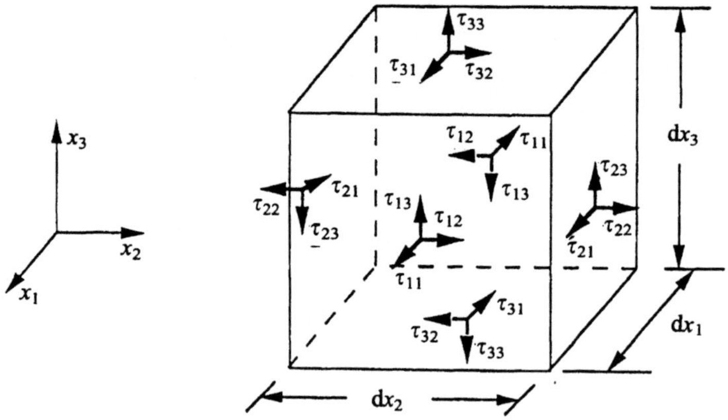

Fig. 1.41 Stresses acting on a volume element in equilibrium.

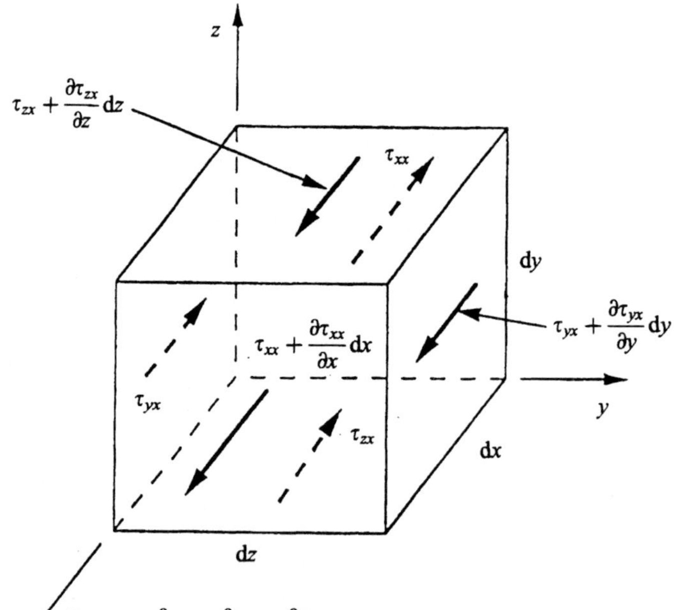

$$\frac{\partial \tau_{xx}}{\partial x} + \frac{\partial \tau_{yx}}{\partial y} + \frac{\partial \tau_{zx}}{\partial z} = \rho \ddot{x}$$

Fig. 1.42 Unbalanced stresses acting on an elemental cube in the *x*-direction.

In cylindrical coordinates, an orthogonal system, the stresses τ_{ij} assume the subscripts r, θ, and z. The relations between the components in rectangular and cylindrical coordinates are given by

$$\begin{aligned}
&\tau_{xx} = \tau_{rr}\cos^2\theta + \tau_{\theta\theta}\sin^2\theta - \tau_{r\theta}\sin 2\theta;\\
&\tau_{yy} = \tau_{rr}\sin^2\theta + \tau_{\theta\theta}\cos^2\theta + \tau_{r\theta}\sin 2\theta;\\
&\tau_{zz} = \tau_{zz};\ \tau_{xy} = (\tau_{rr} - \tau_{\theta\theta})\sin\theta\cos\theta\\
&\qquad + \tau_{r\theta}(\cos^2\theta - \sin^2\theta);\\
&\tau_{zx} = \tau_{zr}\cos\theta - \tau_{z\theta}\sin\theta;\\
&\tau_{zy} = \tau_{rz}\sin\theta + \tau_{z\theta}\cos\theta.
\end{aligned} \tag{1.103}$$

When all stress components in a certain direction, such as z, vanish, the body is in a state of plane stress; this occurs, for example, in a thin membrane. The stress tensor, the left-hand side of eqn (1.99), is now given by

$$\begin{vmatrix} \tau_{xx} & \tau_{xy} & 0 \\ \tau_{xy} & \tau_{yy} & 0 \\ 0 & 0 & 0 \end{vmatrix}. \tag{1.104}$$

The principal stresses σ_k are the roots of the cubic equation

$$-\sigma^3 + I_1\sigma^2 - I_2\sigma + I_3 = 0, \tag{1.105}$$

where I_1, I_2, and I_3, the so-called stress invariants, are defined by

$$\begin{aligned}
I_1 &\equiv \sigma_1 + \sigma_2 + \sigma_3;\\
I_2 &\equiv \sigma_1\sigma_2 + \sigma_2\sigma_3 + \sigma_3\sigma_1;\\
I_3 &\equiv \sigma_1\sigma_2\sigma_3.
\end{aligned} \tag{1.106}$$

The stress invariants are independent of the rotation of the coordinate system; their importance lies in their role in the interpretation of experimental results regarding the behavior of materials when forces are applied.

A set of stress deviators s_{ij} representing the difference in the loading on a surface relative to a mean stress S has been defined in terms of the tensor

$$\begin{aligned}
&s_{ij} \equiv \tau_{ij} - S\delta_{ij};\\
&S \equiv \tfrac{1}{3}(\sigma_1 + \sigma_2 + \sigma_3) = \tfrac{1}{3}I_1 = \tfrac{1}{3}(\tau_{11} + \tau_{22} + \tau_{33}).
\end{aligned} \tag{1.107}$$

A similar set of invariants J_k exists for the stress deviators (repeated indices i, j, or k in products are summed over the index):

$$\begin{aligned}
J_1 &= s_{11} + s_{22} + s_{33} = 0;\\
J_2 &= 3S^2 - I_2 = \tfrac{1}{2}s_{ij}s_{ij};\\
J_3 &= I_3 + J_2S - S^3 = s_{11}s_{22}s_{33}.
\end{aligned} \tag{1.108}$$

It has been established by many tests that yielding or similar behavior of homogeneous materials is not affected, within reasonable limits, by the presence of a hydrostatic pressure S. For this reason, such behavior is sometimes described in terms of the deviatoric stresses, where, in view of the first of eqns (1.108), the hydrostatic components are eliminated.

Forces or moments are categorized by the nature of the deformation that they produce. The ratio of the limiting load or stress—defined as that producing either failure or a specified amount of temporary or permanent deformation—to the applied load or stress is called the factor of safety, which must be greater than unity, and frequently substantially so.

1.3.3 Deformation

Two types of uniaxial (or normal) strain ϵ_{ii} have already been defined for a slender rod loaded by tensile or compressive forces (Fig. 1.37), quantified by means of eqns (1.95) which apply for either large or small (infinitesimal) strains. As a practical matter, strains less than 1 per cent may always be regarded as small; this bound may be increased to 10 per cent in certain cases. Other deformations due to simple loading situations are illustrated in Fig. 1.43. Figure 1.43(a) portrays the distortion of a section due to the application of a set of parallel, oppositely directed forces, or couple, producing shearing; the resulting angle of inclination is a measure of the shear strain. A more general definition of shear will be presented subsequently. In Fig. 1.43(b), a straight bar loaded by a moment M (whose vector direction is normal to the bar axis) produces bending; here, elements at the top are in tension and those at the bottom in compression. Twisting of a cylinder by a torque T (a vector similar to a moment, where, however, the unit vector points along the axis of the cylinder) produces torsion, shown in Fig. 1.43(c); the distortion of the elements is similar to that of Fig. 1.43(a). The compression of a very thin rod (column) due to compressive forces may produce buckling, shown in Fig. 1.43(d), which, when excessive, can lead to collapse.

Although not required conceptually, the rational definition of infinitesimal strain in three dimensions

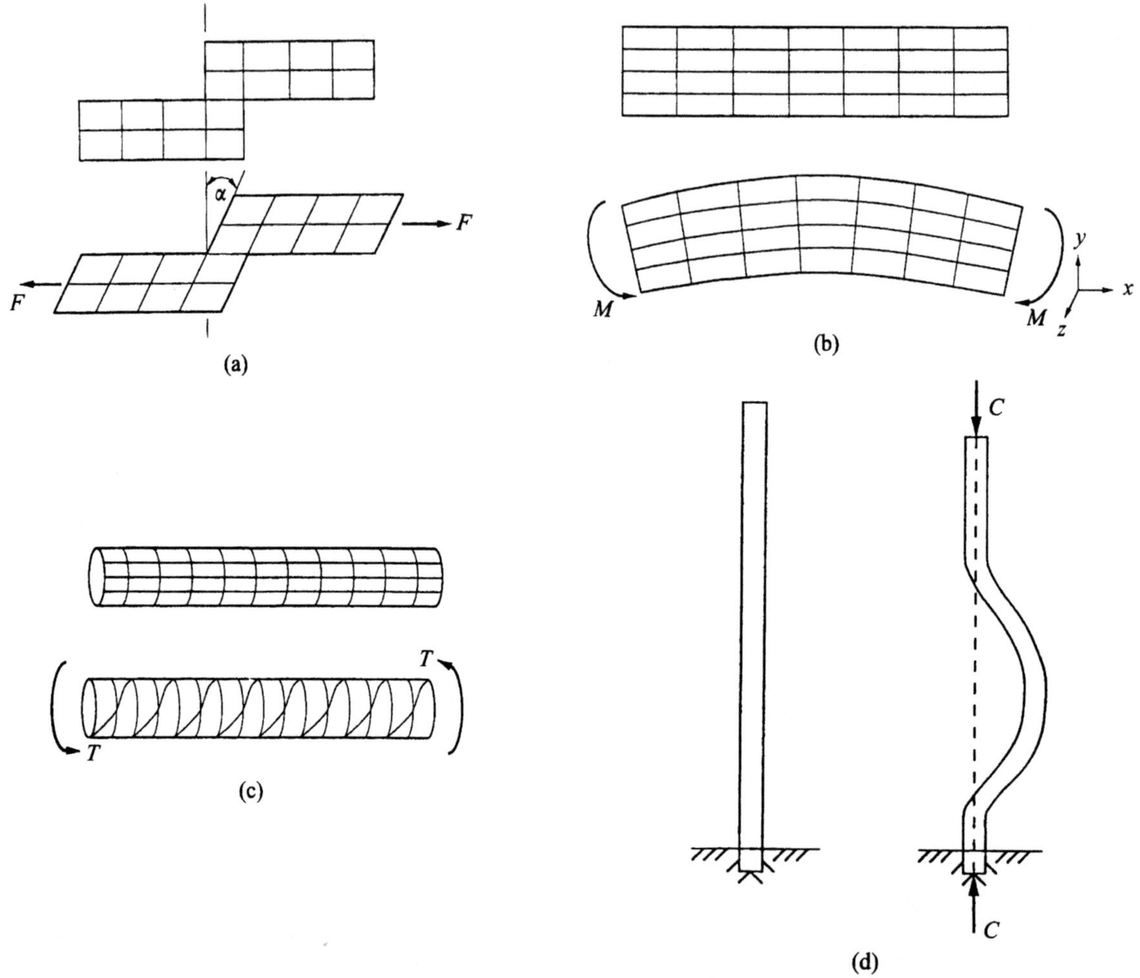

Fig. 1.43 Deformations due to simple loading: (a) shearing due to a couple produced by two parallel and equal forces F; (b) bending of a beam due to moment M; (c) torsion of a bar due to twist or torque T; (d) buckling of a slender column due to axial compressive forces C.

requires consideration of the deformation of a volume element V_0 containing two neighboring points C and D, initially positioned at C_0 and D_0, and separated by infinitesimal displacement d$\boldsymbol{X}$. The position vector $\boldsymbol{X}$ of C_0 has components X_1, X_2, and X_3 with respect to the coordinate origin O of an inertial frame. When deformed, the volume element becomes V_f, the position of C_0 is translated by vector $\boldsymbol{r}$ to C_f whose position from O is $\boldsymbol{x}$, and that of D_0 by $\boldsymbol{r}^*$ to D_f, as illustrated in Fig. 1.44. The original infinitesimal distance $\overline{C_0D_0} = \mathrm{d}\boldsymbol{X}$ has been distorted into line $\overline{C_fD_f} = \mathrm{d}\boldsymbol{x}$; this involves both a change in length (either stretching or compression) and a rotation of the initial element d$\boldsymbol{X}$. The relations between the vector components are

$$x_i = X_i + r_i; \quad r_i + \mathrm{d}x_i = r_i^* + \mathrm{d}X_i \quad (i = 1, 2, 3). \tag{1.109}$$

Differentiation of the first of eqns (1.109) and substitution in the second yields

$$\mathrm{d}x_i = \mathrm{d}X_i + \mathrm{d}r_i; \quad r_i^* = r_i + \mathrm{d}r_i. \tag{1.110}$$

Using only the first term of a Taylor expansion, dr_i may be written as

$$\mathrm{d}r_i = \frac{\partial r_i}{\partial x_j}\mathrm{d}x_j \equiv r_{i,j}\mathrm{d}x_j \quad \text{(in standard tensor notation), or}$$

$$\mathrm{d}r_i = \tfrac{1}{2}(r_{i,j} + r_{j,i})\,\mathrm{d}x_j + \tfrac{1}{2}(r_{i,j} - r_{j,i})\,\mathrm{d}x_j. \tag{1.111}$$

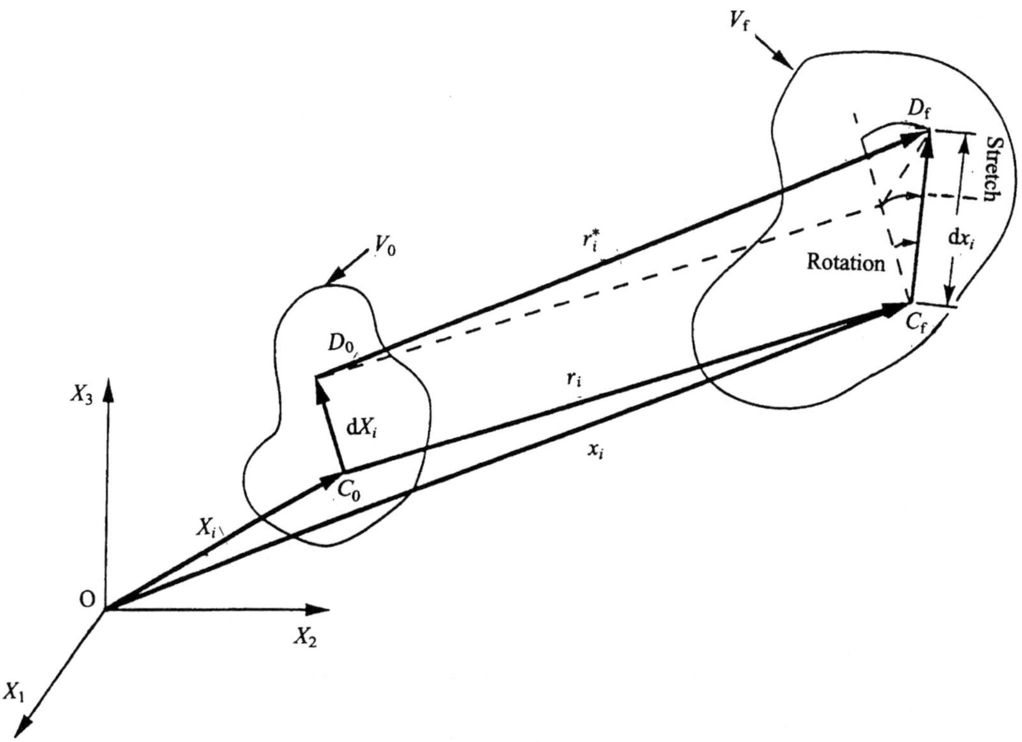

Fig. 1.44 Finite deformation of a volume element.

The first term on the right side in the last of eqns (1.111) represents the deformation due to stretching or compression of the volume element and is defined as the strain tensor $\epsilon_{ij} = \frac{1}{2}(r_{i,j} + r_{j,i})$. The second term defines the rotation tensor $\omega_{ij} = \frac{1}{2}(r_{i,j} - r_{j,i})$. These terms are linear due to the assumed presence of only infinitesimal deformation. For finite deformations, these representations must be significantly modified by addition of higher-order terms in the Taylor expansion, and the results will be nonlinear and complicated. Other measures of strain have been used in the literature. One example is the stretch ratio $\Lambda = L/L_0$ representing the quotient of current and initial lengths of a line element, illustrated for a unit cube in Fig. 1.45. This concept is associated with the true strain given by eqn (1.96) and is particularly useful for large strains that often occur in biological systems. Further information on large strains may be found in advanced books on continuum mechanics.

In accordance with the definitions given above, infinitesimal strains can be grouped into normal components ϵ_{ii} when $i = j$; these characterize the deformations in the direction of loading. For rectangular Cartesian coordinates with displacements u, v, and w in the directions x, y, z, respectively, these normal strains form the set

$$\epsilon_{xx} = \frac{\partial u}{\partial x}; \quad \epsilon_{yy} = \frac{\partial v}{\partial y}; \quad \epsilon_{zz} = \frac{\partial w}{\partial z}. \tag{1.112}$$

The incremental deformations giving rise to eqns (1.112) are presented in Fig. 1.46(a) for the one-dimensional and in Fig. 1.46(b) for the two-dimensional case; this is readily extended to three dimensions. For shear strains ϵ_{ij}, with $i \neq j$, the displacement pattern is shown in Fig. 1.46(c) for a plane element, where the angular deformation is

$$\alpha + \beta = \gamma_{ij} \left[\text{for example, } \gamma_{xy} = \left(\frac{\partial u}{\partial y} + \frac{\partial v}{\partial x}\right)\right]. \tag{1.113}$$

However, to conform to standard matrix notation, both normal and shear strains must be represented by a single

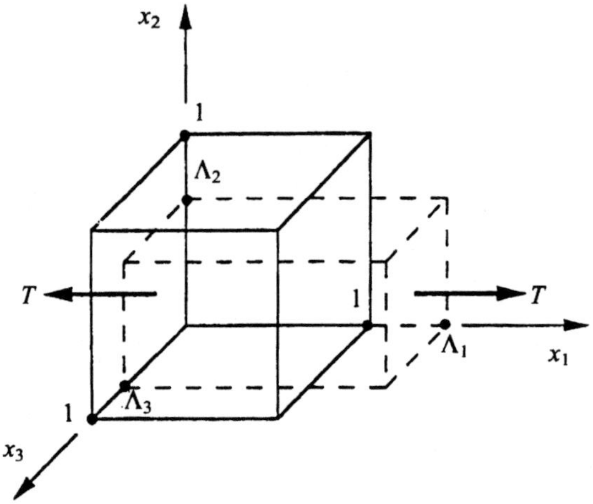

Fig. 1.45 Stretch ratios for an originally cubical volume element: (*solid lines*) original cube; (*dashed lines*) deformed volume.

relation; this is achieved by the definition

$$\epsilon_{ij} = \frac{1}{2}\left(\frac{\partial u_i}{\partial x_j} + \frac{\partial u_j}{\partial x_i}\right) = \epsilon_{ji}. \qquad (1.114)$$

In rectangular coordinates, this leads to the normal strains expressed by eqns (1.112) and yields a typical shear strain $\epsilon_{xy} = \frac{1}{2}\left(\frac{\partial u}{\partial y} + \frac{\partial v}{\partial x}\right)$ when this notation is employed. Thus $\epsilon_{xy} = \frac{1}{2}\gamma_{xy}$ *and this distinction must be carefully observed.*

Analogous to the stress tensor, there exists a strain tensor of the form

$$\begin{vmatrix} \epsilon_{xx} & \epsilon_{xy} & \epsilon_{xz} \\ \epsilon_{yx} & \epsilon_{yy} & \epsilon_{yz} \\ \epsilon_{zx} & \epsilon_{zy} & \epsilon_{zz} \end{vmatrix} = \begin{vmatrix} \frac{1}{2}\gamma_{xx} & \frac{1}{2}\gamma_{xy} & \frac{1}{2}\gamma_{xz} \\ \frac{1}{2}\gamma_{yx} & \frac{1}{2}\gamma_{yy} & \frac{1}{2}\gamma_{yz} \\ \frac{1}{2}\gamma_{zx} & \frac{1}{2}\gamma_{zy} & \frac{1}{2}\gamma_{zz} \end{vmatrix}. \qquad (1.115)$$

Coordinates x_i and x_j may also take on values for any orthogonal system, such as the cylindrical coordinates r, θ, and z. Furthermore, just as in the case of the stress tensor, there exists a rotation of the coordinate system such that for a particular set of orthogonal directions, the shear strains vanish, and the matrix denoting eqn (1.115) is transformed to the principal strain matrix

$$\begin{vmatrix} e_1 & 0 & 0 \\ 0 & e_2 & 0 \\ 0 & 0 & e_3 \end{vmatrix} \quad (e_i = \text{principal strains, } i = 1, 2, 3). \qquad (1.116)$$

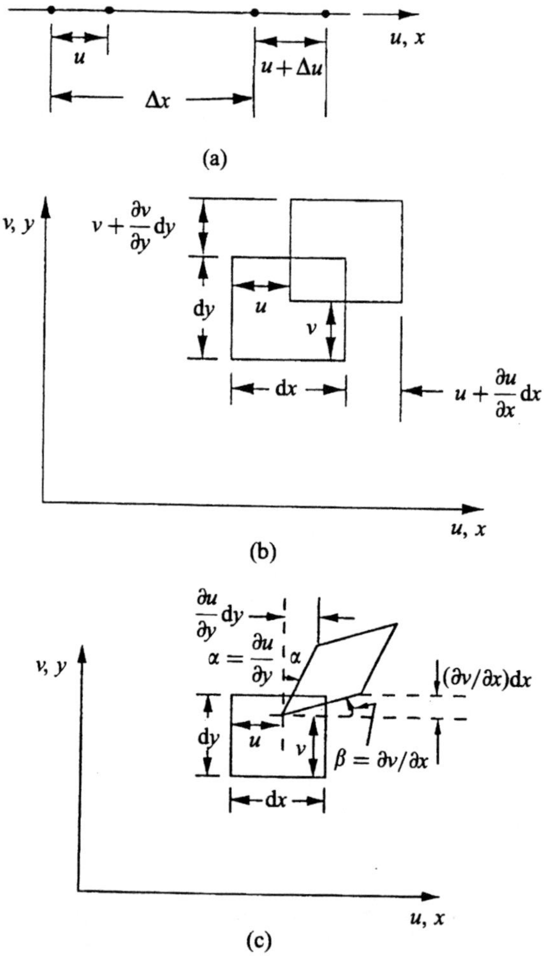

Fig. 1.46 Deformation under loading due to: (a) normal stresses (extension) in one dimension; (b) normal stresses in two dimensions; (c) shear stresses in two dimensions.

This is obtained by substituting strains ϵ_{ij} for the stresses in eqn (1.100).

When a case of plane strain exists, such as in the case of a very thick beam, the strain tensor is given by eqn (1.115) with each term of the third row and column replaced by zero, which may be diagonalized to provide principal strains e_1 and e_2. One may also formulate a set of strain invariants analogous to those for stresses. A fundamental thesis of continuum mechanics, the effect of rigid-body translations or rotations on strains in a deformable medium, is too complex to be discussed here; the reader is referred to the literature.

1.3.4 Models of material behavior

Construction of models for material behavior involves the development of a relation between stress, strain and their derivatives with respect to time which also features some material constants and is called a constitutive equation. Before engaging in this task, it is vital to define two concepts: *homogeneity* and *isotropy*. A material is said to be homogeneous if all of its macroscopic properties in the original (reference) state are independent of location measured with respect to a fixed coordinate system. This concept permits variation of certain properties, such as resistance to deformation, with direction at a given point, as long as that property is the same in the parallel direction at *any* other point. An object is isotropic if the properties along all possible axes emanating from any origin are the same, as occurs to a first approximation in a compact bone; however, a more accurate description indicates it to be otherwise (that is, anisotropic), a condition more strongly manifested in a global description of the scalp and the dural membrane of the head. Isotropy is generally associated with (at least macroscopic) homogeneity; however, materials may be macroscopically homogeneous, but anisotropic, as is the case for cross-woven fiber sheets, such as Kevlar, a substance employed for human-body protection.

A plethora of composite materials, consisting of two or more distinct components, have been developed recently; these are represented by a net of fibers or particles embedded in a matrix. The spinal cord is a biological example of a composite, consisting of intertwined white and gray matter. Skin is another composite, constituted of a layer of vascular tissue, the dermis, on whose surface are located the sensing elements, or papillae, while embedding certain organs such as sweat glands; the dermis is covered by a membrane called the epidermis (Gray 1984). Composites can often be modeled as homogeneous and, depending on the specific structure, as either isotropic or anisotropic.

Many different types of anisotropy exist; for example, if a body exhibits the same properties in any direction at all points of parallel planes, but which differ from those orthogonal to that plane, it is called 'transversely isotropic'. The nature of the anisotropy is governed by the symmetry present in the material and is frequently modeled after some crystal structure. An orthotropic material is one with three mutually perpendicular planes of symmetry, exemplified by wood. An entire long bone is a composite material, with a spongy, cancellous interior, that may be regarded as macroscopically isotropic, and an outer compact part. The skull is also a composite, consisting of two compact bone shells sandwiching an interior cancellous core (the diploë). Soft tissues, such as muscles or ligaments, are incapable of sustaining compression and are pre-tensioned in the *in vivo* state; their behavior is similar to that of cables in the technological environment.

Although other descriptions exist, the most frequent simulations of nonfluid material behavior and matching physical models are: the elastic solid (depicted by a spring); the viscous solid (described by a dashpot); the plastic solid (portrayed by two parallel sliding surfaces); or combinations of these. The first model displays instantaneous and total reversal of any deformation produced by loading when such loads are removed. A viscous material is characterized by a continual slow increase in deformation upon load application and a similar slow reduction of the deformation (recovery) upon load removal; the stress–strain curves depend on rate of deformation. Purely viscous objects are usually fluids. Plastic solids, such as some metals, will deform permanently, but remain intact when loaded and only the elastic component will be recovered when unloaded. Wet bone, an anisotropic substance, exhibits plasticity (as well as viscous effects) (Fondrk *et al.* 1988), but that is likely produced by a different process than in metals, while dry bone is elastic/brittle. Hence the theory of plasticity (Chung 1988) will not be presented here. Continuum descriptions of solids are valid only before fracture or other types of failure occur. Many organic substances may be regarded as viscoelastic, combining elastic and viscous features. Materials can be considered elastic when deformation or full recovery occur contemporaneously with loading or unloading and when the response is independent of loading rate. Stress–strain data is often obtained from experiments on thin uniform specimens loaded uniaxially in a testing machine.

The loading effects discussed here for solids are assumed to occur under isothermal (constant temperature) conditions, so that dependence of the deformations on temperature can be ignored. The quantitative description of solid material behavior, briefly presented below, is paralleled by a corresponding relations for fluids; here, however, temperature variations are more important, particulary in the case of gases, where they are preeminent.

For biological bodies, a stress–strain curve is often replaced by a load–deformation curve as a characterization of material behavior (Yamada 1970). A large difference between the loading and unloading curves is a sign of a dissipation due to either friction (such as rubbing of adjacent fibers) or viscoelasticity. The area between these curves, shown in Fig. 1.47(a) for synthetic skin (Frankel 1981) is called a hysteresis loop. It should be emphasized that all material models only approximate the actual behavior; the most reliable information for the response of a biological or other system to loading is obtained from careful measurement.

Another fundamental concept in the specification of materials is the difference between 'linear' and nonlinear behavior. In the linear case, the final deformation state of a body due to successive applications of two different forces $\boldsymbol{F}_1$ and $\boldsymbol{F}_2$ (or, similarly, moments $\boldsymbol{M}_1$ and $\boldsymbol{M}_2$) will be the same when the sequence of loading is reversed. On the other hand, a nonlinear elastic material such as shown in Fig. 1.47(a), typical of many biological substances, exhibits different instantaneous deformation states when a loading sequence is reversed. In such cases, the principle of superposition fails and the specified order of loading must be strictly observed; this invariably, but not exclusively, occurs when the deformations are large. In linear systems, the stresses and their time derivatives are linearly related to the strains and strain rates.

1.3.5 Elastic materials

Elastic materials exhibit instantaneous deformation response to loading and absence of rate dependence; the second condition can be verified for an unknown substance by obtaining nearly identical results in two tests performed at very different loading rates, such as quasi-static and impact loading, respectively. A uniaxial static tensile (or, when possible, a compressive) test (in a direction denoted by x) of an elastic uniform sample prior to the onset of permanent deformation or failure may display a linear response such as shown in Fig. 1.47(b). The governing equation for this model, also known as Hooke's law, and the corresponding ratio of transverse to axial strains, known as Poisson's ratio ν, are

$$\tau_{xx} = E\epsilon_{xx} \qquad \nu \equiv \left|\frac{\epsilon_{yy}}{\epsilon_{xx}}\right|. \tag{1.117}$$

Quantities E, Young's modulus, and ν have meaning only for uniaxial loading unless the material is isotropic; if nonisotropic, these terms are replaced by more general elastic constants C_{ijkl}. For nonlinear behavior, as shown in Fig. 1.47(a) and (c), the stress–strain relation, at least for limited ranges, is often empirically written in the form (with C and $n > 1$ as constants)

$$\tau_{xx} = C\epsilon_{xx}^{n}. \tag{1.118}$$

Similarly, for pure torsion, the linear relation between shear stress and strain is $\tau_{xy} = 2G\epsilon_{xy}$; G is the shear modulus or modulus of rigidity. A composite skull sample displays the uniaxial radial transverse isotropic behavior shown in Fig. 1.47(d), elastic at first, followed by diploë crushing, especially radially, an example of failure, but not of plastic deformation.

Linear uniaxial stress–strain relations can be extended to three dimensions by means of the generalized Hooke's law which linearly relates the stress tensor, whose terms are given by $\tau_{i,j}$, to the terms of the strain tensor ϵ_{kl} in the form

$$\tau_{ij} = C_{ijkl}\epsilon_{kl}$$
$$(i,j = 1,2,3 \text{ or } x,y,z;\ \ k,l = 1,2,3 \text{ or } x,y,z, \text{ in turn}). \tag{1.119}$$

Here, the C_{ijkl} are the elastic constants and, recall that, in standard tensor notation, a repeated subscript indicates summation over all values of that subscript. Since $\tau_{ij} = \tau_{ji}$ and $\epsilon_{kl} = \epsilon_{lk}$, the constants must exhibit the symmetry property $C_{ijkl} = C_{jikl} = C_{ijlk}$; this reduces the initial number of constants C_{ijkl} from 81 to 36. In all practical cases, symmetry considerations will reduce the number of constants needed to describe elastic material behavior still further. In rectangular components, the normal stress τ_{xx} derived from eqn (1.119) can with a change of notation be expressed by

$$\tau_{xx} = C_{11}\epsilon_{xx} + C_{12}\epsilon_{xy} + C_{13}\epsilon_{xz} + C_{22}\epsilon_{yy} + C_{23}\epsilon_{yz} + C_{33}\epsilon_{zz}, \tag{1.120}$$

and five relations can also be written for the other normal and shear stresses.

A particularly useful symmetry property frequently encountered in many substances is isotropy where the properties of the body at any specified point are identical in all directions. In this case, eqn (1.119) or (1.120), together with the five other corresponding expressions, reduce to the relations

$$\begin{aligned} \tau_{xx} &= \lambda(\epsilon_{xx} + \epsilon_{yy} + \epsilon_{zz}) + 2G\epsilon_{xx}; \\ \tau_{yy} &= \lambda(\epsilon_{xx} + \epsilon_{yy} + \epsilon_{zz}) + 2G\epsilon_{yy} \\ \tau_{zz} &= \lambda(\epsilon_{xx} + \epsilon_{yy} + \epsilon_{zz}) + 2G\epsilon_{zz}; \\ \tau_{xy} &= 2G\epsilon_{xy}; \quad \tau_{xz} = 2G\epsilon_{xz}; \\ \tau_{yz} &= 2G\epsilon_{yz} \end{aligned} \tag{1.121}$$

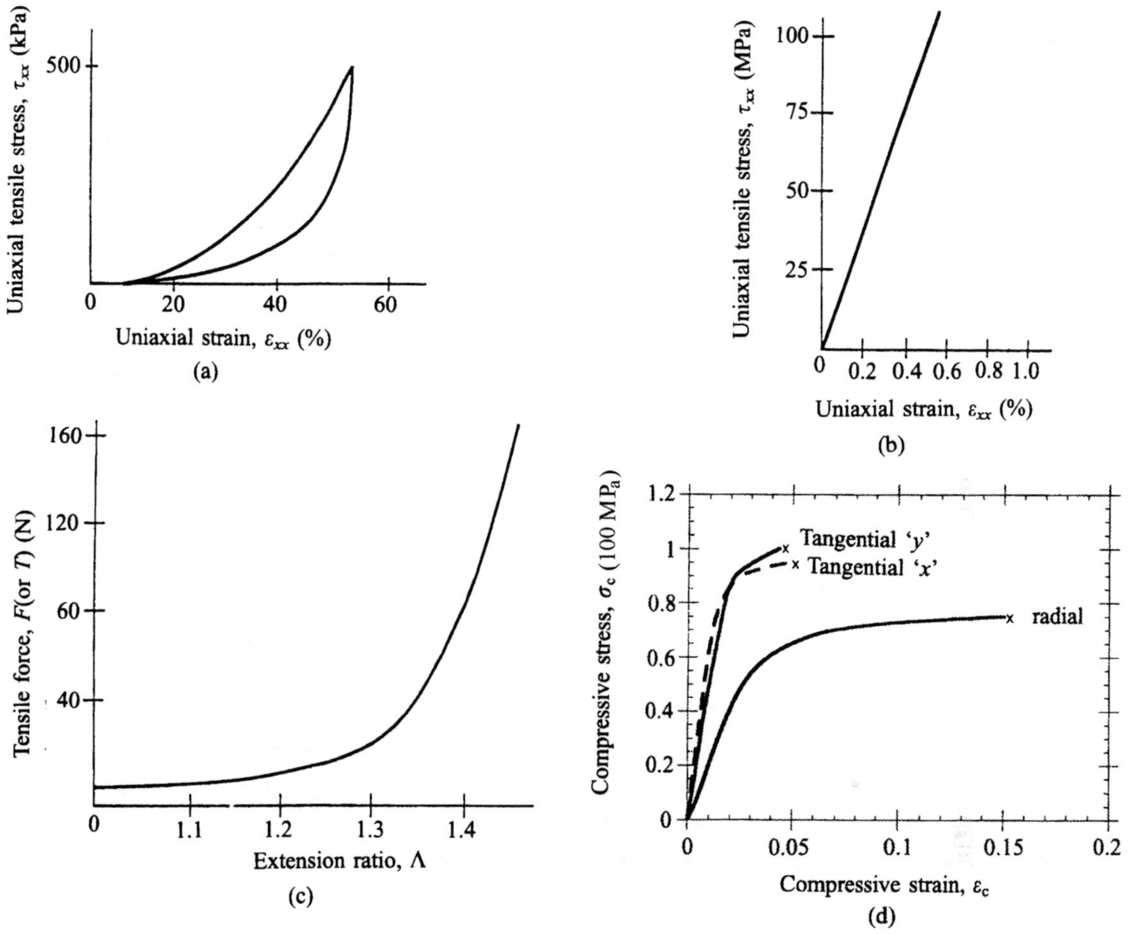

Fig. 1.47 Uniaxial elastic material behavior: (a) tensile stress–strain curve for synthetic skin showing nonlinearity and hysteresis (Frankel 1981); (b) tensile linear stress–strain curve for human compact bone (Yamada 1970; Evans 1973); (c) typical nonlinear force–extension curve for human muscle (Winter 1979); (d) compressive data for a complete skull specimen exhibiting transverse isotropy due to crushing of the diploë when loaded radially (McElhaney *et al.* 1970; Goldsmith 1973).

The elastic parameters reduce to the two contained in eqns (1.121), λ and G, which are called the Lamé constants. These equations can be written in compact form as

$$\tau_{ij} = \lambda\epsilon_{ii}\delta_{ij} + 2G\epsilon_{ij};$$

$$\delta_{ij} = \text{Kronecker delta} = \begin{Bmatrix} 1 \text{ for } i = j \\ 0 \text{ for } i \neq j \end{Bmatrix} \qquad (1.122)$$

The Lamé constants are related to Young's modulus E and Poisson's ratio v by

$$\lambda = \mu E/(1+\mu)(1-2\mu) = G(E-2G)/(3G-E);$$

$$G = \frac{E}{2(1+v)} \qquad (1.123)$$

The bulk modulus $K = p/(\Delta V/V)$, the resistance measure to uniform compression, is the ratio of hydrostatic pressure p to unit volume change; $K = E/3(1-2v)$.

Although most biological tissues, living or dead, exhibit inhomogeneities, anisotropies, and multiple component features, it is still sometimes possible to represent average properties by means of the simple constitutive relations, eqns (1.120), particularly when applied to uniaxial loading states. However, the assumption of small strains and deformations must frequently be abandoned, particularly in the case of soft tissue. This is evident from the simple stress–strain curve of Fig. 1.47(c), and from the many examples given by Yamada (1970) and by Fung (1993). Figure 1.48(a) portrays the slightly viscoelastic uniaxial stretch ratio (τ vs Λ) curve derived from the measurements of Fung (1967) for the rabbit mesentery, the membrane-like connective tissue between the intestines. When Fung reevaluated the data to provide a relation for the slope $d\tau/d\Lambda$ as a function of τ, Fig. 1.48(b), it was found to be nearly a straight line of the form

$$d\tau/d\Lambda = k_1(\tau + k_2) \quad \text{for } \Lambda < \Lambda_{\text{limit}}, \tag{1.124}$$

where Λ_{limit} represents the upper bound of validity for this linear approximation, and k_1 and k_2 are material constants. By integrating eqn (1.124) and requiring that the stress vanish when $\Lambda = 1$, the uniaxial governing equation for the constitutive behavior of the mesentery

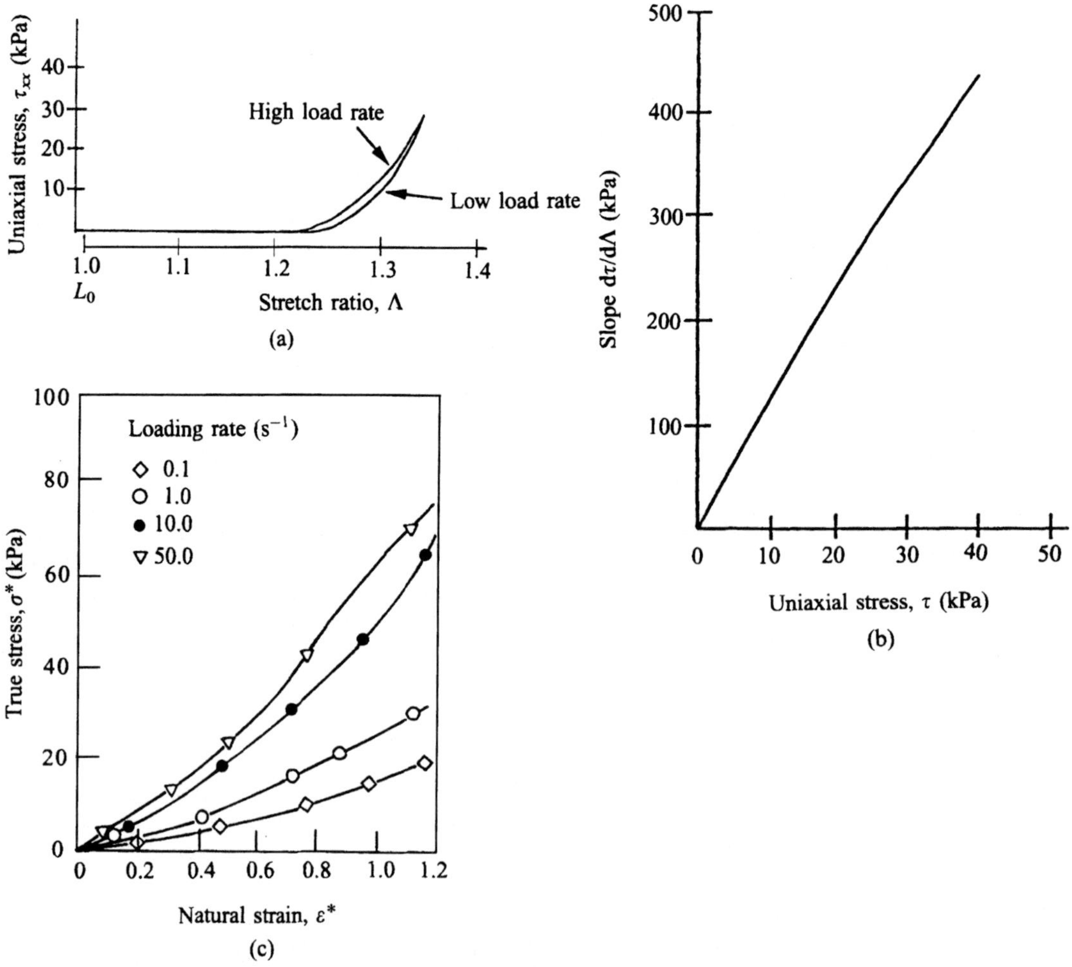

Fig. 1.48 Uniaxial tensile data for the rabbit mesentery (Fung 1967): (a) stress–stretch ratio curve; (b) slope of the data shown in (a) as a function of uniaxial engineering stress. (c) True compressive stress–strain curves for human brain at strain rates from 0.1 to 50 s^{-1} showing viscoelastic behavior (McElhaney *et al.* 1969; Goldsmith 1973).

becomes

$$\tau + k_2 = k_2 e^{k_1(\Lambda-1)} \quad \text{for } \Lambda < \Lambda_{\text{limit}}. \qquad (1.125)$$

Alternatively, this relation can be transformed by choosing as reference some fixed point on the curve, (τ_c, Λ_c); eqn (1.125) then becomes

$$\tau + k_2 = (\tau_c + k_2)e^{k_1(\Lambda-\Lambda_c)}$$
$$\text{with } k_2 = [\tau_c e^{-k_1(\Lambda_c-1)}]/[1 - e^{-k_1(\Lambda_c-1)}], \qquad (1.126)$$

as τ is still 0 when $\Lambda = 1$. Many soft tissues exhibit an exponential uniaxial constitutive equation. The observed uniaxial tensile behavior of soft tissue is more accurately described by a quadratic relation between $d\tau/d\Lambda$ and τ which, upon integration, provides a more complicated exponential relation than eqn (1.125).

1.3.6 Viscoelastic solids

Viscoelastic behavior is characterized by rate-dependence, exemplified by Fig. 1.48(c). Such substances exhibit combined elastic and time-dependent response and are described by some functional relation between stress and its time derivatives and strain and similar time derivatives; in most cases only the first derivative plays a role. For linear isotropic materials, a representation of the constitutive relation is obtained in analogy to the linearly elastic law, eqn (1.119), by an integral operation called a convolution process, expressed as (see Appendix, Section A.2)

$$\tau_{ij}\langle t\rangle = \int_{-\infty}^{t} \phi_{ijkl}\langle t - t'\rangle \frac{d\epsilon_{kl}\langle t'\rangle}{dt'} dt'$$
$$\left(\text{or } \tau_{ij}\langle t\rangle = \int_{-\infty}^{t} \epsilon_{kl}\langle t - t^*\rangle \frac{d\phi_{ijkl}\langle t^*\rangle}{dt^*}\right) dt^* \qquad (1.127)$$

for the stresses τ_{ij}. The tensor relaxation function ϕ_{ijkl} has the property

$$\phi_{ijkl}\langle t\rangle = \phi_{jikl}\langle t\rangle = \phi_{ijlk}\langle t\rangle, \qquad (1.128)$$

which is similar to the properties exhibited by elastic constant C_{ijkl}. If the strain history has a discontinuity at time $t = 0$, eqn (1.127) can be generalized by the addition of a function $\phi_{ijkl}\langle 0\rangle\epsilon_{kl}\langle t\rangle$; this relation then represents a Boltzmann solid. Alternatively, eqn (1.127) can be inverted to provide an integral equation for strain ϵ_{ij} in the form

$$\epsilon_{ij}\langle t\rangle = \int_{-\infty}^{t} \psi_{ijkl}\langle t - t^*\rangle \frac{d\tau_{kl}\langle t\rangle}{dt^*} dt^*, \qquad (1.129)$$

where ψ_{ijkl} is the tensorial creep function. Corresponding expressions for the anisotropic case are currently not available. The complexities arising from the three-dimensional nature of eqns (1.127) and (1.128) can be avoided in most problems by either (a) describing the material behavior directly in terms of experimental uniaxial test results for specified loading conditions, or (b) portraying the comportment of the material by an analytical model composed of elastic and viscous elements.

The description of the material can be obtained by means of a uniaxial relaxation test by observing the reduction of stress with time in a standard specimen when a constant strain is suddenly applied and either maintained for a finite time or held until the end of the test. From eqn (1.127), the uniaxial constitutive equation in terms of the uniaxial relaxation function ϕ is

$$\tau\langle t\rangle = \int_{-\infty}^{t} \phi\langle t - t'\rangle \frac{d\epsilon\langle t'\rangle}{dt'} dt'. \qquad (1.130)$$

If a step function $H\langle t\rangle$ of strain is applied, i.e. $\epsilon = \epsilon_0 H\langle t\rangle$, the relaxation function can be determined from eqn (1.127) as

$$\phi\langle t\rangle = 0, \quad \text{for } t < 0; \quad \phi\langle t\rangle = \frac{\tau\langle t\rangle}{\epsilon_0}, \quad \text{for } t \geq 0. \qquad (1.131)$$

Conversely, the uniaxial equivalent of eqn (1.129) is

$$\epsilon\langle t\rangle = \int_{-\infty}^{t} \psi\langle t - t^*\rangle \frac{d\tau\langle t^*\rangle}{dt^*} dt^*, \qquad (1.132)$$

and a step function of stress $\tau_0 H\langle t\rangle$ yields the creep function

$$\psi\langle t\rangle = 0, \quad \text{for } t < 0; \quad \psi\langle t\rangle = \frac{\epsilon\langle t\rangle}{\tau_0}, \quad \text{for } t \geq 0. \qquad (1.133)$$

The stress and strain histories defined in eqns (1.131) and (1.133) are easily measured. They can often be represented in exponential form, i.e.

$$\phi\langle t\rangle = \sum_{i=0}^{m} a_i e^{-t/t_i'} \text{ and } \psi\langle t\rangle = \sum_{j=0}^{n} b_j e^{t/t_j^*}; \; a_i, b_j, t_i', t_j^* > 1 \qquad (1.134)$$

for the relaxation and creep functions, respectively. Another method of determining viscoelastic material properties is the measurement of the response when harmonic loading is applied to a uniaxial specimen.

Physical models of isotropic linear viscoelastic materials can also be constructed in terms of any desired combination of springs and dashpots. The most general relation between stress and strain for an n-parameter body related to such a model is given in terms of material constants p_i and q_j as

$$\sum_{i=0}^{n} p_i \left(\frac{\mathrm{d}^i \tau}{\mathrm{d}t^i} \right) = \sum_{j=0}^{m} q_j \left(\frac{\mathrm{d}^j \epsilon}{\mathrm{d}t^j} \right) \qquad (1.135)$$

The simplest and most commonly used versions of eqn (1.135) are the representations depicted in Fig. 1.49. They include (a) a Newtonian dashpot; (b) a Kelvin–Voigt solid, featuring a parallel spring and dashpot; (c) a Maxwell model (also a fluid), with a spring and dashpot in series; and (d) the standard linear solid which combines the Kelvin–Voigt and the Maxwell solids. The one-dimensional constitutive relations for these simulations are (Fung 1994), with c as the damping parameter per unit length:

$$\text{(a) } \tau = c\frac{\mathrm{d}\epsilon}{\mathrm{d}t} = c\dot{\epsilon} \qquad \text{(b) } \tau = E\epsilon + c\dot{\epsilon}$$

$$\text{(c) } \tau + \frac{c\dot{\tau}}{E} = c\dot{\epsilon}; \quad \dot{\tau} = E\dot{\epsilon}_1 \quad \text{and} \quad \tau = c\dot{\epsilon}_2;$$

$$\epsilon = \epsilon_1 + \epsilon_2$$

$$\text{(d) } \tau + \frac{c}{E_2}(\dot{\tau}) = E_1\epsilon + \left(\frac{c}{E_2}\right)(E_1 + E_2)\dot{\epsilon}$$

$$\text{since } \epsilon = \epsilon_1 + \epsilon_2 \text{ and } E_2\epsilon_2 = c\dot{\epsilon}_1. \qquad (1.136)$$

Alternatively, eqns (1.136) can be represented by the integral expression

$$\tau = E_1\epsilon + E_2 \int_{-\infty}^{t} \mathrm{e}^{-(E_2/c_2)[t-t']} \frac{\mathrm{d}\epsilon\langle t'\rangle}{\mathrm{d}t'} \mathrm{d}t', \qquad (1.137)$$

which is a special case of eqn (1.127) with the relaxation function as a single exponential. Generalized Kelvin–Voigt models can be constructed by combining a number of single units in series, and a generalized Maxwell model is obtained by connecting individual units in parallel; corresponding analytical expressions are special forms of eqn (1.135). The creep and relaxation curves for these generalized solids assume the exponential forms of eqns (1.134). For the simple models presented in Fig. 1.49, the relaxation and creep responses for step loading of strain $\epsilon_0 H\langle t - t_0\rangle$ or stress $\tau_0 H\langle t - t_0\rangle$ at time t_0 are given by

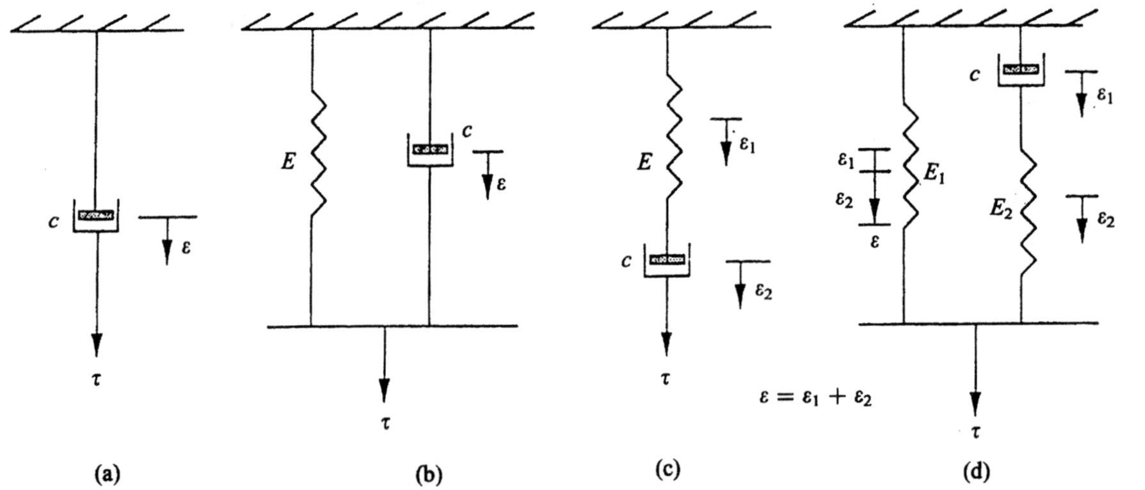

Fig. 1.49 Representations of simple uniaxial linear viscoelastic bodies: (a) dashpot; (b) Kelvin–Voigt (spring and dashpot in parallel); (c) Maxwell (spring and dashpot in series); (d) standard linear solid (three parameters).

Creep *Relaxation*

$$\begin{array}{lll}
\text{(a)} & \epsilon\langle t\rangle = \left(\dfrac{\tau_0}{c}\right) tH\langle t-t_0\rangle & \tau\langle t\rangle = c\epsilon_0\delta\langle t-t_0\rangle \\[2ex]
\text{(b)} & \epsilon\langle t\rangle = \left(\dfrac{\tau_0}{E}\right) H\langle t-t_0\rangle[1-\mathrm{e}^{-(E/c)t}] & \tau\langle t\rangle = c\epsilon_0\delta\langle t-t_0\rangle + E\epsilon_0 H\langle t-t_0\rangle \\[2ex]
\text{(c)} & \epsilon\langle t\rangle = \left[\dfrac{1}{E}+\dfrac{t}{c}\right]\tau_0 H\langle t-t_0\rangle & \tau\langle t\rangle = \epsilon_0 H\langle t-t_0\rangle E\mathrm{e}^{-(E/c)t} \\[2ex]
\text{(d)} & \epsilon\langle t\rangle = \tau_0 H\langle t-t_0\rangle/E_1 & \\
& \quad\times\left[1-\left(\dfrac{E_2}{E_1+E_2}\right)\mathrm{e}^{-E_1E_2t/c(E_1+E_2)}\right] & \tau\langle t\rangle = \epsilon_0 H\langle t-t_0\rangle\left[1+\left(\dfrac{E_1}{E_2}\right)\mathrm{e}^{-E_2t/c}\right]E_1
\end{array} \tag{1.138}$$

The creep and relaxation curves for these materials are shown in Fig. 1.50. If the same functions ϕ and ψ are obtained for several different constant values of suddenly applied strain or stress, respectively, the material is linearly viscoelastic, and superposition, or integral representation in the form of eqns (1.130) or (1.137), applies. It is fortunate that many substances exhibit linear characteristics, because, although several nonlinear theories of viscoelastic behavior have been proposed, there is no concensus on which approach best represents the actual behavior and, further, how the appropriate material properties of such substances can be measured.

Many if not most soft tissues exhibit viscoelastic characteristics to a greater or lesser degree. This includes

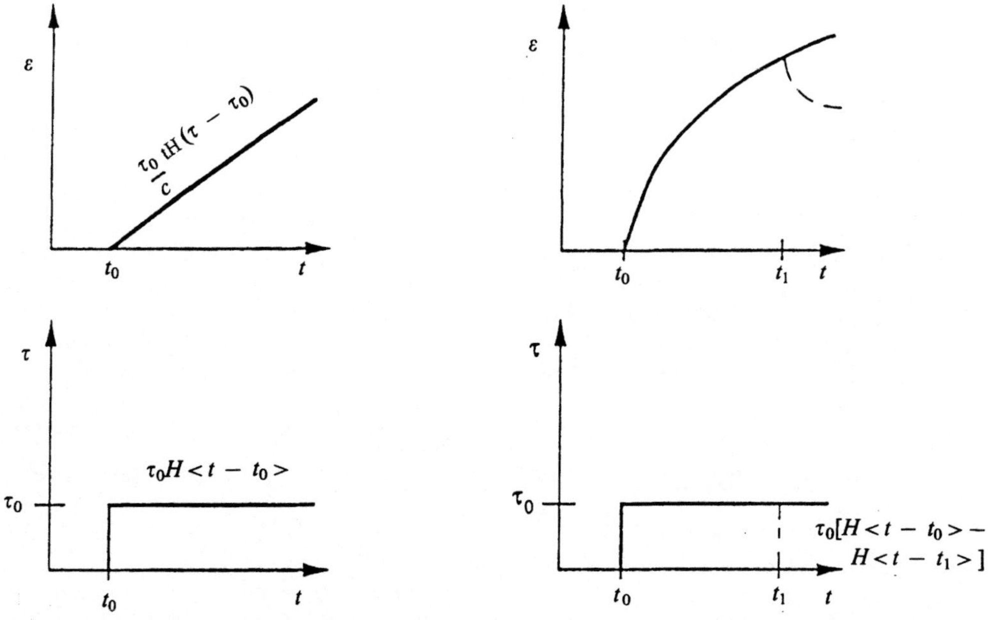

(a) Creep (b)

Fig. 1.50 (Part 1) Creep and relaxation curves for models (a)–(d) of Fig. 1.49 (Fung 1993). Dashed lines represent case of finite load duration. [This figure is in 4 parts spread over the next 4 pages.]

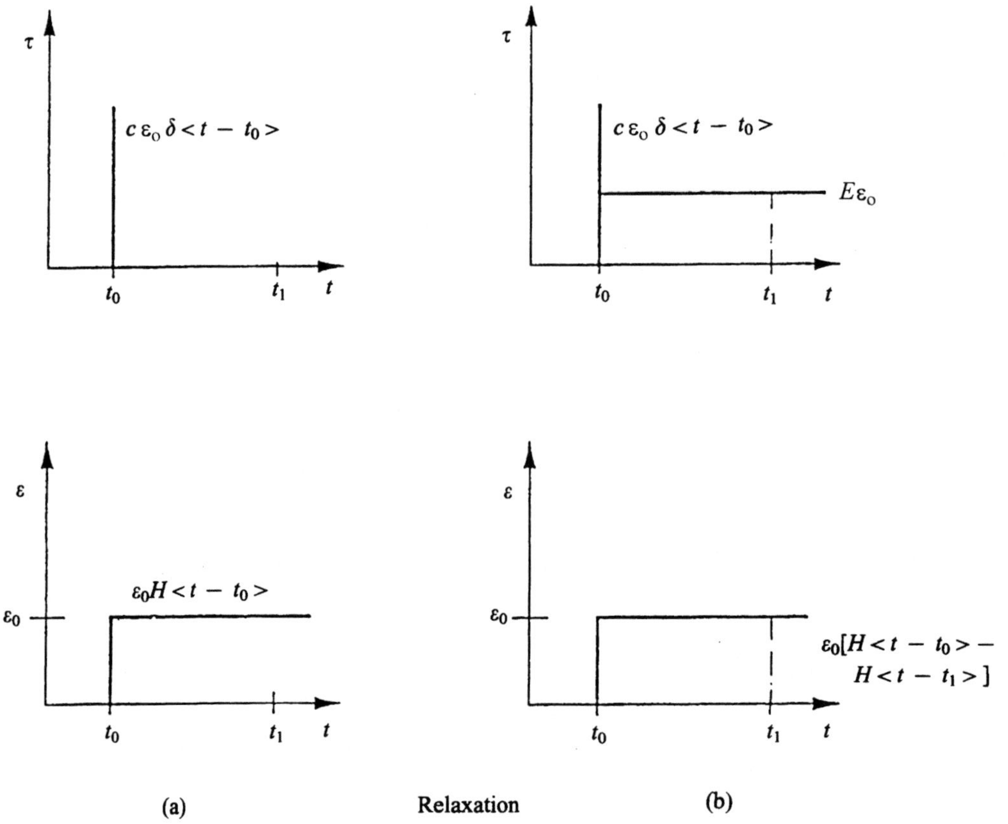

Fig. 1.50 (Part 2)

muscles, tendons, ligaments, connective tissue, skin, organs, and even bone (Sedlin 1965; Fung 1993; Woo *et al.* 1993). Viscous properties affect performance and function, particularly by damping the effects of shocks or repeated loading on the bodies of humans and animals, and thus serve a vital protective purpose.

1.4 Strength of materials—deformation of simple linearly elastic systems

The subject of strength of materials is a simplified modeling approach to the deformation of structural elements under load which frequently provides a sufficiently accurate determination of the strain and deformation for practical purposes (Popov 1976; Gere and Timoshenko 1990; Case *et al.* 1993). A more precise description of this behavior requires the use of continuum mechanics.

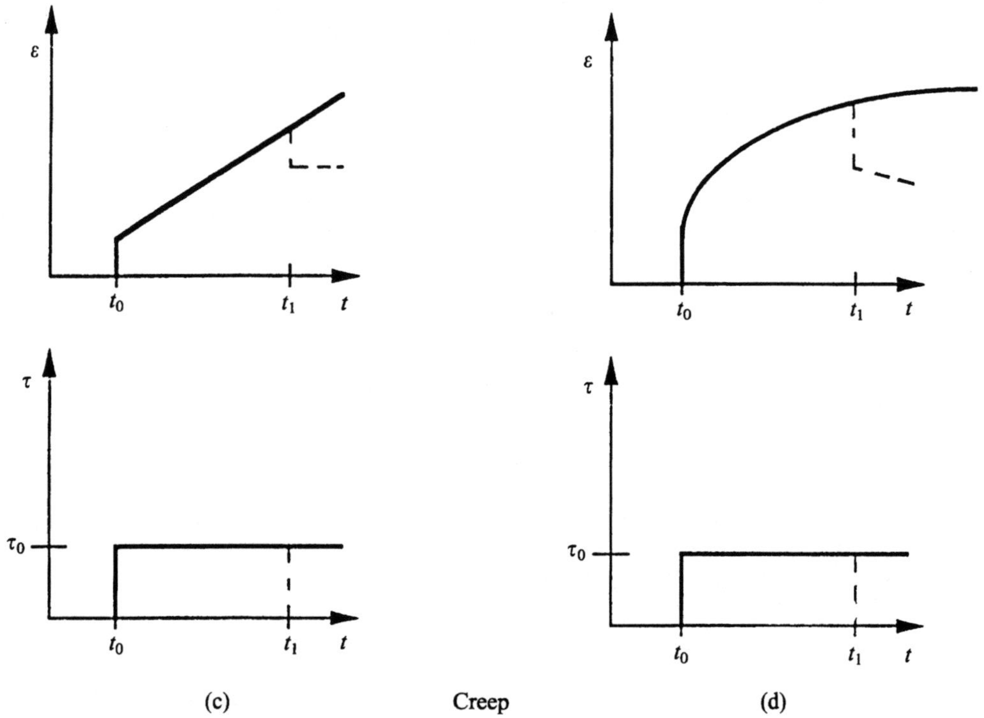

Fig. 1.50 (Part 3)

1.4.1 Bending of beams

Beams are relatively thin and narrow structural members capable of carrying moments and axial and transverse loads; the latter may be concentrated or distributed. The loading considered here is restricted to two dimensions; thus, three equations of equilibrium are available, and, for the beam to be statically determinate, only three unknown reactions can be present at the supports. These may be either fixed (allowing normal, shear, and bending moment reactions at this location, Fig. 1.51(a); or simply supported (where only axial and transverse forces can be transmitted), Fig. 1.51(b); they may be represented by rollers or pins, (where only normal forces can be transmitted), Fig. 1.51(c); or they may be free, Fig. 1.51(d), where the forces and moments must vanish. Other modes of support for the member can be used, such as a linear or torsional spring that can transmit forces or moments proportional to linear or angular displacements, respectively.

It is customary to plot shear and bending moment diagrams for beams. These can be drawn using either the method of sections; that is, by cutting the beam at some section x and writing the equations of equilibrium for either portion, or else by summation of forces and moments on a differential beam element. As an illustration, the action of specified concentrated loads F_1 and F_2 (and known directions) on a beam with a simple support at A and a roller at B is presented in Fig. 1.52(a); the corresponding free-body diagram is shown in Fig. 1.52(b). The analysis begins by solving the equations of equilibrium for this case, given by

$$\begin{aligned} &\sum F_y = 0: \; A_y + B_y - F_1 - F_2 \cos\gamma = 0; \\ &\sum F_x = 0: \; A_x - F_2 \sin\gamma = 0 \quad (B_x = 0) \\ &\sum M_B = 0: \; A_y L - F_1(L-a) - (F_2\cos\gamma)b = 0 \end{aligned} \tag{1.139}$$

for the unknown reaction components A_x, A_y and $B_y = B$. The beam is now cut into several segments, Fig. 1.52(c), so that each successive section includes the action of an additional vertical (shear) force, or couple, if present, as shown in Fig. 1.52(b). The equations of equilibrium are obtained for each segment; the horizontal forces are

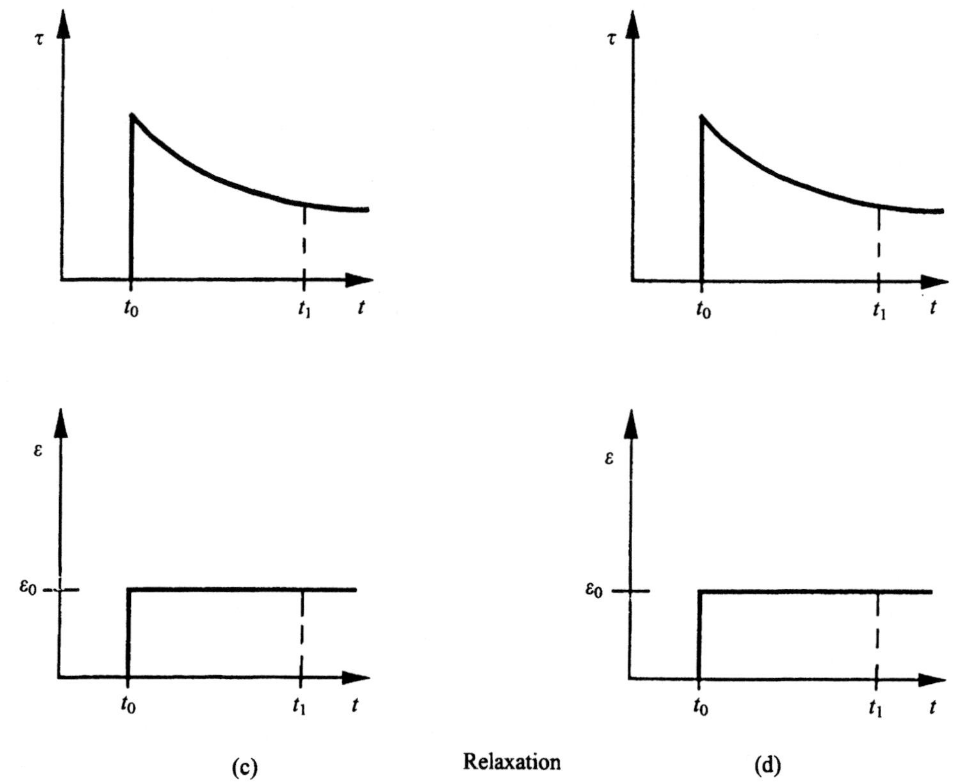

Fig. 1.50 (Part 4)

resisted by the simple support at A. In this case, that involves sections x_1, x_2 and x_3, and yields the relations

$$\begin{aligned}
&\sum F_x = 0:\ F_x = A_x;\quad \sum F_y = 0:\ V = A_y;\\
&\sum M|_{x_1} = 0:\ M = A_y x \quad \langle \text{at } x_1\rangle \quad \text{for } 0 \le x \le a\\
&\sum F_x = 0:\ F_x = A_x = F_2 \sin\gamma;\\
&\sum F_y = 0:\ V = F_1 - A_y;\\
&\sum M|_{x_2} = 0:\ M = A_y x - F_1(x-a) \quad \langle \text{at } x_2\rangle\\
&\text{for } a \le x \le L-b;\\
&\sum F_x = 0:\ F_x = 0;\\
&\sum F_y = 0:\ V = F_1 + F_2\cos\gamma - A_y;\\
&\sum M|_{x_3} = 0:\ M = A_y x - F_1(x-a)\\
&\qquad - (F_2\cos\gamma)(x+b-L) \quad \langle \text{at } x_3\rangle\\
&\text{for } L-b \le x \le L
\end{aligned} \tag{1.140}$$

The resulting shear and moment diagrams obtained from eqns (1.140) are exhibited in Figs 1.52(d) and (e). As will be shown shortly, the bending moment variation can also be obtained from the shear diagram by integration.

A case of uniformly distributed loading w (N m^{-1}) is shown in Fig. 1.53(a), (b), (c), and (d), analogous to the case shown in Fig. 1.52. The vertical reactions at the supports are obtained as $\frac{1}{2}wL$ from the equilibrium equations for the beam, and the corresponding equations for the section shown are

$$\begin{aligned}
&F_y = 0:\ V = w(\tfrac{1}{2}L - x);\\
&\sum M = 0:\ M|_x = \tfrac{1}{2}w(Lx - x^2) \quad 0 \le x \le L.
\end{aligned} \tag{1.141}$$

There are no applied horizontal force components; thus, the reaction force A_x vanishes and the maximum moment $M_{\max} = wL^2/8$ occurs at the center. A combination of concentrated and distributed vertical loading is presented

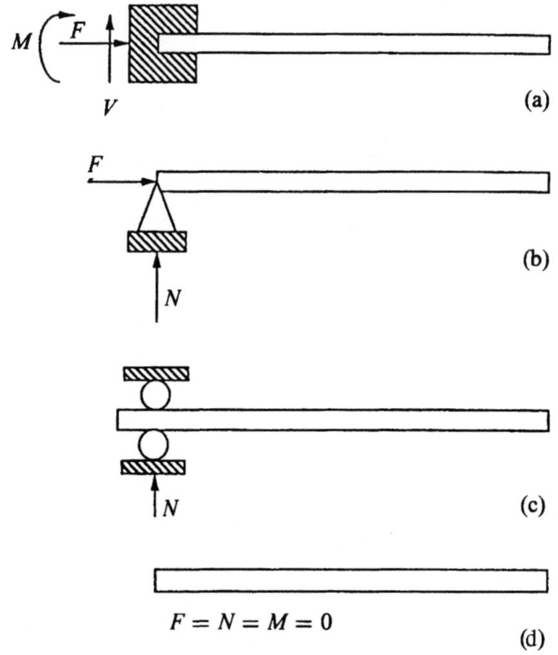

Fig. 1.51 Types of supports for simple beams: (a) fixed; (b) simple; (c) roller; (d) free.

in Fig. 1.54(a), (b), (c), and (d). Again, there are no forces in the x-direction and the governing equations for the equilibrium of the entire beam and for a single section at $b < x < c$ are

$$F_y = 0\colon\ A_y + B_y - F_1 - (c-b)[w_1 + \tfrac{1}{2}(w_2 - w_1)] = 0$$

$$M_A = 0\colon\ B_y L - F_1 a - \tfrac{1}{2}(c-b)[w_1(b+c) + \big(b + \tfrac{2}{3}(c-b)\big)(w_2 - w_1)] = 0$$

for the entire beam, and

$$F_y = 0:\ V = F_1 + w_1(x-b) + \tfrac{1}{2}[(w_2 - w_1)(x-b)^2/(c-b)] - A_y \qquad (1.142)$$

$$M = 0\colon\ A_y x - F_1(x-a) - w_1\frac{(x-b)^2}{2} - \frac{(w_2 - w_1)(x-b)^3}{6(c-b)}$$

for $b \le x \le c$.

For distributed loading, the differential equation involving load $w(x)$ per unit length of the beam can also be obtained from the equilibrium of the element shown in Fig. 1.55, and is given by

$$F_y = 0\colon\ \frac{dV}{dx} = -w(x);\quad M = 0\colon\ \frac{dM}{dx} = -V$$

$$\text{and hence}\quad w(x) = \frac{d^2M}{dx^2}. \qquad (1.143)$$

Integration of these equations, with due regard for discontinuities due to concentrated loads or impulsive moments, will also yield the proper shear and bending moment diagrams.

Beams subjected to loads or moments will bend, producing tension in one part and compression in the other. The dividing boundary, where the stress is zero, is known as the neutral plane; the intersection of this plane with an orthogonal surface in the plane of the beam defines a line known as the neutral axis. A fundamental principle of the present simplified analysis for prismatic beams is that plane cross sections of the beam remain plane after bending deformation; that is, no warping of the cross-section occurs. For beams governed by Hooke's law, the stresses are assumed to vary linearly with distance from the neutral axis, as indicated in Fig. 1.56. Hence, $\tau_{xx} = ky^*$, where k is the slope of the stress variation along cross-sectional area A and y^* is the distance from the neutral axis to elemental area dA.

When the beam is subjected to an external moment M, equilibrium demands that there can not be any resultant axial forces; hence integration over the cross-sectional area of the normal stresses yields

$$\int_A \tau_{xx}\, dA = k \int_A y^*\, dA = 0, \qquad (1.144)$$

which, by comparison with eqn (1.21), shows that the neutral axis passes through the centroid of the section. The second equilibrium condition requires that the external moment M must balance that induced by the stresses about the z-axis, or

$$M + \int_A (\tau_{xx} y^*)\, dA = 0;$$

$$\text{hence } M = -k \int_A y^{*2}\, dA = -kI', \qquad (1.145)$$

where I' is the areal moment of inertia defined in Subsection 1.1.13. Thus,

$$\tau_{xx} = -My^*/I' \quad\text{and}\quad |\tau_{xx}|_{\max} = Mc/I', \qquad (1.146)$$

where c is the distance from the neutral axis to the outer beam fiber, i.e. $c = |y^*_{\max}|$.

In the deformation of the beam as shown in

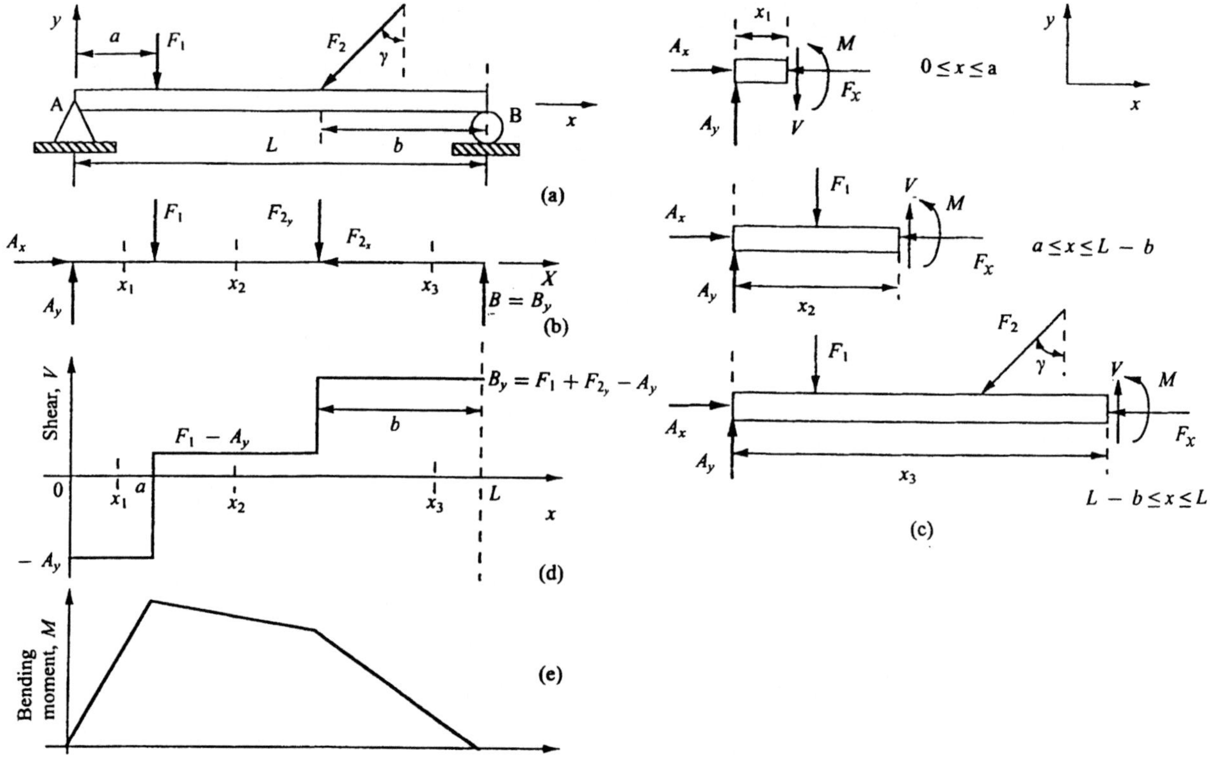

Fig. 1.52 Analysis of a determinate beam loaded by concentrated forces: (a) physical description; (b) free-body diagram of beam; (c) free-body diagram of sections; (d) shear diagram; (e) bending moment diagram.

Fig. 1.57(a), the bending moment generating its concave upward curve is taken so that its vector points along the positive z-axis. By considering this deformation of a prismatic beam of constant cross section, it is evident that the linear extension Δu of a fiber equals $-y^*\Delta\theta$, as indicated in Fig. 1.57(b).

If, further, the differential length element at the neutral axis between two initially parallel beam sections is denoted by Δs, then the strain ϵ_{xx} is

$$\epsilon_{xx} \underset{\text{Lim } \Delta s \to 0}{=} \frac{\Delta u}{\Delta s} = -y^*\left(\frac{d\theta}{ds}\right) = -\frac{y^*}{R}, \quad (1.147)$$

where R is the radius of curvature of the neutral axis. Using eqn (1.147) and Hooke's law, eqn (1.117) and the second of eqns (1.146) yield the flexure formula

$$\frac{1}{R} = \frac{M}{EI'}. \quad (1.148)$$

For a two-dimensional Cartesian coordinate system, the radius of curvature is

$$\frac{1}{R} = (d^2y/dx^2)/[1 + (dy/dx)^2]^{3/2} \approx \frac{d^2y}{dx^2} \quad \text{when } \frac{dy}{dx} \ll 1, \quad (1.149)$$

which often happens. Then, with y as the beam deflection, eqn (1.148) becomes

$$\frac{d^2y}{dx^2} = \frac{M}{EI'}. \quad (1.150)$$

The deflection $y\langle x\rangle$ can be obtained by double integration when $M\langle x\rangle$ is known, subject to the appropriate boundary conditions at end $x = 0$ and/or $x = L$; since eqn (1.150) is stated in terms of M, it may also be expressed in terms

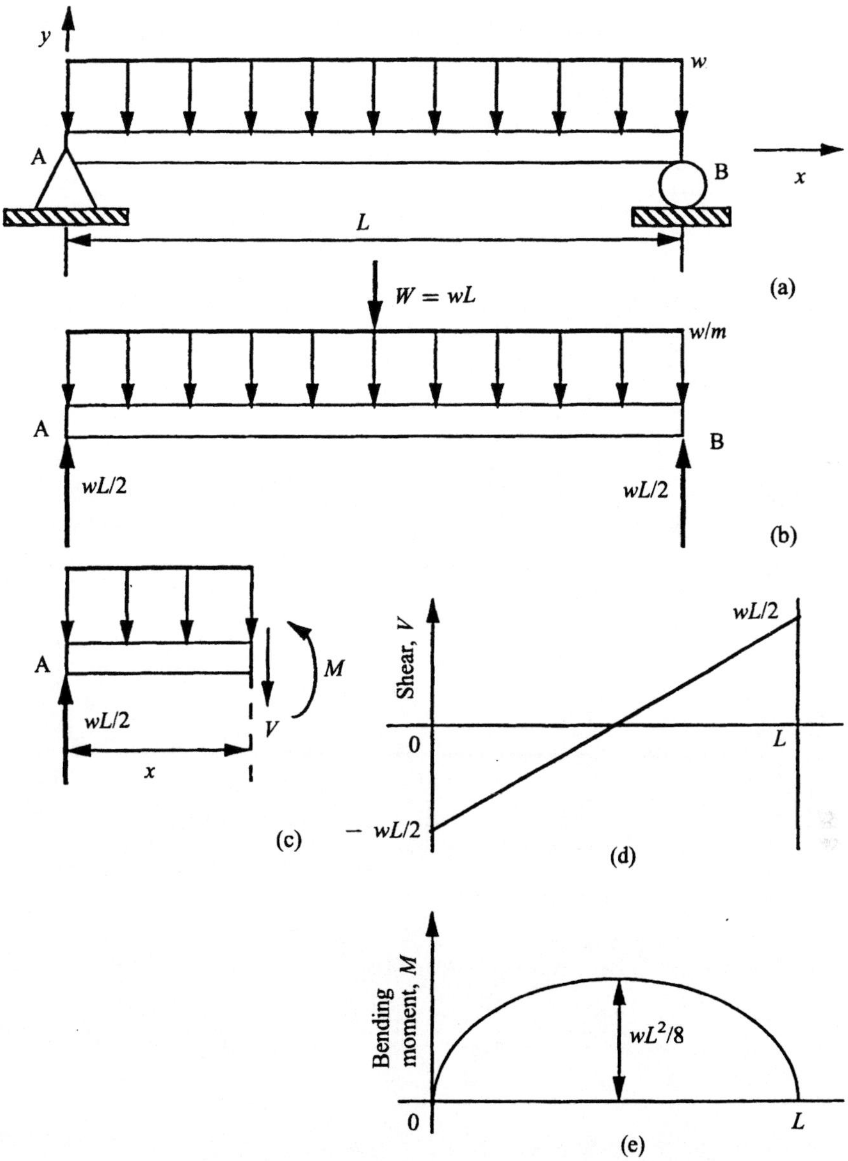

Fig. 1.53 Analysis of a determinate beam subjected to a uniform load: (a) physical description; (b) free-body diagram of beam; (c) free-body diagram of a section; (d) shear diagram; (e) bending moment diagram.

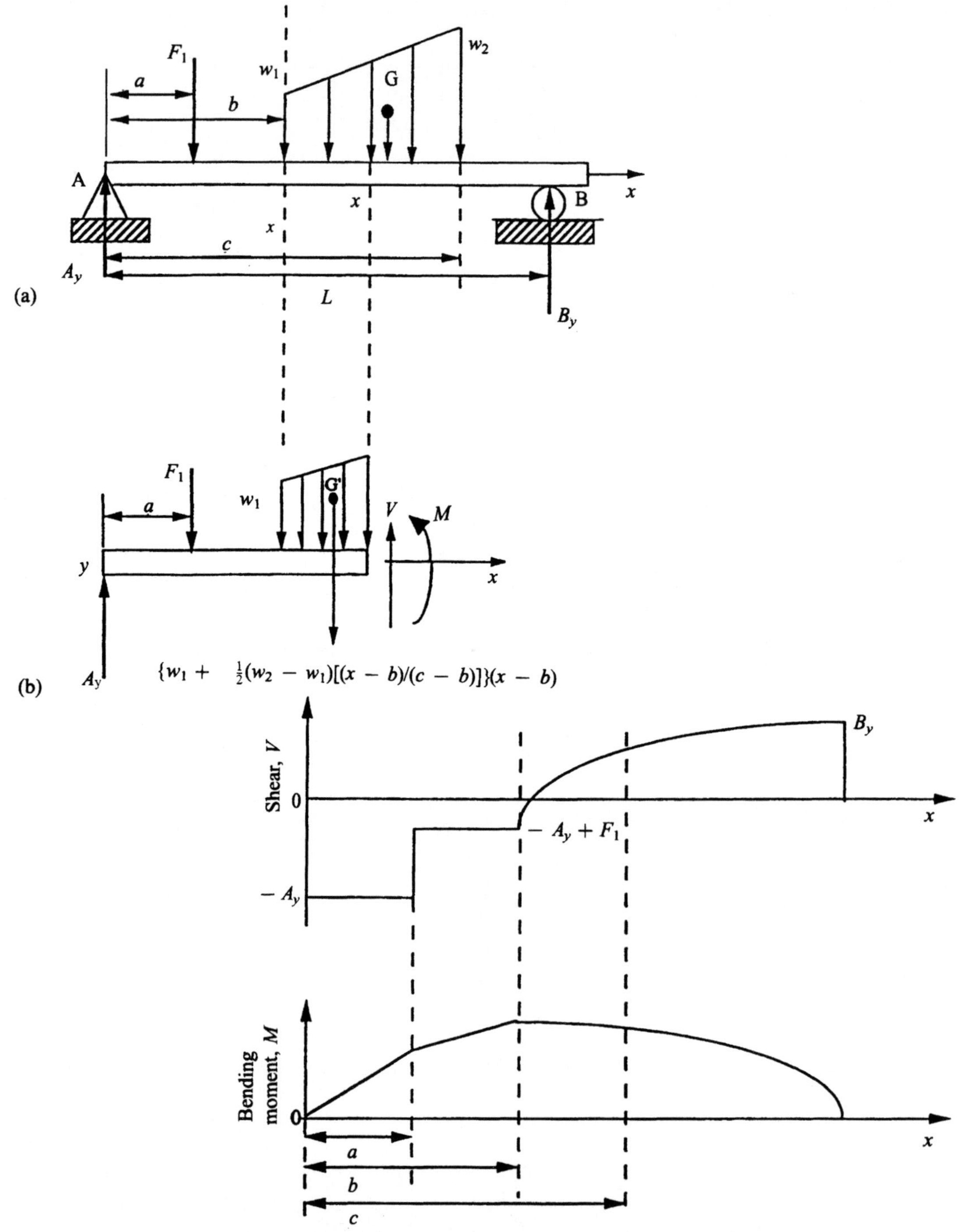

Fig. 1.54 Analysis of a determinate beam subjected to concentrated and distributed loads: (a) free-body diagram of beam; (b) free-body diagram of typical section; (c) shear diagram; (d) bending moment diagram.

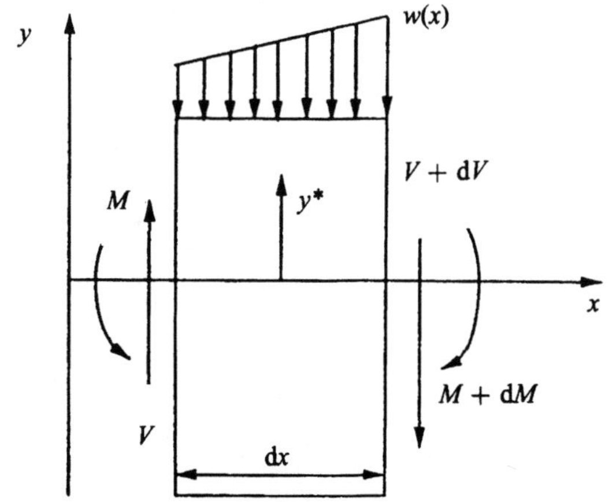

Fig. 1.55 Free-body diagram of a section of a loaded beam. (Other arrow directions are also in use.)

of the shear $V\langle x\rangle$ or load per unit length $w\langle x\rangle$ in the forms

$$V = -\frac{dM}{dx} = -\frac{d}{dx}\left(EI'\frac{d^2y}{dx^2}\right)$$
$$\text{or}\quad w\langle x\rangle = -\frac{dV}{dx} = \frac{d^2M}{dx^2} = \frac{d^2}{dx^2}\left(EI'\frac{d^2y}{dx^2}\right). \qquad (1.151)$$

When either or both the modulus E and area moment I' change along the length, this variation must be taken into account in the solution of eqns (1.151). In the special case when both are constant along the length of the beam, the second of eqns (1.151) becomes much simpler and can be written as

$$w\langle x\rangle = EI'\left(\frac{d^4y}{dx^4}\right). \qquad (1.152)$$

It is necessary to specify four boundary conditions, as eqn (1.152), which is expressed in terms of the load per unit length, is a fourth-order differential equation. For example, a pinned support or a roller support requires that the bending moment M and deflection y are zero at such points. A built-in support requires the vanishing of the deflection y and slope dy/dx, while a free end demands vanishing of both shear and bending moment at that position.

As an example of the application of these equations, consider a bricklayer supporting a load of bricks on his upper arm (humerus), as shown in Fig. 1.58(a), constituting a uniform load of $w = 1\ \text{N mm}^{-1}$ (5.71 lbf in^{-1}); the upper arm will be considered fixed at the shoulder. With an effective upper arm length of 254 mm (10 in), this member will be idealized as a uniform circular cylindrical shell with an outer diameter of $D_0 = 50.8$ mm (2 in) and an inner diameter $D_i = 38.1$ mm (1.5 in), and to possess a Young's modulus of 9 GPa. It is desired to find the variation of bending moment, shear and deflection along the arm and to find the maximum bending moment and deflection. The effect of the forearm and hand will be neglected and the elbow will be considered to be a free support. The

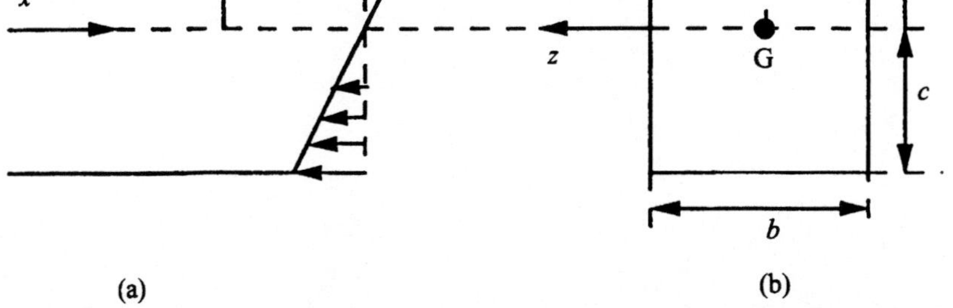

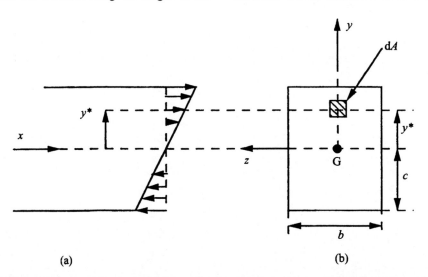

Fig. 1.56 Stresses in a beam: (a) linear distribution about neutral axis through mass center G for a section along beam length; (b) cross-sectional area. Note: y^* represents a variable distance from the neutral axis to a beam elemental area; y is the beam deflection.

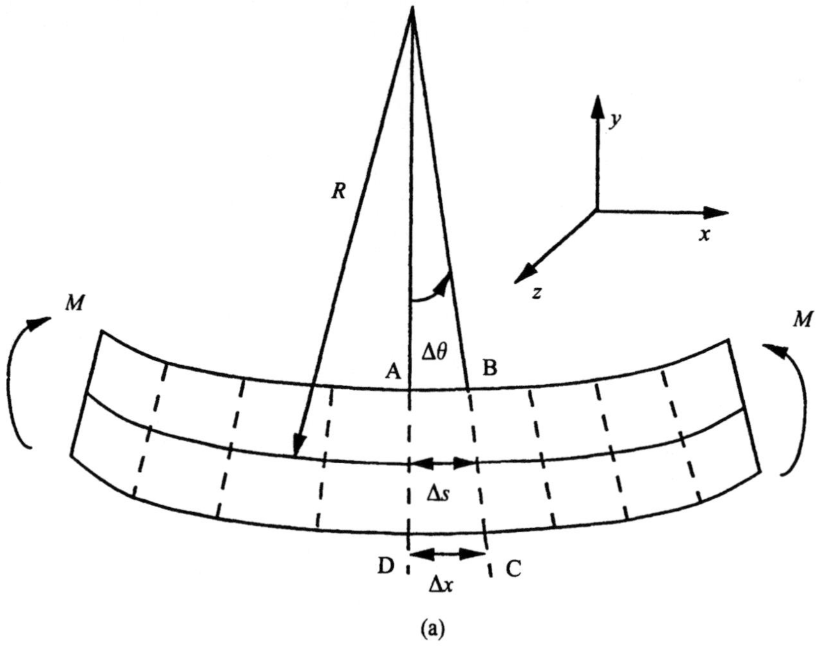

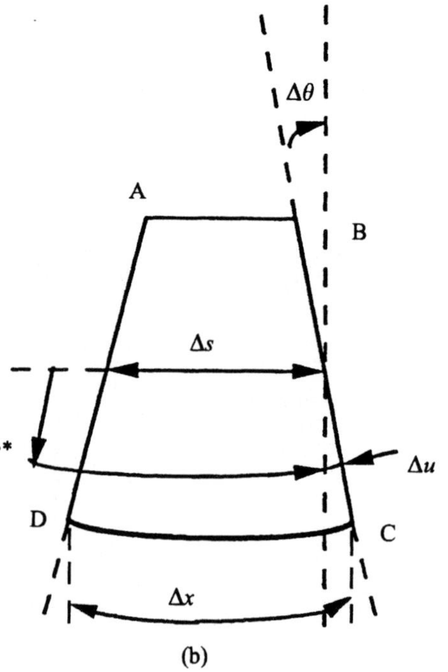

Fig. 1.57 Geometry of beam bending: (a) free-body diagram; (b) distortion of a differential section. Note: the bending moment shown in this diagram producing concave-upward curvature will be assigned a positive value.

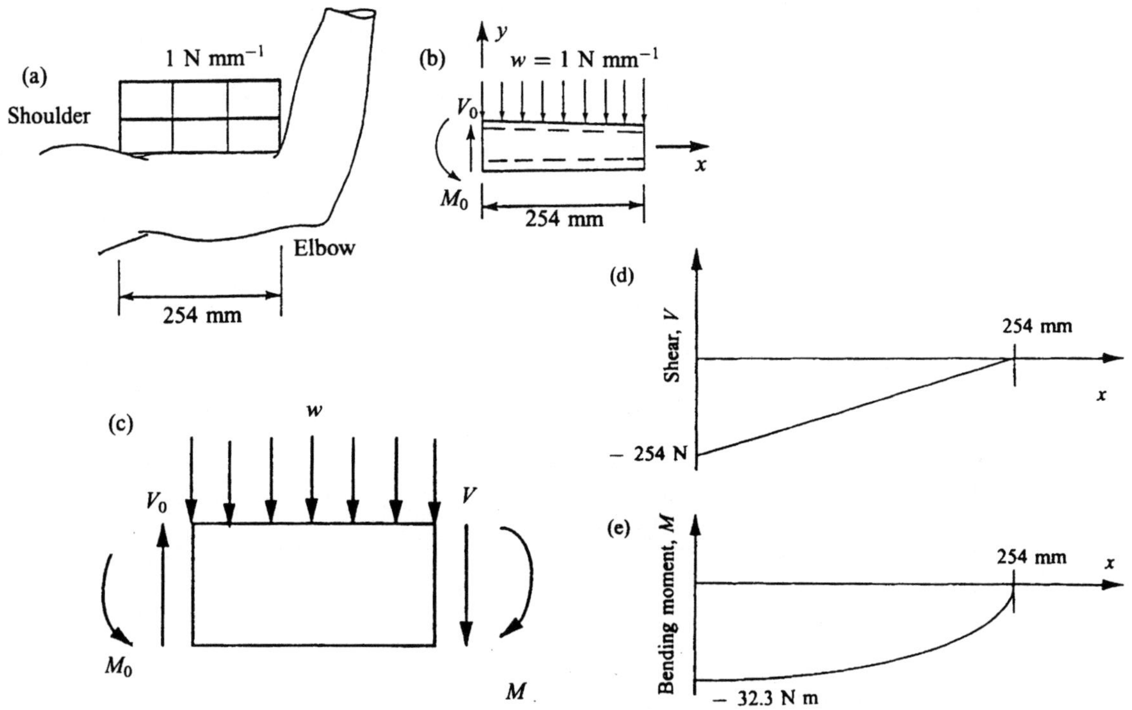

Fig. 1.58 Idealized loading and deflection of the forearm of a bricklayer: (a) schematic; (b) free-body diagram of system; (c) free-body diagram of section; (d) shear diagram; (e) bending moment diagram.

maximum deflection takes place at the elbow; the maximum stress occurs at the shoulder, which is taken as the origin of the coordinate system. The value of c in eqn (1.146) is $\frac{1}{2}D_0 = 25.4$ mm.

The free-body diagram of the system is shown in Fig. 1.58(b); $w\langle x\rangle = -1 \text{ N mm}^{-1}$ (in the negative y-direction). Units are in N and mm. The shear and bending moment relations, from which Fig. 1.58(c) and (d) are obtained, become

$$V = -\int w\langle x\rangle \mathrm{d}x + C = x + C;$$

$$V\langle 254\rangle = 0 = 254 + C; \quad C = -254$$

$$V = -254 + x, N;$$

$$M = -\int V \,\mathrm{d}x + C_1 = 254x - \tfrac{1}{2}x^2 + C_1; \tag{1.153}$$

$$M\langle 254\rangle = 0 = 64\,520 - 32\,260 + C_1;$$

$$C_1 = -32\,260;$$

$$M = -32\,260 + 254x - \tfrac{1}{2}x^2, \text{N mm}.$$

Since the deflection for this case of constant beam properties may be evaluated in the form of the indefinite integral, eqn (1.150) becomes

$$y = \frac{1}{EI'}\int_0^x\int_0^x [M\,\mathrm{d}x]\,\mathrm{d}x + (C_3/EI')x + (C_4/EI'), \tag{1.154}$$

where, for the present example, the conditions of zero moment and shear at the elbow require that the constants of integration C_3 and C_4 are zero. Thus,

$$y = \frac{1}{EI'}\int_0^x\int_0^x \left[\left(-32\,260 + 254x - \frac{1}{2}x^2\right)\mathrm{d}x\right]\mathrm{d}x$$

$$= \frac{1}{EI'}\left(-16\,130x^2 + \frac{127}{3}x^3 - \frac{1}{24}x^4\right). \tag{1.155}$$

For a value of $E = 9$ GPa $= (1.3 \times 10^6 \text{ lbf in}^{-2}) = 9 \times 10^3 \text{ N mm}^{-2}$ for compact bone and $I' = \frac{1}{4}\pi(R_0^4 - R_i^4) = \frac{1}{4}\pi[(25.4)^4 - (19.05)^4] = 0.223 \times 10^6$ mm^4, the maximum deflection becomes

$$y = (9 \times 10^3 \times 0.223 \times 10^6)^{-1}$$
$$\times (-10.41 \times 10^8 + 6.937 \times 10^8 - 1.734 \times 10^8)$$
$$= -0.259 \text{ mm}$$

The maximum stress is

$$|\tau_{xx}|_{max} = \frac{Mc}{I'} = \frac{|(-32\,260 \times 25.4)|}{0.223 \times 10^6}$$
$$= 3.67 \text{ N mm}^{-2}$$
$$= 3.67 \text{ MPa } (=532 \text{ lbf in}^{-2}).$$

1.4.2 Torsion of circular members

In the development of a simple formula expressing the torque applied to a circular member of radius R to its deformation, expressed in terms of an angle of twist ϕ, Fig. 1.59(a), it is assumed that (a) plane sections remain plane, and (b) the shear strain varies linearly with distance from the axis of symmetry to the outer fiber. These conditions, which are very similar to those employed in the beam analysis, are depicted in Fig. 1.59(b). The total torque applied, T, can now be calculated from the diagrams as

$$T = \int_0^R r[|\tau_{xz}|]\, dA = \int_0^R [|\tau_{xz}|_{max}/R]r^2\, dA$$
$$= [|\tau_{max}|/R]\int_0^R r^2\, dA = [|\tau_{xz}|_{max}/R]J', \quad (1.156)$$

where J' is the areal polar moment of inertia and τ_{xz} is the shear stress. For a circular section, $J' = \frac{1}{2}\pi R^4$, so that

$$|\tau_{xz}|_{max} = \frac{TR}{J'} \quad \text{and} \quad \tau_{xz} = \frac{Tr}{J'}. \quad (1.157)$$

By using Hooke's law for the one-dimensional shear stresses,

$$\gamma_{xz} = 2\epsilon_{xz} = \tau_{xz}/G \quad \text{and} \quad \gamma_{xz_{max}} = |\tau_{xz}|_{max}/G, \quad (1.158)$$

and noting for the linear shear strain distribution shown in Fig. 1.58(a) that $\gamma_{xz}\langle x\rangle dx = r\, d\phi$ and hence $\gamma_{max}\, dx = R\, d\phi$, there results the expression

$$\phi = \int_0^x [T\langle x\rangle/(J'\langle x\rangle G\langle x\rangle)]\, dx + D_1, \quad (1.159)$$

where D_1 is the angle of twist at the origin. If this is zero

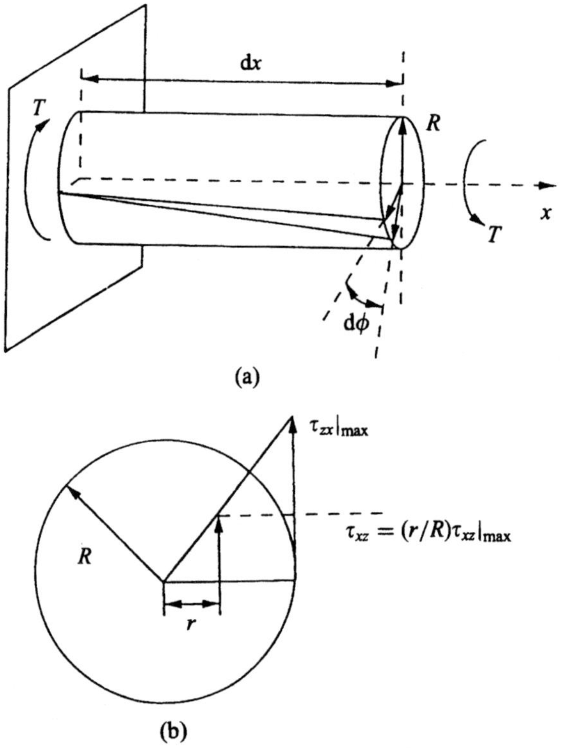

Fig. 1.59 Torsion of a circular bar: (a) deformation pattern; (b) stress distribution for a cross section.

and T and J are constant along the length L of the member, $\phi = TL/J'G$.

Torsion of hollow shafts can be evaluated in a similar manner by using the appropriate value of J'. However, torsion of noncircular members requires a much more complicated analysis. Spiral fractures of the tibia in skiing accidents are invariably produced as the result of a twist applied by ski and boot, which are either fixed or turn opposite to the direction of rotation of the foot and leg. It can be shown by means of a stress transformation using a technique known as Mohr's circle (Popov 1990) that this is equivalent to failure in tension at an angle to the axis of the leg of 45°. Tibial fractures frequently occur at the level of the ski boot due to bending with the boot top acting as a fulcrum.

1.4.3 Buckling of slender columns

When a beam with a length large compared to its cross sectional dimensions is subjected to two equal and oppositely directed compressive axial loads F_a, as shown in Fig. 1.60, the configuration is called a slender column. Large transverse deformation, leading to an instability called buckling, will occur if load F_a is sufficiently large; the initial transverse deflection may, in practice, be due to slight malalignment of the forces, inhomogeneities, or a transverse force F_t which may be very small. A biomechanical example of column action is the support provided by a cane when a patient leans on it heavily.

The analysis will first be carried out without the presence of force F_t. From the beam theory leading to eqn (1.150), the equation of equilibrium is

$$EI'\frac{d^2y}{dx^2} = M = -F_a y \quad \text{or} \quad \frac{d^2y}{dx^2} + \xi^2 y = 0; \qquad (1.160)$$
$$\xi^2 \equiv F_a/EI'; \quad 0 \leq x \leq L.$$

This is a homogeneous differential equation of second order whose solution is

$$y = D_1 \sin \xi x + D_2 \cos \xi x \qquad (1.161)$$

with constants D_1 and D_2 to be determined from the boundary conditions: (a) $y = 0$ at $x = 0$; (b) $y = 0$ at $x = L$. This yields the solutions $D_2 = 0$ and $D_1 \sin \xi L = 0$; however, the case $D_1 = 0$ is trivial, as eqn (1.161) then yields a null identity. Thus, for a nontrivial solution, $\sin \xi L = 0$, implying that $\xi L = L(F_a/EI')^{1/2} = i\pi$, $i = 1, 2, 3\ldots$ The critical force F_a to produce instability, also called the Euler buckling load, is then given by

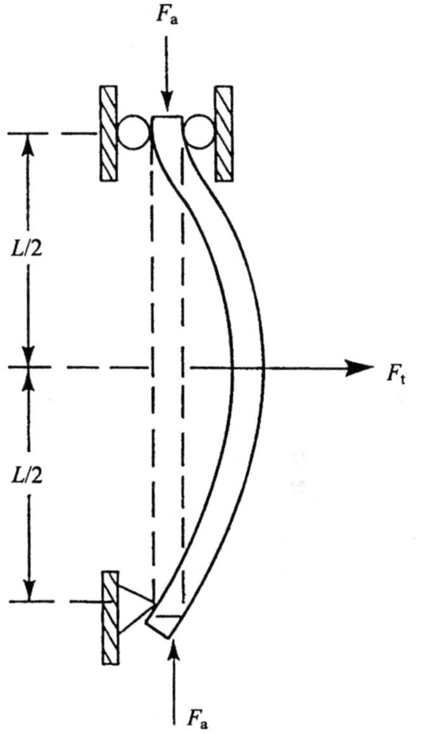

Fig. 1.60 Column action in the presence of a transverse force F_t. Buckling is exaggerated for emphasis.

$$F_c = (i\pi)^2 EI'/L^2, \qquad (1.162)$$

whose minimum value occurs for $i = 1$. The result of eqn (1.162) depends critically on the boundary conditions at the ends, and can, with n as a constant, generally be represented by

$$F_c = n\pi^2 EI'/L^2. \qquad (1.163)$$

For the case of a column clamped at the base and free at end $x = L$, $n = \frac{1}{4}$; for a column clamped at the base and simply supported at the top $x = L$, $n = 2.05$; for a column of length L clamped at both ends, $n = 4$. Further results can be found in Popov (1976, 1990).

Returning to Fig. 1.60, the presence of F_t at midspan adds the term $-\frac{1}{2}F_t x$ to the right-hand side of eqn (1.160), resulting in a nonhomogeneous equation with a particular solution $y_p = -\frac{1}{2}(F_t x)/F_a$, which is added to the homogeneous solution provided by eqn (1.161). The presence of this term changes the lateral deflection, but does not alter the condition of instability, so that eqn (1.162) for the case shown in Fig. 1.60 still applies.

1.4.4 Other loading situations

The stress and deflection analyses for beams, shafts subjected to torsion, and column action before instability were based on the assumption of linearity, and hence superposition applies in the case of combined loading. An example of such a situation occurs in the case of eccentric axial force application to a compression member. It is also evident that such superposition is not permissible when the deflections are large. The accuracy of the results for the simple models presented here can be improved by more rigorous examination using the theories of elasticity (Sokolnikoff 1956; Hearmon 1961; Timoshenko and Goodier 1969; Lekhnitskii 1981; Lur'e 1990; Mal and Singh 1991; Barber 1992; Chou and Pagano 1992), viscoelasticity (Flügge 1975; Christensen 1982; American Chemical Society 1992), and continuum mechanics (Chung 1988; Bowen 1989; Mase and Mase 1992) treated extensively in numerous texts and monographs. More complex geometries, such as plates, shells, tubes, and rings, whose deformation characteristics (Timoshenko and Woinowsky-Krieger 1959; Vinson 1989; Soedel 1993) have been examined in detail, are also suitable for modeling purposes in biomechanical systems.

One other loading condition that generates very high localized stresses can occur in a biological system and requires a brief discussion. Such a situation is called a stress concentration (or stress raiser) and occurs when sudden discontinuities in the otherwise more or less uniform geometry of a structure occur, with particular emphasis on hard tissue or implants. The magnitude of the average stress computed by the techniques described in the preceding section can then be exceeded at such a position by factors as high as five or ten (Petersen 1974; Sih 1981); examples of such a state are represented by a very small hole in a beam or plate, the presence of a screw, or the sudden enlargement of a section without an adequately contoured curved connection.

Obviously, such conditions should be avoided or mitigated, but they may be inadvertently generated by bony components as the result of deformity, impacts, or natural bone loss (osteoporosis), or else may occur in prosthetic devices; it is important that the danger of such an enhanced stress be recognized. The magnitude of the stress concentration factors for many types of discontinuities can be calculated from the theory of elasticity or other continuum mechanical treatments, and they are delineated in the literature concerned with these subjects. Also, finite element procedures have recently permitted the evaluation of the local stress state in small domains of the structure.

However, when the geometry is too complex, or the region of concern is too small compared to the size of the loaded member, analytical procedures may be inadequate for the accurate quantification of this factor. In such cases, experimental methods may be employed for its determination (Kobayashi 1993) such as; for example, the techniques of photoelasticity, holography or caustics (a pattern created by interference of light when passing through a stressed zone). These topics are covered in books or papers dealing with optical stress analysis, a speciality within the realm of experimental mechanics (Dally and Riley 1991; Kobayashi 1993).

Finally, it should be remarked that space constraints prevent the discussion of limiting states of permissible loading that result either in material failure or else in unacceptable deformations. The initiation, propagation, and arrest of cracks has been extensively described in monographs and research papers dealing with the subject of fracture mechanics (Meguid 1989; Freund 1990; Gdoutos 1993; Parton 1992; Kobayashi 1973, 1975); this development applies primarily to hard tissue, principally bone, but also includes nails and teeth. In soft tissue, such as muscles, ligaments, tendons, or skin, as well as in organs, partial or total tearing or puncture constitutes mechanical dysfunction.

Mechanical failure of a material is usually represented by the value of the uniaxial ultimate stress or strain (Yamada 1970; Hastings and Williams 1980) or else by a limiting amount of absorbed energy. When the substance is represented by a constitutive equation, failure can frequently be described by a combination of stress or strain functions whose one-dimensional loading condition produces either an unacceptable permanent deformation or actual separation of tissues in a corresponding uniaxial test. On the other hand, mechanical failure of a biological tissue is an absolute upper bound of an acceptable loading state, since physiological malfunction usually occurs long before the mechanical limit is reached, except perhaps for bone fracture; the latter has been analyzed by the methods of fracture mechanics (Piekarski 1970; Pope and Outwater 1984; Bonfield 1987; Lakes *et al.* 1990). As yet, no correlation exists between these two conceptually diverse failure conditions.

Biomechanical systems are subjected to a very large number of repeated or cyclic loads that, in general, continuously reduce the stress at which the unit will fail, a phenomenon known as fatigue fracture caused by repetitive plastic deformation (Juvinall and Marshek 1991; Suresh 1991; Dowling 1993). Crack propagation initiates at a discontinuity, possibly microscopic, and continues slowly across the surface until the remaining intact area can no longer sustain the load, at which point catastrophic fracture occurs. For some steels and a few other metals, the diminution in resistive strength compared to some fraction of the ultimate strength for a single load application is represented by a linear logarithmic plot of allowable stress (S) as a function of the number of cycles (N) (see problem 1.28); a level is reached, called the endurance limit, beyond which the strength does not decrease further. Other materials do not exhibit such a well-defined endurance limit, and the degree of strength reduction varies enormously with material composition, being generally much less severe in the case of composites. Another method of defining a fatigue strength is to specify the allowable stress at a given number of cycles, say 10^6. This concept has substantial implications for implants, such as heart valves which must withstand tens or even hundreds of millions of cycles without failure; it is also important for harder tissues such as bone, enamel, and teeth. The ramifications of this loss of strength are further discussed in Chapter 9.

1.4.5 Final comments

The present introduction to analytical modeling of biomechanical processes can not provide an appreciation of the vast breadth and depth of the discipline. References and a further reading list are provided that give some of the major sources in solid mechanics, stress concentrations, and fracture for those who would like to expand their horizons in the engineering science area. On the other hand, the biomechanical bibliography presented here is intended to acquaint the reader with some of the major journals, texts, and proceedings of special meetings dealing with this subject, and some of its subspecialties. In addition, a few citations of individual papers are given that provide some insight into the variety of topics that fall within this purview.

Furthermore, experimental measurement of parameters involved in the characterization of the behavior of organic components or in biomechanical processes is crucial for a full understanding of the performance of biological systems. The resulting data are required to assess the efficacy of models intended to represent the activities of organisms and must serve as the sole portrayer of the event when, for whatever reasons, analysis or numerical models can not be constructed.

Bioinstrumentation utilized in such experiments is an independent field of endeavor which has accumulated a literature in its own right (Normann 1988; D. A. Christensen 1988; Wise 1990, 1991). In certain cases, techniques of measurement used to determine the behavior of inanimate systems, such as the mechanical, electrical, or optical measurement of position, velocity, acceleration, and applied force, stress, or strain (Paipetes and Holister 1985; Lagarde 1987; Ostrovsky *et al.* 1991; Kobayashi 1993) can be transferred directly to the field of biomechanics (Norkin and White 1985; Miles and Tanner 1992). However, precautions must be taken to ensure that the transducers employed will faithfully record the relatively large deformations experienced by soft tissue (compared to most inanimate structural members).

Linear and angular accelerometers securely attached to specific points of human or animal anatomy can also be applied to ascertain both motion histories (Dove and Adams 1964) and associated forces; six accelerometers are required to determine the three-dimensional motion of an object treated as a rigid body. Radiography (X-rays), tomography, and magnetic resonance imaging can provide information concerning internal configurations: flash X-ray or motion picture radiographs can extend this type of observation to dynamic situations. A full discussion of the methodology and the limitation of such measurements will require substantially more space than available here, and the reader is referred to the critical literature in this field.

The material presented in the previous sections constitutes the elements of solid mechanics as applied to topics and problems in biomechanics, a subject which is further subdivided into numerous individually recognized disciplines. These include, but are not limited to: locomotion (kinesiology); biomaterials; orthopedic biomechanics; biomechanics of sports; injury biomechanics, including human response involved in vehicular collisions as well as industrial and home accidents; and the biomechanical aspects of the interaction of humans and the work environment (the man–machine interface), constituting a portion of the subject known as ergonomics (Chaffin and Anderssen 1991). Each of these disciplines has acquired its own very substantial literature, selectively represented in the bibliography at the end of this chapter. A thorough portrayal of each of these themes would require a complete individual text. The first three specialties, which exemplify the diversity and the enormous breadth of the field of biomechanics, are discussed in more detail in subsequent chapters. Broad coverage of the field of biomechanics may be found in two preeminent journals covering this subject: the *Journal of Biomechanics* published by Pergamon Press, and the *Journal of Biomechanical Engineering* published by the ASME, as well as in numerous more specialized periodicals.

Worked examples

E1. A 70 kg person is standing rigidly on the tip of the toes at A (Fig. 1.61) so that the sole of the foot makes an angle of 15° with the floor. The mass center, G, in this position is 0.9 m from the back of the sole at C; each arm and leg carries half the total load. Distance $\overline{AC}$ is 0.2 m; the outstretched arms make an angle of 20° with the horizontal, and line $\overline{BG}$, 0.85 m in length, makes an angle of 30° with the horizontal. Friction acts so that all resultant forces pass through G. Determine the force components carried along the axes of the arms and legs.

Solution. The forces acting on the sytem consist of the weight of $70 \times 9.81 = 687$ N passing vertically through G and the reactions at contact points A and B. In a three-force system, the forces must be concurrent and pass through G, by hypothesis, as shown in the free-body diagram, Fig. 1.61(b); F_B is inclined at 30° to the horizontal. $\overline{AG} = [(0.9)^2 + (0.2)^2]^{1/2} = 0.922$ m; thus angle $\phi = \sin^{-1}[0.9/0.922] = 77.46°$. Thus, the angle $\overline{AG}$ makes with the horizontal is $\beta = 180° - 77.46° - 15° = 87.54°$ whose complement is 2.46°. From the concurrent force diagram, $\gamma = 87.5° - 30° = 57.5°$. Thus $\sin\gamma/\sin\delta = \sin 57.5°/\sin 120° = 687/F_A; F_A = 705$ N; $\sin 57.54°/\sin 2.46° = 687/F_B$; $F_B = 34.9$ N. These are the total reaction forces acting at the floor and wall, respectively. They must be resolved in the direction of the extremities in order to determine the components transmitted along the axes of the arms and legs. Thus, as shown in the diagram, Fig. 1.61(c), $F_{legs}/F_A = (0.9)/(0.922)$; $F_{legs} = 688$ N, or each leg carries 344 N. Also, for the arms, Fig. 1.61(d), the right triangle for the arm and F_A gives: $F_{arms} = F_B \cos 10° = 34.4$ N; per arm, this is 17.2 N. Note that the force along the centerlines of the arms and legs is not the total force that is supported; there is a transverse force acting normal to each leg of $(0.2 \times 353)/(0.922) = 76.6$ N and also a transverse force of $\frac{1}{2}(34.9) \sin 10° = 3.0$ N carried by each arm.

E2. [Variation of Hudson and Johnson (1976) and Ghista (1982)]. A rock climber with a weight W climbing up a chimney is shown in equilibrium in Fig. 1.62(a); the feet and shoulders press against the vertical sides whose friction coefficients are f_1 and f_2, respectively. What is the vertical distance, d, between contact points A (foot) and B (shoulder)?

Solution. Again, the climber is a three-force member with total reactions R_A and R_B, each composed of a normal and frictional (tangential) force N_A, f_1N_A and N_B, f_2N_B, respectively, that must meet at a point O on the vertical line through the mass center G. From eqns (1.20), equilibrium requires (a) horizontally: $N_A = N_B = N$; (b) by summing moments about B and A: $Wb + Nd - f_1Nc = 0$ and $Wa - f_2cN - Nd = 0$, and thus $d = [f_1c - (W/N)b]$ and $d = (W/N)a - f_2c$, or $d = af_1 - bf_2 = a(f_1 + f_2) - cf_2$. This shows that the larger is a (for fixed f_1 and f_2), i.e., the more the back is straightened, the larger is d. Also, if $f_1a < f_2b$ (when A is slippery), d may be negative, as shown in the diagram, Fig. 1.62(b), i.e. the support point at B is below that at A. The same equations of equilibrium give: (a) $N_A = N_B = N$; $M_B = 0 = Wb - f_1(a + b)N - Nd$; $M_A = 0 = Wa - f_2Nc + Nd$, so that $d = f_2b - f_1a > 0$, the same actual result as above, as required. Distances a and b may vary, but their sum equals the chimney width c.

E3. The head, Fig. 1.63, considered as a rigid body for the present purpose, rocks on the occipital condyles C where an axial force, F_A, a shear force, V, and a neck torque, T_0, resist motion. A blow (treated as an instantaneous load B at an angle of 63° to the horizontal) is applied to the chin, and the initial linear acceleration, a, of the mass center, G, is photographically determined to be 140g. For a head mass of 3.5 kg, a moment of inertia about an axis perpendicular to the sagittal plane of 0.0356 kg m^2, and the dimensions shown, what are the

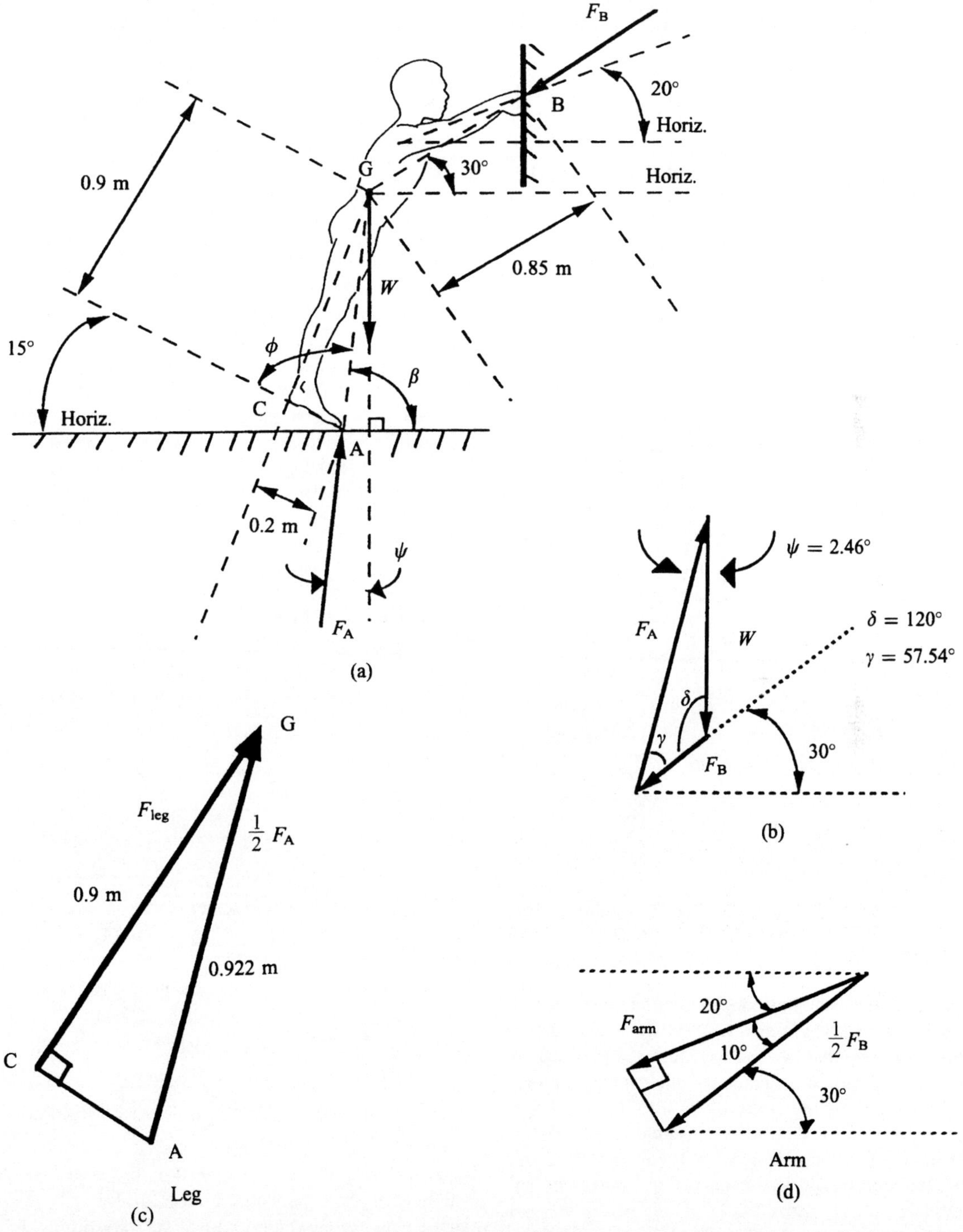

Fig. 1.61 Person pushing against a vertical wall: (a) schematic; (b) vector diagram of external forces; (c) force along leg; (d) force along arm.

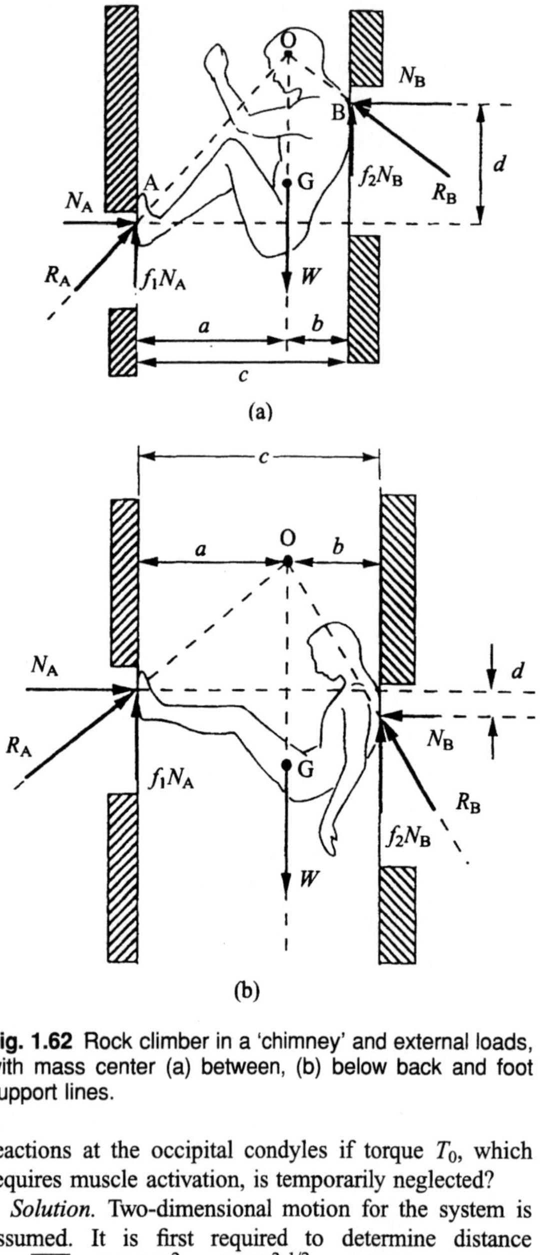

Fig. 1.62 Rock climber in a 'chimney' and external loads, with mass center (a) between, (b) below back and foot support lines.

reactions at the occipital condyles if torque T_0, which requires muscle activation, is temporarily neglected?

Solution. Two-dimensional motion for the system is assumed. It is first required to determine distance c: $\overline{MG} = [(0.08)^2 + (0.13)^2]^{1/2} = 0.1526$ m. Angle DMG $= \tan^{-1}$ 0.13/0.08 = 58.4°. Angle PMG = 63° − 58.4° = 4.6°; $c = \overline{PG} = \overline{MG} \sin 4.6° = 0.0122$ m. The equations of motion, eqns (1.72), in the inertial coordinate system are: (a) $\Sigma F_x = B_x - V = m\ddot{x}$ (b) $\Sigma F_y = B_y - mg - F_A = m\ddot{y}$ (c) $\Sigma M_G = F_A(0.02) + V(0.08) + Bc + T_0 = I_G\alpha$, or, resolving the observed acceleration, which is normal to line $\overline{CG}$, into x and y components: line $\overline{CG}$, of length $[(0.08)^2 + (0.02)^2]^{1/2} = 0.0825$ m, is inclined at $\tan^{-1}(8/2) = 76°$ to the $-x$-axis. Now $\ddot{x} = a \cos(90° - 76°) = 140 \times 0.970 \times 9.81 = 1333$ m s^{-2}; $\ddot{y} = a \sin 14° = 140 \times 0.242 \times 9.81 = 332$ m s^{-2}. Also from eqns (1.20), (a): $B \cos 63° - V = (3.5)(1333) = 4666$ N. Further, for (b): $B \sin 63° - 3.5 \times 9.81 - F_A = (3.5)(332)$ or $0.891B - F_A = 1196$. The moment equation, the last of eqns (1.72), is (c): $0.02F_A + 0.08V + 0.0122B + 0 = 0.0356\alpha = 0.0356(140 \times 9.81/0.0825) = 592.6$ N m, as the initial angular acceleration about pivot point C is the linear acceleration a divided by distance $\overline{CG}$. Solving (a), (b), and (c) gives $V = 2112$ N, $F_A = 12\,110$ N and $B = 14\,930$ N. Note that, if torque T_0 is not negligible, the system is indeterminate and an additional condition must be specified. As a point of interest, accelerations of 250 g due to a gloved blow at 8 m s^{-2} have been observed in boxing simulations (Johnson *et al.* 1975; Ghista 1981). It was advocated early that 4480 N (1000 lbf) should represent the upper tolerance limit for impact forces to the head (Gurdjian 1975), but this has been long exceeded.

E4. Consider 'schussing' in the fall-line in alpine skiing (Fig. 1.64); it is desired to estimate the terminal velocity. If θ is the maximum downhill slope, F_s is the sliding friction with constant coefficient f, F_D is air drag given by eqn (1.25) as $\frac{1}{2}C_DA'\rho v^2$, and L is the lift, then equilibrium along y gives:

$$\Sigma F_y = mg \cos \theta - L - N = 0$$

so that $F_s = f(mg \cos \theta - L)$ and eqn (1.72) for x is:

$$\Sigma F_x = m(\mathrm{d}v/\mathrm{d}t)$$
$$= mg \sin \theta - f(mg \cos \theta - L) - \tfrac{1}{2}C_DA'\rho v^2$$

where C_D is the drag coefficient, A' the projected area of the skier, and ρ the density. Ignoring the lift and assuming C_D constant yields

$$\frac{\mathrm{d}v}{\mathrm{d}t} = g \cos \theta[\tan \theta - f] - (C_D\rho A'v^2)/2m)$$

and substitution in this relation of the third of eqns (1.33), gives

$$v = [(1 - e^{-2Bx})(C/B)]^{\frac{1}{2}};\ B = (C_D\rho A'/2m)$$
$$\text{and } C = g \cos \theta(\tan \theta - f)$$

with the initial conditions $v_0 = 0$ and $x_0 = 0$. For large values of x, the exponential term in the velocity relation approaches zero and the terminal velocity becomes:

$$v_{\text{terminal}} = (C/B)^{\frac{1}{2}} = [(2mg/C_D\rho A')\cos \theta(\tan \theta - f)]^{\frac{1}{2}}.$$

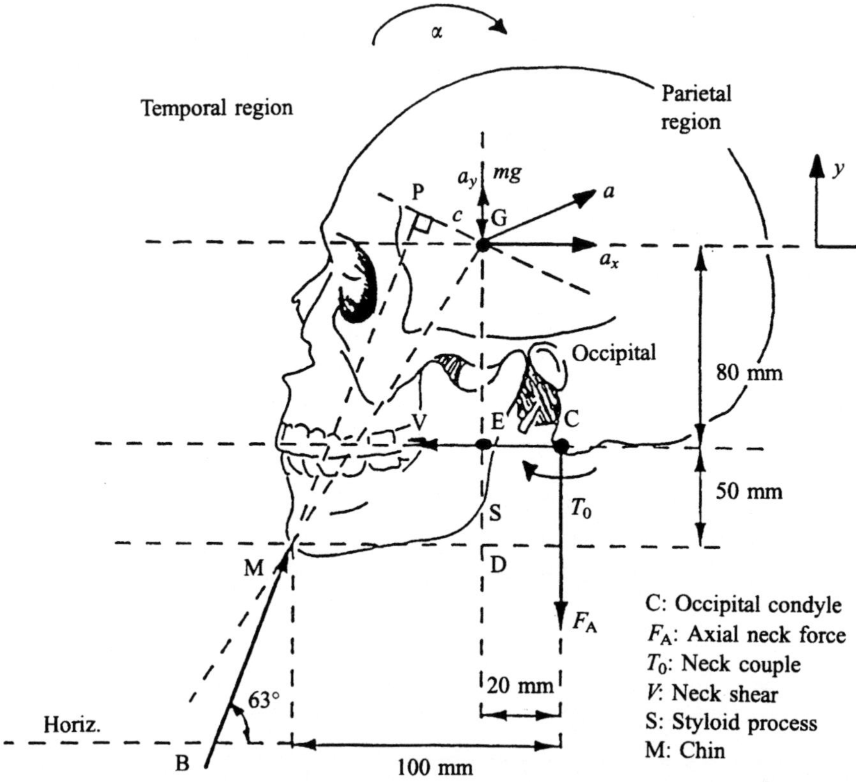

Fig. 1.63 Loading diagram on a human head.

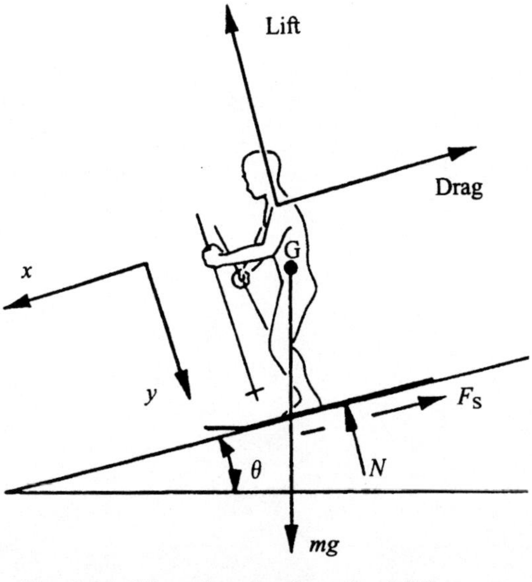

Fig. 1.64 Skier schussing down the fall line.

Minimum values of sliding friction for skiing are about 0.02, while the term $2mg/C_D\rho A'$ has a range from 1000 $m^2\ s^{-2}$ for a slow skier to 8000 $m^2\ s^{-2}$ for a fast skier. For an intermediate value of, say 2500 $m^2\ s^{-2}$, a friction coefficient of 0.08, and a slope of 25°, the above expression gives a theoretical maximum speed value for schussing of 29.6 m s^{-1}, which is very fast indeed.

E5. Assume that a broad jumper can attain a maximum horizontal speed of 10 m s^{-1}. How should he jump so as to attain a maximum horizontal distance (by pushing off correctly) and what will this distance be? The time for pressing on the ground during push-off is 0.1 s and the force is constant. Neglect air resistance.

Solution. Although well-known, the equation for the maximum range of a body moving with a given initial velocity, but variable angle θ to the horizontal under gravity will be developed. The kinematic equations of two-dimensional motion provide separate, independent relations for x and y from eqns (1.33): $x = (v_0 \cos \theta)t$;

$y = (v_0 \sin\theta)t - \frac{1}{2}gt^2 = 0$ (to return to the horizontal level). Thus $t = (2v_0 \sin\theta)/g$ and $x_f = 2(v_0^2/g)\sin\theta\cos\theta$. For maximum range, $dx/d\theta = 0 = \cos^2\theta - \sin^2\theta$; $\tan\theta = 1$, $\theta = 45°$, $x_{max} = (2v_0^2/g)\ \sin 45°\ \cos 45° = v_0^2/g = (10)^2/9.81 = 10.2$ m. During pushoff, with constant force and lasting 0.1 s, the jumper must change horizontal velocity partly to vertical velocity. The vertical component of the force exerted by the ground on the foot (and hence on the Achilles tendon) produces the impulsive change from horizontal velocity v_0 to vertical velocity $0.707v_0$, and the impulse–momentum relation, eqn (1.48), yields $0.1F = m(0.707v_0)$.

For a 75 kg jumper, the force is $F = 75 \times 10 \times 0.707/0.1 = 5303$ N. Assuming a cross-sectional area of 2 cm^2 = 200 mm^2 for the Achilles tendon, which provides the mechanism for this push-off, the stress in this member is $5303 \times 10^6/200 = 26.5$ MPa. Yamada (1970) lists the ultimate strength of this tendon as 5.6 kgf mm^{-2} = 54.9 MPa; thus the factor of safety is 54.9/26.5 = 2.07, a value considered to be dangerously low. On the other hand, the jumping achievement is optimum, and for an experienced jumper, the strength of the tendon is likely to be higher than average. Data by Miller and Nelson (1973) and Hay (1993) indicate peak performances of 6.5 m with a pushoff time of 0.12 s, a vertical force of 4270 N, an initial velocity of 9.45 m s^{-1}, and a take-off angle of 22.2°. The *Guinness book of sports records* (1992) lists the longest jump as 8.9 m, which is not far from the theoretical maximum. At the same take-off velocity, the distance achieved in the broad jump with angles of inclination of 35° and 50° would be 9.52 m and 10.02 m, respectively, indicating that the distance achieved is not greatly sensitive to the actual take-off angle in the vicinity of 45°.

E6. [Adapted from Feller *et al.* (1985)]. A patented device for separating dirt clods from potatoes gathered from the field by a mechanical harvester consists of the feed conveyor shown in Fig. 1.66(a), which moves with a constant horizontal speed of $v_0 = 0.325$ m s^{-1}, a roller with a diameter of 320 mm rotating clockwise with a speed of 50 rev min^{-1}, and a bin with a separator S which divides the rebounding potatoes from the surrounding dirt clods. The separation is achieved by virtue of the difference in the coefficients of restitution for the potatoes, which is 0.55, while it is 0.25 for the dirt clods. To prevent bruising, the center of the roller is placed at the maximum distance of $h_1 = 250$ mm below the conveyor belt and the impact point P is located at the 45° position of the roller. The final tangential velocity after impact is the average of the initial tangential velocity of the objects and the roller tangential velocity; this effect is due to friction. The top of the separator is located at a distance $h_2 = 175$ mm from the roller center.

(a) What is the total velocity of both potatoes and clods at the instant of impact with the roller and its direction relative both to the 45° line and the vertical? What is the normal rebound velocity of the potatoes and the clods?

(b) What is the horizontal position d_1 of the center of the roller relative to the end of the conveyor for the objects to strike at the 45° position, P?

(c) What are the final tangential velocities after impact and the total horizontal and vertical velocities of both potatoes and clods after impact?

(d) What must be the position of the separator, distance d_2, to be located halfway between the points of the trajectories of potatoes and clods 175 mm below the center of the roller?

Solution. Attach an x–z coordinate system to the entire unit as shown in Fig. 1.66(b) and maintain directions throughout the analysis. Let distances d and b represent the horizontal and vertical separations, respectively, from the conveyor belt end and the point of contact P,

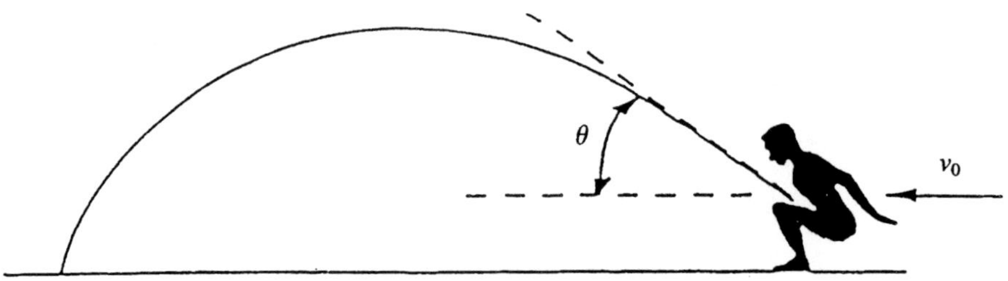

Fig. 1.65 Trajectory of a broad jumper.

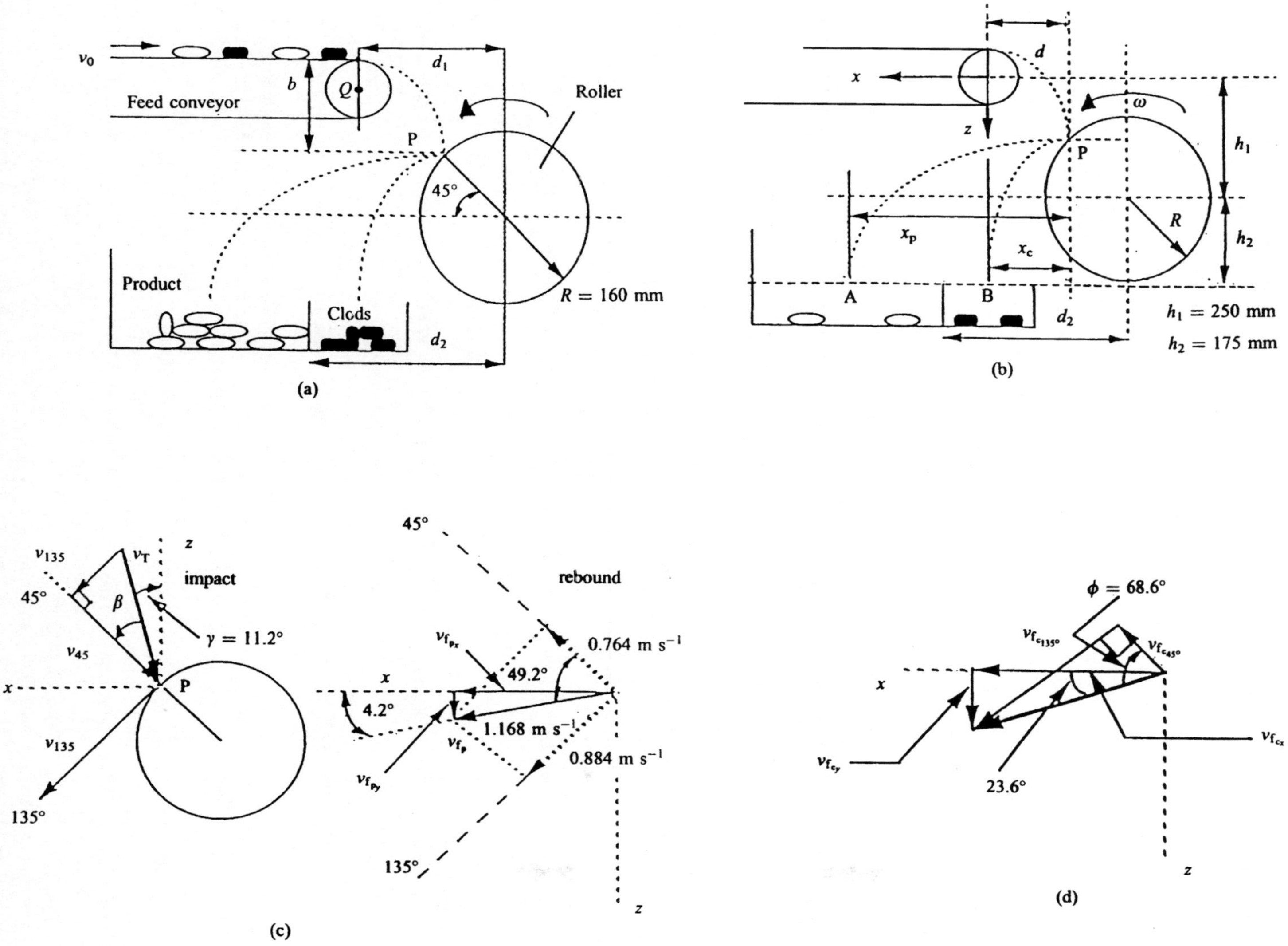

Fig. 1.66 Conveyor belt drop for separating potatoes from dirt clods: (a) schematic; (b) dimensions; (c) velocity components before and after impact for potatoes; (d) final velocity of clods after impact.

respectively, and let x_p and x_c represent the horizontal distances from contact point P to points A and B in the horizontal plane at $h_2 = 175$ mm below the roller center where the trajectories of the rebounding products and clods intersect the top bin surface. Using the laws of motion in the vertical plane under the influence of gravity only, and assuming free vertical motion starts at point Q, there results:

(a) Distance $b = 0.250 - R \sin 45° = 0.1369$ m. From eqn (1.34),

$$b = \tfrac{1}{2}gt^2: \ 0.1369 = \tfrac{1}{2}(9.81)t_1^2;$$

t_1 = time from leaving conveyor to striking roller = 0.167 s with speed $v_z = gt_1 = 9.81 \times 0.167 = 1.639$ m s^{-1}. The total impact speed at P, $v_T = [(1.64)^2 + (0.325)^2]^{\frac{1}{2}} = 1.672$ m s^{-1}. The angle between v_T and the vertical, $\gamma = \tan^{-1}$ (0.325)/(1.64) = 11.2°, as indicated in Fig. 1.66(c). The angle between v_T and the 45° line is thus $\beta = 45° - 11.2° = 33.8°$. The components of the total velocity v_T along the 45° and 135° lines are $v_{45} = -1.672 \cos 33.8° = -1.389$ m s^{-1} (down, to the right) and $v_{135} = 1.672 \sin 33.8° = 0.93$ m s^{-1} (down, to the left), respectively. (The signs for these velocities are arbitrarily taken to be positive if their x-component is positive). After rebound, v'_f, the normal components for the product and the clods are (eqn 1.62)

$$v'_{f_{prod}} = -ev_{45_{prod}} = -0.55 \times (-1.389)$$
$$= 0.764 \text{ m s}^{-1} \text{ (up and to left)},$$

$$v'_{f_{clod}} = -ev_{45_{clod}} = -0.25 \times (-1.389)$$
$$= 0.347 \text{ m s}^{-1} \text{ (up and to left)}.$$

(b) Distance from the conveyor end to impact point P is: $d = v_0 t_1 = 0.325 \times 0.167 = 0.0543$ m. Thus, $d_1 = d + R \sin 45° = 0.0543 + 0.160 \times \sin 45° = 0.1674$ m.

(c) $v_{roller} = 0.160(2\pi)(50/60) = 0.8378$ m s^{-1};

$$v_{tang,_{prod, final}} = v_{tang,_{clod, final}} \equiv v_t$$
$$= \tfrac{1}{2}(0.93 + 0.8378) = 0.8839 \text{ m s}^{-1}.$$

The total final product velocity is $v_{f_p} = [(0.764)^2 + (0.884)^2]^{\frac{1}{2}} = 1.168$ m s^{-1} inclined at angle $\tan^{-1}$ (0.884/0.764) = 49.2° to the 45° line; for the clod, $v_{f_c} = [(0.347)^2 + (0.884)^2]^{\frac{1}{2}} = 0.950$ m s^{-1} inclined at the angle $\phi = \tan^{-1}$ (0.884/0.347) = 68.6° from the 45° line, as shown in Fig. 1.66(d). Their velocity components in the x and z directions are:

$$\text{Product}: \ v_{f_{p,x}} = 1.168 \cos(49.2° - 45°) = 1.165 \text{ m s}^{-1};$$
$$v_{f_{p,z}} = 1.168 \sin 4.2° = 0.0855 \text{ m s}^{-1};$$
$$\text{Clod}: \ v_{f_{c,x}} = 0.950 \cos(68.6° - 45°) = 0.8705 \text{ m s}^{-1};$$
$$v_{f_{c,z}} = 0.950 \sin 23.6° = 0.380 \text{ m s}^{-1}.$$

Hence $v_{f_c} = [(0.8705)^2 + (0.380)^2]^{\frac{1}{2}} = 0.950$ m s^{-1}.

(d) The vertical distance from the impact point to the top of the separator is

$$z = h_2 + R \sin 45° = 0.175 + 0.160 \sin 45°$$
$$= 0.175 + 0.113 = 0.288 \text{ m}.$$

Using eqns (1.34), for the product:

$$z = v_0 t_2 + \tfrac{1}{2}gt_2^2: \ 0.288 = 0.0855t_2 + \tfrac{1}{2}(9.81)t_2^2;$$
$$t_2^2 + 0.0174t_2 - 0.0587 = 0$$

t_2, the time from impact of the product to its arrival at the separator top is 0.234 s. $x_p = v_{f_{p,x}} t_2 = 1.165 \times 0.234 = 0.271$ m. Similarly, for the clod, $0.288 = 0.380t_3 + \frac{1}{2}(9.81)t_3^2$; $t_3 = 0.206$ s. Thus, $x_c = v_{f_{c,x}} t = 0.8705 \times 0.206 = 0.179$ m. Hence, the distance of the separator from the center of the roller is

$$d_2 = \tfrac{1}{2}(0.271 + 0.179) + 0.160 \cos 45° = 0.338 \text{ m}.$$

E7. A man swings a hammer which rotates about the elbow through an angle of 110° to strike a nail. The mass of his forearm is 1.54 kg, with a 0.12 m distance from mass center to elbow, and a transverse centroidal moment of inertia of 0.01 kg m^2, Fig. 1.67(a). His hand, of mass 0.58 kg (that will be treated as a point mass), is located 275 mm from the elbow; the hammer consists of a circular handle with a mass of 0.25 kg and a length of 0.18 m with a 0.9 kg head at its tip that is also treated as a point mass. (a) If the man exerts a constant elbow moment of 50 N m, what is the velocity with which the head hits the nail and the angular acceleration of the arm? (b) If the rotation of the system is stopped in 0.05 s, the mass of the nail is neglected, and the distance from elbow to nail is L, what is the force transmitted to the hammer head by the object struck, assuming no rebound?

Solution. The motion is fixed-axis rotation about the elbow, a noncentroidal point, Fig. 1.67(b).

(a) First, the individual centroidal moments of inertia of the various parts are calculated and transferred to the elbow. From Subsection 1.1.13,

$$I_{\text{hammer shaft}} = \tfrac{1}{12}mL_h^2 = \tfrac{1}{12} \times 0.25 \times (0.18)^2$$
$$= 0.000675 \text{ kg m}^2.$$

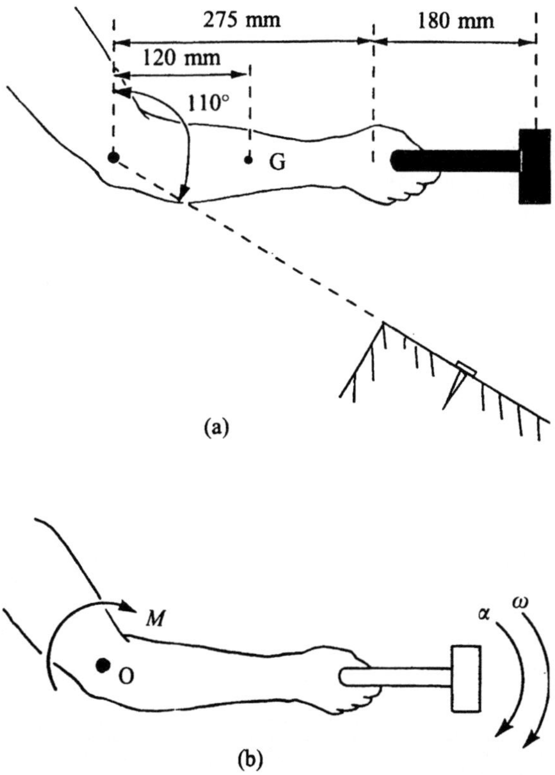

Fig. 1.67 Person striking object with hammer: (a) schematic; (b) motion diagram of lower arm.

Using the parallel axis theorem, the inertial moment about the elbow is:

$$I_{\text{elbow}} = I_{\text{arm}} + I_{\text{hand}} + I_{\text{shaft}} + I_{\text{tip}}$$
$$= [0.01 + 1.54 \times (0.12)^2] + [0.58 \times (0.275)^2]$$
$$+ [0.000675 + 0.25(0.365)^2 + 0.9(0.455)^2]$$
$$= 0.296 \text{ kg m}^2.$$

From the third of eqns (1.74): $M_O = I_O\alpha$: $50 = 0.296\alpha$; $\alpha = 169$ rad s^{-2}; $\omega^2 = \omega_0^2 + 2\alpha\theta$: $\omega^2 = 2 \times 169 \times (110/[360/2\pi]) = 648.9\text{rad}^2 \text{ s}^{-2}$; $\omega =$ 25.5 rad s^{-1}; $v_{\text{tip}} = R\omega = 0.455 \times 25.5 = 11.60$ m s^{-1}.

(b) This solution uses the angular impulse–momentum equation (1.75) applied to a rigid body; further, an assumption is required concerning the shape of the impulse. If this shape is assumed to be a square, i.e. a step function, then, from eqn (1.51), for a constant moment M, there results: $M\Delta t = F \times L\Delta t = I\Delta\omega$, or $0.455F \times 0.05 = 0.296 \times 25.5$, so that $F = 332$ N. The energy transformed into friction and plastic deformation is $(\frac{1}{2}I\omega^2 = (\frac{1}{2})(0.296) \times (25.5)^2 = 96.2$ N m.

E8. A motorcyclist wearing a helmet is riding at a speed of 56 km h^{-1} when he is struck from the side at a right angle by a car travelling at 48 km h^{-1}, Fig. 1.68(a). The total mass of cycle and rider is 364 kg, and that of the car and occupants is 1500 kg. The coefficient of restitution between the vehicles is zero.

(a) If the cycle rider is jarred loose, acquiring the velocity of the cycle immediately after the impact, and is observed to leave the cycle at 20° to the horizontal, where will he land on the pavement relative to the initial impact point? Assume that his mass center is initially 0.75 m above the pavement and that he lands horizontally.

(b) Assuming no loss of speed from air resistance, what is his kinetic energy at impact with the pavement if his total mass is 72.5 kg?

(c) If he does not land horizontally, but strikes the pavement on the helmeted portion of his head, what is the energy that must be absorbed by helmet and head if (i) only the mass of head, neck and helmet, about 7.25 kg, needs to be stopped (with the neck flexibility isolating the effect of the impact of the rest of his body), (ii) if in addition the neck transmits enough force so that the energy due to a portion of his torso with an additional mass of 22.5 kg must also be absorbed? If a superior motorcycle helmet can absorb about 136 J (100 ft lbf) of energy, is the protective capacity of the helmet adequate for this case?

(d) From dummy test results, the force–time curve for such an impact can be approximated by a triangle, Fig. 1.68(b), of 10 ms (0.01 s) duration—about twice that found in controlled laboratory tests. What is the peak force experienced by the head–helmet system if there is no rebound? Compare this with one suggested safe limit (Gurdjian 1975) of 4460 N (1000 lbf).

(e) Prior to use, the helmet worn in the accident, which has a mass of 1.36 kg, was impact-tested according to government standards. The helmet is draped over an inverted headform containing an accelerometer, mounted on an arm; the total mass of the unit is 4.55 kg (10 lb). The entire assembly is then dropped from a height of 1.83 m (6 ft) onto a steel anvil; rebound is negligible. The acceleration history, Fig. 1.68(c), is also found to be approximately triangular with a rise time of 0.001 s and a duration of 0.004 s, Fig. 1.68(f). If this system can be regarded as a rigid body, or where appropriate, as a

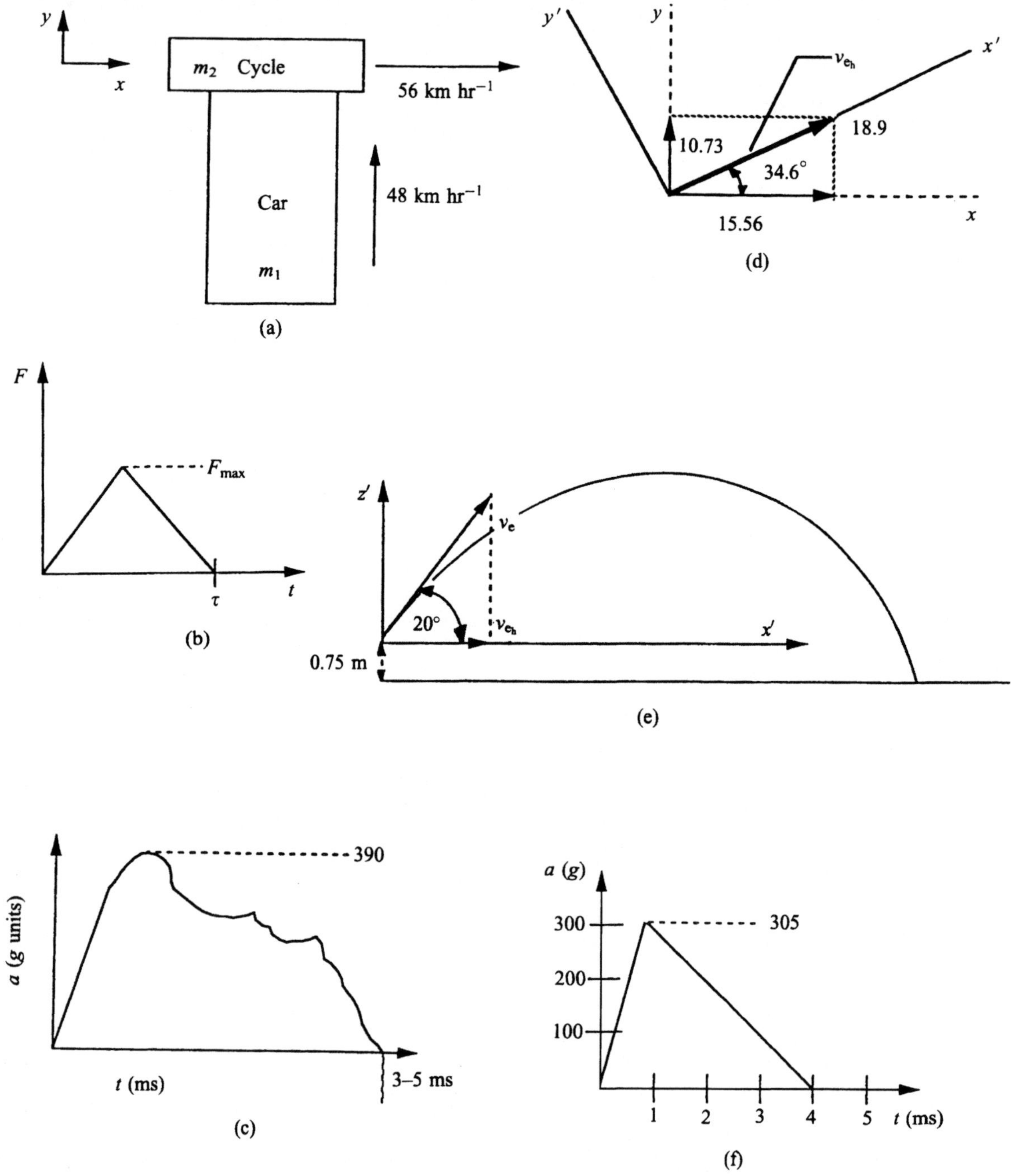

Fig. 1.68 Collision of car and motorcycle: (a) schematic; (b) force history for the impact of rider with pavement; (c) typical motorcycle helmet test record of acceleration history; (d) horizontal ejection velocity diagram of motorcycle rider; (e) rider trajectory in vertical plane; (f) acceleration history for present helmet.

particle, what is the maximum force and maximum acceleration? Compare this with the NHTSA (National Highway Traffic Safety Administration) 1984 rule that an acceleration must (1) never exceed 400 g, (2) not exceed 200 g for more than 2 ms, and (3) not exceed 150 g for more than 4 ms.

(f) Compare the conditions of (e) with another standard, the severity index $SI = \int_0^\tau a^{2.5}\,dt < 1500$, where a is in units of g, time is in units of seconds, and τ is the impact duration. Also compare the result to another suggested limit (Ono *et al.* 1980) of 140g for the threshold of concussion and 200g for skull fracture.

Solution. Consider the collision as shown in the diagram, Fig. 1.68(a): x is the initial direction of the cycle, mass 2, and y is the initial direction of the car, mass 1. The cycle, rider and car will be treated as particles.

(a) $v_{f_{x2}} = v_{o_{x2}} = 56$ km h^{-1} $= 15.56$ m s^{-1}; $v_{o_{y1}} =$ 48 km h^{-1} $= 13.3$ m s^{-1}. From eqn (1.61), conservation of linear momentum in the y direction, with v_f as the common velocity at the point of maximum compression (or the instant when the cyclist is jarred loose), gives for the y-velocity component of the cycle: $m_1 v_{o_{y1}} = (m_1 + m_2)v_{f_y}$: $1500 \times 13.33 = [1500 + 364]v_{f_y}$; hence, $v_{f_y} = 10.73$ m s^{-1}. The rider is thrown off by the joint velocities of v_{f_y} and the cycle. The horizontal ejection velocity component $v_{e_h} = [(10.73)^2 + (15.56)^2]^{1/2}$ $= 18.9$ m s^{-1} at an angle $\beta = \tan^{-1}(10.73/15.56)$ $= 34.6°$ (Fig. 1.68(d)) with respect to the x-axis. For the trajectory, choose z positive up (perpendicular to the pavement) and x' along the direction of v_{e_h}, Fig. 1.68(e); then, eqn (1.34) gives $z = v_{0_z}t - \frac{1}{2}gt^2$: $-0.75 = (18.9 \tan 20°)t - \frac{1}{2}(9.81)t^2$: solving $t^2 - 1.402t - 0.1529 = 0$ gives the flight time $t = t_F$ which is found to be $t_F = \frac{1}{2}(1.402 \pm 1.605) = 1.504$ s, where the plus sign had to be chosen to obtain a positive time. The horizontal travel distance is $x' = (v_{e_h} \times t_F) = (18.9)(1.504) =$ 28.4 m.

(b) The impact velocity of the cyclist with the pavement, $v_I = [v_{I_{x'}}^2 + v_{I_z}^2]^{1/2}$; $v_{I_{x'}} = 18.9$ m s^{-1} with no air drag. From eqn (1.34), $v_{I_z} = v_{o_z} - gt_F =$ $18.9 \tan 20° - 9.81 \times 1.504 = -7.88$ m s^{-1} (downwards); $v_I = [(18.9)^2 + (-7.88)^2]^{1/2} = 20.47$ m s^{-1}. The impact energy of the rider, from eqn (1.45), is $T_I = \frac{1}{2}mv_I^2 = \frac{1}{2}(72.5)(20.47)^2 = 15\,200$ J.

(c) (i) $T = \frac{1}{2}(7.25)(20.47)^2 = 1520$ J; (ii) $T = \frac{1}{2}(29.75)(20.47)^2 = 6230$ J. The energy to be absorbed for case (i) is 1520 J; if the helmet can absorb 136 J, the head must absorb 1384 J. The energy cited by Gurdjian (1975) as a maximum fracture energy for dry cadaver skull fractures is 104 J (920 in lbf). Thus, it appears as though there is not enough energy absorption capability for head and helmet to sustain this impact, and damage, probably serious or lethal, must be expected.

(d) From eqn (1.49), $\int_0^\tau F\,dt = \Delta mv$; $\frac{1}{2}F_{max}(0.01) = mv_I = 7.25 \times 20.47$; $F_{max} = (2 \times 7.25 \times 20.47/(0.01) = 29\,680$ N, more than sixfold greater than the stipulated safety limit of 4460 N (1000 lbf) and greater than the highest tolerance that can be construed from the US Federal Motor Vehicle Standard 218 of the absolute limit of 400g for motorcycle helmets to which a 4.54 kg (10 lb) head could be subjected.

(e) Here, the mass dropped is $(4.55 + 1.36) =$ 5.91 kg, and the impact velocity is $v_I = [2 \times 9.81 \times 1.83]^{\frac{1}{2}} = 5.99$ m s^{-1}. For the triangular impulse of 4 ms duration, Fig. 1.68(f), $\int_0^\tau F\,dt = \frac{1}{2}F_{max}\tau = \Delta(mv) = mv_I$; $F_{max} = (2 \times 5.91 \times 5.99)/(0.004) = 17\,700$ N (3979 lbf). From eqn (1.14), $a_{max} = F_{max}/m = 17\,700/5.91 = 2995$ m s^{-2} $= 305g$. Thus, this helmet conforms to all three acceleration level criteria (1), (2), and (3), set by NHTSA, in that the peak acceleration is less than 400g (305g); 200g is exceeded for less than 2 ms, and the acceleration for a 4 ms duration is zero, hence less than 150g. Thus, the helmet would be acceptable according to US Department of Transportation. However, it exceeds Ono's criterion (Ono *et al.* 1980) for concussion of 140g by a factor of 2.2.

(f) For the severity index, the rising curve is a (in units of g) $= 305\,000t$ for $0 \le t \le 0.001$. The descending portion of the curve is given by a (in units of g) $= 407 - 101\,900t$, $0.001 \le t \le 0.004$. Thus,

$$\begin{aligned} SI &= \int_0^{0.001} (305\,000t)^{2.5}\,dt \\ &\quad + \int_{0.001}^{0.004} (407 - 101\,900t)^{2.5}\,dt \\ &= (1/3.5)(305\,000)^{2.5}t^{3.5}\Big|_0^{0.001} \\ &\quad + [407 - 101\,900t)^{3.5}]/(-3.5)(101\,900)\Big|_{0.001}^{0.004} \\ &= 464 + 1391 = 1855\ \text{ft}^{2.5}\text{s}^{-4}. \end{aligned}$$

This is more than the upper limit of acceptability of 1500 specified for the SI criterion, and thus this motorcycle helmet would not be certified when this standard is employed.

It is instructive to note that the levels of acceleration in the accident were far greater than those generated in the acceptance test. In consequence, the helmet would not provide adequate protection from serious or fatal injury in the accident considered, provided the mathematical model depicts the physics of the collision with reasonable accuracy, which is believed to be the case. The mandated drop test, on the other hand, does not pretend to duplicate any particular real-life impact situation, but simply assures a certain quality in the manufacture of the helmet that would give its wearer a better chance at survivability than if he were to wear a helmet not meeting the stipulated requirements. The physical situation of a helmeted human head striking a pavement, where both helmet and head are capable of some deformation before permanent damage ensues, and that of a helmeted metal headform striking a steel anvil are simply not comparable, and rigid-body analysis for the accident case is applicable only for the mass center of the head. Furthermore, in the accident, the contact duration is longer than in the acceptance test. From the linear impulse–momentum law, a longer contact duration for a specified change of momentum will reduce the peak force level, which is a correlator with degree of trauma. Thus, an increase in the contact duration for direct impacts of or to the head (all other things being equal) should reduce the resulting damage. Another standard, the head injury criterion (Versace 1971; Goldsmith 1979) addresses the most rapidly occurring rate of change of the acceleration during the impact and is based on the greatest value of an integral during any fraction of the total contact duration.

Recently, modern modeling has been applied to the design of both motorcycle and bicycle helmets that promises to significantly enhance their performance (Gilchrist and Mills 1994; Mills and Gilchrist 1991).

E9. A laborer, whose mass is 81.8 kg (180 lb), uses a shovel, of negligible mass but of length $s = 1.5$ m to the pivot point to lift dirt with a weight W_1 from a pile into a truck through an arc of 45° from the horizontal, Fig. 1.69(a). The two primary muscles activating this motion are the deltoid and the erector spinae, both performing in pairs. The force–strain relations for muscles, Fig. 1.69(b), are approximated by $F = Ak\left[\frac{e}{a-e}\right]$, where A is the cross-sectional area of the muscle, k is the constant of the nonlinear spring, e is the extension (as a fraction) over rest length L_0, and a is the extension for physical failure, 70 per cent (Merrill *et al.* 1984). In the relaxed state L_1, when he starts lifting, both muscle pairs are extended 10 per cent, and at the maximum lift point, the deltoid length L_2 is 35 per cent extended and the erector spinae length L_2 is 20 per cent extended. If the areas A, spring constants k, and rest lengths of these two muscles are (1) deltoid: $A = 0.04\ \text{m}^2$, $k = 18$ kPa, $L_0 = 0.2$ m; (2) erector spinae: $A = 0.0850\ \text{m}^2$, $k = 35$ kPa, $L_0 = 0.18$ m, determine the weight W_1 of dirt lifted routinely every 8 s.

Solution. The work done $= W_1 h = W_1 s \sin 45° = W_1 \times 1.5 \times 0.707 = 1.06 W_1$ J. The work is also equal to that done by the two muscle pairs. From eqn (1.44), this is Work $= \int_{L_1}^{L_2} F\,dx = Ak \int_{L_1}^{L_2} \left[\frac{e}{a-e}\right] dx$. The extended length of the muscle is $L = (1 + e)L_0$; thus $e = (L/L_0) - 1$ and $de = dL/L_0$, and hence $dx = L_o\,de$. Thus work for both muscle pairs

$$
\begin{aligned}
&= \sum_{i=1}^{2} 2A_i k_i L_{0_i}[-e_i - a\ln(a - e_i)]\|_{e_{i_0}}^{e_{i_f}} \\
&= (2 \times 0.04 \times 18 \times 10^3 \\
&\quad \times 0.2\{[-0.35 - 0.7\ln(0.7 - 0.35)] \\
&\quad - [-0.1 - 0.7\ln(0.7 - 0.1)]\}) \\
&\quad + (2 \times 0.085 \times 35 \times 10^3 \times 0.18 \\
&\quad \times \{[-0.2 - 0.7\ln(0.7 - 0.2)] \\
&\quad - [-0.1 - 0.7\ln(0.7 - 0.1)]\}) \\
&= 36.7 + 29.7 = 66.4\ \text{J}.
\end{aligned}
$$

Hence, $W_1 = 66.4/1.06 = 62.6\ N$ (14 lbf). The power developed for this task is obtained from Subsection 1.2.6; here, it is the average work done per unit of time $= 66.4/8 = 8.30\ \text{N m s}^{-1}$ (or J s^{-1}) $= 8.30$ W $= 0.011$ hp.

A minimum level of food consumption for a person is about 2000 kcal d^{-1} (or 41 670 calories per half hour) so that in half an hour of shoveling, or 225 lifts, the energy expended will be $225 \times 66.4 = 14\,940$ J $= 3570$ cal, or about 8.5 per cent of this energy intake (since 1 cal = 4.184 J). However, as the power of an 81.8 kg (180 lb weight) person running up two stair steps, each 0.203 m (8 in) high, per second is $P = 81.8 \times 9.81 \times 2 \times 0.203 = 326$ W (0.437 hp), the process of lifting the dirt as described is not as efficient as running up stairs.

E.10 As a tool in head injury severity prediction, a model, Fig. 1.70, has been proposed consisting of that portion of the head mass directly moving under a blow, m_1, connected to the rest of the head, of mass m_2, by a series combination of a linear spring of constant k and linear dashpot of constant c_1, paralleled by another dashpot of constant c_2. Dashpot c_2 appears to be the damping contribution of the brain. What are the governing equations for this system?

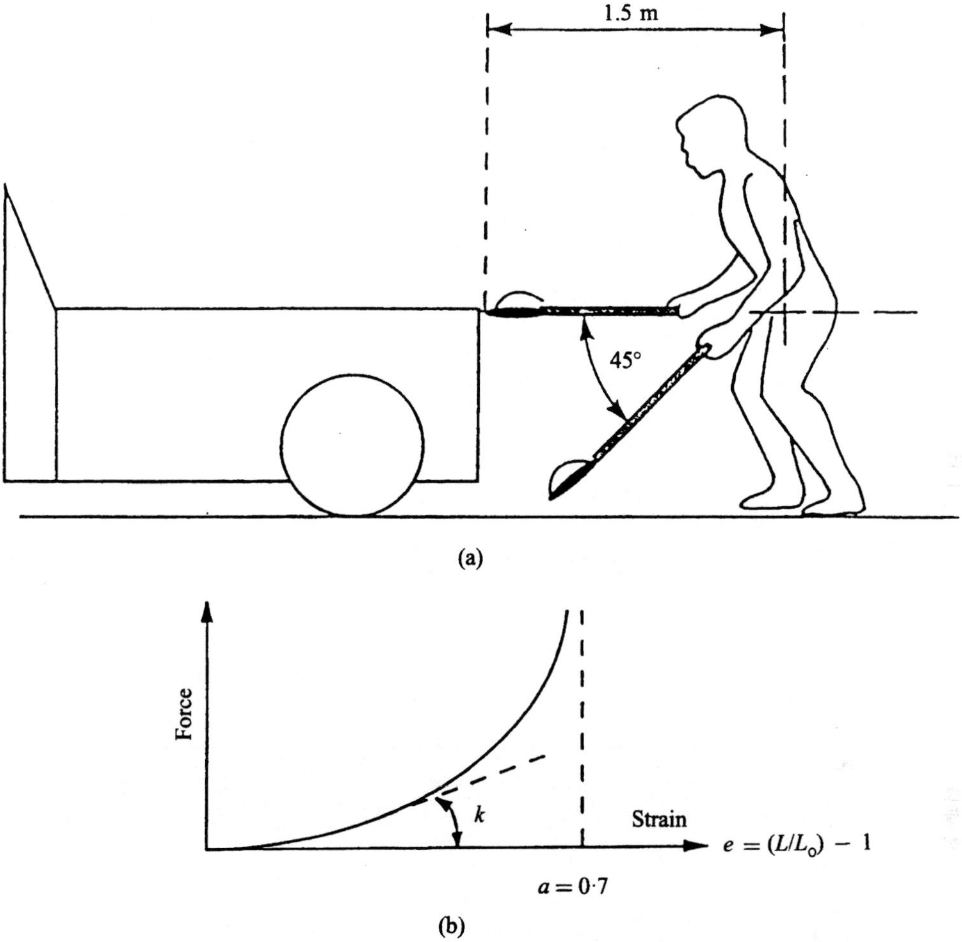

Fig. 1.69 Laborer shoveling dirt into truck: (a) schematic; (b) force–strain curve for muscle.

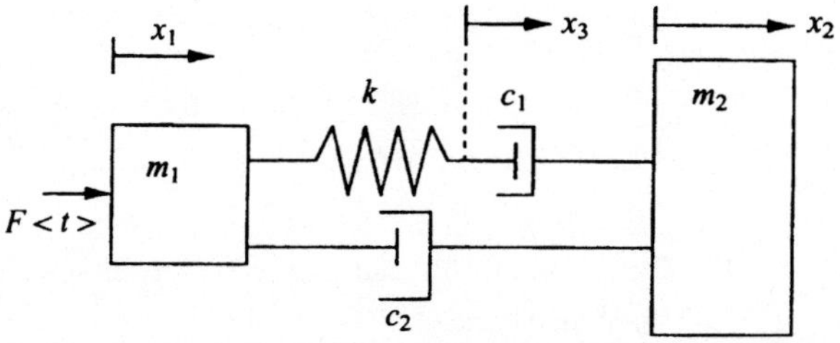

Fig. 1.70 Two-mass model of human head.

Solution. The constraint is that $m_1 + m_2 = M$, the total mass of the head with the brain. The equations of motion, eqn (1.79), for masses m_1 and m_2 are $F\langle t\rangle = m_1\ddot{x}_1 + c_2(\dot{x}_1 - \dot{x}_2) + k(x_1 - x_3)$; $m_2\ddot{x}_2 = c_1(\dot{x}_3 - \dot{x}_2) + c_2(\dot{x}_1 - \dot{x}_2)$. Force continuity: $k(x_1 - x_3) = c_1(\dot{x}_3 - \dot{x}_2)$.

E11. A tree, Fig. 1.71(a), (b), is to be considered as a uniform cylinder with a specific gravity of 0.9 and a diameter of 0.3 m. If Young's modulus is taken as

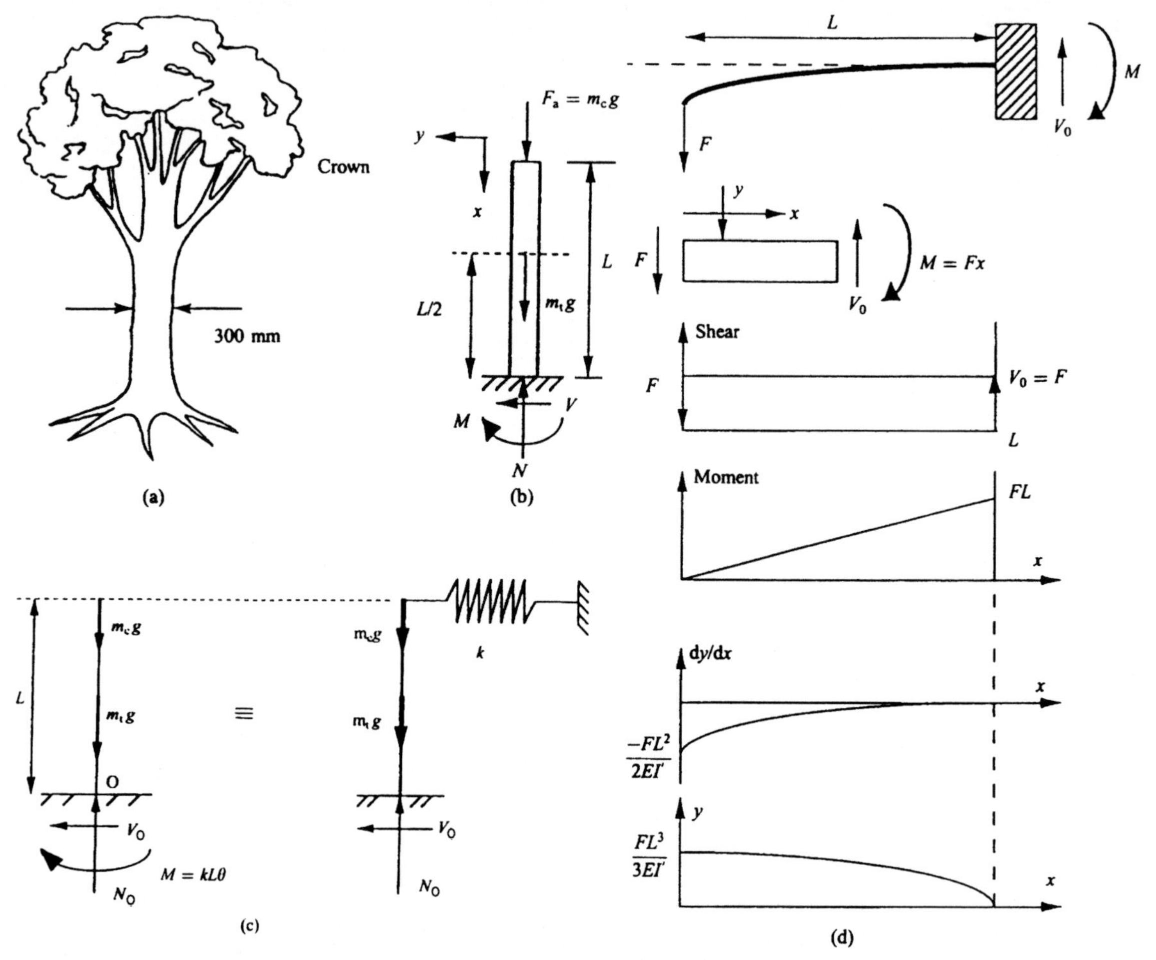

Fig. 1.71 Tree: (a) schematic; (b) loading diagram; (c) free-body diagram of tree and equivalent loading diagram; (d) free-body diagram of a beam and an element, and shear, moment, slope, and deflection diagrams for the tree.

$E = 1.379$ GPa (200 000 lbf in^{-2}) and the crown has a mass of 60 kg, determine (a) the buckling length of the tree ignoring the contributions to the moment due to the trunk mass, (b) the deflection per unit transverse load applied at the top if the tree length is $\frac{3}{8}$ the critical buckling length, and (c) the resulting natural frequency of the system if it is modeled as two concentrated masses with a linear spring as shown acting at the crown (use $L = \frac{3}{8}L_c$ determined in part b).

Solution. The moment of inertia of the circular cross section $I' = \frac{1}{4}\pi R^4 = \frac{\pi}{4}(0.15)^4 = 0.0003976$ m^4.

(a) The Euler buckling equation must be used here for a built-in column,

$$F_{a_{crit}} = \tfrac{1}{4}\pi^2 EI'/L_c^2;$$

$$L_c^2 = \left(\frac{1}{4}\pi^2 \times \frac{1.379 \times 10^9 \times 0.0003976}{60 \times 9.81}\right) = 2298 \text{ m}^2.$$

Thus $L_c = 47.9$ m.

(b) The equation for the transverse deflection of the beam is given by $y = (F/6EI')(2L^3 - 3L^2x + x^3)$ (derived at the end of the problem). Here, $L = \frac{3}{8}L_c = 17.96$ m, but L will be taken as 18 m. The maximum deflection occurs at $x = 0$, and thus

$y\langle 0\rangle = FL^3/3EI'$ and spring constant $k = [y\langle 0\rangle]^{-1}$ for unit force $F = 1$; $k = 3EI'/L^3 = 3 \times 1.379 \times 10^9 \times 0.0003976/(18)^3 = 282$ N m^{-1}. The deflection per unit transverse load is $\delta = 1/k = 3.5$ mm.

(c) The mass of the trunk is $m_t = \pi R^2 L\rho = \pi(0.15)^2 \times 18 \times 0.9 \times 10^3 = 1145$ kg. (1 m^3 of water has a mass of 10^3 kg; 1 cm^3 has a mass of 1 g). This calculation contradicts the assumption made in (a) that the mass of the trunk does not contribute to the buckling load. It is, of course, possible to recalculate the critical length including the trunk mass by integrating along the trunk to find the local bending moment.

For the present model, the tree is assumed to move linearly in accordance with simple harmonic motion theory, eqn (1.80) or its equivalent for angular motion, and deflections are to remain small. The model is shown in the diagram, Fig. 1.71(c). Taking moments about the base O with the fixed-end moment taken as that due to the trunk deflection, the equation of motion, eqn (1.75), is transformed into eqn (1.80), so that

$$\begin{aligned} M_o &= m_t g(\tfrac{1}{2}L\theta) + m_c g(L\theta) - Lk(L\theta) \\ &= I_o\ddot{\theta} = (\tfrac{1}{3}m_t L^2 + m_c L^2)\ddot{\theta}, \end{aligned}$$

$$\text{or } \ddot{\theta} + \theta(kL^2 - \tfrac{1}{2}m_t gL - m_c gL)/(\tfrac{1}{3}m_t L^2 + m_c L^2) = 0$$

(subscripts t and c denote trunk and crown). Here, $kL^2 = 91\,370$; $\frac{1}{2}m_t gL = 101\,100$ N m; $m_c gL = 10\,590$ N m; hence $kL^2 < \frac{1}{2}m_t gL + m_c gL$, and thus the equation reads: $\ddot{\theta} - \omega^2\theta = 0$, and any motion based on the present model would *not* be stable; this is unrealistic, and hence the postulated conditions (such as a cylindrical trunk) and/or assumptions (i.e. neglect of damping) must be changed. If, consistent with the procedure employed above, the moment due to the concentrated trunk mass is ignored, but its inertia is retained, then simple harmonic motion will result: the governing equation is $\ddot{\theta} + [(kL^2 - m_c gL)/(\frac{1}{3}\, m_t L^2 + m_c L^2)]\theta = 0$, where $\omega^2 = 80\,770/143\,100 = 0.564$ rad^2 s^{-2} and $\omega = 0.751$ rad s^{-1}, too fast a sway for realistic motion.

Actually, ignoring the overturning force of the trunk mass is not that poor an assumption, because the trunk does not deform linearly; the major portion of the lateral deflection is close to the crown. A more exact analysis should include the actual position of the mass center of the trunk; however, the model using an equivalent spring is so crude that it would not be more realistic to include such a moment. The proper model incorporating all mass and deflection effects is highly nonlinear and hence very complicated. It is expected that the natural frequency of the tree has a significant influence on the buckling phenomenon and that it will be reduced near the critical load. Palm trees and other slender agricultural products, such as corn stalks, are believed to attain heights very close to their critical length.

For the present problem, the bending moment and shear diagrams of the tree are shown in Fig. 1.71(d); the shear $V = \text{const.} = -F$ except at the end points. Now from eqns (1.143) for concentrated loads, $dM/dx = -V = F$ except at the end points. Hence, $M = Fx + M\langle 0\rangle$; but $M\langle 0\rangle = 0$; thus, $M = Fx$ and $M_{x=L} = FL$; $d^2y/dx^2 = M/EI' = Fx/EI'$ and, further, $y\langle L\rangle = dy/dx = 0$ at $x = L$. Thus, upon integrating,

$$\frac{dy}{dx} = \frac{1}{2}\frac{Fx^2}{EI'} + C_3 \text{ and } C_3 = -\frac{1}{2}\frac{FL^2}{EI'}.$$

A further integration gives

$$y = \frac{1}{6}\frac{Fx^3}{EI'} - \frac{1}{2}\frac{FL^2x}{EI'} + C_4;$$

$$0 = \frac{1}{6}\frac{FL^3}{EI'} - \frac{1}{2}\frac{FL^3}{EI'} + C_4 \quad \text{and } C_4 = \frac{1}{3}\frac{FL^3}{EI'}.$$

Thus:

$$y = \frac{F}{6EI'}(x^3 - 3L^2x + 2L^3).$$

E12. A skier with a mass of 82 kg catches the edge of one ski while making a turn with an angular speed of $\frac{1}{2}$ rev s^{-1} and is stopped fully in a time $t = 0.2$ s. The moment of inertia of the skier about the vertical axis is 1.8 kg m^2 at the moment this accident happened. The minimum cross section of the tibia is approximated by an elliptical shell, Fig. 1.72, whose outer semi-major and semi-minor axes are 0.01 and 0.008 m, respectively, with corresponding semi-major and semi-minor axes for the interior hole of 0.009 m and 0.0072 m, respectively. The ultimate strength in shear is obtained from Yamada (1970) as 58 MPa. Determine whether the tibia will be fractured.

Solution. The moment of inertia of the skier is approximated from the data of Miller and Morrison (1975) by including assumed positions of the extremities and poles in a turn. The average angular acceleration is: $\ddot{\theta} = \frac{1}{2}(\omega_0 + \omega_f)/t = (2\pi \times 0.25)/0.2 = 7.854$ rad s^{-2} (since $\omega_f = 0$). From Newton's law of rigid-body rotation, the rotational moment $M = I\ddot{\theta} = 1.8 \times 7.854 = 14.14$ Nm, which also represents the torque applied to the tibia. The polar moment of inertia of the tibial section is $J_2 - J_1$, where subscripts 2 and 1

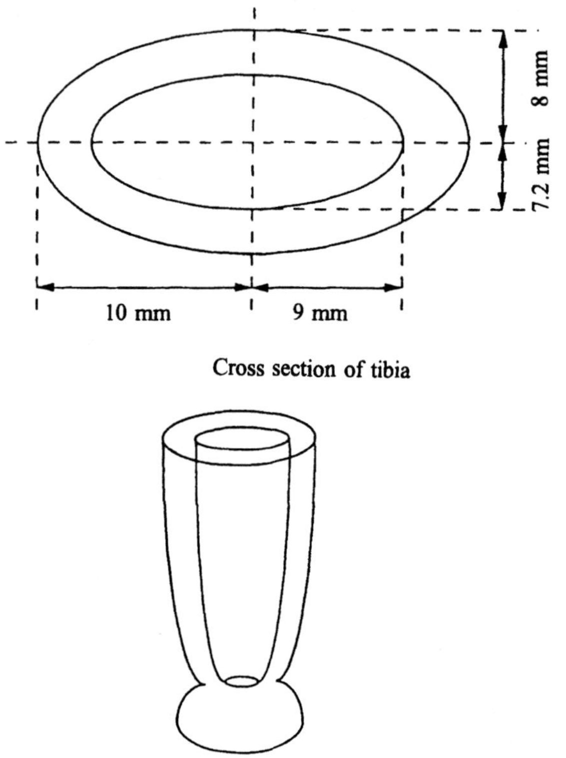

Fig. 1.72 Tibia and tibial cross section.

refer to the outer and inner boundaries of the hollow tibial shaft, respectively. $J = \frac{1}{4}A(a^2 + b^2)$ with $A = \pi ab$, where a and b are the semi-major and semi-minor axes, respectively. Thus,

$$J_1 = \tfrac{1}{4}\pi(0.01)(0.008)[(0.01)^2 + (0.008)^2]$$
$$= 1.03 \times 10^{-8}\text{m}^4;$$
$$J_2 = \tfrac{1}{4}\pi(0.009)(0.0072)[(0.009)^2 + (0.0072)^2]$$
$$= 0.6761 \times 10^{-8}\text{m}^4; \quad J = J_1 - J_2.$$

Thus, $J = (1.03 - 0.6761)10^{-8} = 0.354 \times 10^{-8}$ m^4. Then $\tau_{xy} = Tc/J = (14.14 \times 0.01)/0.354 \times 10^{-8} =$ 39.9 MPa. This approaches the value of the allowable strength of 42.5 MPa, so one would not expect fracture, but it might occur due to the variability of human bone.

Problems

P1.1 An oarsman, of 75 kg mass, is sitting symmetrically in a shell, Fig. 1.73. He pulls on the oars and exerts a total horizontal force of 140 N. His seat is pinned to the shell, but his feet rest on boards that slide back and forth on rollers. His center of gravity G is where shown; what reactions do the seat and board exert on him?

P1.2 A stamp collector views his prize exhibits through a magnifying glass with his head, whose mass is 5.5 kg, in the stance indicated in Fig. 1.74. The weight of his head, W, acts along the vertical passing through the center of mass G. His position is maintained by a muscle force P whose line of action makes an angle $\phi = 30°$ with the horizontal, and by a normal force N and shear force V at the occipital condyles C, whose resultant R makes the angle $\theta = 55°$ with the horizontal. Find R and P.

P1.3 The equilibrium of a person bending forward, Fig. 1.75, is provided by the contraction of the pair of erector spinae muscles, one of the more important tissues of the abdominal and thoracic regions. The effective center of rotation is the articulation of the sacrum with the fifth lumbar vertebrae. These muscles, while in tension, are inactive when a person is in the erect position; the weight supported by the sacrum is then resisted by the articulating processes in pure compression.
(**a**) If the mass of the head and torso above the sacrum is

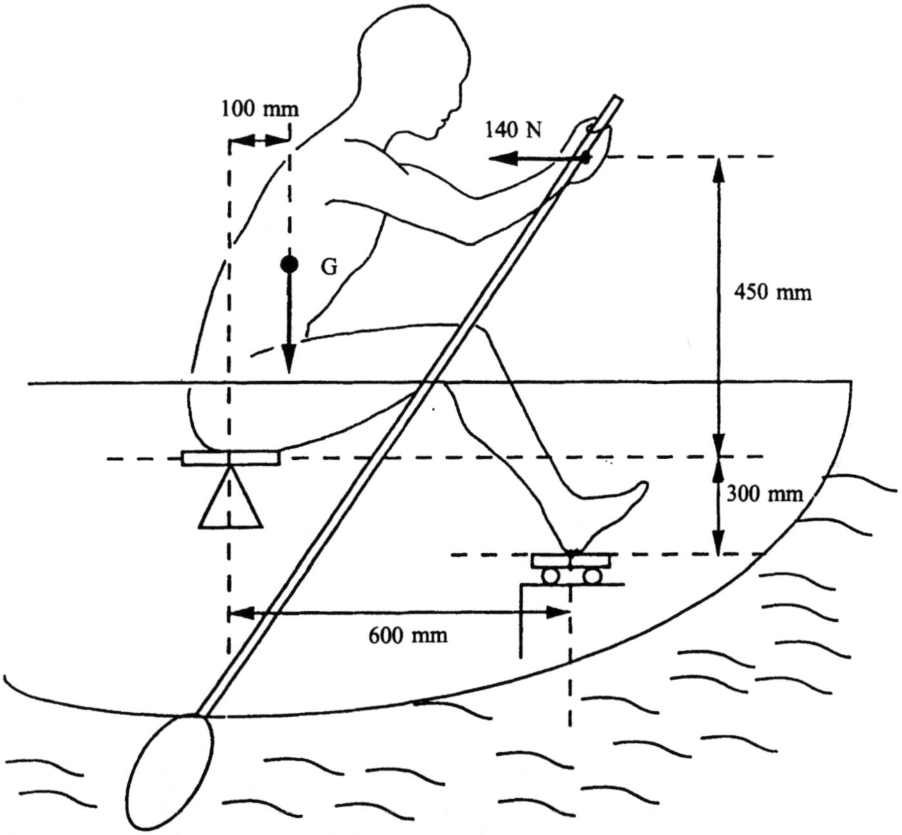

Fig. 1.73 Rower in boat.

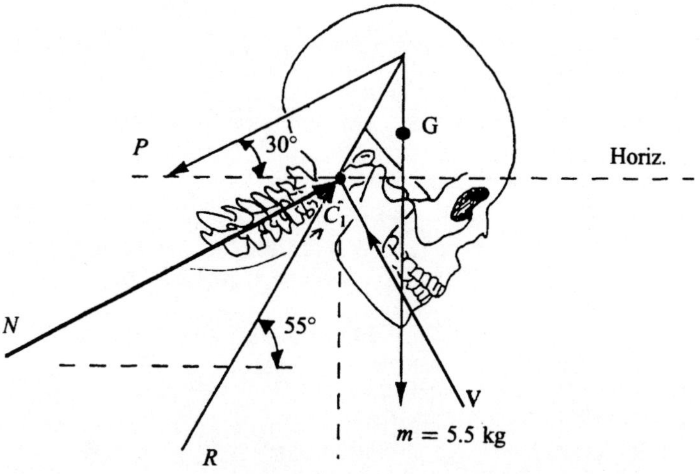

Fig. 1.74 Stamp collector examining exhibit.

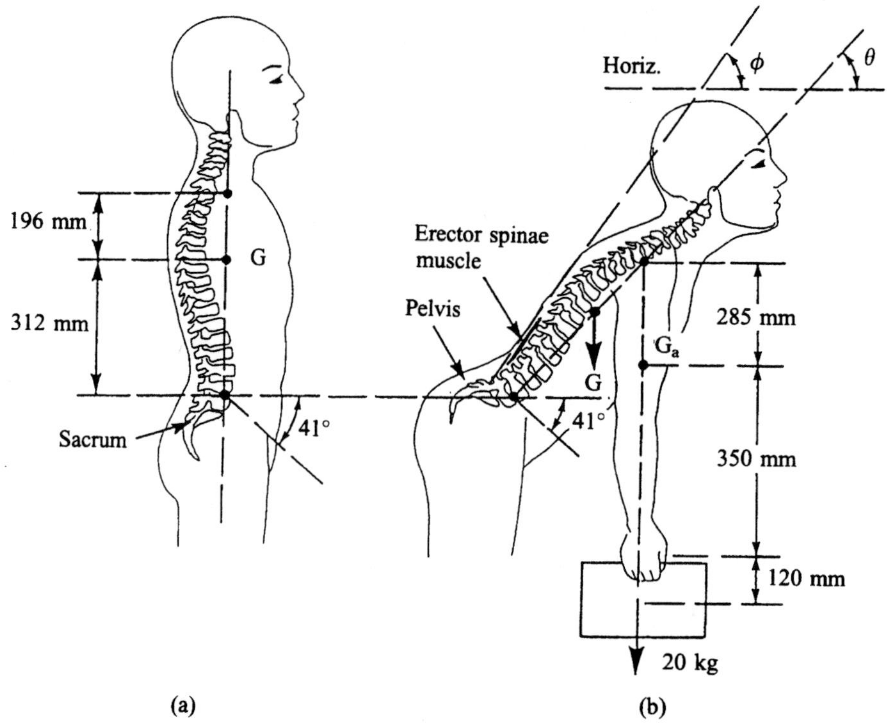

Fig. 1.75 Person carrying load.

45 kg acting at the center of mass G and each of the upper limbs of the person has a mass of 3.75 kg, acting at the limb mass center G_a, and if the angle of the articulating surface is 41° when standing erect, what are the compressive and shear forces transmitted to the sacrum?

(**b**) The person now carries a mass of 20 kg in both hands with both arms vertically down (the load is directly below the shoulder). What are the compressive and shear forces on this joint if his torso (spine) makes an angle of (i) 30°, (ii) 55° with the horizontal? Some dimensions for the person are shown in the diagram. Other anatomical data are given in Table 1.1.

P1.4 Consider the action of the biceps when carrying a load of 8 kg with the forearm making an angle of 20° with the horizontal and the upper arm making an angle of 60° with the horizontal, Fig. 1.76. The forearm has a mass of 1.35 kg acting at G and the hand has a mass of 0.5 kg with the mass center G_h also supporting the load. The locations are shown in the diagram. The biceps is attached to the forearm 50 mm ahead of the elbow joint and the humerus articulates with the radius over a 1200 mm^2 joint surface at the elbow.

(**a**) What is the pull exerted by the biceps, which must act along the muscle centerline, and what is the magnitude and direction of the force exerted by the humerus on the head of the radius?

(**b**) What is the average joint pressure?

P1.5 One method to find the average density of the body is to immerse the body completely in a tub of water filled to the brim and measure the volume V of the water that spills out.

Table 1.1. Data for problem P3

Angle of inclination of torso, θ (°)	Erector spinae angle ϕ (°)	Perpendicular distance of erector spinae to joint, c (mm)
30	33	60
55	35	56

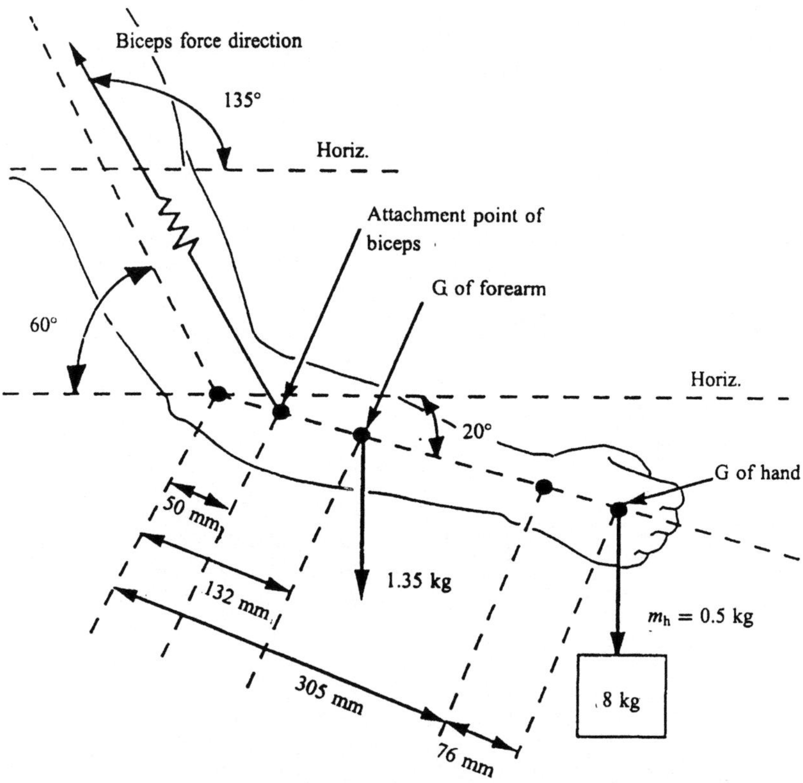

Fig. 1.76 Hand carrying load and attached forearm and upper arm.

(**a**) If a person whose mass is 75 kg as determined by a scale displaced $V = 0.07\ m^3$ of water, what is his average specific gravity and density?

(**b**) If he now immerses one arm up to the shoulder into a full bucket of water, whose overflow is $0.0035\ m^3$, while standing on a scale, what is the weight of one arm and hand, assuming the validity of the average density, and what is the scale reading? (*Hint:* Use the law of buoyancy.)

P1.6 Another simple but not very accurate method of measuring body segment masses is the use of a board pivoted at one end and attached to a scale at the other, as shown in Fig. 1.77(a). The person lies on the board with the position of the top of the head exactly defined, say at pivot P. The scale provides a reading Q_1 for this position and another Q_2 with the limb of concern, say one entire arm and hand, raised to the vertical position, as indicated in Fig. 1.77(b). The mass center of the raised section must be known; this can be found by making a cast, or approximated by considering the segment to be a frustum of a cone with attached rectangular parallelepiped, as shown in Fig. 1.77(c).

(**a**) (Optional) Assuming a uniform density for arm and hand, with the circumference at the shoulder joint as 457 mm and at the hand as 203 mm, an arm length of 533 mm, and a hand with dimensions 178 mm long, 99 mm wide, and 38 mm thick, determine the mass center of the whole arm and hand relative to the shoulder joint, X_a. The mass center of a frustum of a cone is given by $X = \frac{1}{4}d[1 + 2k + 3k^2]/[1 + k + k^2]$; k = ratio of radii $R_1/R_2 < 1$ of the end surfaces, and X is measured on the centerline from the larger end.

(**b**) Q_1 reads 330 N with the arm on the uniform board, which is 2.45 m long and has a mass of 18 kg. If the person is 1.83 m (6 ft) tall and if the mass of the person is 85 kg, find the location of the mass center of the person relative to the bottom of his feet, x_p.

(**c**) Assuming that the distance from the shoulder joint to the mass center of the arm, X_a, is that computed in (a), or if not so computed, that it is 37% the length of the arm and hand, and that the distance from the shoulder

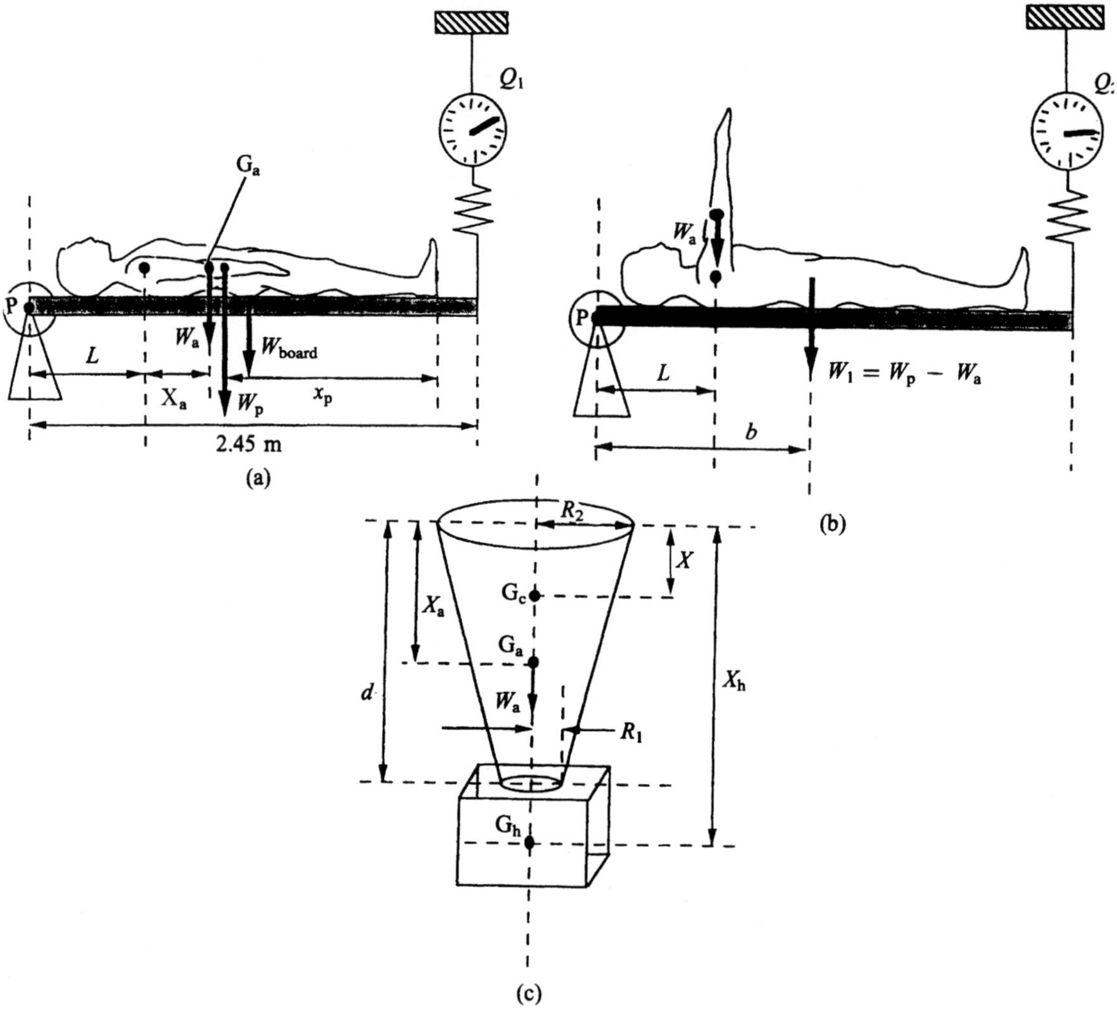

Fig. 1.77 (a) Person lying completely supine on a beam attached to a scale; (b) person lying supine with arm raised; (c) mass center locations of frustrum of a cone and a rectangular parallelepiped.

joint to the top of the head $L = 279$ mm, and if the arm is raised vertically as shown in Fig. 1.77(b) when the scale reads $Q_2 = 330.6$ N, determine the mass (weight) of the segment consisting of the upper arm, forearm, and hand.

(**d**) The upper limb is totally immersed in a bucket of water to the level of the shoulder joint; this displaces 3.33 liters (or 3330 g) of water. Assuming a uniform density for this limb, what is the specific gravity of the arm and hand?

P1.7 In a method to determine the position of the mass center of a person, the subject rests on a board of negligible weight, one of whose ends is supported by a pivot P, the other by a scale Q which measures vertical reaction, Fig. 1.78. The scale is read when the subject is completely prone and when one of his legs is raised, as shown. Assuming that the mass center of the leg and foot is at the shank center, a distance x'_2 from P that is measured, derive the expression for the distance $x_2 - x'_2$ as defined in the diagram. Let W be the total weight of the

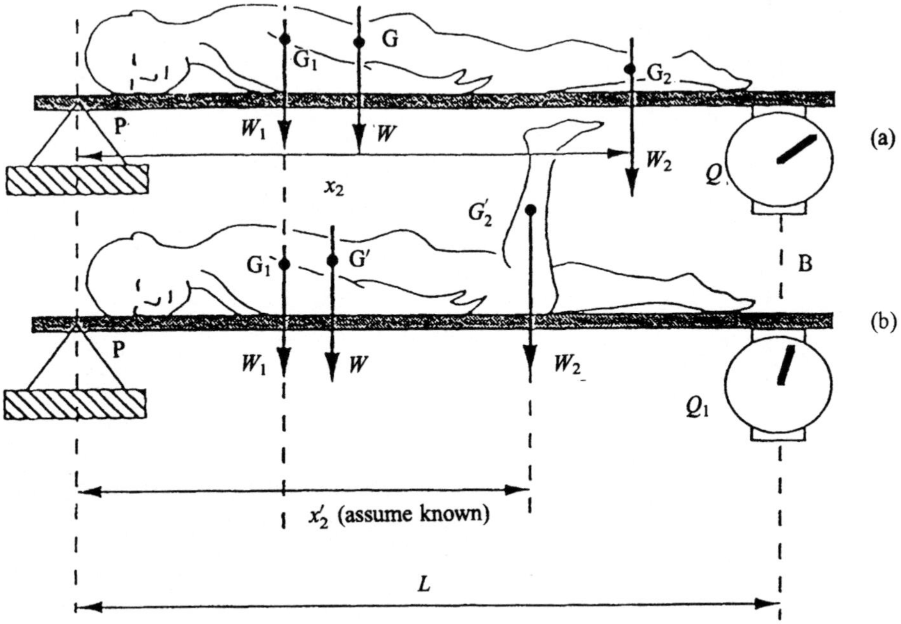

Fig. 1.78 Person lying completely prone on a beam scale (a), and with a raised leg (b).

person, W_2 the weight of one entire lower leg and foot, and $W_1 = W - W_2$. Note that the position of the mass center of W_1 does not change during the raising of the limb, but the mass center position of W does change.

P1.8 The shape of a palm tree, Fig. 1.79, is symmetrical about its axis and its outline is given by the equation: $y = 19.2 - 120x^2$ (y, x in meters). The tree is 18 m high and has a specific gravity of 1.8. On its top sits a crown with a mass of 5000 kg.

(a) Determine the mass of the tree trunk

(b) What is the compressive stress in MPa (1 Pa = 1 N m^{-2}) that the tree experiences at its top and at its base?

(c) If the roots of the tree had been cut with a saw, at what height y_1 would a rope have to be attached which, with a horizontal pull of 5000 N, would tip the tree about its base?

(d) What is the expression for the compressive stress (force per unit area) in the tree as a function of the height y above the base?

P1.9 In the analysis of the motion of a leg, Fig. 1.80, it is assumed that the trajectory remains in a fixed vertical plane, that the axis of rotation between shank and thigh is fixed at the (apparent) knee joint, i.e. at the level of the femoral condyles, and that the leg acts as a rigid body. Considering the inertial frame $OXYZ$ and the moving frame $O'xyz$ attached to the leg at the knee as shown, derive an expression for the knee moment M_k in terms of the necessary geometric, kinematic, and other variables (which should be defined) during the swing phase of the leg when there is no contact with the ground. Express this moment in terms of accelerations a_1 and a_2 in directions parallel to x that can be accurately measured by transducers attached to the leg at the positions shown.

P1.10 A person of 90 kg mass pulls a crate up a 10° slope by means of a rope inclined at 15° to the plane, Fig. 1.81. The coefficient of friction for the crate and plane is 0.2.

(a) If a maximum force of 538 N can be exerted on the rope, what is the maximum crate weight W that can be moved at a velocity of 1 m s^{-1} and what is the power developed? Give results also in hp.

(b) If 0.11 kW can be developed, what is the maximum pulling speed?

P1.11 The total mass of a bicycle, Fig. 1.82, is 14.5 kg and its rider has a mass of 75 kg. The pedal length is

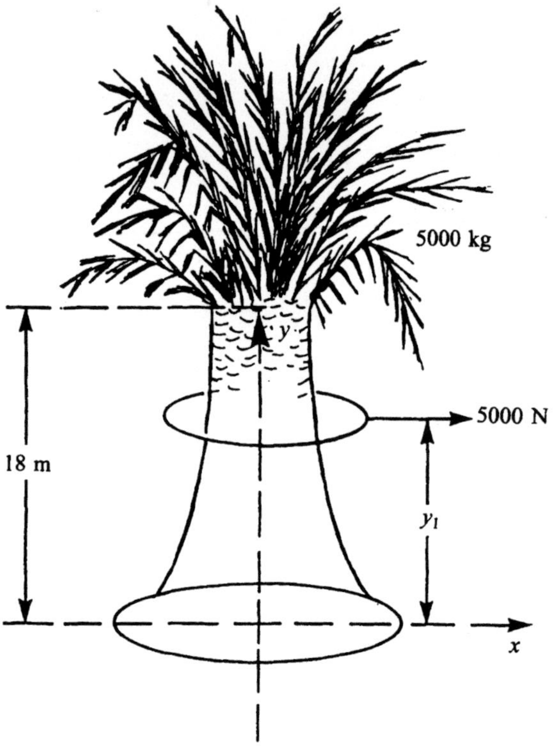

Fig. 1.79 Palm tree configuration.

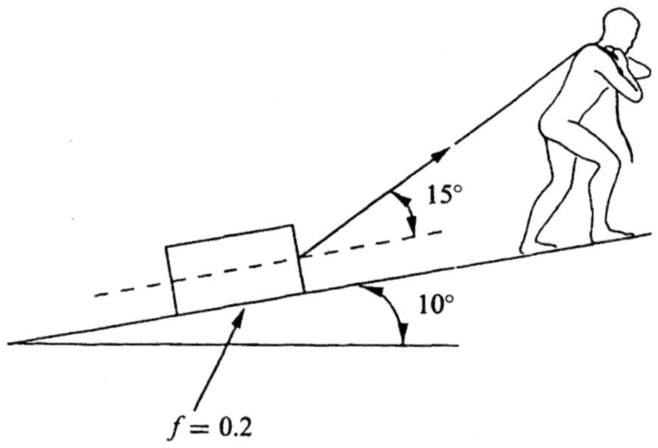

Fig. 1.81 Person pulling a load up a 10° incline.

178 mm, the larger sprocket wheel diameter is 190 mm and the back wheel sprocket of the chain drive has a diameter of 76 mm. The mass of each wheel is 1.36 kg with a radius of 318 mm. If the radius of gyration of the wheel is 292 mm, the mass center of the rider is 762 mm above the ground, and the mass center of the complete bicycle is 406 mm above the ground, determine the forces exerted by the foot of the rider on the pedal to produce a linear acceleration of the system of 0.18 m s^{-2} in the plane of motion. Assume pure rolling of the wheels, that air resistance and mass of sprockets and chain are neglected, and that only one foot produces a constant moment for half a pedal revolution.

P1.12 A baseball player hits a home run over a fence 122 m (400 ft) away that has a height of 3.66 m (12 ft)

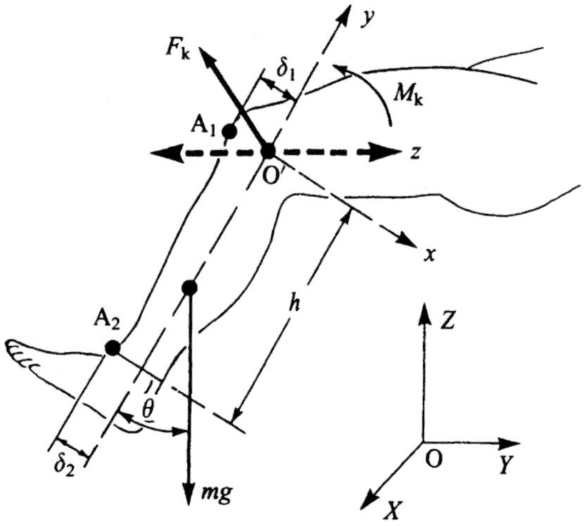

Fig. 1.80 Raised leg and local coordinate system, showing positions of accelerometer transducers A_1 and A_2, and inertial coordinate system at lower right.

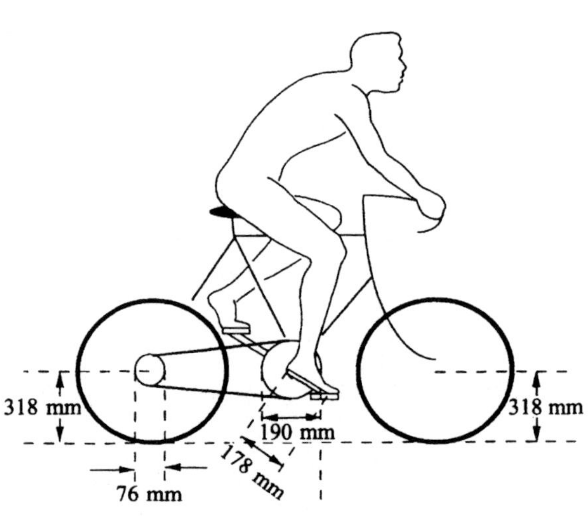

Fig. 1.82 Bicycle rider on a horizontal plane.

above the horizontal plane in which he swings his bat, Fig. 1.83. The bat has a mass of 1.36 kg (3 lb), a length of 0.911 m (36 in) from the grip, and a radius of gyration of 508 mm (20 in), and the ball hits 711 mm (28 in) from the wrist, assumed to be the point of rotation. The throw is clocked at 120 km h^{-1} (75 mph), not a fast ball; the ball leaves the bat at an angle of 30° (in part due to friction) with the horizontal after the batter has swung through an angle of 3 radians. The 'effective' coefficient of restitution between ball and bat, i.e. that relating to their absolute velocities, is known to be 0.45. Assume the bat velocity to be unaffected by the impact and neglect air resistance (a poor assumption, but it simplifies the problem).

(**a**) What wrist moment must the batter use to just clear the fence?

(**b**) Is this within the limit of 163 N m (120 lbf ft) regarded as attainable?

(**c**) If the ball is in contact with the bat for 3 ms, what is the maximum striking force when the force history is known to be a triangle and the ball weighs 142 g (5 oz)?

P1.13 It is desired to ascertain the performance of a bicyclist who competes internationally. His weight is 660 N and that of his bicycle is 90 N. He rides up an incline with a 1:10 grade, Fig. 1.84, starting from rest, to achieve a velocity of 7.5 m s^{-1} in 5 s of uniform acceleration, with an aerodynamic drag equivalent to 16 N.

(**a**) What forwared force must be exerted on the bicycle?

(**b**) Assuming that gearing produces an effective driving wheel diameter of 2 m, and that the applied torque is constant, determine the magnitude of the latter.

(**c**) If the crank radius of the pedal is 0.175 m, determine the required pedal force that the rider must produce.

(**d**) Determine the tractive power developed by the rider after attaining and maintaining the final speed.

P1.14 A few years ago, a stunt team from the British Commonwealth came to San Francisco and shocked the residents by jumping off the Golden Gate Bridge (they were later fined). Of course, they had tied a nylon rope around their waists, with a free length of 50 m, the other being attached to the bridge rail. The bridge deck is 72 m above the water level. The spring constant of the rope was found by attaching a 60 kg mass to one end of a vertical, initially unstretched rope section 25 m long and noting an extension of 1 m. Neglecting air resistance, for a 70 kg jumper,

(**a**) what is the person's speed at the end of the free rope length, i.e. at the point when the rope begins to stretch?

(**b**) will the maximum downward motion keep the person above the water level, and if so, by how much?

(**c**) what is the amplitude and half period of the first oscillation if not immersed in the water?

(**d**) what is the maximum strain on the rope if it could be considered uniformly distributed over the entire length, and what is the maximum stress based on an original rope diameter of 16 mm?

(**e**) Now assume that there is air resistance of $R = -kv^2$, and that the value of k/m is found to be 10^{-2} m^{-1}. What is the terminal velocity without considering the presence of the rope?

P1.15 The senior shotput throw involves a steel sphere with a mass of 7.27 kg (16 lb) and a specific gravity of 7.7. It has been stated that, in competition, the departure

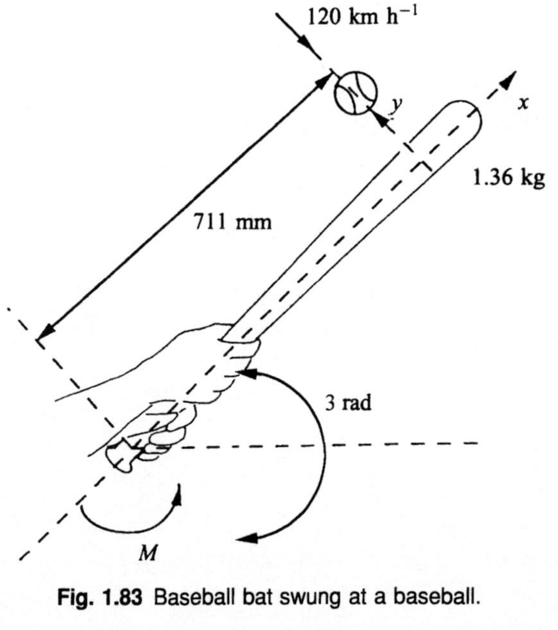

Fig. 1.83 Baseball bat swung at a baseball.

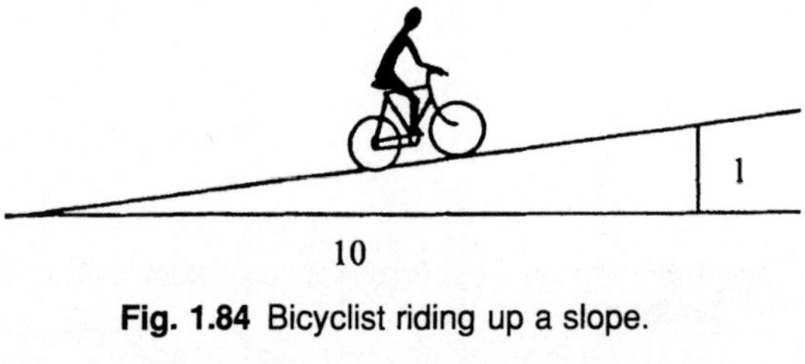

Fig. 1.84 Bicyclist riding up a slope.

angle ranges from 34 to 40 degrees and throw distances just below world record are 20–23 m.

(**a**) If the shot is released at a height of 1.83 m (6 ft) above the ground at an angle of 35° and thrown a distance of 19.8 m (65 ft), Fig. 1.85(a), what must be the initial release velocity, neglecting air resistance?

(**b**) If the motion from rest to release is monitored by a camera and found to take 0.1 seconds, what is the force exerted by the hand to produce this throw? The force remains constant.

(**c**) Assume now that the motion of the upper arm, forearm and hand of the thrower can be considered to occur in a single plane inclined at the appropriate angle to the horizontal, and assume further that the upper and forearm remain at right angles as shown in Fig. 1.85(b), with the locations of the centers of mass of the three segments shown by G_u, G_f, and G_h (the last coincides with that of the sphere mass), determine the angular acceleration, assumed constant, required to produce this result and the moment exerted on the upper limb and sphere by the shoulder joint. Required data for the geometric and kinematic properties are given in Table 1.2. The x-axis is perpendicular to the plane of arm motion, and z is along the body segment centerline. The shoulder joint should be the inertial reference coordinate origin for this case.

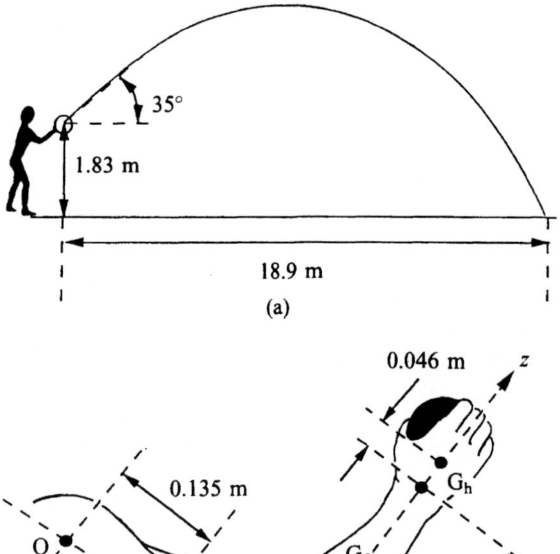

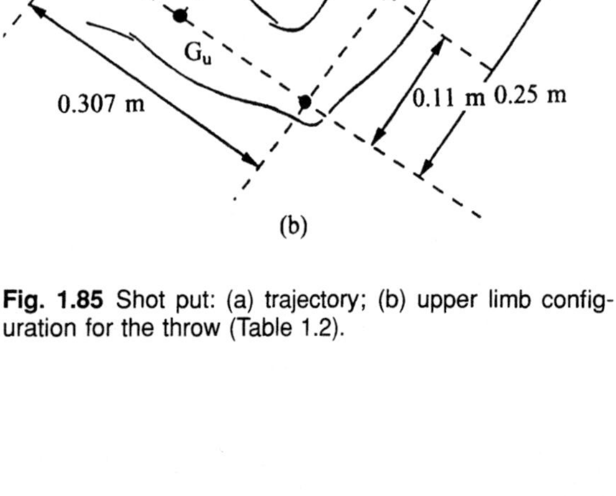

Fig. 1.85 Shot put: (a) trajectory; (b) upper limb configuration for the throw (Table 1.2).

P1.16 The curves in Fig. 1.86 (Lippert 1963) indicate the comfort level of humans when subjected to whole-body vibrations.

(**a**) If a building sways with simple harmonic motion of 3 Hz frequency, what is the maximum amplitude of this motion when human reaction becomes alarming?

(**b**) Repeat this calculation when the frequency is 18 Hz. Also plot the velocity curve for the alarming case.

P1.17 It will be assumed that a ski acts on a skier both as a linear spring of constant k and as a parallel linear dashpot of constant c, Fig. 1.87. When the skier, whose mass is m, traverses regularly shaped harmonic moguls of the same amplitude b and wavelength L, as shown, the forcing function acting on the skier is given by $F/m = 2c_1 b\Omega \cos \Omega t + n^2 b \sin \Omega t$, where $\Omega = 2\pi v/L$, v is the constant forward speed of the skier, L is the spacing of the moguls, b their amplitude, $c_1 = c/2m$, and $n^2 = k/m$.

(**a**) Neglecting the mass of the ski and assuming that the mass center G of the skier always remains the same height h above the slope, write the equation of motion for the skier. The spring–dashpot elements represent the action of the ski.

(**b**) Show that the motion y is given by $y = bQ \sin(\Omega t - \alpha)$, and determine Q which depends on Ω. Plot $Q\langle\Omega\rangle$ for two cases: (a) wooden skis, where

Table 1.2. Data for problem P15(c)

Component	Mass (kg)	Length (mm)	Mass moment of inertia (kg) mm^{2*} $I_{xx} = I_{yy}$	I_{zz}
Sphere	7.27	121.8†	$\frac{2}{5}mr^2$	$\frac{2}{5}mr^2$
Hand	0.52	100	0.0005	0.0005
Forearm	1.30	254	0.010	0.0011
Upper arm	2.15	307	0.030	0.0032

* Contini (1972) † Diameter

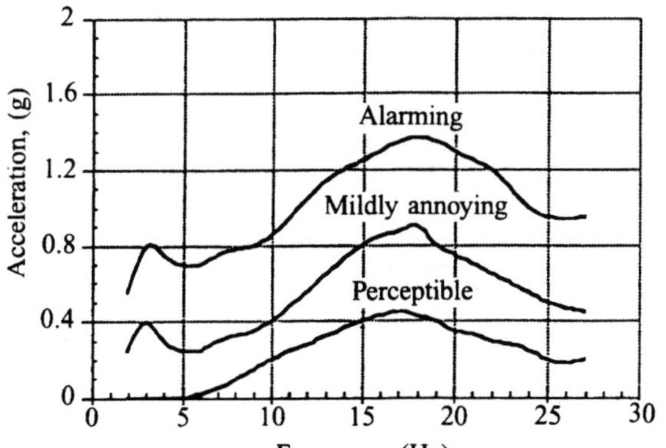

Fig. 1.86 Human reaction to whole-body vibration.

$c = 39$ N s m^{-1} and $k = 19\ 520$ N m^{-1}, and (b) metal skis, $c = 3.9$ N s m^{-1}, $k = 19\ 630$ N m^{-1}. The mass of the skier is 60 kg. (Note that the problem is very much more complicated if the height of G from the slope varies.)

P1.18 A passenger and a car are modeled as a three-mass system with the head and neck, seat and body, and car as the three masses shown in Fig. 1.88, connected by two springs for the head (one linear, one rotational) and another between body and car. At rest, the head is a distance L above the top of mass m_2 at pivot B. If the system is subjected to the acceleration shown, derive the equation of motion for the system. Define any required symbols for quantities not specified.

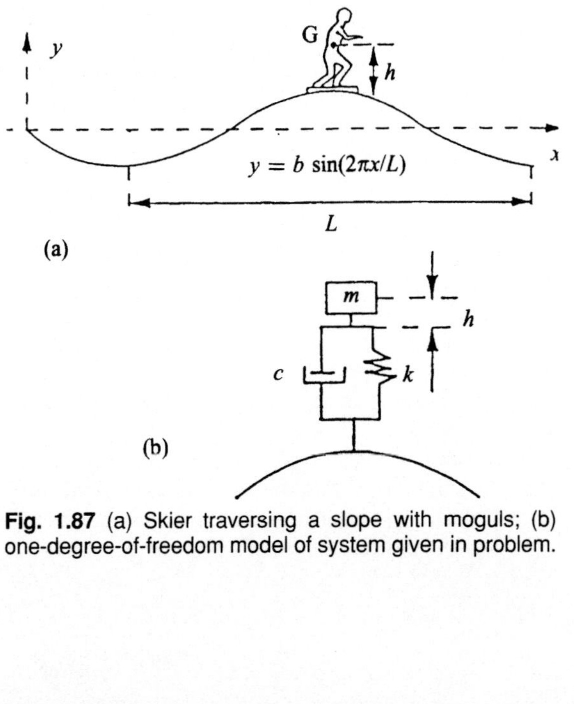

Fig. 1.87 (a) Skier traversing a slope with moguls; (b) one-degree-of-freedom model of system given in problem.

P1.19 Many soft tissue components have been found to be well represented by the uniaxial stress–strain relation $\tau = ke/(1 - e/e_a)$, Fig. 1.89, where e is the extension $(L - L_o)/L_o$ (in tension), L_o is the rest length, k is a spring constant, and e_a is the extension when the tissue breaks. If $e_a = 0.7$ and $k = 33.4$ kPa for the sternocleidomastoid muscle, attached to the head:

(**a**) Plot the stress-extension curve.

(**b**) Obtain the energy per unit volume that can be stored in the muscle at 60 per cent elongation.

(**c**) What is the total energy required to produce this elongation if the unloaded extension in the muscle is 10 per cent?

(**d**) The cross-sectional area of this muscle has an average of 360 mm^2, its rest length (no stress) is 200 mm, the head mass is 5 kg, and its centroidal moment of inertia normal to the plane of symmetry (sagittal plane) is $mk^2 = 0.05$ kg m^2. If the 60 per cent extension is reached as the result of a sudden backward acceleration of the head due to a 500 N blow to the chin, whose moment arm from the occipital condyles is 85 mm, determine the angular acceleration of the head developed if rotation occurs about the occipital condyles (0.1 m from head mass center). (Note that the sternocleidomastoid must resist the blow due to the weight of the head even before the blow.)

(**e**) How does the stress developed in this muscle compare with the ultimate strength of such a muscle, given by Yamada (1970) as 186 kPa? (Note: Other muscles also oppose the rotation).

P1.20 For the broad jumper of worked example E5, assume that the peak force, 5303 N (1192 lbf) as computed is transmitted to the tibia of one leg when he jumps, without reduction. The elastic modulus of compact bone (i.e. the shell) is 8.27 GPa (1.2×10^6 lbf in^{-2}).

(**a**) Determine the compressive stress and strain in this member at the four points indicated if the cross section can be approximately described as an axisymmetric tube with a diameter and thickness varying uniformly as shown in the model of the tibia, Fig. 1.90.

(**b**) Is this load dangerous if the compressive strength of the tibia is of the order of 172 MPa (about 25 000 lbf in^{-2})?

P1.21 Beef leg bone, treated elastically, is tranversely isotropic, but, if approximated as isotropic, has a Young's

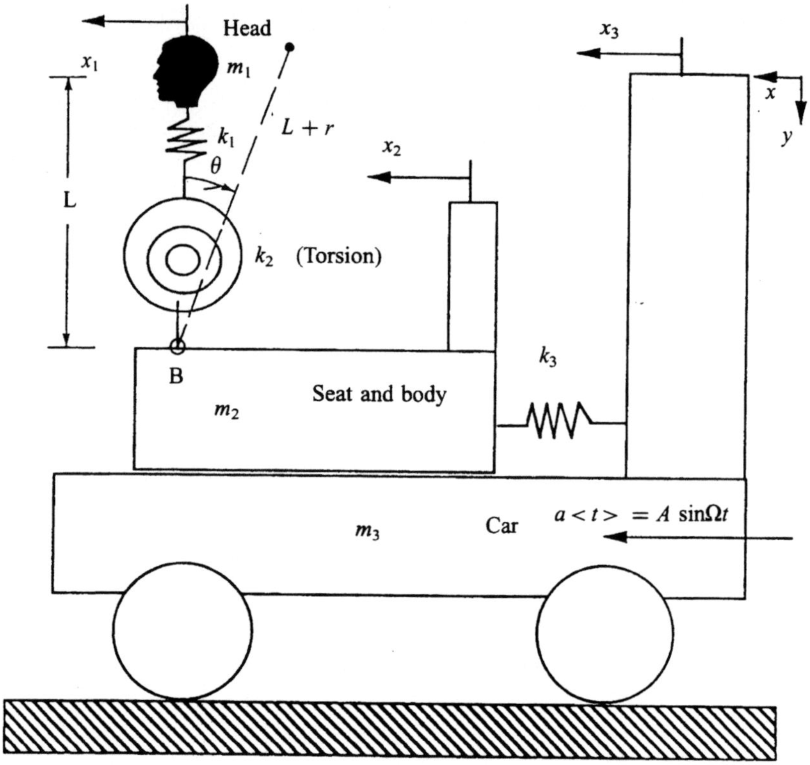

Fig. 1.88 Model of passenger seated in a car.

modulus $E = 8.25$ GPa and a Poisson's ratio $v = 0.33$.

(**a**) Determine the shear modulus G, the Lamé constant λ, and the compressibility modulus $K = \lambda + \frac{2}{3}G$.

(**b**) Compressive loading below the yield point is given by Eqn (1.137), with relaxation function $\phi = [13.63 - 5.38\varepsilon_\phi(1 - e^{-t/32.2})]$ GPa, $t \geq 0$; $\phi = 0$, $t < 0$ replacing the integrand, where t is expressed in microseconds. The bone cross section is a uniform shaft of length 380 mm with respective outside and inside diameters of 63.5 and 45.7 mm. Plot the force-deflection curves for several constant values of time t from 1 to 10 μs.

P1.22 The mechanical behavior of articular cartilage is to be defined in terms of a Kelvin–Voigt rheological model. Instead of creep and relaxation functions, it is possible to use an inverse property description in terms of compliance $J\langle t\rangle$ which is related to a retardation time t' and corresponding function $L\langle t'\rangle$ by

$$J\langle t\rangle = J_u + \int_{-\infty}^{\infty} L\langle t'\rangle(1 - e^{-t/t'})\, d\langle \ln t'\rangle,$$

where $J_u = J\langle t\rangle$ as $t' \to 0$ is the inverse of the instantaneous (unrelaxed) shear modulus and $J\langle t\rangle$ is the inverse of the apparent shear modulus $G\langle t\rangle$ at any instant. The observed behavior of a small animal cartilage is shown in Fig. 1.91. Determine the compliance for the retardation spectrum distribution shown in the diagram, where

$$L\langle t\rangle = \text{constant} = 2\ \text{MPa}^{-1} \text{ for } t'_{\min} < t < t'_{\max};$$
$$L\langle t\rangle = 0 \text{ otherwise}$$

(*Hint:* Let $t' = 1/x$.)

P1.23 US Federal motor vehicle helmet standards require that each helmet be placed on a magnesium headform with a mass of 5.27 kg (11.6 lbf) and dropped onto a steel anvil from a height of 1.83 m (6 ft). An accelerometer at the center of the headform determines the acceptability of the helmet; failure is regarded to occur when any of the following criteria are exceeded: (1) a maximum acceleration of 400g; (2) an acceleration

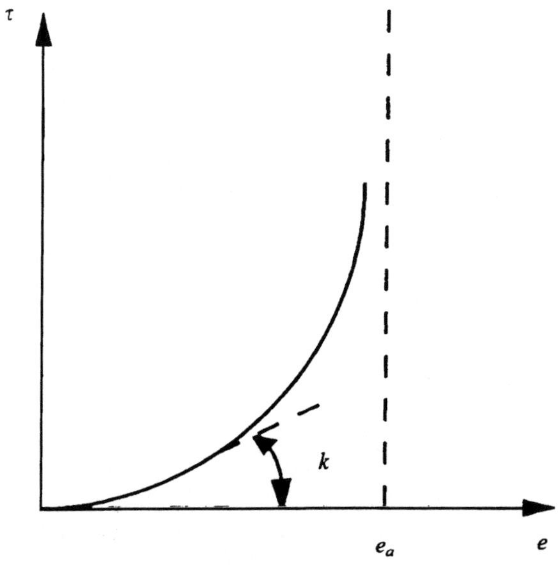

Fig. 1.89 Soft tissue uniaxial stress–extension relation.

of 200g lasting more than 2 ms; or (3) an acceleration of 150g lasting more than 4 ms.

(**a**) What is the impact energy if the helmet has a mass of 1.136 kg (2.5 lbf)?

(**b**) If the unit rebounds to a height of 0.152 m (6 in) and

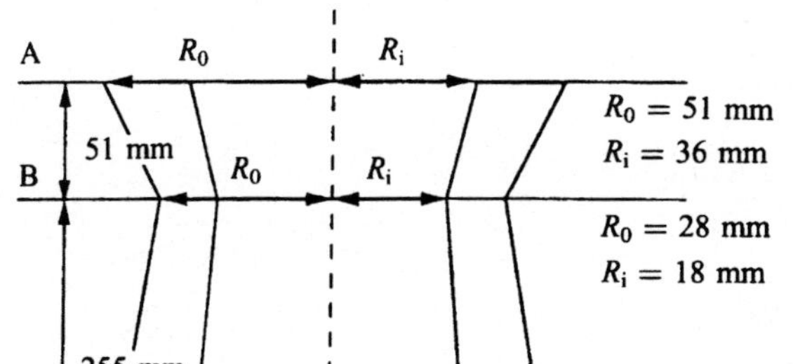

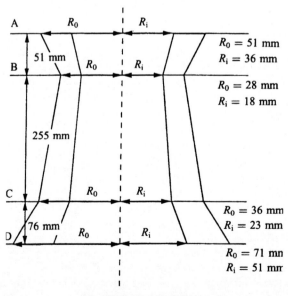

Fig. 1.90 Model of tibia.

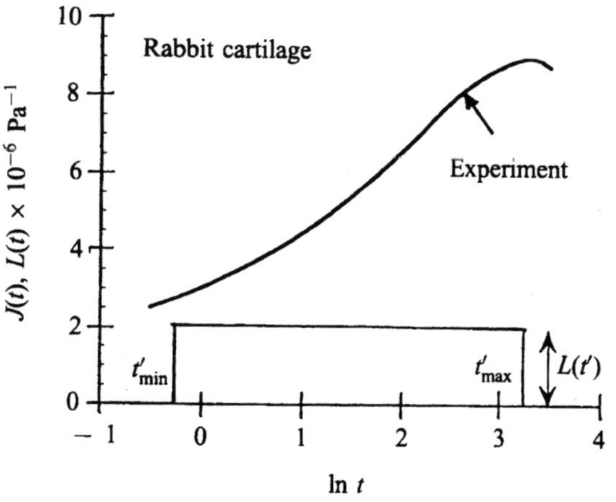

Fig. 1.91 Response of rabbit cartilage to a step load.

is observed to be in contact with the anvil for 10 ms, what is the maximum force on the system for force–time curves consisting of (i) a triangle, (ii) a half-sine wave, (iii) a square pulse?

(**c**) If the system is treated as a rigid body, will the helmet meet federal requirements for any of these force histories?

(**d**) Repeat the above if the observed duration of contact is 5 ms.

1.24 A pole vaulter Fig. 1.92, takes a running start carrying a pole with a mass of 2.273 kg (5 lbf) and a length of 3.05 m (10 ft). When he achieves a speed of 8.54 m s^{-1} (28 ft s^{-1}), he emplaces the pole in the ground and leaps upward. His mass is 77.27 kg (170 lbf) and his mass center in an erect position is 0.915 m (3 ft) from the ground. If it is assumed that the vaulter holds the pole at the end, that 12.5 per cent of the total kinetic energy of man and pole is converted into bending energy at the time when he is in a horizontal position while the pole is angled at 60° to the horizontal, that this bending energy plus his push-off from the pole gains him an additional 1.067 m (3.5 ft) of height, and that his mass center is 76.2 mm (3 in) vertically above the bar when he clears,

(**a**) determine the height at which the bar is set.

(**b**) assuming that he has no vertical speed when clearing the bar, determine the velocity he has when he returns to the ground in an erect fashion.

(**c**) Assuming that it takes 0.25 s for the sand to bring him uniformly to rest, what is the average force exerted on him by the sand?

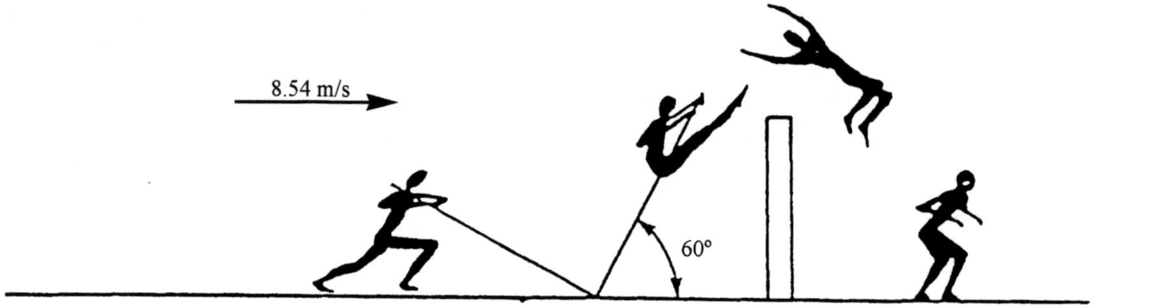

Fig. 1.92 Trajectory of a pole vaulter.

(**d**) If the resistance of the sand is modeled as a linear spring, what is its constant?

P1.25 The resistance of the air to a falling object is given by $F_D = \frac{1}{2}C_D \rho S v^2$, where C_D is the drag coefficient, dimensionless, taken as 1, ρ is the mass density, 1.2 kg m^{-3} (0.00233 slug ft^{-3}) at sea level (which can be taken as constant up to 5000 m), S is the 'presented' area of the body, and v its velocity, all in consistent units.

(**a**) Determine the 'terminal' (that is, the maximum possible) velocity of a 72.73 kg mass (160 lb) person falling (i) vertically, where the presented area is an ellipse with semi-major and semi-minor axes of 0.25 m (9.8 in) and 0.076 m (3 in), respectively; (ii) horizontally, where the presented area is a rectangle, 1.83 m × 0.305 m (6 ft × 1 ft).

(**b**) The parachutist jumps in a vertical position from a plane flying at 1524 m (5000 ft). The full opening of the parachute, which has a circular base radius of 3.81 m (12.5 ft), takes 10 seconds after pulling the cord, including the stretching of the nylon supporting cords. If the parachutist wishes to open the chute so that his body experiences no more than a 450 N (100 lbf) force due to the deceleration from the chute opening, assumed to be constant during this operation, but as close to the ground as possible, what distance must he be in 'free fall' after exiting the plane before pulling the parachute cord? For the purpose of this calculation, neglect air resistance during free fall and assume that he is uniformly decelerated during the 10 seconds of the full parachute opening. Neglect the chute weight.

P1.26 The force–extension curves for skin (Fung 1993) tested in the two directions x and y (with the stretch ratio Λ held constant in the other direction) are shown in Fig. 1.93. Using the Fung relation $d\tau/d\Lambda = a(\tau + b)$, where τ is the engineering stress (based on original area), determine the values of the constants if the cross-sectional areas for the tests along x and y are $A_x = 61.8$ mm^2 and $A_y = 68.5$ mm^2, respectively.

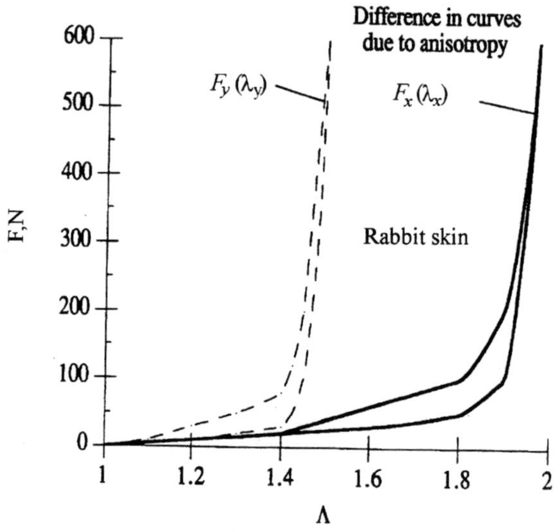

Fig. 1.93 Force–extension curve for rabbit skin (Fung 1993).

P1.27 The 3.05 m (10 ft) long, horizontal branch of a cedar tree may be considered as a uniform circular rod of 0.0762 m (3 in) diameter. It is attached to the tree by a joint that can be considered fixed or 'embedded'. A bird of mass 4.545 kg (10 lbf) is perched on the end of the tree as shown in Fig. 1.94. The specific gravity of cedar is 0.61 and its modulus of elasticity, E, is 10.38 GPa

(1.5 × 10^6 lbf in^{-2}). Considering both the weights of the bird and the uniformly distributed weight of the branch,

(**a**) write the equations for and construct the shear and bending moment diagrams for this situation;

(**b**) obtain the deflection curve;

(**c**) determine the maximum deflection of the branch;

(**d**) determine the maximum stress in the branch.

P1.28 A serious dental problem is bruxism, the unconscious sideways gnashing of teeth which produces major material damage by flattening out sharp teeth as well as producing erosion at the root. In order to quantify this problem, consider the digestion of food by biting involving (i) vertical forces only, Fig. 1.95(a), (ii) sideways chewing involving horizontal loads which act cyclically on the tooth creating this horizontal motion as well as a vertical force, Fig. 1.95(b). Assume that the two force components required to chew the food are the same, 50 N. The root of the idealized tooth has a width of 4 mm and a depth of 3 mm. The root is located 10 mm below the surface where the load is applied.

(**a**) What is the direct stress acting on the root of the tooth for case (i)?

(**b**) What is the maximum bending stress acting on the root of the tooth for case (ii)? The area is the same as in (a).

(**c**) What is the factor of safety with sideways chewing after (i) 1 year and (ii) after 50 years if 200 such motions occur per day? The fatigue curve (that is, the strength remaining after application of specific number of cycles) for the bone material is shown in Fig. 1.95(c).

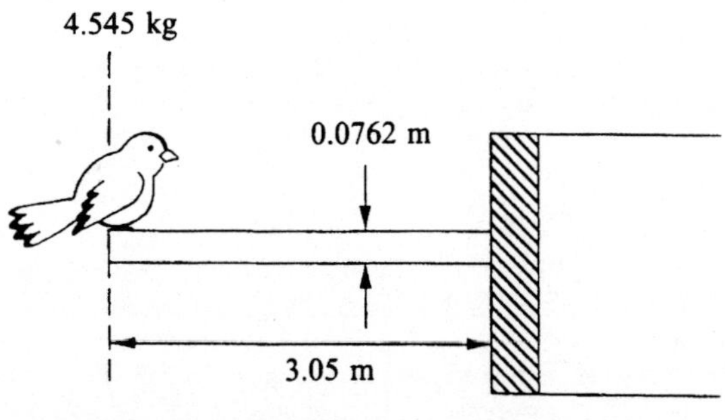

Fig. 1.94 Bird perched on a horizontal tree branch.

P1.29 A bird with a mass of 0.23 kg perches on a vegetable stalk, vertically embedded in the ground, with a height of 0.61 m (2 ft) and a Young's modulus of 138 MPa (20 000 lbf in^{-2}), Fig. 1.96.

(**a**) If the stalk can be regarded as a uniform cylinder of radius R, determine the minimum radius of the stalk (whose mass is neglected) to prevent buckling of the system. Use a column factor of $n = \frac{1}{4}$.

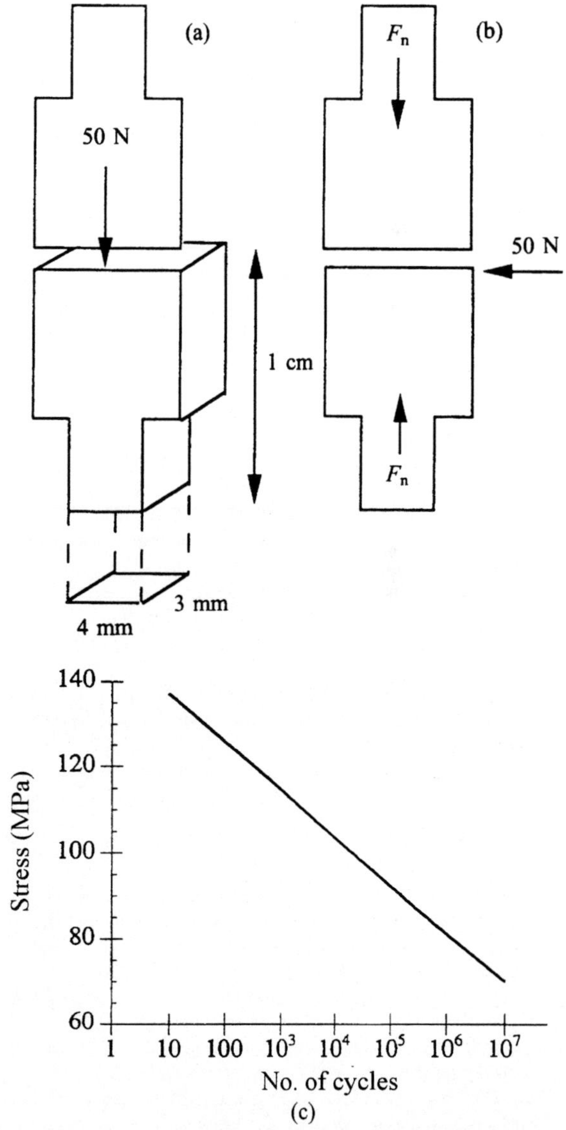

Fig. 1.95 Model of tooth with (a) axial loading, (b) axial and tangential loading; (c) fatigue curve for the system.

(b) If the system is now regarded as a beam with an equivalent spring constant $k = 3EI'/L^3$, determine the frequency at which the stalk will swing back and forth for the case of the radius found in part (a).

P1.30 The cervical disk can be considered as a Kelvin–Voigt solid. In the cervical region, the average cross section of the first six disks is 150 mm^2 and the average thickness is 5 mm. If a mass of 20 kg is suddenly placed and maintained for some time on the head, and if the Young's modulus (spring constant) of the disk is 30 MPa, and the viscous constant of the dashpot is 25 N s m^{-1}, determine (a) the strain in the disks at the end of 10 seconds, and (b) the total change in length of the neck (cervical region) at that instant. Assume that the entire load is carried by the disks.

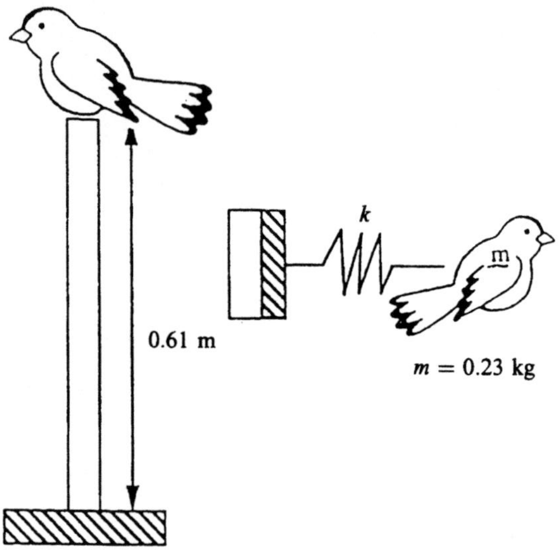

Fig. 1.96 Bird perched on top of a stalk.

References

Adrian M. and Cooper, J. M. (1989). *The biomechanics of human movement.* Benchmark Press, Indianapolis.

Aerts, P. (1992). Fish biomechanics—purpose or means. *Netherlands Journal of Zoology*, **42**, 330–44.

Aldman, B. and Chapon, A. (eds.) (1983). *The biomechanics of impact trauma*, Proceedings of the International Center for Transportation Studies, Amalfi, Italy (series). Elsevier, New York.

Alexander, R. M. (1992). *Exploring biomechanics: animals in motion.* Scientific American Library (distributed by W. H. Freeman), New York.

American Chemical Society (1992). *Viscoelasticity of biomaterials*, ACS Symposium Series **489**. American Chemical Society, Washington, DC.

Barber, J. R. (1992). *Elasticity.* Kluwer, Dordrecht and Boston.

Bonfield, W. (1987). Advances in the fracture mechanics of cortical bone. *Journal of Biomechanics*, **20**, 1071–81.

Bowen, R. M. (1989). *Introduction to continuum mechanics for engineers.* Plenum Press, New York.

Case, J., Chilver, Lord, and Ross, C. T. F. (1993). *Strength of materials and structures: with an introduction to finite element methods*, (3rd edn). Edward Arnold, London.

Chaffin, D. B. and Andersson, G. B. J. (1991). *Occupational biomechanics*, (2nd edn). Wiley, New York.

Chou, P. C. and Pagano, N. J. (1992). *Elasticity: tensor, dyadic and engineering approaches.* Dover, New York.

Christensen, D. A. (1988). *Ultrasonic bioinstrumentation.* Wiley, New York.

Christensen, R. M. (1982). *Theory of viscoelasticity: an introduction*, (2nd edn). Academic Press, New York.

Chung, T. J. (1988). *Continuum mechanics.* Prentice-Hall, Englewood Cliffs, NJ.

Contini, R. (1972). Body segment parameters, II. *Artificial Limbs*, **16**, (Spring), 1–19.

Dally, J. W. and Riley, W. F. (1991). *Experimental stress analysis*, (3rd edn). McGraw-Hill, New York.

Dempster, W. T. (1955). The anthropometry of body action. *Annals of the New York Academy of Sciences*, **63**, 559–85.

Dimarogonas, A. D. and Haddad, S. (1992). *Vibration for engineers.* Prentice-Hall, Englewood Cliffs, NJ.

Dove, R. C. and Adams, P. H. (1964). *Experimental stress analysis and motion measurement: theory, instruments and circuits, techniques.* C. E. Merill Books, Columbus, OH.

Dowling, N. E. (1993). *Mechanical behavior of engineering materials: engineering methods for deformation, fracture and fatigue.* Prentice-Hall, Englewood Cliffs, NJ.

Dowson, D. and Wright, V. (ed.) (1981). *An introduction to the biomechanics of joints and joint replacements.* Mechanical Engineering Publications, London.

Drillis, R. and Contini, R. (1966). *Body segment parameters*, Technical Report No. 1166.03. New York University, School of Engineering and Science, Research Division Technical Report 1166.03.

Easterby, R., Kroemer, K. H. E., and Chaffin, D. B. (ed.) (1982). *Anthropometry and biomechanics.* Plenum Press, New York.

Evans, F. G. (1973). *Mechanical properties of bone*. C. C. Thomas, Springfield, IL.

Feller, R., Margolin, E., Zacharin, A., and Pasternak, H. (1985). Development of a clod separator for potato packing houses. *Transactions of the American Society of Agricultural Engineers*, **28**, 1019–23.

Flügge, W. (1975). *Viscoelasticity*, (2nd edn). Springer-Verlag, Berlin, New York.

Fondrk, M., Bahnuk, E., Davy, D. T., and Michaels, C. (1988). Some viscoplastic characteristics of bovine and human cortical bone. *Journal of Biomechanics*, **21**, 623–30.

Frankel, L. J. (1981). Three-dimensional response of a humanoid head–neck system to dynamic loading, M.S. thesis. University of California, Berkeley.

Freund, L. B. (1990). *Dynamic fracture mechanics*. Cambridge University Press.

Frost, H. M. (1973). *Orthopaedic biomechanics*, Orthopaedic Lectures, Vol. 5. C. C. Thomas, Springfield, IL.

Fung, Y. C. E. (1967). Elasticity of soft tissue in simple elongation. *American Journal of Physiology*, **213**, 1532–44.

Fung, Y. C. (1993). *Biomechanics: mechanical properties of living tissue* (2nd edn). Springer-Verlag, New York.

Fung, Y. C. (1994). *A first course in continuum mechanics: for physical and biological engineers and scientists* (3rd edn.) Prentice-Hall, Englewood Cliffs, NJ.

Gans, C. (1974). *Biomechanics: an approach to vertebrate biology*. Lippincott, Philadelphia.

Gdoutos, E. E. (1993). *Fracture mechanics: an introduction*, Solid Mechanics and its Applications, Vol. 14. Kluwer, Dordrecht and Boston.

Gere, J. M. and Timoshenko, S. P. (1990). *Mechanics of materials*, (3rd edn), PWS-Kent Series in Engineering. PWS-Kent, Boston.

Ghista, D. N. (ed.) (1981). *Biomechanics of medical devices*, Biomedical Engineering and Instrumentation Series, Vol. 7. Dekker, New York.

Ghista, D. N. (ed.) (1982). *Human body dynamics*, Oxford Medical Engineering Series. Clarendon Press, Oxford.

Gilchrist, A. and Mills, N. J. (1994). Modelling of the impact response of motorcycle helmets. *International Journal of Impact Engineering*, **15**, 201–18.

Goldsmith, W. (1973). Biomechanics of head injury. In *Biomechanics: its foundations and objectives*, (ed. Y. C. Fung, N. Perrone, and M. Anliker), pp. 585–634. Prentice Hall, New York.

Goldsmith, W. (1979). Some aspects of head and neck injury and protection. In *Progress in biomechanics*, Proceedings of the NATO Advanced Study Institute, Series E, (ed N. Akkas), pp. 333–77. Sijthoff and Noordhoff, Alphen aan den Rijn, The Netherlands.

Gozna, E. R. and Harrington, I. J. (with special contribution by D. C. Evans) (1982). *Biomechanics of musculoskeletal injury*. Williams & Wilkins, Baltimore.

Gray, H. (1985). *Anatomy of the human body*, (30th American edn, ed. C. D. Clemente). Lea & Febiger, Philadelphia.

Gurdjian, E. S. (ed.) (1970). *Impact injury and crash protection*. C. C. Thomas, Springfield, IL.

Gurdjian, E. S. (1975). *Impact head injury*. C. C. Thomas, Springfield, IL.

Hastings, G. W. and Williams, D. F. (ed.) (1980). *Mechanical properties of biomaterials*. Wiley, Chichester, New York.

Hay, J. G. (1968). Mechanical energy in pole vaulting. *Track Technique*, **33**, 1047–51.

Hay, J. G. (1993). *The biomechanics of sports techniques*, (4th edn). Prentice-Hall, Englewood Cliffs, NJ.

Hay, J. G. and Reid, J. G. (1988). *Anatomy, mechanics and human motion*, (2nd edn). Prentice Hall, Englewood Cliffs, NJ.

Hearmon, R. F. S. (1961). *An introduction to applied anisotropic elasticity*. Oxford University Press.

Hinson, M. M. (1981). *Kinesiology*, (2nd edn.) W. C. Brown Company, Dubuque, IA.

Hubbell, M. (1980). Dynamics of the pole vault. *Journal of Biomechanics*, **13**, 965–76.

Hudson, R. and Johnson, W. (1976). Elementary rock climbing mechanics. *International Journal of Engineering Education*, **4**, 357–68.

Johnson, A. T. (1991). *Biomechanics and exercise physiology*. Wiley, New York.

Johnson, J., Sorecki, J., and Wells, R. P. (1975). Peak accelerations of the head experienced in boxing. *Medical and Biological Engineering*, **13**, 396–404.

Johnson, W., Al Hassani, S. T. S., and Lloyd, R. B. (1975). Aspects of pole vaulting mechanics. *Proceedings of the Institution of Mechanical Engineers, London, Applied Mechanics Group*, **189**, 53–75.

Journal of Bone and Joint Surgery, Inc. (1922–). *Journal of Bone and Joint Surgery, American Edition*. Boston.

Juvinall, R. C. and Marshek, K. M. (1991). *Fundamentals of machine component design*, (2nd edn), pp. 257–296. Wiley, New York.

King, W. F. and Mertz, H. J. (ed.) (1973). *Human impact response, measurement and simulation*. Plenum Press, New York and London.

Kobayashi, A. S. (ed.) (1993). *Handbook on experimental mechanics*, (2nd rev. edn) VCH, Bethel, CT (Society for Experimental Mechanics).

Kobayashi, A. S. (1973, 1975). *Experimental techniques in fracture mechanics*, Society for Experimental Stress Analysis, SESA Monograph No. 1, 2. Iowa State University Press, Ames, IA.

Kreighbaum, E. and Barthels, K. M. (1990). *Biomechanics: a qualitative approach for studying human movement*, (3rd edn). Macmillan, New York.

Lagarde, A. (ed.) (1987). *Static and dynamic photoelasticity and caustics: recent developments*. Springer-Verlag, Vienna, New York.

Lakes, R. S., Nakamura, S., Behiri, J. C., and Bonfield, W. (1990). Fracture mechanics of bone with short cracks. *Journal of Biomechanics*, **23**, 967–75.

Lekhnitskii, S. G. (1981). *Theory of elasticity of an anisotropic elastic body*. Mir Publishers, Moscow.

LeVeau, B. P. (1992). *Williams & Lissner's biomechanics of human motion*, (3rd edn). W. B. Saunders, Philadelphia.

Lippert, S. (ed.) (1963). *Human vibration research. A collection of articles sponsored by the Human Factors Society*. Macmillan, New York; Pergamon, Oxford.

Lur'e, A. I. (1990). *Nonlinear theory of elasticity*, (trans. K. A. Lurie). North-Holland, Amsterdam, New York.

Mal, A. K. and Singh, S. J. (1991). *Deformation of elastic solids*. Prentice-Hall, Englewood Cliffs, NJ.

Marcotte, M. R. (1990). *Biomechanics in orthodontics*. B. C. Decker, Toronto, Philadelphia.

Mase, G. E. and G. T. Mase (1992). *Continuum mechanics for engineers*. CRC Press, Boca Raton, FL.

McElhaney, J. H., Stalnaker, R. L., Estes, M. S., and Rose, L. S. (1969). Dynamic mechanical properties of scalp and brain. In *Proceedings of the 6th Annual Rocky Mountain Bioengineering Symposium*. Laramie, 67–73.

McElhaney, J. H., Fogle, J. L., Melvin, J. W., Haynes, R. R., Roberts, V. L., and Alem, N. H. (1970). Mechanical properties of cranial bone. *Journal of Biomechanics*, **1**, 495–512.

Meguid, S. A. (1989). *Engineering fracture mechanics*. Elsevier Applied Science, London and New York.

Merrill, T. H., Goldsmith, W., and Deng, Y.-C. (1984). Three-dimensional response of a lumped parameter head-neck model due to impact and impulsive loading. *Journal of Biomechanics*, **17**, 81–95.

Miles, A. W. and Tanner, K. E. (ed.) (1992). *Strain measurement in biomechanics*. Chapman & Hall, London.

Miller, D. I. and Morrison, W. E. (1975). Prediction of segmental parameters using the Hanavan human body model. *Medicine and Science in Sports*, **7**, 207–12.

Miller, D. I. and Nelson, R. C. (1973). *Biomechanics of sport—a research approach*. Lea & Febiger, Philadelphia.

Mills, N. J. and Gilchrist, A. (1991). The effectiveness of foams in bicycle and motorcycle helmets. *Accident Analysis and Prevention*, **23**, 153–63.

Minorski, N. (1983). *Nonlinear oscillations*, (2nd edn). R. E. Krieger, Malabar, FL.

Mohsenin, N. N. (1986). *Physical properties of plant and animal materials: structure, physical characteristics, and mechanical properties*, (2nd rev. and updated edn). Gordon & Breach, New York.

Morecki, A. (ed.) (1987). *Biomechanics of engineering: modelling, simulation, control*, Courses and Lectures, No. 291. Springer-Verlag, Vienna, New York.

Mow, V. and Hayes, W. C. (ed.) (1991). *Basic orthopaedic biomechanics*. Raven Press, New York.

Nahum, A. M. and Melvin, J. (ed.) (1993). *Accidental injury: biomechanics and prevention*. Springer-Verlag, New York.

Niklas, K. J. (1992). *Plant biomechanics: an engineering approach to plant form and function*. University of Chicago Press, Chicago.

Niwa, S., Perren, S. M., and Hattori, T. (1992). *Biomechanics in orthopedics*. Springer-Verlag, Tokyo, New York.

Norkin, C. C. and White, D. J. (1985). *Measurement of joint motion: a guide to goniometry*. Davis, Philadelphia.

Normann, R. A. (1988). *Principles of bioinstrumentation*. Wiley, New York.

Ono, K., Kikushi, A., Nakamura, M., Kobayashi, H., and Nakamura, N. (1980). Human head tolerance to sagittal impact reliable estimation deduced from experimental head injury using subhuman primates and human cadaver skulls. In Proceedings of the 24th Stapp Car Crash Conference, pp. 101–60, Paper 801303. Society of Automotive Engineers, Warrendale, PA.

Ostrovsky, Yu. I., Shchepinov, V. P. and Yakovlev, V. V. (1991). *Holographic interferometry in experimental mechanics*, Springer Series in Optical Sciences, Vol. 60. Springer-Verlag, Berlin, New York.

Paipetes, S. A. and Holister, G. S. (ed.) (1985). *Photoelasticity in engineering practice*. Elsevier Applied Science, London and New York.

Parton, V. Z. (1992). *Fracture mechanics: from theory to practice*, (trans. L. Mann). Gordon & Breach, Philadelphia.

Pergamon Press (1969–). *Journal of Biomechanics*, **1**–. Oxford.

Petersen, R. E. (1974). *Stress concentration factors; charts and relations useful in making strength calculations for machine parts and structural elements*. Wiley, New York.

Phipps, L. J. (1983). *Mechanics in agriculture*, (3rd edn). Interstate Printers and Publishers. Danville, IL.

Piekarski, K. (1970). Fracture of bone. *Journal of Applied Physics*, **41**, 215–33.

Plagenhoef, S. (1971). Patterns of human motion. Prentice-Hall, Englewood Cliffs, NJ.

Pope, M. H. and Outwater, J. O. (1984). Fracture characteristics of bone substance. *Journal of Biomechanics*, **5**, 457–65.

Popov, E. P. (1976). *Mechanics of materials*, (2nd edn) Prentice-Hall, Englewood Cliffs, NJ.

Popov. E. (1990). *Engineering mechanics of solids*. Prentice-Hall, Englewood Cliffs, NJ.

Rasch, P. J. (1989). *Kinesiology and applied anatomy*, (7th edn). Lea & Febiger, Philadelphia.

Reznicek, R. (ed.) (1988). *Physical properties of agricultural materials and products*. Hemisphere, Washington, DC.

Schafer, N. C. (1987). *Clinical biomechanics: musculoskeletal actions and reactions*, (2nd edn). Williams & Wilkins, Baltimore.

Sedlin, E. D. (1965). A rheological model for cortical bone: a study of the physical properties of human femoral samples. *Acta Orthopaedica Scandinavica*, Suppl. 83, 1–77.

Sih, G. C. (ed.) (1981). *Experimental evaluation of stress concentration and intensity factors: useful methods and solutions to experimentalists in fracture mechanics*. Nijhoff, The Hague, Boston.

Silver, F. H. (1987). *Biological materials: structure, mechanical properties and modelling of soft tissue*. New York University Press, New York.

Soedel, W. (1993). *Vibrations of shells and plates*, (2nd rev. edn). Dekker, New York.

Sokolnikoff, I. S. (1956). *Mathematical theory of elasticity*, (2nd edn). McGraw-Hill, New York.

Steidel, R. F., Jr (1989). *An introduction to mechanical vibrations*, (3rd edn). Wiley, New York.

Stronge, W. J. (1990). Rigid body collisions with friction. *Proceedings of the Royal Society of London, Series A*, **431**, 169–81.

Suresh, S. (1991). *Fatigue of materials*. Cambridge University Press.

Synge, J. L. and Griffith, B. A. (1959). *Principles of mechanics*, (3rd edn). McGraw-Hill, New York.

Timoshenko, S. P. and Goodier, J. N. (1969). *Theory of elasticity.* (3rd edn.) McGraw-Hill, New York.
Timoshenko, S. and Woinowsky-Krieger, S. (1959). *Theory of plates and shells.* McGraw-Hill, New York.
Vaughan, C. L. (ed.) (1989). *Biomechanics of sport.* CRC Press, Boca Raton, FL.
Vaughan, C. L., Murphy, G. N., and de Toit, L. L. (1987). *Biomechanics of human gait: an annotated bibliography,* (2nd edn). Human Kinematics Publishers, Champaign, IL.
Versace, J. (1971). A review of the Severity Index. In Proceedings of the 15th Stapp Car Crash Conference. SAE Paper No. 710381. Society of Automotive Engineers, Warrendale, PA.
Vinson, J. R. (1989) *The behavior of thin walled structures: beams, plates and shells,* Mechanics of Surface Structures, Vol. 8. Kluwer Academic, Dordrecht, Boston.
Walker, H. S. and Kirmser, P. G. (1982). Biomechanical parametric analysis of pole vaulting and optimization of performance. In *Human body dynamics* (ed. D. N. Ghista), Oxford Medical Engineering Series. Clarendon Press, Oxford.
Webb, P. W. and Weihs, D. (eds.) (1983). *Fish biomechanics,* Praeger Special Studies. Praeger, New York.
Winter, D. A. (1990). *Biomechanics and motor control of human movement,* (2nd edn). Wiley, New York.
Winter, D. A. (1991). *Biomechanics and motor control of human gait: normal, elderly and pathological,* (2nd edn). University of Waterloo Press, Waterloo, Ont.
Wise, D. L. (ed.) (1990). *Bioinstrumentation: research, developments and applications.* Butterworths, Boston.
Wise, D. L. (ed.) (1991). *Bioinstrumentation and biosensors.* Dekker, New York.
Woo, S. L., Johnson, G. A., and Smith, B. A. (1993). Mathematical modelling of ligaments and tendons. *Journal of Biomechanical Engineering,* B, **115**, 468–73.
Yamada, H. (1970). *Strength of biological materials* (ed. F. G. Evans). Williams and Wilkins, Baltimore.
Young, M. (ed.) (1992). *The Guinness book of sports records.* Facts on File, New York.

Further reading

ADIS Press (1984–). *Sports Medicine,* **1**–. Newtown, PA.
Alexander, R. M. (1975). *Biomechanics.* Chapman & Hall, London.
American College of Sports Medicine (1969–). *Medicine and Science in Sports and Exercises,* **1**–. Madison, WI.
American Orthopaedic Society for Sports Medicine (1972–). *American Journal of Sports Medicine,* **1**–. Waltham, MA.
American Society of Mechanical Engineers (ASME), Bioengineering Division (1974–). *Advances in bioengineering.* ASME, BED, New York.
American Society of Mechanical Engineers (ASME) (1975–). *Biomechanics symposium,* **10**–. ASME, New York.
American Society of Mechanical Engineers (ASME), Bioengineering and Design Divisions (1987). *Biomechanics in sport—a 1987 update.* ASME, New York.
American Society of Mechanical Engineers (ASME) (1977–). *Transactions of the ASME, Journal of Biomechanical Engineering,* **99**–. ASME, New York.
Anderson, T. L. (1991). *Fracture mechanics: fundamentals and applications.* CRC Press, Boca Raton, FL.
Backaitis, S. H. (ed.) (1993). *Biomechanics of impact injury and injury tolerances to the head–neck complex.* Society of Automotive Engineers, Warrendale, PA.
Bauld, N. R., Jr (1986). *Mechanics of materials,* (2nd edn). PWS Engineering, Boston.
Beer, F. P. and Johnston, E. R., Jr (1988). *Vector mechanics for engineers: statics and dynamics,* (5th edn). McGraw-Hill, New York.
Biewener, A. A. (ed.) (1992). *Biomechanics—structures and systems: a practical approach.* IRL Press at Oxford University Press.
Bleustein, J. L. (ed.) (1973). *Mechanics and sport,* AMD vol. 4. American Society of Mechanical Engineers, New York.
Caputo, A. A. and Standles, J. P. (1987). *Biomechanics in clinical dentistry.* Quintessence Publishing Company, Chicago.
Cavanagh, P. R. (ed.) (1990). *Biomechanics of distance running.* Human Kinetics Books, Champaign, IL.
Christensen, R. M. (1991). *Mechanics of composite materials.* Krieger, Malabar, FL.
Cowin, S. C. (ed.) (1989). *Bone mechanics,* CRC Press, Boca Raton, FL.
Dally, J. W., Riley, W. F., and McConnell, K. G. (1991). *Instrumentation for engineering measurements* (2nd edn). Wiley, New York.
Den Hartog, J. P. (1956). *Mechanical vibrations* (4th edn.). McGraw-Hill, New York.
Ecker, T. (1985). *Basic track and field biomechanics.* Tafnews Press, Los Altos, CA.
Friedrich, K. (ed.) (1989). *Application of fracture mechanics to composite materials.* Elsevier, Amsterdam, New York.
Frocht, M. M. (1969). *Photoelasticity: the selected scientific papers of M. M. Frocht,* (ed. M. M. Leven). Pergamon Press, Oxford, New York.
Fung, Y. C. (ed.) (1966). *Biomechanics,* Proceedings of a Symposium Sponsored by the Applied Mechanics Division of the ASME. ASME, New York.
Fung, Y. C. (1990). *Biomechanics: motion, flow, stress and growth.* Springer-Verlag, New York.
Fung, Y. C., Perrone, N., and Anliker, M. (ed.) (1972). *Biomechanics, its foundations and objectives.* Prentice-Hall, Englewood Cliffs, NJ.

Gabelnick, H. L. and Litt, M. (1973). *Rheology of biological systems*. C. C. Thomas, Springfield, IL.

Ghista, D. N. and Roaf, R. (ed.) (1978). *Orthopaedic mechanics; procedures and devices*. Academic Press, New York.

Greenwood, D. T. (1988). *Principles of dynamics*, (2nd edn). Prentice-Hall, Englewood Cliffs, NJ.

Gregor, R. J., Broker, J. P., and Ryan, N. M. (1991). The biomechanics of cycling. *Exercise and Sport Sciences Review*, **19**, 127–69.

Haher, T. R., O'Brien, M., Kauffman, C., and Liao, K. C. (1993). Biomechanics of the spine in sports. *Clinics in Sports Medicine*, **12**, 449–64.

Hall, S. J. (1991). *Basic biomechanics*. Mosby Year Book, St. Louis.

Harper and Row Medical Department (1976–). *Spine*, **1**–. Philadelphia. (Formerly published by J. P. Lippincott, Philadelphia.)

Hay, J. G. (1987). *A bibliography of biomechanics literature*, (5th edn). J. G. Hay, Iowa City, IA.

Hay, J. G. (1992). The biomechanics of the triple jump. *Journals of Sports Sciences*, **10**, 343–78.

Hertzberg, R. W. (1989). *Deformation and fracture mechanics in engineering materials* (3rd edn). Wiley, New York.

Hibbeler, R. C. (1992). *Engineering mechanics: statics and dynamics*, (6th edn). Macmillan, New York.

Hirokawa, S. (1993). Biomechanics of the knee joint: a critical review. *Critical Reviews in Bioengineering*, **21**, 79–135.

International Congress of Biomechanics (1967–). *Proceedings*, **1**–. University Park Press, Baltimore; Human Kinetics Publisher, Champaign, IL; Free University Press, Amsterdam.

International Research Committee on the Biomechanics of Impacts (IRCOBI) (1971–). *Proceedings*. IRCOBI Secretariat, Bron, France.

Kenedi, R. M. (ed.) (1980). *A textbook of biomedical engineering*. Blackie, Glasgow.

Komi, P. V. (1990). Relevance of in vivo force measurements to human biomechanics. *Journal of Biomechanics*, Supplement 1, 23–34.

Maiman, D. J. and Yoganandan, N. (1991). Biomechanics of cervical spine trauma. *Clinical Neurosurgery*, **37**, 543–70.

Mero, A., Komi, P. V., and Gregor, R. J. (1992). Biomechanics of sprint running. A review. *Sports Medicine*, **13**, 376–92.

Narasimhan, M. N. L. (1993). *Principles of continuum mechanics*. Wiley, New York.

NATO Advanced Study Institute (1979). *Progress in biomechanics*, (ed. N. Akkas), Nato Advanced Study Institute, Series E: Applied Sciences, Vol. 32. Sijthoff and Noordhoff, Aalphen aan den Rijn, Netherlands.

Nordin, M. and Frankel, V. H. (1989). *Basic biomechanics of the musculoskeletal system*, (2nd edn). Lea & Febiger, Philadelphia.

Park, J. B. (1979). *Biomaterials: an introduction*. Plenum Press, New York.

Parton, V. Z. and Boriskovsky, V. G. (1989–90). *Dynamic fracture mechanics*, (rev. edn), (trans. R. S. Wadhwa; R. B. Hetnarski). Hemisphere, New York.

Pauwels, F. (1980). *Biomechanics of the locomotor apparatus*. Springer-Verlag, Berlin, New York.

Profio, A. E. (1993). *Biomedical engineering*. Wiley, New York.

Reid, J. G. and Jensen, R. K. (1990). Human body segment inertia parameters: a survey and status report. *Exercise and Sport Sciences Reviews*, **18**, 225–41.

Russell, A. (ed. in chief) (1987). *1987 Guinness book of world records*. Sterling, New York.

Sahay, K. B. and Saxena, R. K. (ed.) (1989). *Biomechanics*. Wiley, New York.

Schendel, M. J., Wood, K. B., Butterman, G. R., Lewis, J. L., and Ogilvie, J. W. (1993). Experimental measurement of ligament force, facet force, and segment motion in the human lumbar spine. *Journal of Biomechanics*, **26**, 427–38.

Schmid-Schonbein, G. W., Woo, S.-L. Y., and Zweifach, B. W. (ed.) (1986). *Frontiers in biomechanics*. Springer-Verlag, New York.

Simonian, C. (1981). *Fundamentals of sports biomechanics*. Prentice-Hall, Englewood Cliffs, NJ.

Skalak, R. and Chien, S. (ed.) (1987). *Handbook of bioengineering*. McGraw-Hill, New York.

Society for Experimental Biology (1980). *The mechanical properties of biological materials*. Cambridge University Press.

Society for Sports Sciences (1983–). *Journal of Sports Sciences*. E. & F. N. Spon, London.

Society of Automotive Engineers (1986). *Human tolerance to impact conditions as related to motor vehicle design*, SAE 885. SAE, Warrendale, PA.

Stapp Car Crash Conferences (1955–). *Proceedings*, **1**–. Society of Automotive Engineers, Warrendale, PA.

Stein, J. L. (ed.) (1987). *Biomechanics of normal and prosthetic gait*, BED Series, Vol. 4, DSC Series Vol. 7. ASME, New York.

Thieme Verlag (1980–). *International Journal of Sports Medicine*, **1**–. Stuttgart.

Timoshenko, S. P. (in collaboration with D. H. Young) (1955). *Vibration problems in engineering*, (3rd edn). Van Nostrand, New York.

Tischauer, E. R. (1975). *Occupational biomechanics: an introduction to the anatomy and function of man at work*, Rehabilitation Monograph No. 51. Institute for Rehabilitation Medicine, New York University Medical Center, New York.

United States Department of Transportation, National Highway Traffic Safety Administration (1988). Motor Vehicle Safety Standard Number 218, Docket 72–6, rev. 4/6/88.

Vincent, J. F. V. (ed.) (1992). Biomechanics—materials: a practical approach. IRL Press at Oxford University Press.

Wilder, D. G. (1993). The biomechanics of vibration and low back pain. *American Journal of Industrial Medicine*, **23**, 577–88.

Winters, J. M. and Woo, S. L.-Y. (ed.) (1990). *Multiple muscle systems: biomechanics and movement organization*. Springer-Verlag, New York.

Yettram, A. L. (ed.) (1989). *Material properties and stress analysis in biomechanics*. Manchester University Press.

Zatsiorsky, V. M. and Yakunin, N. (1991). Mechanics and biomechanics of rowing—a review. *International Journal of Sport Biomechanics*, **7**, 229–81.

FUNDAMENTALS OF FLUID MECHANICS

2

Lawrence Talbot

Contents

2.1 Basic concepts in fluid mechanics

Fluid mechanics plays an important role in many areas of physiological functioning, as for example the flows in the circulatory and respiratory systems and the intercellular transport of material, to name a few. Our object here is to present at an elementary level some of the basic concepts and laws which govern fluid motion and how they relate to some common physiological phenomena. At the level of our presentation, we can but scratch the surface of this highly complex subject, but we hope that the information presented may provide previously uninitiated readers with a starting point for a more in-depth study of the subject.

2.1.1 Definition of a fluid

A fluid, either liquid or gas, is defined as a medium which deforms (undergoes motion) continuously under the action of a shearing stress. This definition distinguishes a fluid from a solid, which is able to resist a shear by static deformation. A consequence of this definition is that a fluid at rest can have no shear stresses acting within it or at its boundaries, and we describe such a fluid state as being a state of *hydrostatic equilibrium*. Normal stress, or pressure, p and its variation within a fluid in hydrostatic equilibrium are due solely to gravitational forces. (We exclude consideration of electromagnetic phenomena.)

2.1.2 A fluid as a continuous medium

We will for the most part consider the fluids we deal with as continuous media. Even though we know that both liquids and gases are made up of individual molecules, on a macroscopic scale we ignore this and treat fluid properties such as temperature, pressure, velocity, etc. as point functions which vary continuously and smoothly throughout the fluid, except possibly at interfaces. However, this description of a fluid as a continuous medium will have to be abandoned later on when capillary blood flow is considered, where the discrete nature of the formed elements in blood (the red blood cells, principally) has to be taken into account.

2.1.3 Physical properties

The intrinsic mechanical properties of a fluid of most importance in physiological applications are its density and viscosity, and to a lesser extent its surface tension and compressibility. In describing these properties, we will use primarily SI units, although other units in common usage will be mentioned.

Density

Fluid density, which is its mass per unit volume, is measured in SI units in kilograms per cubic meter (kg m^{-3}), and is here denoted by the symbol ρ. In the mass–length–time (M–L–T) system of fundamental units, the dimensions of ρ are thus $[\rho] = [ML^{-3}]$, where the square brackets indicate that the dimensions of the quantity are being expressed. Fluid densities are also often cited in cgs units, g cm^{-3}, and it is obvious that ρ (g cm^{-3}) $\times$ 1000 = ρ (kg m^{-3}). The density of a fluid at constant pressure varies inversely with temperature due to thermal expansion. The rate of decrease is much greater in gases than in liquids. However, fluid flows in physiological systems are for the most part isothermal, so that except in special circumstances fluid densities in these systems can be taken as constant properties. We term such flows *incompressible*.

A few numerical values may be of interest. Water at a temperature of 15°C has a density $\rho = 999$ kg m^{-3}, whereas air at this temperature and standard atmospheric pressure has a density $\rho = 1.22$ kg m^{-3}. Blood at body temperature (37°C) has a density $\rho = 1060$ kg m^{-3}, about 6 per cent greater than that of water.

The density of a liquid is often cited in terms of its specific gravity, s, which is defined as the ratio $s = \rho(\text{liquid})/\rho(H_2O)$, with $\rho(H_2O)$ being chosen at some reference temperature, commonly 15°C. Thus the specific gravity of blood at 37°C would have the value $s = 1.06$. Note that specific gravity is a dimensionless quantity, and we would indicate this by writing $[s] = [M^\circ\ L^\circ\ T^\circ]$.

The specific weight of a fluid, which is the force exerted by gravity on a unit mass of fluid, is often given a special symbol. Here we will not do so, but instead express it as ρg, where g is the acceleration of gravity (9.81 m s^{-2}).

Viscosity

Before defining fluid viscosity quantitatively it is essential to call attention to a phenomenon common both to liquids and to gases at normal densities, called the *no-slip* condition. This condition, established experimentally, states that the relative velocity of a fluid in contact with a solid boundary is zero. Thus if the boundary is stationary, as for example a pipe wall, the fluid in contact with the wall will have zero velocity, although elsewhere in the fluid there may be motion. If the boundary is moving, the fluid in contact with the boundary will have the same velocity as the boundary.

Having described the no-slip condition, we may now consider the conceptual experiment illustrated in Fig. 2.1(a). Fluid is contained between two infinite parallel plates, separated by a distance h. A tangential force F is applied to the upper plate, causing it to translate at a constant velocity U, while the lower plate is held fixed.[1] Thus the fluid contained between the plates is being sheared at a constant rate. (This topic is also briefly discussed in Section 1.1.12.)

If we were to carry out this experiment for different values of F (see footnote 1), we would observe for many fluids the following facts:

(a) The tangential force per unit plate area, or shear stress, $F/A \equiv \tau$ exerted by the upper plate upon the fluid is directly proportional to the translational velocity U of the plate. Since U is constant, and hence the acceleration is zero, a force balance requires that the shear stress τ exerted on the fluid by the upper plate be balanced by an equal and opposite shear stress exerted by the lower plate upon the fluid. In fact a little thought leads us to conclude that the shear stresses acting on the upper and lower faces of any element of fluid within the gap between the plates must be equal and opposite, as shown in Fig. 2.1(b), as it deforms as indicated by the dashed lines.

[1] An approximation to this conceptual experiment may be achieved in the laboratory by means of containing fluid in an apparatus consisting of two concentric cylinders with annular gap width small compared to their length and radii, and holding one cylinder fixed while applying torque to rotate the other one.

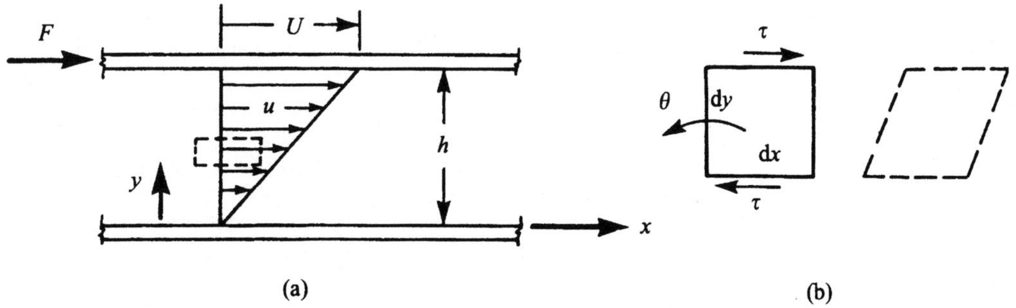

Fig. 2.1 Viscous flow and shearing action produced by relative motion between two parallel plates.

(b) If now we were to repeat this experiment, varying the gap width h while keeping U constant, we would find that the observed shear stress varies inversely with h.

When we combine these two observations, we conclude that

$$\tau \propto \frac{U}{h}, \tag{2.1}$$

that is, the shear stress within the fluid is directly proportional to U and inversely proportional to h.

Because of the no-slip condition, the fluid velocity u at the lower plate must be zero, and at the upper plate must be equal to U, as shown in Fig. 2.1, and in fact u must vary linearly with the transverse distance y. This means that the *velocity gradient*, $\mathrm{d}u/\mathrm{d}y$, must be constant and in fact equal to U/h. Hence eqn (2.1) may be expressed alternatively as

$$\tau \propto \frac{\mathrm{d}u}{\mathrm{d}y}. \tag{2.2}$$

Equation (2.2) is of course only a statement of proportionality. To convert it into an equality, we must insert the constant of proportionality, which we define as the *viscosity*, μ. Thus we write

$$\tau = \mu\frac{\mathrm{d}u}{\mathrm{d}y}. \tag{2.3}$$

Equation (2.3) represents the simplest definition of a Newtonian fluid, the property of which is that the viscous shear stress within the fluid is linearly proportional to the local velocity gradient, with the constant of proportionality being the fluid viscosity.[2] The dimensions of viscosity may be easily inferred from eqn (2.3). We have

$$[\mu] = [\tau]\Big/\left[\frac{\mathrm{d}u}{\mathrm{d}y}\right]. \tag{2.4}$$

Since τ = force/area, and force has by Newton's law the dimension of mass × acceleration, we have $[\tau] = [F]/[A] = [\mathrm{MLT^{-2}}]/[\mathrm{L^2}] = [\mathrm{ML^{-1}T^{-2}}]$. Similarly, we have $[\mathrm{d}u/\mathrm{d}y] = [\mathrm{LT^{-1}}]/[\mathrm{L}] = [\mathrm{T^{-1}}]$. Putting these together, we obtain

$$[\mu] = [\mathrm{ML^{-1}T^{-1}}]. \tag{2.5}$$

In SI units, then, the units of viscosity are kg $\mathrm{m^{-1}s^{-1}}$.[3] However, despite the growing acceptance of SI units in the scientific literature, one still finds cgs units employed. In cgs units, viscosity is measured in g $\mathrm{cm^{-1}s^{-1}}$, and in fact the quantity of 1 g $\mathrm{cm^{-1}s^{-1}}$ has been assigned a special name, the poise, in honor of the French physiologist Jean-Louis-Marie Poiseuille (1799–1869), about whose work more will be said later. It can be easily verified that the conversion factor is 1 poise = 0.1 kg $\mathrm{m^{-1}s^{-1}}$ (or 0.1 Pa s). The continued attractiveness of the poise as a unit of measure of viscosity seems to be that the viscosity of water at normal temperature is approximately 0.01 poise, or 1 centipoise.[4]

[2] This equation for the shear stress has been derived under the assumption that the flow is rectilinear with only one component of velocity u in the x-direction. In the case of a more general planar flow, having as well a velocity component v in the y-direction, the expression for the shear stress is $\tau = \mu\left(\frac{\partial u}{\partial y} + \frac{\partial v}{\partial x}\right)$. We note for future reference that the quantities $\frac{\partial v}{\partial x}$ and $-\frac{\partial u}{\partial y}$ are in fact the angular velocities of the line elements $\mathrm{d}x$ and $\mathrm{d}y$, respectively.

[3] If we had chosen to use force rather than mass as a fundamental dimension, then in SI units with the unit of force being the Newton (N), and with stress having the units of N $\mathrm{m^{-2}}$ or Pascals (Pa), the units of viscosity would be given in Pascal-seconds (Pa s). Numerical values of viscosity, whether in Pa s or kg $\mathrm{m^{-1}s^{-1}}$, will be identical.

[4] The ratio μ/ρ often appears in analytical results, and has been given the name 'kinematic viscosity', and denoted by the symbol ν. The dimensions of ν are $[\mathrm{L^2T^{-1}}]$. In the cgs system water has a kinematic viscosity of about 0.01 $\mathrm{cm^2s^{-1}}$, or one centistoke. The unit 'stoke' (1 $\mathrm{cm^2s^{-1}}$) is named in honor of the English scientist Sir George G. Stokes (1819–1903).

Although we have introduced viscosity as the 'constant' of proportionality relating local fluid shear stress to local velocity gradient, we do not mean to imply that the viscosity of a given fluid is an absolute material constant. In fact, the viscosity of a fluid is strongly dependent on fluid temperature. In gases, viscosity increases as gas temperature increases, while in liquids just the reverse occurs, and a hot liquid will have a smaller viscosity and resistance to shear than it would possess at a lower temperature.

Equation (2.3) provides us with the simplest definition of a Newtonian fluid, in which the shear stress acting within the fluid at any point is linearly proportional to the local velocity gradient or shear rate at that point. Many important fluids, in particular air and water, can be classified as Newtonian. However, for blood the situation is not so simple. It turns out that the viscometric or rheological behavior of blood can be described as Newtonian for sufficiently high shear rates, but more complicated relationships between shear stress and shear rate must be employed at low rates of shear. Some further discussion of this will be given later in Chapter 3; for the present we will assume that all the fluids we deal with are Newtonian in character.

While viscosity is indeed a fundamental physical property of all fluids, this does not mean that viscosity plays a decisive role in governing all fluid motions. In fact, as we shall expand upon later on, there are numerous examples of fluid flow for which the effects of viscosity are confined to very narrow regions of flow (boundary layers) adjacent to solid boundaries, and for which in the vast bulk of the fluid, shear stresses are unimportant and the fluid behaves as if it were not possessed of viscosity. Such fluid flows are commonly termed as *inviscid*. Very many important fluid flow problems can be successfully analyzed under the inviscid flow assumption. When we employ this assumption, we discard the no-slip requirement, allowing a finite tangential fluid velocity to exist adjacent to a solid wall, although we still require that the wall be impermeable and thus that the fluid velocity normal to the wall must be zero, or if the wall is moving, equal to its normal velocity.

Surface tension

A large number of familiar phenomena, such as those associated with soap bubbles, indicate that the surface of a liquid behaves like a stretched membrane. The reasons for this have to do with intermolecular forces, but for our purposes a macroscopic approach is sufficient. Because a liquid surface is in tension, if we imagine a cut made in the surface, we will expose a tensile force. The force per unit length of the cut is called the *surface tension*, which we denote by the symbol σ. The dimensions of σ are Newtons per meter in SI units. For the most common interface, air–water, the value of σ is 0.073 N m^{-1} at 15°C. This is a nominal value; it can be reduced considerably if the surface contains contaminants or surfactants. Surface tension effects are responsible for capillary action, and for determining whether a liquid wets (water on glass) or does not wet (water on paraffin, mercury on glass) a surface. They are likewise responsible for the break-up of liquid jets, and the fact that droplets take on a spherical shape, because that is the configuration of minimum surface energy. Related to this last fact is that the pressure inside a drop exceeds the ambient pressure, which we now examine briefly.

We imagine a cut made through the equator of a spherical drop, as shown in Fig. 2.2, thus exposing the surface tension force acting around the boundary and the force due to the excess pressure Δp acting within the drop, required to balance the surface tension. The surface tension force is the surface tension σ multiplied by the circumference of the drop, while the pressure force is the product of the excess pressure and the cross-sectional area of the drop. Neglecting gravity effects, a force balance gives

$$2\pi R\sigma = \pi R^2 \Delta p$$

or

$$\Delta p = \frac{2\sigma}{R} \tag{2.6}$$

Thus we observe that the excess pressure Δp is directly proportional to σ and inversely proportional to drop

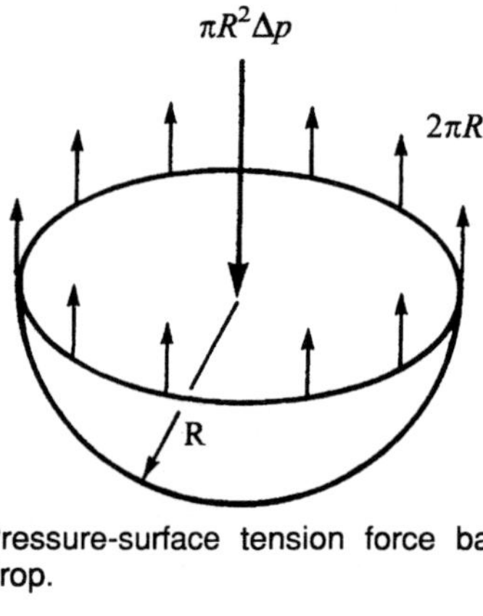

Fig. 2.2 Pressure-surface tension force balance for a spherical drop.

radius. This fact is of importance in pulmonary mechanics.

Compressibility

All substances are to a greater or lesser extent compressible, gases much more so than liquids or solids. The measure of the compressibility of a substance is the bulk modulus E_v, defined by

$$E_v = \rho \frac{dp}{d\rho}. \tag{2.7}$$

E_v obviously has the dimension of pressure or $[ML^{-1}T^{-2}]$. For gases, E_v takes on different values depending on how the compression is accomplished, as for example adiabatically or isothermally. For liquids and solids, the differences are negligible. Water at 15°C has the value $E_v = 2.15 \times 10^9$ N m^{-2}.

The speed of sound propagation c in a substance is directly related to E_v by the equation

$$c = \sqrt{E_v/\rho}. \tag{2.8}$$

Thus for water at 15°C, we have

$$c = \sqrt{\left(\frac{2.15 \times 10^9}{999}\right)} = 1470 \text{ m s}^{-1}.$$

The speed of sound in body tissue and in blood is of importance in the quantitative interpretation of clinical ultrasound imaging (see Appendix 2.B).

2.2 Kinematics of fluid flow

The most important information needed in the analysis of fluid flow problems is knowledge of the velocity field, since if it is known in an incompressible flow, all other quantities of interest such as local pressures and forces on boundaries can be calculated. While in solid mechanics we may be interested in following the motion of individual particles, the so-called Lagrangian approach, this is generally not feasible (or desirable) in fluid mechanics[5] and instead we adopt the Eulerian approach in which we focus on the fluid velocity at a given point x, y, z in space as a function of time t, without concern for the fact that the point will be successively occupied by different fluid 'particles' as the flow proceeds. Thus we write the fluid velocity vector V as

$$V = V(x, y, z, t)$$

where V has velocity components u, v, w, in the x, y, and z directions respectively. The properties of the velocity field V constitute the kinematics of the flow.

2.2.1 Fluid acceleration

When we apply Newton's laws of motion to fluid flow, we will need an expression for fluid acceleration in terms of the Eulerian description of the velocity field. We therefore require the total time derivative of $V(x, y, z, t)$, considering x, y, z, and t as independent variables. By the chain rule of differentiation,

$$\frac{dV}{dt} = \frac{\partial V}{\partial t} + \frac{\partial V}{\partial x}\frac{dx}{dt} + \frac{\partial V}{\partial y}\frac{dy}{dt} + \frac{\partial V}{\partial z}\frac{dz}{dt}. \tag{2.9}$$

Now, dx represents the infinitesimal change in the x-position of a fluid particle during the infinitesimal time interval dt, and thus dx/dt is the local x-component of fluid velocity u, and similarly for dy/dt and dz/dt. Hence eqn (2.9) can be written as

$$\frac{dV}{dt} = \frac{\partial V}{\partial t} + u\frac{\partial V}{\partial x} + v\frac{\partial V}{\partial y} + w\frac{\partial V}{\partial z}. \tag{2.10}$$

Equation (2.10) in fact comprises three scalar equations, with V in turn replaced by each of its Cartesian components u, v, and w. The term $\partial V/\partial t$ represents the local acceleration, which is zero for a steady flow, whereas the remaining terms represent the convective acceleration, which can exist even when the flow is steady, because the velocity of a fluid particle can change as it is convected through the flow field.

It is often convenient to evaluate dV/dt along a particular curve s drawn within the fluid, to which V is instantaneously tangent. If this curve is expressed parametrically by $s = s(x, y, z, t)$, then the tangential acceleration along the curve s is given by

$$\left(\frac{dV}{dt}\right)_{\text{tang.}} = \frac{\partial V}{\partial t} + V\frac{\partial V}{\partial s}, \tag{2.11}$$

[5] An exception to this is in the application of marker particles to track fluid flows, as in the case of dye and thermal dilution techniques for the measurement of blood flow, which are described in Appendix 2.B.

where V is the scalar magnitude of the vector $\boldsymbol{V}$. If the path s has curvature, then along with the tangential acceleration there will be a centripetal acceleration normal to the path in the direction of its center of curvature given by

$$\left(\frac{\mathrm{d}V}{\mathrm{d}t}\right)_{\text{norm.}} = \frac{V^2}{R_c}, \qquad (2.12)$$

where R_c is the local radius of curvature of the particle path.

2.2.2 Steady vs. unsteady flow

Imagine that we visualize a fluid flow by mixing within the fluid light-reflecting particles which are small enough to faithfully follow the fluid motion, and that we take a short time exposure photograph of the flow. Our photograph would display a large number of bright streaks produced by the light-reflecting particles. The lengths δx, δy of these streaks represent the particle displacements in, say, the x–y plane of the photograph during the exposure time δt, and thus an analysis of all of the streaks of the photograph can provide us with an approximate quantitative measure of the x and y components of the velocity field at the time t_1 at which the photograph was taken. The ideas are illustrated in Fig. 2.3. (This method works well only if the z-component of velocity, w, is negligible.)

Let us now suppose that we repeat this process at a subsequent time t_2. If we find that the analysis of our new streak pattern gives us the same velocity field as that obtained at time t_1, then we have a time-invariant *steady* flow. If the velocity field has changed, then we have an unsteady flow. Mathematically, the condition for steady flow is that the term $\partial\boldsymbol{V}/\partial t$ in eqn (2.10) is zero. We emphasize that the vanishing of $\partial\boldsymbol{V}/\partial t$ does *not* mean that the fluid acceleration is zero, since the convective terms $\partial\boldsymbol{V}/\partial x$, etc. in eqn (2.10) remain, and represent fluid accelerations associated with changes in fluid particle velocities as they move throughout the flow field, even when the flow is steady.

2.2.3 Streamlines, pathlines, and streaklines

Fluid flows lend themselves readily to a variety of visualization technique which can provide qualitative and sometimes quantitative information on the flow properties. The most useful indicator of a flow is the streamline pattern, since this flow characteristic is that which is most readily predicted by analysis and the one which provides the easiest understanding of flow behavior. Streaklines and pathlines, while reasonably simple to construct from experimental flow visualizations, are less easy to relate to analytical results, in the general case of unsteady flow.

A *streamline* is a curve which is everywhere tangent to the local velocity vector at a given instant. For a planar flow, as described in Section 2.2.2, we may determine the streamline patterns at successive times t_1, t_2, etc. by constructing curves tangent to the photographically recorded light streaks throughout the plane of the photograph, as suggested in Fig. 2.4.

Pathlines can also be constructed via the photographic procedure described with an assemblage of particles identified by their light streaks at an exposure time t_1. Subsequent exposures at times t_2, t_3, etc. can be used to track the motion of identified particles throughout the flowfield and thus construct the pathlines of these particles.

Streaklines are at the same time the easiest to produce experimentally and the most difficult to interpret

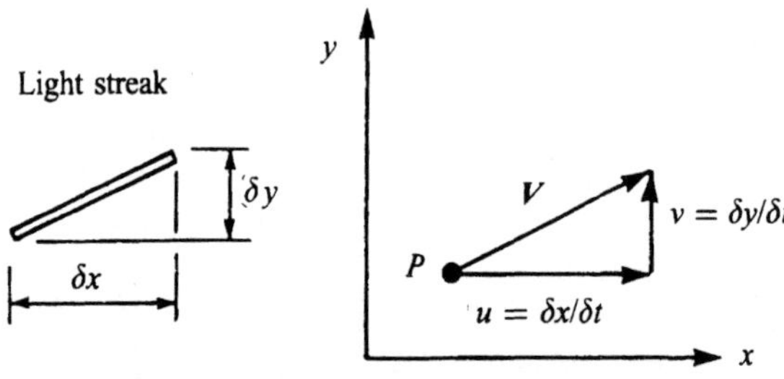

Fig. 2.3 Construction of an instantaneous velocity field from light streaks.

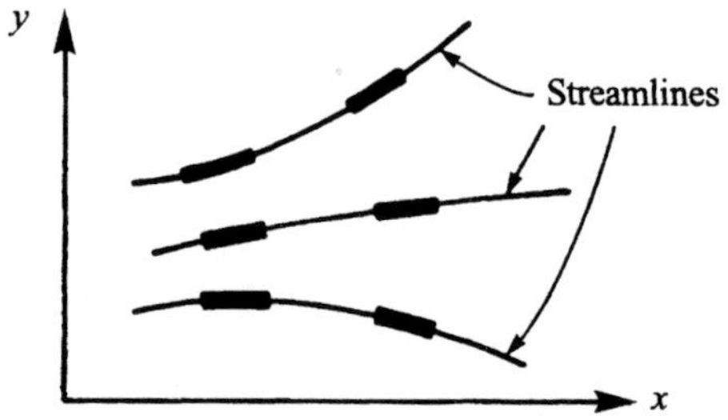

Fig. 2.4 Construction of a streamline pattern.

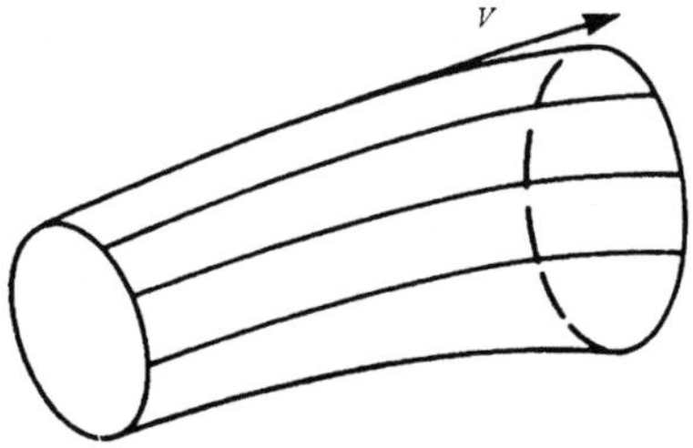

Fig. 2.5 Streamtube composed of a bundle of streamlines.

analytically. A streakline is defined as the locus of positions of particles which earlier have passed through a prescribed point in the flowfield. A common example of a streakline is the instantaneous configuration of a dye filament released from a point in the flow.

Streamlines, pathlines and streaklines will in general be different in an unsteady flow. For a steady flow, however, they are all identical.

In the case of steady flow, it is useful to extend the concept of a streamline to that of a *streamtube*, which may be visualized as a bundle of streamlines forming the lateral boundary of a duct within the flow, as shown in Fig. 2.5. From the definition of a streamline as a curve tangent to the local velocity vector, it follows that since a streamtube is composed of a bundle of streamlines, there can be no flow across a streamtube wall. Thus a streamtube may be constructed from an arbitrarily chosen surface composed of fluid streamlines, or may be a rigid surface such as the wall of a pipe. There is one difference, however. The no-slip condition need not apply at the boundary of a streamtube arbitrarily constructed within a flow, whereas at a pipe wall it does if viscous effects are taken into account.

2.2.4 Vorticity

Before leaving the subject of fluid kinematics, it will be useful to introduce one additional concept, that of fluid vorticity. We present only the results; derivations can be found in most fluid mechanics texts. We restrict our attention to x–y planar flows.

The *vorticity* ζ of a fluid element is equal to twice its angular velocity, and which can be evaluated as the sum of the instantaneous angular velocities of two mutually perpendicular differential line elements drawn in the fluid at a point. In Cartesian coordinates, with the velocity vector given by $V = \{u, v, 0\}$ (cf. footnote 2).

$$\zeta = \left(\frac{\partial v}{\partial x} - \frac{\partial u}{\partial y}\right), \tag{2.13}$$

and in cylindrical coordinates, with the velocity vector given by $V = \{v_r, v_\theta, 0\}$,

$$\zeta = \frac{1}{r}\frac{\partial}{\partial r}(rv_\theta) - \frac{\partial v_r}{\partial \theta}. \tag{2.14}$$

The importance of vorticity as a kinematic quantity is that it provides evidence of the effects of viscous shear. An element of constant density fluid, started in motion from rest, can only acquire vorticity (that is, rotational velocity) if it is acted on by tangential viscous stresses. If viscous stresses are negligible, a fluid element initially without rotation will remain so as it moves. We term such flows as *irrotational*.[6]

Vorticity is most often generated by viscous stresses at solid boundaries, though also in shear layers formed between fluid streams moving parallel to one another at different speeds. The vorticity thus produced diffuses away from its site of generation much as heat and mass diffuse, and in this way increases the extent of the flow region which is rotational. The concepts of vorticity production and diffusion play an essential role in explaining the behavior of viscous and inviscid flows.

It is important not to confuse rotation of a fluid element with curved motion of the fluid in the large. For the fluid element sketched in Fig. 2.1, we see that although the fluid motion is rectilinear, the vorticity everywhere within the fluid has the value $\zeta = -du/dy$ (note the sense of positive angular velocity $\dot{\theta}$). On the other hand, a circular fluid motion whose velocity field in the r–θ coordinate system is given by

$$v_r = 0, \quad v_\theta = \frac{\text{const}}{r}$$

is in fact irrotational, as can easily be shown by evaluating ζ as given by eqn (2.14). The physical explanation for this is that as a fluid element moves along its circular path, the initially chosen line segments dx and dy do indeed both rotate, but at equal angular speeds in opposite directions, and the mean angular velocity remains zero.

[6] The situation is somewhat more complicated when density variations are present, but that need not concern us here.

2.3 Hydrostatics

A fluid at rest in a gravitational field is said to be in a state of hydrostatic equilibrium when no shear stress exists and the pressure variation within the fluid is due solely to the weight of the fluid.

Consider a small element of fluid of cross-sectional area dA and vertical height dz, as sketched in Fig. 2.6. Since there is no acceleration in the horizontal direction, the pressure p can be a function of z only, and the force balance in the z-direction is, taking the positive direction as upward,

$$p\,\mathrm{d}A - \left(p + \frac{\mathrm{d}p}{\mathrm{d}z}\mathrm{d}z\right)\mathrm{d}A - \rho g\,\mathrm{d}A\,\mathrm{d}z = 0,$$

where the last term represents the weight of the fluid element. Hence, we obtain the fundamental hydrostatic relationship

$$\frac{\mathrm{d}p}{\mathrm{d}z} = -\rho g. \qquad (2.15)$$

The negative sign indicates that p decreases as z increases, as is well known. Since we are dealing with a constant density fluid, eqn (2.15) can be integrated immediately and we have

$$\int_{p_1}^{p_2} \mathrm{d}p = -\rho g \int_{z_1}^{z_2} \mathrm{d}z$$

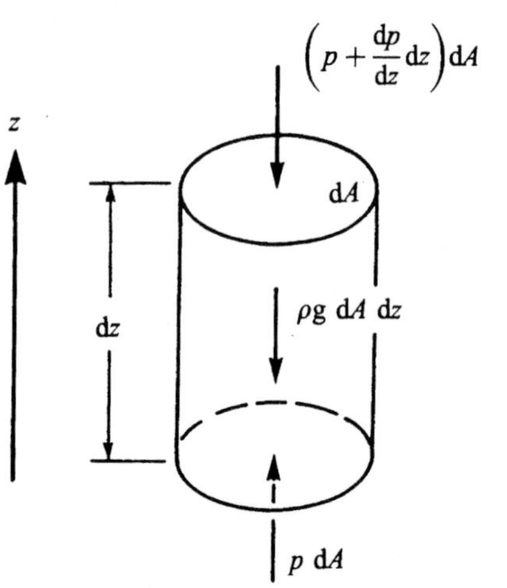

Fig. 2.6 Hydrostatic equilibrium of a fluid element.

or

$$p_2 - p_1 = -\rho g(z_2 - z_1), \qquad (2.16)$$

giving the difference in pressure between two elevations z_2 and z_1 in the fluid.

In hydrodynamics, pressures may be specified either in absolute units, referenced to zero pressure or absolute vacuum, or in 'gage' units, referenced to local atmospheric pressure p_a, which normally is taken to be the standard value $p_a = 1.01 \times 10^5$ N m^{-2}. Thus the pressure at the free surface of a liquid exposed to the atmosphere would be equal to p_a in absolute units, but would be zero in gage units. However, another method of measuring pressure, in widespread use clinically, is to specify gage pressure in equivalent units of height of a column of mercury in millimeters, as is also employed in reporting absolute barometric pressure. If we consider a column of mercury of density ρ_{Hg} and height $h = (z_2 - z_1)$, then the pressure difference $\Delta p = p_1 - p_2$ between the bottom of the column and the top (which is often at atmospheric pressure or zero gage pressure) is

$$\Delta p = \rho_{\mathrm{Hg}} g h. \qquad (2.17)$$

With ρ_{Hg} taken as 1.35×10^4 kg m^{-3} at 20°C the pressure difference equivalent to 1 mm (10^{-3} m) height of a mercury column is

$$(\Delta p)_{1\mathrm{mmHg}} = (1.35 \times 10^4)(9.81)(10^{-3}) = 133 \text{ Pa}.$$

A familiar example of the application of hydrostatics in the clinical context is afforded by the sphygmomanometer, a device used to measure blood pressure. The instrument is sketched in Fig. 2.7. The device is operated as follows. The cuff is pressurized by repeated squeezing of the inflation bulb, the check valve allowing new volumes of air to enter the bulb each time it expands. The pressure p_c in the cuff is the same as the pressure at the surface of the mercury in the reservoir, and this is the same as the pressure at that level in the mercury column where the graduation on the scale is set to zero, since the glass tube is open at its bottom. As the pressure p_c is increased, mercury is forced into the tube from the reservoir. Since the column is exposed to atmospheric pressure at the top, we see from eqn (2.17) that at any

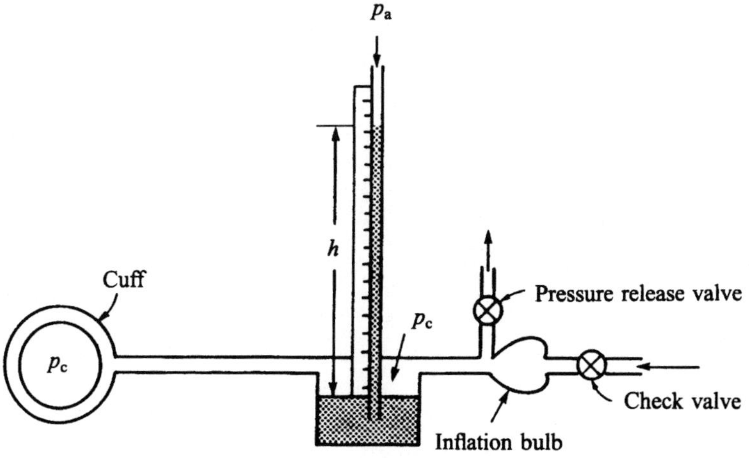

Fig. 2.7 Sphygmomanometer.

instant of time the cuff pressure is given by

$$p_c - p_a = \rho_{Hg} g h$$

or simply, in gage pressure,

$$p_c = \rho_{Hg} g h.$$

The cross-sectional area of the reservoir is made much larger than that of the tube, so that the zero of the scale against which h is measured remains essentially at the liquid level in the reservoir. From observation of the sounds produced by the blood flow (a topic we shall return to later) as the cuff pressure is gradually released, the values of systolic and diastolic pressure are determined, and conveniently recorded in terms of the corresponding values of h in mmHg.

We might note that a mercury barometer is constructed in almost the same fashion, except that the reservoir is exposed to the atmosphere, so that p_c becomes p_a, while the upper end of the tube is closed and evacuated, so that the pressure acting on the top of the mercury column is just the mercury vapor pressure p_v (0.17 N m^{-2} at 20°C). Since atmospheric pressures are in the region of 1×10^5 N m^{-2}, the correction for the vapor pressure effect is negligible, and we have

$$p_a - p_v \approx p_a = \rho_{Hg} g h.$$

The principles of hydrostatics can be used to analyze the U-tube manometer, one of the most common devices used to measure pressure differences. Imagine two closed vessels 1 and 2 filled with, say, water, and connected with a U-tube partially filled with, say, mercury, as shown in Fig. 2.8. We wish to determine the pressure difference $p_1 - p_2$ that exists at the elevations z_1 and z_2 in the two vessels. (For simplicity, we will take $z_1 = z_2$, though because the vessels are closed, p_1 can be different from p_2). To do this, we make repeated use of eqn (2.16).

Starting at point 1, we have for the water occupying the left side of the U-tube

$$p_1 - p_A = -\rho_{H_2O} g (z_1 - z_A), \qquad \text{(a)}$$

where p_A is the pressure existing at the water–mercury interface in the tube, and z is measured from an arbitrary datum. Now p_A is also the pressure existing at the same elevation $z_{A'} = z_A$ in the mercury on the right-hand side of the U-tube, since we may 'move' from A to A′ while remaining within the mercury. Similarly, we may write

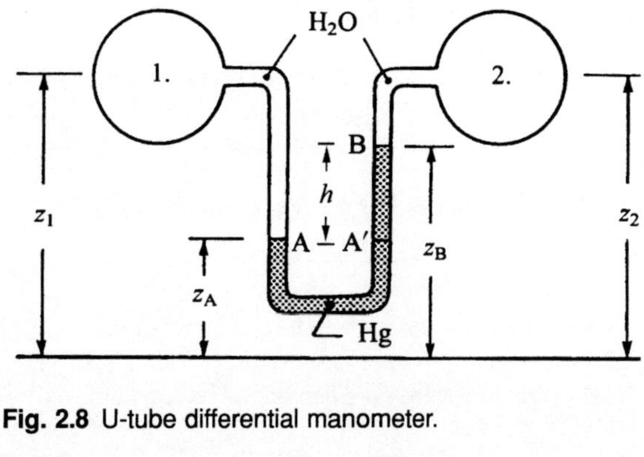

Fig. 2.8 U-tube differential manometer.

for the mercury column A′B

$$p_B - p_{A'} = p_B - p_A = -\rho_{Hg} g (z_B - z_A) \quad \text{(b)}$$

and then finally for the remaining water column

$$p_2 - p_B = -\rho_{H_2O} g (z_2 - z_B). \quad \text{(c)}$$

A little algebra allows us to eliminate p_B and p_A while combining the three equations (a), (b), and (c) and using $z_1 = z_2$, and we obtain, with $z_B - z_A = h$,

$$p_1 - p_2 = (\rho_{Hg} - \rho_{H_2O}) g h. \quad (2.18)$$

Note that it is the *difference* in specific weights of the liquids, $(\rho_{Hg} - \rho_{H_2O})g$ times the manometer deflection h which gives the pressure difference. We may remark that once one obtains some practice in analyzing hydrostatics problems, one can carry out the foregoing calculation in just one step instead of writing separate equations for each liquid column and eliminating terms between them. We have taken this more cumbersome approach in order to illustrate more clearly how the laws of hydrostatics are applied in this situation.

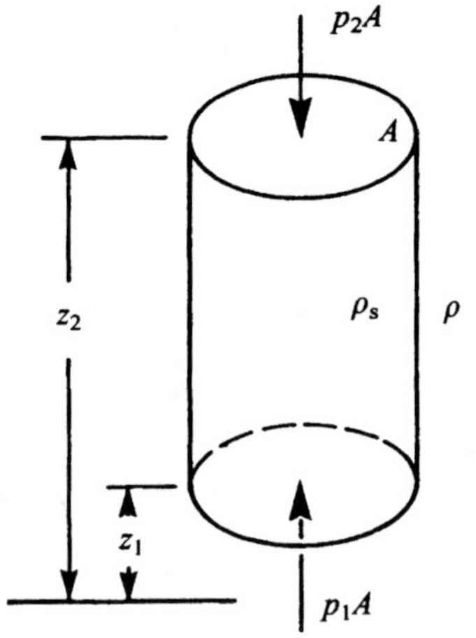

Fig. 2.9 Pressure differences producing buoyancy effect.

As a final example of an application of hydrostatics, we examine the buoyancy effect. We return to a situation similar to that sketched in Fig. 2.6, but this time consider the cylinder to be a solid object, with a density ρ_s, immersed in a fluid of density ρ as illustrated in Fig. 2.9. The net pressure force F_B in the positive (upward) direction acting on the cylinder is

$$F_B = (p_1 - p_2)A,$$

where A is the cylinder cross-sectional area. Since from hydrostatics the pressure difference in the fluid is given by $p_1 - p_2 = -\rho g (z_1 - z_2)$ we have

$$F_B = \rho g (z_2 - z_1) A.$$

But $(z_2 - z_1)A$ is simply the volume $\mathcal{V}$ of the cylinder, so we have

$$F_B = \rho g \mathcal{V}, \quad (2.19)$$

where we have used the subscript B to indicate that this is the buoyancy force. This is the famous law first enunciated by Archimedes, that the buoyancy force on a body is equal to the weight of the fluid it displaces, and it is easily shown that this holds true for bodies of general shape. The body will be in equilibrium when fully immersed only if $\rho_s = \rho$; if $\rho_s \neq \rho$ the body will either rise or sink, depending on whether its weight is less than or greater than the weight of the fluid it displaces.

There are many other interesting and more complex applications of hydrostatics, involving extended surfaces immersed in fluids, as for example the curved wall of a tank containing liquid, for which the basic equation (2.15) must be integrated over the surface in order to obtain the resultant pressure force, and moments calculated to determine its line of action. There are however few if any such applications arising in biological situations, so we leave the subject of hydrostatics at this point. The interested reader can find further information on the subject in any elementary fluid mechanics textbook.

2.4 Conservation relations

Conservation relations form the backbone of the analysis of fluid motion. There are basically three such relations—the conservation of mass, of momentum, and of energy—and in general all must be used together to analyze fluid flow problems. However, in the case of incompressible flows, the conservation of momentum relationship often yields the same information as does the conservation of energy, particularly in the case of inviscid flow, because in incompressible flow fluid thermal energy is essentially uncoupled from fluid mechanical energy. This will become more clear as we proceed, and as we shall see, for the problems we will consider we will not need recourse to a separate energy conservation analysis.

2.4.1 Conservation of mass: the continuity equation

We consider a section of a streamtube, as described earlier in Section 2.2.3, which may either be a bundle of streamlines arbitrarily chosen within a flow, or a section of pipe confining a flow, as sketched in Fig. 2.10. We assume that the fluid density ρ is constant. The mass flux (rate of mass flow) of fluid entering the streamtube across the inlet boundary A_1 is given by

$$\text{influx} = \int_{A_1} \rho V_{1n}\, dA_1,$$

where V_{1n} is the magnitude of the flow velocity normal to the differential element of area dA_1 at the cross section A_1, and the integration is taken over A_1. Similarly, the efflux of fluid crossing the section A_2 of the streamtube is given by

$$\text{efflux} = \int_{A_2} \rho V_{2n}\, dA_2.$$

The difference between the efflux and influx is equal to the rate at which mass is accumulated within the streamtube volume $\mathcal{V}$ defined by the sections A_1 and A_2 and the streamtube lateral walls. Hence, we have for our basic conservation of mass relationship

$$\int_{A_2} \rho V_{2n}\, dA_2 - \int_{A_1} \rho V_{1n}\, dA_1 = -\frac{\partial}{\partial t}\int_{\mathcal{V}} \rho d\mathcal{V} \qquad (2.20)$$

where the term on the right hand side of eqn (2.20) represents the time rate of change of the total mass contained at any instant of time within the streamtube volume, and the negative sign accounts for the fact that if efflux exceeds influx, the total mass contained within $\mathcal{V}$ will decrease with time. Clearly, with our assumption of constant fluid density we could cancel out ρ in each of the terms in eqn (2.20), but we have retained it up to here in order to emphasize the fact that we are dealing with the principle of conservation of mass.

Equation (2.20) simplifies considerably in many flow applications. Consider, for example, a situation where:

(a) the velocities V_1 and V_2 are constant across the sections 1 and 2, and normal to them, so that $V_1 = V_{1n}$, and similarly for V_2;
(b) the flow is steady and the streamtube volume is constant in time.

Then we have simply

$$A_2 V_2 = A_1 V_1 = Q \text{ (a constant)}, \qquad (2.21)$$

where now we make use of the assumption that ρ is constant. The constant Q in eqn (2.21) is the volumetric flow rate, having the dimensions $[\mathrm{L}^3\ \mathrm{T}^{-1}]$ and in SI units is measured in $\mathrm{m^3 s^{-1}}$. Equation (2.21) is often referred to as the *continuity equation*. Note that if the velocities V_1 and V_2 varied across the streamtube cross-sectional area, as they might do in the case of a pipe flow, even if the flow were steady we would still have to carry out the integrals on the left hand side of eqn (2.20) in order to evaluate Q. The two integrals would of course give the same value of Q. If the flow were unsteady but the streamtube indistensible, then at any instant of time we could still write

$$Q = A_1 \bar{V}_1 = A_2 \bar{V}_2,$$

where $\bar{V}_1$ and $\bar{V}_2$ represent velocities averaged over the cross sections A_1 and A_2, i.e.

$$\bar{V}_1 = \frac{1}{A_1}\int_{A_1} V_1 dA,$$

and similarly for $\bar{V}_2$. Finally, we note that if the streamtube is distensible, as for example a segment of

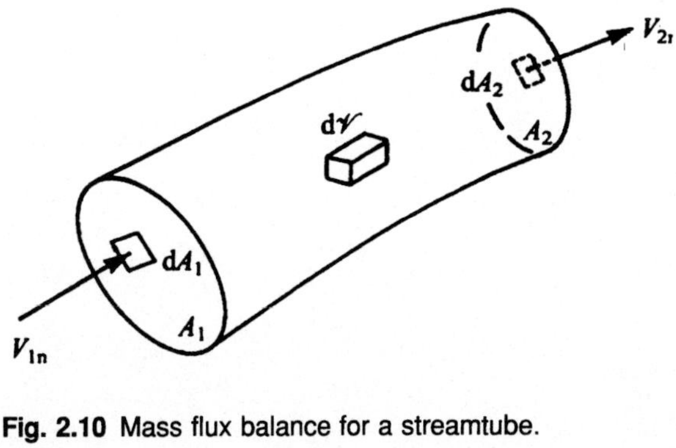

Fig. 2.10 Mass flux balance for a streamtube.

an artery in the circulatory system, then under unsteady flow conditions we must employ the full form of eqn (2.20), treating A_1, A_2, V_{1n}, V_{2n} and $\mathcal{V}$ as functions of time.

One obvious conclusion which may be drawn from the continuity relationship, either in the form of eqn (2.21) or eqn (2.20), is that under steady or unsteady flow conditions, when the area of an indistensible streamtube or vessel varies along its length, the average velocity of flow at any time will also vary along its length.

2.4.2 Conservation of momentum: Bernouilli's equation

The conservation of momentum is an alternative way of expressing Newton's equation of particle motion, force equals mass × acceleration (or more precisely, force equals time rate of change of momentum), since in the absence of an applied force the particle acceleration is zero and its momentum conserved. We wish now to apply Newton's equation of particle motion to a fluid element, and in so doing obtain the famous Bernouilli equation governing an important class of fluid flows.

Consider an element of fluid of cross-sectional area dA and length ds moving along a streamline s with speed V, as illustrated in Fig. 2.11. We assume the flow to be inviscid, so that no shear stresses act on the lateral surfaces of the fluid element. Then the only forces acting on the fluid element in the streamline direction are the two pressure forces exerted by the surrounding fluid on the element's upstream and downstream faces, and the component of its weight in the streamline direction. We have for the sum of these forces

$$\sum F_s = p\,dA - \left(p + \frac{\partial p}{\partial s}ds\right)dA - \rho g \sin\theta\, dA\, ds,$$

where the positive direction is the same as the direction of V. Simplifying, and using the fact that $\sin\theta = dz/ds$, we have

$$\sum F_s = \left(-\frac{\partial p}{\partial s}ds - \rho g \frac{dz}{ds}ds\right)dA.$$

We can now apply Newton's law of motion. The acceleration of the fluid element is given by eqn (2.11) and its mass is $\rho\, dA\, ds$, so we write force equals mass × acceleration along the streamline as

$$(\rho\, dA\, ds)\left(\frac{\partial V}{\partial t} + V\frac{\partial V}{\partial s}\right) = \left(-\frac{\partial p}{\partial s}ds - \rho g\frac{dz}{ds}ds\right)dA,$$

which upon simplification becomes

$$\rho\left(\frac{\partial V}{\partial t} + V\frac{\partial V}{\partial s}\right)ds = -\frac{\partial p}{\partial s}ds - \rho g\, dz. \qquad (2.22)$$

We may now integrate the above expression between any two arbitrary points 1 and 2 along the streamline,

$$\rho\int_1^2\left(\frac{\partial V}{\partial t} + V\frac{\partial V}{\partial s}\right)ds = -\int_1^2\frac{\partial p}{\partial s}ds - \rho g\int_1^2 dz,$$

with the result[7]

$$\rho\int_1^2\frac{\partial V}{\partial t}ds + \rho\left(\frac{V_2^2 - V_1^2}{2}\right) + (p_2 - p_1) + \rho g(z_2 - z_1) = 0. \qquad (2.23)$$

Equation (2.23) is the unsteady form of Bernouilli's equation for incompressible inviscid flow along a streamline. For steady flow, the first term of eqn (2.23)

[7] We have used the fact that $V\frac{\partial V}{\partial s} = \frac{\partial}{\partial s}(V^2/2)$.

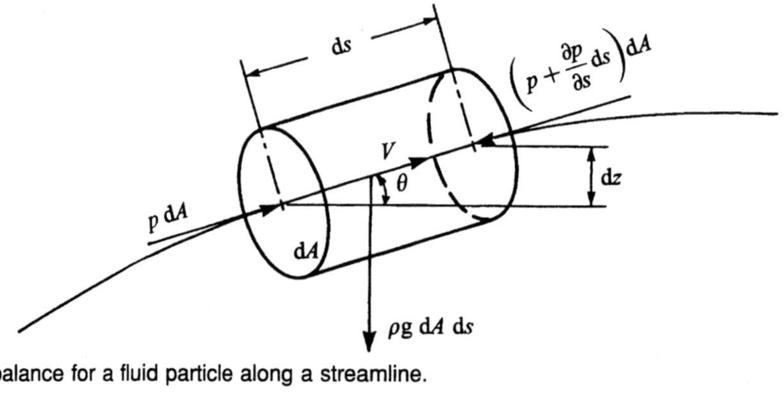

Fig. 2.11 Force balance for a fluid particle along a streamline.

vanishes, and, since stations 1 and 2 are arbitrary points on the streamline, we have

$$p + \rho \frac{V^2}{2} + \rho g z = H \text{ (a constant)} \qquad (2.24)$$

for all points along the streamline. The Bernouilli constant H may vary from streamline to streamline, however. Equation (2.24) is referred to as the steady Bernouilli equation. All the terms in eqn (2.24) have the dimensions of pressure, $[\mathrm{ML}^{-1}\mathrm{T}^{-2}]$, which is the same as the dimensions of energy per unit volume. The term $\rho V^2/2$ is called the *dynamic pressure*, while p is referred to as the *static pressure*.

The Bernouilli equation is in fact a mechanical energy balance for a fluid element (thermal energy terms are absent), although it was derived from a momentum balance. The terms $\rho V^2/2$ and $\rho g z$ are clearly the kinetic and potential energies respectively per unit fluid volume. However, the pressure term is not an energy term. It arises from the fact that the quantity $(p_2 - p_1)$ in eqn (2.23) represents the work done per unit volume by the fluid element upon the adjacent fluid, the so-called 'flow work', and it would be wrong to speak of it as 'pressure energy'. The concept of flow work is treated in detail in elementary texts on fluid mechanics and thermodynamics in connection with the derivation of the steady flow energy equation.

2.4.3 Some examples of the use of the Bernouilli equation

Let us now consider some simple applications of the steady Bernouilli equation. As a first example we take the case of liquid discharging into the atmosphere from a small orifice of area A_2 in the side of a large open tank of area A_1 as shown in Fig. 2.12. We note that from the continuity equation, $Q = A_1V_1 = A_2V_2$ (eqn 2.21). Since $A_1 \gg A_2$ we have $V_1 \ll V_2$ and we may effectively take $V_1 = 0$. Next we note that since the liquid surface at 1 and the jet emerging at 2 are both exposed to atmospheric pressure, we have $p_1 = p_2 = p_a$. We may now write the Bernouilli equation for any streamline 1–2 as

$$\rho \frac{V_1^2}{2} + p_1 + \rho g z_1 = \rho \frac{V_2^2}{2} + p_2 + \rho g z_2,$$

which reduces to

$$0 + p_a + \rho g z_1 = \rho \frac{V_2^2}{2} + p_a + \rho g z_2$$

or

$$V_2 = \sqrt{2g(z_1 - z_2)} = \sqrt{2gh}. \qquad (2.25)$$

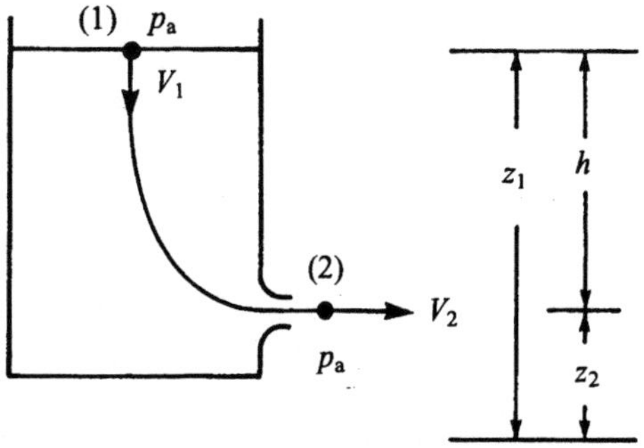

Fig. 2.12 Discharge of fluid from a large tank through a small orifice.

This is the famous Torricelli theorem. Notice that the velocity of efflux V_2 is independent of the area A_2 of the orifice (if its diameter is small compared to h) and of the fluid density. The volumetric flow rate Q out of the orifice will of course depend on A_2, since $Q = A_2V_2$.

As a second example we examine a slight variation of the preceding one. Consider that we have fluid in a large chamber at a pressure p_0, and that at some instant of time we open a valve allowing the fluid to flow into a connecting duct.[8] The situation described is sketched in Fig. 2.13, where for clarity we have not shown the valve. Also, for simplicity we have chosen the arrangement as horizontal, thus eliminating the effects of elevation changes, and we assume that the velocity V_1 is uniform across the duct cross section at station 1. One may in fact imagine this example as an idealization of steady flow issuing from the left ventricle of the heart into the aorta, though in actuality flow through heart valves is unsteady (pulsatile), as we shall discuss later on.

If we write Bernouilli's equation between stations 0 and 1, again as before invoking continuity to take $V_0 \approx 0$, we have simply

$$p_0 = p_1 + \tfrac{1}{2}\rho V_1^2 \qquad (2.26)$$

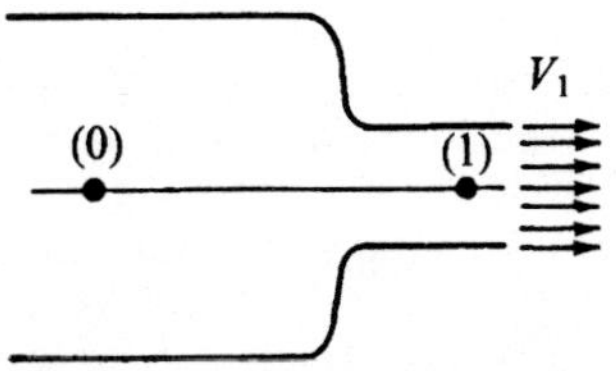

Fig. 2.13 Flow of fluid from a large chamber into a duct.

[8] The reason for our choice of subscript 0 for the chamber pressure will become evident.

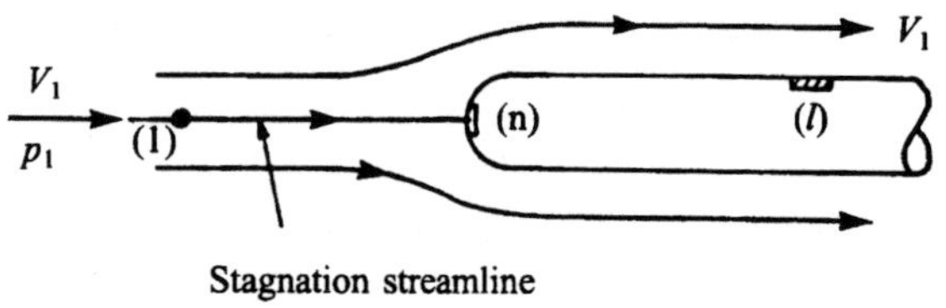

Fig. 2.14 Illustration of the occurrence of a stagnation streamline.

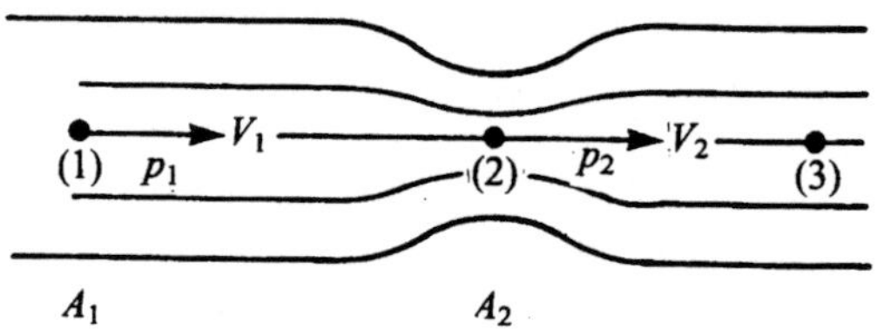

Fig. 2.15 Flow through a constriction in a pipe.

and thus p_0 is the sum of the static and dynamic pressures. We see that if we are able to measure the pressure difference $(p_0 - p_1)$, we can determine the velocity V_1 if the fluid density is known.

The pressure p_0, as defined by eqn (2.26), is called the *stagnation pressure* (sometimes the *total pressure*), for the following reason. Suppose we were to place into the oncoming flow at station 1 a small probe, as sketched in Fig. 2.14. Since the flow must divide around the probe there will be a streamline, the *stagnation streamline*, along which the flow will be brought to rest at one point on the nose of the probe. If a pressure sensor (n) is located at this point, then the Bernouilli equation along this stagnation or dividing streamline gives on comparing with eqn (2.26)

$$\rho\frac{V_1^2}{2} + p_1 = p_n = p_0.$$

On the other hand, if a pressure sensor is located on the lateral surface of the probe (l) sufficiently far downstream from its nose so that the surrounding flow velocity has returned to its upstream value V_1, the pressure p_l recorded by that sensor will be the same as the static pressure p_1 of the oncoming stream. Thus a probe of this design which measures both p_0 and p_1 can be used to determine the velocity V_1, and this is indeed the principle underlying the operation of pitot-static probes used to measure fluid velocities in wind tunnels as well as the velocities of boats and low-speed aircraft. Likewise, in clinical applications, cardiac catheters employing the same principle are available for measuring blood velocity and estimating cardiac output. We might remark that oftentimes when eqn (2.26) is employed in clinical literature, the pressure difference $(p_0 - p_1)$ is measured in equivalent units of mmHg, while ρ and V_1 are in SI units. The result, using the value for the density of blood, is cited as

$$p_0 - p_1 = \Delta p \doteq 4V_1^2,$$

where V_1 is in m s^{-1} and Δp in mmHg and obviously the coefficient 4 is not a dimensionless constant. We might also remark that in clinical literature Δp is usually termed a pressure *gradient* rather than its mathematically proper description as a pressure *difference*. A discussion of clinical fuid dynamic measurements is given in Appendix 2.B.

As a final example of the application of the Bernouilli equation, we consider incompressible flow through a constriction in a pipe, as illustrated in Fig. 2.15. Again for simplicity we take the pipe to be horizontal, and we assume the velocities and static pressures in the pipe to have uniform values V_1, V_2 and p_1, p_2 at the cross-sectional areas A_1 and A_2. As before, we treat the flow as inviscid and steady.

From the equation of continuity we have immediately

$$A_1V_1 = A_2V_2 = Q,$$

which tells us something that is intutively obvious;[9] as the fluid passes from the larger area A_1 to the smaller area A_2, its velocity increases. (This is in fact an example of convective acceleration.) Next we see what the Bernouilli equation, which we may apply because the flow is assumed inviscid, tells us about this flow. We have

$$p_1 + \rho\frac{V_1^2}{2} = p_2 + \rho\frac{V_2^2}{2}.$$

Now, we may eliminate V_2 in favor of V_1 by means of the continuity equation, and thus obtain

$$p_1 - p_2 = \frac{1}{2}\rho V_1^2\left(\frac{A_1^2}{A_2^2} - 1\right). \qquad (2.27)$$

Since the right hand side of eqn (2.27) is positive, $p_1 > p_2$ and the pressure decreases as the fluid velocity increases. The same arguments applied to the section between (2) and (3) show that where the pipe area returns to A_1 the pressure recovers its original value p_1.

We may easily convert this pipe constriction into a flow measuring device. If we were to drill small holes in the pipe wall (pressure taps) at stations (1) and (2) and

[9] Our intuition would lead us astray if this were a compressible supersonic flow, however.

connect a mercury U-tube manometer between them, then from the deflection of the manometer liquid according to eqn (2.18) we would measure $\Delta p = p_1 - p_2$.[10] Since the areas A_1 and A_2 are presumed known, the measurement of Δp determines V_1 according to eqn (2.27), and further, since $Q = A_1 V_1$, we can write eqn (2.27) as a metering equation,

$$Q = A_1 \sqrt{\left(\frac{2\Delta p/\rho}{(A_1^2/A_2^2) - 1}\right)}. \tag{2.28}$$

This is the ideal (inviscid) formula governing the operation of a venturi meter, a device widely used to measure volumetric flow in pipelines.

As we have pointed out several times, Bernouilli's equation applies to an incompressible fluid flow in which viscous effects can be neglected. Viscous effects play a negligible role in some of the flow situations we have discussed, but are dramatic in others and predictions based on the Bernouilli equation fail altogether. In the following sections we examine the effects of viscosity, and will then revisit some of the examples just discussed to see how these effects might alter the results obtained so far.

[10] Note that the pressure varies hydrostatically across the pipe cross section if the streamlines are parallel, so that even if the pressure taps at (1) and (2) were at different elevations, eqn (2.18) would still be correct if $z_1 = z_2$, as it is in this example. If the pipe were inclined, and $z_1 \neq z_2$, then Δp in eqn (2.28) would be replaced by $(p_1 + \rho g z_1) - (p_2 + \rho g z_2)$, as is easily verified.

2.5 Viscous flow

All fluids possess viscosity, the effects of which vary from one flow situation to another. Sometimes viscosity effects can be neglected, as we have supposed in the previous section; sometimes they dominate the flow. The key to determining which condition prevails is often provided by the value of a dimensionless parameter called the Reynolds number, which we shall introduce shortly.

When viscous effects are important in a flow, they may be confined to thin boundary layer regions, or they may extend throughout the entire flow field of interest. Even when they are confined to boundary or shear layers, the behavior of these layers can exert profound influence on the remaining essentially inviscid flow, principally through the phenomenon of flow separation.

In considering viscous flows, in which fluid shear stresses play an important role in determining fluid motion, as contrasted with inviscid irrotational flows in which, by definition, shear stresses are of negligible importance, we must distinguish between *laminar* and *turbulent* flows. Laminar flows are smooth, well-organized flows. They are characterized by well-defined streamline patterns, however complex they might be. For simple configurations, laminar flows are amenable to exact analysis, and even for complicated flow configurations numerical methods of analysis can give quite accurate results.

Turbulent flows on the other hand are characterized by the existence of random time-varying fluctuations of velocity, pressure, and shear superimposed on the mean flow values, even if the mean flow is steady. No complete theory exists for predicting turbulent flows, and fluid dynamicists must resort to a combination of statistical modeling and experimentally determined parameters in order to estimate quantities of technical importance such as wall shear forces, pressure losses in conduits, drag forces on immersed objects, and the like. Turbulent flows are in general characterized by increased values of shear stress, pressure losses, drag, and heat and mass transfer as compared to their laminar flow counterparts. Many aspects of turbulent flows can be explained in terms of vorticity dynamics.

Some flows may be turbulent from their outset. Others may commence as laminar flows, but become unstable and undergo transition to turbulence. They key factor in determining whether such transitions will occur is the flow Reynolds number, suitably defined (see later). For values of this parameter less than a critical value, which will be different for different flows, the flow will remain laminar, while for values exceeding this critical value infinitesimal disturbances which are always present in a fluid flow will amplify. The laminar motion becomes unstable, and turbulence eventuates. Fluid dynamicists have achieved considerable success in predicting the onset of laminar flow instability, but much less success in describing the subsequent events leading to fully developed turbulence. A good example of laminar–

turbulent flow transition is afforded by the behavior of the buoyant plume of hot air arising from a cigarette, as visualized by the streaklines marking the flow produced by the smoke particles. In a quiet room with negligible ambient air currents, the streaklines will initially exhibit laminar behavior, but as they proceed upward soon take on oscillations, and ultimately transition to chaotic turbulent motion takes place.

Most flows occurring in nature are turbulent, as for example flows in water pipes and in rivers, air flow about automobiles, and the like. However, many flows of physiological importance are in fact laminar, which aids immensely in their analysis. In the following we will discuss some elementary examples of such flows.

2.5.1 Laminar flow in a circular tube

As our first example, we consider steady laminar flow of a Newtonian fluid in a straight circular constant diameter tube, at a section far away from the tube's entrance and exit. For simplicity we take the tube to be horizontal, although the results come out exactly the same for an inclined tube. We neglect hydrostatic effects and assume that the pressure is constant across the cross section of the tube, although again this is not necessary for the results we will obtain. Since the flow is assumed steady, and the tube diameter is constant, the acceleration of fluid elements, both local and convective, must be zero, and a force balance on any such element will involve only a balance of pressure and viscous forces.

Such a force balance is illustrated in Fig. 2.16. The fluid element is taken to be an annulus of radius r, thickness dr and length dx, centered in the tube of radius a. The difference in the pressure forces acting on its upstream and downstream faces must be balanced by the difference in the shear forces acting on the inner and outer surfaces of the element. Hence

$$2\pi(r+dr)\left(\tau+\frac{d\tau}{dr}dr\right)dx - 2\pi r\tau\, dx - 2\pi r dr\frac{dp}{dx}dx = 0,$$

which on simplification and dropping higher-order terms gives[11]

$$\frac{d}{dr}(r\tau) = r\frac{dp}{dx}. \tag{2.29a}$$

This result actually applies to both laminar and turbulent flow, and to Newtonian and non-Newtonian fluids alike, since as yet we have not specified the functional form of τ. For laminar Newtonian flow we may write, following eqn (2.3),

$$\tau = \mu\frac{du}{dr}$$

and obtain

$$\mu\frac{d}{dr}\left(r\frac{du}{dr}\right) = r\frac{dp}{dx}. \tag{2.29b}$$

Since (dp/dx) is not a function of r, eqn (2.29b) may be integrated once with respect to r to obtain

$$\mu\frac{du}{dr} = \frac{r}{2}\frac{dp}{dx} + \frac{C_1}{r}. \tag{2.30}$$

Now, at the centerline, $r = 0$, the velocity gradient (du/dr) must vanish because of symmetry, and this requires that $C_1 = 0$. A second integration with respect to r gives

$$u = \frac{r^2}{4\mu}\frac{dp}{dx} + C_2.$$

[11] Both $d\tau/dr$ and dp/dx turn out to be negative quantities.

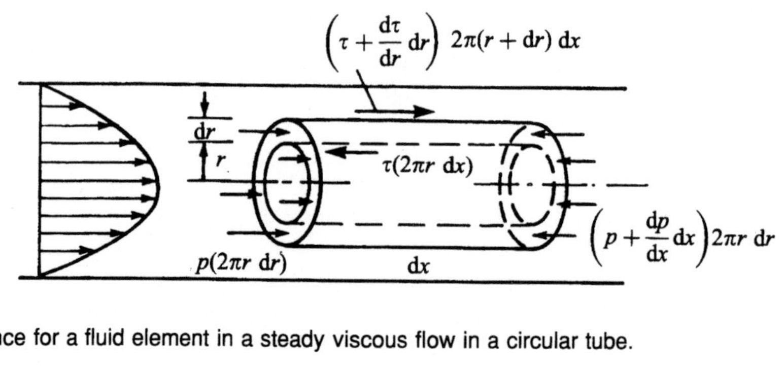

Fig. 2.16 Force balance for a fluid element in a steady viscous flow in a circular tube.

The constant C_2 is evaluated by requiring that u vanish at $r = a$, the no-slip condition, and we obtain

$$u = -\frac{1}{4\mu}\frac{dp}{dx}(a^2 - r^2). \tag{2.31a}$$

Note that u is positive (because dp/dx is negative) and varies parabolically from its maximum value $u_m = -(a^2/4\mu)(dp/dx)$ at $r = 0$ to zero at $r = a$. An alternative form for the velocity profile is obviously

$$u = u_m\left(1 - \frac{r^2}{a^2}\right). \tag{2.31b}$$

We may easily calculate the average flow velocity V from

$$V = \frac{1}{\pi a^2}\int_0^a u 2\pi r \, dr = -\frac{a^2}{8\mu}\frac{dp}{dx} = \frac{1}{2}u_m. \tag{2.32}$$

Also, since dp/dx is constant we may write it as the negative of the pressure difference $\Delta p = p_1 - p_2$ between two locations in the tube divided by the distance L separating them.[12]

$$\frac{dp}{dx} = -\frac{(p_1 - p_2)}{L} = -\frac{\Delta p}{L}.$$

Substituting for dp/dx in eqn (2.32) and employing the tube diameter $d = 2a$, we have, with $Q = \pi d^2 V/4$,

$$\Delta p = \frac{32\mu L V}{d^2} = \frac{128\mu L Q}{\pi d^4}. \tag{2.33}$$

This is the famous Poiseuille law (1840) for steady laminar flow in a section of tube remote from its entrance and exit. It states that the pressure drop is proportional to the length L of the tube section and to the volumetric flow rate Q, and inversely proportional to the fourth power of the tube diameter.

It is revealing to express Δp, as given by the first form of the right hand side of eqn (2.33), in dimensionless form. If we multiply numerator and denominator by ρV, and rearrange slightly, we obtain

$$\frac{\Delta p}{\frac{1}{2}\rho V^2} = \frac{64}{(\rho V d/\mu)}\cdot\frac{L}{d}. \tag{2.34}$$

This interesting result states that the pressure drop Δp ratioed to the dynamic pressure based on the average velocity is proportional to the length ratio L/d, and inversely proportional to a dimensionless group of variables, $\rho V d/\mu$. This group of variables[13] is called the Reynolds number, Re, in honor of the British scientist Osborne Reynolds (1842–1912), who made many important studies of viscous flows. One of the most important of these is his experimental observation that the laminar regime of tube flow which we have just analyzed reliably occurs only if the Reynolds number of the flow does not exceed the critical value

$$Re_{crit} = \left(\frac{\rho V d}{\mu}\right)_{crit} \approx 2000 \tag{2.35}$$

and that for Reynolds numbers exceeding this value the flow can be expected to be turbulent. Our analysis would not apply in this event.

The implications of the observation given by eqn (2.35) are profound. It states that it is the *combination* of variables comprising the Reynolds number which determines whether the flow is laminar or turbulent, not the values of the individual quantities involved. This concept is further explored in Appendix 2.A to this chapter dealing with dimensional analysis and flow similitude.

The Poiseuille result illustrates clearly the restrictions inherent in the Bernouilli equation, eqn (2.23). If it were applied between locations (1) and (2) along a tube containing a viscous flow, since the average velocity V is the same at all cross sections and the tube is assumed horizontal (although this is not essential for our argument), we would have $V_1 = V_2$ and $z_1 = z_2$, from which we would conclude from the Bernouilli equation that $p_1 = p_2$ and hence $\Delta p = 0$, which is obviously inconsistent with the Poiseuille result, eqn (2.33). The resolution of this inconsistency clearly lies in the fact that the Bernouilli equation was derived under the assumption of negligible viscous effects, whereas Poiseuille flow is dominated by such effects, which in fact extend throughout the entire flowfield. The Bernouilli equation is not a total energy balance, but only a 'mechanical' energy balance. The reason the Bernouilli 'constant' H is in fact not a constant in a viscous flow is that thermal energy is not taken into account in the Bernouilli equation. A full energy balance shows that viscous dissipation transforms mechanical energy into irrecoverable thermal energy, which is manifested as a decrease in H in the flow direction but a concomitant increase in the thermal energy of the fluid. Energy is conserved though degraded in availability for mechanical usefulness in a viscous incompressible flow.

[12] The only change that occurs for flow in an inclined pipe is that the quantity $\Delta p = p_1 - p_2$ is replaced by $(p_1 + \rho g z_1) - (p_2 + \rho g z_2)$.

[13] We may easily verify that Re is indeed dimensionless: $[Re] = [\rho V d/\mu] = [(ML^{-3})(LT^{-1})(L)/(ML^{-1}T^{-1})] = [M^\circ L^\circ T^\circ]$.

It has been noted in the derivation of the Poiseuille law that it applies to steady laminar flow in a section of tube far removed from its entrance or exit. The reason for this is that in the latter regions the fluid may undergo convective accelerations or decelerations which influence the tube velocity profiles in the vicinity of its entrance or exit, and thus the flow does not conform to the assumptions used to derive the Poiseuille law. The nature of the flow in the entrance of a tube best illustrates the phenomena involved.

When fluid first enters a tube through a smooth contraction from, say, a large chamber, as illustrated in Fig. 2.17, the velocity profile across the tube is essentially that of a uniform flow. Viscous effects associated with the no-slip boundary condition at the tube wall are confined to a very thin boundary layer adjacent to the wall. As the flow proceeds down the tube, the thickness δ of this boundary layer grows by the action of vorticity diffusion until finally at a distance L_e from the entrance it closes in on itself and the parabolic profile associated with Poiseuille flow is established.[14] At this point the flow is said to be *fully developed*. It is only from this point onward, to some location upstream of a subsequent distortion of the profile produced by an exit effect, that the Poiseuille law is valid. Both theory and experiment yield for the value of L_e, which is termed the *entry length*,

$$\frac{L_e}{d} \approx 0.06\left(\frac{\rho V d}{\mu}\right) = 0.06Re. \qquad (2.36)$$

(Again we note the appearance of the Reynolds number as the governing parameter.) It is ironic that although Poiseuille's investigations were motivated by his interest in the circulation of blood, in fact virtually none of the vessels in the circulatory system meets the requirements inherent in his results. Either they are too short in length,[15] with $L < L_e$, or they are too small in diameter, in which case the assumption that the blood can be treated as a Newtonian fluid fails because the size of the red blood cells becomes comparable to the diameter of the blood vessel. An additional complication is that blood vessels are generally tapered and curved, and undergo frequent branchings, so that even apart from entry-length and non-Newtonian considerations, the Poiseuille law is inapplicable. Nevertheless, the law provides a very useful means for estimating the order of magnitude of the pressure drop in viscous flow through circular tubes, and is a convenient reference value to which pressure losses in more complex flow situations can be compared.

2.5.2 Flow through a constriction revisited

In Section 2.4.3 the flow through a convergent–divergent passage was considered under the assumption that viscosity effects could be neglected. The flow streamlines were assumed to follow faithfully the contours of the pipe wall, converging smoothly into the contraction and diverging smoothly as they exited it. Let us now examine what takes place when viscosity effects play a role.

Suppose that the contraction is located near the entrance of a tube, where flow is in the entry-length region and viscous effects are still confined to a thin boundary layer of the thickness δ at the wall, as sketched in Fig. 2.18. In the convergent section, the pressure gradient (dp/dx) is negative (termed 'favorable') and the inviscid core flow outside the boundary layer is accelerated smoothly, as predicted from consideration of the Bernouilli and continuity equations. The fluid

[14] The thickness of this laminar boundary layer as a function of distance x from its origin can be roughly estimated as $(\delta/x) \approx 5.0\sqrt{\nu/Ux}$, where U is the velocity at the outer edge of the layer, in the present case approximately the mean of the inlet velocity and the centerline velocity u_m for the Poiseuille velocity distribution. Note that this introduces a Reynolds number based on the distance x, $Re_x = Ux/\nu$. The quantity $\nu = \mu/\rho$ is the kinematic viscosity, as previously noted.

[15] To illustrate, for a typical arterial Reynolds number of 1000, $L_e/d \approx 60$.

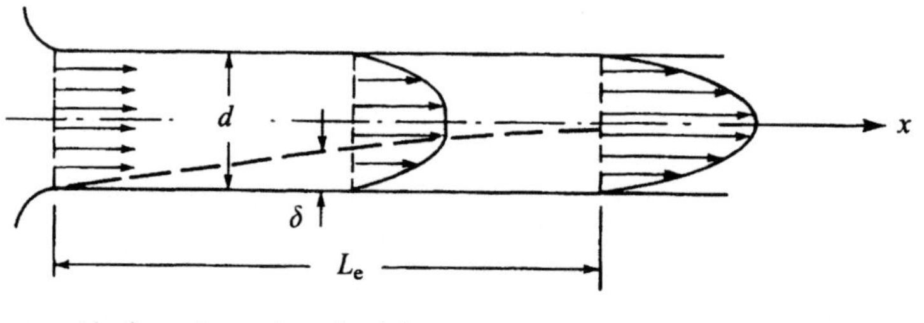

Fig. 2.17 Flow development in the entry region of a tube.

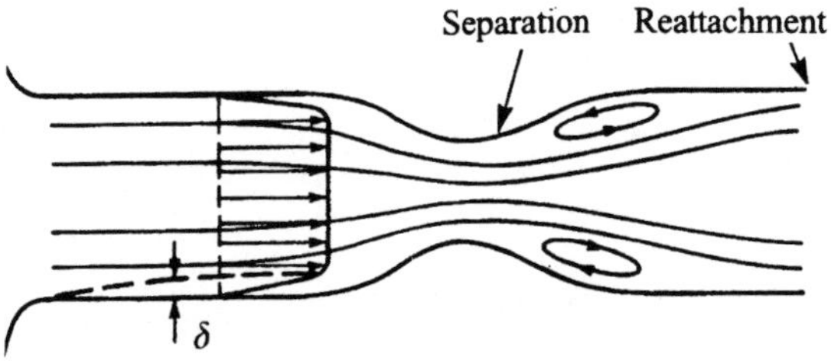

Fig. 2.18 Boundary layer separation in the divergent portion of a pipe constriction.

occupying the thin boundary layer is likewise accelerated and in fact the rate of growth of the boundary layer thickness is diminished. The situation can however be markedly different in the divergent section. If the expansion of the passageway is too rapid, then the phenomenon of *flow separation* will occur. The boundary layer fluid, having less momentum than the core fluid, is more retarded by the rising pressure (termed an 'adverse' pressure gradient) than the core flow, and is unable to follow the contour of the wall. The bulk of the fluid issues from the contraction as a jet-like flow, bounded by a recirculating vortex-like region, with reverse flow occurring at the wall of the tube. Farther downstream, the jet reattaches to the wall, and the flow takes on again an appearance similar to that ahead of the constriction, though with a thicker boundary layer. (Ultimately, if the tube were long enough and the tube Reynolds number less than the critical value, the Poiseuille velocity distribution would be established.)

Boundary layers can be either laminar or turbulent, depending on the value of the Reynolds number $Re_x = \rho Ux/\mu$ (see footnote 14). The critical condition in this case is $Re_x \lesssim 500\,000$ for laminar conditions to prevail. Turbulent boundary layers, it turns out, are more resistant to separation in regions of adverse pressure gradient than laminar layers. There are few if any instances of turbulent wall boundary layers occurring under physiological flow conditions. However, the vortical shear layer formed between the core jet flow and the recirculating flow when separation occurs is highly unstable, and it is here that turbulence is sometimes found in physiological flows.

Separation is often observed at arterial branchings and downstream of arterial stenoses (narrowings due to atherosclerotic plaque formation). At the downstream flow reattachment point, turbulent pressure fluctuations associated with the interaction of the turbulent shear layer with the arterial wall can sometimes be detected, if the stenosis is sufficiently severe. These turbulent pressure fluctuations have been hypothesized to cause fatigue of the vessel wall, resulting in post-stenotic dilatation of the artery. Similar circumstances arise in flow through heart valves. If a valve is stenotic and unable to open fully, the higher-than-normal velocity of the jet issuing from the valve during systolic flow can have a turbulent shear layer bounding it which, when it interacts with a boundary—the ventricular wall in the case of the mitral valve and the aorta in the case of the aortic valve—produces pressure fluctuations which are detected stethoscopically as ejection murmurs. The intensity of these murmurs is correlated with the severity of the stenosis.

Boundary layer separation and reattachment also play a central role in the functioning of the sphygmomanometer, which was described earlier in Section 2.3. When the cuff is fully inflated, the brachial artery monitored stethoscopically is completely collapsed and there is no flow through it. As the cuff pressure is gradually reduced, a point is reached where it falls below the maximum systolic pressure produced in the pulsatile cycle of the heart's output, and flow is established initially as a series of short-duration narrow jets issuing through the arterial constriction during the portion of the heartbeat cycle when the systolic pressure exceeds the cuff pressure. The sounds detected stethoscopically, called the Korotkoff sounds, appear as sharp clicks and are believed to be due to the combined effects of arterial wall flutter and the interaction of the separated turbulent jet with the arterial wall at the reattachment point. As the cuff pressure is further reduced, the sharp clicks become more muffled and of longer duration, probably because the artery is no longer fully collapsed over any portion of the heartbeat cycle, and the sounds are due primarily to jet–wall interaction. Eventually, as the cuff pressure is further reduced, the sounds disappear, and the pressure at which this occurs is identified as the diastolic pressure. The clinical significance of systolic and diastolic blood pressure is discussed in standard physiology textbooks.

2.5.3 Flow through curved tubes

The vessels of the circulatory system are generally curved. Laminar flow in curved tubes differs markedly from flow in straight tubes, and is considerably more complex. We present here a qualitative description of such flows.

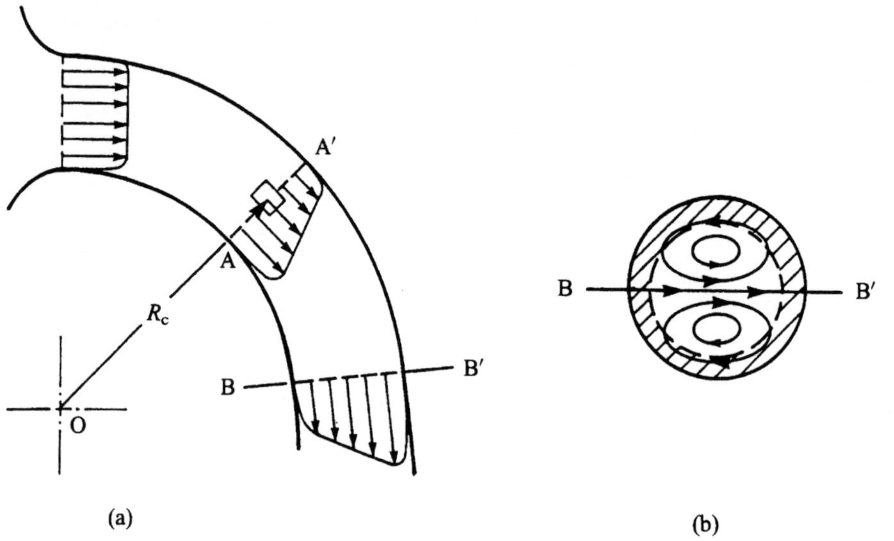

Fig. 2.19 Axial and secondary flow patterns in a curved tube.

Consider the situation of flow entering a curved pipe, as shown in Fig. 2.19. The pipe is assumed to lie in the horizontal plane, so that gravity effects can be ignored. Initially, as the flow enters the bend, the wall boundary layer is very thin, and the velocity profile is essentially uniform. Since the streamlines of the entering flow all originate from a region of constant Bernouilli head H (cf. eqn 2.24) it follows, in the absence of gravitational effects, that the quantity

$$p + \rho\frac{V^2}{2} = H \tag{2.37}$$

is the same along all the entering streamlines. However, since the flow undergoes curvature immediately upon entering the tube, a centripetal acceleration is experienced by the fluid elements, as given by eqn (2.12). This acceleration must be due, according to Newton's law, to an inwardly acting force on the fluid elements, causing them to follow their curved paths. The force is provided by a pressure variation or gradient across the tube section, with the pressure being higher at the outer wall A′ than at the inner wall A. Since H is the same for all streamlines, it follows from eqn (2.37) that the velocity is higher at the inner wall of the bend than at the outer wall. The result is a skewing of the velocity profile, as sketched for section AA′.

As the flow proceeds down the pipe, the wall boundary layer, as indicated by the shaded region in Fig. 2.19(b), grows in thickness and begins to exert its influence on the overall flow. The pressure gradient associated with the curvature of the core flow is more or less uniform across the tube cross section, and all fluid elements including those within the boundary layer experience approximately the same inwardly acting pressure force and hence the same centripetal acceleration, V^2/R_c. Now, since the velocities in the boundary layer are less than those of the core, the paths of the boundary layer fluid elements must take on smaller radii of curvature in order to maintain V^2/R_c uniform. The result is an inwardly directed circumferential motion near the wall of the tube, accompanied by an outwardly directed motion of the faster moving core fluid near the center plane. The overall flow field is composed of a main axial component on which is superimposed a *secondary flow* circulation consisting of two counter-rotating vortices, as shown in Fig. 2.19(b). This flow pattern is often referred to as a Dean-type flow, after the fluid dynamicist who first analyzed it. A consequence of the Dean motion is that the faster-moving fluid is transported toward the outer wall of the bend, and the axial velocity profile is now skewed in the opposite direction, as sketched at plane BB′ in Fig. 2.19(a). This shift in the axial velocity profile from the pattern shown for section AA′ to that for section BB′ actually begins within a distance of between one and two pipe diameters downstream from the pipe inlet, although the fully developed profile of section BB′ is not achieved until much farther downstream.

We have described the establishment of the Dean-type circulation in the context of an entry flow situation. The same situation occurs if a fully developed Poiseuille flow encounters a bend in a pipe, although we cannot invoke arguments based on the Bernouilli equation to explain the skewing of the velocity profile, since H varies across the flow. If the bend is of limited extent and is followed by a second section of straight pipe, then the Dean circulation will gradually decay by viscous dissipation, and far downstream of the bend Poiseuille flow will be reestablished. If on the other hand the pipe consists of a continuous coil, then once the Dean flow is established in the entry region of the pipe it will persist as the steady-state flow. The entry length for the establishment of steady Dean flow appears from both analysis and experiment to be somewhat less than that of Poiseuille flow at the same mean flow Reynolds number, and is correlated by a dimensionless parameter called the Dean number, defined as

$$\kappa = \frac{1}{2}\left(\rho\frac{\bar{V}d}{\mu}\right)\sqrt{\frac{a}{R}}, \tag{2.38}$$

where a $(= d/2)$ is the pipe radius, and R the radius of curvature of the pipe axis.[16] The ratio of the pressure drop per unit length of fully developed curved pipe flow $(\Delta p/L)_c$ to that of straight pipe Poiseuille flow $(\Delta p/L)_s$ is likewise found to be a function of the Dean number, and both theory and experiment yield the result that

$$\frac{(\Delta p/L)_c}{(\Delta p/L)_s} > 1.$$

The reason for the increased pressure drop in curved laminar pipe flow is associated with the increased shear stress and viscous dissipation at the outer part of the bend due to the higher axial velocities which are produced by the flow curvature in this region. Relatively little is known about pressure losses in the entry region of curved pipe flow; this is unfortunate since virtually all curved pipe flows which occur in physiological applications are of the entry-flow variety.

In the case of fully developed turbulent pipe flow, a bend in a pipe also produces secondary Dean-type circulation, but this is of more importance in engineering applications than in physiological ones. It is of interest to note that the transition from laminar to turbulent flow, which occurs at $Re \approx 2000$ for a straight pipe, may be delayed up to $Re \approx 5000$ for a pipe with a curvature ratio $a/R = 0.1$.

2.5.4 Flow past immersed objects

The nature of the flow about an immersed object in an unbounded flow field is of major importance in aerodynamic applications, but is of much less significance in the physiological context. Possibly the most important application is in the transport of aerosols in the airways of the respiratory system. Although the variety of different flow patterns encountered in the flow about immersed objects provides a fascinating kaleidoscope of fluid motions, we restrict our discussion here to the case of flow about a sphere, the immersed-body flow of most physiological importance, and refer the reader to the many texts on fluid mechanics for extended treatment of the subject.

The regimes of immersed-body flows are classified according to the value of the Reynolds number, which for a sphere is defined as $Re = U_\infty d/\nu$, where U_∞ is the velocity of the sphere relative to the ambient fluid, d the sphere diameter and ν the fluid kinematic viscosity. The Stokes flow (or creeping flow) regime is the limiting case $Re \to 0$, which in practical terms turns out to be $Re \lesssim 1$. In this limit inertial forces are negligible compared to viscous forces, and the fluid resistance to the body motion relative to the fluid, the *drag*, is due to the balance of surface viscous shear stresses and pressures. In this limit an analytical result is obtained for the sphere, which is

$$\text{drag} = 3\pi\mu U_\infty d \tag{2.39}$$

or, in terms of the *drag coefficient*, C_D, which is the drag force normalized by the dynamic pressure and the sphere frontal area,

$$C_D = \frac{\text{drag}}{\frac{1}{2}\rho U_\infty^2\left(\frac{\pi d^2}{4}\right)} = \frac{24}{Re}. \tag{2.40}$$

To illustrate the use of these results, we consider two fluids, one of which is water at 20°C ($\rho = 998$ kg m^{-3}, $\mu = 1.01 \times 10^{-3}$ kg m^{-1}s^{-1}), the other being glycerin at 20°C ($\rho = 1260$ kg m^{-3}, $\mu = 1.49$ kg m^{-1}s^{-1}), and

[16] Note that the Dean number involves two dimensionless ratios, the Reynolds number and the curvature ratio a/R. For highly curved flows, correlation of the flow characteristics appears to require the use of both these parameters separately, rather than their combination in the form of the single parameter κ.

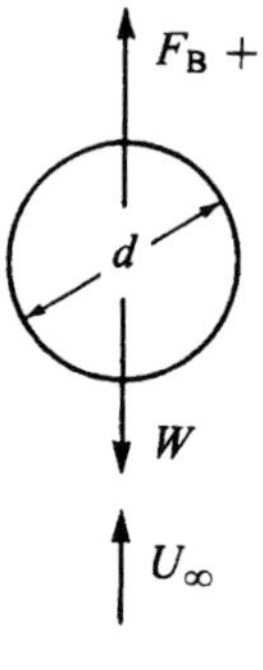

Fig. 2.20 Force balance for a spherical droplet falling in a fluid.

we ask what sphere diameters would correspond to a settling velocity of $U_\infty = 0.01$ m s^{-1}, if in each case the Reynolds number was unity. We find that for the water case $d = 0.101$ mm, whereas for the glycerin case $d = 11.8$ cm. The point of this comparison is that it is the Reynolds number, not the sphere diameter, which determines whether the flow is in the Stokes regime, although for most fluids only very small particles can meet the criterion $Re \leq 1$.

There are many practical instances when the velocity U_∞ is not known *a priori*, but must be found as part of the solution of the problem. A typical case is that of the settling rate in air (density ρ) of a water droplet whose diameter d and density ρ_w are presumed known. The equilibrium speed of fall U_∞ of the droplet (its terminal velocity) will be determined by a balance between the upwardly acting sum of the drag and buoyancy forces $D + F_B$ on the droplet,[17] and its downwardly acting weight, W, as shown in Fig. 2.20. (We choose a frame of reference attached to the droplet, so that its downward speed of fall is replaced by an equivalent upward velocity of the air.) The weight W of the droplet is given by

$$W = \tfrac{1}{6}\pi d^3 \rho_w g$$

and the buoyancy F_B is given by (cf. eqn 2.19)

$$F_B = \tfrac{1}{6}\pi d^3 \rho g.$$

If we were sure that the droplet motion were in the Stokes flow regime, then we could employ eqn (2.39) or its alternative eqn (2.40) for the drag force D to solve for U_∞. However, this would not yield a correct result unless the value of U_∞ obtained corresponded to a value of the Reynolds number less than unity. A reasonable way to proceed is to assume that eqn (2.40) applies, and calculate U_∞ from the vertical force balance $D + F_B - W = 0$, or

$$C_D \rho \frac{U_\infty^2}{2}\left(\frac{\pi d^2}{4}\right) + \frac{1}{6}\rho g \pi d^3 - \frac{1}{6}\rho_w g \pi d^3 = 0. \quad (2.41)$$

Then, having found U_∞ we can calculate the Reynolds number and determine whether it is less than unity. If it is, our problem is solved. But if it is not, then we must look for a relation for the drag coefficient which applies for Reynolds numbers greater than unity. Such a relation, which is essentially empirical, is represented by the plot of sphere drag coefficient vs. Reynolds number shown in Fig. 2.21. As seen in this figure, for $Re < 1$, the drag coefficient curve follows the Stokes value $24/Re$, but thereafter departs from it, because of the occurrence of flow separation and the formation of a wake behind the sphere. We note that in the range $10^3 \leq Re \leq 10^5$ the drag coefficient is roughly constant at about the value 0.5. The abrupt drop in C_D at $Re \approx 3 \times 10^5$ is caused by boundary layer transition from laminar to turbulent, resulting in a narrower wake. It may be remarked that for $Re > 100$, the contribution to the drag due to viscous shear is much less than that due to the pressure difference between the front and back of the sphere and this is the reason that the change in the location of the separation point when transition occurs has such a large effect on C_D.

In the context of the problem of the settling velocity of the spherical droplet, if the value of U_∞ found from solution of eqn (2.41) yields a value of $Re > 1$, we would resort to iteration to obtain the value of U_∞ which is consistent with both the force balance of eqn (2.41) and the C_D vs. Re data given by Fig. 2.21.

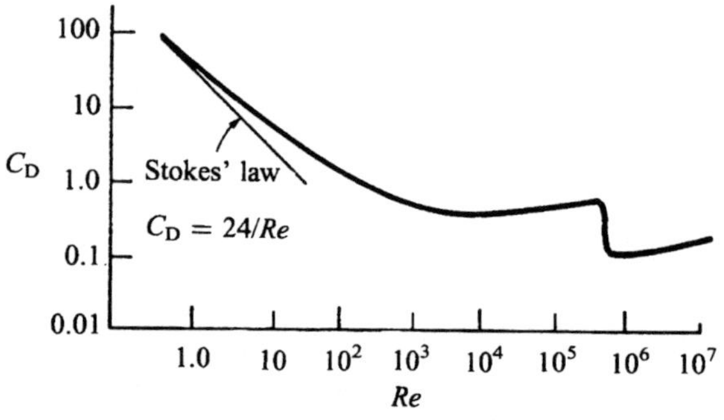

Fig. 2.21 Variation of sphere drag coefficient with Reynolds number.

[17] The buoyancy force turns out to be of negligible importance in this example, since $\rho_w \gg \rho$, but is included for correctness of the force balance.

2.6 Unsteady flows

Flows in portions of the circulatory system are obviously unsteady, because of the pulsatile nature of the pumping action of the heart. Unsteady flows are present in the heart itself and in the major arteries. The combination of compliance and viscous dissipation within the smaller vessels comprising the microcirculation effectively damps flow pulsations, however, and the flow in the venous return to the heart is essentially steady. In this section we examine some elementary characteristics of unsteady flows, and in particular how they relate to the steady flows we have discussed previously.

2.6.1 The concept of the Stokes layer

Consider an infinite wall bounded on one side by an unbounded extent of viscous fluid. The wall is caused to undergo sinusoidal motion in its own plane parallel to itself. Because of the no-slip condition the fluid immediately adjacent to the wall must follow the motion of the wall, undergoing excursions back and forth along with the wall. Intuitively we expect that as we move outward into the fluid, in the direction normal to the wall, we will find that the excursions of fluid elements diminish in amplitude, until at some distance δ_s they have been reduced to a very small fraction of that of the wall. This expectation is borne out by analysis which predicts that the value of δ_s associated with diminution to 1 per cent of the wall motion is given by

$$\delta_s = 5(\nu/\omega)^{1/2}, \qquad (2.42)$$

where $\nu = \mu/\rho$ is the fluid kinematic viscosity and ω is the radian frequency of the sinusoidal oscillation of the wall. The quantity δ_s is sometimes referred to as the Stokes layer thickness or viscous wave penetration depth.[18]

2.6.2 Oscillatory viscous flow in a long straight tube

We illustrate the use of the concept of the Stokes layer by considering the problem of viscous flow in a long straight circular tube, far from its entrance or exit. We know from Section 2.5.1 that for a constant imposed pressure gradient dp/dx we obtain the Poiseuille parabolic velocity profile given by eqn (2.31b). If now we imagine the pressure gradient to vary slowly in a sinusoidal fashion, we may reasonably expect the velocity profile to follow faithfully the prediction of eqn (2.31b), and at each instant of time have a parabolic distribution whose maximum amplitude is directly proportional to the instantaneous value of the pressure gradient. We designate this behavior as *quasi-steady* flow.

As the frequency of oscillation of the pressure gradient is gradually increased, a situation will develop wherein the velocity profile will no longer be able to respond rapidly enough to the changing pressure gradient, and the effects of unsteadiness become manifested. The faithful response of the velocity profile to the time-varying pressure gradient requires that the viscous penetration depth be at least equal to the tube radius a; otherwise the distribution of vorticity in the unsteady flow situation will be different from that in steady flow. This leads us to introduce a parameter α proportional to the ratio a/δ_s,

$$\alpha = a\left(\frac{\omega}{\nu}\right)^{1/2}. \qquad (2.43)$$

This parameter is called the Womersley number. For small values of α, we expect quasi-steady flow, whereas for large values we do not. For large values of α corresponding to $\delta_s < a$ the fluid near the wall of the tube, having lower velocity (hence lower inertia) than that near the axis, responds more rapidly than the core fluid to the changing pressure gradient, with the result that the core fluid lags behind the wall region fluid in following the pressure gradient. This phase lag increases as α increases, and in addition the amplitude of the motion in the core decreases. At very high frequencies, very little motion of the core fluid occurs, and the oscillatory flow is localized within the thin viscosity-dominated Stokes layer adjacent to the tube wall. For the aorta ($a = 0.0125$ m, $\nu = 3.5 \times 10^{-6}$ m^{2}s^{-1}) at the frequency of 1 Hz or $\omega = 2\pi$ (1 Hertz $= 2\pi$ rad s^{-1}) we have

$$\alpha = (0.0125)\sqrt{\left(\frac{2\pi}{3.5 \times 10^{-6}}\right)} = 16.8.$$

[18] We can easily verify that $(\nu/\omega)^{1/2}$ has the dimensions of length.

This is a large value and indicates that even apart from entry-length and curvature effects the flow in the aorta could not resemble quasi-steady Poiseuille flow. The physically relevant problem, pulsatile curved-pipe entry flow, is extremely complicated and as yet only limited experimental and numerical results are available.

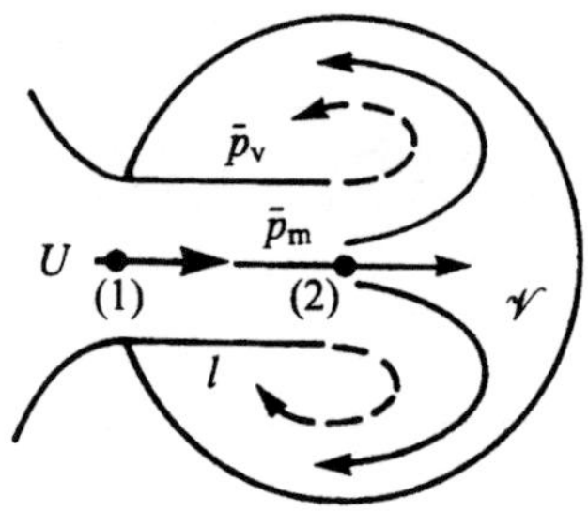

Fig. 2.22 Flow model for mitral valve closure.

2.6.3 Flow through heart valves

Healthy heart valves are remarkably efficient, offering very little resistance to flow, and opening and closing in response to small pressure differences. They are essentially passive organs whose motions are due mainly to fluid-dynamical forces exerted on them. The mechanism of valve closure is of particular interest since it has been found that the initiation of closure of a healthy valve begins during the forward phase of flow through it, and only a very small amount of reverse flow is required to accomplish the final sealing of the valve. This desirable feature of valve behavior has its origin in the unsteady nature of the flow, which we now proceed to examine, considering first the mitral valve and then the aortic valve. A simple analysis based on the unsteady Bernouilli equation describes the essential features of the closure process.

We model the mitral valve when fully open as a cylindrical sleeve of cross-sectional area A_1 and length l projecting symmetrically into a spherical ventricle, as shown in Fig. 2.22. The inflow is taken to be a parallel stream of velocity U, uniform except for a negligibly thin boundary layer adjacent to the surface of the valve. Upon establishment of flow the jet issuing from the valve impinges upon the far wall of the ventricle and turns back to form a ring-like vortex motion. During the part of the cycle when the ventricular volume $\mathcal{V}$ is expanding at an increasing rate so that $d^2\mathcal{V}/dt^2$ is positive, the flow velocity U will also be an increasing function of time, since $UA_1 = d\mathcal{V}/dt$, and $\partial U/\partial t > 0$. When the rate of increase of ventricular volume begins to slow, and $d^2\mathcal{V}/dt^2$ becomes negative, then the rate of filling of the ventricle will also slow, and $\partial U/\partial t < 0$. This is the decelerative phase of flow through the valve, and signals the beginning of valve closure, which we now analyze.

Since the flow is assumed inviscid, except for the thin boundary layer region, we may apply the unsteady form of Bernouilli's equation, eqn (2.23), to the valve flow. Referring to Fig. 2.22, we have

$$\int_1^2 \frac{\partial U}{\partial t}\,dx + \frac{U_2^2 - U_1^2}{2} + \frac{p_2 - p_1}{\rho} = 0. \tag{2.44}$$

But since the valve at this moment has constant cross-sectional area, we have $U_1 = U_2 = U$, independent of x (but a function of time), and eqn (2.44) reduces to

$$p_1 = p_2 + \rho l \frac{\partial U}{\partial t}. \tag{2.45}$$

The average pressure acting on the interior of the valve is then

$$\bar{p}_m = \frac{p_1 + p_2}{2} = p_2 + \frac{1}{2}\rho l \frac{\partial U}{\partial t}. \tag{2.46}$$

The value we assign to the average pressure $\bar{p}_v$ exerted on the ventricular side of the valve will depend on what we assume for the vortex flow behind the valve. If the vortex is weak, then it is reasonable to assume that $\bar{p}_v = p_2$, whereas if the vortex is strong, the velocity on the outside of the valve cusp will vary from zero at the root of the cusp to a value close to U at the tip of the cusp, and with some assumption regarding this variation the unsteady Bernouilli equation could again be applied to this outside flow to evaluate the average pressure $\bar{p}_v$. Under the most conservative assumption that the vortex is weak and $\bar{p}_v = p_2$, we find that the average pressure difference across the valve is given by

$$\bar{p}_v - \bar{p}_m = -\frac{1}{2}\rho l \frac{\partial U}{\partial t}, \tag{2.47}$$

and valve closure begins when $\bar{p}_v > \bar{p}_m$ at the onset of flow deceleration. If a strong vortex is assumed, this simple model predicts that closure would commence earlier in the cycle. The result following from the assumption $\bar{p}_v = \bar{p}_2$ agrees reasonably well with observation.

Although the structure of the aortic valve is quite different from the mitral valve, its dynamics are similar. The aortic valve is comprised of three semicircular leaflets or cusps, attached at their margins to the root of the aorta (radius a). Behind each of the leaflets are outpouchings, called the sinuses of Valsalva. Fig. 2.23 shows the flow configuration in the plane of symmetry of one of the sinuses when the valve is fully open. The flow issuing from the valve is split at the sinus ridge R into two parts. Part of the flow is directed into the sinus where it forms a convectively driven vortex before reemerging, out of the plane of the figure, to join the main part of the flow in the ascending aorta, and this flow configuration is maintained until closure commences.

The closing process can be analyzed in the same way as for the mitral valve. We have from eqn (2.44), with $l \approx 2a$,

$$p_1 - p_2 = 2\rho a \frac{\partial U}{\partial t}, \tag{2.48}$$

so the average pressure $\bar{p}_a$ on the aortic side of the valve cusp is, as for the mitral valve,

$$\bar{p}_a = p_2 + \rho a \frac{\partial U}{\partial t}. \tag{2.49}$$

As in the case of the mitral valve, the value we take for the average pressure $\bar{p}_s$ on the sinus side of the cusp depends on whether the vortex is assumed to be strong or weak. If it is assumed weak, then we may assume the sinus pressure to be uniform, $\bar{p}_s = p_2$, and recover eqn. (2.47) obtained for the mitral valve. If the vortex is assumed strong (implying $U_C \approx U$) we can expect a variation in p_s from the value p_B at the base of the cusp, which will be close to that of the sinus ridge stagnation pressure $p_R = p_2 + (\rho U^2/2)$, to a value p_2 at the distal margin of the cusp. This would yield a value for $\bar{p}_s$ greater than p_2, and would lead to the prediction of an earlier valve closure. In either case, we see that closure will be initiated during the forward phase of flow, as is observed with normal valves.

The sinuses play an essential role in valve closure. As flow deceleration begins and the leaflets move toward apposition, more flow is diverted into the sinuses to fill the increasing volume behind them, while still maintaining forward flow in the aorta. In the absence of sinuses, with the cusps opening up until they contact the aortic wall, the only way fluid can be provided to fill the volume behind the valve cusps when they begin to close is by reverse flow, akin to a door blown shut by wind. From a physiological point of view this is an inefficient or incompetent valve, since a substantial fraction of the blood pumped by the ventricle is returned to it, rather than delivered to the circulatory system. Highly stenosed valves behave in this fashion, and the clinical implications of the additional workload required of the heart to provide adequate circulatory flow with the existence of an incompetent valve are not difficult to imagine.

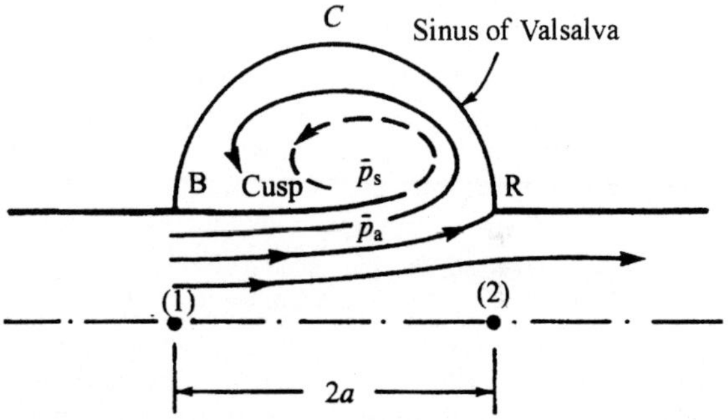

Fig. 2.23 Flow in the plane of symmetry of a sinus of Valsalva, showing flow past the valve cusp and the formation of a vortex in the sinus.

Appendix 2.A: Similitude and dimensional analysis

On several occasions in this chapter we have made a point of examining the dimensionality in units of M, L, and T of certain quantities, to make certain that mathematical expressions are dimensionally homogeneous, and the powers of M, L, and T appearing on the left-hand side of an equation are the same as those appearing on the right-hand side. We also saw that it was possible to rewrite equations so as to convert them into relationships between dimensionless groups of variables, as for instance eqn (2.40) which gives the dimensionless drag coefficient C_D for a sphere in Stokes flow in terms of dimensionless Reynolds number *Re*. The numerical example we described, comparing the size of a sphere falling in glycerin to one falling in water at the same value of Re, is a demonstration of the application of the ideas of similitude. We found in this example that a 12 cm diameter sphere falling in glycerin would have about the same drag coefficient as a 1 mm diameter sphere falling in water if both fell at a speed of 1 cm s^{-1}. They would not experience the same *drag force* D, however, since $D = C_D(\rho U_\infty^2/2)(\pi d^2/4)$, and in fact the drag force acting on the sphere in glycerin would be of the order of 10^6 times that on the sphere in water.

The foregoing example indicates that if we are able to express the equations governing a physical phenomenon in terms of dimensionless groupings of variables, we can investigate *prototype* flows of interest to us by means of *model* experiments, provided we are able to maintain the same values of the relevant dimensionless parameters in the two instances. Thus we may be able to determine the aerodynamic stability characteristics of a new aircraft design by means of wind tunnel measurements on a geometrically similar small-scale model (before constructing and flying the full-scale prototype aircraft!), or on the other hand we may be able to learn certain things about the characteristics of flow in the microcirculation through measurements made within enlarged models which could not easily be accomplished in the tiny actual-size vessels. A word of caution is in order here. Although scaling, either up or down, of prototype dimensions to useful model dimensions often is a practical experimental approach, sometimes it turns out to be altogether unworkable. Later on we shall give such an example.

It thus becomes evident that we require a method for identifying the relevant dimensionless parameters which govern a particular flow situation. When one has acquired enough experience in dimensional analysis, it is often possible to identify by inspection the important dimensionless groups, and to write down immediately the functional *form* (though *not* the precise equation) governing the relationship between them. One rational procedure for identifying the nondimensional groups is the *power-product* method, which we illustrate by considering the case of fully developed viscous flow through a pipe.

We are interested in the magnitude of the pressure drop Δp which occurs in a pipe of length L and diameter d, through which fluid of density ρ and viscosity μ is flowing at an average velocity V. (We do not specify whether the flow is laminar or turbulent.) In so describing the problem, we have in effect chosen the relevant variables, and postulated that

$$\Delta p = f(L, d, \rho, V, \mu). \qquad (2.A.1)$$

Suppose further that instead of the general functional form $f(\ldots)$ we now consider a much more restricted form for the right hand side of eqn (2.A.1), namely a power-product relation of the form

$$\Delta p = \text{const.}\ (L)^{\alpha_1}(d)^{\alpha_2}(\rho)^{\alpha_3}(V)^{\alpha_4}(\mu)^{\alpha_5} \qquad (2.A.2)$$

in which the exponents will be either integers or rational fractions. (This is just a convenient way of obtaining the dimensionless groups; the functional form of eqn (2.A.1) is likely to be more complicated.) For eqn (2.A.2) to be dimensionally correct, the powers of the fundamental dimensions M, L and T on the left-hand side of the equation must be the same as those on the right-hand side. The M–L–T dimensions of each of the quantities are:

$$\begin{aligned}
[\Delta p] &= [\mathrm{ML^{-1}T^{-2}}]\\
[d] &= [\mathrm{L}]\\
[L] &= [\mathrm{L}]\\
[\rho] &= [\mathrm{ML^{-3}}]\\
[V] &= [\mathrm{LT^{-1}}]\\
[\mu] &= [\mathrm{ML^{-1}T^{-1}}]
\end{aligned}$$

and of course, [const.] = $[\mathrm{M^\circ L^\circ T^\circ}]$. Inserting these in eqn (2.A.2), we have dimensionally

$$[\mathrm{ML^{-1}T^{-2}}] = [\mathrm{L}]^{\alpha_1}[\mathrm{L}]^{\alpha_2}[\mathrm{ML^{-3}}]^{\alpha_3}[\mathrm{LT^{-1}}]^{\alpha_4}[\mathrm{ML^{-1}T^{-1}}]^{\alpha_5}$$

We now equate in turn the powers of M, L, and T on the two sides of the above equation:

$$\begin{aligned}
\mathrm{M}:&\quad 1 = \alpha_3 + \alpha_5\\
\mathrm{L}:&\quad -1 = \alpha_1 + \alpha_2 - 3\alpha_3 + \alpha_4 - \alpha_5\\
\mathrm{T}:&\quad -2 = -\alpha_4 - \alpha_5
\end{aligned}$$

giving us three equations for the five alphas, so that we may eliminate any three of them in terms of the remaining two. Suppose we choose to keep α_1 and α_5, and express α_2, α_3, and α_4 in terms of them. Then we find that

$$\begin{aligned}
\alpha_3 &= 1 - \alpha_5\\
\alpha_4 &= 2 - \alpha_5\\
\alpha_2 &= -\alpha_1 - \alpha_5
\end{aligned}$$

These exponents can now be inserted into our proposed power law relationship, eqn (2.A.2), and we obtain

$$\Delta p = \text{const.}\ \rho V^2 \left(\frac{L}{d}\right)^{\alpha_1}\left(\frac{\mu}{\rho V d}\right)^{\alpha_5}, \qquad (2.A.3)$$

or, rearranging slightly to make both sides dimensionless,

$$\frac{\Delta p}{\rho V^2} = \text{const.}\left(\frac{L}{d}\right)^{\alpha_1}\left(\frac{\mu}{\rho V d}\right)^{\alpha_5} \qquad (2.A.4)$$

Comparing eqn (2.A.4) with the result we obtained for Poiseuille flow, eqn (2.34), we see that the two equations agree if the constant has the numerical value of 32, and the exponents α_1 and α_5 are both unity. The functional form given by eqn (2.A.4) applies also to fully developed turbulent flow through smooth pipes, and experiments show that while α_1 remains unity, both the constant and α_5 are found to have values quite different from the laminar case. Dimensionless analysis by way of the power-product method has thus identified three non-dimensional groupings of variables—$\Delta p/\rho V^2$, L/d, and the Reynolds number $Re = (\rho V d/\mu)$—and indicated a possible relationship between them.[19] A more generalized approach called the pi-theorem yields the result that

$$\frac{\Delta p}{\rho V^2} = f\left(\frac{L}{d}, Re\right) \qquad (2.A.5a)$$

or

$$f\left(\frac{\Delta p}{\rho V^2}, \frac{L}{d}, Re\right) = 0 \qquad (2.A.5b)$$

without specifying the functional form of f. For the drag on a sphere, the pi-theorem gives

$$\frac{D}{\frac{1}{2}\rho U_\infty^2 A} \equiv C_D = f(Re) \qquad (2.A.6)$$

which is exactly what Fig. 2.21 depicts. The actual functional dependence of the drag coefficient on the Reynolds number can be found analytically only in the Stokes flow regime, $Re < 1$, but has to be determined experimentally or numerically for the remaining range of Re. Despite the fact that dimensional analysis can at best yield only a suggested functional relationship between dimensionless groupings of variables, it does, as we pointed out earlier, tell us the proper way to interpret the results of scaled-up or scaled-down experiments, when the relevant dimensionless groups are maintained constant. Scaling need not apply only to size changes. For example if we wish to evaluate the drag coefficient of an airplane using a 1/10 scale model in a wind tunnel, while maintaining the same Reynolds number for the model as for the prototype,

$$\frac{\rho_m U_m L_m}{\mu_m} = \frac{\rho_p U_p L_p}{\mu_p},$$

it would appear that we have at the minimum several different choices. With $L_m = (1/10)L_p$, we could obtain Reynolds number similarity by making either $U_m = 10U_p$, $\mu_m = (1/10)\mu_p$, or $\rho_m = 10\rho_p$. Practically, our choices are limited. If we choose the wind tunnel speed to be ten times the flight speed, $U_m = 10U_p$, we will likely introduce into our model tests high-speed Mach number compressibility effects which may be absent in the prototype situation, and complete similarity would be destroyed. The possibility that we could make $\mu_m = 10\mu_p$ is not practical, since it would require substantial heating of the wind tunnel air. The best approach actually turns out to be an isothermal pressurization of the wind tunnel, making $\rho_m = 10\rho_p$, and this is in fact what is done in some wind tunnels.

It is clear that the results we obtain from dimensional analysis will depend on what variables we choose. In relatively simple situations the choices may be fairly obvious. For example, in the case of entry flow in a pipe the important lengths are the boundary layer thickness δ and the distance x from the entrance, rather than the pipe diameter d. Dimensional analysis would then yield Reynolds numbers based on either x or δ, $Re_x = \rho V x/\mu$ or $Re_\delta = \rho V \delta/\mu$, rather than $Re = \rho V d/\mu$ which governs fully developed pipe flow.

For some complicated problems, where the effects of different physical properties and phenomena are not easily anticipated in advance, it is often necessary to include a large number of variables, and after performing the dimensional analysis determine by appropriate experiments which if any of the derived dimensionless groups when changed in value do not alter the primary results of the experiment and can therefore be discarded. On the other hand, in a few simple circumstances, it is possible to show *a priori* that a particular quantity cannot enter the problem. Take for instance the case of a simple pendulum of length l, with a concentrated mass m at its free end, in a gravitational field g. We wish to know how the period τ of the pendulum is related dimensionally to the quantities m, l, and g. Now $[\tau] = [\mathrm{T}]$, $[m] = [\mathrm{M}]$, $[l] = [\mathrm{L}]$, and $[g] = [\mathrm{LT}^{-2}]$. But since only one of the four variables contains the dimension M, we have no way

[19] If we had chosen to evaluate the α-exponents differently, say eliminating α_1, α_2, α_4 in favor of α_3 and α_5, we would have obtained different nondimensional groupings of variables, but they would not be independent of those we did obtain, and could be converted to them. In fact, the pi-theorem states that if we have n variables and m fundamental dimensions, we can have no more than $(n - m)$ independent dimensionless groups. In the present example $n = 6$ and $m = 3$ (M, L, and T), so the number of dimensionless groups is 3, as eqn (2.A.4) and eqn (2.A.5) indicate.

of forming a dimensionless group involving m, and we conclude that τ must be independent of m, which is correct.

Many dimensionless groups may be interpreted as force ratios. We illustrate this in the case of the Reynolds number, which turns out to be proportional to the ratio of inertia to viscous forces acting on a fluid element. The inertia force F_I acting on a typical fluid element will be proportional to its mass times acceleration, or

$$F_I \propto \rho L^3(L/T^2) \propto \rho V^2 L^2,$$

where L is a length characterizing the element and L/T characterizes its velocity. The viscous force acting on the element will be proportional to the product of a viscous shear stress and a representative area,

$$F_V \propto (\mu V/L)(L^2),$$

where V/L is a measure of the velocity gradient across the fluid element and L^2 is proportional to a representative area. The ratio

$$\frac{F_I}{F_V} = \frac{\rho V^2 L^2}{\mu V L} = \frac{\rho V L}{\mu}$$

gives us a Reynolds number. This result provides us with additional insight as to for example why in the limit of very small Reynolds numbers we may be justified in neglecting inertia effects, as is done in the case of Stokes flow about a sphere.

Finally, just as a matter of interest, we give an example of a situation where dynamic similitude is clearly impossible to achieve via model testing. (There are unfortunately many such instances.) Suppose we wished to study the tidal flow patterns in the San Francisco Bay by means of a model. The typical horizontal scale length of the Bay is 10^4 m, and the average water depth is perhaps about 10 m. If we constructed a scale model of the Bay with a horizontal dimension of the order of 10 m (a scale factor of 1/1000) then the average water depth in our model would have to not exceed 1 cm for correct vertical scaling. This would immediately defeat our modeling efforts, since sheets of water this thin are subject to surface tension effects which are of no influence whatsoever in determining tidal motions. What is worse, if we wanted the Reynolds number based on the depth to be the same for the model as for the prototype, we would have to produce flow velocities on the order of 1000 m s^{-1} in our model to simulate typical tidal velocities of 1 m s^{-1}, which is clearly impossible. Models of the Bay have indeed been constructed, but they are *distorted* models in which the vertical scale factor is different from the horizontal one, and geometric similitude is abandoned. Nevertheless, the results obtained from such distorted models when appropriately interpreted have yielded a wealth of valuable information.

Appendix 2.B: Clinical fluid dynamic measurements

Fluid dynamicists have at their disposal a wide array of measurement techniques which can be employed in laboratory or field experimentation. Among them are optical methods, various types of velocity and pressure probes, marker particles, electrical discharges, and various other types of sensing and transduction methods. References listed at the end of Chapter 6 provide a good information source for this subject.

Clinicians, however, are under many more constraints as to the measurement methods which can be used *in vivo*, because of the imperative that patient risk be kept at a minimum. Noninvasive methods are obviously preferable to invasive ones, but unfortunately even the most technologically sophisticated noninvasive measurement techniques currently available cannot provide all the fluid dynamic information needed in the diagnosis and treatment follow-up of cardiovascular diseases, and clinicians must perforce use a combination of invasive and noninvasive measurements. Invasive measurements generally involve the use of some form of catheter, a slender flexible tube which is introduced percutaneously into an internal blood vessel and maneuvered under fluoroscopic control to the site of interest. The book by Netter (1969) contains some excellent illustrations of different catheter placements. Catheters come in a variety of designs, according to the function required, as will be evident in the discussion which follows.

2.B.1 Cardiac output measurement

The amount of blood pumped by the heart per unit time can be determined by application of the Fick principle, which states that if the amount of a tracer material added to the flow is known, and its concentration is determined proximal and distal to the point of introduction, then the volumetric flow can be calculated. In the classical application of the Fick principle, oxygen is the tracer, whereas in other applications dye (indicator dilution method) and temperature (thermal dilution method) are the tracers.

Fick method

Since all the blood pumped by the heart passes through the lungs, the measurement of the blood oxygen content proximal and distal to the lungs, together with the amount of oxygen consumed during a given time interval, provides the necessary information for application of the Fick principle. A catheter positioned in the right atrium or at the entrance to the pulmonary artery in the right ventricle of the patient is used to sample the mixed venous blood. A sample of the arterial blood is extracted from a peripheral artery. The patient's expired air is collected by a spirometer for a period of time, and its volume and oxygen content are used to determine the oxygen consumption. This quantity, together with the measurements of the oxygen concentrations in the venous and arterial blood samples, then determine the cardiac output according to the relation

$$\underset{(\text{l min}^{-1})}{\text{cardiac output}} = \frac{\overset{(\text{ml min}^{-1})}{O_2 \text{ consumption}}}{\underset{(\text{ml l}^{-1})}{\text{arterial } O_2 - \text{venous } O_2}} \qquad (2.B.1)$$

with the units most commonly used indicated.

Dilution methods

As mentioned, both dye and temperature can also be used as tracers in the application of the Fick method. The use of dye, however, requires the continuous sampling and densitometric analysis of the blood dye concentration at the distal observation point, and has largely been replaced by the thermal method, which does not require this. For a discussion of indicator dilution methods, see Bassingthwaite (1974).

In the application of the thermal dilution technique, a bolus of fluid (usually a saline solution) at a temperature different from that of the blood is injected through a catheter into the right atrium. The catheter is designed such that when its injection port is in the right atrium, its tip, at which a temperature-sensing thermistor is located, is positioned at the entrance to the pulmonary artery. With this arrangement, the right ventricle becomes an open system with one inlet, the tricuspid valve opening, and one outlet, the pulmonary artery. In the analysis which follows, it will be assumed that the injected fluid is at a higher temperature than that of the blood, although in actual practice, the converse is the case. This convention is adopted as a convenience to avoid having to deal with negative temperature and energy differences, and the final results come out exactly the same for the two cases.

Consider a thermally insulated open system with a single outlet area A. A thermal indicator of temperature T_i, density ρ_i, total volume $\mathcal{V}_i$ and specific heat c_i is impulsively injected at the inlet of the system. The energy change ΔE produced is

$$\Delta E = \rho_i \mathcal{V}_i c_i (T_i - T_b), \qquad (2.B.2)$$

where T_b is the temperature of the base fluid flow (i.e. the blood). Since the system is assumed to be thermally insulated, this energy increment must also represent the total integral over time of the energy increment flux across the outlet plane A, so that

$$\rho_i \mathcal{V}_i c_i (T_i - T_b) = \int_0^\infty dt \int_A \rho_m c_m V (T - T_b)\, dA, \qquad (2.B.3)$$

where ρ_m, c_m, V, and T are the mixed density, mixed specific heat, velocity, and temperature respectively of the fluid crossing the outlet plane A, all of which can be functions of time and position at the outlet plane. Equation (2.B.3) is exact. However, for clinical applications, where high precision is not required, a number of simplifying assumptions are made (see Roselli *et al.* 1975, for discussion of these assumptions). It is assumed that $\rho_m = \rho_i = \rho_b$, $c_m = c_i = c_b$, and that $T = \bar{T}$, $V = \bar{V}$ represent position-independent averages at the outlet plane A, and in addition that $\bar{V}$ is independent of time. With these assumptions, we obtain for the volumetric flow Q through the system

$$Q = \bar{V} A = \frac{\mathcal{V}_i (T_i - T_b)}{\int_0^\infty (\bar{T}(t) - T_b) dt}. \qquad (2.B.4)$$

We see from this result that if the volume $\mathcal{V}_i$ and temperature T_i of the injected fluid bolus are known, and the temperature 'washout' integral is measured (by recording and processing the catheter-tip thermistor signal), the volumetric flow rate (cardiac output) can be determined.[20] When properly carried out, and with corrections applied if necessary to account for recirculation effects on the integral due to the reappearance of the indicator following its transit through the systemic circulation, thermal dilution measurement of cardiac output yields results within 5 per cent of the classical Fick method. Strictly speaking, the velocity $\bar{V}$ is a spatial average over the outlet plane area but is a function of time in a pulsatile flow. However, the washout process takes place over many heartbeat cycles, and the assumption that $\bar{V}$ is independent of both space and time yields good results, as has been demonstrated by Roselli *et al.* (1975).

2.B.2 Valvular stenosis measurements

Valvular stenosis, which produces high-velocity jet flows, has been traditionally diagnosed through the use of pressure catheters to measure the trans-valvular pressure 'gradient' (difference), and applying a simplified form of the Bernoulli equation, eqn (2.26). For the tricuspid and pulmonic valves, venous catheters are used, whereas for the aortic valve, an arterial catheter is employed, which is passed through the valve to measure the appropriate heart chamber stagnation pressure, and then withdrawn to measure the static pressure of the jet issuing from the valve. In the case of the mitral valve, where a catheter measurement of the left atrium pressure is difficult to obtain with an arterial catheter, the so-called 'pulmonary wedge' pressure is used. This is the pressure measured by a venous balloon-tipped catheter which is maneouvered up the pulmonary artery until it wedges in the narrowing vessel. The pressure measured at this location is taken to be approximately equal to the left atrium pressure, since the pressure drop across the lungs is very small.

Difficulties encountered in pressure catheter procedures are the estimation of the valve cross-sectional area and the accurate positioning of the catheter for measurement of the jet static pressure, particularly in the case of highly stenosed valves, and the use of continuous-wave Doppler measurements, to be discussed next, has largely supplanted pressure catheter measurements for the evaluation of valvular stenosis.

[20] It is evident that if $T_i < T_b$ as in common practice, both the numerator and the denominator of eqn (2.B.4) are negative, and the final result is unaltered.

2.B.3 Doppler ultrasound measurements

Of the several noninvasive flow measurement and imaging techniques available clinically, as described in Chapter 12, the Doppler ultrasound method is the one in most widespread usage because of its low relative cost and ease of application. The method is based on the fundamental equation for the Doppler frequency shift

$$\Delta f_d = \frac{2Vf_o}{c}\cos\theta, \qquad (2.B.5)$$

where V is the velocity of the structure (moving red blood cells or heart wall) reflecting the incident ultrasonic beam of frequency f_o, c the speed of sound in the medium, and θ the angle between the incident beam and the velocity vector.

Doppler ultrasound systems are operated in two modes, continuous-wave and pulsed or range-gated. In the continuous-wave mode, an ultrasound beam (typically of 2–5 MHz frequency) is continuously transmitted by one piezoelectric crystal, and a separate detector crystal is used as the receiver. The Doppler-shifted reflected waves emanating from moving structures within the combined fields of view of the transmitter and receiver are Fourier-analyzed and a Doppler shift frequency spectrum is constructed. While this spectrum provides no spatial information, either positional or directional, it can be useful in indicating the maximum flow velocity within the field of view, and the spectral broadening observed may be at least suggestive in some circumstances of turbulence levels in highly disturbed jets issuing from stenosed heart valves.

In the pulsed mode of operation, the same piezoelectric crystal is used as both transmitter and receiver. Short pulses of high-frequency sound (typically f_o = 5–20 MHz) are transmitted at a much lower pulse repetition frequency f_{PR}. Within the time interval between pulses, a a detection 'window' or gate is selected during which the reflected Doppler-shifted waves are acquired. The gate location within the pulse time interval determines the sampling location, since the gate time delay τ_{TD} following a pulse and the distance l between the transmitter and the sampling location are related by $\tau_{TD} = 2l/c$. To illustrate the idea, suppose that a pulsed Doppler transducer is placed at the chest wall to interrogate the blood velocity in a vessel 12 cm within the body. Using the value $c = 1540$ m s^{-1} for the body tissue sound speed, we have $\tau_{TD} = 2 \times 0.12/1540 = 0.156$ ms for the gate time delay following the pulses. Hence the maximum pulse repetition frequency must be less than

$(f_{PR})_{max} = 1/\tau_{TD} = 6.4$ kHz. This then indicates that the maximum Doppler shift frequency which can be measured is 3.2 kHz, if the Nyquist criterion is applied. According to the Doppler equation, eqn (2.B.5), we find

$$V_{max} = (\Delta f_d)_{max} c/2f_0 \approx 0.5 \text{ m s}^{-1},$$

assuming $\theta = 0$ and taking $f_0 = 5$ MHz. However, the blood velocities associated with many heart diseases, in particular valvular stenosis, are much higher, reaching 4–5 m s^{-1}, and for this reason the use of both pulsed and continuous-wave Doppler in clinical assessments is standard practice. An excellent introduction to the clinical applications of Doppler ultrasound is provided in the book by Hatle and Angelsen (1982), although it should be noted that many technological developments have occurred since its publication. One such recent development in Doppler imaging is the implementation of pulsed phased-array, color-coded flow mapping systems. In such systems, a number of transmitter–receiver transducers are used, and their pulse and receiving gate timing sequences chosen by computer control such that a planar flow field, usually in the shape of a sector, is insonified and Doppler shifts within it detected. The measured shifts are color-coded according to velocity magnitude, providing the clinician with a real-time, easily interpreted video display. The information provided is only semiquantitative, however, because of the limitations on maximum velocities detectable with pulsed Doppler systems, the limited amount of directional sensitivity (forward and reverse flow can be discriminated, but not velocity vector direction), and the loss of signal definition caused by high levels of flow disturbances (turbulence). Nevertheless, color-coded flow mapping provides a wealth of diagnostic information.

Another important recent advance in Doppler ultrasound technology is the development of intracoronary Doppler catheters. These catheters, tipped with piezoelectric transmitter–receiver crystals operated in the pulsed mode, are small enough to be maneuvered using a central guide wire up through the aorta into the coronary ostia, where they 'look' downstream at the coronary blood flow. Through varied range gating, the Doppler shift spectra at locations proximal to, within, and distal to a coronary artery occlusion can be obtained and the degree of stenosis assessed. However, as discussed by Denardo *et al.* (1994), measurement of coronary blood flow by this means is at best semiquantitative due to uncertainties in the velocity profile, insonification pattern, occlusion shape, and catheter position. But despite this, intracoronary Doppler catheter measurements provide valuable clinical information on relative levels of coronary blood flow before and following surgical interventions such as angioplasty and atherectomy.

2.B.4 Electromagnetic flow transducers

Electromagnetic flowmeters have played an important role in clinical measurements for many years, but are now rarely employed intracorporeally except in major surgical procedures, since they require the exposure of an intact vessel. They continue to be used extracorporeally in perfusion systems, and in animal model research.

E-M flowmeters exploit the fact that when a conductor (e.g. blood flow in a vessel) moves through an applied transverse magnetic field, an emf (voltage) is produced perpendicular to the magnetic field and the direction of motion. The basic equation for the emf ε developed is

$$\varepsilon = B\bar{V}d, \tag{2.B.6}$$

where B is the magnetic flux density, $\bar{V}$ the average flow velocity, and d the diameter of the vessel. In a practical clinical application a cuff containing coils which produce a parallel transverse magnetic field and having vessel wall electrode contacts at a right angle to the field is placed around a vessel, and the emf generated is observed. To illustrate the magnitudes of the quantities involved, suppose that B has the value of 0.05 Wb m^{-2} (500 gauss), the average flow velocity $\bar{V}$ is 0.01 m s^{-1}, and the vessel diameter is 0.01 m. Then eqn (2.B.6) yields for the emf developed $\varepsilon = 5.0$ μV, which is easily measured. As stated, the disadvantage of E-M flowmeters is that they are highly invasive when employed intracorporeally on exposed blood vessels. Their attractiveness, particularly when used extracorporeally as integral self-contained metering units, is that they respond to phasic flow and, as can be shown by theoretical analysis, the emf produced by the flow is independent of the shape of the velocity profile, provided only that it is axisymmetric. Thus both laminar and turbulent steady or time-varying flows having the same mean velocities, whether they be of entry type or fully developed, will generate the same steady or time-varying emf's. Miniaturized E-M catheters have also been developed, but are now not much used clinically because they sample only a small portion of the flow field and are difficult to calibrate.

An informative discussion of E-M flowmeter biomedical applications is given by Cobbold (1974). The basic theory of these devices can be found in Shercliff (1962).

References

Bassingthwaite, J. B. (1974). The measurement of blood flows and volumes by indicator dilution, *Medical engineering* (ed. C. D. Ray), Ch. 20. Yearbook Medical Publishers, Chicago.

Cobbold, R. S. C. (1974). Biomedical measurement systems, in *Medical engineering*, (ed. C. D. Ray) ch. 15. Yearbook Medical Publishers, Chicago.

Denardo, S. J., Talbot, L., Hargrave, V. K., Fitzgerald, P. J., Selfridge, A. R., and Yock, P. G. (1994). Analysis of pulsed wave Doppler ultrasound spectra obtained from a model intracoronary catheter. *IEEE Transactions on Biomedical Engineering*, **41**, No. 7, pp. 635–48.

Hatle, L. and Angelsen, B. (1982). *Doppler ultrasound in cardiology*. Lea & Febiger, Philadelphia.

Netter, F. H. (1969). *Heart*, Ciba Collection of Medical Illustrations, vol. 5. Ciba Pharmaceutical Company, Summit, NJ.

Roselli, R. J., Talbot, L., and Abbott, J. A. (1975). Evaluation of the thermal dilution technique for the measurement of steady and pulsatile flows. *Journal of Biomechanics*, **8**, 157–66.

Shercliff, J. A. (1962). *Theory of electromagnetic flow measurements*. Cambridge University Press, Cambridge, England.

PHYSIOLOGICAL FLUID MECHANICS

3

Stanley A. Berger

Contents

3.1 Introduction

We now turn to the application of the ideas and principles of the preceding sections to the flow of fluids or motion within fluids in the human body. Bioengineering would encompass fluid flows within a wider biological setting; for example, flow of sap in plants and trees, motion of microorganisms in various fluid media, but we shall not have the space to discuss these here.

The fluids which are of principal concern in physiological flows are a gas—air—and various liquids, e.g. blood, lymphatic fluids, mucus, urine, and reproductive fluids, the first of these liquids being the most ubiquitous and the flow of which has been most extensively studied. Common to all flows in the body is that they are relatively slow. This has two immediate consequences: (i) the flows can generally be considered to be incompressible, and (ii) the Reynolds numbers are also reasonably low, of the order at most of thousands, and not millions, typical of industrial and aeronautical flows. The flow of air in the respiratory system is typically at the larger values of Reynolds numbers because of differences in density and viscosity between gases and liquids, in general. As a result of (ii) the flows are typically laminar, rather than turbulent, except for parts of the respiratory tract. The fact that most physiological flows of interest are laminar and incompressible is an enormous advantage in analyzing them. Complicating factors are that these flows (i) are often oscillatory or nonsteady (for example, pulsatile blood flow in the arteries, peristaltic urine flow in the ureter, and inspiratory/expiratory airflow in the respiratory tract) (ii) occur in complex geometric configurations; for example, the complex pattern of branching vessels in the circulatory and respiratory systems. Additional complications also arise because some of the fluids, blood, for example, contain suspended material, and for this, or other reasons, exhibit non-Newtonian properties under certain flow conditions.

We begin with the area of physiological fluid mechanics that has received by far the most attention, blood flow in the circulatory system.

3.2 Blood flow in the circulatory system

We assume that the reader has had a basic biology course dealing with the physiology and anatomy of the human circulatory system. We shall just quickly review some of the basic elements and describe features important to the discussion that follows.

Shown in Fig. 3.1 is a schematic of the flow in the systemic and pulmonary circulations, the major blood circulatory systems in the body. In the systemic circulation, contraction of the left ventricle ejects blood through the aortic valve into the aorta, the largest artery in the body, whereupon the blood divides as it moves through many generations of junctions and bifurcations (the principal larger arteries are illustrated in Fig. 3.2), passing through progressively smaller arteries and arterioles, finally reaching the capillaries, the smallest blood vessels in the body, across whose walls the gaseous and mass transfer processes necessary for life occur. The return circuit is through the venules, the veins, the vena cavae, and finally into the right atrium. The deoxygenated blood is then ejected through the tricuspid valve into the right ventricle, then pumped into the pulmonary arteries, ultimately reaching the pulmonary capillaries surrounding the aveoli in the lung where it is reoxygenated, returning then to the left atrium. Ejection through the mitral valve into the left ventricle then completes one cycle of this pulsatile motion.

Table 3.1 lists some properties of the circulation and of the blood. While blood has about the same density as water, it is about three times more viscous, primarily the result of the large number of suspended red blood cells. The average Reynolds numbers are sufficiently low for the flow to be laminar throughout the circulation, and they reach very low values in the microcirculation (the flow in the arterioles and capillaries). The mean pressure falls from a relatively high value of 100 mmHg in the largest arteries to values of the order of 20 mmHg in the capillaries and even lower in the return venous circulation. This means that the arteries and arterioles must be relatively thick-walled, whereas the capillaries, venules and veins can be, and are, thin-walled. This is schematically illustrated in Fig. 3.3, which also shows the mixture of the four basic tissues in each of the blood vessel types. The proportionally large amount of elastic tissue in the larger arteries means that these vessels act rather passively in response to the pressure pulse, whereas the arterioles, with a proportionally much larger fraction of smooth muscle can, by dilating or contracting in diameter as ordered by the autonomic nervous system, actively affect the amount of blood that flows through them. This feature of the circulation is the dominant flow

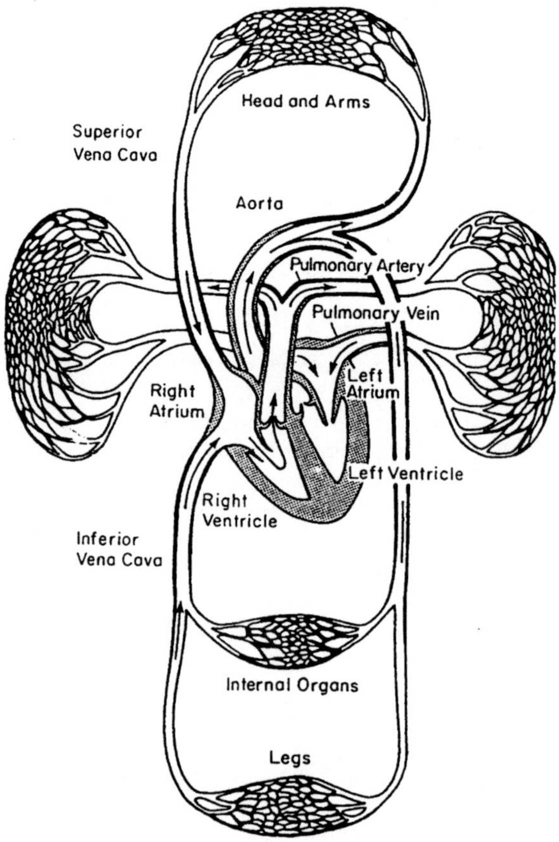

Fig. 3.1 Schematic of the systemic and pulmonary circulations.

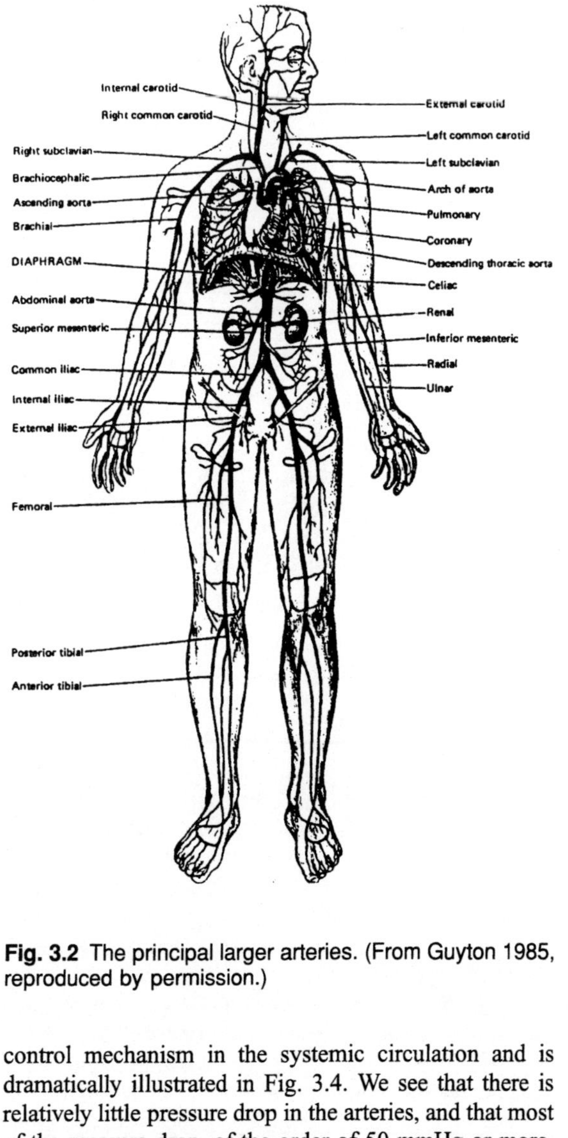

Fig. 3.2 The principal larger arteries. (From Guyton 1985, reproduced by permission.)

control mechanism in the systemic circulation and is dramatically illustrated in Fig. 3.4. We see that there is relatively little pressure drop in the arteries, and that most of the pressure drop, of the order of 50 mmHg or more, occurs in the arterioles. This seemingly poorly designed system, so much of the high pressure the heart muscle develops being lost, is what makes it possible for the body to shunt blood so effectively to different organs—for digestion, physical exertion, etc.—as needed.

That changes in arteriolar diameter can so dramatically affect the pressure can be explained by eqn (2.33), showing that in a Poiseuille flow (a rough approximation to the flow in these vessels) the pressure drop varies with the inverse fourth power of vessel radius, so that a

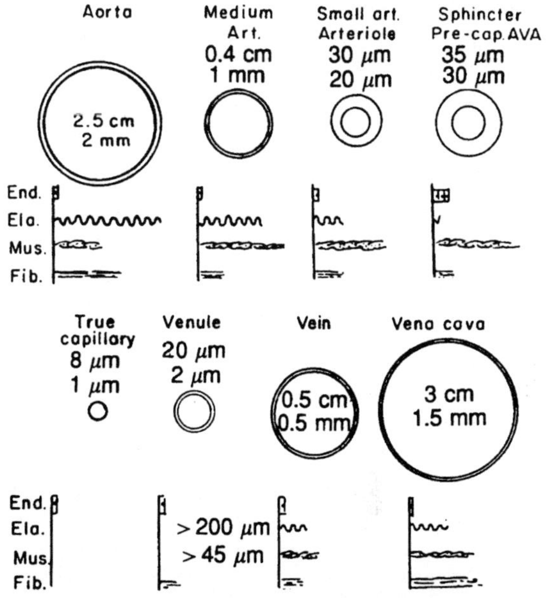

Fig. 3.3 Variety of sizes, thickness of wall and admixture of the four basic tissues in the walls of different blood vessels. The figures directly under the name of the vessel represent the diameter of the lumen; below this, the thickness of the wall. *End.* endothelial lining cells. *Ela.* elastic fibers. *Mus.* smooth muscle. *Fib.* collagenous fibers. (From Burton 1965, by permission of the American Physiological Society).

decrease in diameter by a factor of two increases the resistance to flow sixteenfold, and vice versa. We note, also, that most of the pulsatility in the pressure has died out by the time the capillary level is reached—this holds generally, with possibly some few exceptions, for the systemic circulation, less so for the pulmonary circulation.

We now discuss in turn and in some detail each major part of the circulatory system, starting with the heart and progressing down to the capillaries.

3.2.1 Heart valves

Flows through the mitral and aortic valves are discussed extensively in Section 2.6.3.

3.2.2 The arteries

The most obvious feature of blood flow in the arteries is the pulsatile nature of the flow, the external evidence of this being the existence of a pulse at sites remote from the heart. Thus we begin our discussion of pulsatile blood

Table 3.1. Some properties of the circulation and blood

Number of red blood cells (mm^{-3})	5×10^6	Specific gravity	1.06
Number of white blood cells (mm^{-3})	10^4	Heart rate (min^{-1})	60–70
Blood volume (L)	5–6	Cardiac output (L min^{-1})	5–6
Viscosity of whole blood (mPa s; cP)	3–4*	Stroke volume (mL)	70

Vessels	Diameter (mm)	Length (cm)	Wall thickness (mm)	Contained volume (cm^3 or mL)	Mean pressure (mmHg)	Average velocity (cm s^{-1})	Reynolds number	
							Average	Maximum
Aorta	25.0	40.0	2.0	100	100(av.)	40(av.)	3000	8500
Arteries	15–0.15	15.0	0.8	350	90(av.)	40–10	500	1000
Arterioles	0.14–0.01	0.2	0.02	50	60	10–0.1	0.7	—
Capillaries	0.008	0.05	0.001	300	30–20	< 0.1	0.002	—
Venules	0.01–0.14	0.2	0.002	300	20	< 0.3	0.01	—
Veins	0.15–15	18.0	0.6	2500	15–10	0.3–5	150	—
Vena cava	30.0	40.0	1.5	300	10–5	5–30	3000	—

* In the larger vessels.

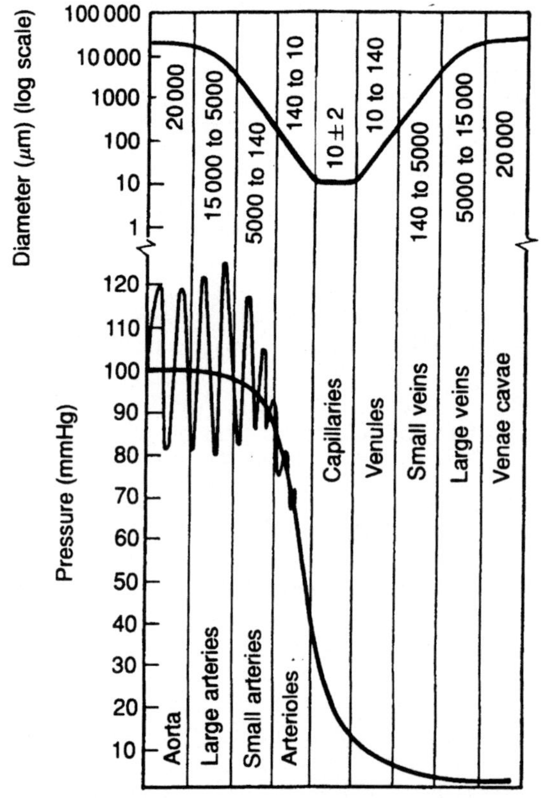

Fig. 3.4 Pressure variation in the systemic circulation.

flow with an analysis of pulse-wave propagation in the arteries.

Inviscid analysis

We assume an idealized artery: an infinitely long, circular, cylindrical elastic tube, isolated from its surroundings. The blood flowing through this tube is assumed to be a homogeneous, incompressible, inviscid fluid. (We note that the assumption of tube elasticity is crucial to the analysis because wave propagation is impossible in an *incompressible* fluid in a *rigid* tube.) The pulse wave is also assumed to be such that the wave amplitude and the tube radius are both small compared with the wavelength. These assumptions allow us to assume that the flow in the tube is one-dimensional, with a single velocity component $u(x, t)$, where x is the coordinate measured in the direction of the axis of the tube, u is the velocity in this direction, and t is time.

The differential form of the mass conservation, or continuity, equation, corresponding to eqn (2.20), is

$$\frac{\partial A}{\partial t} + \frac{\partial (uA)}{\partial x} = 0, \tag{3.1}$$

where A is the cross-sectional area of the tube.

Conservation of momentum is expressed by eqn (2.22), dropping the gravitational term and making the obvious change of notation:

$$\frac{\partial u}{\partial t} + u\frac{\partial u}{\partial x} + \frac{1}{\rho}\frac{\partial p}{\partial x} = 0. \tag{3.2}$$

Finally, we require a relationship governing the dependence of the tube radius, a, on the internal pressure, p, in the tube. The simplest elastic model is one that assumes

$$\mathrm{d}a = \frac{\gamma}{2}\mathrm{d}p, \tag{3.3}$$

where γ is the compliance of the elastic tube (the factor 2 is introduced for later convenience).

We now have three equations for the three unknowns u, p, and A ($= \pi a^2$). Combining these, after linearizing them for small disturbances, leads to the single equation

$$\frac{\partial^2 p}{\partial x^2} = \frac{1}{c^2}\frac{\partial^2 p}{\partial t^2}, \quad c^2 = \frac{a}{\rho\gamma}. \tag{3.4}$$

(The same equation holds for u.) This is the well-known wave equation, with wave speed c.

For blood flow in the arteries an important example of such an elastic tube is a thin-walled tube whose material satisfies Hooke's law. Consider a small change in radius $\mathrm{d}a$ due to the change in pressure $\mathrm{d}p$ in the half cross-sectional element shown in Fig. 3.5. The increase in the circumference is $2\pi\ \mathrm{d}a$, so the associated circumferential strain is $2\pi\ \mathrm{d}a/2\pi a = \mathrm{d}a/a$. Then according to Hooke's law, the circumferential stress is equal to $E(\mathrm{d}a/a)$, where E is the Young's modulus of the wall material (see Section 1.3.5). A force balance for equilibrium on the upper half of the vessel (Fig. 3.5) gives, per unit axial length,

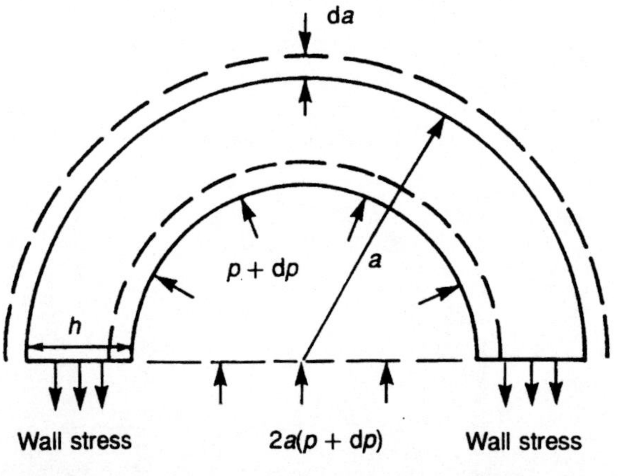

Fig. 3.5 Forces on a half cross-sectional element of an artery.

$$2a\ \mathrm{d}p = 2\left(E\frac{\mathrm{d}a}{a}\right)h,$$

or

$$\mathrm{d}a = \left(\frac{a^2}{Eh}\right)\mathrm{d}p. \tag{3.5}$$

Comparing this with eqn (3.3) implies that the compliance of the tube is

$$\gamma = \frac{2a^2}{Eh} \tag{3.6}$$

and therefore that the speed of pulse-wave propagation, eqn (3.4), is

$$c = \sqrt{\frac{a}{\rho\gamma}} = \sqrt{\frac{Eh}{2\rho a}}. \tag{3.7}$$

This expression for the wave speed in an inviscid incompressible fluid in a thin-walled simple elastic tube, one of the most important formulas in arterial fluid dynamics, is called the Moens–Korteweg formula, although discovered much earlier, in 1808, by Thomas Young.

We shall see shortly that eqn (3.7) is a good approximation to the wave speed in the larger arteries even when the viscosity of blood is accounted for.

Viscous analysis

A more realistic model of blood flow in the arteries considers wave propagation in an arbitrarily *distensible* tube containing a *viscous* incompressible fluid. This is too advanced for treatment here, so we shall briefly outline the analysis. The fluid is assumed to be a Newtonian fluid (eqn (2.3) and discussion below that equation) because, as we shall discuss later, this is an excellent approximation in the arteries. For the tube wall behavior there is a whole hierarchy of assumptions and approximations that can be, and have been, employed. The tube is again generally assumed to be an infinitely long, freely moving circular cylinder. The displacements of the wall due to passage of the wave are assumed to be small, and the wall material to have linear response, and to behave in a purely elastic, viscoelastic, or more complicated manner. Finally, the tube is taken to be either thin-walled or thick-walled. Whichever of these various wall material behavior and geometry assumptions are made, one writes down two equations governing the axial and radial displacements of the wall, respectively. These are *linear* partial differential equations, whose independent variables are time, the axial coordinate, and (in the case of a thick-walled tube) the radial coordinate.

The fluid motion is characterized by three dependent variables: the pressure and two velocity components, the axial and radial velocities. The three governing equations are the continuity equation and the axial and radial momentum equations. For a viscous fluid the momentum equations, the Navier–Stokes equations, are highly nonlinear because of the presence of the convective acceleration terms (see eqns (2.10) and (2.22)). However, because the pulse-wave speed is much larger than typical fluid velocities in the arteries, these nonlinear terms are small and may be dropped.

We then have a total of five governing equations: two for the wall motion, three for the fluid motion; these are linear partial differential equations in three independent variables: time and the axial and radial spatial coordinates. Specification of boundary conditions completes the formulation of the problem. Some of these relate to the fluid only, others require that the motion and state of stress in the fluid and wall match at their interface.

The solution is obtained in terms of a traveling wave solution, of the form

$$v_x = \hat{v}_x(r)\exp[i\omega(t - x/c)]$$
$$v_r = \hat{v}_r(r)\exp[i\omega(t - x/c)] \tag{3.8}$$

where x and r are the axial and radial coordinates, v_x and v_r the corresponding velocity components, and ω and c the frequency and wave speed of the pulse wave, respectively. The system is dissipative because of the fluid viscosity, so c is complex, so also are $\hat{v}_x(r)$ and $\hat{v}_r(r)$, which turn out to be Bessel functions of complex arguments. The details of the analysis are quite elaborate; we highlight here only some features of the solution. (The analysis as we have described it is primarily due to Womersley 1957.)

The solution depends on the Womersley, or frequency, parameter

$$\alpha = a\sqrt{\frac{\omega}{\nu}}, \tag{3.9}$$

where ν is the kinematic viscosity, introduced earlier in Section 2.6.2 (see eqn (2.43)), which also contains a discussion of the physical significance of α and how its value determines the nature of the flow. Typical values of α in the aorta and the femoral arteries of various mammals, including man, are given in Table 3.2. Similitude considerations (see Chapter 2, Appendix 2.A) require that if two such flows as are being considered here are to be 'similar', α in both should have about the same value. From Table 3.2 we see why the dog is commonly used as a model for arterial flows in man, whereas neither the mouse, the rabbit, nor the elephant(!) could play this role.

Table 3.2 Typical values of frequency parameter α in the aorta and femoral arteries of various mammals

Animal	Aorta	Femoral arteries
Mouse	1.5	—
Rabbit	4	1.6
Dog	10	2.4
Human	15	3
Elephant	50	—

Among the quantities of interest is the ratio c/c_o, where c is the actual pulse-wave velocity and c_o is the Moens–Korteweg velocity for an inviscid fluid, defined in eqn (3.7). The variation of c/c_o with the Womersley parameter α, illustrated in Fig. 3.6, shows that for values of $\alpha \geqslant 5$, which is appropriate for the larger arteries, the actual wave speed is very close to the inviscid wave speed (this is about 6 m s^{-1} in the aorta, 8–10 m s^{-1} in the femoral). If one calculates the attenuation of the pulse wave in one wavelength (the only obvious characteristic axial dimension), it is found to be quite high, suggesting a not very efficiently designed system,

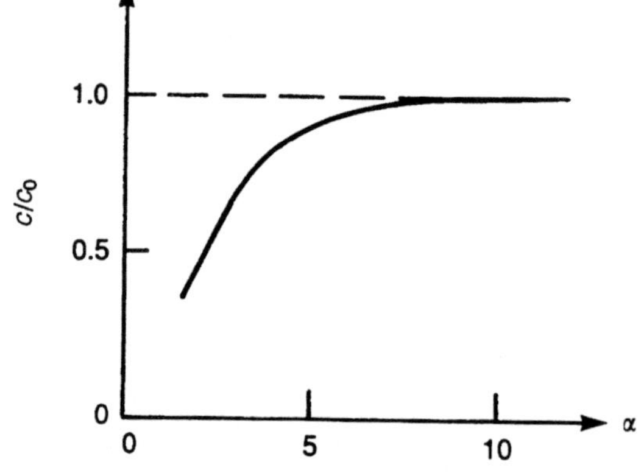

Fig. 3.6 Variation of wave speed c/c_0 with α.

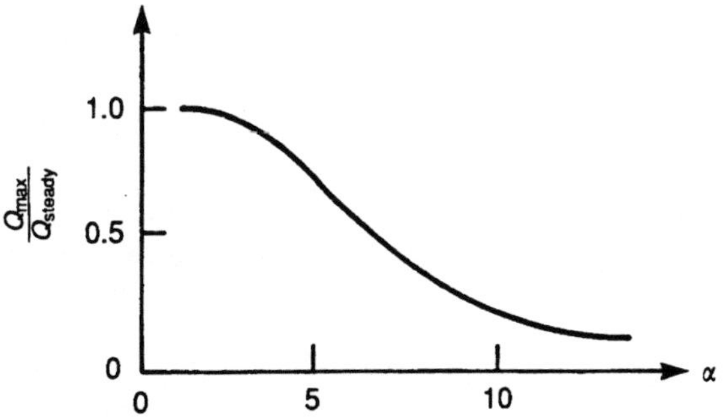

Fig. 3.7 Variation of flow rate ratio Q_{max}/Q_{steady} with α.

until one realizes that corresponding to a fundamental pulse frequency f of 1 Hz the wavelength is $\lambda = c/f = (6 \text{ m s}^{-1})/(1 \text{ Hz}) = 6$ meters, whereas the distance from the heart to the furthest extremity is typically about one-quarter of this distance, about 1.25 meters. In distances of this order there is little attenuation. (The wavelengths for higher values of α are shorter, but it is not until the 4th harmonic of the fundamental that an entire wavelength is encompassed within the body.) Fig. 3.7 shows the ratio of the maximum flow rate in the artery to the steady Poiseuille value for the same pressure gradient. Values of this ratio for larger α are quite small; clearly there is a price that is paid in having a heart that beats so that it can rest over part of the cycle.

Simplified analysis

One can carry out a simplified analysis which contains the essential features of the analysis and solutions of the preceding two subsections. Consider the flow of a Newtonian fluid in a straight tube of circular cross section (Fig. 3.8). As a model of the flow we use the (approximate) axial momentum equation

$$\frac{\partial v_x}{\partial t} + v_x \frac{\partial v_x}{\partial x} = -\frac{1}{\rho}\frac{\partial p}{\partial x} + \nu\left[\frac{\partial^2 v_x}{\partial r^2} + \frac{1}{r}\frac{\partial v_x}{\partial r} + \frac{\partial^2 v_x}{\partial x^2}\right] \quad (3.10)$$

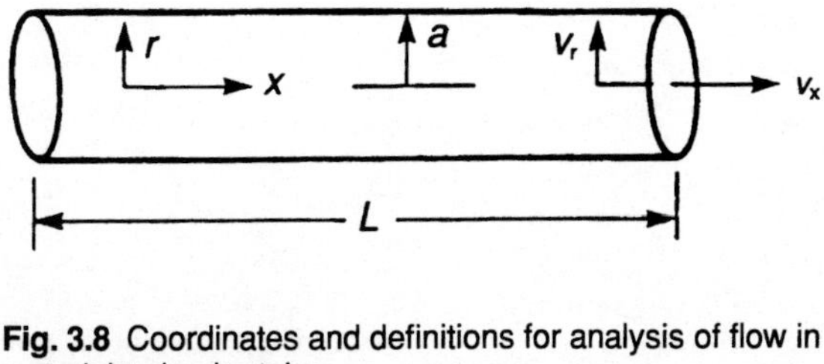

Fig. 3.8 Coordinates and definitions for analysis of flow in a straight circular tube.

for the axial velocity $v_x(r, x, t)$. As before, we assume that $v_x/c \ll 1$ and drop the nonlinear convective inertia term on the left-hand side. We introduce nondimensional variables

$$\hat{r} = \frac{r}{a}, \quad \hat{t} = \omega t, \quad \hat{x} = \frac{x}{L}, \quad (3.11)$$

where L is some representative axial length. Equation (3.10) becomes

$$\frac{\partial v_x}{\partial \hat{t}} = -\frac{1}{\rho\omega L}\frac{\partial p}{\partial \hat{x}} + \frac{1}{\alpha^2}\left[\frac{\partial^2 v_x}{\partial \hat{r}^2} + \frac{1}{\hat{r}}\frac{\partial v_x}{\partial \hat{r}} + \frac{a^2}{L^2}\frac{\partial^2 v_x}{\partial \hat{x}^2}\right] \quad (3.12)$$

using the definition of α (eqn (3.9)).

Limit $\alpha = a\sqrt{\omega/\nu} \to 0$

Taking this limit we drop the term on the left-hand side of eqn (3.12), but not the pressure term on the right-hand side, because this provides the only driving force for the flow, obtaining

$$0 = -\frac{\alpha^2}{\rho\omega L}\frac{\partial p}{\partial \hat{x}} + \left[\frac{\partial^2 v_x}{\partial \hat{r}^2} + \frac{1}{\hat{r}}\frac{\partial v_x}{\partial \hat{r}} + \frac{a^2}{L^2}\frac{\partial^2 v_x}{\partial \hat{x}^2}\right]. \quad (3.13)$$

In the limit $a/L \to 0$, assuming $\partial p/\partial x =$ const., and, consistent with these assumptions, that $v_x = v_x(r)$ only, eqn (3.13) reduces to

$$\frac{d}{d\hat{r}}\left[\hat{r}\frac{dv_x}{d\hat{r}}\right] = \frac{a^2}{\rho\nu}\hat{r}\frac{\partial p}{\partial \hat{x}}.$$

This is exactly of the form of eqn (2.29b) and can be similarly immediately integrated to

$$v_x(r) = -\frac{a^2}{4\mu}\left(1 - \frac{r^2}{a^2}\right)\frac{\partial p}{\partial x}, \quad (3.14)$$

which is the Poiseuille parabolic velocity solution discussed extensively in Section 2.5.1.

Limit $\alpha = a\sqrt{\omega/\nu} \to \infty$

In this limit the bracketed terms on the right-hand side of (3.12) vanish, so the equation reduces to

$$\frac{\partial v_x}{\partial \hat{t}} = -\frac{1}{\rho\omega L}\partial p/\partial \hat{x}. \quad (3.15)$$

If we assume that $-\partial p/\partial \hat{x} = A \cos \omega t = A \cos \hat{t}$, a simple sinusoidal pressure gradient, eqn (3.15) can be integrated to

$$v_x = \frac{A}{\rho\omega} \sin \hat{t} = \frac{A}{\rho\omega} \sin \omega t. \quad (3.16)$$

The velocity distributions given by eqn (3.16) are (i) flat profiles varying in time, uniform across the cross section, and the same at every axial location, and (ii) 90° out of phase with the pressure gradient.

If in this limiting case of $\alpha \to \infty$, instead we assume the tube is not rigid, we must supplement the linearized momentum equation (3.15) with the conservation of mass equation, eqn (3.1). Combination of these two equations leads to, in dimensional terms,

$$\frac{\partial^2 p}{\partial x^2} = \frac{1}{c^2}\frac{\partial^2 p}{\partial t^2}, \quad (3.17)$$

or

$$\frac{\partial^2 v_x}{\partial x^2} = \frac{1}{c^2}\frac{\partial^2 v_x}{\partial t^2}, \quad (3.18)$$

where $c = \sqrt{(A/\rho)\, \mathrm{d}p/\mathrm{d}A}$, the same equations derived earlier under 'Inviscid analysis' and which are the equations for linear wave propagation with speed c.

If the nonlinear convective term in eqn (3.10) is included in eqn (3.15), as well as the nonlinear term in eqn (3.1), the governing equations are

$$\frac{\partial v_x}{\partial t} + v_x\frac{\partial v_x}{\partial x} = -\frac{1}{\rho}\frac{\partial p}{\partial x}$$
$$\frac{\partial A}{\partial t} + \frac{\partial (v_x A)}{\partial x} = 0. \quad (3.19)$$

Associated with these nonlinear equations are the so-called characteristics or characteristic directions, given by

$$\frac{\mathrm{d}x}{\mathrm{d}t} = v_x \pm c, \quad (3.20)$$

along which certain quantities, combinations of the dependent variables, called the Riemann invariants, remain constant. These quantities and ideas can be employed to yield an exact solution, the Riemann solution, to eqns (3.19). An important consequence of this solution is the steepening of compression waves, leading to the formation of shock waves. This has led to some literature concerning the possibility of shock waves developing in arterial blood vessels.

Figure 3.9 shows velocity profiles in a straight rigid tube at various time increments in the cycle for a sinusoidal pressure gradient for a range of values of the Womersley parameter α. These illustrate the theory outlined above, namely that for a low value of α ($\alpha = 3.34$) the flow is quasi-steady, with roughly parabolic Poiseuille velocity profiles, whereas for the larger values of α the profiles are relatively flat, with the flow in the core lagging behind the rapidly responding wall (Stokes) layers. Section 2.6.1 has an extensive

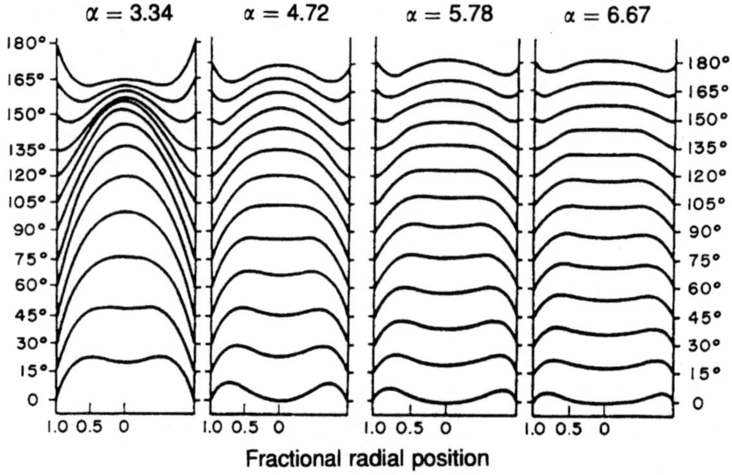

Fig. 3.9 Velocity profiles in a straight rigid tube at various points in the cycle of a sinusoidal pressure gradient for different values of α.

physical discussion of these cases and their interpretation in terms of Stokes layers at the wall.

General comments

Comparison with experiments (McDonald 1974) of the results of the most refined arterial flow analysis, those discussed under 'Viscous analysis', shows significant discrepancies, e.g. that the axial movement of the artery is much greater than is observed. To account for this and other discrepancies Womersley (1957) incorporated into his original analysis additional factors, such as tethering—the attachment of the artery to the surrounding tissue—and internal viscous damping in the artery wall. These have the effect of making the agreement between theory and experiment quite good. There are other factors which play a role in accurately modeling arterial flows, such as taper of the artery, more complex arterial wall behavior, etc. Notwithstanding all these additional complications, it is interesting that the results for pulsatile flow in an infinitely long, uniform, *rigid* circular tube, such as shown in Fig. 3.9, reasonably well reflect the state of flow in the arteries for different values of α.

There are, however, further effects beyond those discussed above that should be considered; these are nonlinear effects, that include (i) convective acceleration terms in the Navier–Stokes momentum equations, (ii) finite strain of the blood vessel wall, and (iii) nonlinear viscoelasticity of the vessel wall. Of these the first is probably the most important. To account for such nonlinearities requires extensive numerical calculations, unlike the linearized theories described earlier.

After obtaining a reliable, accurate model of flow in an artery, one still faces the fact that the arterial system in any mammal is a complex multi-generational system of branching, bifurcating tubes. It would be hopeless, using even the largest computers, to attempt to solve the governing fluid flow and wall motion equations for such a complex system. What is done instead is to use ideas from AC circuit theory. It is an easy exercise to show that the integrated linearized continuity and Navier–Stokes momentum equations are equivalent to the equations for signal transmission through a uniform cable (the 'telegraph' equations), with the correspondence shown in Table 3.3. From human anatomical arterial atlases the corresponding physical characteristics on the left-hand side of this Table, such as resistance, inertance, compliance, etc., are calculated for each arterial segment. This determines their electrical analogues, the entries on the right-hand side. After modeling each segment of the arterial tree using this correspondence, the equations for this complex AC circuit are solved to determine the voltage V and current I in each segment, from which the corresponding values of pressure p and flow Q in each segment are determined (see, for example, Noordergraaf 1978). (See also 'Examples of lumped paths' in Subsection 7.5.3.)

Table 3.3 Corresponding arterial and electrical quantities

Arterial	Electrical
Pressure, p	Voltage, V
Flow, Q	Current, I
Inertance per unit length, ρ/A	Inductance per unit length, L'
Resistance per unit length, $8\mu/\pi a^4$	Resistance per unit length, R'
Compliance† per unit length, dA/dp	Capacitance per unit length, C'
Leakage‡ per unit length, W'	Conductance per unit length, G'

ρ = density of blood.
A = cross-sectional area of vessel of radius a.
† Due to distensibility.
‡ Due to lateral branches.

There is one physiologically important aspect of arterial bifurcation that should be pointed out here. It is a straightforward matter to analyze a one-dimensional model of a simple bifurcation of a parent vessel into two daughter vessels, as shown in Fig. 3.10. When a traveling pressure wave encounters such a bifurcation, part of the wave is transmitted down the daughter vessels and part is reflected. Continuity of volumetric flow and pressure at the junction is enough to determine the

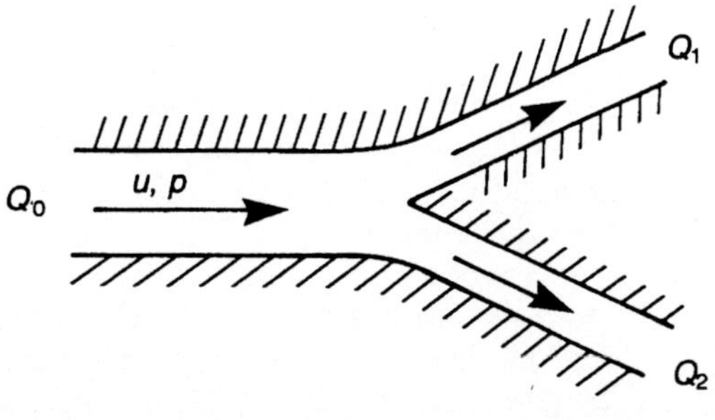

Fig. 3.10 Simple bifurcation of a parent vessel into two smaller vessels.

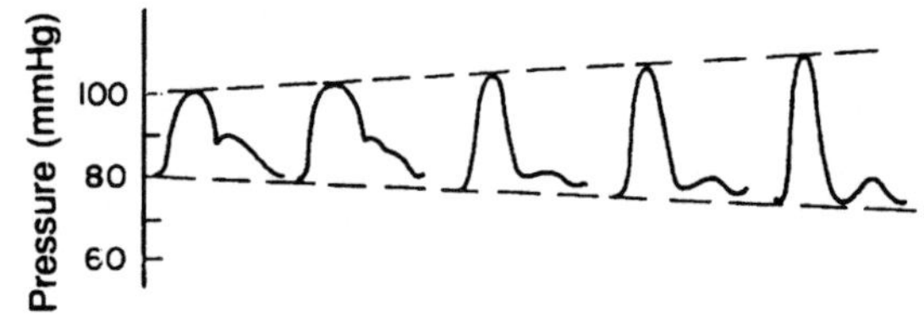

Fig. 3.11 Change in arterial pressure pulse with increasing distance from the heart.

magnitudes of the reflected and transmitted waves. It is the reinforcing effect of all the reflected waves from each of the multitude of bifurcations in the arterial tree that explains why, in the face of the dispersive, dissipative effects of the fluid and wall, the pressure pulse shape becomes 'peakier' as it moves further from the heart into the smaller arteries (Fig. 3.11).

For a general comprehensive discussion of the circulation see Caro *et al.* (1978); for more mathematical treatments see Bergel (1972), Pedley (1980), Fung (1984), and S. A. Berger (1993).

3.2.3 The microcirculation

The microcirculation is the term used for those arterial and venous system vessels smaller than approximately 180 micrometers. This includes the arterioles, the capillaries, and the venules. The arterioles are those very small vessels whose diameters lie between those of the smallest arteries (approximately 180 μm) and the true capillaries (diameter 10 μm and smaller). These are the arterial vessels with the largest amount of smooth muscle and through which vascular control is mediated. The capillaries have diameters between 10 μm and 3 μm and it is here that all the important gaseous and mass transfer processes between the red blood cells and the tissues occur. The venules are the venous side vessels corresponding to the arterioles. The microcirculatory bed is different in different organs and tissues—some of these are shown in Fig. 3.12—with different numbers of other microcirculatory vessels, such as metarterioles, preferential or thoroughfare channels, arterio-venous anastomoses, and precapillary sphincters, the latter three being important flow control agents.

Much of the pulsatility in the arterial system has died out by the time the blood has reached the arterioles (Fig. 3.4). On the other hand the flow in these vessels is exceedingly hard to analyze because these vessels range in size from approximately 20 times to slightly larger than the red cell diameter, so it is no longer quite proper to treat blood as a homogeneous fluid. Further, experimental observations and data are most lacking for this group of blood vessels, for they are too thick-walled to be transparent, and therefore to see into, and too narrow for probes to be inserted without disturbing the flow. Thus, much of what we know about these flows is gleaned from model, *in vitro*, experiments. As just noted, one should analyze blood flow in the arterioles taking account of the individual red cells. This is a formidable task, for they are very flexible, they are so densely packed that interactions between them are very important, and they interact in a nonlinear way with the flow of the plasma in which they are suspended. Thus, as an alternative, continuum models have been utilized, treating blood in the microcirculation, particularly in the arterioles, as a non-Newtonian fluid. It is therefore appropriate to begin this section on the microcirculation with a discussion of blood rheology.

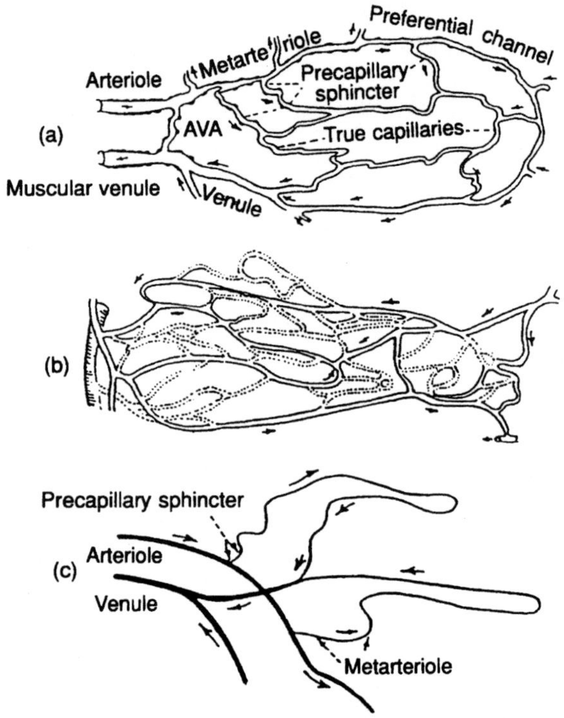

Fig. 3.12 Three types of pattern of terminal vascular beds. (a) muscle, with the preferential channels. The location of smooth-muscle control vessels is indicated. (b) the mesentery. The true capillaries are shown by the dotted lines. (c) the unique hairpin capillary loops of the human nail bed. (Zweifach 1950).

Rheology of blood

We begin with a discussion of instruments to measure the rheological behavior of blood and rheological laws to model this measured behavior.

Viscometry. There are two broad categories of instruments used to measure the viscosity of blood: (i) tube, and (ii) rotational instruments. We discuss each of these briefly in turn.

Tube viscometers. In these viscometers, also known as capillary-tube viscometers, the test fluid is made to flow from a reservoir down a narrow tube of precisely known calibre and either (i) the total volume that passes within a given time, or (ii) the time necessary for a given volume to pass, under a given pressure gradient, is measured. Usually either precision-bore glass or steel tubing is used, with diameters of less than 100 μm[1] so that the flow remains laminar, and to minimize sample volume. A common simple variant of the tube viscometer is the Ostwald viscometer in which the test fluid is partially sucked up into a narrow tube and then allowed to fall under gravity, and a record taken of the time necessary for the meniscus to pass between two fixed points in the tube.

For a Newtonian fluid, from Poiseuille's law, eqn (2.33), it follows that

$$\mu = \left(\frac{\pi d^4}{128LQ}\right)\Delta p, \qquad (3.21)$$

where d is the tube diameter, L the length of the tube, Q the volume flow rate, and Δp the pressure gradient or head. Because of the fourth power of d, the bore of the tube must be known with great precision. Corrections are also needed for end effects. These problems can be avoided by calibrating these instruments with Newtonian fluids of known viscosity. Thus values commonly reported are relative viscosities—usually in terms of that of water.

Note that unlike the rotational viscometers discussed below, the rate of shear is not constant in tube viscometers, varying from a maximum at the tube wall to zero at the axis. Thus they should, strictly speaking, only be used for Newtonian fluids, although this injunction is often breached.

[1] The bore cannot be made too small because, for reasons discussed later, the viscosity then becomes dependent on tube diameter, an anomaly called the Fahraeus–Lindquist effect.

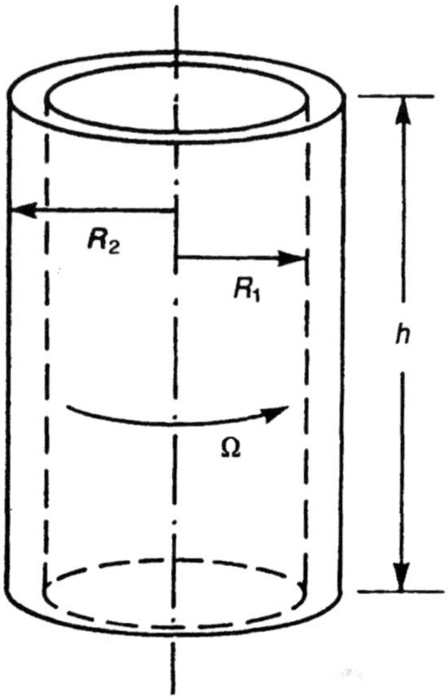

Fig. 3.13 Schematic of a coaxial cylinder viscometer.

Rotational viscometers. This type consists of two elements, separated by the liquid being tested, that rotate relative to one another about a common axis. The viscosity is determined by the relationship between the rotational velocity of the rotating element and the torque developed on the other. Within this broad group of instruments there are two principal subgroups, depending on the geometry: the coaxial cylinder type and the cone–plate type.

The *coaxial cylinder* type consists of two coaxial circular cylinders which rotate relative to one another (Fig. 3.13). The instrument is usually operated with the outer cylinder spinning at a constant angular velocity, say Ω, and the inner one fixed. If the fluid is Newtonian, the torque transmitted to the inner cylinder is given by

$$T = \left(\frac{4\pi R_1^2 R_2^2 h\Omega}{R_2^2 - R_1^2}\right)\mu, \qquad (3.22)$$

so the absolute viscosity μ can be determined directly in terms of the geometry of the instrument and the

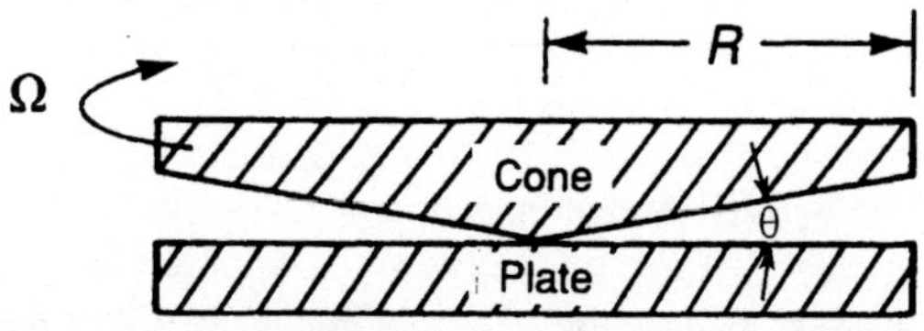

Fig. 3.14 Schematic of a cone–plate viscometer.

measured torque T. To compensate for effects that are not accounted for in eqn (3.22), such as end effects, since eqn (3.22) is based on the assumption of infinitely long cylinders, a common procedure is to calibrate the instrument with a fluid of known viscosity, assuming that the neglected effects cause the same fractional errors in the unknown and calibrating fluids.

In the *cone–plate* type the test liquid is contained in the space between a cone of a very large apex angle and a flat surface normal to the axis (Fig. 3.14). One of these elements is rotated and the torque on the other measured. The perpendicular distance between the cone and plate is proportional to r, so also is the relative linear velocity of cone and plate, thus the rate of shear in the fluid is constant throughout, equal to Ω/θ, where Ω is the rotational speed. The shear stress is $3T/2\pi R^3$. For a Newtonian fluid, then

$$\mu = \frac{3T\theta}{2\pi R^3 \Omega}. \tag{3.23}$$

The advantage of such rotational viscometers is that the volume of test fluid required can be made quite small, an important consideration when dealing with biological samples, such as blood. This is accomplished by making the gaps between the rotating and stationary elements as small as possible. This has another very important consequence. For Newtonian fluids it does not matter if the rate of strain varies across the test fluid, because there is a linear relationship between stress and rate of strain, and the viscosity is constant. This is not the case for non-Newtonian fluids. Equations (3.22) and (3.23) are not valid for such fluids. To obtain the relationship between stress and rate of strain for a non-Newtonian fluid one must make a series of tests in each of which the rate of strain is constant. This property is exactly true for one flow configuration only, planar Couette flow, the flow between infinite parallel plates in relative translational motion, discussed in Section 2.1.3. As mentioned there, a narrow-gap coaxial cylindrical arrangement is an approximation to the physically unattainable planar geometry. Using a range of rotational speeds, and thereby imposing different, approximately constant, rates of strain on the test fluid, one can determine the stress–rate-of-strain relationship for non-Newtonian fluids.

Summary. Tube viscometers are used extensively for blood viscosity measurements because they (i) are inexpensive, (ii) require only a small sample of blood, (iii) are easy to use, and (iv) simulate, *in vitro*, to some degree the geometry through which the blood flows. Important disadvantages are that (i) in each measurement the blood is subject to a wide range of shear rates and (ii) wall effects in tubes that are too narrow may cause significant errors. Coaxial cylinder viscometers have the advantage of nearly constant shear rate. On the other hand, end effects are sometimes difficult to overcome or assess, particularly for non-Newtonian fluids. The instruments are relatively expensive, particularly if designed for small samples at low shear rates (which is often necessary for blood measurements). Cone–plate viscometers possess many of the advantages of coaxial cylinder instruments, and in general require less blood. They are, however, also expensive. Certainly for research purposes, and increasingly more generally, the trend is for the greater use of rotating viscometers. For more routine and clinical measurements of blood and plasma viscosity, however, tube instruments are still widely used.

Stress–rate-of-strain laws. We consider here some simple stress–rate-of-strain relations, beginning with that for a Newtonian fluid, and including wherever possible the usual mechanical models used to represent them. These are the ones most relevant to blood rheology. For a much fuller account of this topic see the extensive discussion of the mechanics of deformable bodies in Chapter 1.

Newtonian fluid. This has been extensively discussed (Section 2.1.3 and immediately above). The relationship between stress τ and rate of strain D is linear,

$$\tau = \mu D. \tag{3.24}$$

The mechanical model is a simple dashpot (Fig. 3.15).

Power-law fluid. There is no simple mechanical model; this is an empirical relationship of the form

$$\tau = \mu D^s, \quad s = \text{const.} \quad (s = 1, \text{Newtonian}). \tag{3.25}$$

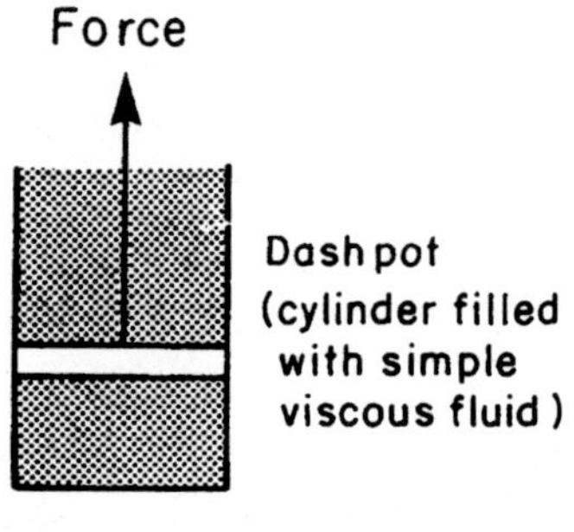

Fig. 3.15 Mechanical model of a viscous fluid.

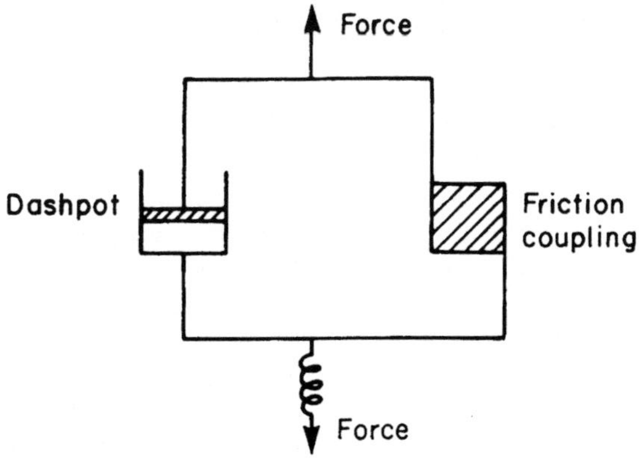

Fig. 3.16 Mechanical model of a Bingham plastic.

Bingham plastic. The mechanical model is shown in Fig. 3.16. Friction between the sliding surfaces in the coupling makes extension impossible unless F is greater than some frictional force. If the rate of extension beyond this value is proportional to the applied stress, the material is called an ideal Bingham plastic; otherwise it is a generalized Bingham plastic. Thus, in the former case

$$\tau - \tau_y = \mu D \quad \text{or} \quad \tau = \tau_y + \mu D, \tag{3.26}$$

where τ_y is called the yield stress.

Casson plastic. There is no mechanical analogue here; this is an empirical extension of the law for a Bingham plastic, given by

$$\sqrt{\tau} - \sqrt{\tau_y} = \mu\sqrt{D}. \tag{3.27}$$

Figure 3.17 shows these various laws for τ vs D.

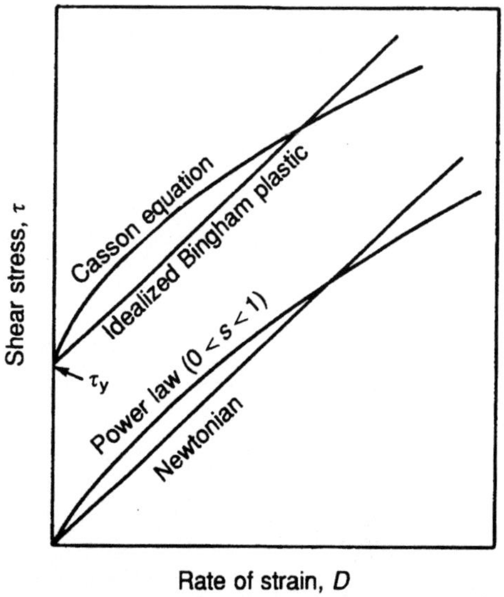

Fig. 3.17 Stress vs. rate of strain for different rheological models.

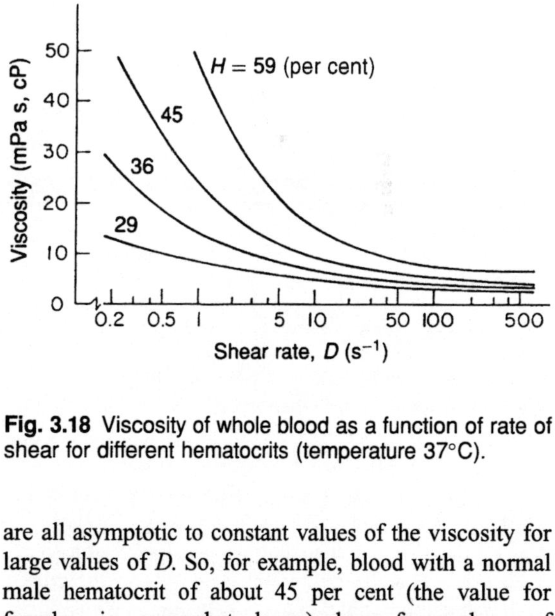

Fig. 3.18 Viscosity of whole blood as a function of rate of shear for different hematocrits (temperature 37°C).

Blood viscosity; stress–rate-of-strain laws. Blood is a complex heterogeneous suspension and although we know much about its rheological behavior, some of it is still controversial and much remains yet to be investigated satisfactorily. Of the suspended material, the numbers of white cells and platelets being so small, the principal contributors to blood rheological behavior are the red blood cells, measured in terms of the hematocrit, H, the volume percentage of red cells.

Using one of the rotating viscometers described above to measure the viscosity of whole blood at different shear rates, one would typically obtain the results shown in Fig. 3.18 for samples of different hematocrit. Not surprisingly, the viscosity increases significantly as the percentage of red cells increases. What is most significant about these curves is that they are all asymptotic to constant values of the viscosity for large values of D. So, for example, blood with a normal male hematocrit of about 45 per cent (the value for females is somewhat lower) has, for values of $D \geqslant 100\ \text{s}^{-1}$, a constant viscosity of about 3–4 mPa s (3–4 cP). The mean rates of strain or shear in the arteries are generally much higher than $100\ \text{s}^{-1}$, so it is a very good approximation to assume, as we have done in our previous analysis, that blood flows in the arteries as a Newtonian fluid with a constant viscosity.

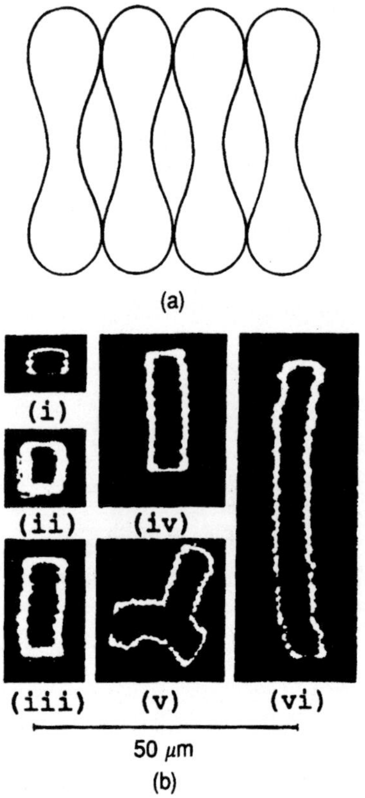

Fig. 3.19 Rouleaux. (a) schematic; (b) photomicrographs of human cells, showing single linear and branched aggregates (left part). The number of cells in linear arrays 2, 4, 9, 15 and 36 in (i), (ii), (iii), (iv) and (vi), respectively. (Goldsmith 1972, reproduced by permission).

Although nominal rates of shear in the arterioles may also be much larger than $100\ \mathrm{s}^{-1}$, the rheological behavior of blood in the arterioles is not well modeled by the Newtonian law. Of the various models discussed above, the two main ones used are the power-law equation (3.25) and the Casson equation (3.27). Investigators have found that the power-law equation (3.25) with s lying in the range $0.68 \leqslant s \leqslant 0.80$ provides a reasonable approximation to the behavior of whole blood for shear rates between 5 to $200\ \mathrm{s}^{-1}$. Of the Casson form, two well-known expressions are that of Whitmore (1968),

$$\sqrt{\frac{\tau}{\mu_o}} = 1.53\sqrt{D} + 2.0, \qquad (3.28)$$

where μ_o is the plasma viscosity, and that of Charm and Kurland (Whitmore 1968),

$$\sqrt{\tau} = 0.166\sqrt{D} + 0.33. \qquad (3.29)$$

The rheological formulas (3.26) and (3.27) both include a yield stress. Most evidence suggests that whole blood exhibits a yield stress, at least above a certain minimum hematocrit ($\sim$5–8%) and that $\tau_y < 0.01$ Pa ($0.1\ \mathrm{dyn\ cm}^{-2}$). There is also general agreement that this yield stress is due to the formation at very low velocity conditions of rouleaux, aggregations of two or more red blood cells, shown in Fig. 3.19. Electrostatic forces due to net charges on the red cell membranes have been suggested as the mechanism leading to these aggregates, but the picture is still unclear. In any event the yield stress τ_y is the (small) value of stress necessary to break up these aggregates before flow commences.

Anomalies in the viscosity of blood.

Effect of tube size: Fahraeus–Lindquist effect. As long as the diameter of the capillary in a tube viscometer is more than 1 or 2 mm, the relative viscosity of whole blood remains the same. If, however, tubes of narrower diameter are used, the value of viscosity decreases, this is known as the Fahraeus–Lindquist effect. Two different explanations have been offered for this effect.

(1) *Plasma skimming:* Next to the tube wall, because no cell can be closer to the wall than one cell radius, there is a relatively cell-free layer which is primarily plasma (see Fig. 3.20) and so has a lower viscosity than whole blood; the smaller the tube, the greater the proportion of the cross-sectional area that consists of this cell-poor 'sleeve' and so the lower the total effective viscosity. By carrying out a simple analysis of such a flow, assuming an axial core of a given viscosity surrounded by a sheath of lower viscosity, with a sharp boundary between the two, one can estimate the size of the sheath. Results of

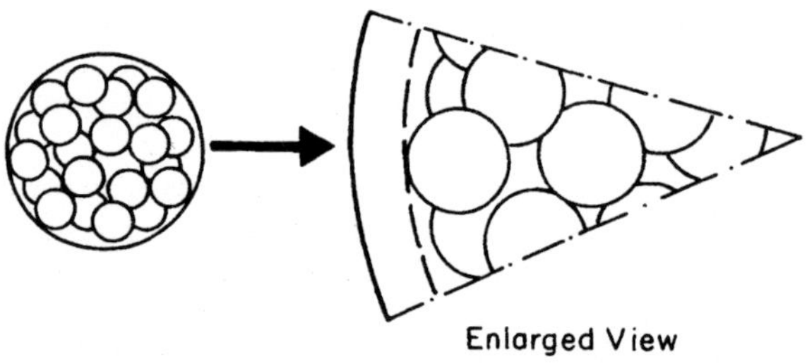

Fig. 3.20 Schematic of plasma skimming due to finite cell size.

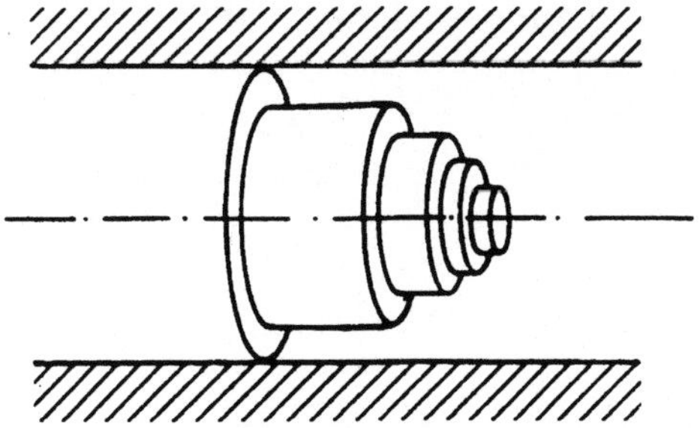

Fig. 3.21 Annuli of constant shear due to finite cell size.

such analysis suggest that in blood with a hematocrit between 40 and 50 per cent the width of this cell-free region lies between 1 μm and 3–5 μm.

(2) *Finite-summation correction:* Since the red cells are not of negligible size compared with the tube radius, one should use summation, rather than integration leading to Poiseuille's law, to obtain the relation between flow rate and pressure gradient. Assuming that shear occurs in layers of thickness δ, separated by unsheared layers (Fig. 3.21), and summing over these finite intervals we find

$$Q = G_{\mathrm{p}}\left[1 + \frac{2\delta}{a} + \frac{\delta^2}{a^2}\right]\frac{\mathrm{d}p}{\mathrm{d}x}, \qquad (3.30)$$

where a is the tube radius and $G_{\mathrm{p}} = \pi a^4/8\mu_{\text{plasma}}$. The corresponding Poiseuille result is (eqn (2.33)):

$$Q_{\text{Poiseuille}} = G_{\mathrm{p}}\frac{\mathrm{d}p}{\mathrm{d}x}. \qquad (3.31)$$

Comparing eqns (3.30) and (3.31) we see that the apparent viscosity of blood decreases as the radius a decreases, i.e. as δ/a increases for fixed δ. Unfortunately, to explain the data with eqn (3.30) the minimum lamina thicknesses are, for different hematocrits, either unrealistically large or small.

We have therefore two quite different, plausible explanations for the Fahraeus–Lindquist effect. Neither is acceptable, as regards the experimental data, as the sole explanation over the whole range of hematocrits, and perhaps a proper combination of the two theories would prove satisfactory. Nevertheless, the decrease of effective blood viscosity with blood vessel size is accepted as a real physiological phenomenon.

Effect of tube size: Fahraeus effect. When whole blood flows from a well-stirred reservoir into a narrow tube, there is a 'screening' effect which reduces the hematocrit that discharges from the tube, H_{D}, to less than that in the feed reservoir H_{F}. This screening effect is a consequence of interaction and collisions of the red cells with the entrance rim and with each other. This discharge hematocrit will not, however, represent the average hematocrit in the tube, H_{T}; generally $H_{\mathrm{T}}/H_{\mathrm{D}} < 1$ for vessels narrower than about 500 μm (the Fahraeus effect) because the red cells are not uniformly distributed across the lumen of the vessel (see below, for instance), the hematocrit being lowest at the vessel wall where the velocity is least and highest nearer the center where the velocities are larger, so high-hematocrit blood flows rapidly, and low-hematocrit blood slowly, through the vessel.

Cell migration. Even in rigid glass tubes, blood exhibits non-Newtonian properties, the apparent viscosity decreasing as the flow velocity increases. The explanation for this effect involves the phenomenon of cell migration: as the velocity of flow increases, there is a greater and greater tendency of the red cells to move toward the axis of the tube, increasing the hematocrit there and lowering it near the wall. Figure 3.22 schematically shows observed migratory paths of rigid and deformable particles in slow viscous flow at very low Reynolds numbers, for which inertial effects are negligible, and in viscous flows at larger Reynolds numbers, for which intertial effects are not negligible. In most cases the particles drift to the axis or to some annular station between the axis and tube wall. The migration of neutrally buoyant rigid spheres to an annulus due to inertial effects is given a special name, the Segre–Silberberg effect, after its discoverers. These observations have been made in model experiments and are only suggestive of the behavior of red cells. The reason why these various cases are shown in Fig. 3.22 is because red cells, depending on the flow state and vessel geometry, may exhibit various degrees of flexibility and inertial effects may be more or less important. In addition to this migration, red cells will also rotate and deform in sheared flows.

Flow in the blood vessels of the microcirculation

Arterioles. It would be a hopeless task to attempt to analyze anything as complex as the motion of a very concentrated suspension of real, deformable, red cells in narrow tubes whose diameter is of the order of $1\frac{1}{2}$ to 20 times the diameter of these cells. A common and more

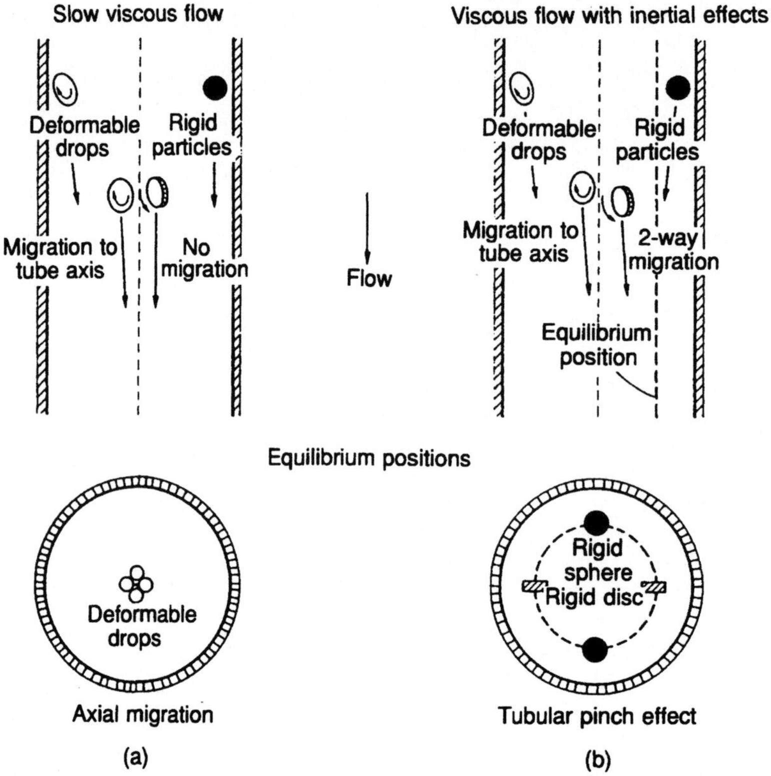

Fig. 3.22 Schematic representation of the differences in the observed migration of rigid and deformable particles in the median plane of a tube for (a) very low and (b) moderately low Reynolds number, $Re \geqslant 0(1)$ and higher. The lower diagrams show the tube end-on with the equilibrium positions reached in migration due to (a) particle deformation (b) inertia of the fluid—the tubular pinch effect. (Goldsmith 1971, as reproduced in Caro *et al.* 1978, p. 389, reproduced by permission).

tractable, but much idealized, model of such a flow is that of a suspension of rigid, equally sized, neutrally buoyant, dispersed spheres in a simple Newtonian liquid flowing at a slow, constant velocity in an infinitely long, straight cylindrical tube. Consistent with the ideas and concepts discussed above, a two-fluid shearing-core model is assumed, wherein an axial core of fluid of one viscosity is surrounded by an annulus containing liquid of a different (lower) viscosity.

The model is fairly simple to analyze but a number of important factors must be taken into account. One of these is that when a suspension of particles flows down a tube, the volume concentration may be different in the tube than in the feeding or collecting vessels or reservoirs because, as discussed earlier, the particles may be displaced radially in the tube, and consequently move at different velocities. This in turn affects how long it takes them to transit the vessel and their concentration in the vessel.

The shearing-core model can be used to predict theoretically the mean concentration of spherical particles in the tube $\bar{c}$, in terms of the feed concentration c_f, the two viscosities, and the geometric parameters of the tube and particles. In interpreting the results, attention must be given to the exact nature and the position of the boundary between the core and the annulus. Care must also be taken because the concentration of spherical particles cannot be constant to the outer edge of the core, as assumed by the theory, if the particles are all contained in the core, but must fall to zero in successive narrow annuli within a sphere diameter of the core boundary. In any event, the expression for $\bar{c}$ does predict that the mean concentration is always less than the feed concentration c_f, and that $\bar{c}$ decreases with decreasing tube diameter, which would lead to a decrease in apparent viscosity of the suspension, a result consistent with the Fahraeus–Lindquist effect.

At higher Reynolds numbers, when inertial effects, ignored in the theory described above, become important, the picture changes very significantly. Particle migrations are much different (Fig. 3.22), and the sphere-free annular zone adjacent to the wall is very dependent on velocity and concentration. The thickness of this annulus, for a given velocity of flow, decreases rapidly with particle concentration, essentially disappearing at normal hematocrit concentrations.

Applying the above model theoretical ideas and results to whole blood is problematic because red cells are very deformable biconcave discoids and not rigid spheres, and the question of the correct choice of a radius for the shearing-core of the model remains open. This is compounded by the lack, due to the difficulties in carrying out experiments in this area, of good experimental data and the contradictory nature of much of what has been measured. It does seem in fact that no one model is yet capable of explaining the flow behavior—concentration and viscosity changes, the effect of changing velocity, etc.,—in blood vessels the size of arterioles. There is agreement, however, between theory and experiment on some of the effects on the flow of some of the most important parameters. For hematocrits less than about 45 per cent the mean concentration of red cells drops appreciably near the tube wall, leading to a decrease in mean tube hematocrit and a concomitant drop in apparent viscosity. The overall flow velocity has only a small effect, suggesting that cell and wall interactions, rather than inertial effects, play the dominant role. Less certain is the effect of tube diameter on cell distribution near the wall. It is not definitely known yet how the thickness of the peripheral annular zone varies with changes in diameter nor whether the radial distribution of red cells within the shearing core is constant or not. In connection with the latter we recall the variety of radial migratory patterns just discussed.

In summary, the best theoretical model of flow in these narrow tubes, at normal hematocrits of 40 per cent or so, is the shearing-core, two-fluid model, with a core of a high constant viscosity determined by the concentration of red cells, surrounded by a very thin annulus of fluid, of thickness half a red cell diameter or less, and viscosity that of plasma. This annular peripheral zone around the wall of the tube can be regarded as an idealization of the unavoidable fall in cell concentration that must occur at the wall. At lower hematocrits a thicker annular layer of suspending fluid develops at the wall, leading to an appreciable decrease in apparent blood viscosity.

This constitutes but the briefest of discussions of the exceedingly complex subject of blood flow in narrow vessels. For a much more complete treatment see Whitmore (1968), which provided the substance of some of the material above. For a more comprehensive discussion of the physiology of the microcirculation see Wiedman *et al.* (1981).

Capillaries. At the capillary level, vessels ranging in size from 10 to 3 μm, the motion of individual cells must be considered. We recall that a typical red blood cell is a biconcave disk of diameter 8 μm and thickness 2 μm, filled with a viscous Newtonian-like fluid, hemoglobin, and surrounded by a thin flexible membrane. The blood flow is no longer pulsatile at this level. The extreme flexibility of red cells is most evident in vessels of this diameter. For example, a red cell can readily move through a micropipette as narrow as 3 μm in diameter by folding about an axis through its center into a so-called 'crêpe Suzette' shape. In somewhat wider tubes, say 5–7 μm, it typically takes on the so-called 'parachute' shape. For the largest capillaries the red cells move down undeformed, probably with their broad face perpendicular to the flow direction. In all these cases they move down in single file, one following the other.

The study of the motion of undeformed individual red cells in the larger capillaries is perhaps the easiest and has received the most attention. A typical such configuration is shown in Fig. 3.23. The usual analysis idealizes the red cells as rigid circular disks moving steadily with constant velocity in a rigid circular tube surrounded by plasma on all sides. Because of the very small dimensions and low cell velocities, the Reynolds numbers for capillary flow are exceedingly small (of the order of 10^{-3}) and these flows fall into the category of Stokes or 'creeping flows', discussed extensively in Section 2.5.4. The governing differential equations for

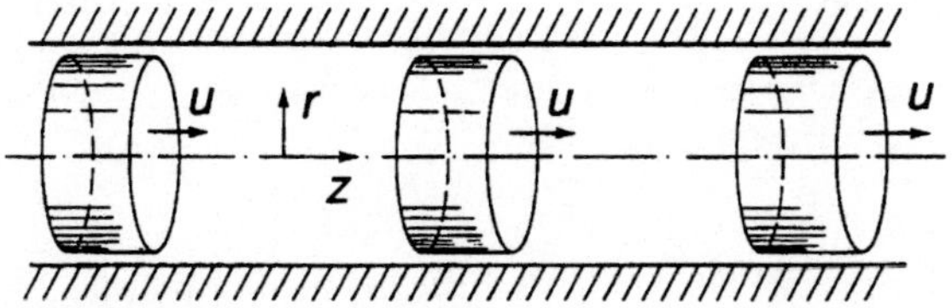

Fig. 3.23 Rigid disks in a tube as a model of red cell flow in a capillary.

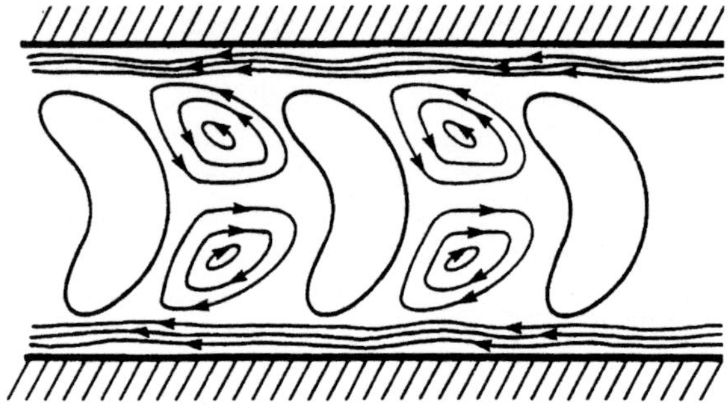

Fig. 3.24 Bolus flow; streamlines shown are those in a frame moving with the red cells.

such flows are linear and may, for simple configurations such as that shown in Fig. 3.23, be solved in closed form.

In somewhat narrower capillaries, the cells must deform to fit into the capillary, say into a parachute shape. The flow in the narrow gap thus created between the cell and the capillary wall can be analyzed as a lubrication flow, such as that in a journal and bearing, with due account taken of cell deformability. This is a special kind of linearized creeping flow which can also be readily analyzed.

Little, if any, analysis has been carried out of the 'crêpe Suzette' motions in the narrowest capillaries, for this situation is unlike those discussed immediately above, in that it is not axisymmetric and is therefore much more difficult to analyze.

Theory and experiment are in general agreement that, because of the presence of individual red cells, resistance to flow depends strongly on the capillary diameter and in the larger capillaries is of the order of 5 to 10 times greater than that of Poiseuille flow of a fluid of the overall whole blood viscosity in the same-sized tube.

An important aspect of red cell motion in the capillaries is the nature of the flow in the plasma region between cells. The streamlines of this so-called 'bolus flow', in the frame of the red cells, are illustrated in Fig. 3.24. At one time it was speculated that this recirculatory flow pattern significantly facilitated the mass transfer of gases between the cells and the tissue surrounding the capillary, but a detailed quantitative analysis showed that this was not the case.

3.2.4 The venous system

The venous system is free of any heart-induced pulsations, the slight unsteadiness in the largest veins, near the heart, being due primarily to respiratory pressure changes. Pressures on the venous side of the circulation are low; blood is forced to the heart by the squeezing action on the thin-walled veins of surrounding tissues (e.g. muscles), with the assistance of valves in the veins to prevent backflow. This motion is much less regular than arterial flows and plays a much lesser role in the dynamics of the circulation; it has thus received much less attention.

Another important difference between arterial flow and flow in the veins, because the latter is a low-pressure system, is that the transmural pressure in the veins is sometimes negative, due to hydraulic pressure gradients, muscular action, intrathoracic pressure changes, etc. When the pressure outside the vein exceeds that inside, there is a tendency for the vessel to collapse. There is accordingly a sizable literature on flow in collapstible tubes (see, for example, Shapiro 1977), a problem where the structural behavior and geometry of the tube play a central role.

3.2.5 Pulmonary circulation

In the pulmonary circulation the entire output of deoxygenated blood from the right ventricle passes through the pulmonary arteries to the alveolar capillaries and returns, oxygenated, via the pulmonary veins, to the left atrium. There is also a secondary, much smaller, circulation, the bronchial circulation, off the thoracic aorta, which nourishes the lung itself (in the same way as the coronary circulation, off the aortic sinuses, nourishes the heart muscle).

Although the pulmonary circulation is like the systemic circulation in the types of vessels—arteries, capillaries, veins–there are important differences between these two major circulations:

(1) the pulmonary circulation is a low-pressure, low-resistance system; the time-average pressure in the arteries is 15 mmHg or 1/6 that in the systemic arteries;
(2) the pulmonary arteries have much thinner walls than the systemic arteries (this is associated with (1)); they have the same Young's modulus as the latter but are more distensible because they are thinner;
(3) the pulmonary vascular bed is not regionally specialized;
(4) vasomotor control is relatively unimportant under normal conditions; unlike the systemic arteries, the

pulmonary vessels do not undergo large active changes in their dimensions; the arteries and arterioles have primarily elastic tissue, little smooth muscle.

The mechanics of the pulmonary circulation are affected to a major degree by the mechanics of the lung and the thoracic pressure changes over the respiratory cycle. The branching of arteries is very similar to that in the systemic circulation. The shape of the pressure wave in the main pulmonary arteries is like that in the aorta, but there is no significant change in shape along the pulmonary arterial system because it is so short—the total length from the beginning of the pulmonary trunk to the capillaries being only 20 cm—that the pulse occurs simultaneously along the entire length. The velocity profiles in the arteries are relatively flat because $Re \gg 1$ and $\alpha = a\sqrt{\omega/\nu} \gg 1$, and the vessels are so short that new boundary layers begin on each flow divider and never fill the vessels.

The pulmonary microcirculation has several features that constrast sharply with the systemic microcirculation:

(1) there is little vasomotor activity in pre-capillary vessels (or elsewhere) and most of the resistance to flow occurs in the capillaries themselves;
(2) the capillaries have lengths comparable to their diameters, are flattened, and are arranged like two-dimensional channels passing between pairs of alveoli, the walls of which are flexible and held apart by posts (typical thickness ≈ 4 μm), so the blood flows as a sheet of fluid between almost parallel alveolar membranes held apart by frequently occurring posts of connective tissue;
(3) the flow is pulsatile.

Like in the systemic microcirculation, $Re \ll 1(\approx 0.005)$, typically velocities being about 0.005 m s^{-1}, and $\alpha \ll 1$ (≈ 0.0002 for $\omega = 10$ Hz). These two characteristics of the flow imply that fluid inertia is negligible, the force balance in the capillary sheet being between pressure and viscous forces, and also that the flow is quasi-steady (see Section 2.6.2), that is, the flow rate is in phase with the pressure gradient at all parts of the sheet at all times. For a Newtonian fluid flowing inside a rigid sheet they would also imply that $Q \propto \mathrm{d}p/\mathrm{d}x$, i.e. the flow rate is directly proportional to the pressure gradient, but here the sheet is elastic, so the sheet thickness, h, depends on $\mathrm{d}p/\mathrm{d}x$ (actually the transmural pressure gradient). Since h affects the fluid resistance, this implies that resistance depends on $\mathrm{d}p/\mathrm{d}x$. Using these ideas and assumptions one can develop a simple theory for flow in such a sheet. Since the pressure outside the sheet, the alveolar pressure, p_{alv}, plays a major role in determing the width of the sheet, the nature of the flow depends on the relative magnitudes of three pressures: the arterial, the venous, and the alveolar pressure. An increase in p_{alv}, leading to a decrease in sheet width, particularly to values as small as 2 to 3 μm, has a profound effect on the rate of flow, for even in a Poiseuille flow of a Newtonian fluid in such a channel the resistance to flow increases as h^{-4}, where h is the channel width, and the increase is so much greater in the case of whole blood with 8 μm red cells that must be deformed to flow through the sheet.

The dependence of the flow in the pulmonary microcirculation on the relative sizes of the arterial pressure, p_{a}, the venous pressure, p_{v}, and the alveolar pressure, p_{alv}, combined with the hydrostatic variation of pressure (eqn (2.16)), has important implications for the zonal distribution of blood flow in the lungs, in particular, as it leads to nonuniform perfusion over the height of the lung. Consider the schematic of a lung shown in Fig. 3.25. In a person when upright the height of the lung is about 30 cm, so the total hydrostatic head between the top and bottom of the lung is 30 cm H_2O or 22.5 mmHg. The pulmonary arteries enter the lung about half-way up, and since p_{a} is on the average about 15 mmHg, the arterial pressures are approximately 26 mmHg at the bottom and 4 mmHg at the top of the lung. Since there is a 7–8 mmHg drop in pressure across the pulmonary circulation, the mean pressure is normally subatmospheric in the microcirculation at the top of the lung! Whenever the pressure in the tissue surrounding a blood vessel exceeds that in the vessel (that is, the transmural pressure is negative), there will be a force tending to close the vessel. Because the pressure external to the pulmonary capillaries, p_{alv}, is approximately atmospheric, the relative magnitudes of p_{a}, p_{v}, p_{alv} at different levels in the lung will be as shown in Fig. 3.25, and these vessels will be fully open near the bottom of the lung, partially open in the middle, and fully closed at the top. The flow rate, Q, in the lowest, fully open capillaries (zone III) is proportional to $p_{\mathrm{a}} - p_{\mathrm{v}}$, while that in zone II is proportional to $p_{\mathrm{a}} - p_{\mathrm{alv}}$. (The latter zone behaves as a so-called Starling resistor, or waterfall, because like a waterfall the flow rate does not depend on the total height of fall, in this case the overall pressure drop $p_{\mathrm{a}} - p_{\mathrm{v}}$.) The levels at which these zones merge into one another depends on the alveolar pressure, and therefore varies over the breathing cycle, and with the arterial pressure, which increases, for example, during exercise.

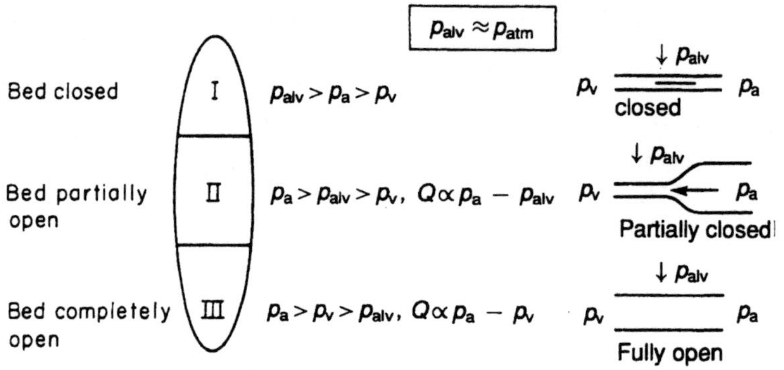

Fig. 3.25 Schematic of a human lung showing relationship of p_a, p_v, p_{alv} at different heights.

Air enters the lung nearer the top and considerable effort is required to get it to the deepest recesses of the lung (as anyone who has taken a lung function test knows!), whereas the above considerations imply that the blood flow is greatest at the bottom and least (or none at all) at the top, suggesting grossly unequal perfusion and ventilation in the lung. Partly because of the highly deformable lung structure, however, the situation is not as bad as it seems, and there is a much closer balance of air and blood over the length of the lung.

3.3 Respiration

In the respiratory system, air passes from the trachea to increasingly smaller air passages, the bronchi, the bronchioles and finally flows into the alveolar ducts and sacs. Table 3.4 lists characteristic diameters and lengths of some of the 20 generations of airways, together with average velocities ($\bar{u}$) and Reynolds numbers (*Re*) for two overall flow rates, corresponding to quiet and vigorous breathing. What distinguishes the respiratory system from the circulatory system is that because the fluid is a gas—air—rather than a liquid—blood—the density and absolute viscosity are much lower, and the flow velocities are much higher. The typical Reynolds numbers are generally much higher, large enough for the flow to be turbulent in the larger air passages. Another, equally important, difference is that over the breathing cycle, from inspiration to expiration, the flow changes direction completely.

As for flow in the larger air passages, the bronchi, their typical length is about 3.5 times their diameter, too short for fully developed flow, so the flow in them always has an entry-length character (see Section 2.5.1 for a discussion of these concepts). A consideration of the development of the boundary layers on the inside walls of the flow dividers at the bifurcations of these vessels plus its enhancement by secondary flow at these junctions (like that in curved tubes, as discussed in Section 2.5.3) leads to the result that in these vessels

$$\frac{\text{viscous dissipation}}{\text{dissipation in Poiseuille flow}} \propto \left(\frac{Re\, d}{l}\right)^{1/2}, \qquad (3.32)$$

where d is the diameter of the vessel and l the vessel length. Most of the energy dissipation in these flows occurs in these boundary layers. The $Re^{1/2}$ dependence in eqn (3.32) is seen to be significant when one observes that in light breathing (10 L min^{-1}) *Re* varies from 800 in the trachea, to 100 in the 5th generation, to about 10 in the 10th generation, whereas in heavy breathing (100 L min^{-1}) the values of *Re* are 10 times as great! The implication of this is that, very much unlike the systemic circulatory system, the loss of total pressure, or the flow resistance, occurs primarily in the early generations of the lung airways. Also, because the resistance is proportional to $U^{1/2}$, where U is the average flow velocity, the loss of total pressure, equal to resistance times U, is proportional to $U^{3/2}$ (in constrast to the linear dependence on U of Poiseuille flow).

Table 3.4 Characteristics of various generations of airways in the human respiratory system

Generation	Diameter (cm)	Length (cm)	Quiet breathing		Vigorous breathing	
			$\bar{u}$ (cm s^{-1})	*Re*	$\bar{u}$ (cm s^{-1})	*Re*
Trachea	1.80	12.0	197	2325	790	9300
1	1.22	4.76	215	1719	859	6876
2	0.83	1.90	235	1281	941	5124
3	0.56	0.76	250	921	1002	3684
4	0.45	1.27	202	594	809	2376
5	0.35	1.07	161	369	643	1476
10	0.13	0.46	38	32	151	127
15	0.066	0.20	4.4	1.9	17.8	7.6
20	0.045	0.083	0.3	0.09	1.2	0.37

The above considerations are based on the assumption of laminar flow in the airways; however, the flow in the trachea is normally turbulent and in heavy breathing remains so in the first three to four generations of airways. To the question of how the above conclusions are modified, the answer is 'not much', because the thin boundary layers on the flow dividers are still expected to be laminar, implying that the total pressure drop still varies as $U^{3/2}$. In any case, for these Reynolds numbers the pressure drop due to turbulence itself varies as $U^{3/2}$, implying that the pressure drop across the whole bronchial tree obeys this dependence on velocity.

As an example of the more complex flows that occur in the airways because the flow goes in both directions over the breathing cycle, consider Fig. 3.26, which shows the expiratory flow in a single bifurcation, two daughter vessels joining a parent vessel, and velocity profiles in the parent tube. Note the secondary motions, as well as the decidedly nonaxisymmetric velocity profiles in the cross section of the parent. A similar schematic for inspiration is shown in Fig. 3.27.

For an excellent summary of work in this area, and the source of much of what appears above, see Pedley (1977), and the references that appear therein. For a fuller discussion of the physiological and bioengineering aspects of the lung as a whole, see West (1977).

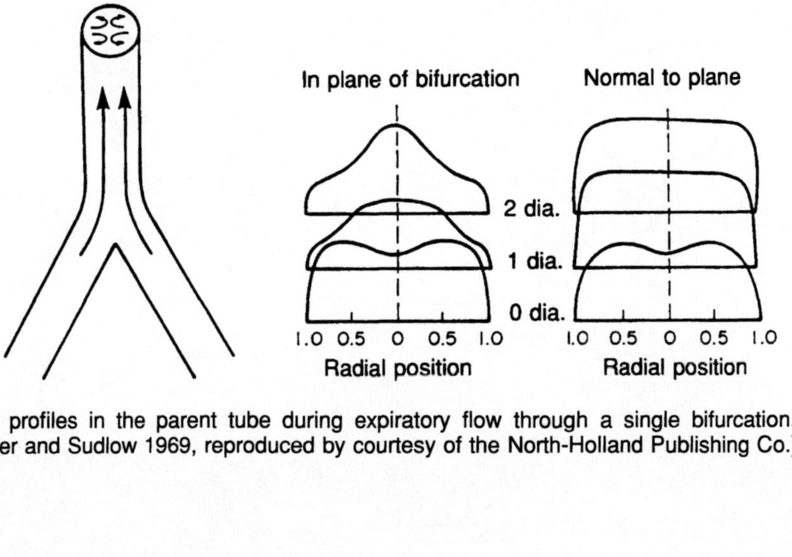

Fig. 3.26 Velocity profiles in the parent tube during expiratory flow through a single bifurcation. Flat entry profile; *Re* = 700. (Schroter and Sudlow 1969, reproduced by courtesy of the North-Holland Publishing Co.)

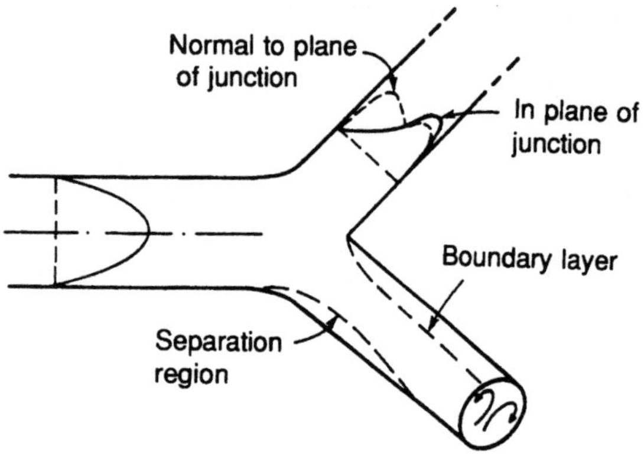

Fig. 3.27 Qualitative picture of flow downstream of a single junction with Poiseuille flow in the parent tube, as during inspiration. Direction of secondary motions, new boundary layer, and separation region are indicated in the lower branch; velocity profiles in and normal to the plane of the junction are shown in the upper branch. (Pedley 1977, reproduced by permission of Annual Reviews Inc.).

Clearly, the much higher Reynolds numbers and the bidirectional flows make analysis of the fluid motion in the airways considerably more difficult than blood flow in the larger arteries. Well-founded theories require a solid foundation of experimental data, and it is the gathering of the latter that has attracted most attention recently.

The terminal passages of the respiratory system are shown in Fig. 3.28. Flow in the smallest passages, particularly the alveolar sacs, is very slow and can be modeled as a creeping or Stokes flow (Section 2.5.4). The alveolar sacs are surrounded by the pulmonary capillaries with which they exchange gases: oxygen and carbon dioxide. Fig. 3.29 shows the various membranes, fluid and cellular layers, and interstitial spaces that these gases must pass through, if life is to be sustained, in the fraction of a second that it takes for a red cell to traverse a capillary. Fortunately, the diffusional rates through each of these zones are sufficiently high for all of this to occur.

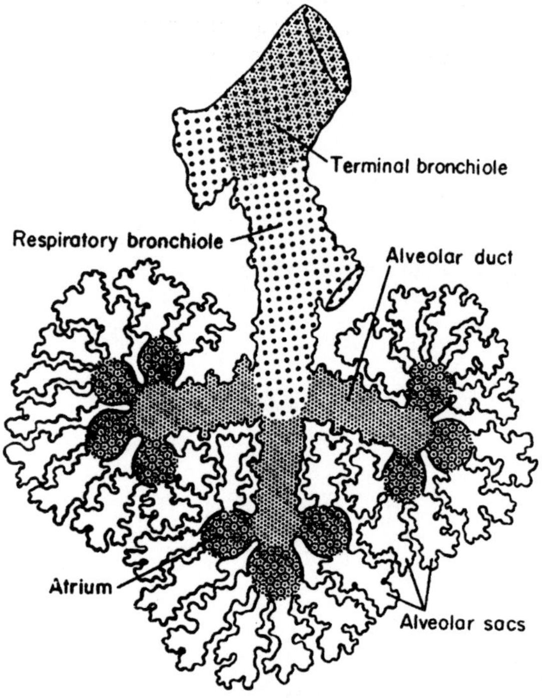

Fig. 3.28 The respiratory lobule. (From Guyton 1977, after Miller 1947, reproduced by permission of the publishers.)

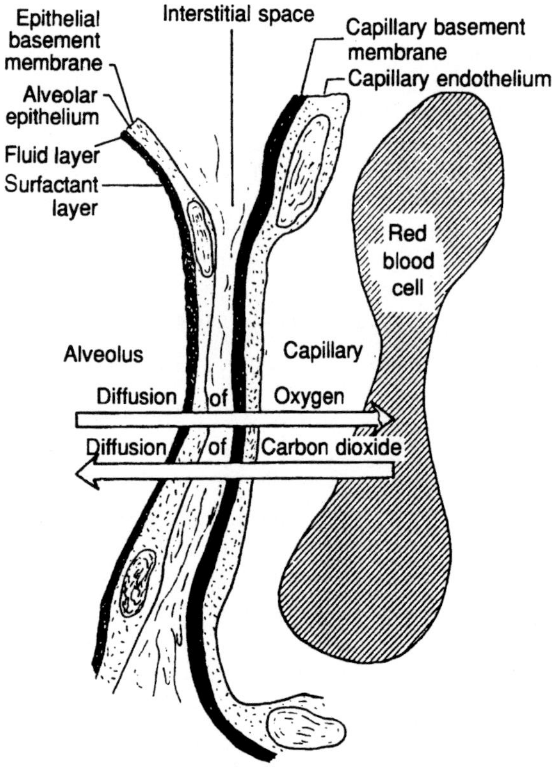

Fig. 3.29 Ultrastructure of the respiratory membrane. (From Guyton 1977, reproduced by permission.)

3.4 Peristaltic motion

A wave of area contraction propagating down a liquid-filled distensible tube will cause the fluid to move in the direction of the wave. We refer to the driving mechanism of this fluid motion as peristaltic pumping. This peristaltic pumping action is an inherent neuromuscular property of any tubular smooth muscle, and is used in the body to propel or mix the fluid or contents of a distensible tube, such as the ureter, the gastro-intestinal tract, the bile duct, and other glandular ducts. It is also used in man-made devices, such as roller pumps, much used as part of heart–lung machines.

To understand how peristaltic pumping works, it is convenient to consider a long tube closed at both ends, filled with a viscous fluid on which a peristaltic contractile wave moves to the right at speed c, and examine the flow in both the laboratory and wave frames (Jaffrin and Shapiro 1971) as shown in Fig. 3.30. To preserve mass flow the velocity profile must appear as in Fig. 3.30(b), associated with which is a pressure gradient, such that the pressure is higher on the right than the left. Since this same pressure drop exists in the laboratory frame, we see immediately that the peristaltic wave produces a rising pressure in the direction of the wave. (If the ends are open any additional flow will superpose on this and not change the basic picture.) From these simple considerations two conclusions result: (i) dissipation due to viscosity is an essential feature, without which the flow in the contraction would be plug flow and no pressure rise ahead of the contracted section would arise, and (ii) in the laboratory frame the fluid in the contracted section moves in a direction opposite to the wave, while the fluid in the uncontracted section moves in the wave propagation direction.

For peristaltic pumping in a two-dimensional channel or tube there are three characteristic length scales: a, the mean half-width or mean radius; b, the amplitude of the peristaltic wave; and λ, the wavelength of the wave. From these we can form two independent parameters, $2\pi a/\lambda = \beta$, a dimensionless wavenumber, and $b/a = \phi$, a dimensionless amplitude ratio. The appropriate Reynolds number is $Re = \beta(ac/\nu)$, where c is the speed of the wave. (It can be shown, since the wave frequency $\omega = 2\pi c/\lambda$, that $Re = \alpha^2$, where α is the Womersley or frequency parameter, defined and discussed in Section 2.6.2. It then follows that Re has a similar physical interpretation as α, and that the limits $Re \to 0$ and

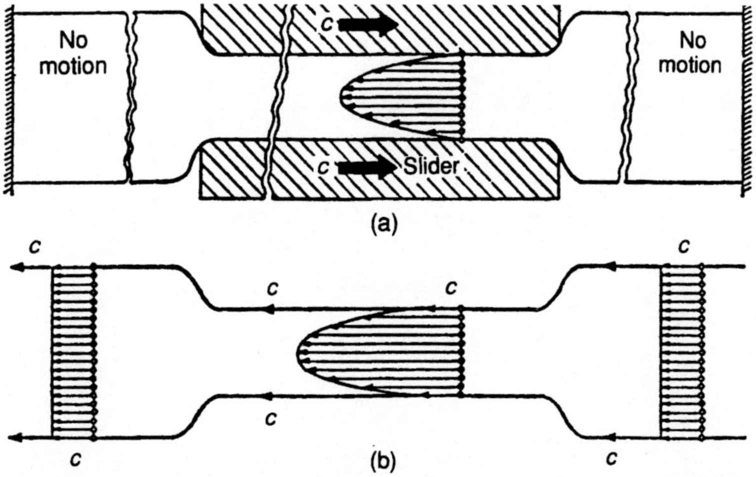

Fig. 3.30 Schematic of the flow due to peristalsis: (a) laboratory frame; (b) wave frame. (From Jaffrin and Shapiro 1971, reproduced by permission of Annual Reviews Inc.)

$Re \to \infty$ lead to instantaneous velocity profiles similar to those in the limits $\alpha \to 0$, $\alpha \to \infty$, discussed in Section 3.2.2).

Various investigators have studied this problem for plane and axisymmetric flow for a wide range of choices of β, ϕ, and Re. In general, one or more of these must be assumed to be small or zero, the other(s) arbitrary if any progress is to be made analytically. (See Jaffrin and Shapiro (1971) for a discussion of this work.) Table 3.5 suggests that no single such analysis is likely to apply to all physiologically or biomedically interesting problems. We shall concentrate in what follows on peristaltic flow in the ureter, as treated by Shapiro and his students in an attempt to explain how bacteria sometimes travel, in times (hours) too short to be explained by molecular diffusion or bacterial self-propulsion, and in the absence of retrograde peristaltic waves, from the bladder to the kidney against the mean urine flow.

The physiological data for the ureter is not well established and varies greatly among individuals and from time to time in the same individual. Representative values are a length of 30 cm, an inside diameter roughly from 0.01 to 0.5 cm (the ureter is almost closed when no urine is flowing), peristaltic wave speed 1 to 6 $\mathrm{cm\,s^{-1}}$, frequency several waves per minute, wavelength 1 to 15 cm.

Shapiro *et al.* model the ureter as a two-dimensional channel, and assume that (i) $a/\lambda \ll 1$ (that is, very large wavelengths), and (ii) $Re \ll 1$, implying that inertial effects are negligible. The first of these assumptions is consistent with the characteristic values for the ureter quoted above and has as consequences that transverse velocities and pressure gradients are smaller than axial values and the pressure may be taken as constant over the cross section. Unfortunately, the second assumption is not a very good one, because for the values above, $Re \approx 1$ rather than being very small. However, creeping or Stokes flow solutions, strictly valid only for vanishingly small Reynolds numbers, often give reasonable results for $Re \approx 1$.

In the wave frame, in which the flow is steady, the reduced axial momentum equation, using the assumptions above, is

$$\frac{\mathrm{d}p}{\mathrm{d}\bar{x}} = \mu \frac{\partial^2 \bar{u}}{\partial \bar{y}^2} \quad (\bar{p} = p(\bar{x}), \text{ only}), \tag{3.33}$$

subject to the boundary conditions

$$\frac{\partial \bar{u}}{\partial \bar{y}} = 0 \quad \text{at} \quad \bar{y} = 0 \tag{3.34a}$$

$$\bar{u} = -c \quad \text{at} \quad \bar{y} = h(x, t), \tag{3.34b}$$

where barred quantities are those measured in the wave frame, $\bar{x}$ and $\bar{y}$ are rectangular coordinates along and perpendicular to the axis of the channel, u and v are velocities in those directions, respectively, and h is the amplitude of the peristaltic wave measured from the axis. Equation (3.33) is the planar analogue of eqn (2.29b).

Table 3.5 Geometry and parameter values of peristaltic flow analyses

Authors*	Geometry	*Re*	β	ϕ	Other restrictions
Shapiro (1967)	plane	0	0	arb.	none
Shapiro, Jaffrin and Weinberg (1969)	plane and axi-symm.	0	0	arb.	none
Jaffrin (1971)	plane	small	small	arb.	none
Zien and Ostrach (1970)	plane	small	small	arb.	zero mean flow
Hanin (1968)	plane	arb.	0	$\ll 1$	zero mean pressure gradient
Fung and Yih (1963)	plane	arb.	arb.	$\ll 1$	none
Yin and Fung (1969)	axi-symm.	arb.	arb.	$\ll 1$	none
Burns and Parkes (1967)	plane and axi-symm.	0	arb.	$\ll 1$	zero mean pressure gradient
Barton and Raynor (1968)	axi-symm.	0	arb.	$\ll 1$	zero mean pressure gradient

* For cited works, see Jaffrin and Shapiro (1971), from which this table is taken, by permission of Annual Reviews Inc.

The solution of eqn (3.33) satisfying eqns (3.34) is

$$\frac{\bar{u}}{c} = -1 - \frac{a^2}{2\mu c}\frac{\mathrm{d}p}{\mathrm{d}\bar{x}}\left[\left(\frac{h}{a}\right)^2 - \left(\frac{\bar{y}}{a}\right)^2\right], \tag{3.35}$$

so $\bar{u}$ is Poiseuille-like instantaneously at each cross section. With the following nondimensional quantities,

$$\xi \equiv 2\pi\frac{x}{\lambda}, \quad \eta \equiv \frac{y}{a}, \quad \tau \equiv \frac{2\pi ct}{\lambda}, \quad H = \frac{h(x,t)}{a}, \tag{3.36}$$

and the transformation back to the laboratory frame,

$$u = \bar{u} + c$$
$$x = \bar{x} + ct \tag{3.37}$$
$$y = \bar{y},$$

eqn (3.35) becomes, in the laboratory frame,

$$\frac{u}{c} = \frac{2\pi a^2}{\lambda\mu c}\left(\frac{\mathrm{d}p}{\mathrm{d}\xi}\right)\frac{(\eta^2 - H^2)}{2}. \tag{3.38}$$

We define $P_\xi = (2\pi a^2/\lambda\mu c)\ (\mathrm{d}p/\mathrm{d}\xi)$, a dimensionless pressure gradient. The nondimensional flow rate, in half of the channel, is then

$$Q(\xi,\tau) \equiv \frac{1}{2}\frac{\text{(instantaneous total flow rate)}}{ac}$$
$$\equiv \frac{1}{ac}\int_0^h u\,\mathrm{d}y = -P_\xi\frac{H^3}{3}. \tag{3.39}$$

A brief calculation gives

$$Q(\xi,\tau) = \bar{Q}(\tau) + H, \tag{3.40}$$

where $\bar{Q}(\tau)$ is the flow rate in the wave frame, which, by continuity, cannot vary with ξ. Combining eqns (3.39) and (3.40) yields

$$P_\xi = \frac{-3[\bar{Q}(\tau) + H]}{H^3}. \tag{3.41}$$

Note that when $H \ll 1$, i.e. $h \ll a$, the contractions are severe, that very large pressure gradients result. Integrating eqn (3.41) over a dimensionless tube length L yields

$$\Delta P_L(\tau) = -3\int_0^L \frac{\mathrm{d}\xi}{H^2} - 3\bar{Q}(\tau)\int_0^L \frac{\mathrm{d}\xi}{H^3}, \tag{3.42}$$

from which one can determine $\bar{Q}(\tau)$. These expressions can also be used to determine the conditions required for the flow to be steady in the wave frame. These are that H, P_ξ and $\bar{Q}$ are periodic in $(\xi - \lambda)$. With this assumption, the train of waves may be regarded as infinite for a finite-length tube if it contains an integral number of waves.

Integrating eqn (3.39) over one wavelength, assuming a wave form $H = 1 + \phi\cos(\xi - \tau)$, yields for the time-mean flow, $\langle Q\rangle$,

$$\langle Q\rangle \equiv \frac{1}{2\pi}\int_0^{2\pi} Q(\xi,\tau)\,\mathrm{d}\tau = \bar{Q} + 1. \tag{3.43}$$

Now integrating eqn (3.42), setting $L = 2\pi$, and eliminating $\bar{Q}$ using eqn (3.43), leads to

$$\langle Q\rangle = \frac{3\phi^2}{2+\phi^2} - \frac{1}{3\pi}\frac{(1-\phi^2)^{5/2}}{2+\phi^2}\Delta P_\lambda, \tag{3.44}$$

where ΔP_λ is the pressure rise per wavelength. The first term in eqn (3.44) represents the flow pumped by the peristaltic waves in the absence of a pressure gradient, the second is a back 'leakage' Poiseuille flow induced by the mean pressure gradient. Note also that the pumping effectiveness is at a maximum when there is total occlusion of the tube ($\phi = b/a = 1$), since then the leakage vanishes and the fluid is pumped by positive displacement with $\langle Q\rangle = 1$.

The pumping range, defined as the range in which ΔP_λ and $\langle Q\rangle$ are both positive, extends from $\langle Q\rangle = 0$, for which the pressure rise is, from eqn (3.44),

$$(\Delta P_\lambda)_{\max} = \frac{9\pi\phi^2}{(1-\phi^2)^{5/2}}, \tag{3.45}$$

which physically corresponds to a tube whose exit is closed off, to maximum flow, given by eqn (3.44) with $\Delta P_\lambda = 0$,

$$\langle Q\rangle_{\max} = \frac{3\phi^2}{2+\phi^2}, \tag{3.46}$$

which physically corresponds to a tube whose inlet and outlet reservoirs are at the same head. Interesting limiting cases are: $\phi \ll 1$, very small squeeze, for which eqns (3.45) and (3.46) reduce to $(\Delta P_\lambda)_{\max} \approx 9\pi\phi^2$ and $\langle Q\rangle_{\max} \approx 3\phi^2/2$; and $\phi \to 1$, near occlusion, for which $(\Delta P_\lambda)_{\max} \approx 9\pi/(1-\phi^2)^{5/2} \to \infty$, and $\langle Q\rangle_{\max} \to 1$.

From eqns (3.39) and (3.43) one can write the velocity profile in the laboratory frame as

$$\frac{u}{c} = \frac{3}{2H^3}(\langle Q\rangle + H - 1)(H^2 - \eta^2), \tag{3.47}$$

which implies that at a fixed cross section the fluid moves alternatively back and forth as the phase of the wave changes, but this tells us nothing about the time-average direction of fluid elements in this unsteady flow. A calculation of the Lagrangian trajectories (Section 2.2) of fluid particles does this, and shows that even for a zero time-mean flow, fluid particles near the axis experience a net positive displacement (that is, in the direction of the wave), while particles near the wall have a net displacement in the negative direction for small mean flows and positive for large mean flows. This net retrograde motion of fluid near the wall under certain conditions is called *reflux*. Unfortunately, while reflux seems like a plausible explanation for the retrograde motion of bacteria from the bladder to the kidneys, the conditions, under which reflux is likely probably do not occur in the healthy normal ureter.

There is another interesting consequence of the analysis, although the physiological implications are not obvious. Whereas the streamlines, in the wave frame, generally are similar to the wall shape, with decreasing amplitude as the axis is approached, under certain conditions the center streamline bifurcates and encloses a bolus of fluid; in the laboratory frame this bolus moves as a whole at the wave speed, much higher than the surrounding fluid, as if trapped by the wave.

One can obtain corrections for moderate inertia, $Re \neq 0$, and finite wavelength, $\beta \lll 1$. These modify the regimes of the parameters for which reflux and trapping occur.

Table 3.6 Characteristic values of peristaltic parameters for various applications

	Re	β	ϕ
Ureter	1	0.02	0–1(?)
Roller pumps	10	0.5	0.9–1
Gastrointestinal tract	10	2	0.3

Source: Jaffrin and Shapiro (1971), by permission of Annual Reviews Inc.

Physiological applications of peristalsis will occur in tubes rather than two-dimensional channels, as analyzed above. One reason for considering channels is that experimental data is primarily available for this geometry. Fortunately the results for an axisymmetric tube do not differ qualitatively from those for a channel. The solution methods for both geometries are also similar.

Finally we note from Table 3.6 that whereas the inertia-free ($Re = 0$), infinite-wavelength ($\beta = 0$) analysis presented above may serve as a reasonably good approximation for the ureter, it would be much less so for the gastrointestinal tract.

3.5 Ciliary and flagellar transport

Our previous discussion has concerned mainly human physiological fluid mechanics. We now turn briefly to a topic that involves human as well as other living organisms. Contractile elements of various kinds and sizes have evolved as essential or important elements of many life functions including ingestion, digestion, excretion, circulation, respiration, reproduction, and locomotion. These elements can be grouped into four classes: (i) prokaryotic (cells without nuclei) flagella; (ii) cytoplasmic filaments or microtubules; (iii) eukaryotic (cells with nuclei) cilia and flagella; and (iv) smooth and striated muscle. We shall consider only classes (i) and (iii), which have in common that the contractile elements, cilia or flagella, are slender oscillating structures that are responsible for the propulsion of the organism or propulsion of a fluid. Flagella vary in length from a few hundred down to less than ten micrometers; cilia are generally much shorter, of the order of ten to twenty micrometres. While we do not distinguish in using the term flagella between prokaryotic and eukaryotic cells, the flagella for these cells are evolutionarily unrelated, nor do they utilize the same energy source. Cilia and eukaryotic flagella, however, have essentially the same structure; both also utilize ATP (adenosine triphosphate) as their primary energy source, while that of prokaryotic cells is unknown. Examples of ciliary propulsion in the body are the upward movement of irritating foreign matter in the mucus lining the trachea and upper respiratory airways by cilia lining these passages, and cilia lining the oviduct helping to propel the ovum toward the uterus. Many simple organisms, such as bacteria, use the cilia lining their bodies both to propel themselves and to stir up their surroundings so as to enrich the nutrients in their immediate environment. Other self-propelling organisms use one or more flagella as their propulsive agents. In the human, as well as in other higher animals, flagellar propulsion is the principal mechanism for sperm transport.

The fluid mechanics of the microscopic world of cilia and flagella can be a lot different than that of our more familiar macroscopic world, in spite of apparent similarities. Thus, for example, sperm under a microscope look very much like tadpoles, and the gross similarity of their moving patterns apparent in a cursory examination of a moving film of their motions would suggest that they propel themselves in similar fashions. This is, however, far from the case, because the Reynolds numbers, the measure of the relative size of inertia to viscous forces, are so much different for the two. In the large Reynolds number world of the tadpole, propulsion is like that of all the larger animals, and little different from that of modern jet planes: conservation of momentum, the imparting of which to the surrounding fluid results in an equal and opposite momentum propelling the object. In the very small Reynolds number world of sperm or ciliated organisms or structures, little inertia can be imparted to the surrounding fluid (imagine your trying to swim in molasses or treacle!) and propulsion must occur through the utilization of viscous and pressure forces only. This is accomplished by cilia or eukaryotic flagella by the propagation of a bending wave; generally this is from the base to the tip. The cause of this bending wave is the sliding, without change of length, with respect to one another, of the nine microtubules that run along the length of the cilium or flagellum around the periphery of the cross section. (Electron micrographs show that the cross section has a total of 11 tubules, 9 around the circumference and 2 in the center; this theory of the origin of the bending wave is called the *sliding filament model.*)

Because of the exceedingly small lengths, diameters, and velocities of propulsion of cilia and flagella, an analysis of propulsion or transport by these elements involves the study of the fluid mechanics of slender bodies in Stokes or creeping, inertia-less flows (Section 2.5.4). There are two characteristic Reynolds numbers measuring the relative importance of inertia and viscous forces: a 'steady' one based on the translational velocity U, and an 'unsteady' one based on a typical frequency of beating of the cilium or flagellum, ω. Defining $Re_{\text{transl}} = UL/\nu$, and $Re_{\text{oscill}} = \omega l^2/\nu$, where L and l are characteristic lengths (not necessarily the same), and ν is the kinematic viscosity of the surrounding fluid, we find bounds on these are

$$10^{-6} \leqslant Re_{\text{transl}} \leqslant 10^{-2}, \quad Re_{\text{oscill}} \approx 10^{-3}.$$

The lower bound for Re_{transl} is for bacteria, the upper bound for spermatozoa. It is clear that these flows are inertia-less. The governing equations are then the so-called Stokes equations, reduced forms of the full Navier–Stokes equations, which are linear partial differential equations for the velocity vector and the pressure. Although the linearity of these equations makes their analysis much simpler than would be the case otherwise, it is still far beyond the level of this chapter, so we shall just broadly highlight some aspects of the analysis. The governing Stokes equations have fundamental or singular solutions which correspond to a point force or moment at any point on the slender body being considered (i.e. the cilium or flagellum). These fundamental solutions, the most important of which are called a stokeslet, a rotlet, and a potential doublet, can be superposed, by integration along the slender body, in terms of unknown source strengths. These unknown source strengths are determined by imposing the conditions that the resulting total force and moment on the slender body are zero (otherwise these essentially massless bodies would have imparted to them infinite linear and angular velocities). For a self-propelling body these conditions also determine the velocity of propulsion. This, and the solution as a whole, depend strongly on the *normal* and *tangential resistive force coefficients*, C_n and C_s, a measure of resistance to motion perpendicular and parallel to the slender body, and, in particular, to their ratio, which is given approximately by $C_s/C_n \approx 1/2$. In carrying out this analysis, account should properly also be taken of the force, or drag, of the cell or head to which the cilia or flagella are attached.

There are a great number of eukaryotic organisms that move by flagellar propulsion, and a variety of possible configurations and wave motions. Generally there are one or two flagella along which is propagated either a planar or helical wave, or some combination of the two, typically two wavelengths long. The wave is most often propagated from base to tip, with propulsion normally occurring in the direction opposite to the direction of the wave.

Even a cursory discussion of ciliary motions is more demanding than that for flagella because of the greater variety of beating patterns. Cilia are essentially short flagella, often occurring in large arrays, such as on self-propelling eukaryotic organisms or ciliated epithelia, which may beat or oscillate at different rates in a manner indistinguishable from eukaryotic flagella, or in *metachrony*, a collaborative motion with a definite, slightly advanced or retarded, phase relationship between the beats of neighboring cilia. Each cilium has a fairly regular beat pattern, consisting of a bend propagating from the base to the tip. The phase of the beat in which the cilium is moving so as to propel the organism, generally straightening out as it does so, is called the *effective stroke*. The *recovery stroke* encompasses the remainder of the beat, during which the cilium returns to its original bent configuration, and through most of which it is moving tangentially to the fluid, rather than normally, as in the effective stroke, suggesting that the cilium is taking advantage of the difference between C_n and C_s. This motion is often three-dimensional, some recovery taking place out of the plane of the effective stroke. In metachrony, the direction of wave propagation may have almost any orientation relative to the direction of the effective stroke; it is called symplectic, antiplectic, or diaplectic when the metachronal wave propagation and the effective stroke are, respectively, in the same direction, in the opposite direction, or normal to one another.

The hydrodynamic analysis of ciliary motion of eukaryotic cells is usually easier than mammalian ciliated tissues because the fluid in the former is usually Newtonian, whereas in the latter it is often highly non-Newtonian (mucus, for example).

The hydrodynamics bears a great similarity to that for flagellar motion but is complicated by the great number of closely packed cilia arrayed on a possibly complex finite-sized geometric surface and undergoing complex bending motions. To make the analysis tractable, most studies assume that the surface to which the cilia are attached is infinite and flat and that the ciliary motions are spatially and temporally periodic, forming metachronal waves. Two principal models have been employed to treat the cilia–fluid interaction: an *envelope* and a *sublayer* model. In the envelope model the cilia are assumed to be sufficiently densely packed so that the surrounding fluid effectively experiences an oscillatory, impenetrable, material surface. Solutions of the Stokes equations are then sought directly, without resorting to the integral representation of distributed fundamental solutions, with unknown source strengths. Envelope models have been used to analyze, for example, the propulsion of fluid in ciliated tubes of mammalian reproductive systems and propulsion of mucus in the respiratory airways. The sublayer models take into account the interactions between individual cilia as well as their interaction with the surrounding fluid. This is done by using the integral fundamental solution representation, necessarily taking into account the image systems of the fundamental solutions in the wall to which the cilia are attached. There have been various attempts to apply these infinite-sheet models to finite-length organisms.

As mentioned above, the difficulty in analyzing internal fluid flows where cilia are the propelling mechanisms is that the fluids are usually non-Newtonian, generally a mucus or other colloidal suspension of long-chain glycoproteins. For concentrations greater than about 1% the suspension exhibits viscoelastic, or other non-Newtonian (for example, shear-thinning) properties. At higher concentrations, gel networks form in the fluid, which break up at high-enough shear rates, the gel becoming liquid again. The mucus forms a blanket of highly viscoelastic fluid which may be either separated from the ends of the cilia by a much less viscous (serous) fluid or just barely in contact, during the effective stroke; the mechanism of propulsion is different in these two cases.

Most of the above discussion has considered flagella or cilia beating in unbounded fluids. This may not always be the case, as, for example, when spermatozoa, *in vivo*, swim near walls or in narrow passages, or are constrained, *in vitro*, between a slide and a coverslip on a microscope stage. Effects of nearby boundaries are more important for Stokes flows than for flows at large Reynolds numbers because they decay less rapidly with separation distance. Therefore they must be taken into account in such situations.

Our present understanding of the hydrodynamics of ciliary systems is still rather limited; many ciliary systems never have been analyzed at all from this point of view. This is particularly true for internal ciliary flows, which are also the ones most difficult to study experimentally.

For excellent reviews of flagellar and ciliary hydrodynamics and propulsion see Brennen and Winet (1977) and Lighthill (1976).

3.6 Other fluid mechanics problems

We now discuss briefly other fluid systems in the human body.

3.6.1 Fluid mechanics of the cochlea

Since there are chemicals that are responsive to light (for example, those which make photography possible), there was a natural path to the development of the sense of sight in higher organisms. The situation is different for hearing, where, if there are any chemical substances responsive to sound, there are certainly none that could provide the exquisite sensitivity and acuity to frequency and volume of the human ear, and those of other animals.

Figure 3.31(a) shows a schematic of the auditory system, the outer, middle and inner ear. The outer and middle ear serve primarily as an impedance-matching transducer of acoustic energy into motion of the fluid, perilymph, contained within the cochlea of the inner ear. Mechanically, the airborne acoustic signal enters the external ear and sets the eardrum into motion, in turn causing motion of the middle-ear bone. Figure 3.31(a), and 3.31(b) in an enlarged schematic, show the cochlea, which is a small fluid-filled chamber that forms part of the inner ear, and is the structure that converts the acoustic signals into neural signals. The cochlea is tapered and normally coiled into a spring; it is shown unrolled in Fig. 3.31. The cochlea is divided into the three ducts shown; the central or cochlear duct is separated from the upper duct, the vestibular canal, by a very thin membrane (Reissner's membrane, not shown in the figure) and from the lower one, the tympanic canal,

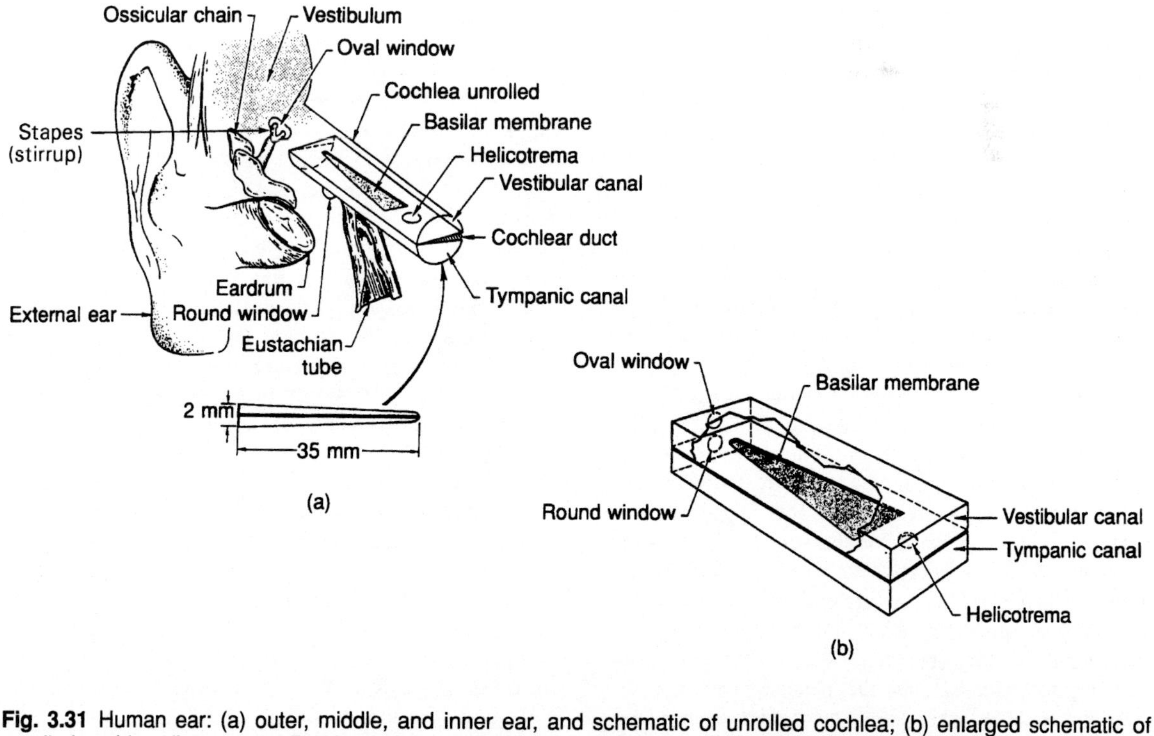

Fig. 3.31 Human ear: (a) outer, middle, and inner ear, and schematic of unrolled cochlea; (b) enlarged schematic of unrolled cochlea. (Lesser and Berkley 1972, reproduced by permission of Cambridge University Press.)

by a bony shelf and another membrane, the basilar membrane. The upper and lower ducts or canals communicate with the middle ear via the membrane-covered oval and round windows, respectively; there is also, at the apical end of the cochlea, an opening, the helicotrema, which allows fluid to pass between the upper and lower ducts. The basilar membrane is narrowest and stiffest near the oval window and widens and becomes less stiff as the apex of the cochlea is approached. Sitting on the basilar membrane is the organ of Corti, richly endowed with sensory hair cells to which are attached neurons which run to the brain.

The theory of hearing most generally accepted today is the 'place' theory, first scientifically proposed and analyzed by Helmholtz in the nineteenth century and validated by the experiments of von Bekesy early in this century. The transduction of acoustic signals into neural signals is initiated when the stapes, or stirrup, one of the middle ear bones, vibrates the oval window. This creates a wave motion in the fluid in the cochlea, which develops a sharp peak in amplitude at a particular location as it passes over the basilar membrane from the oval window toward the apex and then decays rapidly. The place on the basilar membrane where the amplitude of the wave, and therefore the displacement of the membrane, is greatest depends on the frequency, each frequency mapping onto a single point in a unique one-to-one fashion. A neural signal, an action potential, is induced in those neurons attached to the hair cells at that point on the organ of Corti where the maximum excursion of the basilar membrane occurs. The brain associates signals from different points on the basilar membrane uniquely with different frequencies. High-frequency tones are communicated to the brain from points on the basilar membrane near the oval window, where the membrane is stiffest, and low-frequency tones from points nearer the apex, where the membrane is most compliant.

There is a very large literature on cochlear mechanics, too vast and too far beyond the level of this chapter to consider. What is perhaps the most dramatic aspect of the dynamics of the cochlea, and certainly intrigues many researchers, is the sharpness of the peak amplitude of the wave. This clearly depends on the fluid mechanics of the wave motion in the ducts of the cochlea, and there have been many analyses of this flow, generally one or two-dimensional potential flow solutions. But just as important are the geometry, and the structural and rheological properties, of the basilar membrane. So, this is a problem where solid mechanicians have been as active and have made as important contributions as fluid mechanicians. We refer the interested reader to a number of comprehensive surveys, consolidations, and extensions of theoretical and experimental studies of cochlear mechanics (Inselberg 1978; Lighthill 1981; Steele 1976).

3.6.2 Fluid mechanics of the eye

A sectional view of the human eye is shown in Fig. 3.32. An early application of fluid mechanics to the human eye arose in connection with the use of indentation tonometers to measure corneal pressure for the diagnosis of glaucoma. The indentation of the cornea causes an elevation of pressure, which in turn causes fluid, the water-like aqueous humor, to flow out of the anterior chamber, between the iris and the cornea, into the canal of Schlemm, a drainage canal around the periphery of the cornea. The nature of this flow is important also from the standpoint of the pathogenesis of glaucoma, since aqueous humor is continuously secreted and forced into the anterior chamber. The volume of this chamber is very small, so the functioning of the outflow system is critical to maintaining normal pressure levels in the eye. This is accomplished by having the aqueous humor continuously drained from the anterior chamber into the trabecular meshwork through Schlemm's canal into collector channels and finally into the veins in the scleral tissue of the eye. This flow can be modeled as a low-Reynolds-number porous media problem, the analysis of which is able to predict conditions which can lead to ocular hypertension.

The cornea obtains the oxygen necessary for the metabolic maintenance of its cells directly from the air via dissolved oxygen in the tears when the eye is open, and from the aqueous humor and the capillaries on the back side of the lids when the eye is closed. Major factors determining the availability of oxygen to the cornea are the dimensions of the tear reservoir, the rate of blinking, and the efficiency of tear replacement by a blink. This question of the adequacy of oxygen delivery to the cornea is a critical aspect affecting material choice and design of a corneal contact lens. Here again, blinking is important because it is by this action that oxygen dissolved in the tears is supplied to the part of the cornea covered by the lens. The adherence and tendency for centralization of a contact lens on the corneal surface are also central to the efficacy of this prosthesis. This requires an analysis of the physical forces applied to the lens. In turn, this requires, as does the analysis of the

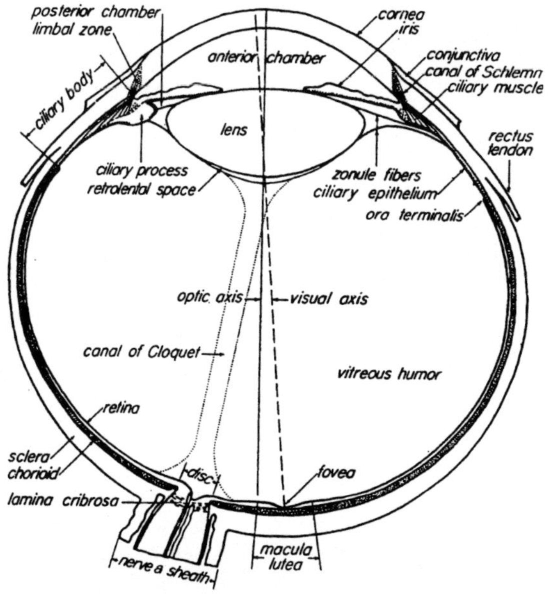

Fig. 3.32 Section of the human eye. (From Walls 1942, reproduced by permission.)

post-blink tear film, a fluid mechanical analysis. The relevant Reynolds number is very low, of the order of 10^{-3}, so the creeping flow, or even more simply, the lubrication, equations are the governing equations (Hayashi 1977). In the post-blink pulling by the upper eyelid of a tear film over the cornea, as well as in the adherence and centralization of a contact lens, surface tension and/or surface tension gradients appear to be the major driving forces (R. E. Berger 1973).

A comprehensive mathematical model for the elastic and fluid mechanical behavior of the human eye has also been carried out (Weinbaum 1965). The model simulates the aqueous flow and intraocular pressure behavior of the eye, and though conceptually simple and idealized, is still sufficiently realistic to predict correctly the overall intraocular pressure phenomena of the eye. The integrated governing differential equations yield algebraic relationships closely related to some of the widely used empirical formulae in ophthalmology. The intraocular pressure behavior of the eye is analyzed and related to the aqueous production and outflow. Transient phenomena in an externally disturbed eye, such as when a tonometer is applied, are also considered; one result is an equation for the mean intraocular pressure curve obtained in a tonographic tracing. The overall model, while exploratory, is nevertheless a good example of the results that may flow from relatively simple, but physically sound, theoretical mathematical models of complex biological systems.

3.6.3 Other body fluid systems

The above discussion hardly exhausts the important fluid systems in the human body. Some others are:

1. *Lymphatic system.* This system plays the role of a drainage system, a pathway for fluids from the interstitial spaces to reenter the circulatory system. An additional important function of the lymphatic system is to carry protein and large particulate matter, neither of which can be absorbed directly into the blood capillaries, away from the tissue spaces.
2. *Cerebrospinal fluid system.* This refers to the fluid system enclosing the brain and spinal cord, a major function of which is to cushion the brain.
3. *Pleural fluid system.* This, and the following three fluid systems, pertain to fluids within the so-called 'potential' spaces of the body. They all normally contain only a few milliliters of fluid and provide the lubrication for the cavities within which important internal organs move. (They are called potential spaces because the spaces, otherwise small, can swell to contain very large amounts of fluids.) Typically these spaces are lined with membranes that offer little resistance to the passage of fluids. The pleural cavity is that surrounding the lungs.
4. *Pericardial fluid system.* The pericardial fluid occupies the pericardial cavity within which the heart moves.
5. *Peritoneal fluid system.* The peritoneal cavity is that which surrounds the abdomen.
6. *Synovial fluid.* Synovial fluid lubricates the joint cavities and bursae.

All the fluid systems mentioned above share similar characteristics and functions. They provide crucial drainage and lubrication functions, maintenance of fluid pressure, etc. How well they perform these functions depends, among other factors, on rates of fluid production, diffusion and filtration rates, outflow paths, absorption rates, exchanges with other fluid systems, osmosis, chemical and electric potentials, physical and chemical properties of surrounding and interposing tissue, etc. Compared to the flows discussed in earlier sections, the driving forces due to these effects are small and the resulting fluid motions are much slower, inertial effects being negligible, and the flows being determined primarily by the balance between all the driving forces.

Problems

P3.1 Beginning with the general expression for conservation of mass, eqn (2.20), derive eqn (3.1) for the special case of a variable area tube filled with an incompressible fluid.

P3.2 Derive the wave equation, eqn (3.4), governing pulse-wave propagation in an elastic artery, using the procedure suggested.

P3.3. For steady laminar flow in a *rigid* circular tube

$$Q = -\frac{\pi a^4}{8\mu}\frac{dp}{dx}, \tag{1}$$

where Q is the volumetric flow rate, dp/dx is the pressure gradient, a the radius of the tube, and μ the dynamic viscosity. Since dp/dx is a constant for this flow, this shows that $Q \propto a^4$.

We now wish to consider laminar flow in an *elastic* tube, a more realistic model of a blood vessel.

(a) Assuming Q is constant, but the vessel radius a is now a function of x because of the elastic deformation, show that the pressure at any point in the vessel is given by

$$p(x) = p(0) - \frac{8\mu}{\pi}Q\int_0^x \frac{1}{[a(x)]^4}dx, \tag{2}$$

where $p(0)$ is the pressure at the entrance to the tube.

(b) Assume that the blood vessel radius obeys the linear pressure–radius relation

$$a(x) = a_0 + \alpha p/2, \tag{3}$$

where a_0 is the tube radius when the pressure p is zero, α is the compliance of the tube, and p is the pressure in the tube. (This should really be the transmural pressure, the difference in pressure between that inside and outside the tube. We are assuming zero external pressure here.) Using (1) and (3) show that for a tube of length L

$$\left(\frac{20\mu\alpha L}{\pi}\right)Q = [a(0)]^5 - [a(L)]^5, \tag{4}$$

where $a(0)$, $a(L)$ are the value of $a(x)$ at $x = 0$ and L.

Thus now $Q \propto a^5$. If the tube at its end $x = L$ is so narrowed that $a(L)/a(0) < \frac{1}{2}$, what does eqn (4) say, approximately, about the relationship between flow rate and the radius at the entrance?

(c) A more *realistic* elastic model of a blood vessel, assuming that the tube-wall material obeys Hooke's law, leads to an alternative expression to (3), of the form

$$a(x) = a_0\left(1 - \frac{a_0}{Eh}p(x)\right)^{-1}, \tag{5}$$

where a_0 has the same definition as in part (b), h is the wall thickness and E is the Young's modulus of the wall material. Show that if the elastic deformation is small; that is,

$$\frac{a_0}{Eh}p(x) \ll 1 \text{ for } 0 \le x \le L,$$

that expression (5) reduces to expression (3). What then is the relationship between α of part (b) and a_0, h, and E?

(d) Substituting eqn (5) into eqn (1) or (2) and integrating show that the pressure $p(x)$ at any point in the vessel can be found from the relationship:

$$\frac{Eh}{3a_0}\left\{\left[1 - \frac{a_0}{Eh}p(x)\right]^{-3} - \left[1 - \frac{a_0}{Eh}p(0)\right]^{-3}\right\} = \left(\frac{-8\mu}{\pi a_0^4}Q\right)x.$$

P3.4 Consider a viscous unsteady (periodic) flow in an infinitely long circular tube. Use the ideas of Dimensional Analysis (see Appendix 2.A) to show that the pressure drop over a length L is given, non-dimensionally, by

$$\frac{\Delta p}{\rho V^2} = f\left(\frac{L}{a}, \frac{\mu}{\rho V a}, \frac{a^2\omega}{\nu}\right), \tag{1}$$

where a = tube radius, μ = dynamic viscosity, ν = kinematic viscosity, ω = frequency of unsteadiness, V = average velocity in the tube, and f is an arbitrary function of its arguments. If we define

$$Re = \text{Reynolds number} = \frac{\rho V a}{\mu}$$

$$\alpha = \text{Womersley parameter} = a\sqrt{\frac{\omega}{\nu}},$$

then (1) can be written as

$$\frac{\Delta p}{\rho V^2} = f\left(\frac{L}{a}, Re, \alpha\right).$$

We see that the pressure drop of this model of arterial flow, for a given tube, depends on Re and α. Why then do you think we kept on emphasizing the dependence of this flow on α rather than Re when we discussed flow in the arteries?

P3.5 To obtain an idea of what happens to the propagation of waves in arteries at junctions consider the following simple bifurcation of a vessel into two branches. (See Fig. 3.33.) A wave travelling down the parent artery will be partially reflected and partially transmitted at the junction. The simplest analysis assumes the flow to be one-dimensional, satisfying at the junction the conditions:

(i) the pressure is single-valued,
(ii) the flow is continuous.

In a one-dimensional model of (longitudinal) wave propagation one can easily show that the flow velocity u is related to the pressure p by the expresssion

$$u = \pm\frac{1}{\rho c}p,$$

where ρ is the density of the blood, c is the speed of the wave, and the sign is + or − depending on whether the wave travels in the + or − direction of the axis of propagation. The flow rate Q in the vessel is then Au,

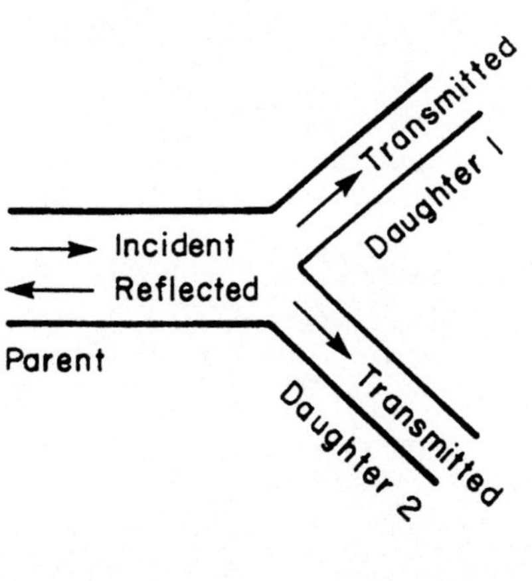

Fig. 3.33

where A is the cross-sectional area. It then follows that

$$Q = Au = \pm\frac{A}{\rho c}p = \pm\frac{1}{Z}p,$$

where $Z = \rho c/A$ is the characteristic impedance of the tube. The above relation has the form of an Ohm's law, $I = V/R$, where I = current, V = voltage, R = resistance; thus Z is analogous to a resistance, or more properly an impedance, since this flow corresponds to an ac, not a dc, circuit.

Use conditions (i) and (ii) at the junction and the expressions cited above to show that the amplitudes of the reflected and transmitted pressure waves at the junction are given in terms of the incident pressure by the expressions

$$p_r = \left\{\frac{Z_0^{-1} - (Z_1^{-1} + Z_2^{-1})}{Z_0^{-1} + (Z_1^{-1} + Z_2^{-1})}\right\}p_i = \mathrm{R}p_\mathrm{i},$$

$$p_{t_1} = p_{t_2} = \left\{\frac{2Z_0^{-1}}{Z_0^{-1} + (Z_1^{-1} + Z_2^{-1})}\right\}p_i = \mathrm{T}p_\mathrm{i},$$

where the subscripts represent: i = incident; r = reflected; $\mathrm{t}_{1,2}$ = transmitted in daughter vessels 1 and 2, respectively; 0 = parent vessel; 1 and 2 = daughter vessels 1 and 2, respectively. R and T denote the reflection and transmission coefficients.

The rate of working by such a wave is given by

$$\dot{W} = \mathrm{p}Au = pQ = \frac{p^2}{Z}.$$

Use this to show that the ratio of energy transmission of the reflected wave to that of the incident wave is R^2, so R^2 is the energy reflection coefficient. Show, similarly, that the rate of energy transfer in the two transmitted waves compared with that in the incident wave, is

$$\left(\frac{Z_1^{-1} + Z_2^{-1}}{Z_0^{-1}}\right)T^2,$$

which is called the energy transmission coefficient.

P3.6 The diameter of the aorta is approximately 2.5 cm or 1 in. This and subsequent blood vessels must bifurcate a sufficient number of times so as to provide nourishment to and carry away waste products from every part of the body. The number of bifurcations is about 28 to 30. In each bifurcation the ratio of total daughter to parent area is about 1.2. The number of generations could be smaller if the ratio of total daughter vessel area to parent vessel area were larger than 1.2. Can you think of a fluid-mechanical reason why this ratio cannot be much bigger than 1.2? (*Hint*: Consider Bernoulli's equation and how the pressure changes when the area increases, and the effect of such a pressure change on flow separation at bifurcations in the larger vessels of the circulation.)

P3.7 Consider the flow in a slighly compliant, cylindrical, infinitely long tube, as a model of a blood vessel. If the flow is unsteady the equation analogous to eqn (2.33) (governing the steady Poiseuille flow in a rigid tube) is

$$-\frac{\partial p}{\partial x} = \frac{\rho}{A}\frac{\partial Q}{\partial t} + \frac{8\mu}{\pi r_0^4} Q, \qquad (1)$$

where $r_0 = d/2$, the radius of the tube, A is the total cross-sectional area, and Q is the flow rate.

Consider a small segment of such a tube. Carry out a one-dimensional mass balance by equating the difference between inflow and outflow of fluid into this segment to the sum of the storage of fluid in the segment (due to its distensibility) and the leakage flow which escapes from the segment through lateral branches. Obtain the resulting continuity equation in the form

$$-\frac{\partial Q}{\partial x} = \frac{dA}{dp}\frac{\partial p}{\partial t} + W'p, \qquad (2)$$

where dA/dp is the compliance of the tube, and W' is the leakage per unit length.

Signal transmission through a uniform cable is governed by the following equations, the so-called telegraph equations,

$$-\frac{\partial V}{\partial x} = L'\frac{\partial I}{\partial t} + R'I, \qquad (3)$$

$$-\frac{\partial I}{\partial x} = C'\frac{\partial V}{\partial t} + G'V, \qquad (4)$$

where V is voltage, I is current, R' is resistance, L' is inductance, C' is capacitance, and G' conductance, all per unit length. The transmission line is schematically shown below. Making an analogy between (1) and (3) and between (2) and (4) show that there immediately follow the correspondances given in Table 3.3.

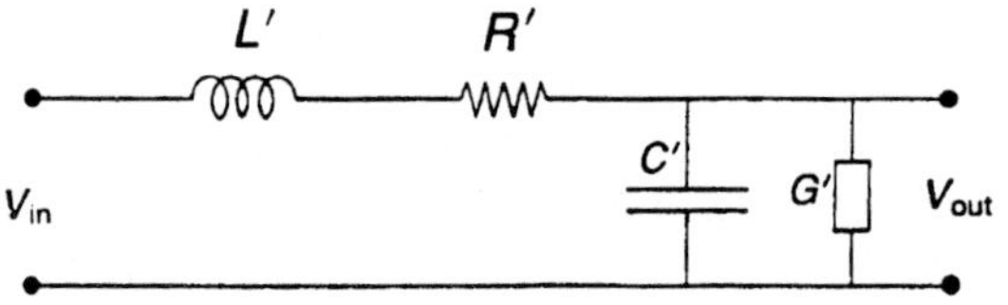

Fig. 3.34

P3.8 Expression (3.22) for the torque in a coaxial cylinder type viscometer is valid for arbitrary values of the radii of the coaxial cylinders, R_1, R_2. Generally these instruments are built with a narrow gap (that is, $R_2 \approx R_1$), so as to minimize specimen size and also to approximate as nearly as possible the flow between infinite parallel plates in relative translational motion, Couette flow, for which the rate-of-strain is constant. Expand expression (3.22) for small gap, $(R_2 - R_1)/R_1 = \varepsilon \ll 1$, and show that this does reduce (3.22) to the expression for Couette flow.

P3.9 From elementary principles derive expression (3.23) for the viscosity of a fluid in a cone-plate type of viscometer.

P3.10 Consider an ideal Bingham plastic, for which

$$\tau = \tau_y + \mu D, \quad \text{for } \tau > \tau_y$$

$$D = 0, \quad \text{for } \tau < \tau_y,$$

where τ_y is the yield stress and D is the rate of shear. Use these expressions and the theory of Section 2.5.1 to analyze the laminar flow of such a fluid in a circular tube. Since there is no flow in the tube if $\tau_w < \tau_y$, assume that $\tau_w > \tau_y$. Calculate the velocity profile across the cross section. (*Hint*: It consists of two regions, a variable-velocity annular region surrounding a central core of uniform velocity.) Show that the average velocity across the entire tube cross section is

$$u_{av} = \frac{\tau_w a}{4\mu}\left[1 - \frac{4\tau_y}{3\tau_w} + \frac{1}{3}\left(\frac{\tau_y}{\tau_w}\right)^4\right],$$

where τ_w is the shear at the tube wall. (This expression is called the Buckingham–Reiner equation.) Plot the velocity profiles for $\tau_y/\tau_w = 0$, 0.25, 0.5, 0.75, and 1.0.

P3.11 In analyzing the flow of power-law fluids it is convenient to use, instead of $\tau = \mu D^s = \mu(du/dy)^s$, when there is only one shear stress component, the expression

$$\tau = \mu\left|\frac{du}{dy}\right|^{\alpha-1}\frac{du}{dy}.$$

For $\alpha = 1$ the fluid is Newtonian and the constant μ is the ordinary Newtonian viscosity. If $\alpha < 1$, the shear stress increases less than linearly with the rate of shear and the fluid is called a pseudoplastic; if $\alpha > 1$, so the shear stress increases more than linearly with the rate of shear, the fluid is called a dilatant fluid.

Using the theory of Section 2.5.1 with the expression above, obtain the velocity profile for laminar flow of a power law fluid in a circular tube. Calculate and plot the profiles for $\alpha = 0$ (plug flow), $\frac{1}{2}$, 1 (Newtonian), 2, and ∞. What can you conclude about the shapes of the profiles of pseudoplastic and dilatant fluids compared to a Newtonian fluid?

P3.12 In the section on flow in the arterioles we discussed the two-fluid shearing core model, wherein an axial core of fluid of one viscosity is surrounded by an annulus of a fluid of a different (lower) viscosity. The use of a thin outer layer of a less viscous fluid, such as water, to decrease the power to pump a more viscous fluid or suspension, such as crude oil or a slurry, through a long pipe is an old and widely implemented idea in industry.

Assume fluid 2 occupies the region from $r = 0$ to $r = \zeta a$, where a is the radius of the tube and $\zeta < 1$, and fluid 1 from $r = \zeta a$ to $r = a$. Integrating the general expression (2.29b) for the regions containing fluids 1 and 2, and requiring that the velocity be continuous at $r = \zeta a$, obtain the velocity profiles in these two fluid regions.

Show that the ratio of the rate of volume flow of fluid 2 to that in the absence of fluid 1 (that is, $\zeta = 1$) is

$$\frac{Q_2}{Q_{20}} = \zeta^2\left(\zeta^2 + 2\frac{1-\zeta^2}{\xi}\right),$$

where $\xi = \mu_1/\mu_2$. From this last expression show that the maximum of Q_2/Q_{20} occurs when $\zeta^2 = (2-\xi)^{-1}$ and that this maximum value is given by

$$\left(\frac{Q_2}{Q_{20}}\right)_{\max} = \frac{1}{\xi(2-\xi)}.$$

P3.13 Using the procedure suggested, derive eqn (3.30), expressing the finite-summation correction for flow in narrow tubes filled with finite-sized particles.

P3.14 Consider the flow of red blood cells moving single file down a capillary. The cells are to be approximated as rigid disks.

h = width of RBC
d = diameter of RBC
L = separation between RBCs
V = average velocity of cells and plasma
ε = thickness of thin layer of plasma separating RBC and capillary wall.

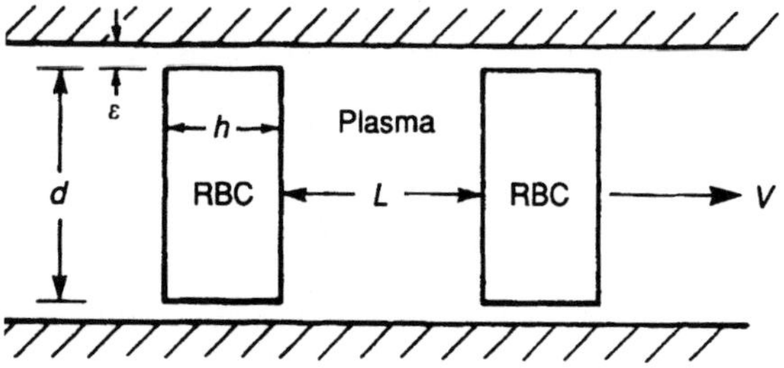

Fig. 3.35

We wish to calculate the pressure drop across the capillary. Do this by considering as a single basic unit one RBC and the plasma region separating it from the next RBC (see above). Assume that the total pressure drop across this unit is the sum of (i) the Poiseuille pressure drop across the plasma region, plus (ii) that in the thin layer between the RBC and the capillary. Calculate (i) from eqn (2.34) and (ii) from the Couette flow formula, eqn (2.3).

Assume the velocity of the RBC and plasma are the same, and $L \approx d$ and $\varepsilon/d \ll 1$. Show that the ratio of (ii) to (i) is $h/8\varepsilon$.

Calculate this ratio for a typical RBC ($h \approx 2$ μm) and a typical flow case with $\varepsilon = 0.1$ μm. *What* can you conclude about the relative sizes of the pressure drop due to the plasma region between cells compared to that due to the 'lubricating' layer between the periphery of the RBC and the capillary wall?

P3.15 Consider the schematic of the pulmonary circulation shown in Fig. 3.36.

Data: (i) gage pressure (pressure above atmospheric) in the right ventricle is 2×10^3 N m^{-2} (15 mmHg or 20 cmH_2O);

(ii) lung has a height of 30 cm, and pulmonary aorta enters the lung about half-way up;

(iii) there is a pressure drop of 0.7×10^3 N m^{-2} in the arteries between the pulmonary aorta and the capillaries.

(a) Show that the hydrostatic gradient of pressure in the pulmonary arteries is about 0.1×10^3 N m^{-2} per centimeter vertical distance.

(b) Use the result of part (a) plus the given data to calculate the pressure at the entrance to thc capillaries at the very top, the very bottom, and half-way up the lung (that is, at the level of the right ventricle).

(c) Assuming dimensions and velocity of flow in the

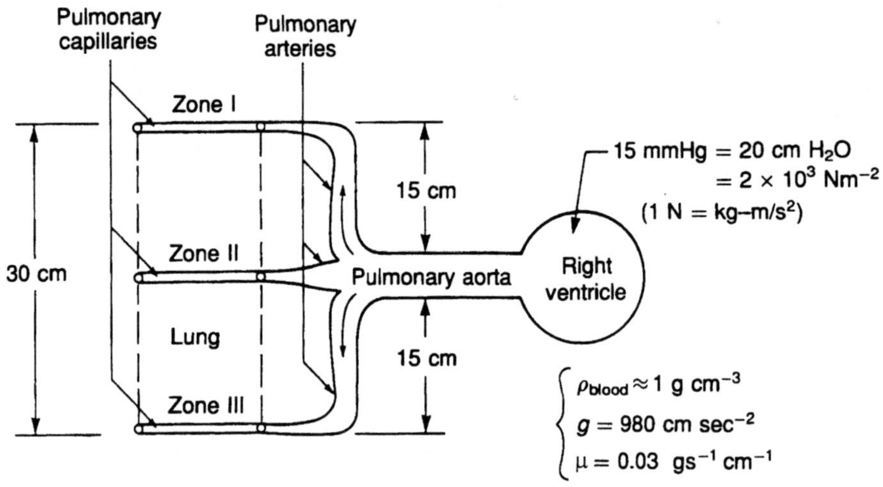

Fig. 3.36

pulmonary capillaries shown in Fig. 3.37 use the Poiseuille flow expression $u_{\text{mean}} = a^2/8\mu(-\mathrm{d}p/\mathrm{d}x)$ to calculate the pressure drop across the capillary.

(**d**) Assuming the actual pressure drop (since the flow in the capillary is *not* a Poiseuille flow) is 5 times what you calculated in part (c), calculate the pressure that exists at the exit of the capillaries at the very top, very bottom and half-way up the lung.

(**e**) If you've done parts (b) and (d) correctly you should have found that at the very top of the lung the pressure at the entrance is subatmospheric, whereas at the bottom and at the center of the lung the pressure is above atmospheric even at the exit. Assuming for any pulmonary capillary (see Fig. 3.38) that the capillary will close if the external pressure, here the alveolar pressure, which is equal to atmospheric pressure, exceeds the internal pressure at any point of the capillary, what can you conclude about capillaries being closed or open at the top and bottom of the lung? (These are zones I and III in Fig. 3.36 at the beginning of this problem.) Do you expect there to be a region where the capillaries are only partially closed, and if so, why?

(**f**) Calculate the location of the highest point in the lung where the capillaries are fully open along their entire length.

P3.16 Calculate approximately the power, or rate of work, of the heart under resting conditions (the left ventricle only). (You can ignore the kinetic energy imparted to the blood as it is ejected.) Express your answer in watts and horsepower. How many human hearts equal one horsepower? (*Note*: 1 mmHg pressure = 1.33×10^3 dynes cm^{-2}.)

The mechanical efficiency of the heart is very low, of the order of 10 per cent. This makes the heart's resting metabolic rate about 10 times what you calculated above. If a normal individual consumes about 2000 kcal per day what percentage of that is therefore used to 'power' the heart?

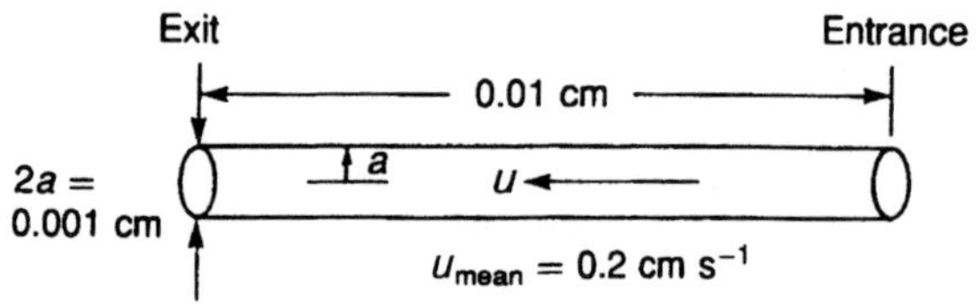

Fig. 3.37

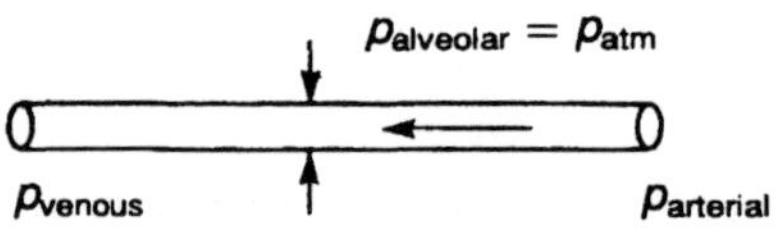

Fig. 3.38

References

Bergel, D. H. (ed.) (1972). *Cardiovascular fluid dynamics*, Vols. 1 and 2. Academic Press, New York.

Berger, R. E. (1973). Pre-corneal tear film mechanics and the contact lens. Unpublished Ph.D. thesis. Johns Hopkins University, Baltimore, MD.

Berger, S. A. (1993). Flow in large blood vessels. In *Contemporary mathematics* (ed. A. Y. Cheer and C. P. van Dam), Vol. 141, pp. 479–518. American Mathematical Society, Providence, RI.

Brennen, C. and Winet, H. (1977). Fluid mechanics of propulsion by cilia and flagella. *Annual Review of Fluid Mechanics*, **9**, 339–98.

Burton, A. C. (1965). *Physiology and biophysics of the circulation*. Year Book Medical Publishers, Chicago.

Caro, C. G., Pedley, T. J., Schroter, R. C., and Seed, W. A. (1978). *The mechanics of the circulation*. Oxford University Press.

Fung, Y. C. (1984). *Biodynamics: circulation*. Springer-Verlag, New York.

Goldsmith, H. L. (1971). Red cell motions and wall interactions in tube flow. *Proceedings of the Federation of American Societies of Experimental Biology*, **30**, 1578–88.

Goldsmith, H. L. (1972). The microrheology of human erythrocyte suspensions. In *Theoretical and applied mechanics*, Proc. 13th IUTAM Congress, (ed. E. Becker and G. K. Mikhailov), pp. 85–103. Springer, New York.

Guyton, A. C. (1977). *Basic human physiology*, (2nd edn). W. B. Saunders, Philadelphia.

Guyton, A. C. (1985). *Anatomy and physiology*. Saunders College Publishing, Philadelphia.

Hayashi, T. T. (1977). Mechanics of contact lens motion. Unpublished Ph.D. thesis. University of California, Berkeley, CA.

Inselberg, A. (1978). Cochlear dynamics: the evolution of a mathematical model. *Society of Industrial and Applied Mathematics Review*, **20**, 301–51.

Jaffrin, M. Y. and Shapiro, A. H. (1971). Peristaltic pumping. *Annual Review of Fluid Mechanics*, **3**, 13–36.

Lesser, M. B. and Berkley, D. A. (1972). Fluid mechanics of the cochlea. Part 1. *Journal of Fluid Mechanics*, **51**, 497–512.

Lighthill, M. J. (1976). Flagellar hydrodynamics—the John von Neumann lecture 1975. *Society of Industrial and Applied Mathematics Review*, **18**, 161–230.

Lighthill, M. J. (1981). Energy flow in the cochlea. *Journal of Fluid Mechanics*, **106**, 149–213.

McDonald, D. A. (1974). *Blood flow in arteries*, (2nd edn). Williams & Wilkins, Baltimore.

Miller, W. S. (1947). *The lung*, (2nd edn.) Charles C. Thomas, Publisher, Springfield, Illinois.

Noordergraaf, A. (1978). *Circulatory system dynamics*. Academic Press, New York.

Pedley, T. J. (1977). Pulmonary fluid dynamics. *Annual Review of Fluid Mechanics*, **9**, 229–74.

Pedley, T. J. (1980). *The fluid mechanics of large blood vessels*. Cambridge University Press.

Schroter, R. C. and Sudlow, M. F. (1969), Flow patterns in models of the human bronchial airways. *Respiratory Physiology*, **7**, 341–55.

Shapiro, A. H. (1977). Steady flow in collapsible tubes. *Transactions ASME, Journal of Biomechanical Engineering*, **99(K)**, 126–47.

Steele, C. R. (1976). Cochlear mechanics. In *Handbook of sensory physiology*, Vol. V, *Auditory system*, Part 3, *Clinical and special topics*. Springer-Verlag, Berlin.

Walls, G. L. (1942). *The vertebrate eye*. Cranbrook Institute of Science, Bloomfield Hills, MI.

Weinbaum, S. (1965). A mathematical model for the elastic and fluid mechanical behavior of the human eye. *Bulletin of Mathematical Biophysics*, **27**, 325–54.

West, J. B. (ed.) (1977). *Bioengineering aspects of the lung*. Dekker, New York.

Whitmore, R. L. (1968). *Rheology of the circulation*. Pergamon Press, Oxford.

Wiedman, M. P., Tuma, R. F. and Mayrovitz, H. N. (1981). *An introduction to microcirculation*. Academic Press, New York.

Womersley, J. R. (1957). *An elastic tube theory of pulse transmission and oscillatory flow in mammalian arteries*. WADC Technical Report TR-56-614. Wright Air Development Center, Dayton, OH.

Zweifach, B. W. (1950). Basic mechanisms in peripheral vascular hemostasis. In *Transactions of the Third Conference on Factors Regulating Blood Pressure*. pp. 13–52. Josiah Macy Foundation, New York. ...

Further reading

An overall view of biomechanics, the proceedings of a symposium on this topic, can be found in:

Fung, Y. C., Perrone, N., and Anliker, M. (ed.) (1972). *Biomechanics: its foundations and objectives*. Prentice-Hall, Englewood Cliffs, NJ.

An extremely comprehensive handbook covering all aspects of bioengineering is:

Skalak, R. and Chien, S. (1987). *Handbook of bioengineering*. McGraw-Hill, New York.

Older but still useful books on the circulation, the former also on respiration, are:

Wolstenholme, G. E. W. and Knight, J. (ed.) (1969). *Circulatory and respiratory mass transport*. Churchill, London.
Burton, A. C. (1965). *Physiology and biophysics of the circulation*. Year Book Medical Publishers, Chicago.

The mechanical properties of living tissues affect almost every aspect of bioengineering. We have seen how important the properties of blood and blood vessel walls are in determining flow in the circulatory system. An excellent source of information and mathematical treatment about these topics, as well as bone, muscle, and other living tissues can be found in:

Fung, Y. C. (1993). *Biomechanics: mechanical properties of living tissues*, (2nd edn). Springer-Verlag, New York.

The general application of the principles of mechanics to the study of living things, including a large amount of material dealing with biofluiddynamics, can be found in this same author's excellent third book on biomechanics.

Fung, Y. C. (1990). *Biomechanics: motion, flow, stress, and growth*. Springer-Verlag, New York.

The fluid dynamics of the circulation, respiration, and a number of other topics are discussed, with the author's characteristic flair and enormous insight, in:

Lighthill, Sir James. (1975). *Mathematical biofluiddynamics*. Society for Industrial and Applied Mathematics, Philadelphia.

Hemodynamics in all its aspects is discussed at great length in:

Milnor, W. R. (1989). *Hemodynamics*, (2nd edn). Williams & Wilkins, Baltimore.

MASS TRANSFER

Michael C. Williams

4

Contents

Symbols

a_i — activity of component i in mixture, related to mole fraction $(= \gamma_i x_i)$

A, B — components of binary mixture (1 = A, 2 = B)

AK — artificial kidney

b — breadth of membrane in planar-surface AK

c — concentration, moles/vol, for entire mixture $\left(= \sum_i c_i\right)$, or for component in a mixture (c_i), or for pure substance

C_p — specific heat capacity at constant pressure, enthalpy/mass–degree

D — diameter of circular channel (for example, capillary)

D_{ij} — diffusion coefficients in multicomponent mixtures (eqn (4.7))

$\mathscr{D}_{ij}$ — diffusion coefficients in binary mixtures (i = A, j = B) (eqn (4.9))

$\boldsymbol{e}$ — surface normal unit vector ($\boldsymbol{e}_x$ normal to surface x = constant)

$\boldsymbol{f}$ — external body force/mass, either $\boldsymbol{g}$ or $\boldsymbol{f}_i$ acting uniquely on component i in a mixture

$\boldsymbol{g}$ — gravitational acceleration vector

G — Gibbs free energy of a mixture (extensive)

h — thickness of narrow planar channel

$\boldsymbol{j}_i$, $\boldsymbol{J}_i$ — diffusion flux of component i in a mixture, mass/area–time (see $\boldsymbol{j}_i$, Table 4.2) or moles/area–time ($\boldsymbol{J}_i$, see Table 4.2). For $\boldsymbol{j}_i^*$ and $\boldsymbol{J}_i^*$, see Table 4.2.

j_M, j_H — jay-factors for mass or heat transfer across an interface (eqn (4.9) *et seq.*)

k_n — reaction rate coefficient ($n = f$ or r)

k_{ix}, k_{ic} — mass transfer coefficient for transport of component i between phases, defined for molar concentration difference (Δx_i or Δc_i) as driving force; see eqn (4.17). Subscripts dropped when obvious, to simplify notation.

$k_{i\omega}$, $k_{i\rho}$ — mass transfer coefficient for transport of component i between phases, defined for mass concentration difference ($\Delta\omega_i$ or $\Delta\rho_i$) as driving force; relationships analogous to eqn (4.17) apply

K, K_x — overall mass transfer coefficient for transport across several resistances (general), or its value for transport of component X

K_{eq} equilibrium constant for a reacting mixture, representative of the final bulk concentration of all components at a given temperature and pressure see eqn (4.5b)

L length of a flow channel

m total mass within a specified volume (m_i is mass of component i)

$\mathcal{M}$ total moles within a specified volume ($\mathcal{M}_i$ is moles of component i)

M average molecular weight of a mixture $\left(= \rho/c = \sum_i x_i M_i\right)$

M_i molecular weight of component i, mass/mole

$\boldsymbol{n}_i$, $\boldsymbol{N}_i$ total flux of component i in a mixture, mass/area–time ($\boldsymbol{n}_i$, see Table 4.2) or moles/area–time ($\boldsymbol{N}_i$, see Table 4.2)

p pressure

Pr Prandtl number

Q volumetric flow rate

$\boldsymbol{r}$ position vector (e.g., $\boldsymbol{e}_x x + \boldsymbol{e}_y y + \boldsymbol{e}_z z$)

R resistance to mass transfer, overall; also the contributions of separate phases (e.g. R_B) to the total

$\boldsymbol{R}$ universal gas constant

r_i, R_i reaction rate by which component i is produced, mass/volume–time (r_i) or moles/volume–time (R_i), with $r_i = M_i R_i$

$\mathcal{R}_i$ total rate of chemical production of component i in a given volume, moles/time

Re Reynolds number; depends on geometry (for capillary flow, $Re = \rho\langle v\rangle D/\mu$)

S surface area of a body, scalar (ds = differential surface area, e.g. $dx\, dy$)

Sc Schmidt number ($= \mu/\rho\mathcal{D}_{AB}$ in binary solutions), a purely material property representing the relative effectiveness of the material for transmitting linear momentum (μ/ρ) and mass ($\mathcal{D}_{AB}$) by diffusive mechanisms

Sh Sherwood number; depends on geometry (for binary mass transfer in a capillary, $Sh = k_{Ax}D/c\mathcal{D}_{AB}$)

t time

T temperature (absolute)

$\boldsymbol{v}$, $\boldsymbol{v}^*$ local velocity of the bulk fluid motion; for solutions, both represent averages over mixture components: mass average $\boldsymbol{v}(\boldsymbol{r}) = \sum_i \omega_i \boldsymbol{v}_i$ or molar average $\boldsymbol{v}^*(\boldsymbol{r}) = \sum_i x_i \boldsymbol{v}_i$

$\boldsymbol{v}_i$ local velocity vector of component i in a mixture, measured relative to fixed laboratory coordinates

V volume of a body (dv = differential volume, e.g. $dx\, dy\, dz$)

w, W bulk material flow rate, mass/time (w) or moles/time (W)

x rectangular coordinate (not subscripted), in flow direction if appropriate

x_i mole fraction of component i in a mixture ($= c_i/c$)

y rectangular coordinate, in direction of velocity or concentration gradient if appropriate

z rectangular coordinate

α_{jk} coefficient in eqn (4.7)

γ_j activity coefficient of component j in a mixture

δ thickness of membrane or other thin layer

Δ difference between values of a quantity evaluated at two locations (for example, $\Delta c = c_2 - c_1$)

ζ_j mobility of component j (see eqn (4.10) and associated footnote)

κ thermal conductivity

μ viscosity

ν number of chemical components in a mixture

ρ mass concentration of bulk material $\left(= \sum_i \rho_i\right)$, mass/vol

$\sum_j$ sum, over components j ($j = 1, 2, \ldots \nu$)

ϕ interface equilibrium coefficient

ω_j weight (mass) fraction of component j in a mixture ($= \rho_j/\rho$)

∇ spatial derivative operator, vector $\left(= \boldsymbol{e}_x \frac{\partial}{\partial x} + \boldsymbol{e}_y \frac{\partial}{\partial y} + \boldsymbol{e}_z \frac{\partial}{\partial z}\text{, if in rectangular coordinates}\right)$

$\langle\ \rangle$ spatial average, taken over a surface (usually a cross-sectional surface in space, through which fluid is flowing)

$(\hat{\ })$ dimensionless quantity

$(\bar{\ })$ partial molar quantity; e.g. for component j, $= [\partial(\ )/\partial N_j]_{i = 1,2, \ldots \nu \neq j}$

$(\bar{\ })$ bulk values (of concentration), far from a wall

Superscripts

o limiting case for concentration span, either for pure component ($x_j \to 1$) or dilute component ($x_j \to 0$), as is clear in context

$^\circ$ quantity having constant value (uniform in space and time)

* molar-average quantity for a mixture, or corresponding component quantity $(\ )_j^*$ measured relative to $\boldsymbol{v}^*$

f related to the forward rate of reaction (reactants → products)

g gas-phase quantity at interface

i interface value

r related to the reverse rate of reaction (products → reactants)

s solid-phase quantity at interface

Subscripts

b body

B blood

c capillary

d drug component

D depletion zone, in a drug delivery device

D dialysate

f forward, for chemical reaction rate coefficient

H hydrodynamic region, in fluids of body tissue just outside a drug delivery device

m	matrix	S	saturation level for the concentration of a component in a mixture
M	membrane	t	tissue
r	reverse, for chemical reaction rate coefficient		

4.1 Introduction

The human body might be envisioned as a chemical reactor of great complexity. Even the simplest cell in the body could be characterized in this way. As in industrial reactors, the body or the cell has inlet streams composed of chemical reactants (carbohydrates, proteins, fats, salts, oxygen, etc. in this case) and one or more outlet streams composed of the products and byproducts of the reactions (carbon dioxide, water, urea, other salts, etc.). Unlike most industrial reactors, these biological reactors operate nearly isothermally. The heat evolved in biological reactions (metabolism, for example) escapes rapidly from its site of generation by a diffusive process known as thermal conduction. Then, it is immediately picked up by flowing body fluids and distributed by the convection process to all parts of the body, including to the skin and other heat-discharge locations, thus keeping the body temperature everywhere at about 37°C.

This heat transfer process of diffusion and convection has an analogy in mass transfer, which also proceeds by diffusion and convection. The inlet and outlet chemical 'streams' mentioned above must penetrate cell walls and cross regions of cytoplasm by diffusion, and the longer-distance transport is handled by convection in flowing blood, lymph, urine, and air. With these biological operations in mind, we will focus in this chapter on the two fundamental mass transport processes of diffusion and convection, with equal emphasis on understanding the processes and applying our understanding to the design of medical devices and artificial organs.

Engineers have used their understanding of mass transfer operations, for many years, to help medical practitioners explain how body organs function and also to invent, produce, and maintain a variety of medical devices. There is a rich literature on this subject, at a range of levels of sophistication. The books by Seagrave (1971) and by Lightfoot (1974) were among the earliest to direct a chemical engineering treatment of mass transfer toward problems of biomedical character, and they are still highly recommended today.

4.2 Mass transfer fundamentals

4.2.1 Conservation of mass

Underlying all mass transfer modeling is the basic concept of mass conservation, about which no model is necessary. This conservation principle is valid at the macroscopic level, where engineers refer to 'material balances', and also at the microscopic level, where we need the power of differential equations. In the latter case, we obtain the so-called equation of continuity, which often yields important information directly. However, the greatest utility for the differential equation approach is for solving problems related to diffusion, which requires coupling the conservation-of-mass equation with another equation describing the physics of diffusion. While the macroscopic material balances are certainly fundamental and useful, we will bypass them for the moment because of our larger goal to solve diffusion problems.

Therefore, we begin here with a derivation of the equation of continuity for a multicomponent chemical system composed of ν species. First, we address the conservation of one of these, say the ith one. (For example, in blood we could identify i as albumin and ν as a very large number, with the component in greatest concentration being water.) The vector velocity of component i at position $\boldsymbol{r}$ is $\boldsymbol{v}_i$, while the mass density of i at $\boldsymbol{r}$ is ρ_i. The mass *flux* of i, designated by $\boldsymbol{n}_i$, is given as $\boldsymbol{n}_i = \rho_i \boldsymbol{v}_i$, so the flux (an 'intensity' of transport, being in units of kg s^{-1} m^{-2}) always bears the same mass units as density and has the vector direction of the velocity.

The flux concept is illustrated in Fig. 4.1, where an arbitrary volume in space (V, the 'control volume') is fixed at $\boldsymbol{r}$, relative to some arbitrary origin of laboratory coordinates. The control volume has an arbitrary surface, S, which is also fixed. This means that the local surface orientation vector—a unit *outward* vector, $\boldsymbol{e}$—is constant too. At any point on S, the local flux of i in space is $\boldsymbol{n}_i$ but not all of $\boldsymbol{n}_i$ is penetrating S. Only the portion of $\boldsymbol{n}_i$ that is perpendicular to S is crossing that boundary; the component of $\boldsymbol{n}_i$ tangent to S is moving away in space. In symbolic fashion, we represent the penetration component by $-\boldsymbol{e} \cdot \boldsymbol{n}_i$ (the minus sign is required because $\boldsymbol{e}$ is defined to have an orientation opposite to the penetration direction).

A flux-related concept is the *rate* of transport (say, in kg s^{-1}), which is a scalar designated by w_i. A rate is always given by a flux times an area; in differential form (see Fig. 4.1) where the surface differential area is the vector $\boldsymbol{ds} = \boldsymbol{e}\,\mathrm{d}s$, the differential rate of transport *into* the control volume at point $\boldsymbol{r}$ is represented by $\mathrm{d}w_i = (-\boldsymbol{ds}) \cdot \boldsymbol{n}_i = (-\mathrm{d}s\,\boldsymbol{e} \cdot \boldsymbol{n}_i)$.

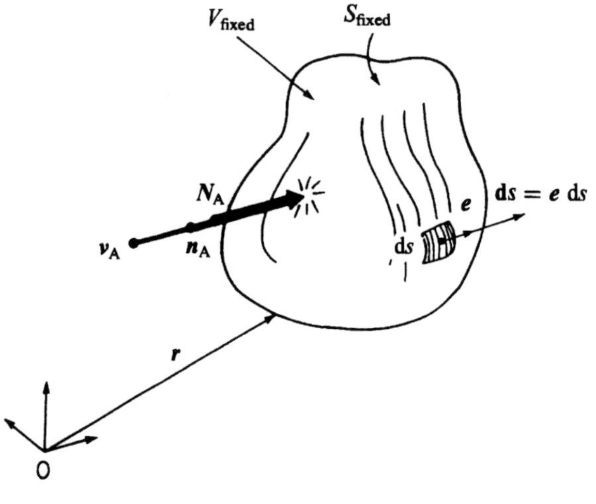

Fig. 4.1 Control volume, fixed at position $\boldsymbol{r}$ and having fixed volume V and surface S, is penetrated by a flow of component A with velocity $\boldsymbol{v}_A$. Colinear with the velocity is the flux of A, $\boldsymbol{n}_A = \rho_A \boldsymbol{v}_A$ (mass flux) and $\boldsymbol{N}_A = c_A \boldsymbol{v}_A$ (molar flux). The outward-normal surface unit vector is $\boldsymbol{e}$, so a differential surface of magnitude ds corresponds to a vector d$\boldsymbol{s}$ = $\boldsymbol{e}$ ds.

The last concept needed in the derivation involves chemical reaction. While such reactions cannot create or destroy mass, they can change its chemical identity (i.e. the label i is changed to j). Thus, we need to account for such conversions by using the scalar reaction rate (per unit volume) by which component i is *created*, namely r_i. We know that r_i depends in general on the concentrations of all reactants (ρ_i and ρ_k and ρ ...) and also on any catalysts that may be present (e.g. enzymes), but for now this concentration dependence need not concern us.

From these elementary concepts, the conservation of mass in V can be stated in terms of the *rates* (mass/time):

Rate of i-mass *accumulating* in V	=	Net rate of i-mass being transported *in* by flux	+	Net rate of i-mass being *created* by chemistry

$$\frac{\mathrm{d}}{\mathrm{d}t}\left(\iiint_V \rho_i \,\mathrm{d}v\right) = \iint_S (-\mathrm{d}s\,\boldsymbol{e}) \cdot \boldsymbol{n}_i + \iiint_V r_i \,\mathrm{d}v \qquad (4.1)$$

We now show that all three terms in eqn (4.1) can be grouped within the same integrand. In the left-hand term, we can move the d/dt operator inside the integral because V is fixed—i.e., not $V(t)$. The flux term, involving a surface integral, can be converted to a volume integral by using the Gauss divergence theorem (totally rigorous)[1]. Then, all terms can be grouped as

$$\iiint_V \left[\frac{\partial \rho_i}{\partial t} + \nabla \cdot \boldsymbol{n}_i - r_i\right] \mathrm{d}v = 0. \qquad (4.2)$$

Since V was chosen arbitrarily, it is unlikely that the integral limits will just happen to give us a zero-valued integration. In general, the integrand must be zero, which gives us directly the equation of continuity for component i. Table 4.1 displays this equation in three coordinate systems, as well as the analogous form on a *molar* basis involving molar concentration c_i, molar flux $N_i = c_i \boldsymbol{v}_i$ and molar reaction rate (per unit volume) R_i.

[1] $\iint_S \mathbf{ds} \cdot \boldsymbol{P} = \iiint_V (\nabla \cdot \boldsymbol{P})\,\mathrm{d}v$, where $\boldsymbol{P}$ can be any vector or tensor (or scalar, if the dot multiplication is removed).

Table 4.1. The equation of continuity

For component i in multicomponent system (mass basis)

Rectangular coordinates:

$$\frac{\partial \rho_i}{\partial t} + \left[\frac{\partial n_{ix}}{\partial x} + \frac{\partial n_{iy}}{\partial y} + \frac{\partial n_{iz}}{\partial z}\right] = r_i \qquad (4.1A)$$

Cylindrical coordinates:

$$\frac{\partial \rho_i}{\partial t} + \left(\frac{1}{r}\frac{\partial}{\partial r}(rn_{ir}) + \frac{1}{r}\frac{\partial n_{i\theta}}{\partial \theta} + \frac{\partial n_{iz}}{\partial z}\right) = r_i \qquad (4.1B)$$

Spherical coordinates:

$$\frac{\partial \rho_i}{\partial t} + \left(\frac{1}{r^2}\frac{\partial}{\partial r}(r^2 n_{ir}) + \frac{1}{r \sin\theta}\frac{\partial}{\partial \theta}(n_{i\theta}\sin\theta) + \frac{1}{r\sin\theta}\frac{\partial n_{i\phi}}{\partial \phi}\right) = r_i \qquad (4.1C)$$

Note: $\mathbf{n}_i = \rho_i \mathbf{v}_i$

For component i in multicomponent system (molar basis)

Equations have same structure as above, replacing n_i by N_i, ρ_i by c_i, r_i by R_i, and using $\mathbf{N}_i = c_i\mathbf{v}_i$. These equations will be designated eqns (4.1D), (4.1E), and (4.1F), for rectangular, cylindrical, and spherical coordinates, respectively.

For total mass

Rectangular coordinates:

$$\frac{\partial \rho}{\partial t} + \left(\frac{\partial}{\partial x}(\rho v_x) + \frac{\partial}{\partial y} + (\rho v_y) + \frac{\partial}{\partial z}(\rho v_z)\right) = 0 \qquad (4.1G)$$

Cylindrical coordinates:

$$\frac{\partial \rho}{\partial t} + \left(\frac{1}{r}\frac{\partial}{\partial r}(\rho r v_r) + \frac{1}{r}\frac{\partial}{\partial \theta}(\rho v_\theta) + \frac{\partial}{\partial z}(\rho v_z)\right) = 0 \qquad (4.1H)$$

Spherical coordinates:

$$\frac{\partial \rho}{\partial t} + \left(\frac{1}{r^2}\frac{\partial}{\partial r}(\rho r^2 v_r) + \frac{1}{r\sin\theta}\frac{\partial}{\partial \theta}(\rho v_\theta \sin\theta) + \frac{1}{r\sin\theta}\frac{\partial}{\partial \phi}(\rho v_\phi)\right) = 0 \qquad (4.1I)$$

For total moles

Equations have same structure as for total mass, replacing ρ by c, v by v^*, and the right-hand zeros by $\sum_k R_k$. These equations will be designated eqns (4.1J), (4.1K), and (4.1L), for rectangular, cylindrical, and spherical coordinates, respectively.

Exercise 1. Since *total* mass is always conserved, another equation of continuity must apply to the entire system. Without repeating the entire derivation, show how the equation can be obtained directly from eqn (4.2).

Solution. We know that total density is $\rho = \sum_i \rho_i$ and total flux $\boldsymbol{n} = \rho\boldsymbol{v}$ (where $\boldsymbol{v}$ is the mass-average velocity, by definition) with $\boldsymbol{n} = \sum_i \boldsymbol{n}_i$ and thus $\rho\boldsymbol{v} = \sum_i \rho_i\boldsymbol{v}_i$. Then, an equation like eqn (4.1) can be written for all components ($i = 1, 2, \ldots \nu$) and the whole set can be added to give

$$\frac{\partial}{\partial t}\left(\sum_i \rho_i\right) + \sum_i \nabla\cdot\boldsymbol{n}_i - \sum_i r_i = 0 \qquad (4.3a)$$

or

$$\frac{\partial \rho}{\partial t} + \nabla\cdot\boldsymbol{n} = 0. \qquad (4.3b)$$

The term $\sum_i r_i$ vanishes because the total mass is unchanged even though the chemical identities of the mass components change; no mass can be created by chemistry. Equation (4.3b) appears in Table 4.1, in three coordinate systems; its counterpart in molar terms is also displayed.

Exercise 2. A polymeric membrane separates an aqueous solution of a medicinal drug from body fluids, also aqueous. If concentrations of the drug are constant in both liquid phases, describe the time dependence of the flux across the membrane and also the distribution ('profile') of flux across the membrane. See Fig. 4.2.

Solution. In Table 4.1, let $i = d$ to designate $c_d(y, t)$ as drug concentration and $\boldsymbol{n}_d(y, t)$ as the drug flux at any position y across the membrane. Since there is no time dependence in c_d anywhere, $\partial c_d/\partial t = 0$. There is also no chemical reaction, so the r_d term vanishes. Then, the equation is reduced to merely $\nabla \cdot \boldsymbol{n}_d = 0$, where $\boldsymbol{n}_d$ is also *independent of time*. The equation states that $\partial n_{dx}/\partial x + \partial n_{dy}/\partial y + \partial n_{dz}/\partial z = 0$, but with only n_{dy} being nonzero we have merely $dn_{dy}/dy = 0$. Thus, we find immediately that $n_{dy} = n_d^\circ$ = constant, *independent of position*. This situation is represented in Fig. 4.2(a). However, we cannot yet predict $c_d(y)$ or the magnitude of n_d°; a preview is given in Fig. 4.2(b).

4.2.2 Constitutive models or 'laws'

While eqns (4.2) and (4.3) are completely general and often useful, few problems can be solved by using them alone. Usually we need to introduce information of the form $r_i(\rho_1, \rho_2, \ldots \rho_j, \ldots)$ and/or $\boldsymbol{n}_i(\rho_1, \boldsymbol{v}_1, \rho_2, \boldsymbol{v}_2, \ldots \rho_j, \boldsymbol{v}_j \ldots)$.

(a) Chemical reaction models, and equilibrium

Choosing a specific function for $r_i(\rho_j \ldots)$ or $R_i(c_j \ldots)$ is the province of chemical kinetics. We illustrate this first for reactions that are 'irreversible' (go only in the forward direction, R^f, forming products from reactants until the reactants are fully consumed). Notation, for now, replaces the component indices 1, 2, ... by A,

Reaction	*Observed c-dependence*	*Kinetic model*	
$A \rightarrow E$	first order	$R_E^f = k_f c_A = -R_A^f$	(4.4a)
$A + B \rightarrow E + F$	second order	$R_E^f = R_F^f = k_f c_A c_B$	(4.4b)
$2A \rightarrow E + F$	second order	$R_E^f = R_F^f = k_f c_A^2$	(4.4c)
$aA + bB \rightarrow eE + fF$	$(a + b)$ order	$R_E^f = e k_f c_A^a c_B^b = k_f' c_A^a c_B^b.$	(4.4d)
$aA + bB \leftarrow eE + fF$	$(e + f)$ order	$R_E^r = -e k_r c_E^e c_F^f = -k_r' c_E^e c_F^f.$	(4.4e)

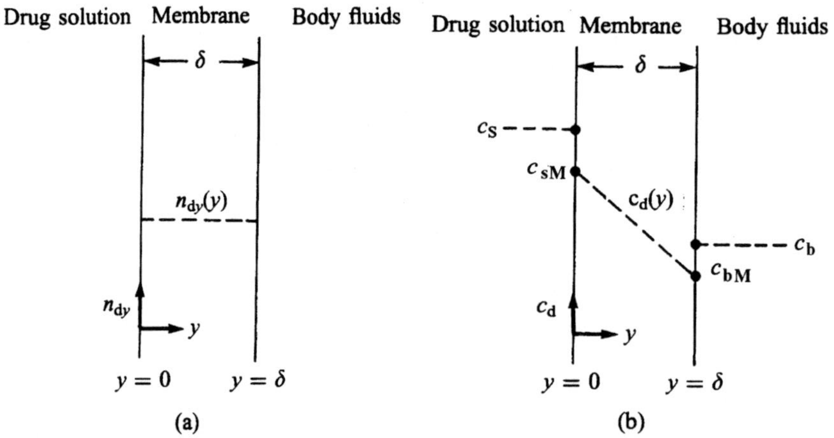

Fig. 4.2 Transport of a drug through a planar membrane. (a) Mass flux profile, $n_{dy}(y)$ = constant; molar flux has a similar profile. These correspond to fixed-concentration boundary conditions. (b) Molar concentration profile, $c_d(y)$, for fixed boundary values. Not shown are deviations from c_d-uniformity outside the membrane, adjacent to its surface; these are addressed later.

B, ..., and the 'law of mass action' is assumed to be correct in governing the reaction mechanisms in single-phase media.

However, some degree of reversibility always prevails, so that the reverse reaction eventually become important, too, as product concentrations increase. The reverse-reaction analogue to eqn (4.4d) is

Since the general case is given by

$$R_E = R_E^f + R_E^r = k_f' c_A^a c_B^b - k_r' c_E^e c_F^f, \qquad (4.4f)$$

we see that chemical equilibrium ($R_E = 0$, reaction 'complete') is characterized not by $c_A = 0$ but rather by a balance of all concentrations that is unique at a given temperature and pressure. This balance is represented by an equilibrium constant, K_{eq}, defined by convention as the ratio of product concentration factors divided by reactant concentration factors and obtained from $R^f + R^r = 0$. In the illustration of eqn (4.4f),

$$K_{eq} \equiv \left[\frac{c_E^e c_F^f}{c_A^a c_B^b}\right]_{eq} = \frac{k_f'}{k_r'} = \frac{k_f}{k_r}. \qquad (4.4g)$$

Note that values of K_{eq} obtained from equilibrium measurements of c_i, contain kinetic information; this is sometimes useful when seeking rate constants (e.g. $k_r = k_f/K_{eq}$). Because K_{eq} is generally not a dimensionless number, its value can depend on both the 'molecularity' of the reaction ($a, b, \ldots e, f$) and the concentration units used (c_i, x_i, and p_i for gases).

When R_i^r is very small, then knowledge of $R_i^f \approx R_i$ may be sufficient for problem-solving. Another useful approximation arises in the case of one reactant being present in great excess; then, its concentration changes by such a small fraction that it may be assumed constant while other components change their concentrations. For example, in eqn (4.4b), an excess of B would give $R_E^f = (k_f c_B) c_A \approx k_f'' c_A$ and pseudo-first-order behavior.

Two final observations are relevant to biological and other complex reacting systems. First, the law of mass action may not apply; stoichiometry may not be relevant to the actual reaction path. Then, empiricism of various kinds may be needed to represent R_i accurately. One example is

$$-R_i = k_1 c_i/(1 + k_2 c_i), \qquad (4.5)$$

which is known as Michaelis–Menten kinetics in the biology literature and as Hougen–Watson kinetics in the chemical engineering literature. Its primary utility is its ability to shift from first-order kinetics at low c_i to zero-order kinetics at high c_i, in a functionally realistic manner. Second, if catalysis by a nonreactant is involved (e.g. by an enzyme), both k_f and k_r are enhanced by the same factor and always in some proportion to catalyst concentration. Since a true catalyst is not consumed in the reaction, the effective k values remain at their high levels throughout the process. However, the equilibrium concentration state is not affected by catalysis, so K_{eq} is unchanged from the noncatalyzed value.

Another aspect of chemical equilibrium is the condition encountered at interfaces: gas–liquid, gas–solid, liquid–liquid, and liquid–solid interfaces. It is common to assume (with apparently negligible error) that all species exist at an interface in equilibrium with their counterpart on the other side of the interface. Thus, molecules of oxygen in lung tissue at the tissue–air interface are in equilibrium with O_2 in the gas phase at that interface (even though, obviously, their concentrations differ). Designating the interface position as I, we can write $c_{O_2}^{(s)}\big|_I = K_L c_{O_2}^{(g)}\big|_I$, where (s) means 'solid phase' (lung tissue) and (g) means 'gas phase' (air), with K_L signifying conditions at a lung interface. This type of relationship, reflecting the affinity of a component (here, O_2) for one chemical environment relative to another, proves to be immensely important in many problems of biomedical engineering as well as in other areas of engineering and chemistry.

(b) Diffusion flux

The total mass flux or molar flux of component i can always be separated into two portions, one convective and the other diffusive. By definition, the convective flux is the fraction carried along with the *average* motion, and the diffusive flux is the remainder (i.e. the fraction moving *relative* to the average motion). In mass units, using $\boldsymbol{v}$ = mass-average velocity and $\rho\boldsymbol{v}$ = total mass flux,

$$\boldsymbol{n}_i \equiv \rho_i \boldsymbol{v}_i = \underbrace{\left(\frac{\rho_i}{\rho}\right)\rho\boldsymbol{v}}_{\text{convection}} + \underbrace{\rho_i(\boldsymbol{v}_i - \boldsymbol{v})}_{\text{diffusion}} = \omega_i \sum_k \boldsymbol{n}_k + \boldsymbol{j}_i, \qquad (4.6a)$$

where ω_i is mass (or weight) fraction and $\boldsymbol{j}_i$ is the mass diffusion flux relative to the mass-average velocity. The molar analogue, with $\boldsymbol{v}^*$ = molar-average velocity and $c\boldsymbol{v}^*$ = total molar flux, is

$$\boldsymbol{N}_i \equiv c_i \boldsymbol{v}_i = \underbrace{\left(\frac{c_i}{c}\right)c\boldsymbol{v}^*}_{\text{convection}} + \underbrace{c_i(\boldsymbol{v}_i - \boldsymbol{v}^*)}_{\text{diffusion}} = x_i \sum_k \boldsymbol{N}_k + \boldsymbol{J}_i^*, \qquad (4.6b)$$

Table 4.2. Mass and molar fluxes in multicomponent systems

With respect to stationary axes		With respect to $\mathbf{v} \equiv \sum_i \omega_i \mathbf{v}_i$		With respect to $\mathbf{v}^* \equiv \sum_i x_i \mathbf{v}_i$	
$\mathbf{v}_i$ (velocity)	(4.2A)	$\mathbf{v}_i - \mathbf{v}$ (relative velocity)	(4.2D)	$\mathbf{v}_i - \mathbf{v}^*$ (relative velocity)	(4.2G)
$\mathbf{n}_i = \rho_i \mathbf{v}_i$	(4.2B)	$\mathbf{j}_i = \rho_i(\mathbf{v}_i - \mathbf{v})$	(4.2E)	$\mathbf{j}_i^* = \rho_i(\mathbf{v}_i - \mathbf{v}^*)$	(4.2H)
$\mathbf{N}_i = c_i \mathbf{v}_i$	(4.2C)	$\mathbf{J}_i = c_i(\mathbf{v}_i - \mathbf{v})$	(4.2F)	$\mathbf{J}_i^* = c_i(\mathbf{v}_i - \mathbf{v}^*)$	(4.2I)
$\sum_i \mathbf{n}_i = \rho \mathbf{v}$	(4.2J)	$\sum_i \mathbf{j}_i = 0$	(4.2L)	$\sum_i \mathbf{j}_i^* = \rho(\mathbf{v} - \mathbf{v}^*)$	(4.2N)
$\sum_i \mathbf{N}_i = c\mathbf{v}^*$	(4.2K)	$\sum_i \mathbf{J}_i = c(\mathbf{v}^* - \mathbf{v})$	(4.2M)	$\sum_i \mathbf{J}_i^* = 0$	(4.2O)
$\mathbf{n}_i = \mathbf{N}_i M_i$	(4.2P)	$\mathbf{j}_i = \mathbf{n}_i - \omega_i \sum_k \mathbf{n}_k$	(4.2T)	$\mathbf{j}_i^* = \mathbf{n}_i - x_i \sum_k \mathbf{n}_k \mathbf{M}_i / M_k$	(4.2X)
$\mathbf{N}_i = \mathbf{n}_i / M_i$	(4.2Q)	$\mathbf{J}_i = \mathbf{N}_i - \omega_i \sum_k \mathbf{N}_k \mathbf{M}_k / M_i$	(4.2U)	$\mathbf{J}_i^* = \mathbf{N}_i - x_i \sum_k \mathbf{N}_k$	(4.2Y)
$\mathbf{n}_i = \rho_i \mathbf{v} + \mathbf{j}_i$	(4.2R)	$\mathbf{j}_i = \mathbf{J}_i M_i$	(4.2V)	$\mathbf{j}_A^* = \mathbf{j}_A M / M_B$	(4.2Z)†
$\mathbf{N}_i = c_i \mathbf{v}^* + \mathbf{J}_i^*$	(4.2S)	$\mathbf{J}_A = \mathbf{J}_A^* M_B / M$	(4.2W)†	$\mathbf{J}_i^* = \mathbf{j}_i^* / M_i$	(4.2AA)

† This applies only to binary systems; there is no simple multicomponent analogue. Note that $M = \rho/c = x_A M_A + x_B M_B = \bar{M}_n$.

where x_i is mole fraction and $\boldsymbol{J}_i^*$ is the molar diffusion flux relative to the molar-average velocity. (Notationally, we use upper-case letters with molar flux and lower-case letters with mass flux; the asterisk indicates referral to mole-average velocity and its absence implies referral to mass-average velocity.) Note that these basic definitions of diffusion flux require that the nature of the average flux (or velocity) be specified. It is possible, within this framework, to employ 'mixed' concepts. Actually, one of these is quite widely used:

$$\boldsymbol{N}_i \equiv c_i \boldsymbol{v}_i = \underset{\text{convection}}{c_i \boldsymbol{v}} + \underset{\text{diffusion}}{c_i(\boldsymbol{v}_i - \boldsymbol{v})} = \frac{\omega_i}{M_i} \sum_k \boldsymbol{n}_k + \boldsymbol{J}_i, \tag{4.6c}$$

where $\boldsymbol{J}_i$ is the *molar* diffusion flux relative to the *mass*-average velocity (and is obviously equal to $\boldsymbol{j}_i/M_i$), and $M_i = \rho_i/c_i$ is the molecular weight of component i. These and other definitions of fluxes are given in Table 4.2 and discussed more extensively by Bird *et al.* (1960), by Cussler (1984), and by Lightfoot (1974).

Exercise 3. Evaluate the sums $\sum_k \boldsymbol{j}_k$ and $\sum_k \boldsymbol{J}_k$.

Solution.

$$\sum_k \boldsymbol{j}_k = \sum_k \rho_k(\mathrm{v}_k - \mathrm{v}) = \sum_k \rho_k \mathrm{v}_k - \mathrm{v} \sum_k \rho_k = \rho \mathrm{v} - \mathrm{v}\rho = 0.$$

$$\sum_k \boldsymbol{J}_k = \sum_k c_k(\mathrm{v}_k - \mathrm{v}) = \sum_k c_k \mathrm{v}_k - \mathrm{v} \sum_k c_k = c\mathrm{v}^* - \mathrm{v}c \neq 0.$$

One can show also that $\sum_k \boldsymbol{J}_k^* = 0$, as is true for all diffusion fluxes defined with internal unit consistency, while sums over diffusion fluxes defined with mixed concepts are never zero.

(c) Transport laws

The three fundamental molecular transport processes involve transport of linear momentum, energy, and mass. These are embedded in the engineering subjects of fluid/solid mechanics, heat transfer, and mass transfer, with the corresponding molecular transport *properties* of shear viscosity (μ) and various kinds of elastic moduli, thermal conductivity (κ), and diffusion coefficients (diffusivities)[2]. One complication encountered with diffusion that is not shared by the other processes (besides the multiplicity of flux definitions) is that there is no such thing as an 'average' diffusivity in a multicomponent mixture. Thus, while such a mixture can be characterized

[2] The properties of bulk viscosity, elongational viscosity, and thermal diffusion coefficient are other transport properties that can be important in some cases but rarely in bioengineering problems.

by μ_{mix} there is no analogue in diffusion. Instead, a large number of independent diffusivities may be involved in such a case, and rarely are their values all known. We illustrate this or the rather frightening general case of mechanically driven diffusion flux:

$$\boldsymbol{j}_i = \boldsymbol{J}_i M_i = \left(\frac{c}{RT}\right)\frac{M_i}{M}\sum_{j=1}^{\nu} M_j D_{ij} x_j \times \left[\sum_{\substack{k=1\\ \neq j}}^{\nu} \alpha_{jk}\nabla x_k + \left(\bar{V}_j - \frac{M_j}{\rho}\right)\nabla p + M_j\left(\boldsymbol{f}_j - \sum_{k=1}^{\nu}\omega_k \boldsymbol{f}_k\right)\right], \tag{4.7}$$

where the D_{ij} are multicomponent diffusion coefficients, $M = \rho/c$ is the average molecular weight of the mixture, $\bar{V}_j$ is the partial molar volume of component j, $\boldsymbol{f}_j$ is the externally imposed force/mass acting on component j (for example, the gravitational acceleration $\boldsymbol{g}_1 = \boldsymbol{g}_2 = \cdots = \boldsymbol{g}$ is such a force), p is the pressure, and $\alpha_{jk} = (\partial\bar{G}_j/\partial x_k)_{T,p,x_m\, m\neq j,k}$ with $\bar{G}_j$ = partial molar Gibbs free energy of mixing $= RT \ln a_j$, thus defining also the activity a_j of component j.

While there are some mathematical restrictions on the values of the D_{ij} (e.g. $D_{ii} = 0$), it is clear that we usually will not have values for all of them. However, what useful information can be extracted from eqn (4.7)?

1. There are three types of 'mechanical' driving forces: concentration gradients, pressure gradients, and net external body force. These correspond, respectively, to *ordinary* diffusion, *pressure* diffusion, and *forced* diffusion.
2. We don't need separate diffusion coefficients to characterize responses to different kinds of mechanical driving forces.
3. When gravity is the only body force, forced diffusion drops out. However, when electromagnetic forces are involved with ionic species (e.g. electrolytic cells, electrophoresis), forced diffusion can be important.
4. Thermodynamics is coupled to ordinary diffusion. Information on solubility of all species in the mixture could be found and expressed as $a_j = \gamma_j x_j$ (where γ_j is the activity coefficient), so $x_j \alpha_{jk}$ multiplying the concentration gradient becomes $(RT/\gamma_j)\cdot(\partial x_j/\partial x_k + \partial\gamma_j/\partial x_k)_{x_m\, m\neq j,k}$. If γ_j = constant or is close to unity ($\gamma_j = 1$ in idealized solutions), then further simplifications become apparent.
5. Thermodynamics is also coupled to pressure diffusion. Using data on $\bar{V}_j$ and ρ tells us, for most mixtures, that the coefficient of ∇p is usually very small. So, except for extremely high pressure gradients (e.g. in an ultracentrifuge), pressure diffusion can be neglected.

Because of complications related to the D_{ij}, we will turn our attention next to the simplest situation: *binary* diffusion ($\nu = 2$). To avoid confusion with the multicomponent case, subscripts 1 and 2 will be replaced here by A and B. Then, eqn (4.7) can be written as

$$\boldsymbol{j}_A = \boldsymbol{J}_A M_A = -\left(\frac{c}{RT}\right)\frac{M_A M_B}{M} \times D_{AB}\left[RT\left(\frac{\partial \ln a_A}{\partial \ln x_A}\right)_{T,p}\nabla x_A + x_A\left(\bar{V}_A - \frac{M_A}{\rho}\right)\nabla p - x_A M_A \omega_B(\boldsymbol{f}_A - \boldsymbol{f}_B)\right] \tag{4.8}$$

with $\boldsymbol{j}_B = -\boldsymbol{j}_A$ (see Exercise 3). Because biomedical systems are not binary ones, eqn (4.8) might seem to have little relevance to bioengineering. However, that is not true; in most cases eqn (4.8) is applied with the strategy that the system is 'pseudo-binary'—that is, component A is the one whose diffusion interests you and component B represents an 'average' environment for A, sort of a 'pseudo-component'. One example is the diffusion of oxygen from an oxygenated red cell through blood plasma to tissue low in oxygen; oxygen is A, and the blood plasma (a complex soup of salts, proteins, metabolites, water, etc.) is viewed as B. This concept is used most frequently when the environment is dominated by one component whose molar or mass concentration is far in excess of the others' concentrations. In the example above, blood plasma as a medium for oxygen diffusion would have properties much like those of water, since the water fraction of plasma is about 0.93 by weight, and much higher on a molar basis.

Thus motivated to examine binary diffusion more closely, we proceed to the classic model for *ordinary* binary diffusion,

$$\mathbf{j}_A = -\rho\mathscr{D}_{AB}\nabla\omega_A \qquad (4.9a)^3$$

[3] Another approach to multicomponent systems that utilizes a binary-diffusion approximation is that leading to the Stefan–Maxwell equations, representing an inversion of the flux–force relationship:

$$\nabla x_i = \sum_{j=1} x_i x_j(\mathbf{v}_j - \mathbf{v}_i)/\mathscr{D}_{ij} = \sum_{j=1}(x_i N_j - x_j N_i)/c\mathscr{D}_{ij}, \tag{4.8'}$$

where the binary coefficients are incorporated. The theoretical justification is discussed by Hirschfelder *et al.* (1954).

or

$$\mathbf{J}_A^* = -c\mathscr{D}_{AB}\nabla x_A, \tag{4.9b}$$

both of which represent Fick's law (and $\mathscr{D}_{AB}$ has the same value in both). Sometimes other approximate forms are used, such as when c or ρ is nearly independent of position:

$$j_A = -\mathscr{D}_{AB}\nabla\rho_A \tag{4.9c}$$

or

$$J_A^* = -\mathscr{D}_{AB}\nabla c_A. \tag{4.9d}$$

Table 4.3 contains Fick's law in three coordinate systems.

The question often arises as to whether $\mathscr{D}_{AB}$ can be treated as a constant. In general, it cannot. We expect that $\mathscr{D}_{AB}$ depends on T, p, and composition. However T-dependence may be unimportant in medical applications because the body is isothermal. Likewise, because pressure variations within the body are so small, p-dependence can be neglected. Composition is usually the biggest problem. For gas-phase diffusion, fortunately, we know (from theory and experiment) that $\mathscr{D}_{AB}$ is independent of composition; this could be relevant to lung behavior and bubble oxygenators. However, in liquids there is definite composition dependence. We can sometimes take refuge with the case of dilute solutions, assuming that the dilute A-molecules will not collide with each other often enough to produce x_A-dependence in $\mathscr{D}_{AB}$; here, one hopes that the limiting dilute value $\mathscr{D}^\circ_{AB}$ holds throughout the system. In other cases, we know that $\mathscr{D}_{AB}(x_A) \neq \mathscr{D}^\circ_{AB}$ but the molecule is diffusing under a small ∇x_A (from x_{A1} to x_{A2}) so that $\mathscr{D}_{AB}$ can be evaluated at an average composition [say, $\bar{x}_A = (x_{A1} + x_{A2})/2)$] and—if $(x_{A1} - x_{A2})/\bar{x}_A \ll 1$—then $\mathscr{D}_{AB}(\bar{x}_A)$ can be used as a constant in calculations.

Exercise 4. Suppose you know $\mathscr{D}^\circ_{AB}$ for a liquid system but plan to work with compositions in the range of $x_A \approx 0.1$. No data on $\mathscr{D}_{AB}$ (x_A) are available. How might you proceed to estimate $\mathscr{D}_{AB}$?

Solution. One possibility, if thermodynamic data are available, is to invoke the form suggested by a comparison of eqns (4.8) and (4.9b): $\mathscr{D}_{AB}(0.1) = \mathscr{D}^\circ_{AB}(\partial \ln a_A/\partial \ln x_A)_{x_A=0.1}$. Alternatively, if viscosity

Table 4.3. Fick's law for binary ordinary diffusion

Vector forms:

$$\boldsymbol{j}_A = \boldsymbol{n}_A - \omega_A(\boldsymbol{n}_A + \boldsymbol{n}_B) = -\rho\mathscr{D}_{AB}\nabla\omega_A \tag{4.3A}$$

$$= -c\mathscr{D}_{AB}(M_A M_B/M)\nabla x_A \tag{4.3B†}$$

$$\mathbf{J}_A^* = \boldsymbol{N}_A - x_A(\boldsymbol{N}_A + \boldsymbol{N}_B) = -c\mathscr{D}_{AB}\nabla x_A \tag{4.3C}$$

$$= -\rho\mathscr{D}_{AB}(M/M_A M_B)\nabla\omega_A \tag{4.3D†}$$

Rectangular coordinates (molar basis):‡

$$J^*_{Ax} = -c\mathscr{D}_{AB}\frac{\partial x_A}{\partial x}, \quad J^*_{Ay} = -c\mathscr{D}_{AB}\frac{\partial x_A}{\partial y}, \quad J^*_{Az} = -c\mathscr{D}_{AB}\frac{\partial x_A}{\partial z} \tag{4.3E}$$

Cylindrical coordinates (molar basis):‡

$$J^*_{Ar} = -c\mathscr{D}_{AB}\frac{\partial x_A}{\partial r}, \quad J^*_{A\theta} = -c\mathscr{D}_{AB}\left(\frac{1}{r}\frac{\partial x_A}{\partial \theta}\right), \quad J^*_{Az} = -c\mathscr{D}_{AB}\frac{\partial x_A}{\partial z} \tag{4.3F}$$

Spherical coordinates (molar basis):‡

$$J^*_{Ar} = -c\mathscr{D}_{AB}\frac{\partial x_A}{\partial r}, \quad J^*_{A\theta} = -c\mathscr{D}_{AB}\left(\frac{1}{r}\frac{\partial x_A}{\partial \theta}\right), \quad J^*_{A\phi} = -c\mathscr{D}_{AB}\left(\frac{1}{r\sin\theta}\frac{\partial x_A}{\partial \theta}\right) \tag{4.3G}$$

† Note that $M = \rho/c = x_A M_A + x_B M_B = \bar{M}_n$.
‡ For the equations on a mass basis, replace J_A^* by j_A, c by ρ, and x_A by ω_A. These equations will be designated as eqns (4.3H), (4.3I), and (4.3J) for rectangular, cylindrical, and spherical coordinates, respectively.

data are available, the Stokes–Einstein relationship for liquids might be used:

$$\mathscr{D}^{\circ}_{AB} = \frac{kT}{\zeta_{AB}} \propto \frac{kT}{\mu^{\circ}_{B}\bar{V}_{A}^{1/3}}, \tag{4.10}$$

where ζ_{AB} is the mobility[4] of molecule A diffusing through an excess of B and μ°_{B} is the viscosity of pure B. Then, presuming $\mu_{B}(x_{A})$ to represent the viscous resistance to the motion of an A molecule, one could estimate $\mathscr{D}_{AB}(0.1) = \mathscr{D}^{\circ}_{AB}\mu^{\circ}_{B}/\mu_{B}(0.1)$. Note, for liquids, that diffusivity and viscosity are *inversely* proportional (though not always as simple as eqn (4.10) suggests, especially when the solute molecules are of polymeric size).

4.2.3 The diffusion equation

In principle, for multicomponent systems, eqns (4.2)/Table 4.1, (4.4), (4.6), and (4.7) should be solved simultaneously. In practice, as indicated above, this is rarely possible. However, for the binary case, progress can be made. Because the ordinary-diffusion expression of $\boldsymbol{j}_{A}$ (eqn (4.9a)) is so much simpler than eqn (4.7) for $\boldsymbol{j}_{i}$ (even when eqn (4.7) is specialized for ordinary diffusion), it is reasonable to make some direct substitutions. The consequence is a more detailed form of the binary equation of continuity, which can be expressed in several ways. For example,

$$\frac{\partial \rho_{A}}{\partial t} = -\boldsymbol{\nabla}\cdot(\rho_{A}\boldsymbol{v}) + \boldsymbol{\nabla}\cdot(\rho\mathscr{D}_{AB}\boldsymbol{\nabla}\omega_{A}) + r_{A}, \tag{4.11a}$$

or,

$$\frac{\partial c_{A}}{\partial t} = -\boldsymbol{\nabla}\cdot(c_{A}\boldsymbol{v}) + \boldsymbol{\nabla}\cdot(c\mathscr{D}_{AB}\boldsymbol{\nabla}x_{A}) + R_{A}, \tag{4.11b}$$

which states that the concentration of A increases with time if the sum of the convected transport and diffusive transport and chemical generation is positive. Equation (4.11) is sometimes described as a *generalized* diffusion equation (as opposed to the transport law for diffusion given previously). The term 'Fick's second law' has also been used, inappropriately. For the case of constant properties, equation (4.11) is displayed in Table 4.4 in three coordinate systems. Its use often involves a series of approximations to obtain mathematical tractability or to represent special (simple) systems. We have already mentioned the ρ = constant and c = constant cases in connection with simplifying Fick's law. Using $\mathscr{D}_{AB}$ = constant (where appropriate) also leads to major simplifications here. Many problems involve no chemical reaction, so $r_{A} = 0$ and $R_{A} = 0$. Finally, convection (bulk flow) may be negligible in diffusion problems—e.g., consider the solid phase or diffusion through a stationary liquid or the migration of species of extremely low concentration; then $\boldsymbol{v} = 0$ and $\boldsymbol{v}^{*} = 0$ are excellent approximations. The result of all of these is

$$\frac{\partial \rho_{A}}{\partial t} = \mathscr{D}_{AB}\nabla^{2}\rho_{A}, \tag{4.12a}$$

$$\frac{\partial c_{A}}{\partial t} = \mathscr{D}_{AB}\nabla^{2}c_{A}, \tag{4.12b}$$

which is known in mathematical circles as the classical diffusion equation. Many analytical solutions to eqn (4.12) have been obtained over the past 150 years, corresponding to a variety of boundary conditions; many of these are described by Crank (1956), and, for the heat-transfer analogue where one obtains $T(\boldsymbol{r}, t)$, also by Carslaw and Jaeger (1959).

Exercise 5. We return to the membrane diffusion problem of Fig. 4.2, first described in Exercise 2 where it was found that the mass flux of drug in the y-direction at steady state is a constant, n°_{d}. It is also true that the molar flux is constant, $N^{\circ}_{d} = n^{\circ}_{d}/M_{d}$. Find the drug concentration profile $c_{d}(y)$ within the membrane, and also evaluate the magnitude of the flux, when external concentrations at the solution side (c_{S}) and body side (c_{b}) are fixed.

Solution. Neglect of convection gives $N^{\circ}_{d} \approx J^{*}_{dy} \approx -\mathscr{D}_{dm}\,dc_{d}/dy$, where $\mathscr{D}_{dM}$ refers to diffusivity of the drug in the membrane. Integration gives $c_{d}(y) = C_{1}y + C_{2}$, where $C_{1} = N^{\circ}_{d}/\mathscr{D}_{dM}$ is taken as constant because $\mathscr{D}_{dM}(y)$ varies so slightly with y. The integration constants C_{2} and C_{1} are both eliminated by using the boundary conditions $c_{d}(0) = c_{SM}$ and $c(\delta) = c_{bM}$, with those concentrations evaluated just *within* the membrane. Thus, $c = c_{SM} - y(c_{SM} - c_{bM})/\delta$. Use of the equilibrium interface relationships $c_{SM} = \phi c_{S}$ and $c_{bM} = \phi c_{b}$ (with ϕ assumed to have the same value at both interfaces of the membrane with aqueous media) leads finally to: $c_{d}(y) = \phi[c_{S} - y(c_{S} - c_{b})/\delta]$. Figure 4.2(b) shows the $c_{d}(y)$ profile.

The flux is obtained from Fick's law by differentiating $c_{d}(y)$: $N^{\circ}_{d} \approx +\mathscr{D}_{dM}\phi(c_{S} - c_{b})/\delta$. Note that drug concentration levels, membrane thickness, drug solubility in the membrane, and drug diffusivity in the membrane all affect N°_{d}. These are all design features that can be

[4] Mobility is defined by $\boldsymbol{F}_{A} = -\zeta_{AB}(\boldsymbol{v}_{A} - \boldsymbol{v}^{*}_{A})$ or $-\zeta_{AB}\boldsymbol{v}_{A}$ when the bulk fluid is stationary. An example is Stokes' law for the motion of a rigid sphere, giving $\zeta_{AB} = 6\pi R\mu^{\circ}_{B}$.

Table 4.4. Relationships for constant properties

Equations of continuity:

$$\text{Constant } \rho: \quad \nabla \cdot \boldsymbol{v} = 0 \tag{4.4A}$$

$$\text{Constant } c: \quad \nabla \cdot \boldsymbol{v}^* = \frac{1}{c}\sum_i R_i \tag{4.4B}$$

Generalized binary diffusion equation (constant ρ, $\mathscr{D}_{AB}$):

Rectangular coordinates:

$$\frac{\partial c_A}{\partial t} + v_x \frac{\partial c_A}{\partial x} + v_y \frac{\partial c_A}{\partial y} + v_z \frac{\partial c_A}{\partial z} = \mathscr{D}_{AB}\left(\frac{\partial^2 c_A}{\partial x^2} + \frac{\partial^2 c_A}{\partial y^2} + \frac{\partial^2 c_A}{\partial z^2}\right) + R_A \tag{4.4C†}$$

Cylindrical coordinates:

$$\frac{\partial c_A}{\partial t} + v_r \frac{\partial c_A}{\partial r} + v_\theta\left(\frac{1}{r}\frac{\partial c_A}{\partial \theta}\right) + v_z \frac{\partial c_A}{\partial z} = \mathscr{D}_{AB}\left(\frac{1}{r}\frac{\partial}{\partial r}\left(r\frac{\partial c_A}{\partial r}\right) + \frac{1}{r^2}\frac{\partial^2 c_A}{\partial \theta^2} + \frac{\partial^2 c_A}{\partial z^2}\right) + R_A \tag{4.4D†}$$

Spherical coordinates:

$$\frac{\partial c_A}{\partial t} + v_r \frac{\partial c_A}{\partial r} + v_\theta\left(\frac{1}{r}\frac{\partial c_A}{\partial \theta}\right) + v_\phi\left(\frac{1}{r\sin\theta}\frac{\partial c_A}{\partial \phi}\right) = \mathscr{D}_{AB}\left[\frac{1}{r^2}\frac{\partial}{\partial r}\left(r^2\frac{\partial c_A}{\partial r}\right) + \frac{1}{r^2\sin\theta}\frac{\partial}{\partial \theta}\left(\sin\theta\frac{\partial c_A}{\partial \theta}\right) + \frac{1}{r^2\sin^2\theta}\frac{\partial^2 c_A}{\partial \phi^2}\right] + R_A \tag{4.4E†}$$

Generalized binary diffusion equation (constant c, $\mathscr{D}_{AB}$):

Equations have the same structure as eqns (4.4C)–(4.4E), replacing v by v^* and adding $-x_A(R_A + R_B)$ to the right side. These will be designated as eqns (4.4F), (4.4G), (4.4H).

† Note that mass units can be recovered from these molar units by multiplying through by M_A. Thus, ρ_A replaces c_A, and r_A replaces R_A.

selected or manipulated by the bioengineer when dealing with synthetic membrane systems in drug delivery devices. These important factors emerged even with all the approximations that were made in the analysis.

4.3 Engineering methods

4.3.1 Material balances

These are commonsense statements of the principle of mass conservation, on a *macroscopic* basis. Typically, only scalar quantities are involved.

Given some kind of macroscopic system, e.g., Fig. 4.3(a), with several inlet streams and outlet streams, each with its own composition, one can write equations representing the conservation of total mass, the conservation of mass of each chemical element, and the balance of mass for each of the molecular species, taking account of chemical reactions. In reference to Fig. 4.3(a), which shows a system with multiple inlet streams (labeled α = a, b, c, ...) and outlet streams (π = p, q, r, ...), one can make overall material balances on both mass and molar bases:

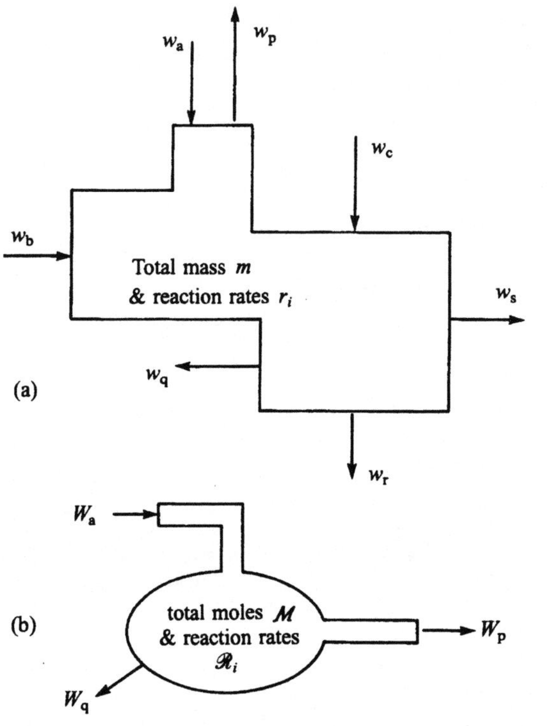

Fig. 4.3 Schematic representations of two classical processing units, for purposes of total material balances: (a) mass units; (b) molar units. Subscripts label the streams (a, b, c, ... = inlet and p, q, r, ... = outlet), except for i serving to identify a specific chemical component.

(a) Total material

$$\textit{mass} \qquad \frac{dm}{dt} = \sum_{\alpha=a,b,\ldots}^{\text{in}} w_\alpha - \sum_{\pi=p,q,\ldots}^{\text{out}} w_\pi \tag{4.13a}$$

$$\textit{moles} \qquad \frac{d\mathcal{M}}{dt} = \sum_{\alpha=a,b,\ldots}^{\text{in}} W_\alpha - \sum_{\alpha=p,q,\ldots}^{\text{out}} W_\pi + \sum_{i=1}^{\nu} \mathcal{R}_i, \tag{4.13b}$$

where m and $\mathcal{M}$ represent the total mass and total moles in the system, respectively; w_α and W_α are the total mass and total molar flow *rates* into the system through stream α, respectively; and $\mathcal{R}_i$ is the total rate of chemical production of moles of component i in the system, so $\mathcal{R}_i = \iiint R_i \, dv$. Note the close resemblance of eqn (4.13a) to the microscopic statement of total mass conservation, eqn (4.3b).

(b) Elements and compounds

$$\textit{mass} \qquad \frac{dm_i}{dt} = \sum_{\alpha=a,b,\ldots}^{\text{in}} w_{i\alpha} - \sum_{\pi=p,q,\ldots}^{\text{out}} w_{i\pi} + \mathcal{R}_i M_i \tag{4.14a}$$

$$\textit{moles} \qquad \frac{d\mathcal{M}_i}{dt} = \sum_{\alpha=a,b,\ldots}^{\text{in}} W_{i\alpha} - \sum_{\pi=p,q,\ldots}^{\text{out}} W_{i\pi} + \mathcal{R}_i, \tag{4.14b}$$

where subscript i designates the element or compound (with $\mathcal{R}_i$ not applicable to elements). Equation (4.14) is structurally and physically similar to eqn (4.2) for the equation of continuity of component i. It should be understood that $m_i = \iiint \rho_i \, dv$, $\mathcal{M}_i = \iiint c_i \, dv$, and similarly for m and $\mathcal{M}$.

Often, the flow rates are expressed in terms different from w or W; the *volumetric flow rate* Q and *average velocity* $\langle v \rangle$ are often used. Here, $\langle v \rangle$ represents an average taken across a conduit cross section, using a velocity component normal to the surface that the fluid is crossing. In a rigorous sense, there is a defining relationship

$$Q \equiv \langle v \rangle S = \iint_S (\boldsymbol{v} \cdot \boldsymbol{e}) \, ds. \tag{4.15}$$

Then, $w = \rho Q = \rho \langle v \rangle S$ and $W = cQ = c\langle v \rangle S$; likewise, $w_i = \rho_i Q$ and $W_i = c_i Q$.

Exercise 6. In Fig. 4.4 we see a commonly used model (The 'Krogh cylinder') for analyzing tissue oxygenation accompanying the flow of blood through a capillary. One can invoke this model for several purposes, but first we want to calculate the flow rate of the blood and also the rate at which oxygen is being delivered to the tissue 'cylinder'. Flow visualization with a microscope determines that erythrocytes are moving at a speed of 0.050 cm s^{-1}, and also that the capillary diameter is 6.0 μm. Arterial blood approaching the capillary has an oxygen concentration of 0.020 g ml^{-1} blood, corresponding (approximately) to saturation of the hemoglobin being carried by red cells; venous blood is only about 80 per cent saturated.

Solution. Red cells move with a speed very close to $\langle v \rangle$; the intervening plasma is carried along between the cells with nearly the same velocity because Q must be the same at all cross sections of the capillary we might consider. Thus, from eqn (4.15), $Q = (500\ \mu\text{m s}^{-1})\ \pi(3.0\ \mu\text{m})^2 = 14\,200\ \mu\text{m}^3\text{s}^{-1}$. Next, eqn (4.14a) is

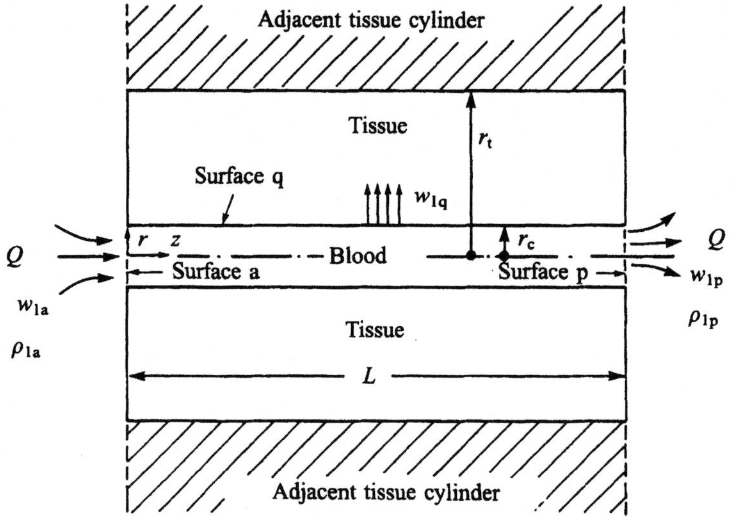

Fig. 4.4 Krogh cylinder model of capillary and surrounding tissue; it is assumed that these cylinders, assembled closely together, account for all the tissue surrounding the capillaries. The model was devised to characterize oxygen transport from blood into the tissue, with no diffusive transport between tissue cylinders (surface at r_t). Dimensions in the microcapillary bed are such that $L \approx 20r_t \approx 200r_c$, with r_c in the range 2–4 μm.

employed for the oxygen balance. The *inlet* stream is the one entering the capillary ('a'), and the two *outlet* streams are the capillary exit flow ('p') and the diffusion 'stream' through the capillary wall into the tissue ('q'). Identifying hemoglobin-bound oxygen as component 1, we use eqn (4.14a) to write (over the capillary)

$$0 = w_{1a} - w_{1p} - w_{1q} = \rho_{1a}Q - \rho_{1p}Q - w_{1q} \quad (4.16a)$$

or

$$\begin{aligned} w_{1q} &= Q(\rho_{1a} - \rho_{1p}) \\ &= (1.42 \times 10^4\ \mu m^3 s^{-1})\,(2 \times 10^{-8}\ g\ \mu m^{-3}) \\ &\quad (1 - 0.8), \end{aligned} \quad (4.16b)$$

giving $w_{1q} = 5.7 \times 10^{-5}\ \mu g\ s^{-1}$. Notice that the chemical reaction term in eqn (4.14a) makes no contribution, even though the oxyhemoglobin is dissociating to release its bound oxygen, because the *bound* oxygen is defined to be our component 1 and the hemoglobin is merely conveying it through the capillary or releasing it; it is not being generated from another chemical species.

4.3.2 Mass transfer coefficients

Near a solid surface, mass transfer between that surface and a fluid flowing past it can be very complex. This is especially true if the fluid is in turbulent flow or if particulate matter is carried by the fluid. In these cases, there is almost no hope of solving differential equations for mass transfer between the wall and the fluid. Instead, the engineer defines an empirical parameter called a *mass transfer coefficient* (k_i for transport of component i) which can be used to represent experimental data. Then, with enough data taken in a wide variety of conditions, the general behavior of k_i can be ascertained and used by other workers who are interested in other molecules.

Figure 4.5 illustrates the situation, with a fluid flowing past a surface at a known Q (or $\langle v \rangle$) which characterizes its macroscopic kinematics. A diffusing species (i) leaves the surface, where its mole fraction (in the *fluid* phase) is x_i°, and joins the bulk flow which has a characteristic bulk composition $\bar{x}_i$, at that cross section. The surface flux due to diffusion, from eqn (4.6b), is used as the context to define the mass transfer coefficients k_{ix} and k_{ic}. For transport in the y-direction (but omitting the y subscript), we write for conditions at $y = 0$,

$$\begin{aligned} J_{io}^* &\equiv N_{io} - x_i^\circ \sum_k N_{ko} \equiv k_{ix}(x_i^\circ - \bar{x}_i) \\ &\equiv k_{ic}(c_i^\circ - \bar{c}_i), \end{aligned} \quad (4.16c)$$

where the subscript x or c on the coefficient reminds us that the latter is tied to a specific type of concentration driving force. Definitions for k_{ip} (used with a partial-

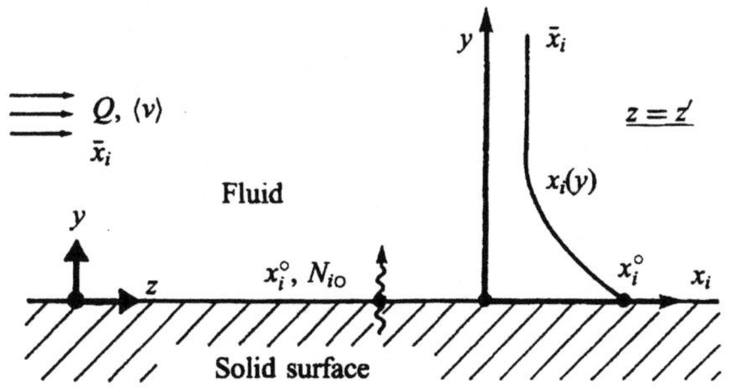

Fig. 4.5 Framework for defining the mass transfer coefficient k_{ix} to correlate interphase transport into (or from) a fluid flowing parallel to a stationary surface at a specific location $z = z'$. Component i is being transported, and a driving force in terms of mole fraction (x_i) differences is employed here. The concentration profile $x_i(y)$ could look different at other z, thereby changing the driving force, but the transport mechanics would hopefully still be characterized by the same k_{ix}, which reflects the state of flow, the velocity profile, geometrical factors, and the diffusion coefficient(s).

pressure driving force) are sometimes used when dealing with molar transport, especially gas, and when using mass flux it is sensible to define and use $k_{i\omega}$ or $k_{i\rho}$. Measurements of J^*_{io} or N_{io} and knowledge of the driving force allow k_{ix} etc. to be calculated and to represent the transport process, empirically. In the Krogh cylinder,

$$w_{1q} = J^*_{1o} 2\pi r_c L \text{ and } J^*_{1o} = k_{1p}(\bar{p} - p^\circ)_{avg}$$

However, the next step also needs to be taken: putting k_{ix} into a *generalized* form where it can be used in a variety of circumstances. Details of the strategy for doing this are found in a number of chemical engineering texts—e.g. Bird *et al.* (1960)—but the scheme will be summarized here in the context of *binary* systems. If the bulk fluid is composed of species A and B, with A diffusing from the wall, then eqn (4.16c) can be interpreted with Fick's law,

$$J^*_{Ao} = \left[-c\mathscr{D}_{AB}\frac{\partial x_A}{\partial y}\right]_{y=0} = k_{Ax}(x_A^\circ - \bar{x}_A), \quad (4.17)$$

which relates k_{Ax} to the basic material property $\mathscr{D}_{AB}$ and the fluid composition profile $x_A(y)$ (which is not possible to obtain analytically at that position z).

Next, we begin to use the tools of dimensional analysis. Reduced (dimensionless) variables will be employed, leading to generalizations which are independent of specific numerical values of variables in experiments. First, a *length scale* characteristic of the fluid kinematics is chosen; in Fig. 4.5 this would be a dimension measured *outward* from the wall, in the direction of the opposite wall (not shown). For flows in pipes, tubes, and capillaries of circular cross section, we choose the diameter D. From this, a dimensionless coordinate can be defined: $\hat{y} \equiv y/D$. Substitution of $y = \hat{y}D$ into eqn (4.17) doesn't change the physics, but allows us to rearrange that equation as

$$\left[\frac{k_{Ax}D}{c\mathscr{D}_{AB}}\right] \equiv Sh = \frac{x_A^\circ - \bar{x}_A}{(-\partial x_A/\partial\hat{y})_{\hat{y}=0}}, \quad \hat{z} \text{ fixed}, \quad (4.18)$$

where the left-hand and right-hand groupings are dimensionless. The Sherwood number (Sh) is, in a sense, independent of the individual magnitudes of k_{Ax}, D, c, $\mathscr{D}_{AB}$, and can be *correlated* with other dimensionless groups—e.g. the flow Reynolds number ($Re \equiv \rho\langle v\rangle D/\mu$ for circular pipes and capillaries) and Schmidt number ($Sc \equiv \mu/\rho\mathscr{D}_{AB}$, a purely material property) and relevant geometrical ratios (e.g. for circular pipes of length L, the ratio L/D emerges). See Bird *et al.* (1960).

The connection of Sh with Re and Sc is made because $x_A(\hat{y}, \hat{z})$—which appears on the right side of eqn (4.18)—is in principle the consequence of simultaneously solving a dimensionless version of the equation of motion (see Chapter 2), in which Re appears, and the generalized diffusion equation (see eqn (4.11)), in which both Re and Sc appear. Applying boundary conditions to get these solutions also introduces geometrical ratios such as L/D.

When fluid and/or wall conditions change with z, as they must here because $\bar{x}_A(z)$ increases as diffusing A merges with the bulk stream, then averages of $Sh(\hat{z})$ must be taken over $\hat{z}$. Alternatively, or equivalently (for correlation purposes), conditions at inlet and outlet of a wall section (inlet and outlet of a capillary, for example) can be used and averaged as appropriate. This eliminates $\hat{z}$ as a variable and produces correlation schemes that characterize the whole system and its operation.

Only one example of such a correlation will be cited here, though many others are available. For flow inside circular tubes, theoretical analysis of *laminar* flow suggests that the grouping

$$j_M \equiv (Sh/Re\, Sc^{1/3})_{avg} \equiv (k_{Ax}/c\langle v\rangle)_{avg} Sc_{avg}^{2/3} \quad (4.19)[5]$$

should be a function of only Re and L/D. Thus, j_M (the 'jay-factor for mass transfer') data are plotted against Re for both laminar and turbulent flow, giving curves as shown in Fig. 4.6. The correlation thus succeeds and can

[5] The 'average' implied here can be specified in different ways, and moreover can be tied to how the driving force was selected in the definition of k_{Ax}.

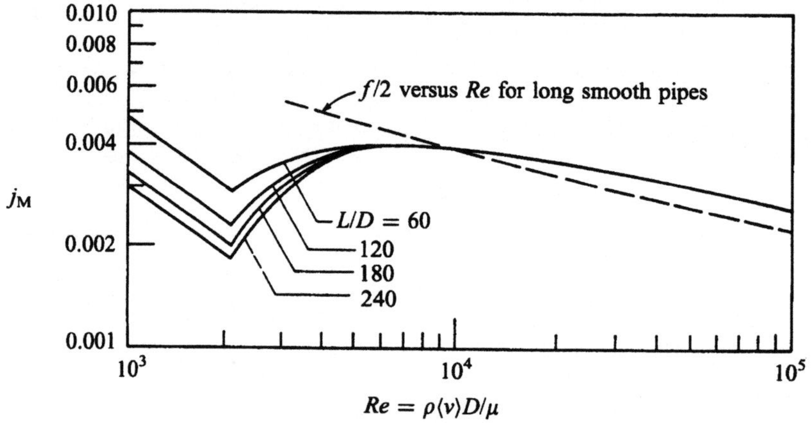

Fig. 4.6 Correlation of data on mass transfer in straight circular tubes with smooth walls and a nearly constant concentration of the transported species at the wall. Various forms of j_M are given in eqn (4.19). The flow is nonlaminar when $Re > 2100$. For flows of almost any kind, correlations of $j_M(Re, L/D)$ give curves very similar to the analogous heat-transfer correlations $j_H(Re, L/D)$, which have been compiled to a much greater extent and may be used with some confidence in mass transfer situations. The same relationship is true between $Sh(Re, Sc, L/D)$ correlations and $Nu(Re, Pr, L/D)$ correlations. For turbulent flow; j_M and j_H are quite close to the function $f/2$ where $f(Re)$ is the Fanning friction factor used to correlate pressure drops ('friction losses') with velocity in Newtonian fluid flows through circular conduits: $f \equiv (\Delta p)D/2L\rho\langle v\rangle^2$. This is represented by the dashed line above. No such similarity is shown between the j-factors and $f/2$ in laminar flow.

be used by engineers working with a wide variety of systems, biomedical and otherwise. Engineers familiar with heat transfer problems will recognize this correlation as being identical to that for $j_H \equiv Nu/Re\,Pr^{1/3}$ in circular pipes used as heat exchangers, where the Prandtl number is $Pr = \mu C_p/\kappa$ and Nusselt number is $Nu = hD/\kappa$ (with κ = thermal conductivity and C_p = heat capacity/mass for the fluid). This analogy between heat and mass transport is important, as it may be possible to solve the latter by using correlations developed for the former. A broad acquaintance with both literatures is therefore recommended.

Exercise 7. A plastic capillary designed for implantation purposes will cause some damage to erythrocytes when blood flows through it. To shield the plastic from cells, it is proposed to deposit a passivating layer of protein on the interior wall by passing a 5 per cent aqueous solution of albumin through the tube. A pump produces a flow rate of $Q = 6\ \text{cm}^3\,\text{s}^{-1}$. The tube diameter is 1.0 mm and its length is 50 cm. If the goal is to deposit a layer with surface density $\rho^{(s)} = 2.0\ \mu\text{g mm}^{-2}$, what should be the duration of flow?

Solution. Although the problem is specified in mass units, we shall solve it using the foregoing equations which were cast in molar units. The task is to calculate the albumin flux to the wall, N_{Ao} (and then n_{Ao}), since blood proteins are strongly absorbed onto foreign surfaces and the deposition rate is thus flux-limited. Because this is effectively a steady process, we can find the necessary flow time from $t_f = \rho^{(s)}/n_{Ao}$. Since $N_{Ao} \approx k_{Ax}(\bar{x}_A - x_A^\circ)$—notice how the driving force in eqn (4.17) is reversed for flow *to* the wall—the major need is to obtain k_{Ax}. Use of Fig. 4.6 requires Re which is calculated to be $\rho\langle v\rangle D/\mu = 4\rho Q/\pi D\mu = (4/\pi)$ $(1.05\ \text{g cm}^{-3})$ $(6\ \text{cm}^3\,\text{s}^{-1})/(0.1\ \text{cm})$ $(0.01\ \text{g cm}^{-1}\,\text{s}^{-1})$ ≈ 8000. Flow is turbulent, in a range where L/D is not a factor. Figure 4.6 then gives $j_M \approx 0.004$. For extracting k_{Ax} from j_M, we need to use $\mathscr{D}_{AW}$ (albumin in water) $= 6 \times 10^{-7}\ \text{cm}^2\,\text{s}^{-1}$—the experimental value in a 5 per cent solution at room temperature—in two places. First, $Sc = (0.01\ \text{g cm}^{-1}\,\text{s}^{-1})/(1.05\ \text{g cm}^{-3})\ (6 \times 10^{-7}\ \text{cm}^2\,\text{s}^{-1}) = 16 \times 10^3$. Then, from eqn (4.19), $Sh = j_M\,Re\,Sc^{1/3} = (0.004)(8000)(16 \times 10^3)^{1/3} \approx 806$. Finally, eqn (4.18) is used with $c = \rho/M \approx (1.05\ \text{g cm}^{-3})/(18\ \text{g mol}^{-1}) \approx 0.058\ \text{mol cm}^{-3}$. [Note that c is dominated by water, since the albumin molecular weight $M_A = 65\,000$ is so high that $c_A = \rho_A/M_A \approx (0.05\ \text{g cm}^{-3})/(65\,000\ \text{g mol}^{-1}) = 0.77 \times 10^{-6}\ \text{mol cm}^{-3}$ is totally negligible in the total.] Thus, $k_{Ax} = Sh\cdot c\cdot\mathscr{D}_{AW}/D = (806)\ (0.058\ \text{mol cm}^{-3})\ (6 \times 10^{-7}\ \text{cm}^2\,\text{s}^{-1})/(0.1\ \text{cm}) = 2.80 \times 10^{-4}\ \text{mol cm}^{-2}\,\text{s}^{-1}$. The calculation of N_{Ao} will proceed by using $x_A^\circ = 0$ (we *assume* the albumin is fully removed from the liquid phase at the wall, even though thermodynamic equilibrium requires that some small nonzero value of x_A° prevails). We also simplify by removing z-dependence from the problem, with the argument that during each pass through the capillary the solution loses so little albumin that $\bar{x}_A = \bar{c}_A/c = 0.77 \times 10^{-6}/0.058 = 1.33 \times 10^{-5} \neq x_A(z)$. Thus, $N_{Ao} \approx (2.80 \times 10^{-4}$ mol cm^{-2} s^{-1}) $(1.33 \times 10^{-5} - 0) = 3.72 \times 10^{-9}$ mol cm^{-2} s^{-1}. Converting to molar units of the targeted $\rho^{(s)}$, we know $\rho^{(s)} = 0.308 \times 10^{-8}$ mol$_A$ cm^{-2}, so finally $t_f = (0.308 \times 10^{-8}$ mol$_A$ cm^{-2})/(3.72 $\times 10^{-9}$ mol$_A$ cm^{-2}s^{-1}) ≈ 0.83 s.

The student should now be able to evaluate the proposal for passivating the plastic surface. Is it a good idea?

4.4 Applications

The physical concepts, differential equations, macroscopic material balances, and transport correlations presented above are useful for solving an enormous range of problems, which includes the biomedical engineering problems of interest here. We will illustrate the latter by first considering the operation of controlled-release drug delivery devices (diffusion-controlled) and then the more complex artificial kidney (diffusion and convection together). Both technologies have had major contributions already from engineers, perhaps to the extent where one could say they were invented by engineers.

4.4.1 Controlled-release drug delivery

(a) Background

Administration of a medication to a specific part of the body can often be done in several ways. For example, the substance can be delivered directly to the desired site or can be placed in the body more generally (orally, or by injection) with the expectation that the blood circulation would convect the agent to that site. While the latter strategy has been conventional in many cases, it is clearly not efficient, because the medication is diluted by the whole body; moreover, there may be undesirable effects in some parts of the body if the drug concentration at the desired site is to be high enough for optimal effectiveness. Therefore, the local-delivery strategy is best if it can be done; examples are the delivery of pilocarpine directly to the eyeball (from a wafer under the lower eyelid) in treating glaucoma and delivery of progesterone to the uterus for preventing conception.

Another important factor is the time dependence of the delivery, which is relevant to both local and systemic treatments. Several time programs are illustrated in Fig. 4.7. For reasons of effectiveness, comfort, and safety, a constant level of drug concentration (locally, or in the whole body) would be better than the time-varying levels resulting from periodic applications. Examples include insulin maintained in diabetic patients and scopolamine released from impregnated bandages on the skin to combat motion sickness, as well as the two other cases cited above.

We will discuss here the operation of several devices (with no moving parts) designed to deliver liquid medication to a local body site at a near-constant rate. Typically, these devices are fabricated from biocompatible polymers and operate by diffusional processes.

(b) Membrane device

A polymeric membrane encapsulates the drug, which is carried in the form of a saturated aqueous solution containing undissolved particles of the drug that maintain the saturation (see Fig. 4.8(a)). Prediction of the steady-state operation of this device has been demonstrated in Exercises 2 and 5, with the boundary conditions being set as follows: c_S = saturation level of drug in solution within the device, and c_b = level in body just outside the device. The molar delivery rate from a spherical capsule of radius R is approximated by

$$W_d^\circ = AN_d^\circ \approx (4\pi R^2)\mathscr{D}_{dM}\phi(c_S - c_b)/\delta, \qquad (4.20)$$

which remains constant—after the initial transient upon placement—until all the drug particles have been

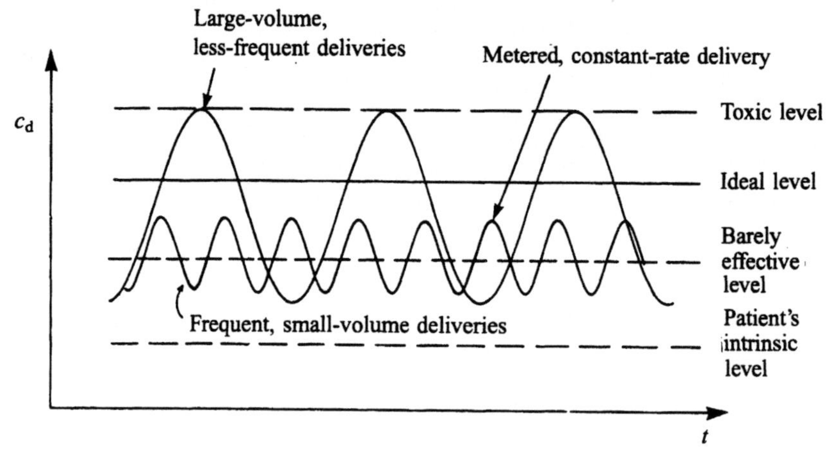

Fig. 4.7 Time dependence of drug concentrations in the body, for various strategies of drug delivery. The controlled-release devices described in this chapter attempt to achieve the ideal-level–constant-rate strategy shown here.

dissolved and drug liquid-phase concentration drops below the saturation level c_S. It is presumed that the body's metabolic and other processes function at a constant rate so that the medication is consumed or cleared at a constant rate too, thus maintaining c_b at a fixed level.

Finally, note that eqn (4.20) uses a form for N_d° that was obtained from a *planar* geometry in Exercise 2, even though the present problem uses a spherical geometry (Fig. 4.8) that we must employ when calculating A. This planar approximation to a curvilinear membrane is justified here because curvature effects do not have much effect on *thin* membranes; as a practical matter, such curvature can be neglected when $\delta/R \ll 1$.

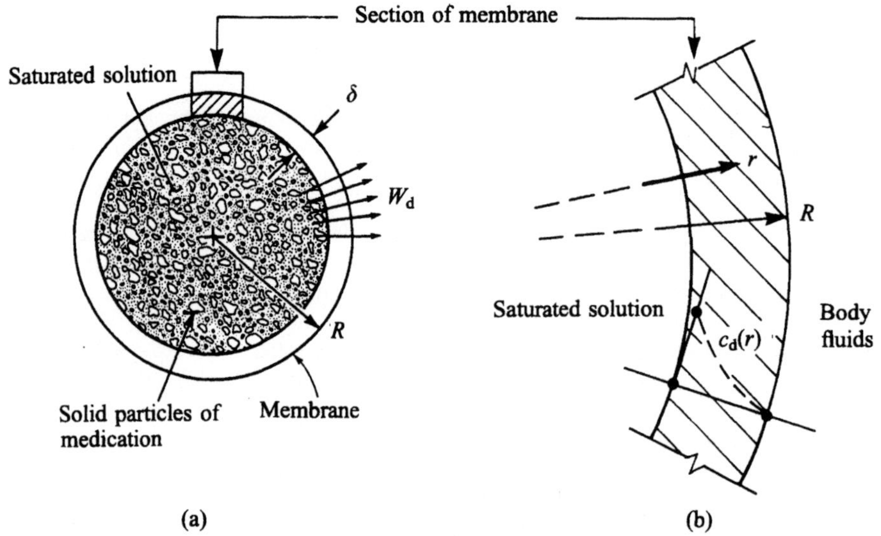

Fig. 4.8 Membrane device for controlled-release delivery of medication at constant rate: (a) overview of the device, represented as a sphere with drug release rate W_d; (b) expanded view of the membrane region, showing curvature of geometry and nonlinear concentration profile $c_d(r)$. Comparison can be made to Fig. 4.2(b).

(c) Matrix device

Here, no membrane is used; the device is analogous to a sponge, with the structurally uniform polymer matrix being initially loaded with a drug-saturated solution (maintained by drug particles, as before). The initial concentration of drug—including the solids—is c_o, throughout the matrix volume V as shown in Fig. 4.9(a). When such a device is implanted, the drug molecules in solution diffuse across the polymer–body interface and a *drug-depleted* region (thickness δ_D) forms on the matrix side of the interface. As diffusion continues, $\delta_D(t)$ grows, while the corresponding 'hydrodynamic' region of enhanced drug concentration (thickness δ_H) outside the device remains at constant thickness because of the continuous washing effect of body fluids and constant clearing processes in the body.

This system does not deliver the drug at a constant rate, but it nearly does so under certain conditions which will now be examined.

Again, we replace the curvilinear system in Fig. 4.9(a) by a planar one (justified if $\delta_D/R \ll 1$) and analyze regions close to the interface as modeled in Fig. 4.9(b). Because we are analyzing a process that is almost a steady-state one, we involve an analytical strategy called 'pseudo-steady state'. Here we set $\partial c_d/\partial t = 0$ in the generalized diffusion equation in order to simplify the mathematics, but we really mean that rates of change are small. Thus, some weak time dependence in $N_d(t)$ can be permitted, and this is introduced through a boundary condition that has time dependence. In our problem, we will incorporate all time dependence through $\delta_D(t)$.

For the planar geometry of Fig. 4.9(b), for flux in the y-direction,

$$N_d = \text{constant} = N_{dm} = N_{db}, \tag{4.21}$$

which equates the steady flux of drug through the matrix depletion zone (N_{dm}) to that through the body's hydrodynamic layer (N_{db})[6]. Again using $N_d \approx J_d$ for these nonconvecting dilute systems, we write eqn (4.21) in terms of Fick's law (dropping the subscript 'd' on c_d to avoid later complexity:

$$\frac{\mathscr{D}_{dm}}{\delta_D}(c_m - c_m^i) = \frac{\mathscr{D}_{db}}{\delta_H}(c_b^i - c_b). \tag{4.22}$$

These expressions are only rigorous if $c(y)$ is linear, as drawn in Fig. 4.9(b). While this is not strictly true (see

[6] Actually, the fundamental equality is between the *rates*, $W_{dm}|_{y=y_1} = W_{db}|_{y=y_2}$. In curvilinear systems, where surface areas change with y, this distinction is important.

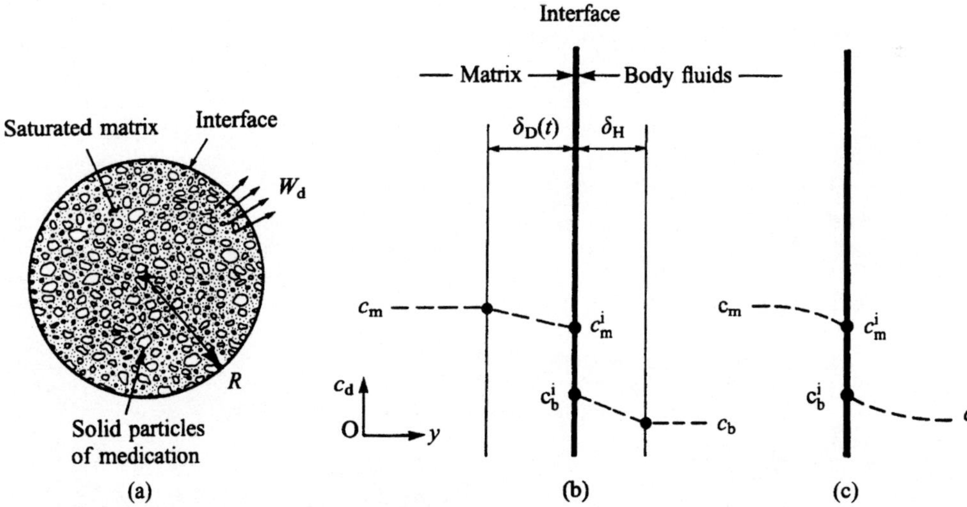

Fig. 4.9 Matrix device for controlled-release delivery of medication at near-constant rate: (a) overview of the device, represented as a sphere with drug release rate W_d; (b) expanded view of the interface region, approximated with planar geometry and sharp boundaries (discontinuities in dc/dy) at the outer edges of the films; (c) a more realistic representation of $c(y)$, without discontinuities (note that film thickness is more difficult to define, rendering the analysis more difficult than in (b), with questionable improvement in results or understanding.

the more realistic Fig. 4.9(c)), the major physical features of the problem are incorporated in eqn (4.22), so we shall retain it for now. Note that the biomedical objective is to control c_b at a steady level, while time dependence is occurring in δ_D, c^i_m, and c^i_b but c_m remains constant (at saturation level).

If c^i_b is eliminated by the interphase equilibrium relationship $c^i_b = Kc^i_m$, then eqn (4.22) can be rearranged to give an explicit expression for c^i_m. This, in turn, can be used to predict flux from the left side of eqn (4.22):

$$N_{dm} = \frac{\mathscr{D}_{dm}}{\delta_D}\left[c_m - \frac{c_m\delta_H\mathscr{D}_{dm} + c_b\delta_D\mathscr{D}_{db}}{\delta_H\mathscr{D}_{dm} + K\delta_D\mathscr{D}_{db}}\right]. \quad (4.23)$$

Note that the flux time dependence lies entirely with $\delta_D(t)$, in this approximate analysis, and the nature of it is controlled by the magnitude of the partition coefficient K as well as by the functional form of $\delta_D(t)$.

The number of moles of drug released from $t = 0$ to time t, $\mathscr{M}_d(t)$, is obtained from a material balance. As shown in Fig. 4.9(b),

$$\begin{aligned}\mathscr{M}_d &= (\text{moles originally present in volume } A_m\delta_D) \\ &\quad - \text{moles remaining in this volume}) \\ &= (A_m\delta_D)c_o - (A_m\delta_D)\langle c\rangle \quad (4.24a) \\ &= A_m\delta_D c_o - A_m\delta_D(c_m + c^i_m)/2 \quad (4.24b)\end{aligned}$$

A modified version of Fig. 4.9(b) is given in Fig. 4.9(c), to represent more realistically the relative magnitudes of parameters in eqn (4.24b). Because the solids loading of the matrix is so much greater than the liquid saturation level (i.e. $c_o \gg c_m > c_{mi}$), we can approximate eqn (4.24b) as $\mathscr{M}_d \approx A_m\delta_D c_o$. Therefore, we obtain

$$\textit{rate} \qquad W_d \equiv \frac{d\mathscr{M}_d}{dt} \approx A_m c_o \frac{d\delta_D}{dt} \quad (4.25a)$$

$$\textit{flux} \qquad N_{dm} \equiv W_d/A_m \approx c_o\frac{d\delta_D}{dt}. \quad (4.25b)$$

Equating eqn (4.25b) and eqn (4.23) leads eventually to a simple differential equation for $\delta_D(t)$:

$$\left(\frac{\delta_H}{\mathscr{D}_{db}} + K\frac{\delta_D}{\mathscr{D}_{dm}}\right)\frac{d\delta_D}{dt} = \frac{c_m K - c_b}{c_o}. \quad (4.26)$$

Integration, using the initial condition $\delta_D(0) = 0$, and rearrangement gives

$$\delta_D\left[\delta_D + 2\frac{\delta_H\mathscr{D}_{dm}}{K\mathscr{D}_{db}}\right] = 2\mathscr{D}_{dm}\frac{c_m}{c_o}\left(1 - \frac{c_b}{Kc_m}\right)t. \quad (4.27)$$

Equation (4.27) is further simplified by recognition that $c_b/Kc_m \equiv (c_b/c^i_b)\,(c^i_m/c_m) \ll 1$, as determined by inspection of Fig. 4.9(c). Thus, the right side of eqn (4.27) becomes $2\mathscr{D}_{dm}(c_m/c_o)t$.

Conditions for near-constant release rates can now be determined. Equation (4.25a) shows that $d\delta_D/dt$ = constant is required, so $\delta_D(t)$ must be linear with t. Such a state can be approached, according to eqn (4.27), if $\delta_D \ll 2\delta_H\mathscr{D}_{dm}/K\mathscr{D}_{db}$. This can be assured in general only if K is very small, meaning that the drug has far more affinity for the matrix than for body fluids. The results are

$$\delta_D \approx K\mathscr{D}_{db}\left[\frac{c_m/c_o}{\delta_H}\right](1 - c_b/Kc_m)t \quad (4.28a)$$

and

$$W_d \approx A_m K\mathscr{D}_{db}c_m(1 - c_b/Kc_m)/\delta_H. \quad (4.28b)$$

This condition is termed 'partition-controlled' and, interestingly, does not involve $\mathscr{D}_{dm}$ or c_b explicitly. The choice of the polymer matrix material, to assure a small K, is one of the tasks of the engineer, who thus finds another materials criterion in addition to biocompatibility and mechanical properties.

(d) Other devices

While the major aspects of diffusion analysis have been illustrated above, it is worth mentioning that other devices have also been proposed and used. For example, there are 'erodable' devices which simply dissolve in body fluids and release their medication at the dissolution rate. These can be formed from a polymer whose chain backbone hydrolyses in the body (for example, polylactic acid) and have the advantage that the implant totally disappears. Still other devices may inherit their constant-rate-of-release properties from a cleverly chosen geometric form, such as illustrated in Problem 8 at the end of the chapter.

4.4.2 Artificial kidney

(a) Background

Kidney disease is a major medical problem, affecting about 5 per cent of the population and accounting for over 60 000 deaths per year in the United States. While the ideal solution is to give the patient a kidney transplant, this is not always possible (lack of suitable donor, cost). To sustain the patient until a transplant is possible, the artificial kidney can be used. It can also be employed during the transplant operation itself or in various emergency situations. Over 85 000 Americans are currently being sustained on such machines; perhaps 15 000 of them have conditions suited to very long-term support.

It is important to understand, first, the function of the natural kidney and the composition of blood. The kidney is the major organ for regulating the composition and volume of body fluids, which constitute about 60 per cent of body weight. Basically, the kidney is a complex chemical processing unit, extracting waste materials (for example, metabolites such as urea) and altering the plasma concentration of other components (salts, water) as needed to achieve balance.

Blood consists of formed elements (cells) and plasma. Among the formed elements are erythrocytes (volume fraction in blood = hematocrit, $H = 0.40$–0.45), white cells, and platelets, with the latter two in very low concentration. Plasma is about 93 per cent water by weight and 7 per cent 'solids', these being solutes of various kinds:

1. *Salts (electrolytes)*: cations Na^+, K^+, Ca^{2+}, Mg^{2+}, and others; anions Cl^-, HCO_3^-, PO_4^{3-}, SO_4^{2-}. The highest concentration belongs to Na^+ and Cl^- because of dietary NaCl consumption, both being on the order of 100 ppm.
2. *Proteins*: albumin (about 5 per cent of plasma weight), fibrinogen, and several globulins. These are all large polymeric molecules, with molecular weights of the order of 10^4–10^6. For example, fibrinogen has $M = 330\,000$ and human albumin $M = 65\,000$; structures tend to be rigid rather than flexible, and most of the conformations are globular.
3. *Metabolites*: Some of the common ones are shown in Fig. 4.10, along with their molecular weights. Note that these are larger than electrolytes but smaller than proteins, so separating them from the rest of the plasma solutes is physically possible by a sequence of filtration processes. Indeed, this is one of the kidney functions. Perhaps the most important metabolite to remove is urea, which has the highest concentration (normally about 25 ppm) and also the smallest size, creating problems for a purely filtration separation.

Intermediate in size between the metabolites cited above and the proteins are the 'middle molecules' with M from 300 to 2000. These are also metabolites but are not completely identified yet; they should also be removed.

When the kidney becomes insufficient, the resulting changes in the body tend to follow a typical pattern. Total water is retained, perhaps by as much as 20 per cent. The plasma concentration of all metabolites increases, which is essentially poisoning the body in its own wastes. Plasma electrolytes may rise or fall: Na^+, Ca^{2+}, HCO_3^-, Cl^- generally are too low in concentration, while K^+, Mg^{2+}, PO_4^{3-}, SO_4^{2-} generally are too high. The increased PO_4^{3-} and SO_4^{2-} levels also upset the acid balance in the blood, reducing the pH.

(b) The natural kidney

Blood purification in the natural kidney is accomplished within specialized cells called nephrons—about a million of them. We can look at the kidney from both an overall viewpoint and the single nephron viewpoint.

Overall, the continuous flow of blood through the kidney represents a huge amount of chemical processing: $Q = 1400\ \mathrm{L\ d^{-1}}$ in the average adult male. The waste stream (urine), consisting of excess water, excess electrolytes, and metabolites, amounts to only about 0.1 per cent of Q. However, to achieve this final separation, the nephrons must execute a highly complex series of operations.

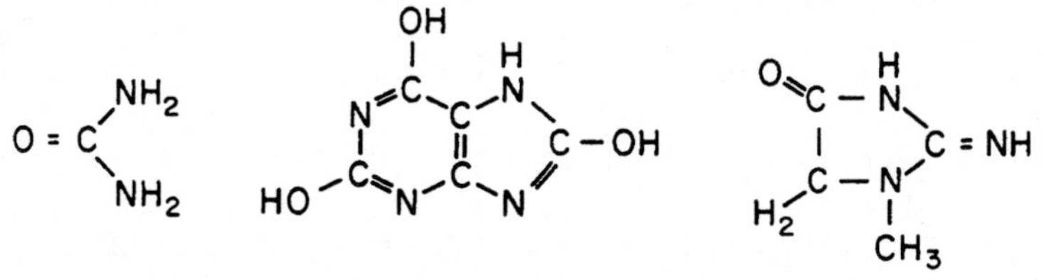

Urea $M = 60$ Uric acid $M = 168$ Creatinine $M = 112$

Fig. 4.10 Chemical structures and molecular weights of three metabolites present in high concentration in blood plasma.

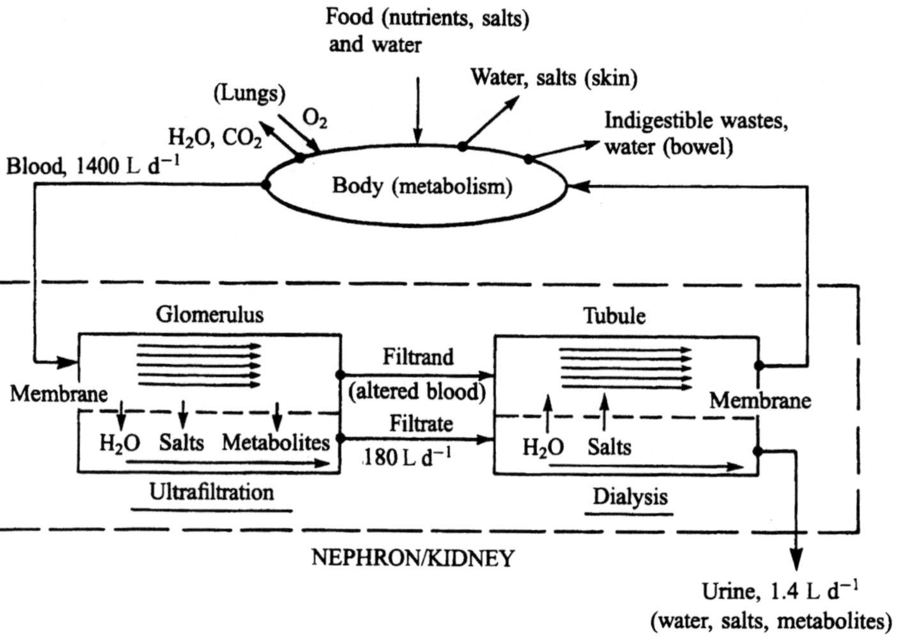

Fig. 4.11 Qualitative material balances on the body, with details of the kidney and how its nephrons function. Two very different membrane processes are involved, with their coupled processing of the blood responsible for fine-tuning of the body's water balance and salt balance and for all elimination of metabolic byproducts.

The duties of a single nephron are diagrammed in Fig. 4.11. To understand this, one must recognize that *two* types of separation process are involved.

1. *Dialysis*. This resembles the ordinary diffusion process discussed earlier, with each diffusing species moving at a different rate through a membrane, from high c_i to low c_i. (When $i = H_2O$, the process is called osmosis.) However, the membrane is 'selective' or 'semipermeable', not allowing large molecules to enter, while permitting small ones to pass through. A typical membrane in the nephron is about 50 nm thick and will admit molecules only of size smaller than about 5 nm; clearly, proteins are completely excluded, while water (0.4 nm) passes easily. Electrolytes can also permeate, and so can metabolites including the middle molecules.

2. *Ultrafiltration*. Here, filtration—cited above—is achieved by the application of a higher pressure on one side of the membrane than the other. All molecules in the plasma are subjected equally to this pressure differential, Δp, but their responses are very different. Proteins cannot pass, because they are too large to be accommodated within the semipermeable membrane. The smallest molecules move through rapidly and the larger ones more slowly, all roughly in proportion to Δp; this is an example of pressure diffusion, discussed previously. In the nephron, the high-pressure side of a membrane can be as much as 50 mmHg (0.066 atm) above atmospheric pressure.

Sometimes dialysis and ultrafiltration can be opposed, as eqn (4.7) clearly shows (when ∇p and ∇c_k have different signs). It is also possible that the *net* Δp can be reduced if osmotic pressure π works in the opposite direction, so $\Delta p_{\text{net}} = \Delta p - \pi$. This can still lead to molecular transport against a concentration gradient, as is termed 'reverse osmosis' when water is being transported.

Figure 4.11 shows that the natural kidney employs both processes. In the glomerulus, ultrafiltration forces salts, metabolites, and much water through a fairly coarse but still selective membrane. The remaining filtrand, depleted in water, clear of metabolites, and partially balanced in electrolytes, moves on to the tubule, as does the filtrate. Here, dialysis is used to achieve the final balances: the proper amount of water is returned by osmosis and the salt levels receive adjustment by transport in one direction or the other.

(c) The artificial kidney (AK)

The AK being used today employs primarily dialysis, with little ultrafiltration being accomplished. It consists of the mass-transfer unit (dialyser) and auxiliary equipment, as shown in Fig. 4.12. The patient's bloodstream is passed through the dialyser in one direction, and a cleansing salt-balanced water stream (dialysate) is passed through in the other direction; this 'countercurrent' operation is very common in many kinds of engineering operations, as it keeps the local concentration differential between the blood and dialysis streams, $\Delta\bar{c}_i(z) \equiv \bar{c}_{iB} - \bar{c}_{iD}$ (for a given i) approximately constant from one end to the other. The two streams are separated by a semipermeable membrane, usually cellulose acetate which has been produced by special methods. Dialysate can be either a standard solution or one freshly prepared to suite the electrolyte imbalance in the patient. Blood is recycled to the patient, while dialysate is discarded.

Home dialysis, often administered by the patient, became feasible with the advent of inexpensive, small dialysers of the sort shown in Fig. 4.13. These units consist of hollow fibers (200 μm i.d., 30 μm wall) through which the blood passes; about 10^4 of these are bundled together in parallel to handle the whole flow. Dialysate washes around the outside of each fiber, removing metabolites and either adding or removing electrolytes. The entire unit is fabricated from polymeric materials, including the cellulose acetate fibers, polymethylethacrylate casing, and polyurethane (rubber) sealant. Fabricated by mass production methods and sterilized before shipping, these units are designed and priced for disposal after one use. There is no need for a blood pump, as the blood flows through the dialyser sufficiently fast with the patient's own blood pressure; typically, the flow causes a pressure loss of only about 15 mmHg.

The operation of the dialyser can be understood by inspecting the mean longitudinal concentration profiles $\bar{c}_i(z)$ in both streams, as shown in Fig. 4.14 for a typical metabolite (for example, i = urea). Each side of the dialyser is identified by number (1 = side of blood entry, 2 = side of blood exit), and the two streams are identified by letter (B = blood, D = dialysate). The four end-point concentrations of this component are thus $\bar{c}_{B1}$,

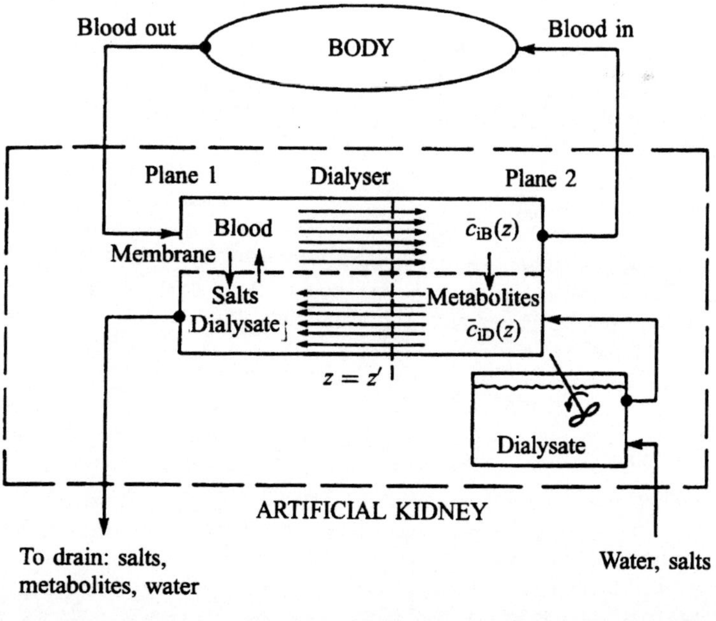

Fig. 4.12 Function of the AK. It operates primarily by dialysis, though a small amount of ultrafiltration may occur (not shown). Salts may be exchanged across the membrane in either direction, depending on composition of the dialysate. A typical dialysate composition (mmol L^{-1}) for treating a patient in renal failure (plasma levels in parentheses) would be: 132 (128) Na^+, 1 (6) K^+, 2.0 (1.6) Ca^{2+}, 1.6 (0.5) Mg^{2+}, 102 (92) Cl^-, 36 (12) HCO_3^-, 0 (1.2) PO_4^{-3}, 0 (0.8) SO_4^{2-}, the corresponding normal levels are approximately 132, 4, 2.0, 1.5, 102, 26, 0.7, and 0.5, respectively.

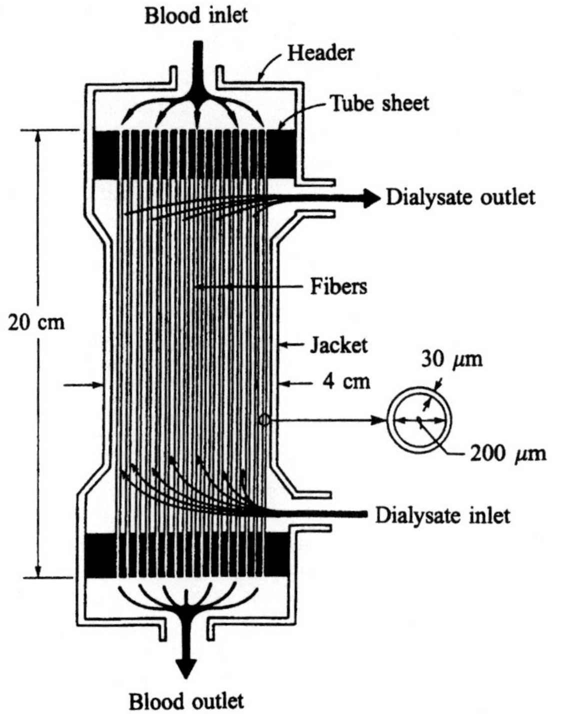

Fig. 4.13 Schematic representation of a portable, disposable dialyser unit operating with hollow-fiber technology. The drawing, with approximate dimensions, grossly exaggerates the spacing between fibers and fiber size in order to show the dialysate flow path; actually, the 10 000 fibers are extremely close to each other.

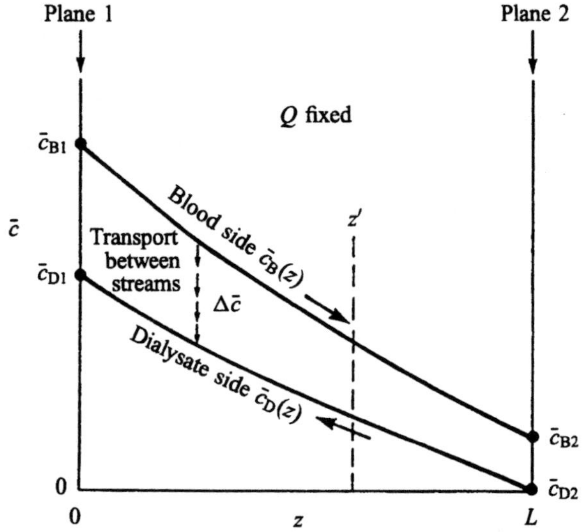

Fig. 4.14 Longitudinal composition profiles $\bar{c}(z)$ for the two streams (blood and dialysate) in countercurrent flow. Because the sketch represents the behavior of a dissolved metabolite, always $\bar{c}_B > \bar{c}_D$ at any z and the dialysate entering is free of that component ($\bar{c}_{D2} = 0$). Note that these profiles will change as Q changes.

$\bar{c}_{B2}$, $\bar{c}_{D1}$, and $\bar{c}_{D2}$ ($= 0$). Figure 4.14 displays how $\bar{c}_B(z)$ and $\bar{c}_D(z)$ behave for countercurrent flow; note that the trans-membrane concentration driving force for transport, $\Delta\bar{c} = \bar{c}_B - \bar{c}_D$, is about the same at all z. The profiles $\bar{c}_B(z)$ and $\bar{c}_D(z)$ need not be linear, though they are often close to linear.

(d) Convective mass transfer

In analyzing the performance of a given AK, we need to be able to predict $\Delta\bar{c}(z)$—or, to put it another way, we must predict how $\bar{c}_{B2}$ responds to changes in $\bar{c}_{D2}$ and Q_D. The problem is even more challenging for design of the basic AK itself, as we must be concerned also with geometrical factors, pressure drop, Q_B, and materials problems.

To understand $\Delta\bar{c}(z)$, we must first analyze the nature of transverse transport at a fixed position $z = z'$. Using the planar schematic of Fig. 4.12, we investigate the y-dependence of $c(y, z')$. An expanded view of the z' region is given in Fig. 4.15, together with the superimposed profile $c(y)$. This profile makes it clear that there are *three* regions in which significant gradients of c occur: the two 'film' regions in the two fluids, of thickness δ_B and δ_D, and the membrane of thickness δ. All three regions must be involved in the overall transport between $\bar{c}_B(z')$ and $\bar{c}_D(z')$. Each region provides a *resistance* to the transport, and—as we shall see—the resistances are additive ($R = R_B + R_M + R_D$) just as for electrical resistances.

We use two basic transport relationships: Fick's law for the membrane and the defining equation for mass transfer coefficients in the liquid phases. In all regions, the flux for a given component is the same (for this planar geometry) at a value J. Thus, we can write three equations for J and invert them as follows:

$$\textit{blood} \qquad \bar{c}_B - c_B^i = \frac{1}{k_{cB}} J, \tag{4.29}$$

$$\textit{membrane} \quad c_{MB}^i - c_{MD}^i = \frac{\delta}{\mathscr{D}} J, \tag{4.30a}$$

or, using the interfacial condition $c_M^i = \phi c^i$,

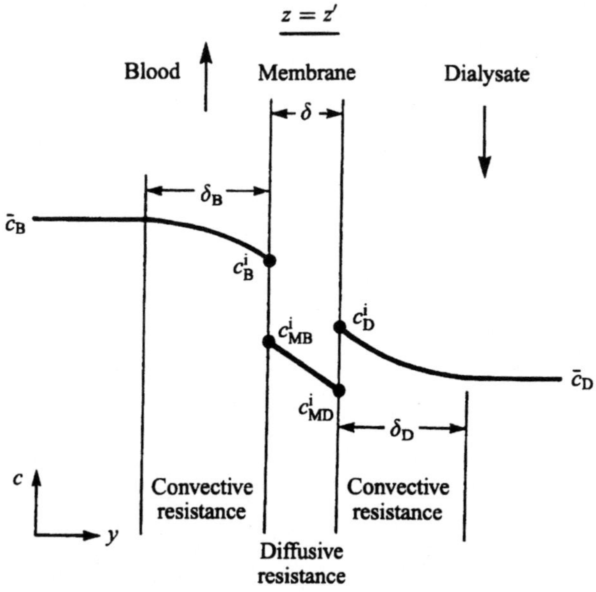

Fig. 4.15 Concentration profile $c(y)$ for a component moving from the blood side to the dialysate side of the membrane, at any given longitudinal position ($z = z'$).

$$\textit{membrane} \quad c^i_{\mathrm{B}} - c^i_{\mathrm{D}} = \frac{\delta}{\phi\mathscr{D}}J, \tag{4.30b}$$

$$\textit{dialysate} \quad c^i_{\mathrm{D}} - \bar{c}_{\mathrm{D}} = \frac{1}{k_{\mathrm{cD}}}J. \tag{4.31}$$

Summation of eqns (4.29), (4.30b), and (4.31) gives

$$\Delta\bar{c} = \bar{c}_{\mathrm{B}} - \bar{c}_{\mathrm{D}} = \left[\frac{1}{k_{\mathrm{cB}}} + \frac{\delta}{\phi\mathscr{D}} + \frac{1}{k_{\mathrm{cD}}}\right]J, \tag{4.32}$$

which is a relationship of the form:

driving force = resistance × flow rate,

and we can see that $R_{\mathrm{B}} = 1/k_{\mathrm{cB}}$, $R_{\mathrm{M}} = \delta/\phi\mathscr{D}$, and $R_{\mathrm{D}} = 1/k_{\mathrm{cD}}$. In most artificial kidneys a good approximation would be $R_{\mathrm{M}} \approx R_{\mathrm{D}} \approx R_{\mathrm{M}} \approx R/3$, though in hollow-fiber dialysers $R_{\mathrm{M}} \approx R/2$ would be closer.

It should be noted that liquid film thicknesses were not used in this analysis. In effect, they were replaced by k_{cB} and k_{cD} as empirical tools for correlating data and making calculations. Equation (4.32) can be inverted to give an explicit expression for flux,

$$J = \frac{\Delta\bar{c}}{\left[\dfrac{1}{k_{\mathrm{cB}}} + \dfrac{\delta}{\phi\mathscr{D}} + \dfrac{1}{k_{\mathrm{cD}}}\right]} \equiv K\,\Delta\bar{c}, \quad \text{local, } z = z', \tag{4.33}$$

which defines the *overall* mass transport coefficient K, which is another way of representing flux relationships. An analogy to the overall heat transfer coefficient in heat exchanger design would be accurate.

Symbols used above did not include designation of the chemical component being transported, but these can be added at will. We shall now do so, considering component X, and write the *rate* of mass transfer of X across a planar membrane section of length dz and breadth b at position z':

$$\mathrm{d}W'_X \approx J'_X(b\,\mathrm{d}z) = K_X(z')\Delta\bar{c}(z')b\,\mathrm{d}z. \tag{4.34}$$

If K_X is independent of z, as is usually true, and if $\Delta\bar{c}$ proves to be independent of z (as is approximated in countercurrent operation), then eqn (4.34) can be integrated trivially to give the total transport rate across the entire membrane:

$$W_X = K_X\,\Delta\bar{c}\,bL = K_X(\bar{c}_{\mathrm{B}} - \bar{c}_{\mathrm{D}})bL, \tag{4.35}$$

where the membrane area is bL. In this case, $(\bar{c}_{\mathrm{B}} - \bar{c}_{\mathrm{D}})$ could be evaluated at any z-position, the end positions ($z = 0$ or L) being common choices.

However, in general, we cannot expect $\Delta\bar{c}$ to be independent of z, even in countercurrent flow. This situation is usually addressed by defining various kinds of *average* $\Delta\bar{c}$ and corresponding *average* K_X. All must predict the same W_X, of course. For example, suppose we have the variations $\bar{c}_{\mathrm{B}}(z)$, $\bar{c}_{\mathrm{D}}(z)$, and even $K_X^{\mathrm{local}}(z)$. Then,

$$W_X = \int_0^L K_X^{\mathrm{local}}(z)[\bar{c}_{\mathrm{B}}(z) - \bar{c}_{\mathrm{D}}(z)]b\,\mathrm{d}z \tag{4.36a}$$

$$\equiv K_X^{\mathrm{avg}}(\Delta\bar{c})_{\mathrm{avg}}bL \tag{4.36b}$$

$$\equiv K_X^{\mathrm{lm}}(\Delta\bar{c})_{\mathrm{lm}}bL, \tag{4.36c}$$

where the average driving forces (averages of the inlet and outlet values) are

arithmetic mean:

$$(\Delta\bar{c})_{\mathrm{avg}} = \tfrac{1}{2}[(\Delta\bar{c})_1 + (\Delta\bar{c})_2] = \tfrac{1}{2}[\bar{c}_{\mathrm{B1}} - \bar{c}_{\mathrm{D1}} + \bar{c}_{\mathrm{B2}} - \bar{c}_{\mathrm{D2}}] \tag{4.37a}$$

logarithmic mean[7]:

$$(\Delta\bar{c})_{\mathrm{lm}} = \frac{(\Delta\bar{c})_1 - (\Delta\bar{c})_2}{\ln[(\Delta\bar{c})_1/(\Delta\bar{c})_2]}$$
$$= \frac{\bar{c}_{\mathrm{B1}} - \bar{c}_{\mathrm{D1}} - \bar{c}_{\mathrm{B2}} + \bar{c}_{\mathrm{D2}}}{\ln[(\bar{c}_{\mathrm{B1}} - \bar{c}_{\mathrm{D1}})/(\bar{c}_{\mathrm{B2}} - \bar{c}_{\mathrm{D2}})]}. \quad (4.37\mathrm{b})$$

Because the numerical values of $(\Delta\bar{c})_{\mathrm{avg}}$ and $(\Delta\bar{c})_{\mathrm{lm}}$ differ, so also do the values of K_X^{avg} and K_X^{lm}. An independent prediction of K_X^{avg} or K_X^{lm} can be made by using the form of eqn (4.33) but replacing the k_{cB} and k_{cD} by either $k_{\mathrm{cB}}^{\mathrm{avg}}$ and $k_{\mathrm{cD}}^{\mathrm{avg}}$ or $k_{\mathrm{cB}}^{\mathrm{lm}}$ and $k_{\mathrm{cD}}^{\mathrm{lm}}$, obtained from correlations similar to those discussed earlier but developed from data analyzed in terms of the respective mean driving force (see footnote 7).

We next abandon the planar geometry used in our discussion thus far and generalize the treatment of transport to apply to a curvilinear system—namely, the cylindrical system used in hollow-tube dialysers. While the physics is no different, it is now important to recognize that the cross-sectional *area* through which the radial flux moves varies with radial position. These factors are automatically accommodated in the differential equations of Tables 4.3 and 4.4, but we will not begin there. Instead, we will refer to Fig. 4.15 and display the analogues of eqns (4.29)–(4.32):

blood
$$\bar{c}_{\mathrm{B}} - c_{\mathrm{B}}^i = \frac{1}{k_{\mathrm{cB}}}\left(\frac{W}{2\pi r_{\mathrm{B}} L}\right), \quad (4.38)$$

membrane
$$c_{\mathrm{B}}^i - c_{\mathrm{D}}^i = \frac{r_{\mathrm{B}} \ln(r_{\mathrm{D}}/r_{\mathrm{B}})}{\phi\mathscr{D}}\left(\frac{W}{2\pi r_{\mathrm{B}} L}\right)$$
$$= \frac{r_{\mathrm{D}} \ln(r_{\mathrm{D}}/r_{\mathrm{B}})}{\phi\mathscr{D}}\left(\frac{W}{2\pi r_{\mathrm{D}} L}\right), \quad (4.39)$$

dialysate
$$c_{\mathrm{D}}^i - \bar{c}_{\mathrm{D}} = \frac{1}{k_{\mathrm{cD}}}\left(\frac{W}{2\pi r_{\mathrm{D}} L}\right), \quad (4.40)$$

where eqn (4.39) represents a solution to the trans-membrane diffusion problem in cylindrical coordinates. Note that the fluxes W/A_{B} and W/A_{D} at the membrane inner and outer surfaces are clearly different. Addition of eqns (4.38)–(4.40) gives

$$\bar{c}_{\mathrm{B}} - \bar{c}_{\mathrm{D}} = \left[\frac{1}{k_{\mathrm{cB}} r_{\mathrm{B}}} + \frac{\ln(r_{\mathrm{D}}/r_{\mathrm{B}})}{\phi\mathscr{D}} + \frac{1}{k_{\mathrm{cD}} r_{\mathrm{D}}}\right]\frac{W}{2\pi L}. \quad (4.41)$$

If we choose to put this in the form of eqn (4.33), we must arbitrarily select one of the cross-sectional areas as a reference surface:

$$J|_{r_{\mathrm{B}}} = \frac{W}{2\pi r_{\mathrm{B}} L} = \frac{\Delta\bar{c}}{\left[\frac{1}{k_{\mathrm{cB}}} + \frac{r_{\mathrm{B}} \ln(r_{\mathrm{D}}/r_{\mathrm{B}})}{\phi\mathscr{D}} + \frac{r_{\mathrm{B}}}{k_{\mathrm{cD}} r_{\mathrm{D}}}\right]}$$
$$\equiv k^{\mathrm{B}}\Delta\bar{c}. \quad (4.42)$$

(e) Design equations for the AK

Three interrelated equations can be written, basically material balances, with one transport expression involved. On a molar basis, these are (for component X excessively concentrated in the blood)

blood side: $\mathrm{d}\mathscr{M}_X/\mathrm{d}t = 0$:

$$0 = (W_{\mathrm{B1}} - W_{\mathrm{B2}})_{\mathrm{flow}} - W_{\mathrm{memb}} \quad (4.43\mathrm{a})$$
$$= Q_{\mathrm{B}}(\bar{c}_{\mathrm{B1}} - \bar{c}_{\mathrm{B2}}) - K_X^{\mathrm{B}}(2\pi r_{\mathrm{B}} L)\Delta\bar{c}, \quad (4.43\mathrm{b})$$

dialysate side:

$$0 = (W_{\mathrm{D2}} - W_{\mathrm{D1}})_{\mathrm{flow}} + W_{\mathrm{memb}} \quad (4.44\mathrm{a})$$
$$= Q_{\mathrm{D}}(\bar{c}_{\mathrm{D2}} - \bar{c}_{\mathrm{D1}}) + K_X^{\mathrm{B}}(2\pi r_{\mathrm{B}} L)\Delta\bar{c}, \quad (4.44\mathrm{b})$$

where it should be recognized that W_{memb} could have been represented equally well by $K_X^{\mathrm{D}}(2\pi r_{\mathrm{D}} L)\Delta\bar{c}$. Overall,

$$0 = (W_{\mathrm{B1}} + W_{\mathrm{D2}} - W_{\mathrm{B2}} - W_{\mathrm{D1}})_{\mathrm{flow}} \quad (4.45\mathrm{a})$$
$$= Q_{\mathrm{B}}(\bar{c}_{\mathrm{B1}} - \bar{c}_{\mathrm{B2}}) + Q_{\mathrm{D}}(\bar{c}_{\mathrm{D2}} - \bar{c}_{\mathrm{D1}}). \quad (4.45\mathrm{b})$$

Note that eqn (4.45) could have been obtained by adding eqns (4.43) and (4.44), showing that only two of these three equations are independent and only two unknown quantities can be obtained by solving them.

---

[7] This peculiar but commonly used average emerges from the solution to an idealized problem: surface concentration c_{B}° (e.g. at the membrane, but in the liquid phase) does not vary at all with z. A material balance in section dz then gives

(Phase B):

$$\mathrm{d}W_X = Q\,\mathrm{d}\bar{c}_{\mathrm{B}} = k_{\mathrm{cB}}(\bar{c}_{\mathrm{B}} - c_{\mathrm{B}}^{\circ} b\,\mathrm{d}z, \quad (\mathrm{F1})$$

or, rearranging,

$$\frac{\mathrm{d}(\bar{c}_{\mathrm{B}} - c_{\mathrm{B}}^{\circ})}{(\bar{c}_{\mathrm{B}} - c_{\mathrm{B}}^{\circ})} = \frac{k_{\mathrm{cB}} b}{Q}\mathrm{d}z \quad (\mathrm{F2})$$

and integrating:

$$\ln[\Delta\bar{c}_{\mathrm{B2}}/\Delta\bar{c}_{\mathrm{B1}}] = k_{\mathrm{cB}} bL/Q. \quad (\mathrm{F3})$$

Direct integration of the left side of eqn (F1) and use of eqn (F3) gives:

$$W_X = Q(\bar{c}_{\mathrm{B2}} - \bar{c}_{\mathrm{B1}}) = k_{\mathrm{cB}} bL \frac{[(\bar{c}_{\mathrm{B2}} - c_{\mathrm{B}}^{\circ}) - (\bar{c}_{\mathrm{B1}} - c_{\mathrm{B}}^{\circ})]}{\ln[(\bar{c}_{\mathrm{B2}} - c_{\mathrm{B}}^{\circ})/(\bar{c}_{\mathrm{B1}} - c_{\mathrm{B}}^{\circ})]} \quad (\mathrm{F4})$$

Data on W_X can be correlated using this logarithmic mean driving force $\Delta(\bar{c}_{\mathrm{B}} - c_{\mathrm{B}}^{\circ})_{\mathrm{lm}}$, in which case the coefficient is labelled $k_{\mathrm{cB}}^{\mathrm{lm}}$.

One typical design question might be: what transfer surface area ($2\pi r_B L$) is required to achieve a desired outlet blood X-concentration ($\bar{c}_{B2}$), when Q_B and Q_D are fixed, and $\bar{c}_{B1}$ and $\bar{c}_{D2}$ are known or controlled, respectively, and only one type of hollow fiber is available (r_B and r_D are given)? This is equivalent to seeking L, but the less obvious feature of the problem is that $\bar{c}_{D1}$ must also be found; $\bar{c}_{D1}$ appears both explicitly and implicitly (examine $\Delta\bar{c}$) in eqns (4.43)–(4.45). Note that the answer will depend on the stipulated value of Q_D, which is arbitrary at the beginning because it can be so easily changed (while Q_B is restricted rather narrowly by the patient's own blood flow rate and medical considerations). The solution must be an iterative one, though it converges very rapidly in most cases. There are further subtleties in this problem as well. For example, if L proves to be large, will the stipulated Q_B be possible, given the limits on Δp imposed by the patient's blood pressure? Should a blood pump be used, or should the outlet objective blood level ($\bar{c}_{B2}$) be altered in the direction of $\bar{c}_{B1}$?

Design work can be very complex, clearly. Somewhat easier is analysis of an existing process or piece of equipment, since there are fewer variables. One example of such analysis is encountered with a given dialyser (surface area fixed, etc.) when one asks: what Q_D should be used in order to achieve a desired $\bar{c}_{B2}$, when $\bar{c}_{B1}$ and $\bar{c}_{D2}$ are known and Q_B is fixed?

In both these two problems, certain information was omitted. It would be necessary to know the components of K^B (or K^D)—k_{cB}, k_{cD}, ϕ, $\mathscr{D}$—or calculate them; in the latter case, the k_c's would be obtained from correlations in terms of Reynolds number (requiring μ and flow-channel spacing). Questions of biocompatibility and strength of materials would have to be addressed semiquantitatively in design work.

Finally, one should observe that the AK has been analysed here as a pseudo-steady-state problem. If blood has had its X-component adjusted to a level $\bar{c}_{B2}$ at the AK exit, it returns to the body and serves to alter the blood composition in the body, so that $\bar{c}_{B1}$ is progressively changing. This dilution influence is, of course, countered by other processes in the body that are producing the adverse concentrations in the first place. None the less, the net result is the systematic decrease (or increase) of $\bar{c}_{B1}$ with time—as must be true if the AK is doing any good at all. The pseudo-steady-state approach would proceed by making an independent calculation of $\bar{c}_{B1}(t)$ by assessing the whole system (body and AK) and using this as an input variable everywhere in the equations above. If this t-variation is slow, in the sense that the convective and diffusive processes in the AK are much faster (i.e. transients in these processes are not apparent), then the approach will be successful.

Additional information about the AK can be found in the treatise by Drucker *et al.* (1978).

Problems

P4.1 Derive the molar equation of continuity for component i and also for total moles in the system.

P4.2 Demonstrate that $\sum_k J_k^* = 0$, using basic definitions of flux.

P4.3 For a binary system (say, albumin in water) in which ordinary diffusion is occurring, show that $\mathscr{D}_{AB} = \mathscr{D}_{BA}$—i.e. the diffusivity of A in an A–B mixture is the same as that of B.

P4.4 The number-average molecular weight of a molecular mixture (e.g. blood plasma) is defined by $\bar{M}_n = \rho/c = \sum_i N_i M_i / \sum_i N_i$, where N_i is the number of molecules of molecular weight M_i in a given sample.

(a) Show that $1/\bar{M}_n = \sum (\omega_i/M_i)$, where ω_i is the weight fraction of component i.

(b) Show that the same result emerges from the mixed (molar and mass) flux relationship in eqn (4.6c). *Hint:* the results of Exercise 3 could be useful.

(c) Calculate $\bar{M}_n$ for blood plasma. How does it compare with M for pure water?

P4.5 Normal blood plasma contains a low level of free hemoglobin (though most hemoglobin resides in erythrocytes), so it could be of interest to know the hemoglobin diffusivity in plasma $\mathscr{D}^{\circ}_{HP}$. Estimate $\mathscr{D}^{\circ}_{HP}$ at 20°C by assuming that this globular protein can be modeled as an impervious sphere of 6.0 nm diameter.

P4.6 The property of *permeability* (P) for a membrane, with respect to a diffusing component X, can be defined by: $J_x^* \equiv P_{XM}(\bar{c}_{X1} - \bar{c}_{X2})$. Here, J_X^* is the scalar (y-directed) molar flux of X across the membrane, while molar concentrations of X in the bulk fluids on the two sides of the membrane are $\bar{c}_{X1}$ and $\bar{c}_{X2}$. Obtain an analytical expression for P_{XM} in terms of fundamental membrane and membrane/diffusant properties. Explain and justify any approximations you make.

P4.7 (a) For the matrix-type drug delivery device discussed in the text, obtain a pseudo-steady-state solution for $\delta_D(t)$ that is valid for all K.
(b) From this $\delta_D(t)$, predict the rate of drug release $W_d(t)$.
(c) Suppose that $\mathscr{D}_{dm} = \mathscr{D}_{db} = 10^{-5}\ cm^2\ s^{-1}$, $\delta_H = 100\ \mu m$, and $c_o/c_m = 100$. Also, take c_b/Kc_m as negligible compared to unity. Then, plot curves on logarithmic graph paper of W_d vs. t, for the cases $K = 10^{-1}$, 10^{-2}, and 10^{-3}. Produce these curves for the time period running from $t = 10$ seconds through $t = 1$ month (roughly 2.6×10^6 s). Over what period is W_d a constant, for each K?
(d) For a device designed as a rectangular wafer, with long dimensions L and B and a small thickness H, obtain an analytical prediction for the useful lifetime of the device.
(e) Use your prediction in (d) to evaluate those lifetimes when $L = B = 5$ mm and $H = 0.5$ mm, given the other parameters in (c), for each of the three K in (c).

P4.8 Most matrix devices for controlled drug release are in the form of a rectangular slab. One major drawback of these systems is that they generally do not display zero-order (constant-rate) release kinetics. To overcome this drawback, various matrix geometries have been considered, and an *inwardly* releasing hemisphere has been proposed to achieve essentially zero-order kinetics. Refer to Fig. 4.16 for detailed structure of such a device, and the geometrical parameters to be used in your analysis. *Show that the theoretical release rate is indeed zero-order* after an initial transient vanishes.

You may assume that the drug concentration at the interface of the matrix and body solution is nearly zero, and also the initial drug loading is much greater than the solubility of the drug in the matrix, i.e. $c_o \gg c_m$. In addition, the geometry of the hemisphere is such that $a_o \gg a_i$.

P4.9 For a *membrane*-type device containing a saturated solution inside, 'dry storage' prior to implantation allows the membrane itself to become saturated at a value c_{SM} (interior solution is at c_S). Consider the initial performance of the device upon implantation.
(a) Without solving any equations, sketch a series of $c(y)$ profiles after the implantation. That is, display $c(y, 0)$, $c(y, t_1)$, $c(y, t_2)$, ... to demonstrate qualitatively the behavior of $c(y, t)$. Carry this out until steady-state operation is reached.
(b) Also sketch $W_d(t)$ over the same period of time covered in (a). Neglect convection.
(c) Using the full partial differential equation needed here, obtain an analytical solution for $c(y, t)$. [*Hint:* assume a solution of the form $c = c_\infty(y) - c_t(y, t)$, where c_∞ is the steady-state result and c_t is the

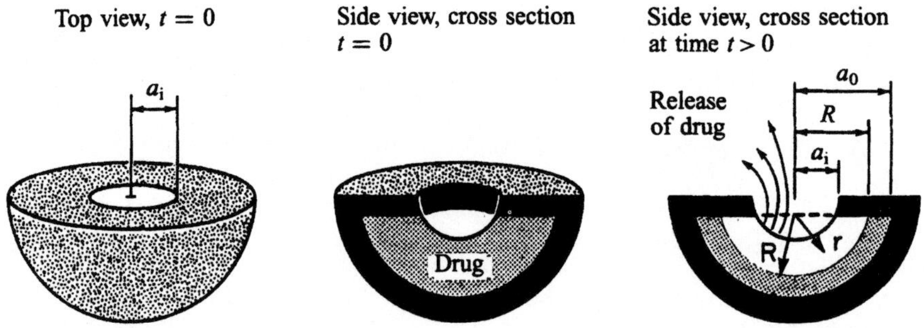

Fig. 4.16 Schematic diagram of an inwardly releasing hemispherical matrix device. Matrix boundaries a_i (inner radius) and a_o (outer radius) are fixed, while the depletion zone boundary $R(t)$ is moving. The white region represents the depletion zone and the shaded region the drug-saturated zone. No diffusion can occur through the impervious shell, which covers the device everywhere except at $r = a_i$.

transient that dies out at long times. Since $c_\infty(y)$ is known, only c_t needs to be found. Then, assume further that the method of separation of variables will work: $c_t = A(t) \cdot B(y)$. Some simplification is achieved by using reduced variables, such as $c^* = c/c_{SM}$, $y^* = y/\delta$, and $t^* = t\mathscr{D}/\delta^2$.]

(d) What is the duration, approximately, of the transient period? Express your result in units of t^* first, and then in terms of t using the parameters $\mathscr{D} = 10^{-5}$ cm^{2}s^{-1} and $\delta = 0.1$ mm.

P4.10 A variety of chemical concentration units are used in connection with the compounds in body fluids. Two of these are 'milligram per cent' (mg%) and, for salts, 'milli-equivalents per liter' (meq L^{-1}):

$$\text{mg\%} = \text{mg of compound}/100 \text{ mL of solution};$$

$$\text{meq L}^{-1} = (\text{mmol L}^{-1}) \times \text{charge of ion};$$

$$= \left[\frac{\text{mg of ion}/\text{M}_{\text{ion}}}{\text{liters of solution}}\right] \times \text{valence of ion}.$$

(a) Suppose you dissolved 0.65 g of H_3PO_4 in water to make 6 liters of solution. Find the H_3PO_4 concentration in mg% and the PO_4 concentration in meq L^{-1}.

(b) An artificial kidney is to be used to reduce a patient's blood levels of PO_4 (from 105 down to 20 meq L^{-1}) and urea (from 3050 down to 1200 mg%). If his blood volume is 4600 mL, how much PO_4 and urea are being removed?

(c) The patient's blood flow rate through the artificial kidney is $Q_B = 140$ mL min^{-1}, and the countercurrent dialysate flow rate is $Q_D = 2200$ mL min^{-1}. Obtain an equation relating the outlet stream concentrations c_{B2} and c_{D1} at the *beginning* of treatment. (Do this for both urea and PO_4.)

P4.11 The resistance R to mass transfer in the parallel-membrane artificial kidney is $R = R_B + R_M + R_D$, where $R = 1/K$ and K is the overall mass transfer coefficient. Given the following data, for the transport of urea in a specific device involving a cellulose acetate membrane,

$$\delta = 0.45 \text{ mm} \qquad k_B = 0.9 \times 10^{-6} \text{ m s}^{-1}$$

$$\phi = 0.28 \qquad k_D = 1.6 \times 10^{-6} \text{ m s}^{-1}$$

$$\mathscr{D}_M = 4 \times 10^{-7} \text{ cm}^2 \text{ s}^{-1},$$

calculate the percentage of R which can be attributed to R_B, R_M, and R_D. As an engineer assigned to reduce R, what would these results suggest to you about directions to take in a device improvement program?

P4.12 The effect of curvilinear geometry on transport through hollow-fiber membranes is shown in eqn (4.39). Obtain the concentration profile $c(r)$ in a cylindrical membrane and use it to predict the transmission rate W.

P4.13 Two types of mean concentration driving forces are defined in eqn (4.37a) and (4.37b). Show that these become the same as the end-point driving forces $\Delta\bar{c}_1$ and $\Delta\bar{c}_2$ approach each other.

P4.14 In the Krogh cylinder model for capillary transport (Fig. 4.4), oxygen in the blood at concentration $c(z)$ crosses the capillary wall and diffuses radially through the tissue. At each z, this is fundamentally the same process, so that in principle the problem need be solved only once.

(a) If the metabolic consumption of oxygen in the tissue occurs at a constant rate $R_{O_2} = C$, obtain an analytical solution for $c(r)$ in the region between r_c and r_t.

(b) Solve the analogous problem when the oxygen kinetics are first-order, eqn (4.4a). (*Hint:* your answer should involve Bessel functions.)

(c) More realistic than (a) or (b) would be the use of Michaelis–Menten kinetics.], eqn (4.5). Although an analytical expression may be unobtainable, the solution form $c(r)$ must lie between the limiting cases for (a) and (b) (note: the Michaelis–Menten constants must be related to C and k_f by $k_1 = k_f$ and $k_2 = k_f/C$). Draw a qualitative graph of c vs. r, showing your solutions from (a) and (b) and also how the Michaelis–Menten result should look.

(d) Use a numerical analysis to validate your proposed result in (c). You may take $k_1 = 19$ s^{-1} and $k_2 = 1.2$ cm^3 nmol^{-1}.

P4.15 Some design considerations will be examined in the context of a flat-plate dialyser modeled as in Fig. 4.12. The operational equation for total transport rate of component X across the membrane is eqn (4.36c), with eqn (4.37b).

(a) Equation (4.37b) can be simplified for the usual case where X is very low everywhere in the dialysate, so that $\bar{c}_{D1}$ and $\bar{c}_{D2}$ can be neglected in comparison to blood-side concentrations. For this approximation, obtain an explicit expression for $\bar{c}_{B2}$ as a function of $\bar{c}_{B1}$ and blood flow rate Q_B. (*Hint:* make an independent material balance on the bloodstream to get another expression for W_X.)

Also important is the relation between Q_B, device dimensions (L, b, h, where h = blood channel height), and pressure drop Δp on the blood side between planes 1 and 2. If Q_B is viewed as the blood flow rate also in the whole body—i.e. if the body and dialyser are in series—then one can derive

$$Q_B = \left[\frac{\pi D^4 bh^3}{12\pi D^4 L_c + 128bh^3 L}\right]\frac{\Delta p}{\mu}, \qquad (4.P15)$$

where laminar flow is assumed for flow in circular tubes of equivalent length L_c and equivalent diameter D (the circulatory system) and also for flow in flat rectangular channels (the dialyser).

(b) Derive eqn (4.P15).

(c) Suppose, in designing such a device, you are interested only in how the variation of h affects performance. Make two qualitative sketches, plotting Q_B vs. h from eqn (4.P15) and ($\bar{c}_{B1}-\bar{c}_{B2}$) vs. h from your answer to (a).

(d) Note that the product $Q_B(\bar{c}_{B1}-\bar{c}_{B2})$ represents the net rate of X-removal from the blood. Using the two curves in (c), sketch their product vs. h and thereby show that an optimal h exists.

(e) In (c) and (d), you were permitted to hold K constant while varying h at constant Q_B and Q_D. Discuss the validity of this, and describe any influences upon K that you think might be caused by h-variation.

(f) Next, assuming K is constant at a value computed in Problem 4.11, and using L_c = 750 cm, μ = 5 mPa s, Δp = 50 mmHg, and D = 0.3 cm, select the dimensions h, b, and L in order to optimize device effectiveness but still be reasonable by standards of common sense. Note that there may be no single correct answer for this, so be sure to explain your logic and rationale carefully.

P4.16 Biochemical agents (say, A) are used in the *gas* phase for a variety of applications—insect pheromones, fragrances, etc. To control their release rate, it has been proposed to use hollow fibers filled with the *liquid* agent (see Fig. 4.17). Vaporization occurs within the fiber, at the liquid–gas interface (where mole fraction x_{AL} is known), and the A-vapor diffuses outward into the environment where $x_A \approx 0$. Perceived advantages of this system include: device useful lifetime is dependent on fiber length (easily controlled); liquid is safely held in by surface tension (since diameter D is small); total release rate W_A is easily controlled by the number of fibers in a fiber bundle and/or D (i.e. governs liquid surface area/fiber).

Your task is to explain how this device works, ultimately predicting W_A (mol h^{-1}) as a function of D and t.

(a) A pseudo-steady-state analysis is employed, and the system is also considered pseudo-binary (agent A + air). Use the full Fickian expression for N_{Az} to guide a *sketch* of the profile $x_A(z)$ at some instant t (long after an initial transient has damped out).

(b) Show several other $x_A(z)$ curves, for other values of t. Be sure to indicate how the curves are progressing with time.

(c) Obtain an analytical expression for $x_A(z)$. Do *not* assume $x_A \ll 1$.

(d) From $x_A(z)$, develop expressions for N_{Az} and W_A.

(e) Write another expression for W_A, in terms of the movement of the liquid level $L(t)$.

(f) Combine your answers from (d) and (e) to determine $L(t)$. From this, answer: Would this system function approximately as a constant-rate release device? *Explain.*

P4.17 Dialysis treatment of a patient suffering from renal insufficiency is actually a time-dependent process,

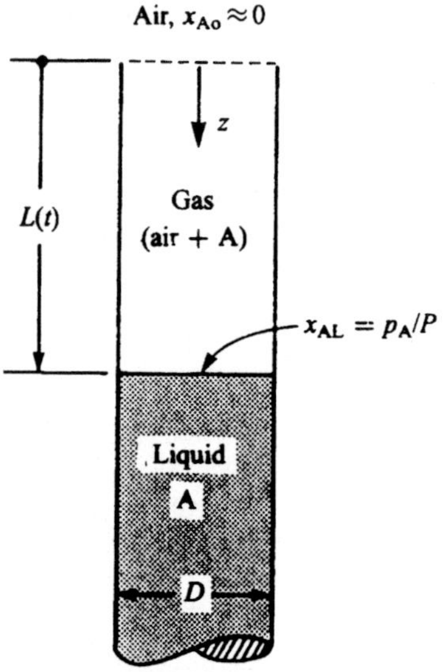

Fig. 4.17 Capillary containing agent A in liquid form, with A-vapor forming at the gas interface $z = L$ at an equilibrium concentration characterized by the liquid vapor pressure p_A. Air in the capillary does not dissolve in liquid A.

though the text did not explicitly examine this. You are now asked to consider the unsteady state.

(a) With a sketch, show how $\bar{c}_B(z)$ and $\bar{c}_D(z)$ in Fig. 4.14 change with time during a single treatment.

(b) From appropriate material balances, develop equations to predict $\bar{c}_{B1}(t)$ in terms of other parameters. Some factors will have to be newly identified and characterized within your equations, so be sure to define all terms carefully.

(c) Solve the equation(s) in (b), predicting $\bar{c}_{B1}(t)$. What is its limiting value, if any, at long times?

References

Bird, R. B., Stewart, W. E., and Lightfoot, E. N., Jr (1960). *Transport phenomena*. Wiley, New York.

Carslaw, H. S. and Jaeger, J. C. (1959). *Conduction of heat in solids*, (2nd edn). Oxford University Press.

Crank, J. (1956). *The mathematics of diffusion*. Oxford University Press.

Cussler, E. L. (1984). *Diffusion: Mass transfer in fluid systems*. Cambridge University Press.

Drucker, W., Parsons, F. M., and Maher, J. F. (1978). *Replacement of renal function by dialysis*. Martin Nijhoff Medical Division, Boston.

Hirschfelder, J. O., Curtiss, C. F., and Bird, R. B. (1954). *Molecular theory of gases and liquids*. Wiley, New York.

Lightfoot, E. N., Jr (1974). *Transport phenomena and living systems*. Wiley, New York.

Seagrave, R. C. (1971). *Biomedical applications of heat and mass transfer*. Iowa State University Press, Ames.

BIOHEAT TRANSFER

Takeshi K. Eto and Boris Rubinsky

5

Contents

Symbols

- A area (m^2)
- Bi Biot number (hL_c/k_{solid})
- c heat capacity (J $kg^{-1}K^{-1}$)
- C capacitance (J K^{-1}); constant
- d diameter (m)
- dx differential length (m)
- E power (W)
- F shape factor
- Fo Fourier number
- h heat transfer coefficient (W $m^{-2}K^{-1}$)
- H enthalpy (J kg^{-1}); height (m)
- I modified Bessel function
- k thermal conductivity (W $m^{-1}K^{-1}$)
- K conductance (W $m^{-2}K^{-1}$)
- L latent heat (J kg^{-1}); length (m)
- m mass (kg); or as defined in Example 5
- n number density of vessel pairs (m^{-2})
- N number of transfer units
- Nu Nusselt number (hL_c/k_{fluid})
- P perimeter (m)
- q heat (J)
- r radius (m); cylindrical coordinate (m); spherical coordinate (m)
- R resistance (K W^{-1})
- t time (s)
- T temperature (°C or K)
- u velocity (m s^{-1})
- U overall heat transfer coefficient (W $m^{-2}K^{-1}$); Darcy velocity (m s^{-1})
- V voltage (V); volume (m^3)
- W weight (kg)
- x spatial Cartesian coordinate (m); exponential equilibration length (m)
- y spatial Cartesian coordinate (m)
- z spatial Cartesian coordinate (m); cylindrical coordinate (m)
- α absorptivity; thermal diffusivity (m^2 s^{-1})
- β directional cosine
- δ tissue thickness (m)
- Δx material thickness (m)
- ε emissivity
- E emissivity function
- ζ dimensionless length; or as defined in eqn (5.62c)
- η dimensionless length
- θ dimensionless temperature; cylindrical and spherical coordinate
- κ thermal equilibration parameter
- λ defined in eqn (5.62e)
- Λ defined in eqn (5.50b)
- μ defined in eqn (5.62f)
- ξ dimensionless length
- ρ reflectivity; density (kg m^{-3})
- σ Stefan–Boltzmann constant (5.67×10^{-8} W $m^{-2}K^{-4}$); or as defined in eqn (5.65c)
- τ transmissivity; dimensionless time

v	defined in eqn (5.62b)
ϕ	conduction shape factor
Φ	defined in eqn (5.63b)
ψ	spherical coordinate
ω	blood perfusion rate (s^{-1}); solid angle
Ω	total solid angle

Subscripts

a	material a; arterial
avg	average
b	blackbody; material b; blood
c	convective; material c; characteristic; deep body core
cond	conduction
conv	convective
D	Darcy velocity
e	exponential equilibration length
eff	effective
fluid	fluid
g	gas
gen	generation
i	interstitial
i	material i; tensor indices
ij	tensor indices
in	inflow
inner	inner
j	jth blood vessel; tensor indices
l	liquid
met	metabolic heat generation
o	initial; no blood perfusion
out	outflow
outer	outer
P	blood perfusion
r	radiative
rad	radiation
ref	reference
s	surface; skin; solid
st	storage
t	thermal; tissue
total	total
v	venous
x	control surface x
$x + dx$	control surface $x + dx$
1	surface 1
2	surface 2
3	surface 3
4	surface 4
1–2	surface 1 to surface 2
∞	environment, far from surface
ε	emissivity function

Superscripts

$'$	spatial derivative
*	j*th blood vessel

Overlines

$\dot{}$	temporal derivative
$\bar{}$	average

5.1 Introduction

The transport of heat plays a vital role in life processes. Homeotherms, or warm-blooded animals (including humans), must maintain their internal temperature relatively stable although they may experience a variety of environmental conditions (Houdas and Ring 1982; Rowell 1983; Bligh 1985). To achieve this thermoregulation, homeotherms have developed very complex control mechanisms. Better knowledge of heat transfer in biological systems can help us in our understanding of life processes.

Knowledge of heat transport is important for thermal comfort. To survive in harsh environments such as in the deep sea or out in space, it is important to design special gear for safety. In addition, thermal comfort is an important design component for the home and workplace (ASHRAE 1985).

Heat transfer analysis can be used to obtain information about the properties of tissue. For example, the flow of blood can be evaluated using a thermal dilution technique. In this procedure, heat is either injected or generated locally and the thermal clearance is monitored. With knowledge of initial thermal conditions and the thermal clearance rate, it is possible to estimate blood flow rates. A review of thermal dilution techniques can be found elsewhere (Bowman 1985).

A person may be subjected to heat/cold exposure as the result of accident or existing medical technology. Accidental heat and cold exposure can have detrimental effects such as during skin burns and frostbite. Heat transfer analysis may help in the development of therapeutic modalities for these injuries (Diller 1985, 1992). Techniques such as ultrasound (Chato 1985; Shiina and Sito 1988) and magnetic resonance imaging (MRI) (Sergiadis *et al.* 1988) can cause heating in tissue. Furthermore, it is known that anesthesia can affect the thermoregulation of the homeotherm (Lipton 1985). In order to ensure patient safety, knowledge of heat transfer process is important.

Understanding of bioheat transport is important in the beneficial applications of heat and cold for medical treatment. Recent advances in the application of heat (hyperthermia), radiation (laser therapy), and cold (cryosurgery), as a means to destroy undesirable tissue, such as cancer, has stimulated much interest in the study of thermal modeling in tissue. In the case of hyperthermia, it is observed that tissue can be destroyed when heated to 42–45°C (Field and Franconi 1987). In hyperthermia, laser treatment, and cryosurgery, thermal modeling of tissue is very critical to aid in the optimization of existing protocols in order to ensure controlled destruction of the cancerous growth (Welch 1985; Baish *et al.* 1986a; Onik and Rubinsky 1988). Another interesting topic that should be mentioned is the area of cell and tissue preservation at low temperatures, both above the tissue's freezing point (hypothermia) and below the tissue's freezing temperature (cryopreservation). Again, an understanding of heat transfer is important to aid in the successful preservation of cells, tissue, and organs at low temperatures (Pegg and Karow 1987).

Finally, not only can we increase our fundamental understanding of life processes, but it may be possible to develop improved protocols through the study of animal survival under extreme thermal environments such as hibernators, deep-sea organisms living in thermal vents, and cold-blooded animals. For example, the wood frog (*Rana sylvatica*) is freeze-tolerant (Storey 1990). That is, rather than migrating to a more thermally favorable environment, the wood frog will actually hibernate in a frozen state. Studying how the wood frog survives can aid researchers in their quest to preserve organs in a cryogenic state.

All of the above examples demonstrate the importance of bioheat transport and thermal modeling. However, the study of heat transfer in tissue is not a trivial task. The analysis of heat transfer in living tissue differs from usual thermal models mainly because of the existence of the vascular system and tissue metabolism. The accounting for blood flow in the thermal modeling of tissue creates considerable difficulties and has been the subject of numerous investigations (Chato 1981, 1987; Shitzer and Eberhart 1985a, b; Cho 1992).

Heat transfer is a complex topic and, in this chapter, only a brief overview will be presented. In the first section, some of the fundamental concepts and techniques in heat transfer are discussed. In the second section, more specific discussion is directed toward how researchers model bioheat transport in living tissue.

5.2 Fundamental concepts

5.2.1 What is heat transfer?

Heat transfer involves the transfer of energy in the presence of a temperature difference and is an extension of the science of thermodynamics. Classical thermodynamics teaches that energy can be transported across a system boundary as work or as heat. The process of heat transfer is driven by the presence of temperature gradients and is associated with changes in entropy. While thermodynamics concerns itself with studying the *magnitude* of heat transfer between two states of thermodynamic equilibrium, the field of heat transfer deals with the study of the *rates* of energy transport and the related temperature distributions in a system.

5.2.2 Modes of heat transfer

Heat transport depends on the nature of complex interactions between the system and its environment. To facilitate analysis it is typical to identify different modes or processes in which heat is transferred in a particular problem.

Conduction

In *conduction*, a temperature difference serves as the driving force for energy transport from high to low temperatures through a medium or two media in contact. Microscopically, this phenomena involves the diffusion of energy through molecular interactions. That is, molecules with higher kinetic energy (i.e. higher temperature) impart some of their energy to neighboring molecules possessing lower kinetic energy (i.e. lower temperature). As a result, 'heat' is transferred. A constitutional relation can be established between the energy flow and the temperature gradient of the form

$$\dot{q} = -kA\frac{\partial T}{\partial x}, \tag{5.1}$$

where $\dot{q}$ is the heat transferred in the direction of the temperature gradient and has units of watts, A is the area normal to the heat flow, T is the temperature, x is the direction of heat flow, and the constant of proportionality k is the *thermal conductivity*. The negative sign in eqn (5.1) is inserted to satisfy the *second law of thermodynamics*, that is, the flow of energy always occurs from high to low temperature. Equation (5.1) is known as *Fourier's law*. Furthermore, $\dot{q}/A$ is known as the *heat flux* or is specified as $\dot{q}''$, where each prime indicates a reciprocal length unit.

The thermal conductivity will vary for different materials. Because of the proximity of neighboring molecules, solids tend to have higher thermal conductivities than liquids and gases. Similarly, liquids tend to possess higher conductivities in comparison to gases. Table 5.1 lists the thermal conductivities of various materials. The thermal conductivities of biological materials vary with water content (Chato 1987) and the magnitude generally increases with both decreasing temperatures and increased blood perfusion (for *in vivo* cases).

Convection

In *convection*, heat transport occurs in a fluid from a combination of molecular diffusion and the fluid's bulk motion or flow. Convection heat transfer is further classified into two categories. During *forced* convection, the fluid's flow is due to an external source such as a fan or pump. During *free* or *natural* convection, the fluid's motion occurs from buoyancy forces due to density variations which arise from temperature differences. It is also possible to have a situation where there is a combination of forced and free convection, often referred to as *mixed* convection.

Convection can also be classified into *exterior* and *interior* convective heat transport. In exterior-type problems, the fluid flow occurs over a surface, for example, hot air blown over the skin. In interior-type problems, the fluid flow occurs in an enclosed area, such as the flow of warm blood in a blood vessel.

Without regard to the classification of convection, the heat flow between a solid and a fluid is evaluated using *Newton's law of cooling*, which is

$$\dot{q} = hA(T_s - T_\infty), \tag{5.2}$$

where h is the *heat transfer coefficient*, A is the wetted area, T is the temperature, and the subscripts s and ∞ stand for surface and fluid far from the surface, respectively.

The heat transfer coefficient depends on many parameters including geometry, surface properties, fluid properties, fluid velocity, and temperature difference. Furthermore, h can vary with spatial location and, thus, it is also common to use an average heat transfer coefficient $\bar{h}$ in eqn (5.2). Table 5.2 lists typical value ranges for the convective heat transfer coefficient. The prediction of the heat transfer coefficient is a complex topic and is the subject of numerous books (Kays and Crawford 1980; Holman 1981; Incropera and DeWitt 1990).

Some empirical relations have been developed to estimate the heat transfer coefficient as listed in Table 5.3.

Table 5.1 Thermal conductivities of various materials

Material	Temperature (°C)	k ($\mathrm{W\,m^{-1}K^{-1}}$)
Solids		
Copper	0	410
Ice	0	2.21
Brain tissue	37	0.628
Skin (*in vitro*)	20–37	0.2–0.4
Skin (*in vivo*)	37	0.5–2.8
Fat (*in vitro*)	20–37	0.1–0.4
Muscle (*in vivo*)	37	0.7–1.0
Bovine liver	0.1	4.17
Bovine muscle	0.1	4.25
Bovine fat	0.1	1.93
Tooth enamel	28	0.653
Bone (*in vivo*)	37	0.4–2.15
Liquids		
Water	0	0.556
Glycerin	0	0.28
Blood	37	0.51–0.53
Gases		
Air	0	0.024
Oxygen	0	0.015
Carbon dioxide	0	0.015

(Sources: ASHRAE 1985; Chato 1985; Shitzer and Eberhart 1985a, b; Yang 1989.)

Table 5.2 Typical convective heat transfer coefficient values

Process	Medium	h ($\mathrm{W\,m^{-2}K^{-1}}$)
Free convection	Gases	2–25
	Liquids	50–1 000
Forced convection	Gases	25–250
	Liquids	50–20 000

(Source: Incropera and DeWitt 1990.)

Table 5.3 Empirical equations to estimate average heat transfer coefficient

Condition	Equation for h ($W\ m^{-2}K^{-1}$)	Comments
Seated	$h = 8.3u^{0.6}$	u = air velocity ($m\ s^{-1}$)
Walking	$h = 8.6u^{0.53}$	u = walking speed ($m\ s^{-1}$)
Treadmill	$h = 6.5u^{0.39}$	u = treadmill speed ($m\ s^{-1}$)

(Source: ASHRAE 1985.)

Radiation

In *radiation* heat transfer, energy is transported through electromagnetic waves (photons) which can be emitted from solids, liquids, and gases. In contrast to conduction and convection, radiation heat transfer can occur in the absence of a medium (i.e. in a vacuum).

The *Stefan–Boltzmann law* describes the thermal radiation transferred from an ideal body E_b as

$$E_b = \sigma A T_s^4, \tag{5.3}$$

where σ is the *Stefan–Boltzmann constant* ($\sigma = 5.67 \times 10^{-8}\ W\ m^{-2}K^{-4}$), A is the area, and T_s is the surface temperature in kelvins. A real surface will only emit a fraction of the energy predicted with eqn (5.3). Thus, the thermal radiation emitted from an actual body is

$$\dot{q} = \varepsilon \sigma A T_s^4 = \varepsilon E_b, \tag{5.4}$$

where the parameter ε is the *(total) emissivity*, which varies from 0 to 1. When the emissivity is equal to 1, the body is an ideal radiator and is referred to as a *black body*. The emissivity is the ratio of actual emitted energy to the emitted energy from a black body at the same absolute temperature. It should be noted that the emissivity is a function of the photon wavelength, i.e. spectral in nature. However, for the scope of this discussion only the *total* emissivity is considered, which means that this parameter has been integrated over all wavelengths. Table 5.4 contains a list of the typical total emissivities of different materials. It is interesting to note that although the skin's reflectivity may vary with skin color within the visible spectrum (wavelength range 0.35–0.75 μm), the total emissivity for skin in the thermal radiation range (wavelength range 0.1–100 μm) is observed at about 0.98 regardless of pigmentation (Love 1985).

Table 5.4 Total emissivities of various materials

Material	Emissivity	Temperature (K)
Copper (polished)	0.023	390
Steel (polished)	0.066	373
Glass (smooth)	0.94	296
Water	0.95–0.963	273–373
Skin	0.97–0.99	305–307

(Source: Shitzer and Eberhart 1985a.)

The net thermal radiation exchange between a black-body surface and its environment is

$$\dot{q}'' = \sigma(T_s^4 - T_\infty^4), \tag{5.5}$$

where the subscript ∞ stands for the environment (also considered a black body). The net thermal radiation exchange between two nonideal surfaces is

$$\dot{q}'' = F_{1-2} E_\varepsilon \sigma (T_1^4 - T_2^4), \tag{5.6}$$

where the subscripts 1 and 2 denote surfaces 1 and 2, respectively. The parameter F_{1-2} is the *shape* or *view factor*, which is the fraction of energy emitted from surface 1 which actually reaches surface 2. The value of the view factor will vary from 0 to 1. The parameter E_ε is the *emissivity function*, which is the ratio of actual energy exchange to the thermal radiation exchange of two black-body surfaces possessing similar absolute temperatures. The value of the emissivity function will also vary from 0 to 1.

A surface with incident thermal radiation can absorb, transmit, or reflect this incoming energy. The fraction of energy absorbed, transmitted, and reflected is the absorptivity α, the transmissivity τ, and the reflectivity ρ, which are related as

$$\alpha + \tau + \rho = 1, \tag{5.7}$$

A material is considered *opaque*, if τ is equal to zero. A black body is also an ideal absorber and, therefore, $\alpha = 1$ and $\rho = \tau = 0$. Based on the definitions stated above,

$$\alpha = \varepsilon, \tag{5.8}$$

which is known as *Kirchhoff's identity*. It should be noted that the parameters α, ρ, τ, and ε represent the *total* properties integrated over all photon wavelengths for the scope of this discussion.

It is also typical to collect all of the effects of thermal radiation between two bodies into an effective heat transfer coefficient in a similar form to Newton's law of cooling for convection, or

$$\dot{q}'' = h_r(T_1 - T_2), \qquad (5.9)$$

where the subscript r stands for radiation. This procedure is performed when the temperature differences between two surfaces is small relative to their absolute value, such as between the surface temperature of an animal and its normal environment.

Change of phase

Heat transfer problems can involve the change of phase of a material. If the phase change is from a gas to a liquid state, the process is called *condensation*. If a liquid is changed into a gas, then the process is called *boiling*. *Solidification* involves a liquid-to-solid phase transition, and the opposite is referred to as *melting*.

In biological systems, phase change is important when considering cooling through evaporation of sweat from the skin. In medical applications, intense heating (e.g. by lasers) and cooling (e.g. by cryoprobes) can cause the tissue to change phase. Furthermore, accidental burning and frostbite will fall under this category.

During phase change the heat flux is associated with the change in enthalpy ΔH between the two phases (for example, $[H_l - H_s]$ during liquid–solid phase transformation or $[H_l - H_g]$ during liquid–gas phase transition). The difference in enthalpy is known as the latent heat L of vaporization or solidification, and the heat flux is

$$\dot{q}'' = \dot{m}''L, \qquad (5.10)$$

where $\dot{m}''$ is the mass flux. The latent heat of solidification ΔH_{sl} for water is 333.8 kJ kg^{-1} (at 0°C and 1 atm) and the latent heat of vaporization for water ΔH_{lg} is 2257 kJ kg^{-1} (at 100°C and 1 atm) (ASHRAE 1985).

Combined modes of heat transfer

The various modes of heat transport discussed above can occur simultaneously. For example, a skin surface exposed to wind and sun can experience convective heat transfer from the moving air and radiation from the sun. The combined effect of the various modes of heat transfer is often linearly superimposed. In this instance, it is typical to define an overall heat transfer coefficient U as

$$\dot{q}'' = \dot{q}''_c + \dot{q}''_r = h(T_s - T_\infty) + h_r(T_s - T_\infty) \qquad (5.11a)$$

$$= (h + h_r)(T_s - T_\infty) \qquad (5.11b)$$

$$= U(T_s - T_\infty). \qquad (5.11c)$$

At times when sweating occurs, the heat transfer can also involve evaporation (i.e. phase transformation).

Example 1. A radiation incubator heats a newborn. The radiator emits energy as a black body at a temperature of 550 K. The emissivity of the newborn skin is one and its body temperature is 36.5°C. There is a small draft in the incubator and the air is at 22°C. The heat transfer coefficient is 15 W m^{-2}K^{-1}. Calculate the amount of heat delivered to the newborn, assuming that its exposed skin surface area is 200 cm^2. Assume that the view factor between the infant and the radiative source is 1.

Solution. The convection and the radiation heat transfer operate together. The radiation heat transfer per surface area is

$$\dot{q}''_r = \varepsilon\sigma(T_\infty^4 - T_s^4)$$
$$= (5.67 \times 10^{-8}\text{W m}^{-2}\text{K}^{-4})[(550\text{K})^4 - (310\text{K})^4]$$
$$= 4.664 \times 10^3 \text{ W m}^{-2}.$$

The convection heat transfer is

$$\dot{q}''_c = h(T_\infty - T_s) = (15 \text{ W m}^{-2}\text{K})(22°\text{C} - 36.5°\text{C})$$
$$= -2.18 \times 10^2 \text{ W m}^{-2}.$$

The total heat transfer is $\dot{q}''_{total} = \dot{q}''_r + \dot{q}''_c = 4.446 \times 10^3$ W m^{-2}. Therefore, $\dot{q}_{total} = (4.446 \times 10^3 \text{ W m}^{-2})(200 \times 10^{-4} \text{ m}^2) = 93.3$ W.

5.2.3 Electrical analogy

Heat transfer through a slab-like configuration in a situation where the temperature is invariant in time is known as one-dimensional steady-state conduction. In steady-state one-dimensional conduction through a material i, the heat flow in rectangular coordinates is obtained through integration of eqn (5.1) with limits $T(0) = T_1$ and $T(\Delta x) = T_2$

$$\dot{q} = -\frac{k_i A}{\Delta x}(T_2 - T_1), \qquad (5.12)$$

where T_1 and T_2 are surface or wall temperatures and Δx is the material thickness. Equation (5.12) can be rewritten as

$$\dot{q} = \frac{T_1 - T_2}{\dfrac{\Delta x}{k_i A}} = \frac{\Delta T}{R_t}, \tag{5.13}$$

where R_t is the thermal resistance. The form of eqn (5.13) is reminiscent of *Ohm's law* in electrical circuit analysis, that is

$$\text{current} = \frac{\text{potential difference}}{\text{resistance}} = \frac{\Delta V}{R}, \tag{5.14a}$$

$$\text{flow of heat} = \frac{\text{temperature difference}}{\text{thermal resistance}} = \frac{\Delta T}{R_t}, \tag{5.14b}$$

where V is voltage.

Extending this electrical analogy, a circuit diagram can be constructed to describe the heat flow between a composite structure as depicted in Fig. 5.1. The circuit diagram will resemble Fig. 5.2 and the thermal resistances are

$$R_{ti} = \frac{\Delta x_i}{k_i A}, \tag{5.15}$$

where i = a, b, c. In a steady-state situation, the heat flow $\dot{q}$ remains constant through each material and, therefore,

$$\dot{q} = \frac{T_1 - T_4}{\sum R_t}. \tag{5.16}$$

For this geometry, an equivalent statement using the concept of the overall heat transfer coëfficient U is

$$\dot{q} = UA(T_1 - T_4), \tag{5.17a}$$

where

$$U = \left(\sum \frac{\Delta x_i}{k_i}\right)^{-1}, \tag{5.17b}$$

and i = a, b, c.

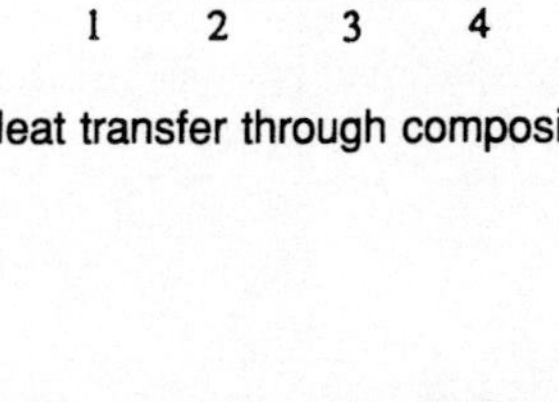

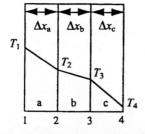

Fig. 5.1 Heat transfer through composite structure.

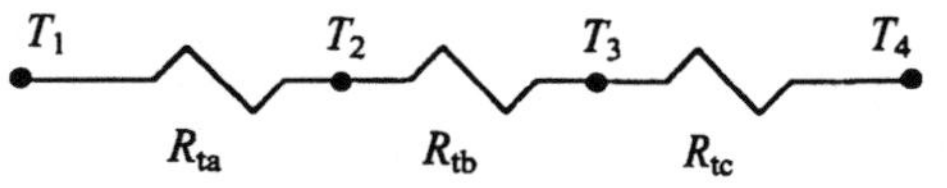

Fig. 5.2 Electrical circuit analogy for heat transfer through a composite structure.

In the case of a series and parallel composite structure as depicted in Fig. 5.3, the electrical analogy of series and parallel resistors can be used as modeled in Fig. 5.4. Here $R_{tb} = \dfrac{\Delta x_b}{k_b(A/2)}$ and $R_{tc} = \dfrac{\Delta x_c}{k_c(A/2)}$, since the area normal to heat flow for the 'parallel' material is $A/2$. The resistances R_{ta} and R_{td} have a similar form as eqn (5.15).

The electrical analogy can also be used to analyze composite cylindrical (Fig. 5.5(a)) and spherical (Fig. 5.5(b)) geometries whose cross-sectional view is depicted in Fig. 5.5(c). The thermal resistances for these cases are

$$\text{cylinder: } \quad R_{ti} = \frac{\ln\left(\dfrac{r_{outer}}{r_{inner}}\right)}{2\pi k_i L}, \tag{5.18a}$$

$$\text{sphere: } \quad R_{ti} = \frac{\dfrac{1}{r_{inner}} - \dfrac{1}{r_{outer}}}{4\pi k_i}, \tag{5.18b}$$

where L is the length of the cylinder, and r_{outer} and r_{inner} are the outer and inner radii of the annulus (for cylindrical geometry) or shell (for spherical geometry), respectively.

It is also simple to incorporate a convection boundary condition into the electrical analogy approach for problems as illustrated in Fig. 5.6. Newton's law of cooling can be rewritten as

$$\dot{q}_{conv} = \frac{T_s - T_\infty}{\dfrac{1}{hA}} = \frac{T_s - T_\infty}{R_{t\,conv}}, \tag{5.19}$$

where 'conv' stands for convection.

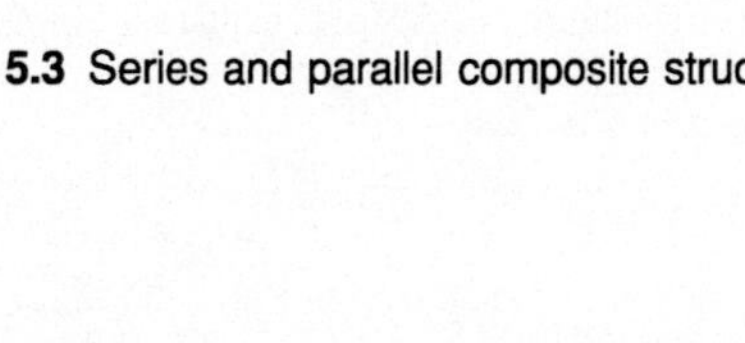

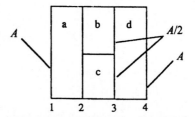

Fig. 5.3 Series and parallel composite structure.

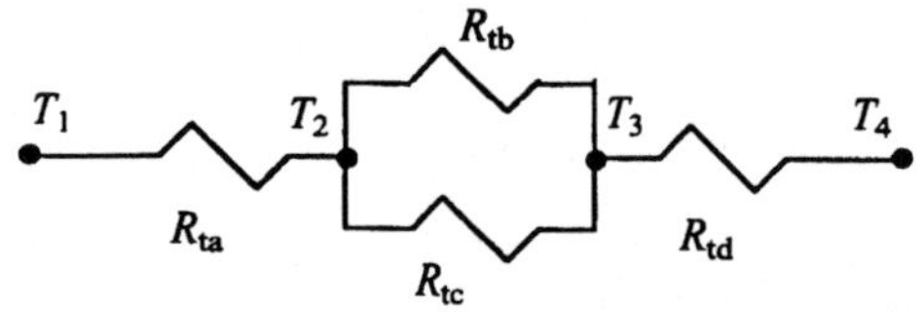

Fig. 5.4 Electrical circuit analogy for heat transfer through series and parallel composite structure.

It should also be noted that it is possible to cast radiation problems into an analogous electrical circuit diagram. This topic will not be covered here but can be explored further in other references (Holman 1981; Siegel and Howell 1981; Incropera and DeWitt 1990).

5.2.4 Conservation of energy

Similar to thermodynamics, heat transfer analysis involves the application of the *first law of thermodynamics*, or the conservation of energy requirement applied to a *control volume* bounded by a *control surface*. The first law applied to a control volume on a rate basis is

$$\dot{q}_{\text{in}} - \dot{q}_{\text{out}} + \dot{q}_{\text{gen}} = \dot{q}_{\text{st}}, \tag{5.20}$$

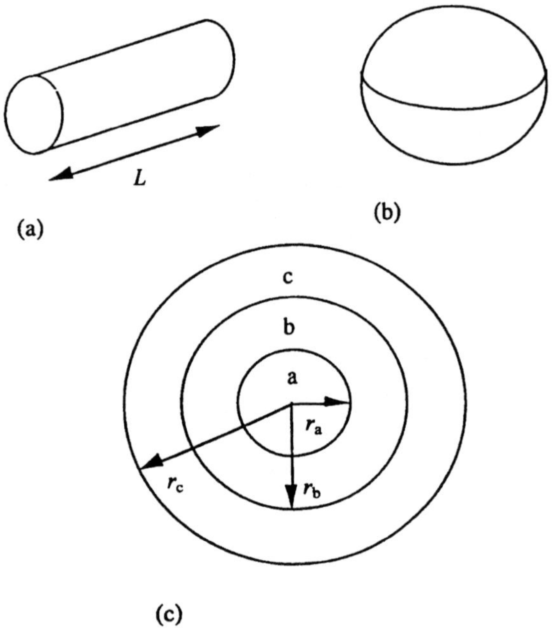

Fig. 5.5 (a) Cylinder. (b) Sphere. (c) Cross-sectional view of composite cylinder and sphere.

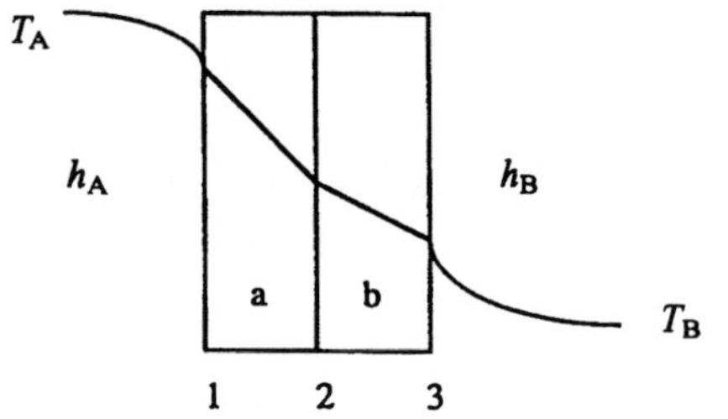

Fig. 5.6 Heat transfer through composite material with convective boundary conditions.

where the subscripts stand for rate of energy entering, leaving, generated, and stored, respectively. The term $\dot{q}_{\text{st}}$ is the rate of change of the material's internal energy and, when included, the expression is considered *transient*. In *steady state*, the energy storage term $\dot{q}_{\text{st}}$ is set to zero. While $\dot{q}_{\text{in}}$ and $\dot{q}_{\text{out}}$ are surface phenomena, $\dot{q}_{\text{gen}}$ is a volumetric phenomenon. That is, the former are associated with the inflow and outflow of energy across the control surface and the latter is energy generated within the control volume.

One can integrate eqn (5.20) over time to derive a similar expression for a time interval:

$$q_{\text{in}} - q_{\text{out}} + q_{\text{gen}} = \Delta q_{\text{st}}. \tag{5.21}$$

An energy balance can be performed on a surface. In this case, only the first two terms of eqn (5.20) are relevant and it simplifies to

$$\dot{q}_{\text{in}} - \dot{q}_{\text{out}} = 0. \tag{5.22}$$

Equation (5.22) is valid for both steady-state and transient cases. For example, the skin surface energy balance for Example 5.1 is

$$\dot{q}''_{\text{cond}} - (\dot{q}''_{\text{rad}} - \dot{q}''_{\text{conv}}) = 0. \tag{5.23}$$

5.2.5 Lumped capacity analysis

Transient problems of heat transfer with convection can also be solved in simplified form under conditions in which the temperature gradients within the solid are assumed negligible relative to those in the surrounding fluid. For example, a small piece of tissue quenched in a cold liquid may fall into this category. From the instant the material is immersed in the fluid, the temperature of the solid will decay until the temperature eventually reaches that of the liquid. In the *lumped capacity formulation*, the material is assumed to be instantaneously at a spatially uniform temperature during the entire transient period and the temperature will vary only in time. As a criterion, this assumption can be made if the Biot number Bi is much smaller than unity or

$$Bi = \frac{hL_c}{k_{solid}} \ll 1, \tag{5.24}$$

where h is the heat transfer coefficient, k_{solid} is the solid's thermal conductivity, and L_c is a characteristic length of the solid (for example, diameter). The Biot number is a dimensionless ratio which can be rewritten as

$$Bi = \frac{\dfrac{L_c}{k_{solid}A}}{\dfrac{1}{hA}} = \frac{R_{t\,cond}}{R_{t\,cond}}, \tag{5.25}$$

where 'cond' and 'conv' stand for conduction and convection, respectively. It is easy to see now that the Biot number is small when the thermal resistance due to convection is much greater than the thermal resistance due to conduction. Thus, temperature gradients due to convection are much greater than those due to conduction.

Taking the space occupied by the solid as the control volume and applying the conservation of energy, eqn (5.20) reduces to

$$\dot{q}_{st} = -\dot{q}_{out}. \tag{5.26}$$

Energy flow out of the control surface is due to convection, and the energy storage term is the internal energy's rate of change. Equation (5.26) can be expanded as

$$-hA(T - T_\infty) = \rho c V \frac{dT}{dt}, \tag{5.27}$$

where ρ is the material's density, c is the material's heat capacity (at constant pressure), V is the solid's volume, t is time, and A is the surface area of the material.

Let $\theta = \dfrac{T - T_\infty}{T_o - T_\infty}$ and $\tau = t/t_c$, where T_o is the temperature of the solid at $t = 0$, T_∞ is the temperature of the liquid, and t_c will be defined at a later stage. Substituting these new nondimensional parameters into eqn (5.27),

$$-\theta = \frac{\rho c V}{hAt_c}\frac{d\theta}{d\tau}. \tag{5.28}$$

If t_c is defined as $\rho cV/hA$, then eqn (5.28) becomes

$$-\theta = \frac{d\theta}{d\tau}. \tag{5.29}$$

The general solution to this ordinary differential equation is simple. Separating variables and integrating,

$$\ln\theta = -\tau + C, \tag{5.30}$$

where C is a constant. Applying the initial condition, i.e. at $t = \tau = 0$, $T = T_o$ or $\theta = 1$, the constant C is identically zero. Therefore, the solution can be written as

$$\theta = e^{-\tau} = \exp(-t/t_c). \tag{5.31}$$

To gain more insight, let us analyze t_c, which can be written as

$$t_c = \left(\frac{1}{hA}\right)\rho cV = R_t C_t. \tag{5.32}$$

Note the use of the thermal resistance of convection in eqn (5.32). The parameter C_t can be thought of as a lumped thermal capacitance. Thus, the lumped method is equivalent to the circuit diagram in Fig. 5.7.

A more comprehensive discussion of this lumped parameter approach and examples of its use are given in Chapter 7.

5.2.6 Governing equations

The equations which describe the heat transfer of a particular problem are known as the *governing equations*. For example, eqn (5.27) is the governing equation for the case when lumped analysis is valid. Under different conditions, different governing equations must be derived and then solved with appropriate initial/boundary conditions.

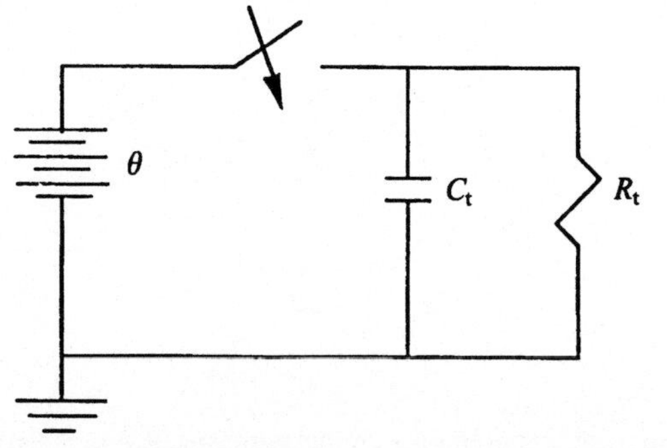

Fig. 5.7 Electrical analogy of lumped capacity method.

Derivation of governing equations

To derive the governing equations, the conservation of energy is applied to a differential control volume. For example, let us consider a problem in which transient conduction heat transfer occurs in a one-dimensional slab with heat generation as depicted in Fig. 5.8. The first law of thermodynamics applied to a control volume, eqn (5.20), with a unit surface area and an infinitesimal thickness dx, yields

$$\dot{q}|_x - \dot{q}|_{x+\mathrm{d}x} + \dot{q}'''\,\mathrm{d}x = \rho c \frac{\partial T}{\partial t}\mathrm{d}x. \qquad (5.33\mathrm{a})$$

In the limit as dx becomes infinitesimally small,

$$\lim_{\mathrm{d}x \to 0} -\frac{\dot{q}|_{x+\mathrm{d}x} - \dot{q}|_x}{\mathrm{d}x} + \dot{q}''' = \rho c \frac{\partial T}{\partial t}, \qquad (5.33\mathrm{b})$$

$$-\frac{\mathrm{d}\dot{q}}{\mathrm{d}x} + \dot{q}''' = \rho c \frac{\partial T}{\partial t}. \qquad (5.33\mathrm{c})$$

Introducing Fourier's law into eqn (5.33c),

$$\frac{\mathrm{d}}{\mathrm{d}x}\left(k\frac{\mathrm{d}T}{\mathrm{d}x}\right) + \dot{q}''' = \rho c \frac{\partial T}{\partial t}. \qquad (5.33\mathrm{d})$$

If the thermal conductivity is constant, the governing equation becomes

$$k\frac{\mathrm{d}^2 T}{\mathrm{d}x^2} + \dot{q}''' = \rho c \frac{\partial T}{\partial t}. \qquad (5.34)$$

In a more general form,

$$\nabla(k\,\nabla T) + \dot{q}''' = \rho c \frac{\partial T}{\partial t}, \qquad (5.35\mathrm{a})$$

or, if the thermal conductivity is constant,

$$\nabla^2 T + \frac{\dot{q}'''}{k} = \frac{1}{\alpha}\frac{\partial T}{\partial t}, \qquad (5.35\mathrm{b})$$

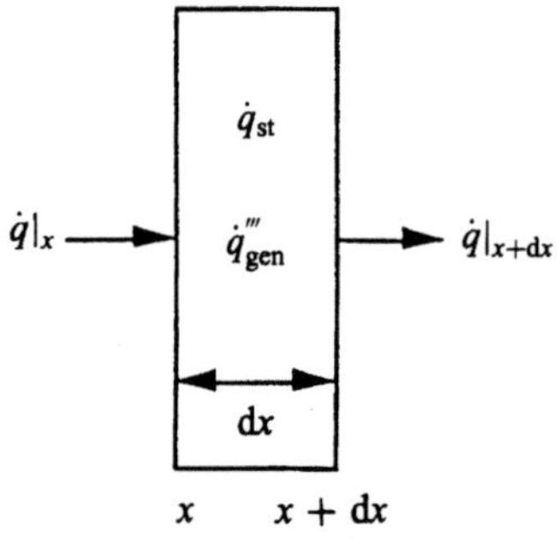

Fig. 5.8 Control volume for transient one-dimensional conduction through a slab with heat generation.

where ∇^2 is the *Laplacian operator* and α is a parameter called the *thermal diffusivity*. Table 5.5 lists the thermal properties of various materials.

In three dimensions for different geometries, $\nabla^2 T$ expands to

$$\text{rectangular:}\quad \nabla^2 T = \frac{\partial^2 T}{\mathrm{d}x^2} + \frac{\partial^2 T}{\mathrm{d}y^2} + \frac{\partial^2 T}{\mathrm{d}z^2}, \qquad (5.36\mathrm{a})$$

$$\text{cylindrical:}\quad \nabla^2 T = \frac{1}{r}\frac{\partial}{\partial r}\left(r\frac{\partial T}{\mathrm{d}r}\right) + \frac{1}{r^2}\frac{\partial^2 T}{\mathrm{d}\theta^2} + \frac{\partial^2 T}{\mathrm{d}z^2}, \qquad (5.36\mathrm{b})$$

$$\text{spherical:}\quad \nabla^2 T = \frac{1}{r^2}\frac{\partial}{\partial r}\left(r^2\frac{\partial T}{\mathrm{d}r}\right) + \frac{1}{r^2 \sin\theta}\frac{\partial}{\partial\theta}\left(\sin\theta\frac{\partial T}{\mathrm{d}\theta}\right) + \frac{1}{r^2 \sin\theta}\frac{\partial^2 T}{\mathrm{d}\psi^2}, \qquad (5.36\mathrm{c})$$

where (x, y, z) are rectangular Cartesian coordinates, (r, θ, z) are cylindrical coordinates, and (r, θ, ψ) are spherical coordinates.

Table 5.5 Thermal properties of various materials

Material	ρ (kg m^{-3})	c (J kg^{-1} K^{-1})	α (m^2 s^{-1})
Air (at 27°C)	1.1774	1.0057	2.22×10^{-5}
Water (at 0°C)	999.8	4225	1.34×10^{-7}
Blood	1060	3889	1.56×10^{-7}
Copper	8954	383.1	1.12×10^{-4}
Aluminium	2707	896	8.42×10^{-5}
Tooth enamel	2800	711.3	4.69×10^{-7}
Bovine liver (at 0.1°C)	—	—	1.05×10^{-7}
Muscle	1070 (human)	3471 (human)	1.05×10^{-7} (bovine at 0.1°C)
Fat	937 (human)	3258 (human)	5.9×10^{-8} (bovine at 0.1°C)

(Sources: Holman 1981; Chato 1985; Yang 1989.)

Initial and boundary conditions

The governing equations in heat transfer are partial or ordinary differential equations. To completely formulate a problem it is necessary to specify the initial and boundary conditions.

The initial condition specifies the system's state at time equal to zero. In one-dimensional rectangular coordinates,

$$T(x, t = 0) = T_o. \qquad (5.37)$$

Boundary conditions have many forms, and a partial listing is presented in Table 5.6. For example, if the case of a solid–solid in tight contact is considered, each solid will possess its own governing equation. To completely formulate the problem, the boundary conditions at their interface must match. More specifically, the temperature and the heat flux at the solid–solid contact should be equivalent as listed in Table 5.6.

Nondimensionalization

It is useful to nondimensionalize the governing equations and their initial/boundary conditions. (This procedure was performed in the previous example of lumped capacity analysis.) The following nondimensional variables are typical parameters that are used in heat transfer analysis:

$$\theta = \frac{T - T_{ref}}{T_0 - T_{ref}}; \quad Fo = \frac{\alpha t}{L^2}, \qquad (5.38a, b)$$

$$Bi = \frac{hL}{k_{solid}}; \quad Nu = \frac{hL}{k_{fluid}}, \qquad (5.38c, d)$$

$$\xi = \frac{x}{L}; \quad \eta = \frac{y}{L}; \quad \zeta = \frac{z}{L}, \qquad (5.38e, f, g)$$

where Fo is known as the *Fourier number* and represents dimensionless time (e.g. τ in eqn (5.28) is $\tau = Fo\,Bi$), and Bi is the previously discussed Biot number. The parameter Nu is the Nusselt number and has a similar form to the Biot number but uses the thermal conductivity of the fluid, k_{fluid}. The subscripts 'o' and 'ref' denote initial and reference temperatures, respectively. The parameter L is a characteristic length defined by the problem (e.g. material thickness, material diameter). The parameters ξ, η, and ζ are dimensionless spatial variables. It is important to note that there is no one correct way of nondimensionalizing equations (e.g. $\theta = \frac{T - T_{ref}}{T_o - T_{ref}}$ and $\theta = \frac{T - T_{ref}}{T_{ref}}$ are both valid dimensionless temperatures). However, it may be possible to facilitate obtaining a solution, depending on how one chooses dimensionless parameters.

Table 5.6 Typical boundary conditions encountered in heat transfer

Condition	Mathematical form[a]
Prescribed temperature	$T(L, t) = T_L(t)$
Prescribed heat flux	$-k\frac{\partial T(L, t)}{\partial x} = \dot{q}''_L$
Insulation (adiabatic)	$\frac{\partial T(L, t)}{\partial x} = 0$
Surface convection	$-k\frac{\partial T(L, t)}{\partial x} = h[T(L, t) - T_\infty]$
Solid (1)–solid (2) interface	$T_1(L, t) = T_2(L, t)$
	$k_1\frac{\partial T_1(L, t)}{\partial x} = k_2\frac{T_2(L, t)}{\partial x}$

[a] Boundary location $x = L$ in each case.

Example 2. Nondimensionalize the surface convective boundary condition in Table 5.6.

Solution. Let $\theta = \frac{T - T_\infty}{T_o - T_\infty}$ and $\xi = x/L$. Substituting into the equation

$$-k\frac{\partial T(L, t)}{\partial x} = h[T(L, t) - T_\infty],$$

we obtain

$$-\frac{k}{L}\frac{\partial \theta}{\partial \xi} = h\theta,$$

$$\frac{\partial \theta}{\partial \xi} = -\frac{hL}{k}\theta = -Bi\,\theta.$$

It is interesting to note the appearance of the Biot number. As previously discussed, when the Biot number is small, the lumped capacity formulation states that the temperature gradients within the material is negligible. Nondimensionalization of the surface boundary condition demonstrates the validity of this assumption. That is, when the Biot number is small, the temperature gradient $\partial\theta/\partial\xi$ is negligible.

Solution techniques

The general equations of heat transfer are partial differential equations that can be solved through a variety of techniques that are well established (Rohsenow and Choi 1961; Arpaci 1966; Holman 1981; Wylie and Barrett 1982; Incropera and DeWitt 1990). Under certain conditions the equations can be simplified, which yields solutions that are more accessible. For instance, in a problem in which the temperature distribution does not change in time (steady-state conditions), the governing equation, eqn (5.35a), becomes

$$k\,\nabla^2 T + \dot{q}''' = 0, \qquad (5.39)$$

subject to appropriate boundary conditions only. When the temperature distribution and the heat transfer are known to change only in one direction, the heat transfer equation becomes

$$k\frac{\mathrm{d}^2 T}{\mathrm{d}x^2} + \dot{q}''' = 0, \qquad (5.40)$$

in Cartesian coordinates.

Example 3. In a bioreactor, cells are maintained in a long test tube with diameter of 20 mm. The test tube's outer surface is maintained at 25°C. Microwave heating warms the test tube's contents uniformly with an intensity $\dot{q}'''$. What microwave intensity will ensure that the temperature at the center of the tube does not rise above 37°C?

Solution. The governing equation in one-dimensional steady-state cylindrical coordinates is

$$\frac{1}{r}\frac{\mathrm{d}}{\mathrm{d}r}\left(kr\frac{\mathrm{d}T}{\mathrm{d}r}\right) + \dot{q}''' = 0.$$

The boundary conditions are $\frac{\mathrm{d}T(0)}{\mathrm{d}r} = 0$ (from radial symmetry) and $T(r_0) = T_s = 25°\mathrm{C}$. The solution is calculated integrating twice:

$$T(r) = -\frac{\dot{q}'''}{4k}r^2 + C_1 \ln r + C_2,$$

where C_1 and C_2 are constants. Applying boundary conditions,

$$T(r) = \frac{\dot{q}'''}{4k}r_o^2\left(1 - \frac{r^2}{r_o^2}\right) + T_s.$$

The temperature at the test tube center is

$$T(0) = \frac{\dot{q}'''}{4k}r_o^2 + T_s.$$

Solving for the intensity,

$$\begin{aligned}\dot{q}''' &= [T(0) - T_s]\frac{4k}{r_o^2}\\ &= (37°\mathrm{C} - 25°\mathrm{C})\frac{(4)(0.62\ \mathrm{W\,m^{-1}\,K^{-1}})}{(0.01\ \mathrm{m})^2}\\ &= 3.0 \times 10^5\ \mathrm{W\,m^{-3}}.\end{aligned}$$

Here, the thermal conductivity of water at the average temperature 31°C = (37°C + 25°C)/2 was used in the calculation (ASHRAE 1985). It should also be noted that the end-effect (i.e. a test tube has a capped bottom end) was neglected.

Example 4. A sea diver's suit has a thickness of 5 mm and is made of a foam with thermal conductivity of 0.05 W $\mathrm{m^{-1}K^{-1}}$. If the diver's skin is at 30°C and the outer temperature of the suit is 20°C, derive an equation for the temperature distribution through the suit. How much heat does the diver lose per unit surface area?

Solution. The steady-state conduction equation in one dimension is

$$\frac{\mathrm{d}^2 T}{\mathrm{d}x^2} = 0,$$

and the boundary conditions are T(0 mm) = 30°C and T(5 mm) = 20°C, taking the coordinate system's origin at the skin surface. Integrating the governing equation twice and applying the boundary conditions, the temperature distribution is

$$T(x) = T(0) + \frac{x}{L}[T(5) - T(0)],$$

where $L = 5$ mm. The heat flux is calculated from Fourier's law:

$$\begin{aligned}\dot{q}'' &= -k\frac{\mathrm{d}T}{\mathrm{d}x} = -k\frac{T(5) - T(0)}{0.005\ \mathrm{m}}\\ &= -(0.05\ \mathrm{W\,m^{-1}\,K^{-1}})\frac{(20°\mathrm{C} - 30°\mathrm{C})}{0.005\ \mathrm{m}}\\ &= 100\ \mathrm{W\,m^{-2}}.\end{aligned}$$

5.3 Mathematical models of bioheat transfer

The modeling of heat transfer processes in biological systems is a topic of great interest, as indicated in the introduction. In this section, we examine the various models developed for analyzing bioheat transport in animals and humans. Unlike previously discussed problems, the difficulty of this subject is the nonhomogeneous nature of tissue. More specifically, tissue is different from other materials because of the presence of the vascular system and its associated blood flow, metabolic heat generation, and heterogeneous thermal properties. The effect of blood flow is most difficult to account for, and researchers have proposed many different models. More detailed reviews of this topic can be found elsewhere (Chato 1981; Shitzer and Eberhart 1985a, b; Charny 1992).

5.3.1 Effective thermal conductivity

Early work on bioheat transfer focused on the heat transport between the skin and the environment. The heat transport was modeled as the product of measured temperature gradients between the body and its surroundings and an effective thermal conductivity k_{eff} which incorporated all the effects of vasculature and heterogeneous properties in a single value (Bazett and McGlone 1927; Charny 1992). The effective thermal conductivity was estimated from experimental data. More complex models were developed, estimating the effect of clothing modeled as additional layers of conduction (Gagge *et al.*, 1938; Charny 1992). Typically, the thermal conductivities were assumed to be constant.

Taking blood flow into account within the tissue is a more complicated problem. A simple method to include blood flow is to separately add the conductance of tissue alone (i.e. without blood flow) and the conductance due to blood flow. The equation is

$$\dot{q}'' = \frac{k_t}{\delta}(T_b - T_s) + \omega\rho_b c_b \delta (T_b - T_s), \quad (5.41a)$$

$$= K_{\text{eff}}(T_b - T_s), \quad (5.41b)$$

where k_t is the tissue thermal conductivity, δ is the thickness of tissue from the outer surface to the depth at which the tissue reaches its deep body temperature, ω is the blood perfusion rate (blood volume flow rate per unit tissue volume), ρ is the density, c is the heat capacity, and the subscripts 'b' and 's' stand for deep body blood and skin, respectively. The parameter K_{eff} is the effective thermal conductance. If K_o is the effective conductance when there is no blood perfusion, then

$$\frac{K_{\text{eff}}}{K_o} = 1 + \frac{\omega\rho_b c_b \delta^2}{k_t}. \quad (5.42)$$

The main problem with this approach is that the temperature difference between the deep body core and the skin surface (i.e. the maximum temperature difference) drives the flow of heat. Therefore, local heat transfer is not taken into account. In addition, since the vascular geometry varies throughout the tissue, the blood perfusion term is an effective parameter and does not describe local blood flow. Finally, this maximum temperature differential is assumed to exist at a small distance δ from the skin surface that is arbitrarily defined.

5.3.2 Bioheat equation

The most widely used thermal model for tissue is commonly known as the bioheat equation (Pennes 1948). This formulation includes a source/sink term which accounts for the heat transferred from blood perfusion. The general statement is

$$\rho_t c_t \frac{\partial T_t}{\partial t} = \nabla(k_t \nabla T_t) + \dot{q}'''_{\text{met}} + \omega\rho_b c_b (T_a - T_v), \quad (5.43)$$

where $\rho_t c_t$ is the volumetric specific heat of tissue and $\dot{q}'''_{\text{met}}$ is the metabolic heat generation. The subscripts 't' and 'b' stand for tissue and blood, respectively. The parameters T_a and T_v are the temperature of arterial blood entering the tissue control volume and that of the outflowing venous blood, respectively. The left-hand side and the first two terms of the right-hand side of eqn (5.43) are familiar from the previous section. The final term is an additional quantity which accounts for the heat transferred from blood flow.

The bioheat equation assumes that all of the blood flow effects can be absorbed into one additional source/

sink term. Typically, the heat exchange is considered complete. That is, the blood enters the control volume at the deep body arterial temperature and leaves the control volume at the venous temperature, which is assumed to equal the tissue temperature. The formulation was originally justified with the assumption that the heat transfer occurs in the capillaries, which are in good thermal equilibrium with the surrounding tissue. If the outflowing venous temperature is different from the tissue temperature, the last term in eqn (5.43) can be modified with a thermal equilibration parameter κ:

$$T_v = T_t + \kappa(T_a - T_t). \quad (5.44)$$

When $\kappa = 0$, the heat transfer is complete, and when $\kappa = 1$, the outflowing blood temperature is at the same as that of the inflowing arterial blood.

Although it has had wide application, the bioheat equation has shortcomings and many of its assumptions have been discredited. One difficulty in the use of the bioheat equation is the determination of the blood perfusion rate. Investigators cite the problems of the empirical nature of this term and claim that it is possible to adjust the perfusion rate (Weinbaum *et al.* 1984). Some of the other shortcomings will be discussed in later sections.

Despite of all its deficiencies, the bioheat equation's simplicity and its ease in implementation in analytical and numerical studies have made it the most widely used thermal model of living tissue. The thermal conductivity is usually chosen as an isotropic quantity (i.e. having the same value in each spatial direction). Therefore, the addition of a single scalar term, which accounts for blood perfusion, facilitates mathematical analysis tremendously. Many applications of the bioheat equation can be found in other references (Shitzer 1973; Shitzer and Eberhart 1985a, b).

Example 5. What is the temperature distribution between the core body temperature T_c and the skin temperature T_s at steady state, if the distance from the outer surface at which the core temperature is measured is δ (usually 8–10 mm)? Assume no metabolic heat generation.

Solution. The steady-state one-dimensional bioheat equation without metabolic heat generation in rectangular coordinates is

$$0 = k_t \frac{d^2 T_t}{dx^2} + \omega \rho_b c_b (T_c - T_t).$$

Setting the origin of the x-axis within the tissue, the boundary conditions are $T(0) = T_c$ and $T(\delta) = T_s$. Let $\theta = \dfrac{T_t - T_c}{T_s - T_c}$ and $\eta = x/\delta$. Substituting into the governing equation,

$$\frac{d^2\theta}{d\eta^2} - m^2\delta^2\theta = 0,$$

where

$$m^2 = \frac{\omega_b \rho_b c_b}{k_t},$$

and the boundary conditions become $\theta(0) = 0$ and $\theta(1) = 1$. The general solution is

$$\theta(\eta) = C_1 \exp(m\delta\eta) + C_2 \exp(-m\delta\eta),$$

where C_1 and C_2 are constants. Applying the first boundary condition at $\eta = 0$, $C_1 = -C_2$. Applying the second boundary condition at $\eta = 1$, $C_1 = [\exp(m\delta) - \exp(m\delta)]^{-1}$. Therefore,

$$\theta(\eta) = \frac{\exp(m\delta\eta) + \exp(-m\delta\eta)}{\exp(m\delta) - \exp(m\delta)}.$$

5.3.3 Tissue as a porous medium

While the bioheat equation's use increased in popularity, other investigators were not satisfied with the thermal model. Wulff (1974) was probably the first investigator to openly question the validity of the bioheat equation. In a brief communication, he argued that it was incorrect to combine local transport phenomena with global convective transport. More specifically, Wulff stated that the blood perfusion term in the bioheat equation was a 'global' term, while the other parameters described local heat transport. Since there are actually three different unknown temperatures (i.e. T_t, T_a, and T_v), there should be three equations which must be solved simultaneously to completely describe the heat transfer process. Another source of dissatisfaction with the bioheat equation came from the fact that blood does not necessarily flow in the direction of the temperature gradient. Wulff proposed a new equation which accounts for the fluid motion in tissue modeled as a porous medium. He utilized an apparent overall (Darcy) blood velocity $\bar{U}_D$ through a porous tissue control volume, defined as

$$\rho_b H_b \bar{U}_D = \frac{1}{4\pi} \int_\Omega \rho_b H_b u \, d\omega, \quad (5.45)$$

where H is the specific enthalpy, u is the actual blood velocity in the capillaries, ω is the solid angle, and Ω is the total solid angle (i.e. $\Omega = 4\pi$). Using this Darcy velocity, a modified convective term in the governing equations was introduced to replace the blood perfusion term in the bioheat equation, or

$$\rho_t c_t \frac{\partial T_t}{\partial t} = \nabla(k_t \nabla T_t) + \dot{q}'''_{\text{met}} - \rho_b c_b \bar{U}_D \nabla T_t. \quad (5.46)$$

It was assumed that the equilibrium of the blood temperature to tissue temperature was complete. Therefore, only one unknown temperature appears in eqn (5.46).

Despite the apparent simplicity of Wulff's modified bioheat equation, this formulation has not received as wide use as the original bioheat equation, most likely due to the difficulty in evaluating the Darcy velocity. There are other shortcomings to this formulation which will be discussed in subsequent sections.

5.3.4 Single-vessel analysis

In the past decade, research interests have shifted toward the understanding of localized heat transport in living tissue. New technological advances and their applications in medicine have stimulated this shift in research effort. For example, both hyperthermic (Overgaard and Nielsen 1980; Baish *et al.* 1986a; Charny and Levin 1988; Roemer *et al.* 1989) and cryosurgical (Onik and Rubinsky 1988) procedures are being used in the treatment of cancer. Ultrasound (Chato 1985) and lasers (Sagi *et al.* 1992; Whiting *et al.* 1992) produce localized heating. In all these examples, it is important to understand the local heat transfer. This interest in local analyses has led investigators to examine the theoretical basis of energy transport in living tissues in relation to the vasculature's architecture and function with more physiological rigor. To this end, investigators addressed how a single blood vessel contributes to the local heat transfer.

Chen and Holmes (1980) were among the first to perform a simple energy balance for a single blood vessel embedded in a tissue as depicted in Fig. 5.9. The convected energy through the blood vessel was equated with the energy due to radial conduction, or

$$u_{\text{avg}} A \rho_b c_b \frac{dT_b}{dz} = UP(T_t - T_b), \quad (5.47)$$

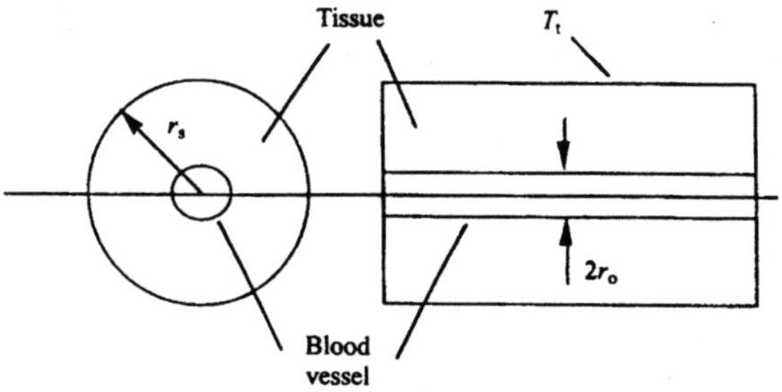

Fig. 5.9 Transverse and longitudinal cross sections of a blood vessel embedded in a tissue cylinder.

where u_{avg} is the average blood velocity in the blood vessel, A is the cross-sectional area of the blood vessel, U is the overall heat transfer coefficient, P is the perimeter of the blood vessel, z is the coordinate system in the axial direction, T_t is the tissue temperature, and T_b is the blood temperature. Equation (5.47) can be written as

$$x_e \frac{dT_b}{dz} = (T_t - T_b), \quad (5.48a)$$

where

$$x_e = \frac{u_{\text{avg}} A \rho_b c_b}{UP}. \quad (5.48b)$$

It is immediately observed that the solution of eqn (5.48a) will involve an exponential decay. Chen and Holmes (1980) introduced the parameter x_e as the exponential equilibration length, that is, the length over which the temperature will decrease by 36.6 per cent. Therefore, the magnitude of x_e is a measure of how fast the temperature decays along the blood vessel.

Let us examine the limits of the exponential equilibration length. If x_e is small relative to the characteristic length of the blood vessel, the blood will exit the vessel at essentially the tissue temperature. If the equilibration length is large relative to the characteristic blood vessel length, then the blood temperature does not decay and will leave the control volume at the same inflow temperature.

Chen and Holmes (1980) estimated the overall heat transfer coefficient as the summation of the thermal resistances due to the tissue and blood, or

$$U^{-1} = R_{tt} + R_{tb}, \quad (5.49a)$$

where 't' and 'b' stand for tissue and blood, respectively.

They estimated the thermal resistances as

$$R_{tt} \approx \frac{r}{k_t} \ln\left(\frac{L_c}{r}\right), \tag{5.49b}$$

$$R_{tb} \approx \frac{r}{k_b\, Nu}, \tag{5.49c}$$

where

$$Nu = \frac{hL_c}{k_b}, \tag{5.49d}$$

and r is the radius of the blood vessel and L_c a characteristic length of the problem, which was taken to be one-half of the typical distance between two blood vessels. After comparing both thermal resistances, the authors concluded that the conductive resistance was dominant. They simplified eqn (5.49a) as

$$U \approx \frac{k_t}{\Lambda r}, \tag{5.50a}$$

where

$$\Lambda = \ln\left(\frac{L_c}{r}\right) + \frac{k_t}{k_b\, Nu}. \tag{5.50b}$$

Incorporating the above expressions into the exponential equilibration length,

$$x_e = \frac{\Lambda}{2} \frac{u_{avg}\rho_b c_b r^2}{k_t}. \tag{5.51}$$

Chen and Holmes (1980) used the vascular data from a 13 kg dog and estimated the exponential equilibration lengths for various blood vessels. Their results are listed in Table 5.7. To identify the vessels listed, Fig. 5.10 illustrates a typical architecture of a vascular bed. Before discussing the results, we briefly describe single-vessel studies performed by other investigators.

Other single-vessel studies

While Chen and Holmes (1980) used a 'semiquantitative' approach to estimate the exponential equilibration length, others employed a more quantitative methodology. For example, Chato (1980) separated the energy flow in the blood vessel and the conduction in the tissue and treated them as heat exchangers. Weinbaum *et al.* (1984) approached the problem as a thermal entry length problem. Therefore, they analyzed the special situation which exists at the vessel entrance where the velocity and temperature profiles are not fully developed (e.g. the velocity is not independent of axial distance). Baish *et al.* (1986a) made use of a conduction shape factor. His formulation of the problem was in similar form as Chen and Holmes (1980), or

$$u_{avg}\rho_b c_b \pi r^2 \frac{dT_b}{dx} = k_t \phi (T_t - T_b), \tag{5.52a}$$

where

Table 5.7 Exponential equilibration lengths

Vessel	r (μm)	x_e (m)	L_c/x_e
Aorta	5000	190	0.002
Large artery	1500	4	0.05
Arterial branch	500	0.3	0.3
Terminal branch	300	0.08	0.1
Small artery	175	0.009	1
Arteriole	10	5×10^{-6}	400
Capillary	4	2×10^{-7}	6000
Venule	15	2×10^{-6}	800
Terminal vein	750	0.1	0.1
Venous branch	1200	0.3	0.3
Large vein	3000	5	0.04
Vena cava	6250	190	0.002

(Source: Chen and Holmes 1980.)

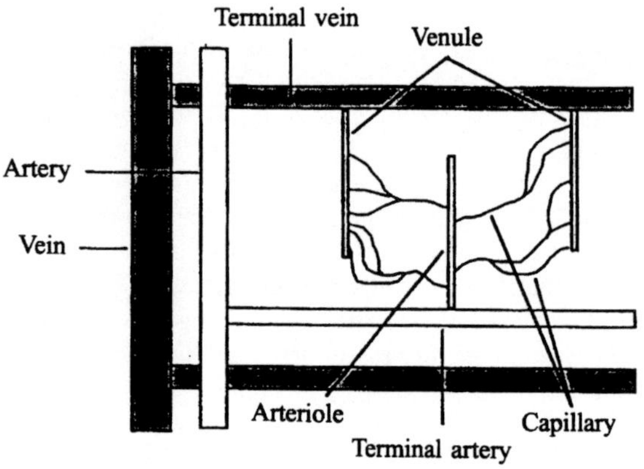

Fig. 5.10 Schematic representation of arterial and venous blood vessels of a terminal vascular bed (Wiedeman 1962).

$$\phi = 2\pi\left[\ln\left(\frac{r_{\text{outer}}}{r_b}\right) + \frac{2k_t}{k_b\,Nu} + \frac{1}{2}\right]^{-1}, \qquad (5.52b)$$

and r_b is the blood vessel radius, r_{outer} the outer radius of the tissue cylinder, and ϕ the conduction shape factor. The exponential equilibration length has the form

$$x_e = \frac{u_{\text{avg}}\rho_b c_b \pi r^2}{k_t \phi}. \qquad (5.53)$$

The solution to eqn (5.52a) will also be an exponential decay as previously discussed.

It should be noted that all of these single-vessel studies make assumptions about the flow and physiology which are not wholly valid. For example, the investigations neglect pulsatile blood flow. Furthermore, the models assume that the equations hold for both large and small blood vessels.

Thermally significant blood vessels

The results obtained from the various single vessel investigations are all similar. The data presented in Table 5.7 shows that the blood in large vessels (radius > 200 μm) will experience little precooling, as the ratio of the characteristic length to the exponential equilibration length is small. In contrast, blood in small vessels (radius < 25 μm) will equilibrate with tissue temperature rapidly, since the ratio L_c/x_e is very large. Recent experimental work showed that the lower limit may be higher at about radius < 50 μm (Lemons *et al.* 1986). Therefore, most of the heat transport actually occurs in the terminal branches and arterioles. Vessels which contribute to local energy transfer are referred as 'thermally significant' blood vessels, and those that equilibrate with tissue temperature instantaneously are called 'thermally insignificant'.

These results demonstrate a major theoretical shortcoming of the bioheat equation. Since capillaries will instantaneously equilibrate with the tissue temperature and there is a range of blood vessels that will actually experience precooling, the assumption of a single volumetric source/sink term is fundamentally and physiologically incorrect. Similarly, Wulff's (1974) porous media model is also not precise, since the model ignores heat transfer in larger blood vessels.

5.3.5 Paired-vessel models

With the identification of the thermally insignificant and significant vessels, it is appropriate to examine where these vessels are located in the tissue. Typically, the thermally significant vessels are found at a depth of 6–10 mm below the skin surface (Weinbaum *et al.* 1984). The vascular architecture of the thermally significant vessels is also of interest; the artery and vein occur close together as a countercurrent pair. This pattern persists until the vasculature reaches the precapillary arterioles, where, in subsequent generations, they form a capillary network.

The vascular architecture of the countercurrent vessel pair aids in the maintenance of the internal temperature and helps in the reduction of environmental heat loss (Eberhart 1985). More specifically, heat can be transferred from the artery to the vein and, since the flows in the vessels are in opposite directions, the heat loss is decreased. Many animals that experience cold thermal conditions employ this type of thermoregulatory mechanism to reduce heat loss (Mitchell and Myers 1968; Bligh 1985).

Mitchell and Myers (1968) analyzed the heat transfer in countercurrent blood vessel pairs. They performed two separate energy balances for control volumes in the artery and in the vein, as depicted in Fig. 5.11. They assumed that the temperature depended only on axial direction, the thermal conductivity was constant, and

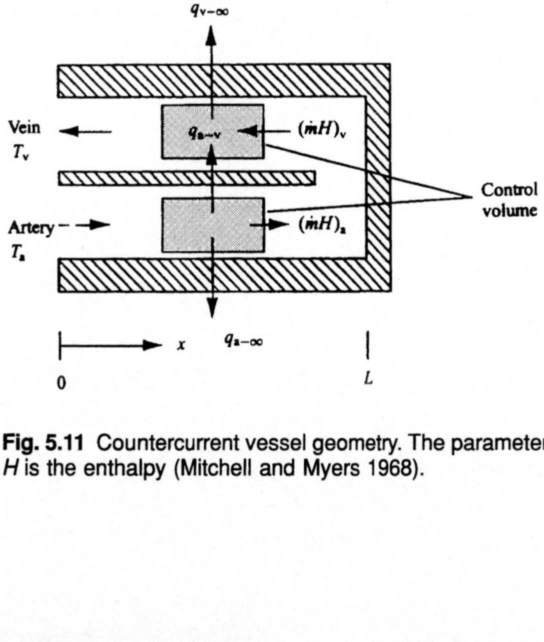

Fig. 5.11 Countercurrent vessel geometry. The parameter H is the enthalpy (Mitchell and Myers 1968).

there was no metabolic heat generation. The conservation of energy gives

$$\text{artery}: \quad 0 = \dot{m}c_b \frac{dT_a}{dx} - (UA')_i(T_v - T_a) - (UA')_a(T_\infty - T_a), \tag{5.54a}$$

$$\text{vein}: \quad 0 = \dot{m}c_b \frac{dT_v}{dx} + (UA')_i(T_a - T_v) + (UA')_v(T_\infty - T_v), \tag{5.54b}$$

where $\dot{m}$ is the blood mass flow rate, U is the overall heat transfer coefficient, and A' is the surface area per unit blood vessel length. The subscripts 'b', 'a', 'v', and 'i' stand for blood, arterial, venous, and interstitial region which lies between the artery and vein, respectively. The parameter T_∞ is the temperature of the surrounding tissue. The first term on the right-hand side of eqns (5.54) is the energy transported by convection in the blood vessels. The following two terms represent the heat transfer in the direction perpendicular to the flow and account for the heat transfer between the interstitial region and the heat transfer between the vessel and surrounding tissue.

The boundary conditions imposed were

$$\text{at } x = 0, \quad T_a = T_c, \tag{5.55a}$$

$$\text{at } x = L, \quad T_a = T_v = T_\infty, \tag{5.55b}$$

where T_c is a specified deep-body arterial temperature, and the authors assumed that all the blood flow will return in the vein and the heat transfer is complete (i.e. at $x = L$).

Mitchell and Myers (1968) nondimensionalized eqns (5.54) and were able to obtain an analytical solution for the coupled differential equations. The solutions for the arterial and venous temperature distributions are

$$\theta_a = \exp(N_v - N_a) \times \frac{\zeta}{2} \frac{C_2 \cosh[C_1(1-\zeta)] + \sinh[C_1(1-\zeta)]}{C_2 \cosh C_1 + \sinh C_1}, \tag{5.56a}$$

$$\theta_v = \exp(N_v - N_a) \times \frac{\zeta}{2} \frac{C_2 \cosh[C_1(1-\zeta)] - \sinh[C_1(1-\zeta)]}{C_2 \cosh C_1 + \sinh C_1}, \tag{5.56b}$$

where

$$C_1 = \frac{1}{2}\sqrt{[(N_a + N_v)(N_a + N_v + 4N_i)]}, \tag{5.56c}$$

$$C_2 = \sqrt{\left(\frac{N_a + N_v + 4N_i}{N_a + N_v}\right)}. \tag{5.56d}$$

The nondimensional parameters are

$$\zeta = \frac{x}{L}; \quad \theta_a = \frac{T_a - T_\infty}{T_o - T_\infty}; \quad \theta_v = \frac{T_v - T_\infty}{T_o - T_\infty}, \tag{5.57a, b, c}$$

$$N_a = \frac{(UA')_a L}{\dot{m}c_b}; \quad N_v = \frac{(UA')_v L}{\dot{m}c_b}; \quad N_i = \frac{(UA')_i L}{\dot{m}c_b}, \tag{5.57d, e, f}$$

where N is called the number of transfer units (NTU) and is a parameter which appears in heat exchanger problems (Holman 1981; Incropera and DeWitt 1990).

After deriving the solution for the temperature distribution, the authors analyzed two flow representations of the countercurrent vessels: the single artery–vein pair and the rete structure, where the artery is completely surrounded with many veins.

Keller and Seiler (1971) investigated heat transfer in countercurrent vessel geometry, including both capillary bleed-off and metabolic heat generation, for the geometry illustrated in Fig. 5.12. In their formulation, three equations were used to describe the heat transport. The additional equation comes from an energy balance performed on a control volume of tissue. The equations are

$$\text{tissue}: \quad k_t \frac{d^2T}{dx^2} + \{(hA''')_a + \rho_b c_b \omega\}(T_a - T_t) - (hA''')_v(T_t - T_v) + \dot{q}'''_{met} = 0 \tag{5.58a}$$

$$\text{artery}: \quad \frac{d}{dx}(\dot{m}_a c_b T_a) + \rho_b c_b \omega T_a + (hA''')_a(T_a - T_t) = 0, \tag{5.58b}$$

$$\text{vein}: \quad -\frac{d}{dx}(\dot{m}_v c_b T_v) + \rho_b c_b \omega T_t + (hA''')_v(T_t - T_v) = 0, \tag{5.58c}$$

where A''' is the average area for heat transfer per unit volume, h is the average heat transfer coefficient, and ω is the capillary perfusion rate. The subscripts 't', 'a', and 'v' stand for tissue, artery, and vein, respectively. In the tissue region, the heat transfer effects from blood are accounted for in the second and third terms in eqn (5.58a). The second term includes both the heat transferred from the artery to tissue and that from capillary perfusion. The third term represents heat that the tissue loses to the vein. The governing equations for the blood in the artery and vein are similar. In contrast to the previously discussed model, these equations contain

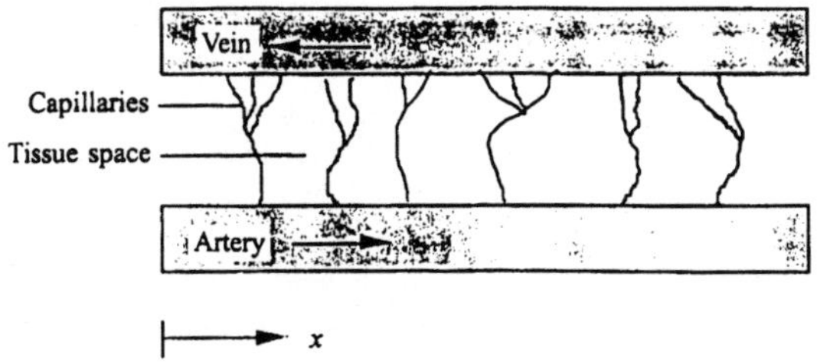

Fig. 5.12 Countercurrent vessel pair with capillary bleed-off (Keller and Seiler 1971).

an extra capillary bleed-off term in the form of $\rho_b c_b \omega T$.

The authors were interested in the heat transfer near the skin surface and, therefore, the boundary conditions imposed were in the form

$$\text{at } x = 0, \quad T_t = T_a = T_c, \tag{5.59a}$$

$$\text{at } x = \delta, \quad T_t = T_v = T_s, \tag{5.59b}$$

where T_c is the specified deep body core temperature and T_s is the specified skin surface temperature.

Keller and Seiler (1971) simplified the problem assuming that the overall heat transfer coefficients were approximately equal, that is,

$$hA''' \approx (hA''')_v \approx (hA''')_a. \tag{5.60}$$

They solved the three coupled differential equations to derive an equation for the heat flux between the skin and the core temperature, using an effective conductivity, in the form

$$\dot{q}'' = \frac{k_{\text{eff}}}{\delta}(T_c - T_s). \tag{5.61}$$

For the case when metabolic heat generation was neglected and the mass flow was assumed constant, the effective conductivity was determined to be

$$k_{\text{eff}} = \frac{-2k_t \sin(\upsilon\pi)}{\pi f(\zeta_0, \mu)} \frac{\lambda\delta\sqrt{2\mu+1}}{\mu(\mu+1)}, \tag{5.62a}$$

where

$$\upsilon = \mu + \frac{1}{2}, \tag{5.62b}$$

$$\zeta = \lambda\delta\sqrt{2\mu+1}\left(1 - \frac{x}{\delta}\right), \tag{5.62c}$$

$$f(\zeta, \mu) = \int_0^{\zeta} \left(\sqrt{\zeta'} I_\upsilon(\zeta') \int_0^{\zeta'} \zeta^{-3/2} I_{-\upsilon}(\zeta)\, d\zeta - \sqrt{\zeta'} I_{-\upsilon}(\zeta') \times \int_0^{\zeta'} \zeta^{-3/2} I_\upsilon(\zeta)\, d\zeta \right) d\zeta' + \int_0^{\zeta'} C\sqrt{\zeta'} I_\upsilon(\zeta')\, d\zeta' \tag{5.62d}$$

$$\delta\lambda = \delta\sqrt{\frac{\rho_b c_b \omega}{k_t}}, \tag{5.62e}$$

$$\mu = \frac{hA'''}{\rho_b c_b \omega}, \tag{5.62f}$$

and I is the modified Bessel function and C a constant. Their effective thermal conductivity k_{eff} absorbed the blood flow effects into a single quantity, which has the general form

$$k_{\text{eff}} = f(\delta\lambda, \mu, \Phi), \tag{5.63a}$$

where

$$\Phi = \frac{\dot{q}'''_{\text{met}}}{(T_c - T_s)\rho_b c_b \omega}. \tag{5.63b}$$

The parameter $\delta\lambda$ is associated with the ratio of heat transfer from capillary perfusion to that of conduction. The parameter μ is the ratio of the blood–tissue heat transfer to that from capillary bleed-off, and Φ is the ratio of metabolic heat to the maximum heat transfer due to capillary perfusion.

Other paired-vessel studies

Numerous investigators have presented many other countercurrent paired-vessel studies. Charny and Levin (1988) used the finite-element numerical solution technique to investigate heat transfer of countercurrent artery and vein pair embedded in muscle tissue. Baish *et al.* (1986b) compared various thermal models applied to the countercurrent vascular architecture with and without the inclusion of capillary bleed-off. Finally, Zhu *et al.* (1988) analyzed the countercurrent vascular geometry in greater detail, pointing out that the vessel pairs are usually of unequal size.

5.3.6 Recent general bioheat transport models

On the basis of new revelations of thermally significant blood vessels, investigators stress the importance of a more physiologically rigorous approach in the accounting for blood flow effects on the heat transfer in tissue. To this end, some new bioheat equations have been proposed. Although these models are superior theoretically and are derived from a more physiologically rigorous foundation, the detailed information needed for these formulations is not usually available. As a result, the evaluation and use of these new thermal models are limited at this stage. With this in mind, we briefly present two recent formulations.

Chen and Holmes model

Chen and Holmes (1980) presented one of the first general models of a bioheat transfer which accounts for thermally significant blood vessels. Their formulation is

$$\rho_t c_t \frac{\partial T_t}{\partial t} = \nabla(k_t \nabla T_t) + \rho_b c_b \omega_j^*(T_a^* - T_t) - \rho_b c_b \bar{U}_D \nabla T_t + \nabla(k_p \nabla T_t) + \dot{q}_{met}'''. \quad (5.64)$$

Several new terms appear in eqn (5.64) and each will be discussed. The term on the left-hand side and the first term on the right-hand side of eqn (5.64) are the usual storage and conduction terms. The second term on the right-hand side of eqn (5.64) is reminiscent of the blood perfusion term in the classic bioheat equation. Here, the parameter ω_j^* is the total blood perfusion rate delivered to the j*th vessel. The asterisk indicates that these quantities are to be evaluated at the j*th vascular generation. Therefore, this term will be important only for thermally significant blood vessels. The third term on the right-hand side of eqn (5.64) is reminiscent of the convective term in Wulff's (1974) porous tissue model discussed earlier. Again, the Darcy velocity appears which accounts for the direction of blood flow. The fourth term on the right-hand side of eqn (5.64) is a new term which represents an enhanced conduction due to capillary perfusion.

Chen and Holmes's modified bioheat equation seems promising, incorporating the blood flow effects into three new terms which are reminiscent of previously discussed bioheat transfer models. Much work must be done to measure the parameters to be used in the governing equation. Arkin *et al.* (1987) have recently attempted to differentiate between convective and perfusive heat transfer modalities in tissue through experimentation.

Weinbaum and Jiji model

Weinbaum and Jiji (1985) proposed an alternative bioheat equation which accounts for countercurrent heat transfer in thermally significant blood vessels. In a series of papers, the authors developed a new detailed three-layer formulation for peripheral bioheat transport, based on vascular casts of a rabbit's hind leg (Jiji *et al.* 1984; Weinbaum *et al.* 1984). In a subsequent study, the authors derived a simplified bioheat equation using local average tissue temperatures and a tensorial thermal conductivity (Weinbaum and Jiji 1985); their formulation has the form

$$\rho_t c_t \frac{\partial T_t}{\partial t} = \nabla[(k_{ij})_{eff} \nabla T_t)] - \frac{\pi^2 \rho_b^2 c_b^2 n r^4 u_{avg}^2}{k_t \sigma} \frac{\partial \beta_i}{\partial x_i} \beta_j \frac{\partial T_t}{\partial x_j} + \dot{q}_{met}''', \quad (5.65a)$$

where

$$(k_{ij})_{eff} = k_t \left(\delta_{ij} + \frac{\pi^2 \rho_b^2 c_b^2 n r^4 u_{avg}^2}{k_t^2 \sigma} \beta_i \beta_j \right), \quad (5.65b)$$

$$\sigma = \frac{\pi}{\cos^{-1}\left(\dfrac{d}{2r}\right)}, \quad (5.65c)$$

and d is the centerline distance between the countercurrent vessels, r is the constant blood vessel radius, δ_{ij} is the Kronecker delta (that is, $\delta_{ij} = 1$, if $i = j$; $\delta_{ij} = 0$, if $i \neq j$), n is the number density per unit area of vessel pairs crossing the control surface, u_{avg} is the average blood velocity, and β_i and β_j are the blood vessel's directional cosines. This new formulation incorporates blood flow effects into an effective thermal conductivity tensor which offers a compact form to describe bioheat transport.

For a one-dimensional geometry with all the blood vessels in the same direction as the temperature gradients, eqn (5.65) simplifies to

$$\rho_t c_t \frac{\partial T_t}{\partial t} = \nabla(k_{eff} \nabla T_t) + \dot{q}_{met}''', \quad (5.66a)$$

where

$$k_{eff} = k_t \left(1 + \frac{\pi^2 \rho_b^2 c_b^2 n r^4 u_{avg}^2}{k_t^2 \sigma} \right). \quad (5.66b)$$

Table 5.8 lists estimated magnitudes of the effective thermal conductivities for various blood vessel dimensions. The large values of the ratio k_{eff}/k_t for larger blood vessels ($r > 150$ μm) show that the countercurrent heat transfer dominates conduction heat transport. In contrast, for blood vessels with small radii, conduction dominates

Table 5.8 Estimated effective thermal conductivities

Radius (μm)	Average blood velocity (cm s^{-1})	Countercurrent pairs (cm^{-2})	k_{eff}/k_t
150	6	1	3.5
100	5	2	1.7
50	3	16	1.2
25	2	64	1.05

(Source: Weinbaum and Jiji 1985.)

countercurrent heat exchange, as the ratio k_{eff}/k_t is of the order of 1.

The use of the Weinbaum and Jiji (1985) bioheat equation presumes knowledge of the vascular system geometry. This physiological information is generally not available and, thus, the use of this new formulation is limited. Other investigators have not been satisfied with some of the assumptions made in the development of the thermal model (Wissler 1987a, b; Weinbaum and Jiji 1987). Still other researchers have used this new formulation to model heat transfer of peripheral tissue for various physiological conditions (Song *et al.* 1987; Song and Jiji 1988).

5.4 Conclusion

In this chapter we introduced the interdisciplinary topic of bioheat transfer. First, a brief overview of some of the fundamental concepts and analyses of heat transfer was presented. Then we discussed various thermal models used to describe heat transport in biological materials.

One of the major difficulties in modeling heat transfer in living tissues is the presence of the vascular system. We reviewed various methods to account for blood perfusion and discussed some of the models' shortcomings. The most widely used model is the Pennes (1948) bioheat equation. Although many of the model's assumptions have been discredited, the bioheat equation remains popular because of the simplicity of its form and ease in implementation for both analytical and numerical studies.

Recent analyses of heat transfer for single and paired vessels have revealed thermally significant and insignificant blood vessels. More physiologically rigorous general governing equations describing heat transfer in living tissue have been developed. However, these models are more complex and their use is limited because of the lack of relevant physiological information.

Taking into account the limitations of the various thermal models, some researchers have attempted to identify which governing equation best describes heat transport in different regions of tissue. To this end, some have developed hybrid models (Charny *et al.* 1990) and others demonstrated that the original bioheat equations actually describe tissue regions with thermally significant blood vessels reasonably well (Baish *et al.* 1986c).

The area of bioheat transport has witnessed increased activity in the past two decades. Some of this interest is driven from many of the beneficial applications of heat and cold in clinical medicine. A better understanding of local heat transfer is important to help optimize protocols for the destruction of undesirable tissues, such as cancer. Modeling of heat transport is also necessary to learn how to preserve tissues and organs at low temperatures.

Although there have been many gains in the understanding of bioheat transfer, much more research remains to be done. Methods for increased accuracy of intrinsic property data (e.g. thermal conductivity) must be developed. Methodologies to measure and evaluate living tissue response (e.g. blood perfusion) under various thermal conditions are needed. Physiological information, such as the blood vessel architecture, is necessary to better apply and test the recently proposed thermal models.

Problems

P5.1 Show that if the net radiation heat transfer is $\dot{q}'' = \varepsilon\sigma(T_1^4 - T_2^4)$ and it is desirable to recast this formulation into the form of eqn (5.9), then $h_r = \varepsilon\sigma(T_1 + T_2)(T_1^2 + T_2^2)$.

P5.2 Construct an electrical network model of Fig. 5.6. What is the form of the overall heat transfer coefficient?

P5.3 The deep body temperature of seals is 36.5°C and is comparable to that of humans. Seals employ a 2.0 cm thick layer of fat to reduce heat loss to their environment. What is the thickness of a diver's suit made of urethane ($k = 0.026$ W m^{-1}K^{-1}) to achieve the same level of insulation as the layer of fat in the seal?

P5.4 The surface area of an adult (human) can be estimated using the DuBois and DuBois (1961) equation,

$$\text{surface area} = 0.203W^{0.425}H^{0.725},$$

where W is the weight in kilograms, H is the height in meters, and the surface area is in units of m^2. To feel thermal comfort, the skin temperature must be approximately 22°C. Roughly estimate the heat loss from a 70 kg, 170 cm tall person when the air temperature is 15°C and the person is (a) seated quietly, (b) walking at 0.89 m s^{-1}.

P5.5 On a clear winter night, the effective sky temperature can drop to −20°C. Assuming that the nurse forgot to pull the shades in the newborn infant care nursery, what is the heat loss from the newborn in Example 1? Assume that the view factor between the newborn and the radiator is 0.2, the view factor between the newborn and the window is 0.4, and all the emissivities are close to 1.

P5.6 A seated person generates 58.2 W m^{-2} of metabolic heat per skin surface area. A person walking at 0.89 m s^{-1} generates 116.4 W s^{-2}. For a person who weighs 75 kg and is 180 cm tall, how much does he have to sweat for the evaporation of sweat to remove the difference in energy generated between the states in which the person is sitting and walking? [You will need the DuBois equation given in P5.4.]

P5.7 The tissue depth at which deep body temperature can be assumed (that is, the core of the body) is about 8 mm. Using the resistance network technique, calculate the equivalent thermal resistance between the outer surface of the skin and the core of the body, if the thickness of the skin is 0.5 mm, the thickness of fat is 3 mm, and the rest is muscle.

P5.8 During heat exhaustion, it is desired to rapidly cool a patient. A doctor in your hospital came up with the idea of cooling a copper disk (1 mm thick, 20 cm in diameter) to 0°C in ice water and then cooling the heat-exhausted patient by putting the disk on his chest. If the skin temperature is 30°C and the equivalent heat transfer coefficient between the disk and the patient skin is 10 W m^{-2}K^{-1}, calculate how long it will take until the copper disk will reach the patient's skin temperature. (Ignore the heat loss from the back of the disk.)

P5.9 If the patient in Problem 5.8 has tissue as described in Problem 7, how long will it take for the disk to reach deep-body temperature?

P5.10 Formulate the boundary conditions in Table 5.6 in a nondimensional form.

P5.11 Derive the boundary conditions for a skin surface that experiences convection, radiation, and evaporation of sweat.

P5.12 Solve Example 4 for a situation in which the water temperature is 15°C and the heat transfer coefficient between the water and the diver is 15 W m^{-2} K^{-1}.

P5.13 Calculate the effective thermal conductance for the case described in problem 7, when the volumetric blood flow rate $(\rho\omega)_b$ is assumed to be 0.02 g ml^{-1} min^{-1}.

P5.14 Derive a dimensionless formulation of eqn (5.43).

P5.15 Using the technique of Chen and Holmes,

demonstrate through calculations whether it is possible to cool biological organs to cryogenic temperatures by perfusing the organs with gaseous helium at $-190°C$. (*Hint:* calculate the thermal decay length using thermal properties of gaseous helium and the technique of Chen and Holmes.)

P5.16 Derive eqns (5.54) from an energy balance.

P5.17 Derive eqns (5.58) from an energy balance.

References

Arkin, H., Holmes, K. R., and Chen, M. M. (1987). Theory on thermal probe arrays for the distinction between the convective and the perfusive modalities of heat transfer in living tissues. *Journal of Biomechanical Engineering*, **109**, 346–52.

Arpaci, V. S. (1966). *Conduction heat transfer*. Addison-Wesley, Massachusetts.

ASHRAE (1985). *ASHRAE handbook: 1985 fundamentals*. ASHRAE Inc. Atlanta, Georgia.

Baish, J. W., Ayyaswamy, P. S., and Foster, K. R. (1986a). Small-scale temperature fluctuations in perfused tissue during local hyperthermia. *Journal of Biomechanical Engineering*, **108**, 246–50.

Baish, J. W., Ayyaswamy, P. S., and Foster, K. R. (1986b). Heat transport mechanisms in vascular tissues: a model comparison. *Journal of Biomechanical Engineering*, **108**, 324–31.

Baish, J. W., Foster, K. R., and Ayyaswamy, P. S. (1986c). Perfused phantom models of microwave irradiated tissue. *Journal of Biomechanical Engineering*, **108**, 239–45.

Bazett, H. C. and McGlone, B. (1927). Temperature gradients in the tissues in man. *American Journal of Physiology*, **82**, 415–28.

Bligh, J. (1985). Regulation of body temperature in man and other mammals. In *Heat transfer in medicine and biology: analysis and applications*, Vol. 1, (ed. A. Shitzer and R. C. Eberhart), pp. 15–52. Plenum Press, New York.

Bowman, H. F. (1985). Estimation of tissue blood flow. In *Heat transfer in medicine and biology; analysis and applications*, Vol. 1, (ed. A. Shitzer and R. C. Eberhart), pp. 193-230. Plenum Press, New York.

Charny, C. K. (1992). Mathematical models of bioheat transfer. In *Bioengineering heat transfer: advances in heat transfer*, Vol. 22, (ed. Y. I. Cho), pp. 19–155. Academic Press, Boston.

Charny, C. K. and Levin, R. L. (1988). Heat transfer normal to paired arterioles and venules embedded in perfused tissue during hyperthermia. *Journal of Biomechanical Engineering*, **110**, 277–82.

Charny, C. K., Weinbaum, S., and Levin, R. L. (1990). An evaluation of the Weinbaum–Jiji bioheat equation for normal and hyperthermic conditions. *Journal of Biomechanical Engineering*, **112**, 80–7.

Chato, J. C. (1980). Heat transfer to blood vessels. *Journal of Biomechanical Engineering*, **102**, 110-18.

Chato, J. C. (1981). Reflections of the history of heat and mass transfer in bioengineering. *Journal of Biomechanical Engineering*, **103**, 97–101.

Chato, J. C. (1985). Measurement of thermal properties of biological materials. In *Heat transfer in medicine and biology: analysis and applications*, Vol. 1, (ed. A. Shitzer and R. C. Eberhart), pp. 167–92. Plenum Press, New York.

Chato, J. C. (1987). Thermal properties of tissues. In *Handbook of bioengineering*, (ed. R. Skalak and S. Chien), pp. 9.1–13. McGraw Hill, New York.

Chen, M. M. and Holmes, K. R. (1980). Microvascular contributions in tissue heat transfer. *Annals of the New York Academy of Science*, **325**, 137–50.

Cho, Y. I. (ed.) (1992). *Bioengineering heat transfer: advances in heat transfer*, Vol.. 22. Academic Press, Boston.

Diller, K. R. (1985). Analysis of skin burns. In *Heat transfer in medicine and biology: analysis and applications*, Vol. 2, (ed. A Shitzer and R. C Eberhart), pp. 85–134. Plenum Press, New York.

Diller, K. R. (1992). Modeling of bioheat transfer processes at high and low temperatures. In *Bioengineering heat transfer*, (ed. Y. I. Cho), pp. 157–357. Academic Press, Boston.

DuBois, D. and DuBois, E. F. (1961). A formula to estimate the approximate surface area if height and weight are known. *Archives of Internal Medicine*, **17**, 863–71.

Eberhart, R. C. (1985). Thermal models of single organs. In *Heat transfer in medicine and biology: analysis and applications*, Vol. 1, (ed. A. Shitzer and R. C. Eberhart), pp. 261–324. Plenum Press, New York.

Field, S. B. and Franconi, C. (ed.) (1987). *Physics and technology of hyperthermia*. Martinus Nijhoff Publishers, Dordrecht.

Gagge, A. P., Winslow, C. A., and Herrington, L. P. (1938). The influence of clothing on the physiological reactions of the human body to varying environmental temperatures. *American Journal of Physiology*, **124**, 30–44.

Holman, J. P. (1981). *Heat transfer*, (5th edn). McGraw-Hill, New York.

Houdas,Y. and Ring, E. F J. (1982). *Human body temperature: its measurement and regulation*. Plenum Press, New York.

Incropera, F. P. and De Witt, D. P. (1990). *Fundamentals of heat and mass transfer*, (3rd edn). Wiley, New York.

Jiji, L. M., Weinbaum, S., and Lemons, D. E. (1984). Theory and experiment for the effect of vascular microstructure on surface tissue heat transfer—part II: model formulation and solution. *Journal of Biomechanical Engineering*, **106**, 331–41.

Kays, W. M. and Crawford, M. E. (1980). *Convective heat and mass transfer*, (2nd edn). McGraw Hill, New York.

Keller, K. H. and Seiler, L. Jr (1971). An analysis of peripheral heat transfer in man. *Journal of Applied Physiology*, **30**, 779–86.

Lemons, D. E., Weinbaum, S., and Jiji, L. M. (1986). The role of the micro and macrovascular system in tissue heat transfer. In *Heat and mass transfer in the microcirculation of thermally significant vessels*, (ed. K. R. Diller and R. B. Roemer), ASME Winter Annual Meeting Proceedings, HTD Vol. 61, pp. 41–8. ASME, New York.

Lipton, J. M. (1985). Thermoregulation in pathological states. In *Heat transfer in medicine and biology: analysis and applications*, Vol. 1, (ed. A Shitzer and R. C. Eberhart), pp. 79–105. Plenum Press, New York.

Love, T. J. (1985). Analysis and application of thermography in medical diagnosis In *Heat transfer in medicine and biology: analysis and applications*, Vol. 2, (ed. A. Shitzer and R. C. Eberhart), pp. 333–52. Plenum Press, New York.

Mitchell, J. W. and Myers, G. E. (1968) An analytical model of the countercurrent heat exchange phenomena. *Biophysical Journal*, **8**, 897–911.

Onik, G. and Rubinsky, B. (1988). Cryosurgery: new developments in understanding and technique. In *Low temperature biotechnology: emerging applications and engineering contributions*, (ed. J. J. McGrath and K. R. Diller), pp. 57–89. ASME, New York.

Overgaard, J. and Nielsen, O. S. (1980). The role of tissue environmental factors on the kinetics and morphology of tumor cells exposed to hyperthermia. *Annals of the New York Academy of Science*, **335**, 254–79.

Pegg D. E. and Karow, A. M. (1987). *The biophysics of organ cryopreservation*. Plenum Press, New York.

Pennes, H. H. (1948). Analysis of tissue and arterial blood temperatures in the resting forearm. *Journal of Applied Physiology*, **1**, 93–122.

Roemer, R. B., McGrath J. J., and Bowman, H. F. (ed.) (1989). *Bioheat transfer, applications in hyperthermia, emerging horizons in instrumentation and modeling*, ASME Winter Annual Meeting Proceedings, BED Vol. 12, HTD Vol. 126. ASME, New York.

Rohsenow, W. M. and Choi, H. (1961). *Heat, mass, and momentum transfer*. Prentice-Hall, New Jersey.

Rowell, L. B. (1983). Cardiovascular aspects of human thermoregulation. *Circulation Research*, **52**, 367–79.

Sagi, A., Shitzer, A., Katzir, A., and Akselrod, S. (1992). Heating of biological tissue by laser irradiation—theoretical model. *Optical Engineering*, **31**, 1417–24.

Sergiadis, B., Sakellaris, J., and Kriezis, E. E. (1988). Electromagnetic field distribution and developed losses inside a finite cylinder and sphere under magnetic resonance imaging. *IEEE Transactions on Medical Imaging*, **7**, 381–5.

Shiina, T. and Saito, M. (1988). Hyperthermia by low-frequency synthesized ultrasound. In *IEEE Engineering in Medicine and Biology Annual Conference*, Part 2, pp. 879–80. IEEE, New York.

Shitzer, A. (1973). Studies of bioheat transfer in mammals. In *Topics in transport phenomena*, (ed. C. Gutfinger), pp. 211–343. Halstead Press.

Shitzer, A. and Eberhart, R. C. (ed.) (1985a). *Heat transfer in medicine and biology: analysis and applications*, Vol. 1. Plenum Press, New York.

Shitzer, A. and Eberhart, R. C. (ed.) (1985b). *Heat transfer in medicine and biology: analysis and applications*, Vol. 2. Plenum Press, New York.

Siegel, R. and Howell, J. R. (1981). *Thermal radiation heat transfer*, (2nd edn). Hemisphere Publishing Corp., New York.

Song, W J. and Jiji, L. M. (1988). Peripheral tissue freezing in cryosurgery. *Cryobiology*, **2**, 153–63.

Song, W. J., Weinbaum, S., and Jiji, L. M. (1987). A theoretical model for peripheral tissue heat transfer using the bioheat equation of Weinbaum and Jiji. *Journal of Biomechanical Engineering*, **109**, 72–8.

Storey, K. B. (1990). Life in a frozen state: adaptive strategies for natural freeze tolerance in amphibians and reptiles. *American Journal of Physiology*, **258**, (Regulatory Integrative Comparative Physiology, 27), R559–68.

Weinbaum, S. and Jiji, L. M. (1985). A new simplified bioheat equation for the effect of blood flow on local average tissue temperature. *Journal of Biomechanical Engineering*, **107**, 131–9.

Weinbaum, S. and Jiji, L. M. (1987). Discussion of papers by Wissler and Baish *et al.* concerning the Weinbaum–Jiji bioheat equation. *Journal of Biomechanical Engineering*, **109**, 234–37.

Weinbaum, S., Jiji, L. M., and Lemons, D. E. (1984). Theory and experiment for the effect of vascular microstructure on surface tissue heat transfer—part I: anatomical foundation and model conceptualization. *Journal of Biomechanical Engineering*, **106**, 321–30.

Welch, A. J. (1985) Laser irradiation of tissue. In *Heat transfer in medicine and biology: analysis and applications*, Vol. 2, (ed. A. Shitzer and R. C. Eberhart), pp. 135–84. Plenum Press, New York.

Whiting, P., Dowden, J. M., Kapadia, P. D., and Davis, M. P. (1992). A one-dimensional mathematical model of laser induced thermal ablation of biological tissue. *Lasers in Medical Science*, **7**, 357–8.

Wiedeman, M. P. (1962). Dimensions of blood vessels from distribution artery to collecting vein. *Circulation Research*, **12**, 375–8.

Wissler, E. H. (1987a). Comments on the new bioheat equation proposed by Weinbaum and Jiji. *Journal of Biomechanical Engineering*, **109**, 226–33.

Wissler, E. H. (1987b). Comments on Weinbaum and Jiji's

discussion of their proposed bioheat equation. *Journal of Biomechanical Engineering*, **109**, 355–6.

Wulff, W. (1974). The energy conservation equation for living tissue. *IEEE Transactions on Biomedical Engineering*, **21**, 494–5.

Wylie, C. R. and Barrett, L. C. (1982). *Advanced engineering mathematics*, (5th edn). McGraw-Hill, New York.

Yang, W-J. (1989). *Biothermal-fluid sciences: principles and applications*. Hemisphere Publishing Corp., New York.

Zhu, M., Weinbaum, S., Jiji, L. M., and Lemons, D. E. (1988). On the generalization of the Weinbaum–Jiji bioheat equation to microvessels of unequal size: the relation between the near field and local average tissue temperatures. *Journal of Biomechanical Engineering*, **110**, 74–81.

THE MODELING APPROACH TO THE STUDY OF PHYSIOLOGICAL SYSTEMS

6

E. L. Keller

Contents

Symbols

DELτ	transport delay of time τ (s)	$L[x(t)]$	Laplace transform of $x(t)$	ϕ	phase angle
$\dot{f}(t)$, $\ddot{f}(t)$	first and second time derivatives of $f(t)$	$L^{-1}[X(s)]$	inverse Laplace transform of $X(s)$	ω	imaginary part of s (rad s^{-1})
$f(t) * h(t)$	convolution of the functions $f(t)$ and $h(t)$	s	complex variable of the Laplace transform		
j	square root of -1	η	real part of s (Np s^{-1})		

6.1 Introduction

Bioengineers and biophysicists have used their mathematical and analytical skills to develop quantitative models of many physiological or biological systems that they and their medical or biological colleagues wish to study. These models, whether computer simulations or explicit sets of mathematical formulae, often provide the best quantitative picture of how complex biological systems function in the healthy or possibly diseased state. Some examples of these types of models are:

- cardiovascular system models
- ecological models of parasite population dynamics
- biochemical reaction dynamics
- neuromuscular control system models
- sensory signal processing models.

The overall objective of the modeling approach to the study of physiological systems is to provide better understanding of the system under study. Specific aims often include some or all of the following list:

(1) to provide a concise summary of the present knowledge about the operation of a particular system;
(2) to predict the outcomes of modes of operation of the system that have not been previously studied or that are impossible to test in a living system;
(3) to provide diagnostic tools to test theories about the possible site of suspected pathology or the action of various drug treatments;
(4) to clarify and simplify complex experimental data or observations that would otherwise be difficult to comprehend;
(5) to suggest new, crucial experiments on the system that are necessary for further advances in understanding, i.e. the model often makes what we do not know about the system very explicit.

In this chapter we will present some examples of the modeling approach as it has been applied to the study of the operation of the nervous system. The style will be to first introduce a system and then discuss some of the typical types of medical or physiological data and measurements that have been made on that system. Later we shall develop an initial model of the system, usually a set of mathematical equations that help to explain this data. We shall proceed by developing the mathematical tools that are necessary for the further analysis of these types of systems. With these analytical tools we shall be able to make predictions about the operation of systems under conditions not yet studied. The development of the mathematical tools will be brief and heuristic, but the reader can consult more extensive expositions given in the texts of mathematical system theory as applied to biological examples listed below for a more rigorous development of the systems theory. Further examples that develop skills in applying systems theory to medical and biological systems are given in the problems set at the end of the chapter. The textbooks by Milhorn (1966), Talbot and Gessner (1973), MacGregor and Lewis (1977), Marmarelis and Marmarelis (1978), and Bahill (1981) are good reference texts for the reader interested in more extensive developments of mathematical systems theory applied to biology. The books by Zuber (1981), Kuffler *et al.* (1984), and Wurtz and Goldberg (1989) are good introduction into the operation of the nervous system for the reader who has had no previous background in the terms used in this chapter.

6.2 Examples of physiological systems

6.2.1 Vestibulo-ocular reflex

This is a very primitive reflex found in almost all animals. The purpose of the reflex is to move the eyes to compensate for movements of the animal's head. When the reflex operates properly, each time the head is rotated, the eyes rotate by an equal but opposite amount, thus leaving the position of the eyes fixed in space. Without this reflex, every head movement would be accompanied by large movements of visual images on the retina. This situation would result in very blurred vision and loss of visual acuity.

Studies of this system have been made by placing humans or animals on a rotating platform which applies carefully controlled rotations to the subject's head. At the same time the positions of the eyes are measured with any one of several highly accurate transducers that have been developed for this purpose. Carpenter (1977) has a discussion of examples of several of these types of biomedical instrumentation devices used to measure eye movements.

Actual measurements in the clinic and in the laboratory have shown that the vestibulo-ocular reflex (VOR) operates very accurately so that eye position in space does remain approximately fixed during head turns, as illustrated in Fig. 6.1. These measurements have been made in total darkness, which indicates that visual inputs are not necessary for the proper operation of the reflex. In the experiment shown in this figure the head is suddenly accelerated to a constant velocity by rotation of the platform. After a short latency period the eyes begin to move in the opposite direction, and after a brief transient period of acceleration, they are moving at the same speed as the head but in the opposite direction. In this system, gaze velocity is defined as the sum of head and eye velocity. Proper operation of the reflex should insure that gaze velocity remains very close to zero whenever the head is turned.

Figure 6.1 also shows that eye velocity only matches head velocity for short intervals of time and then undergoes a short-duration, high-speed resetting motion in the opposite direction. These resets are necessary because the eyes can only rotate through a few degrees in the orbit (about 40° in humans) before reaching the mechanical limits imposed by orbital tissues. Thus the VOR is able to compensate for the head movement for a short period of time and then the eyes are reset to be able to compensate again about a new position in space. These two oppositely directed movements of the eyes are called, respectively, slow phases and quick phases (or

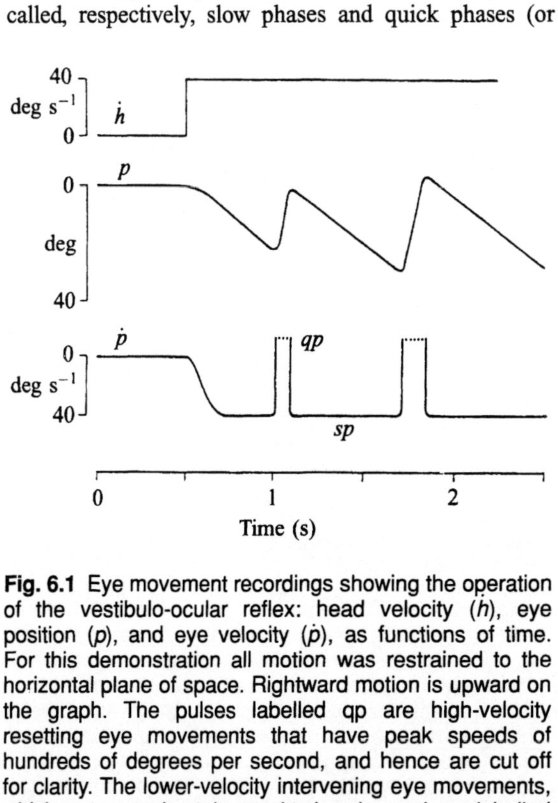

Fig. 6.1 Eye movement recordings showing the operation of the vestibulo-ocular reflex: head velocity ($\dot{h}$), eye position (p), and eye velocity ($\dot{p}$), as functions of time. For this demonstration all motion was restrained to the horizontal plane of space. Rightward motion is upward on the graph. The pulses labelled qp are high-velocity resetting eye movements that have peak speeds of hundreds of degrees per second, and hence are cut off for clarity. The lower-velocity intervening eye movements, which are approximately equal to head speed, are labelled sp.

saccades). We shall return to a system description of saccade generation below. Here a system description of the generation of the slow phases will be made.

Anatomical and physiological studies have shown that the inputs that activate the VOR originate from motion receptors in the inner ear (called the vestibular organ). Furthermore, there are separate receptors for linear accelerations and for angular accelerations of the head about each of the three Cartesian axes in space. When the head is rotated about any axis in space, compensatory eye movements similar to those illustrated for simple horizontal rotation in Fig. 6.1 are recorded, in which the eyes move about the same axis in space to compensate for the head movement.

Figure 6.2 shows a simple model of this system. This level of model merely contains the system inputs and outputs connected through a block or *black box* to indicate a functional relationship between the set of inputs and the outputs. This relationship could just as easily be given by a set of equations of the following form:

$$p_i(t) = F(\ddot{h}_x, \ddot{h}_y, \ddot{h}_z, \ddot{l}_x, \ddot{l}_y, \ddot{l}_z), \tag{6.1}$$

where p_i may be any one of the outputs p_x, p_y, p_z, and $\ddot{h}_x, \ddot{h}_y, \ddot{h}_z, \ddot{l}_x, \ddot{l}_y, \ddot{l}_z$ are system inputs from the vestibular organs in the inner ear.

Equation (6.1) or the corresponding block diagram in Fig. 6.2 states that the outputs of the system could be functions of all the inputs. Notice that each of the inputs and the outputs is a function varying in time. They are called *signals* in systems notation. The form of the equation also tells us that we need to design experiments that establish the form of this functional relationship. A number of such experiments have been conducted on this system which indicate that a close approximation of the

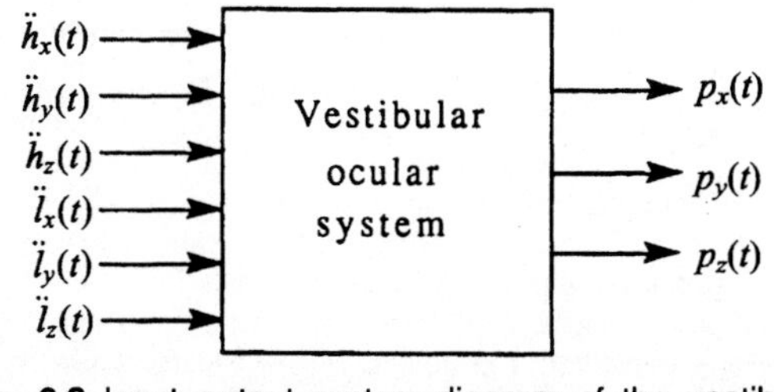

Fig. 6.2 Input–output system diagram of the vestibulo-ocular system. Inputs include angular accelerations of the head ($\ddot{h}_{x,y,z}$) and linear accelerations of the head ($\ddot{l}_{x,y,z}$), both in three-dimensional space. Outputs are rotational positions of the eye ($p_{x,y,z}$) in three dimensions.

system is obtained by considering that the output eye rotation is only dependent on head angular acceleration inputs (the inputs $\ddot{h}_x$, $\ddot{h}_y$, $\ddot{h}_z$). When we further restrict the inputs to the horizontal plane we reduce the system to a simple single-input–single-output system. For this initial introduction to the biological applications of systems theory we will only consider systems of this type (single-input–single-output systems).

From further anatomical and physiological study of this system it has been possible to divide the single function block of Fig. 6.2 into several identified separate blocks as illustrated in Fig. 6.3. The upper portion shows a schematic diagram of the anatomical structures that make up this system. Various signals are placed on the diagram where they exist in the system. The lower part of the figure represents each structure of the actual biological system by a block with appropriate input and output signals. Notice the convention followed in this block diagram. The input to the leftmost block is still head acceleration but its output is now vestibular nerve discharge, $v(t)$. This signal becomes the input to the next functional block labeled vestibular neurons. These neurons are part of the central nervous system and are located in the brainstem. They transform the vestibular nerve discharge (this block's input) into a new form of neural discharge $n(t)$, the output of this block. This signal is the input to another block further to the right named motor neurons. This block's output is motor nerve discharge, $m(t)$. Finally $m(t)$ is the input to the most rightward block labelled orbital mechanics. This block transforms the motor nerve discharge into the system output, $p(t)$, which in this case is eye position.

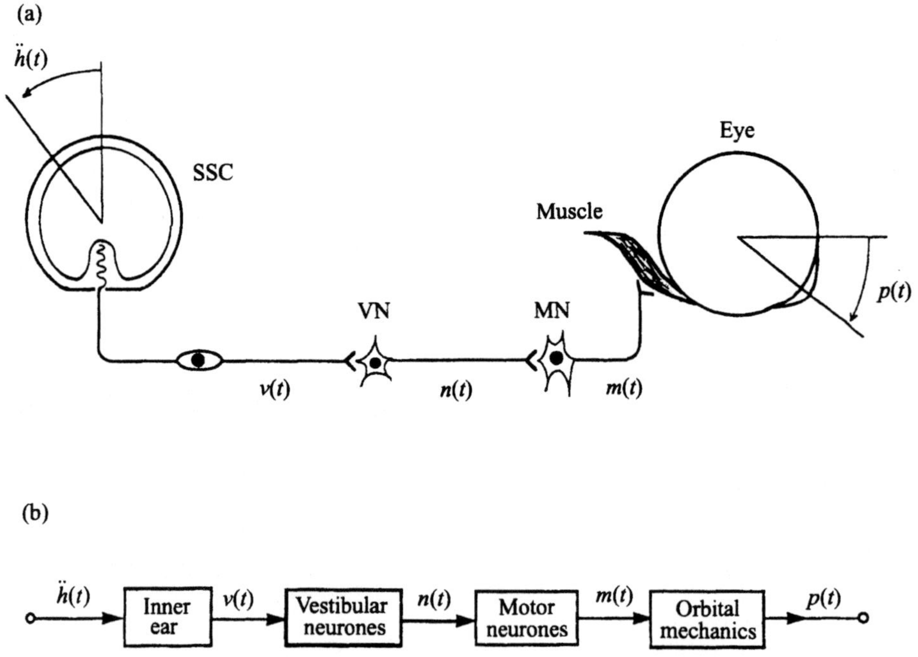

Fig. 6.3 Anatomical arrangement and model of the vestibulo-ocular reflex. (a) Anatomical structures that constitute this reflex. Signals that flow between these structures are indicated with lower-case variables of time. The three irregularly shaped structures with internal dark-filled circles represent the cell bodies of neurons. SCC, semicircular canal; VN, vestibular neurons; MN, motor neurones. The dart-tipped lines represent nerves and synapses, contact points where a neurone can affect the activity of the following neuron. Neurones are directed devices in that signals flow from the cell body toward the synapse; that is, from left to right in this figure. (b) Block diagram, single-input–single-output model of the reflex. There is a separate block for each anatomical structure shown above, $\ddot{h}(t)$, angular acceleration of the head in the plane of the canal; $v(t)$, activity of the vestibular nerve going from the canal to the brain; $n(t)$, activity of vestibular neurones going to motor neurones; $m(t)$, activity of motor neurones going from the brain to an eye muscle; $p(t)$, position of the eye in the horizontal plane.

This system is an example of an open-loop system. We will return to this point after the next example, where we consider an example of a closed-loop or feedback system. Quantitative models of the VOR have been developed by Kamath and Keller (1974), Robinson (1981), and Optican and Zee (1984).

6.2.2 Spinal cord control of posture

Another simple model of a physiological system is illustrated in Fig. 6.4. The upper portion of the figure shows, in abbreviated form, the anatomical elements that comprise the reflex operation of the neural organization

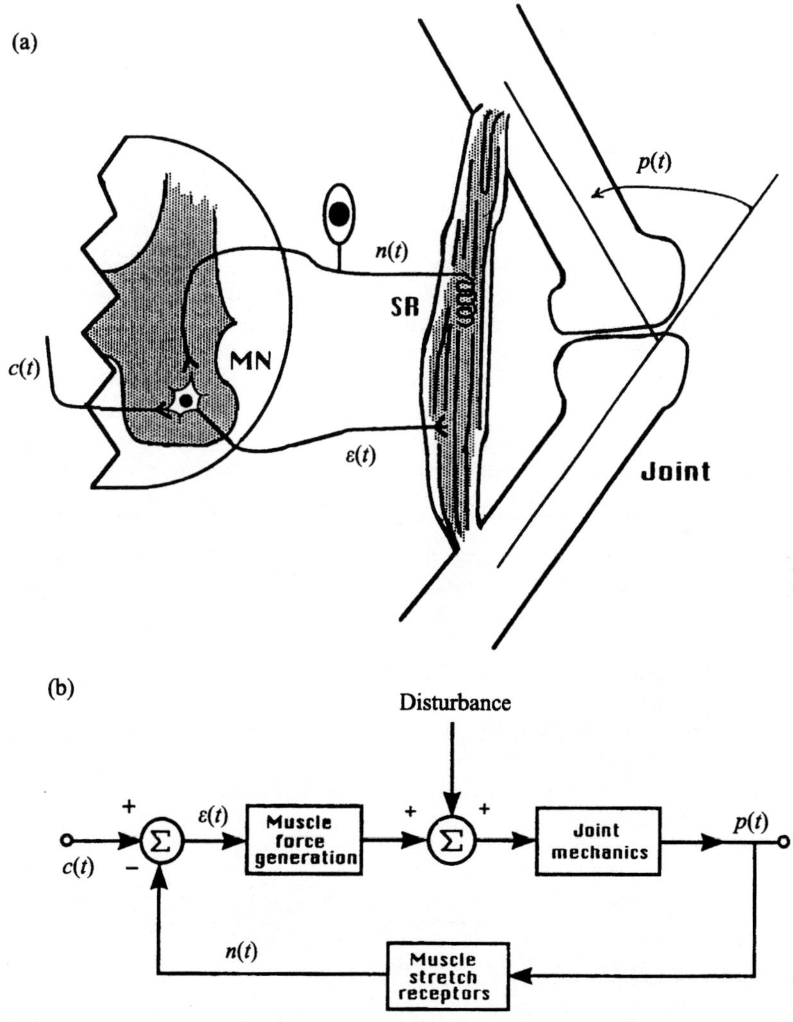

Fig. 6.4 Anatomical arrangement and model of the spinal cord reflex that maintains steady posture of the body. (a) Anatomical elements. Symbols as in Fig. 6.3. The structure on the right shows a single joint and one muscle that can change the angle of this joint. The structure on the left is a hemisection of the spinal cord. SR, muscle stretch receptor that transduces muscle length (joint angle) into neural activity that is returned to the spinal cord. (b) Block diagram model of the reflex. $p(t)$, joint position; $c(t)$, neural activity from the central nervous system specifying desired joint position; $\epsilon(t)$, error signal; $n(t)$, activity of muscle stretch receptors coding joint angle.

in the spinal cord that stabilizes upright posture in the face of unexpected disturbances such as load changes or muscle fatigue. The lower portion (Fig. 6.4b) is a block diagram system model of the operation of this reflex. The input to this system, $c(t)$, shown on the left, is the desired postural position of the body. This signal represents a selectable set point determined by the central nervous system. Just as for eye position in the previous example, the postural input signal represents a desired three-dimensional orientation of the body in space. The components of this central command could easily be given as a vector of signal inputs, but for simplicity we shall consider it as a single-dimensional signal in the following discussion. The output of the postural control system is the actual position of the body (simplified to a one-dimensional angle of a single joint) and labelled $p(t)$ on the right of the figure.

The model for this system has several new structural features not encountered in the previous example. The circles with the sigma labels represent nodal points in the system where two or more signals are algebraically summed to form a single new output signal. The signs next to the input signal arrow heads indicate whether the signals are added or subtracted at this point in the system. This system has a second input, labelled *disturbance*, which is also a signal representing, for example, a suddenly applied change in load.

The output position is sensed by muscle sensors called stretch receptors, which transform the joint position (muscle length) to the nerve signal, $n(t)$, which is returned to the spinal cord. Within the spinal cord the sensory nerves from stretch receptors make direct contact with the same motor neurons that activate their muscle. Therefore the summing junction on the left in the system diagram represents these motor neurons which receive two input signals, one the central command, $c(t)$, and the other the feedback signal from the muscle sensors telling the motor cell the current length of the muscle that it innervates. It has not been made clear at this point why the input from muscle receptors is assigned a negative sign at the summing junction when it also activates (increases the discharge) of the motor neuron. We will return to this point later when we consider this system in more detail. For now we merely assure the reader that the net effect of the stretch receptor input is negative in sign. Notice that the operation performed by the summing junction is comparison: the desired position as coded by $c(t)$, is compared to the actual position, as coded by $n(t)$. When they are the same, the input to the system, $\epsilon(t)$, is zero and no change in output is produced. If either the desired position or the output changes (due to an input disturbance), $\epsilon(t)$ will become nonzero, and the system will respond to this input to again move the output toward the desired position. The signal $\epsilon(t)$ is called the error because it is the difference between the desired position and the actual position. The summing junction on the left is called the *comparator*. This type of system is an example of a feedback system, because the output or some transduced signal representing the output is fed back and combined with the input to continuously control the output. The operation at the summing junction need not be negative or even algebraic summation. Feedback systems exist where the operation is multiplicative. However, the form shown in this example has many advantages, as we shall discuss later, and occurs very frequently in both physiological and man-made feedback systems. It is called a *negative feedback system* or *servo system* by systems engineers.

These two examples have developed the concept of the block diagram as a concise way to represent a complex physiological system. They also introduced some forms of model topology and operator notation such as signal summation and signal flow direction. We shall now develop some analytical tools that will allow us to further examine the operation of these example systems in detail and also others of similar type.

6.3 Systems analysis techniques

6.3.1 Mechanical model of the eye's orbital mechanics

Consider the subsystem block shown on the far right in Fig. 6.3(b). It is labelled orbital mechanics and represents the mechanical action of the eye muscles, muscle suspensory and connective tissues, globe inertia, and orbital viscous tissues. Its input is muscle nerve discharge rate, $m(t)$, and its output is eye position $p(t)$. We want to develop a model for just this portion of the overall system that represents the mechanics at the system's output. One way to do this is to break up this final block into a series of anatomically distinct tissues for which appropriate mechanical models exist (see Chapter 7). For example, a passive muscle ($m(t)$ is set to zero) will have both elastic and viscous characteristics. Although the actual relationship between the passive muscle length (or eye position in the model) and the force it exerts on the globe is nonlinear, we shall develop a linear model that is approximately correct. The definition of linearity (Chapter 7) includes additivity and homogeneity. Consider a system whose input signal is $f(t)$ and whose output signal is $g(t)$. The system is linear if, and only if, for all input signals, $f_1(t)$ and $f_2(t)$, that are applied separately to the system to yield respectively the separate output signals, $g_1(t)$ and $g_2(t)$: (1) when $f_1(t)$ and $f_2(t)$ are applied together (i.e. $f(t) = f_1(t) + f_2(t)$) the response is $g_1(t) + g_2(t)$ (i.e. $g(t) = g_1(t) + g_2(t)$; and (2) when $f(t) = a\, f_1(t)$, where a is any real constant, the response is $g(t) = a\, g_1(t)$. The first condition states that that same output will be obtained whether the input signals are applied separately and the separate outputs are then summed or applied simultaneously to yield a composite output. The second condition states that if the input is scaled by any factor, the output will be scaled by the same factor. This definition of linearity is often called the principle of superposition.

All the parallel tissue elements in the eye muscle are lumped into two separate mechanical elements to represent the total muscle elasticity (k) and viscosity (r) as shown in Fig. 6.5. The elastic and viscous forces summate at the tendon (terminal labelled as $f(t)$), giving the following differential equation as a model:

$$f(t) = kp + r\dot{p}. \tag{6.2}$$

The homogeneous solution (f set to zero) for this equation is:

$$p(t) = Ae^{-(k/r)t} \tag{6.3}$$

where $A = p(0)$ is the initial condition on eye position. Notice that when $t = r/k$, $p(t) = Ae^{-1}$ and that e^{-1} is approximately 1/3. Let this value of t be called T.

A plot of the solution, eqn (6.3), is shown in Fig. 6.6. Equation (6.2) and its solution tell us that the complete operation of this simple linear system can be summarized by two constants, A and T, for system inputs $f(t) = 0$. But in practice we usually want to know the system output for specific, nonzero, inputs.

A general solution to this problem is considered next.

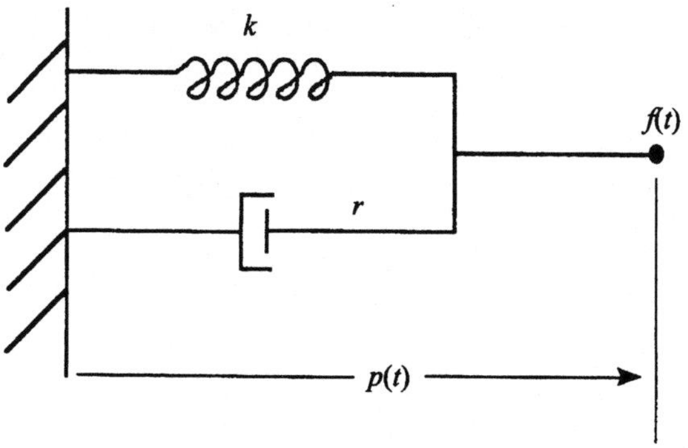

Fig. 6.5 Lumped-element linear model of a passive muscle. The ground structure on the left represents the attachment of one end of the muscle to a nonmoving bone. $f(t)$, signal representing the force at the tendon of the muscle; $p(t)$, signal representing the length of the muscle (or eye position); k, coefficient of muscle elasticity; r, coefficient of muscle viscosity.

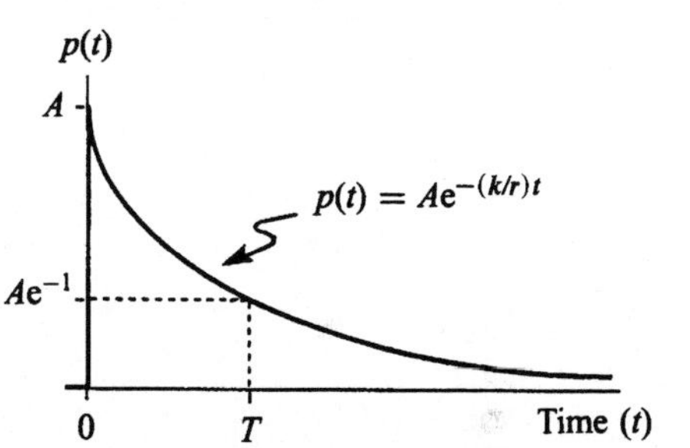

Fig. 6.6 Solution of the model in Fig. 6.5 for initial position A. Solution is a function of time with $T = r/k$ as time constant.

Choose as a specific input:

$$\begin{aligned} \delta(t - t_i) &= A, \quad t_i < t < t_i + \Delta t, \\ \delta(t - t_i) &= 0, \quad t_i + \Delta t < t < \infty, \\ A\Delta t &= 1, \end{aligned} \tag{6.4}$$

where t_i is a constant and Δt is very small (see Appendix, p. 491). The signal $\delta(t)$ is called the *unit impulse function*. It is said to have a weight (area) of unity. It is a very short-duration, large-amplitude pulse that is not strictly defined in the limit as Δt becomes infinitesimally small, but in concept it is a very useful signal. When $\delta(t - t_i)$ is applied to the input of any system, an output, say $h(t - t_i)$, is obtained. By definition, $h(t - t_i) = 0$ when its argument, $t - t_i$, is less than zero. The response, $h(t)$, to $\delta(t)$ is called the system's unit impulse response. Now apply two weighted impulses to the system, one at time t_1 and scaled by a factor of 5, and another at time t_2 scaled by 2. Because of the principle of superposition for a linear system, the output obtained can be shown graphically as in Fig. 6.7, or mathematically expressed as: $5h(t-t_1)+2h(t-t_2)$. The expression $(t-\tau)$ in parentheses means 'shift the signal to the right along the time axis by τ units'. Now consider representing an arbitrary input signal, $f(t)$, as a series of scaled impulse functions (Fig. 6.7c). The slices of $f(t)$ are approximate impulses of width Δt and variable amplitude (the amplitude of the ith such pulse is shown as a_i). By superposition, the output $y(t)$ at time t may be written as:

$$y(t)=\sum_{t=0}^{t/\Delta t} f(i\Delta t)\,h(t-i\Delta t)\,\Delta t=\sum a_i\,h(t-i\Delta t)\,\Delta t. \tag{6.5}$$

Now let Δt become infinitesimally small and in the limit the summation above becomes the integral equation:

$$y(t)=\int_0^t f(\tau)\,h(t-\tau)\,d\tau, \tag{6.6}$$

$$y(t)=f(t)*h(t). \tag{6.7}$$

The right-hand side of eqn (6.6) is called the *convolution integral*. The equation itself is a very powerful result. Notice that it states that if the unit impulse response, $h(t)$, of a linear system is known, then the output of the system can be calculated by convolution for any input $f(t)$. Equation (6.7) is a shorthand way of writing eqn (6.6). The symbol $*$ stands for convolution.

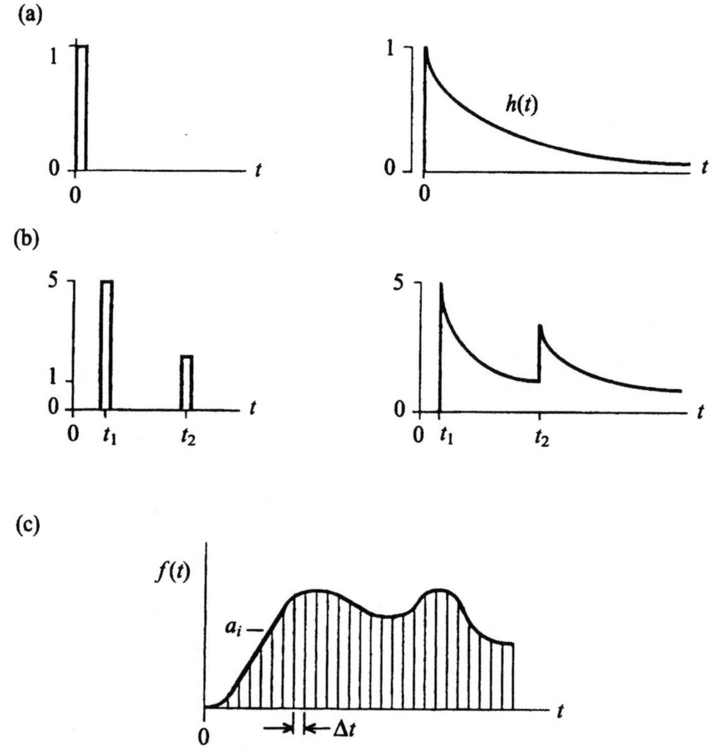

Fig. 6.7 Graphical demonstration of linearity or superposition. (a) System input is a unit impulse shown on the left, and the response is the system's unit impulse response shown on the right. (b) System response to consecutive impulse of weights 5 and 2 at times t_1 and t_2. (c) Continuous input, $f(t)$, conceptually broken up into an infinite sum of very narrow, impulse-like inputs with widths Δt and weights a_i, $i=1, 2, \ldots$, where each a is the amplitude of $f(t)$ in each consecutive interval.

6.3.2 Laplace transform

Although the convolution result is powerful, and the calculations involved in evaluating the integral can be carried out rather easily with digital computers (some digital signal processing circuits have been designed specifically for that purpose), it does not provide much insight into the operation of the system unless $f(t)$ and $h(t)$ are very simple functions.

As an alternative solution we will use the Laplace transform technique (Appendix, Section A.3, p. 494). Return now to the differential equation for the system shown in Fig. 6.5. Take the Laplace transform of each side of eqn (6.2):

$$L[f(t)] = L[kp + r\dot{p}]. \quad (6.8)$$

Tables A.3, A.4, and A.5 in the Appendix give short compilations of the Laplace transforms of some frequently encountered functions of time. Table A.2 gives some basic properties of the transform. More details about this technique and proofs for the basic properties are available in systems-theory or circuit-theory texts, such as Desoer and Kuh (1969). The following notation is used in this chapter. Let $x(t)$ be a time function, then $X(s)$ is its Laplace transform. Let $\dot{x}(t)$ be the time derivative of $x(t)$, then from property 6 in Table A.2, $L[\dot{x}(t)] = s\,X(s)$, i.e. under zero-state conditions the Laplace transform of the derivative of a time function is the Laplace variable, s, times the Laplace transform of the time function. This can be repeated so that

$$L[\ddot{x}(t)] = s^2X(s). \quad (6.9)$$

In many cases an inverse transform exists so that, if $X(s)$ is the Laplace transform of $x(t)$, then $L^{-1}[X(s)] = x(t)$.

The variable, s, of Laplace functions is complex:

$$s = \eta + \mathrm{j}\omega = \eta + \mathrm{j}2\pi f, \quad (6.10)$$

where the variable ω is in units of rad s^{-1} (f being in Hz), and the variable η is in units of Np s^{-1} (e-foldings per s).

Thus, Laplace functions are often called *frequency functions* in contrast to time functions. Functions of time will be denoted by lower-case symbols, $x(t)$, while the signal after being transformed into the corresponding frequency function will be given in capitals; for example, $X(s)$. Both $x(t)$ and $X(s)$ represent the same signal. The Laplace form for the equation of the orbital mechanics, eqn (6.8), is:

$$F(s) = k\,P(s) + rs\,P(s) + (k + rs)P(s), \quad (6.11)$$

$$P(s) = F(s)\left[\frac{1}{k + rs}\right]. \quad (6.12)$$

This equation is now in the operator or transfer function form, since the expression in brackets (the *transfer function*) operates on (multiplies) its input $F(s)$ and produces $P(s)$ as the output. Notice that simple multiplication replaces convolution in the time domain type of solution. This is a big advantage of using the frequency (Laplace) domain solution. The solution is indicated graphically in Fig. 6.8 by a system block diagram. Note that this is equivalent to solving the differential equation. The expression for $P(s)$ can be manipulated into several other forms:

$$P(s) = F\left[\frac{1/k}{1 + Ts}\right] = F\left[\frac{1/r}{s + a}\right] = F[H(s)] = FH, \quad (6.13)$$

where $T = r/k$ and $a = k/r$. Notice that we occasionally drop the arguments from the transformed functions (i.e. $F = F(s)$ and $H = H(s)$), which is a common practice to simplify notation.

The first form is called the *time-constant form* and is a most convenient representation when working with several blocks in cascade (series). The second form often is most convenient when looking up Laplace transforms in tables to reconvert them to time functions (e.g. see Tables A.3, A.4, and A.5 in the Appendix). As in eqn (6.11), the expression in brackets in eqn (6.13) is called the *transfer function* of the system. A system with the particular transfer function shown in this equation is called a *first-order system*, and the function itself is sometimes called a *first-order lag*. Note that zero initial conditions (zero state) are assumed in this operator notation. Nonzero initial conditions can very easily be built into Laplace notation, but in this chapter only the zero-state form will be used.

For the passive orbital mechanics, the transfer function is $H(s) = (1/k)/(1 + Ts)$. The value of frequency, $s = -1/T$, is called a *natural frequency* or *system pole*.

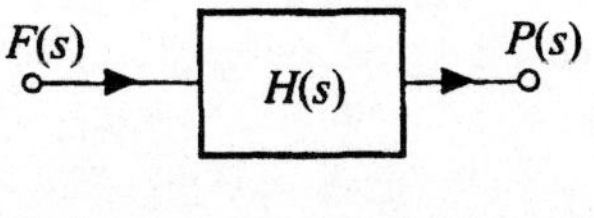

Fig. 6.8 Block diagram of system with transfer function *H*.

Notice that for this value of frequency, the denominator of the transfer function, $H(s)$, becomes zero and the expression for it is not defined.

Consider the situation when the input $f(t)$, to the system is the following time function called the *unit step function*:

$$u(t) = 1, \quad t \geqslant 0,$$
$$u(t) = 0, \quad t < 0. \tag{6.14}$$

Table A.4 (Appendix, p. 496) indicates that the transform of this function is $1/s$. To be more general, consider a step of amplitude A with transform A/s (from the homogeneity property, Table A.2). This can be written as:

$$L[A\, u(t)] = A/s. \tag{6.15}$$

In eqn (6.12), replace the input F with A/s, then the output becomes:

$$P(s) = A/s \cdot \frac{1/r}{s+a} = A/k\left[\frac{1}{s} - \frac{1}{s+a}\right]. \tag{6.16}$$

The final form of the right-hand side of eqn (6.16) is accomplished by the partial-fraction expansion technique (see subsection A.4.3 of the Appendix, p. 496). We can now inverse-transform each of the terms in eqn (6.16) back to time functions using Tables A.3 and A.4 along with the properties in Table A.2:

$$\begin{aligned} p(t) &= L^{-1}[P(s)] \\ &= A/k[1 - e^{-t/T}], \quad t \geqslant 0, \\ &= 0, \qquad\qquad\qquad t < 0. \end{aligned} \tag{6.17}$$

Try plotting $p(t)$. It should appear as a time function starting at $t = 0$ and $p = 0$ which then rises exponentially to a steady-state value of A/k. It should be about two-thirds of the way to A/k at a time value of $t = T$.

Notice that from the properties of Laplace transforms given in Table A.2 we may also interpret the transform $1/s$ as an operator that performs integration in the time domain. Consider a system with a transfer function of just $1/s$. It is usually called an integrator. If the input to this simple system is $F(s)$, then its output, $Y(s)$, is F/s. In the time domain, $y(t)$ is the integral of $f(t)$. If $f(t)$ is an impulse, then $y(t)$ is a steady level for all time $t > 0$. Such a system is said to have infinite memory and is called a perfect integrator. Clearly, it is not physically realizable. In particular, all biological systems with integrator-like properties will have the form $1/(s + a)$, where a may become very small. Such systems are called *leaky integrators*. You should plot the time domain form of the impulse response of a leaky integrator to see why they fit this name.

The transfer function, $H(s)$, given above converts muscle force, $f(t)$, to eye position, but where does the muscle force arise? The usual approach is to include an internal force generator (an active element or source) in the model of orbital mechanics. For simplicity here, we will assume that an internal biochemical mechanism in the muscle converts nerve discharge, $m(t)$, coming into the muscle directly to muscle force. The dynamics of this process is so fast that we can ignore it compared to the muscle time constant. Thus the transfer function from nerve input, $m(t)$, to muscle force, $f(t)$, is just a constant A (newtons per spike per second). Then the overall orbital mechanics can be described as shown in Fig. 6.9. The eye position output as a time function is given by eqn (6.17) if $m(t)$ is a unit step function.

A more realistic model of the orbit is given as $P(s) = M\, G(s)$, where

$$G(s) = \frac{A}{(1 + T_1 s)(1 + T_2 s)}, \tag{6.18}$$

where A is given in degrees per spike per second. This is known as a second-order system because it has two time constants. If $m(t) = u(t)$ and $T_1 = 0.2$ and $T_2 = 0.02$, we can now find the form of $p(t)$ by using Table A.5. In order to use this transform pair, first convert $G(s)$ to the system pole form and multiply by the input, $1/s$, to obtain:

$$P(s) = \frac{A/T_1 T_2}{s(s + 1/T_1)(s + 1/T_2)}. \tag{6.19}$$

We now want to find the inverse transform of $P(s)$ and plot is as a time function. The form of $P(s)$ is exactly the form of the third entry in the inverse table except that it is multiplied by the constant A/T_1T_2, which property 1 in Table A.2 (homogeneity) shows us how to handle. Notice that $a = 1/T_1 = 5$ and $b = 1/T_2 = 50$ are the two system poles (natural frequencies in Np s^{-1}). The plotting of $p(t)$ can be simplified by observing that the coefficient of the first exponential term, $AT_1/(T_2 - T_1)$, is approximately equal to A because $T_1 \gg T_2$, and the coefficient of the second exponential term, AT_2/T_1, is

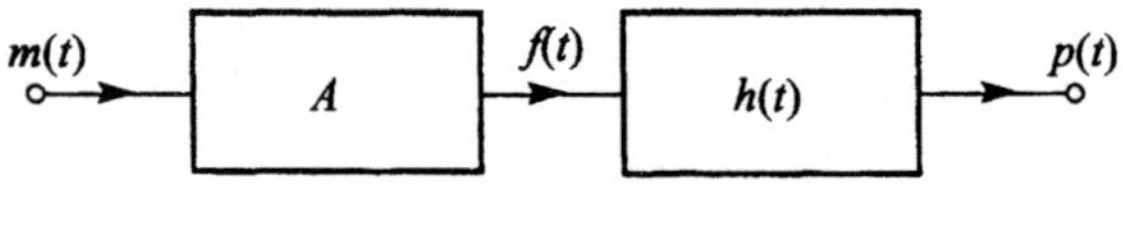

Fig. 6.9 Block diagram of the orbital mechanics.

much smaller than A. Therefore:

$$\begin{aligned} p(t) &\approx A[1 - e^{-5t} + 0.1e^{-50t}] \\ &= A[1 - e^{-t/0.2} + 0.1e^{-t/0.2}], \quad t \geqslant 0 \quad (6.20) \\ &= 0, \quad t < 0. \end{aligned}$$

Try plotting this last function on graph paper and observe that $p(t)$ is dominated by the longer time constant (0.2) except at very short times, where the effect of the shorter time constant (0.02) is to round off the curve to zero slope as t approaches 0. This situation occurs frequently in second-order biological systems; namely, the system time constants are separated by an order of magnitude or more.

In many cases we are only interested in the steady-state value of the system's output. For the case we have been studying we want to know the value of $p(t)$ after it has approached close to its final steady value. In other cases we are only interested in the initial value of the output (at $t = 0$). These values can be found without transforming the system equations back to the time domain by using the following theorems called the *final-value theorem* and the *initial-value theorem* (see Problem 9 of this chapter for constraints on these theorems):

$$\lim_{t\to\infty} f(t) = \lim_{s\to 0} s\,F(s) \tag{6.21}$$

$$\lim_{t\to 0} f(t) = \lim_{s\to\infty} s\,F(s), \tag{6.22}$$

where $F(s) = L[f(t)]$. For the oculomotor mechanics, if $m(t) = u(t)$,

$$P(s) = 1/s \cdot \frac{A}{(1 + T_1 s)(1 + T_2 s)}, \tag{6.23a}$$

then

$$\begin{aligned} \lim_{t\to\infty} p(t) &= \lim_{s\to 0}\left[s \cdot \frac{1}{s} \cdot \frac{A}{(1 + T_1 s)(1 + T_2 s)}\right] \\ &= A, \end{aligned} \tag{6.23b}$$

$$\begin{aligned} \lim_{t\to 0} p(t) &= \lim_{s\to\infty}\left[s \cdot \frac{1}{s} \cdot \frac{A}{(1 + T_1 s)(1 + T_2 s)}\right] \\ &= 0, \end{aligned} \tag{6.23c}$$

6.4 Frequency analysis

It has already been pointed out in the previous section that Laplace transforms can be visualized as functions of the complex frequency variable $s = \eta + j\omega$. In order to plot this Laplace function graphically, it is advantageous to calculate the values of the transform for η set equal to zero; that is, along the imaginary axis $j\omega$.

Consider the transfer function $G(s)$. Let $s = j\omega$, then for a given value of the variable ω (frequency in rad s^{-1}), $G(\omega)$ is, in general, a complex number. This complex number can be represented in two ways:

$$G(\omega) = X(\omega) + j\,Y(\omega), \tag{6.24a}$$

$$G(\omega) = |G(\omega)|e^{j\phi(\omega)}. \tag{6.24b}$$

$|G|$ and ϕ are two real functions of ω called the magnitude and the phase of the transform, respectively. We now want to plot the magnitude and the phase as functions of frequency ω for a variety of transforms that occur often in biological systems.

The transform $G(\omega)$ is often called the phasor transform of the time function $g(t)$. It could be derived directly from $g(t)$, but as we shall see, it is usually analytically simpler to work with the Laplace transform and only convert to the $G(\omega)$ form when it is desired to make a plot of the function. This transform is also called a truncated or single-sided ($\omega > 0$) Fourier transform.

Example 1. Suppose

$$\begin{aligned} G &= \frac{A}{Is^2 + rs} \\ &= \frac{A/r}{s(sT_1 + 1)}, \quad T_1 = 1/r, \end{aligned} \tag{6.25}$$

where I, A and r are real positive constants. Letting $s = j\omega$,

$$G(\omega) = \frac{A}{j\omega(j\omega T_1 + 1)}. \tag{6.26}$$

Notice that for each value of frequency, ω, the denominator is in general a complex number. We would have to rationalize the function first in order to plot it as the real and complex parts of the first form of a complex number above. Instead, use the second form with the function expressed in the magnitude, $|G|$, and phase, ϕ, form. Both $|G|$ and ϕ are real functions of ω which we can plot as separate graphs. For the plot of $|G|$ we use the identities $|a|\ |b| = |ab|$, $|\log a| = \log |a|$, and $\log(a/b) = \log a - \log b$ to get $\log |G| = \log(A/r) - \log |\omega| - \log |j\omega T_1 + 1|$. We usually plot $|G|$ in decibels (dB), which is $20 \log |G|$. Thus for magnitude we have:

$$20 \log|G| = 20 \log(A/r) - 20 \log |\omega| - 20 \log |j\omega T_1 + 1|. \tag{6.27}$$

In theory, this function (the magnitude of the phasor transform) can be plotted for all values of ω (positive and negative), but we will be interested only in the plot for positive or zero values of ω. These are the physically realizable values of frequency that we can use to test a real system. Negative values of frequency have no physical meaning. We proceed by plotting each term ($20 \log(A/r)$, $-20 \log |\omega|$ and $-20 \log |j\omega T_1 + 1|$) on the right-hand side of eqn (6.27) separately and then adding all three plots by inspection. The first term is a constant value, say $M = 20 \log(A/r)$, for all values of ω. That is, its effect is to add a constant offset (in dB) to the plot of the other two terms.

The second term, $-20 \log |\omega|$, will decrease (notice the minus sign in front of this term) linearly (on a log plot) with increasing values of ω. Therefore it will be convenient to plot it as a function of $\log \omega$, as shown in Fig. 6.10(a). Notice that the slope of the plot has a constant slope of -20 dB per decade (dec) of omega increase for all ω and that the value of $-20 \log |\omega|$ at $\omega = 1$ is zero dB. Notice also that we can not represent $-20 \log |\omega|$ as ω goes to zero on this log plot. It is easy to see that the value of this term approaches infinity as the value of ω approaches zero (thus ω is a pole of the system).

The third term, $-20 \log |j\omega T_1 + 1|$, will have two regions of behavior, one for very small values of ω:

$$\omega T_1 \ll 1$$
$$\therefore \quad |j\omega T_1 + 1| \approx 1. \tag{6.28}$$

In this region the third term is approximated by $-20 \log 1 = 0$; that is, a constant value of zero along the $j\omega$ axis. For large values of ω:

$$\omega T_1 \gg 1$$
$$\therefore \quad |j\omega T_1 + 1| \approx \omega T_1 \tag{6.29}$$

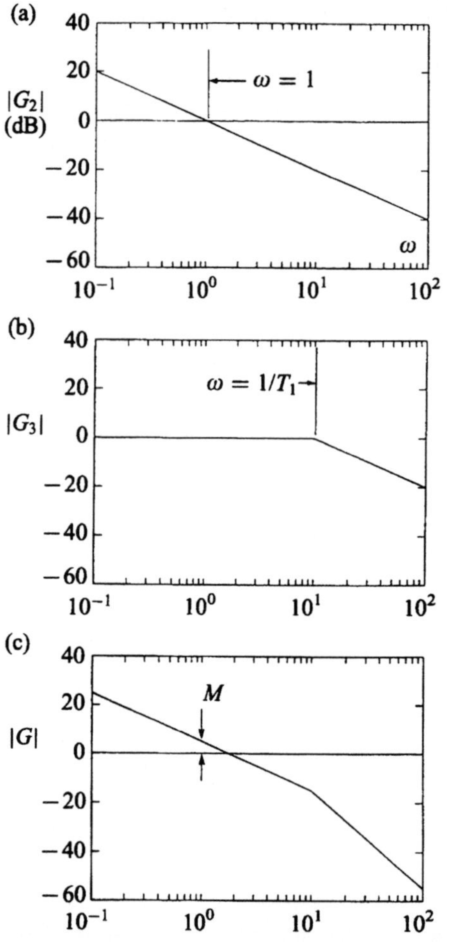

Fig. 6.10 (a), (b) Asymptotic Bode plots of the magnitude of the second and third terms in eqn (6.27). (c) Composite magnitude plot for the whole system.

Thus for large values of ω this term is approximated by $-20 \log \omega$; that is, a line with constant slope -20 dB/dec. The curves representing the asymptotic behavior at low and high values of ω must meet where $\omega T_1 = 1$ (or $\omega = 1/T_1$). This is shown in Fig. 6.10(b).

Now sum the three terms graphically to get the composite plot of $|G|$ as seen in Fig. 6.10(c). The overall curve for the magnitude of $G(j\omega)$ has a low-frequency range where $|G|$ falls at 20 dB/dec and a higher-frequency region of 40 dB/dec decrease. These two regions of differing behavior meet at the frequency of the second system pole, $1/T_1$. Notice that $|1 + jT_1\omega|$ is a smooth function of ω and therefore the actual curve for

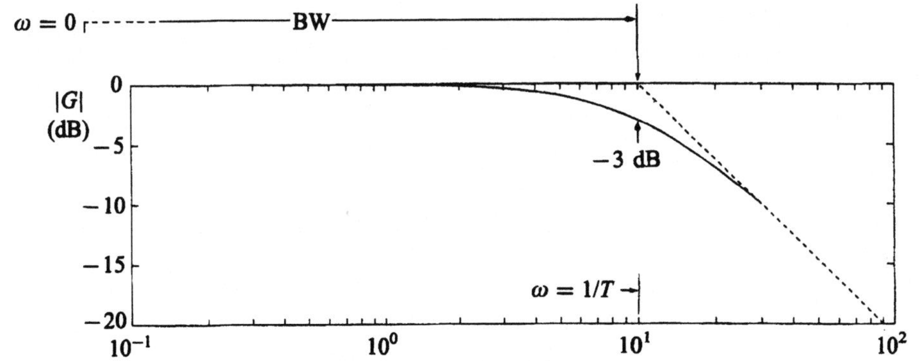

Fig. 6.11 Definition of system bandwidth (BW) for a first-order lag system. The dashed line shows the asymptotic magnitude plot and the solid curve the actual function.

this function only approaches asymptotically the straight-line approximations shown in Fig. 6.10(b). The actual magnitude curve for a first-order lag is shown in Fig. 6.11 for $G = 1/(1 + \mathrm{j}T\omega)$.

The magnitude, $|G|$, of the transfer function is often loosely called the gain of the system. The gain function for a system with the shape shown in Fig. 6.11 is also called a low-pass filter function. Notice that the function is constant for low frequencies, but decreases rapidly for high frequencies. The frequency region of approximately constant gain from $\omega = 0$ to $\omega = 1/T$ is called the system bandwidth (BW). The actual value of $|G|$ at $\omega = 1/T$ (shown by the smooth curve) is $-20 \log|1 + \mathrm{j}(1/T)T| = -20 \log[1 + 1]^{1/2} = -20 \log 2^{1/2} = -3$ dB. Notice that the difference (error) between the asymptotic approximation shown by the two straight lines and the actual smooth curve is greatest at this value of ω. The value $\omega = 1/T$ is also called the break frequency (or corner frequency), for obvious reasons.

Consider the situation when the input signal to a first-order lag is a unit step function. Then the ratio of the gain of the system output at the break frequency to the gain at low frequency is just 0.707. If the output is considered to have a physical interpretation like voltage or force (see Section 7.5), then the square of this ratio has the physical analogy of power, and the ratio at the −3 dB frequency is called the one-half power point.

The slope of the high-frequency portion of the gain curve is often called the roll-off and is usually given in dB/dec. In biological systems the roll-off is frequently given in decibels per octave where one octave is a doubling of frequency. Since 20 log 2 is approximately 6 dB a slope of 20 dB/dec is equal to 6 dB/octave. Thus a slope of −40 dB/dec is −12 dB/octave.

Now consider the plot for the phase angle ϕ of $G(\omega)$ in eqn 6.26 as a function of ω. From the trigonometric representation of a complex number we can show that:

$$\phi(\omega) = \tan^{-1}\left(\frac{\mathrm{Im}[G]}{\mathrm{Re}[G]}\right) \tag{6.30}$$

where Im[G] and and Re[G] are the imaginary and real parts of G respectively. First express each factor in G in the magnitude/phase form as shown below:

$$G(\omega) = \frac{A/r}{(\mathrm{j}\omega)(\mathrm{j}\omega T_1 + 1)}$$

$$= \frac{m_1 \mathrm{e}^{\mathrm{j}\phi_1}}{m_2 \mathrm{e}^{\mathrm{j}\phi_2} \cdot m_3 \mathrm{e}^{\mathrm{j}\phi_3}} = \frac{m_1}{m_2 m_3} \mathrm{e}^{\mathrm{j}(\phi_1 - \phi_2 - \phi_3)}. \tag{6.31}$$

Thus the composite phase angle ϕ is the algebraic sum of the three phase angles in eqn (6.31): $\phi = \phi_1 - \phi_2 - \phi_3$. Phase angles in the numerator are summed and those from the denominator are subtracted. In order to plot these phase angles as functions of ω it is easy to see that:

$$\phi_1(\omega) = 0, \tag{6.32a}$$

since A/r is a purely real number;

$$\phi_2(\omega) = \tan^{-1}\left[\frac{\omega}{0}\right] = +90^{\mathrm{deg}}, \tag{6.32b}$$

a constant phase angle not dependent on ω and

$$\phi_3 = \tan^{-1}\left[\frac{\omega T_1}{1}\right] = \tan^{-1}[\omega T_1]. \tag{6.32c}$$

The shape of the approximate plot for ϕ_3 can be seen from the following points: ϕ_3 approaches zero as ω becomes very small; ϕ_3 approaches 90° as ω becomes very large; and $\phi_3 = +45°$ when $\omega = 1/T_1$; $\phi_3(\omega)$ is a smooth function of ω, with maximum slope at $\omega = 1/T_1$. With these facts in mind, we plot ϕ_3 as shown in Fig. 6.12, again on a log axis for ω. Notice that an approximate plot for ϕ_3 can be made by plotting a straight line between the values $\phi_3 = 0$ at 0.1 of the break frequency, $\phi_3 = 45°$ at the break frequency, and $\phi_3 = 90°$ at ten times the break frequency. The complete curve for ϕ can now be plotted by combining the plots for ϕ_1, ϕ_2, and ϕ_3 with their appropriate signs as shown in Fig. 6.13.

The two plots, the magnitude of G (as shown in Fig. 6.10(c)) and the phase of G (as shown in Fig. 6.13), are normally plotted together for a system, $G(\omega)$, and are called the *Bode* or *frequency plots* for that system.

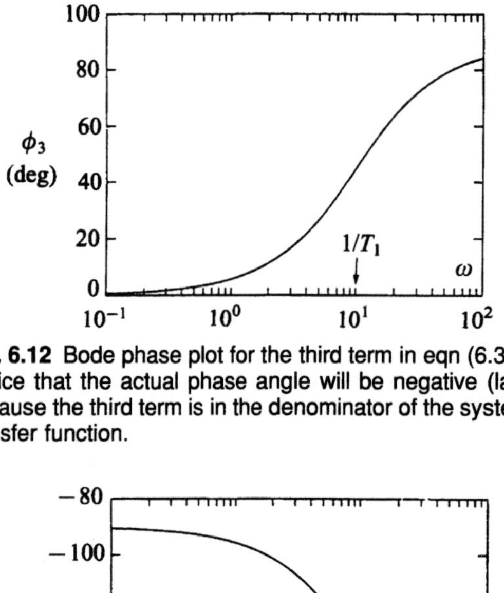

Fig. 6.12 Bode phase plot for the third term in eqn (6.31). Notice that the actual phase angle will be negative (lag) because the third term is in the denominator of the system transfer function.

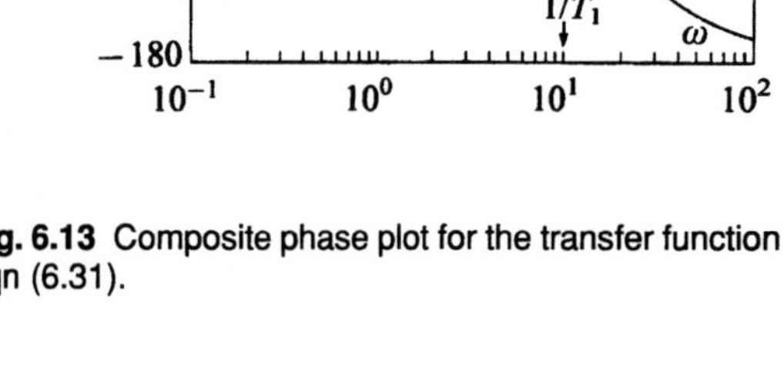

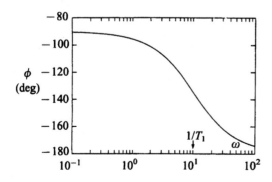

Fig. 6.13 Composite phase plot for the transfer function of eqn (6.31).

Example 2. Consider a different system with transfer function:

$$G(s) = \frac{1 + sT_1}{1 + sT_2}, \quad T_1 > T_2. \tag{6.33}$$

As before, let

$$G(\omega) = \frac{1 + j\omega T_1}{1 + j\omega T_2}. \tag{6.34}$$

The value $\omega = -1/T_1$, where the numerator vanishes, is called a *zero* for the system. This system has one zero and one pole. The zero term will contribute a graph of the form $+20 \log |1 + j\omega T_1|$ to the Bode plot. This graph will have a break frequency at $\omega = 1/T_1$ and then will increase at a steady 20 dB/dec at higher frequencies. The first-order pole term can be plotted as shown in the previous example. The overall Bode diagram (log–log plots of $|G|$ and phase angle of G against ω) is shown in Fig. 6.14. This combination of a zero at a lower frequency than the pole is called a *lead/lag* transform. Notice that the effect of the zero is to create a system with phase lead over a midrange of frequencies. Gain shows a positive slope over this same range of frequencies. The phase lead can be a very useful characteristic, a point we will return to below.

If the system in this example had contained a second pole at a frequency $\omega = 1/T_3$ above the other pole, then the shape of the Bode gain curve would have a flat region of approximately constant gain between frequencies $1/T_2$ and $1/T_3$. Try plotting it. Observe that there is now a low-frequency break point or cut-off (at $\omega = 1/T_2$) and a high-frequency cut-off (at $\omega = 1/T_3$). This type of gain curve is called a *bandpass curve* and its bandwidth is the range of frequencies from $\omega = 1/T_2$ to $\omega = 1/T_3$.

Example 3. Suppose the input to a system with transfer function $G(s) = A/(1 + Ts)^2$ is a unit step signal. Find the frequency form of the system's output response to this input and plot its Bode diagram.

Let $R(s)$ be the system output. From eqn (6.12), we write:

$$R(s) = \frac{A}{s(1 + Ts)^2}$$
$$R(\omega) = \frac{A}{(j\omega)(1 + jT\omega)^2} \tag{6.35}$$

Then

$$|R|_{\text{dB}} = 20 \log A - 20 \log |\omega| - 20 \log |1 + jT\omega|^2$$
$$= 20 \log A - 20 \log |\omega| - 40 \log |1 + jT\omega| \tag{6.36a}$$

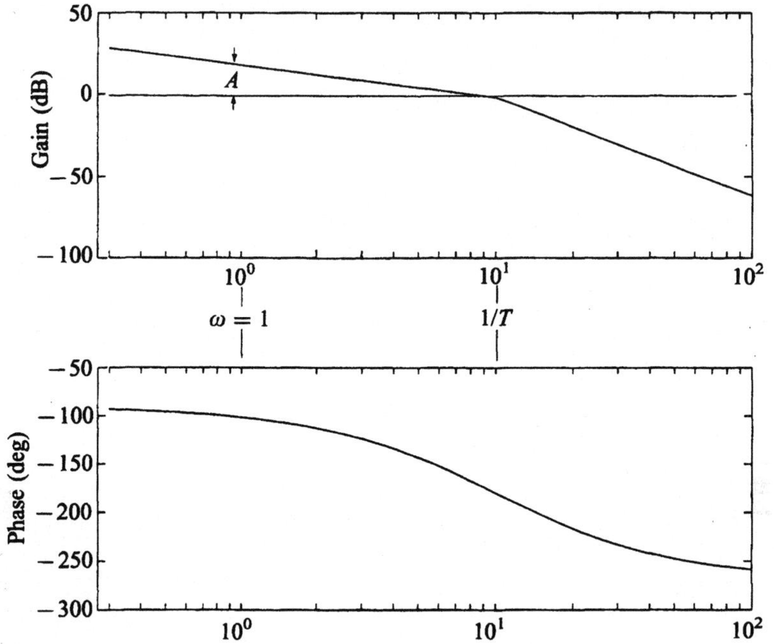

Fig. 6.14 Bode gain and phase plots for the transfer function in Example 2.

and

$$\begin{aligned}\phi &= -\tan^{-1}[\omega/0] - 2\tan^{-1}[T\omega/1] \\ &= -90° - 2\tan^{-1}[T\omega]. \end{aligned} \quad (6.37)$$

Notice that the power of 2 from the quadratic factor becomes a multiplier of 2 due to the properties of logarithms and the rules for the arithmetic of complex numbers. The Bode diagram is shown in Fig. 6.15. This type of transfer function is called a second-order system.

It may be apparent from the consideration of these transform examples that there is a close relationship between the gain and phase plots. If you have not noticed this relationship, re-examine the Bode diagrams for each of the examples again. Notice the relationship, particularly at high frequencies. Each first-order pole leads to a gain roll-off of 20 dB/dec at high frequencies and this is accompanied by 90° of phase lag at this same range of frequencies. A second-order pole, as illustrated in Example 3, has a gain slope of −40 dB/dec at high frequencies and 180° of phase lag. A first-order zero in the transfer function leads to a gain slope of +20 dB/dec and 90° of phase lead. It can be proved that this observation is general—the high-frequency slope of the gain function allows one to specify the phase angle of the transfer function at high frequencies. Each decrement of 20 dB/dec leads to −90° of phase angle and each increment of 20 dB/dec yields +90° of phase angle. In fact, the result is even more general. From integration of the slope of the gain curve over all frequencies, we can predict the phase angle at any given value of frequency. In this chapter we will only use the relationship in the high-frequency, asymptotic region. We need one other condition to ensure that the statement about the gain and phase relationships is always true. So far we have considered transfer function factors of the form: $(\mathrm{j}\omega + a)^n$, $n = \pm 1, \pm 2, \ldots$, where a is a positive real number. Factors of this form vanish at a frequency $\omega = -a$. If we plot the location of poles or zeros of this type in the complex plane, $s = \eta + \mathrm{j}\omega$, we can see that they all lie in the left-half plane. This is the additional condition we need to ensure the validity of the gain and phase relationship, i.e. all poles or zeros of the transfer function must be in the left-half complex frequency plane. If this is true, the transfer function is said to be of the *minimum phase type*. We shall discuss further the meaning of this name below.

Example 4. Consider the transfer function $G(s) = 1/(1 + \beta s + \alpha s^2)$. Plot its frequency response. This is a

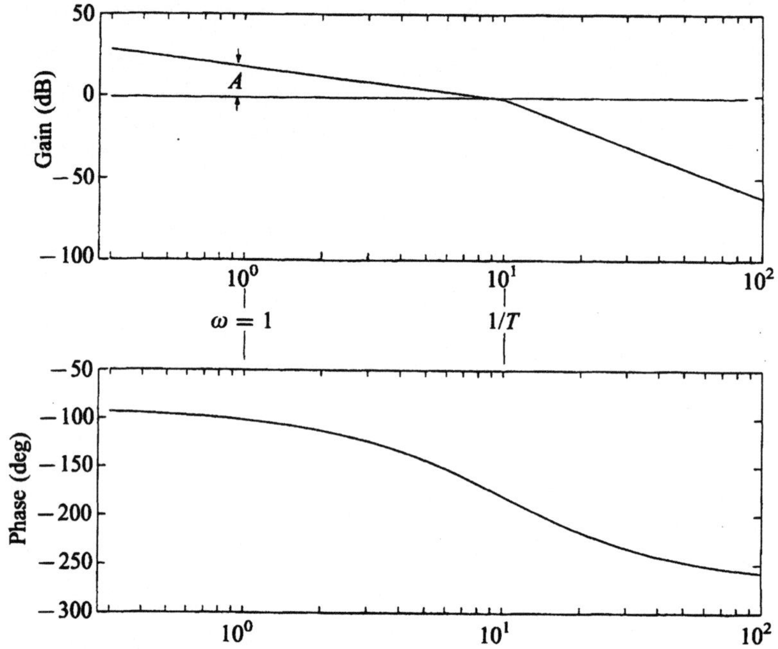

Fig. 6.15 Bode diagram for the transfer function in Example 3.

second-order system, since the highest denominator power of s is 2, but the denominator now appears as a quadratic factor. Letting $s = j\omega$ yields:

$$|G(\omega)|_{\text{dB}} = -20 \log |1 + j\beta\omega - \alpha\omega^2| \quad (6.38)$$

At very large ω, the ω^2 dominates. Thus, at high frequencies the gain is approximately $-20 \log(\alpha\omega^2)$. At high frequencies, this function has a slope of -40 dB/dec because $-20 \log(\alpha\omega^2) = -40 \log(\alpha^{1/2}\omega)$. At very small ω, $|G|$ is approximately $-20 \log(1) = 0$. At ω in between these extreme ranges of frequency, $|G| = -20 \log|(1 - \alpha\omega^2) + j\beta\omega|$. In particular, at $\omega = 1/\alpha^{1/2}$, the real part will vanish and $|G| = -20 \log(\beta\omega)$. Notice that if β is made very small, the gain becomes very large and positive near $\omega = 1/\alpha^{1/2}$. Thus the frequency response over all ω can be plotted as shown in Fig. 6.16 for several different values of β. The peak in the gain curves when β is small is called a *resonance*. The high- and low-frequency asymptotes for both gain and phase are the same as for the second-order factor in Example 3. The shape of Bode curves for gain and phase near the resonance are not easy to approximate, since they are a function of β, so reference has to be made to a set of curves like those shown in Fig. 6.16, which are given in various texts, such as that by Saucedo and Schiring (1968), for a wide range of values of β. (For linear plots of such curves see Figs 1.35 and 1.36.)

When $\beta = 0$, the resonance peak goes to infinity, and it is at a value of frequency exactly equal to $1/\alpha^{1/2}$. This frequency is called the *undamped natural frequency* of the system. By differentiating $|G(\omega)|$ with respect to ω, setting the derivative equal to zero, and solving for ω_p, one can show that for small positive values of β, the peak of the gain function is at a frequency:

$$\omega = \omega_p = \frac{1}{\sqrt{\alpha}}\sqrt{1 - \beta^2/2\alpha}. \quad (6.39)$$

The value of ω_p will be close to, but less than, the undamped natural frequency.

The transfer function can be rearranged to give:

$$G(s) = \frac{1/\alpha}{s^2 + (\beta/\alpha)s + 1/\alpha}. \quad (6.40)$$

The denominator can be factored to yield the values of the system poles

$$s_{1,2} = -\beta/2\alpha \pm \sqrt{(\beta^2 - 4\alpha)/4\alpha^2}. \quad (6.41)$$

If $\beta = 0$, the locations of the system poles are at $\pm j1/\alpha^{1/2}$. That is, they are located on the imaginary ($j\omega$) axis at the positive and negative value of $1/\alpha^{1/2}$, the undamped natural

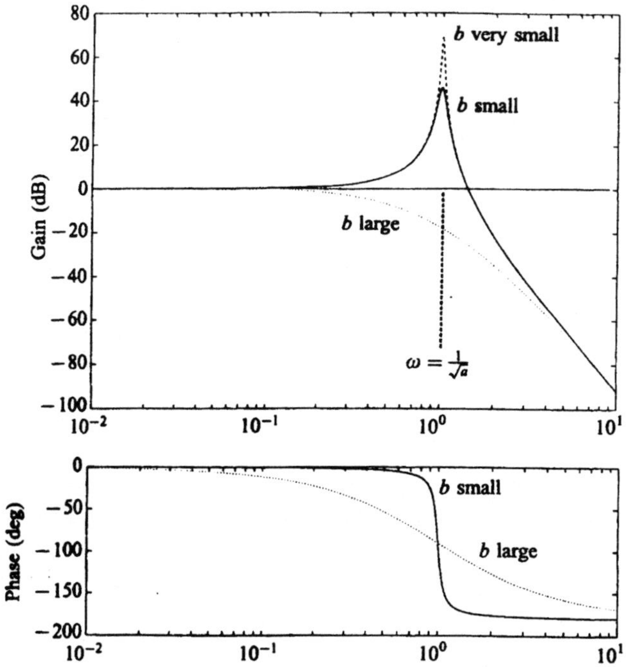

Fig. 6.16 Bode diagram for the transfer function in Example 4.

frequency of the system. If $\beta < 2\alpha^{1/2}$, then s_1 and s_2 are complex conjugate locations in the left half of the s-plane and eqn (6.40) can be restated as follows:

$$G(s) = \frac{1/\alpha}{(s+a)^2 + b^2}, \quad a = \beta/2\alpha, \quad b = \sqrt{(4\alpha - \beta^2)/4\alpha^2} \tag{6.42}$$

From this form, and by reference to the fifth entry in Table A.3, it is easy to see why the constant $a = \beta/2\alpha$ is often called the *damping factor* for the system. Notice that for a constant α, the scalar β determines how fast the time function, $g(t)$, goes exponentially toward an asymptotic value of zero, and thus it is also loosely called the damping factor.

From eqn (6.39) it is easy to show for damping, $\beta > 1.4\alpha^{1/2}$, that no peak (resonance) exists in the gain curve, $|G(\omega)|$ vs. ω. When $\beta = 2\alpha^{1/2}$, the transfer function becomes a second-order pole factor, as in Example 3, and the system is *critically damped.* If $\beta > 2\alpha^{1/2}$, then the poles are located on the negative real axis in the left half of the s-plane and eqn (6.40) can be restated as

$$G(s) = \frac{1}{(s + 1/T_1)(s + 1/T_2)}. \tag{6.43}$$

You should try plotting these poles and understand how they move as functions of the constants α and β in eqn (6.41).

Example 5. Consider again the last element in the vestibular system (the right-hand block in Fig. 6.3(b)). Since this is the mechanical output of this neuromuscular system, it is often called the *system plant.* Its transfer function is given by eqn (6.18). Plot the frequency response of the plant, if $T_1 = 0.2$ s (200 ms), $T_2 = 0.02$ s (20 ms) and $A = 10$ deg/spike per s. Then plot the Bode diagrams for the system's output when the input signal is a unit-step function. Also calculate the time domain form of the output when the input signal is a unit sinewave, sin $(2\pi f_0 t)$, and f_0 is a fixed frequency in Hz. The Bode plot for the transfer function is shown in Fig. 6.17. The plot of the output when the input is a unit-step function is obtained from $O(s) = (1/s)G(s)$, where $O(s)$ is the output signal (in the frequency form), $1/s$ is the input signal, and $G(s)$ is the plant transfer function. Notice that the input, $1/s$, will add a curve of constant slope of -20 dB/dec to all the values shown in Fig. 6.17. When the input is a sinewave of frequency f_0, the output is $[\omega_0/(s^2 + \omega_0^2)]G(s)$, where $\omega_0 = 2\pi f_0$. By partial-fraction expansion (Section A.4.3) or by reference to a more complete table of inverse transforms, it can be shown that the output time signal in this case is the sum of two transient terms that have $e^{-(1/T_1)t}$ or $e^{-(1/T_2)t}$ factors (and thus, decay to zero value as t approaches infinity) and another term of the form

$$A(\omega_0)\sin[\omega_0 t + \psi(\omega_0)], \tag{6.44}$$

where $A(\omega_0)$ is a frequency-dependent gain and $\psi(\omega_0)$ is a frequency-dependent phase. Because the first two terms vanish after some time, the term given in eqn (6.44) is called the *sinusoidal steady-state response.* This is another powerful result. When a sinusoidal signal of frequency f_0 is the input to the system, the system's steady-state output will be a sinusoid of the same frequency, but modified in amplitude and phase. Using partial-fraction expansion (see Section A.4.3) and the second and fourth transform pairs in Table A.3, it can be

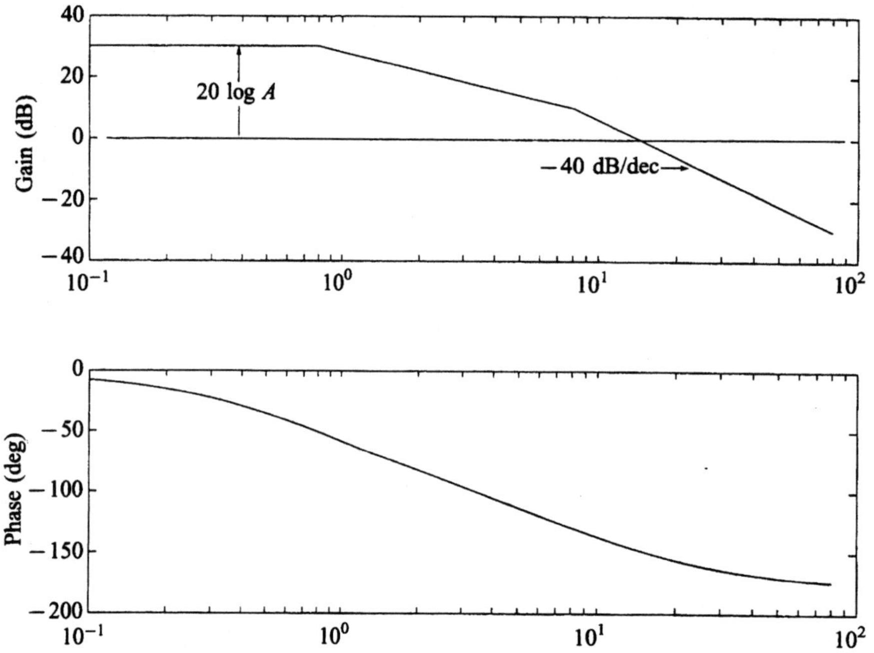

Fig. 6.17 Bode diagram for the unit-step response of the orbital mechanics (the system plant).

shown that

$$A(\omega_0) = |G(j\omega_0)| \tag{6.45}$$

and

$$\psi(\omega_0) = \tan^{-1}\left(\frac{\text{Im}[G(j\omega_0)]}{\text{Re}[G(j\omega_0)]}\right). \tag{6.46}$$

Thus, we can use the unit sinusoidal signal at various frequencies as the input to any linear system, H, to find the magnitude and phase angle of H at these same frequencies and then plot the Bode plots for H. Furthermore, once H has been estimated analytically we can predict the output of the system for any sinusoidal input of arbitrary magnitude and phase angle:

$$\begin{aligned} |O| &= |I|\,|H(j\omega_0)|, \\ \phi_O &= \phi_I + \phi_H(j\omega_0), \\ \phi_H(j\omega_0) &= \tan^{-1}\left(\frac{\text{Im}[H(j\omega_0)]}{\text{Re}[H(j\omega_0)]}\right), \end{aligned} \tag{6.47}$$

where $|I|$ and $|O|$ are the amplitudes of the input and output sinewaves, respectively, and ϕ_I and ϕ_O are their respective phases (see Appendix, Section A.7).

So far, most of the examples have assumed that the analytical form of the system transfer function is known. When this is the case, the usual problem is to determine the form of the output for various input signals. This is called *system analysis* and the steps involved are straightforward, as shown by Example 5. First, change the input signal into its frequency-domain representation by reference to a table of Laplace transforms. Next, multiply the system transfer function by this input. Finally, rearrange the resulting function and convert it to the output time function by reference to a table of inverse transforms.

How do we obtain the system transfer function for a physiological system when it is not given? In one approach the system is broken down into simple network elements for which we know the transform, as illustrated by the muscle example from Fig. 6.5. These network element transforms are combined by the methods given in Chapter 7 to yield the composite system transfer function. This method is a form of *system synthesis*.

A second method is to apply to the system's input a known signal such as a step or a series of sinusoidal signals with different frequencies. The output of the system is then measured experimentally and the form of the system transfer function is estimated from the experimental data. This approach is known as *system identification*.

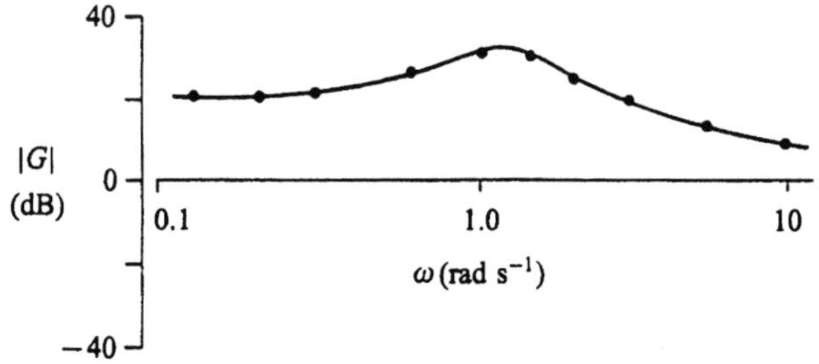

Fig. 6.18 Bode gain plot for the experimentally measured responses of the system in Example 6. Unit-sinewave inputs at various frequencies were applied to the system's input and the steady-state response as shown by the dots was measured at each applied frequency. The smooth curve shown by the solid line was then fit through the measured points.

Example 6. A system was tested with a series of unit sinewave inputs of various frequencies, and the output responses that were obtained are plotted in Fig. 6.18. Estimate the form of the transfer function.

Each of the points shown for various frequencies in Fig. 6.18 represents one experimental measurement on the system at that frequency. From eqn (6.47), it can be seen that these points define $H(j\omega)$, the system transfer function. The gain slope appears to be zero at low frequencies and has a value of 20 dB. Therefore, a scalar factor of 10 must be present in $H(s)$. There is a region beginning at about $\omega = 0.3$ where the slope becomes approximately +20 dB/dec. Therefore, there must be a zero factor, $(s + 0.3)$, present in $H(s)$. The slope becomes −20 dB/dec at about $\omega = 1.0$. This suggests a pole factor $(s + 1)^2$. Thus the overall transfer function is estimated to be

$$H(s) = \frac{10(s + 0.3)}{(s + 1)^2}. \tag{6.48}$$

Now consider the inverse problem. Given that we know the system transfer function and have measured the system's output under various conditions of natural behavior, can we estimate what the input to the system was for a particular output?

Example 7. Consider the response of the saccadic eye movement system as shown in Fig. 6.19. (*Saccade* is a French word meaning a *jerk*. Thus, saccades are the very jerky, rapid eye movements that are made when an individual inspects a new visual surround). In this control situation a target (T), on which the eye (p) is fixated, suddenly steps to a new position. After a short period of time (L) the eye makes a saccade to the new target position. Notice that the peak eye velocity $\dot{p}$ is very high during the saccade (> 600° s^{-1}). This is a much higher velocity than the slow phase vestibularly induced eye movements (compare Fig. 6.19 to Fig. 6.1). In fact, saccades are very much like the quick phase movements in this earlier figure. Also notice that the duration of the 20° saccade is only about 40 ms. Suppose that we accept that an approximate model of the orbital mechanics is $H(s) = A/(1 + 0.2s)(1 + 0.02s)$. The problem is to estimate the time form of the input (neural discharge signal) that created the saccade shown in Fig. 6.19 as an output.

This is an important problem because it allows us to infer the form of a physiological signal (in this case, a nerve discharge pattern) inside the body that we cannot measure directly. Suppose we guess that the nerve signal $m(t)$ is a unit step function, $u(t)$. Then the output P is given as:

$$\begin{aligned} P(s) &= \frac{1}{s} \cdot \frac{A}{(1 + T_1 s)(1 + T_2 s)} \\ &= \frac{a}{s} + \frac{b}{1 + T_1 s} + \frac{c}{1 + T_2 s} \\ T_1 &= 0.2, \quad T_2 = 0.02, \quad A = 20, \end{aligned} \tag{6.49}$$

where the coefficients a, b, and c can be estimated by partial-fraction expansion (see Appendix, Section A.4.3). Applying this technique gives $a = 20$, $b = -22.2$, $c = 2.2$. Thus,

$$\begin{aligned} p(t) &= 20 - 22.2e^{-(1/T_1)t} + 2.2e^{-(1/T_2)t}, & t \geqslant 0, \\ &= 0, & t < 0, \end{aligned} \tag{6.50}$$

after applying the inverse transform. This can be diagrammed as shown in Fig. 6.20.

This predicted eye movement doesn't look at all like the saccade shown in Fig. 6.19, which is over in 40 ms and doesn't look exponential, like the response in Fig. 6.20. Either the system model, $H(s)$, is wrong or the input signal is not a step. Because we think $H(s)$ is correct, we need to try another type of input signal. As a second guess for the probable input to the system, try the signal shown in Fig. 6.21. We can estimate the form of the output graphically by the use of superposition. Consider that the input is made up of the sum of two signals: (1) a positive step of amplitude 300 at $t = 60$ ms; and (2) a negative step of amplitude 260 40 ms later. The

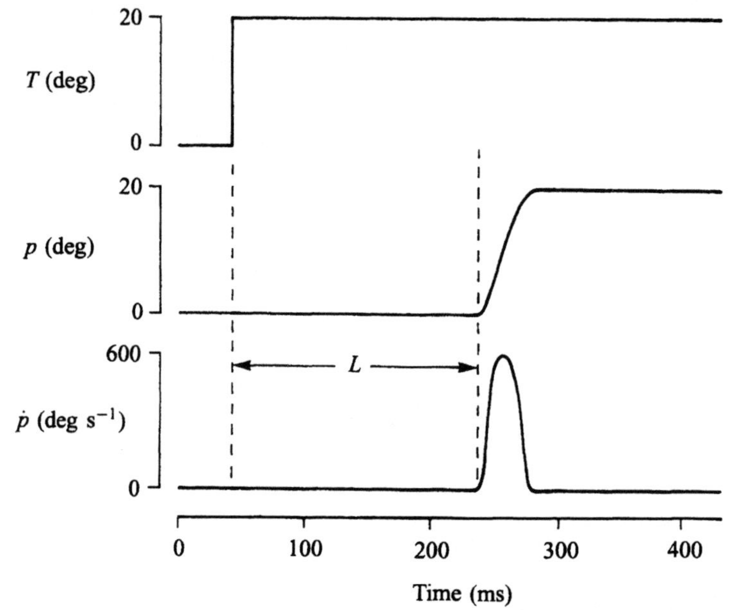

Fig. 6.19 Time-course of a saccadic eye movement. The upper curve shows the target (*T*) as it steps out to a 20° eccentric point. The saccade is shown on a plot of eye position (*p*) in the middle trace after a delay of *L* ms. The eye velocity during the saccade is shown by the lower trace ($\dot{p}$).

temporal sum of these two inputs gives the pulse–step signal seen in this figure. Now it is easy to graph the two outputs to the individual step inputs and then sum these two to arrive at the actual system output. This result is shown as the lower curve in Fig. 6.21. This signal looks much more like an actual saccade. The pulse-step type of control signal, illustrated by $m(t)$ in Fig. 6.21, works well when one wishes to make very rapid movements with no overshoot of the final position for a viscoelastic system which is dominated by viscosity. The orbital mechanics, $H(s)$, in this example is this type of system.

Although the example given above on solving the inverse problem used a graphical approach to estimate the shape of the input signal, a rigorous solution could also have been carried out analytically:

$$P(s) = M \cdot H(s), \tag{6.51}$$

given that the form of $H(s)$ is known and $P(s)$ is measured. Then

$$M(s) = P \cdot H^{-1}(s), \tag{6.52}$$

where $H^{-1}(s)$ is the inverse operator and $M(s)$ is the desired input signal. For the simple transforms with which we have been dealing, the inverse operator is obtained by interchanging the numerator and the denominator of the transfer function.

Notice that the timing of the pulse duration, and also the amplitude of pulse to a lesser extent, are the important factors in determining how far the eye will move during the pulse. Thus the control of pulse duration is crucial in this type of system.

It has now been possible to record the electrical input signals in human eye muscles during saccades, and these signals very closely approximate the pulse–step form predicted on the basis of the inverse analysis. You should now be able to speculate on the proper form of the neural control signal that will produce a rapid limb movement in a system, H, dominated by inertia, e.g. $H(s) = 1/(10^2 + s + 0.1)$.

The inverse problem can also be solved in the time domain using an inverse operation called *deconvolution*. Suppose you have a system, H, with unit impulse response, $h(t)$, input time signal, $x(t)$, and output signal, $y(t)$. You want to find $x(t)$ given the specific form of

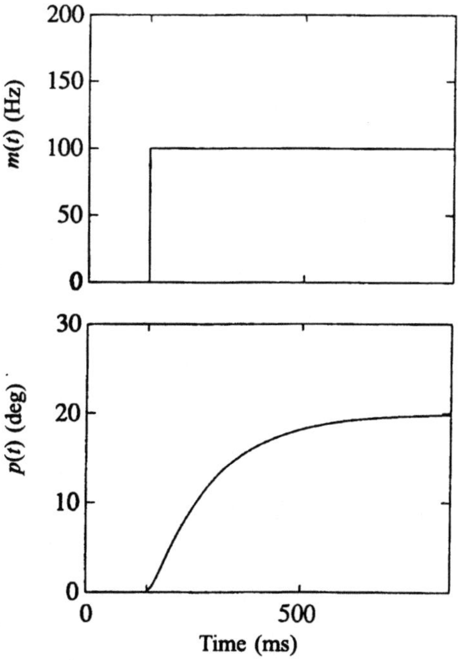

Fig. 6.20 Time response of the eye movement system if the input neural signal (m) is a unit-step function as shown in the upper trace.

$y(t)$. First find $H(s) = L[h(t)]$ and then look up $h^{-1}(t) = L^{-1}[1/H(s)]$ in a table of inverse transforms or use partial-fraction expansion. From convolution we know

$$\begin{aligned} x(t) * h(t) &= y(t), \\ h(t) * h^{-1}(t) &= \delta(t), \quad \text{unit impulse}, \\ x(t) * \delta(t) &= x(t). \end{aligned} \tag{6.53}$$

Then by deconvolution

$$\begin{aligned} [x(t) * h(t)] * h^{-1}(t) &= y(t) * h^{-1}(t), \\ x(t) * [h(t) * h^{-1}(t)] &= y(t) * h^{-1}(t), \\ x(t) &= y(t) * h^{-1}(t). \end{aligned} \tag{6.54}$$

Delay elements as a non-minimum-phase system. Consider a system element that imparts a pure delay to its input signal, but that has a gain of 1 for all frequencies. For example, consider the delay involved in nerve impulse travel from the brain to the spinal cord. If the nerves conduct impulses at 20 m s^{-1}, and the distance is about 1 m, then a delay of 50 ms is added to

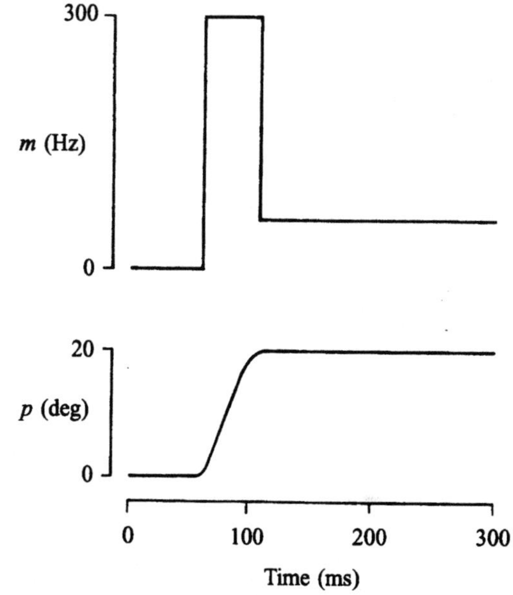

Fig. 6.21 Time response of the eye movement system if the system input is a pulse-step neural signal as shown in the upper trace.

input signals traveling from the head to the base of the spinal cord.

We will use the symbol DEL τ to represent an operator with a delay time of τ. In the example, we have DEL τ, where $\tau = 0.05$ s. The Laplace transform of DEL τ is $e^{-\tau s}$ (see fourth transform pair in Table A.2), and if a signal $f(t)$ is the input to this operator, then its output is $f(t - \tau)$, i.e. $f(t)$ delayed by τ s. In the Laplace domain we have

$$R(s) = F(s)e^{-\tau s}, \tag{6.55}$$

where $F(s)$ is the input and $R(s)$ is the output of the delay operator. The phasor transform of DEL τ is $e^{-j\tau\omega}$, which has an amplitude (gain) of 1 (0 dB) for all values of ω, and a phase lag given by

$$\phi(\omega) = -\tau\omega. \tag{6.56}$$

But this phase angle is in radians and we want the phase angle in degrees. We can convert to degrees by multiplying by $360/2\pi = 57.3$.

Example 8. Construct the Bode plot for a pure time delay operator with $\tau = 50$ ms. We have $\tau = 50$ ms =

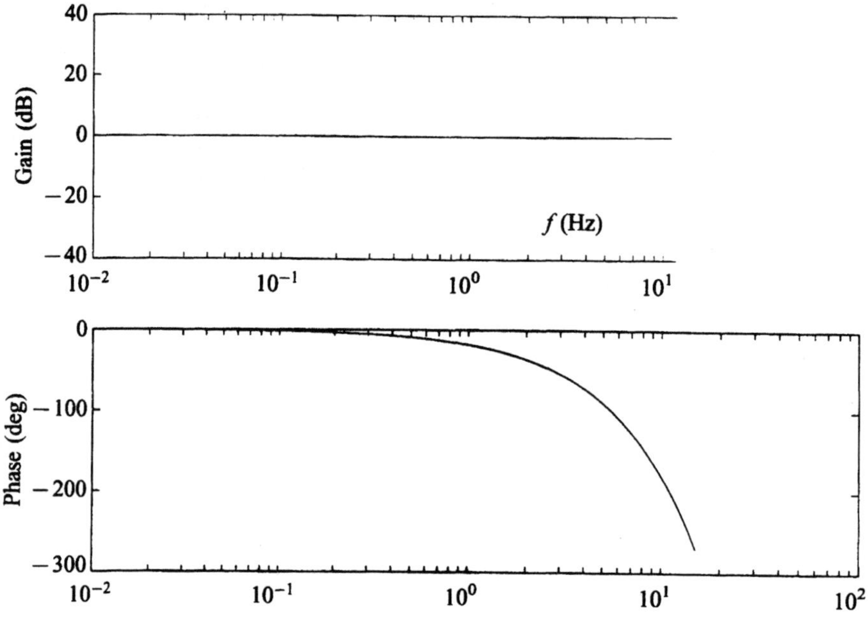

Fig. 6.22 Bode diagram for the non-minimum-phase time delay operator.

0.05 s. The gain = 1 or 0 dB for all ω. To plot the phase for frequency in Hz,

$$f = 0.1, \quad \phi = -2\pi(0.1)(0.05)(57.3) = -1.8°,$$
$$f = 1.0, \quad \phi = -18°,$$
$$f = 10, \quad \phi = -180°. \tag{6.57}$$

Now plot this on a Bode diagram as shown in Fig. 6.22. Notice that phase lag actually increases linearly with frequency, but that on a semilog Bode plot this leads to a very steep slope of the phase curve at higher frequencies.

Next consider a system with a pure delay in series (cascade) with a first-order lag. The overall transfer function is

$$H(s) = e^{-\tau s}/(1 + Ts). \tag{6.58}$$

For a unit impulse input the output may be written as

$$|H(j\omega)|_{dB} = 20 \log |e^{-j\tau\omega}| - 20 \log |1 + jT\omega|$$
$$= 0 - 20 \log |1 + jT\omega|, \tag{6.59}$$
$$\phi = -\tau\omega - \tan^{-1}(T\omega).$$

Notice that gain curve is the same as a first-order lag with a high-frequency roll-off of −20 dB/dec, but the phase curve has the phase lag expected for the lag operator and an additional phase lag of $\tau\omega$. The lag phase curve is thus the *minimum-phase curve* and the delay adds more *non-minimum phase lag*.

Another example of a non-minimum phase transfer function that occurs in physiological systems is $F(s) = 1/(s - a)$. Notice that the pole for this transfer function is located on the positive real axis in the s-plane. Its time domain response will thus have a positive exponential. The impulse response of this system will grow with increasing time and thus is *unstable*.

6.5 Feedback systems

6.5.1 The closed-loop transfer function

A standard negative feedback system is shown in Fig. 6.23 in block diagram form. The block labelled H (shorthand for $H(s)$ or $H(j\omega)$) is called the feedback transfer function or loosely the feedback gain. In many cases it is simply a scalar value of unity or a small positive number. The system then has *proportional*

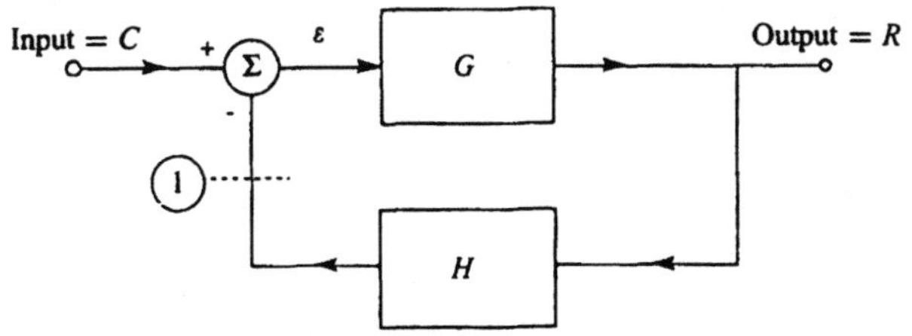

Fig. 6.23 Block diagram of a negative feedback system.

feedback. In other cases the system block H approximates a transfer function equal to s (or $j\omega$). The system then has *derivative* or *velocity feedback*. The transfer function, G, is called the forward loop gain. The signal labelled ϵ is the error signal.

For example, in the physiological system that controls the position of the eyes in response to changes in the position of visual targets in the external world, the position of the target would be the input, C. The position of the eyes would be the output, R. The difference between these two signals would be retinal error. This system is a servo or feedback control system that tries to make the error close to zero. Then the eyes will point (look at) the target.

A second example is the system regulating the level of carbon dioxide in the blood. The sensor for this system is located in the carotid artery in the neck. The input in this case is a steady, desired level of carbon dioxide in the blood. The output is the actual level present. The difference between desired level and actual level is the system error. This type of control system that tries to maintain a steady level or set point is called a regulator.

Break or cut the feedback loop at point 1. The input signal to block G now becomes just $C(s)$, since the output can no longer affect the input to block G. Let the output of cascade system GH be $Q(s)$. Then $Q/C = GH$, which is defined to be the *open-loop transfer function*. The open-loop transfer function is also often loosely called the *open-loop gain*. We now wish to derive an expression for the closed-loop gain of the system. We have

$$R(s) = \epsilon\ G(s) \tag{6.60}$$

and

$$\epsilon(s) = C(s) - R\ H(s). \tag{6.61}$$

Substituting eqn (6.61) for ϵ into eqn (6.60), we get

$$\begin{aligned} R(s) &= (C - RH)G = CG - RHG, \\ R(1 + HG) &= CG, \end{aligned} \tag{6.62a}$$

$$R/C = G/(1 + GH), \tag{6.62b}$$

where $G/(1 + GH)$ is the *closed-loop transfer function*. Equation (6.62b) is a very important result. Rearranging slightly, we have

$$R(s) = C\frac{G}{1 + GH}. \tag{6.63}$$

Thus, we can predict the system's output, $R(s)$ if we know the input, C. If $H = 1$, for unity negative feedback, the expression for the closed-loop transfer function becomes $G/(1 + G)$, and

$$R(s) = C\frac{G}{1 + G}. \tag{6.64}$$

Notice that if the sign of the feedback in Fig. 6.23 is made positive at the comparator, then the expression for the closed-loop transfer function becomes $G/(1 - GH)$.

The expression for the open-loop gain, GH, is often derived from experimental data, as was done for the transfer function in Example 6. It will have a set of open-loop poles and zeros located in the complex s-plane. Notice that the location of the poles of the closed-loop transfer function will not, in general, be the same as those of the open-loop transform. The systematic prediction of how the poles of the closed-loop form change from those of the open-loop transfer function is a method called the *root locus method*. This method is beyond the scope of the present chapter, but can be found in systems theory texts such as Saucedo and Schiring (1968).

A very large number of physiological systems that have been studied have a form, at least when simplified to operate in a linear range, very similar to the standard negative feedback system of Fig. 6.23. We now consider some of the characteristics of this type of system that make it very advantageous for biological as well as human-engineered systems. For simplicity, consider that $H = 1$ in the following discussion.

6.5.2 Characteristics of negative feedback systems

Lack of sensitivity to internal component changes

Define sensitivity (σ) as the absolute value of the relative change in system output, R, to the relative change in the system G. Then

$$\sigma = \left|\frac{dR/R}{dG/G}\right| = \left|\frac{dR}{dG}\cdot\frac{G}{R}\right|. \tag{6.65}$$

Using the expression for R in eqn (6.64) and taking the derivative with respect to G yields

$$\frac{dR}{dG} = C\frac{-1}{(1+G)^2}. \tag{6.66}$$

Then substituting eqns (6.64) and (6.66) into eqn (6.65),

$$\begin{aligned}\sigma &= |C[-1/(1+G)^2]\cdot[G(1+G)/CG]| \\ &= |-1/(1+G)| = 1/|1+G| \\ &\approx 1/|G|\end{aligned} \tag{6.67}$$

if the loop gain $|G|$ is large.

Thus, the sensitivity to changes in $|G|$ (internal system changes) will be very small if $|G|$ is large. To see how this works, consider a similar open-loop system with a transfer function of G. If some parameter in G changes so that $|G|$ goes up by a factor of 2 (i.e. a 200% change in $|G|$), then the system output will also double for the same input. Now consider a similar change in G in a feedback system. Let $|G|$ be approximately 100. Then if $|G|$ doubles, $|R|$ changes only by about 2/100, a 2% change (for C fixed). Thus, it is clear that feedback systems show much less sensitivity to internal parameter changes than open-loop systems, if $|G|$ is high. If the feedback gain, $|H|$, is not unity, then

$$\sigma = 1/(1+|HG|) = 1/|HG| \tag{6.68}$$

if $|HG|$ is large.

Increased response speed

To demonstrate this property of feedback systems we need to invert a Laplace transform: $L^{-1}[H(s)] = h(t)$. Consider an open-loop system with transfer function $H(s) = 10/(1 + Ts)$. Let the input to this system be a unit step function. Then the system response is $R(s) = 10/s(1 + Ts)$. If the system is a muscle with a time constant (T) of 100 ms, the system response will be very sluggish. It will take several hundred milliseconds to approximate a steady-state response level.

Now consider putting $H(s)$ in a unity negative feedback system as shown in Fig. 6.24. The new closed-loop system transfer function is

$$H/(1+H) = \frac{10/(1+Ts)}{1+10/(1+TS)} = (10/11)/(1+T'S) \tag{6.69}$$

where T' is approximately 10 ms. Thus we have shortened the system time constant by a factor of 10 and steady-state response will now be approximated in several tens of milliseconds. But, also notice the effect of feedback on the system's frequency response as shown in Fig. 6.25. The closed-loop system's gain is decreased, but its bandwidth is increased by a factor of 10.

Increased linear range

Consider an open-loop system (G) with a steady-state input–output relationship as shown in Fig. 6.26. Notice that the system response is linear for input values up to an absolute value of 1, but the response then shows a soft saturation for input values of larger absolute value. The

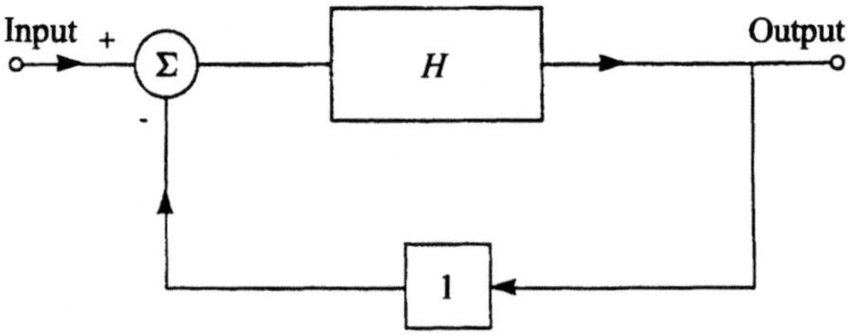

Fig. 6.24 Negative feedback system with unity gain in the feedback path.

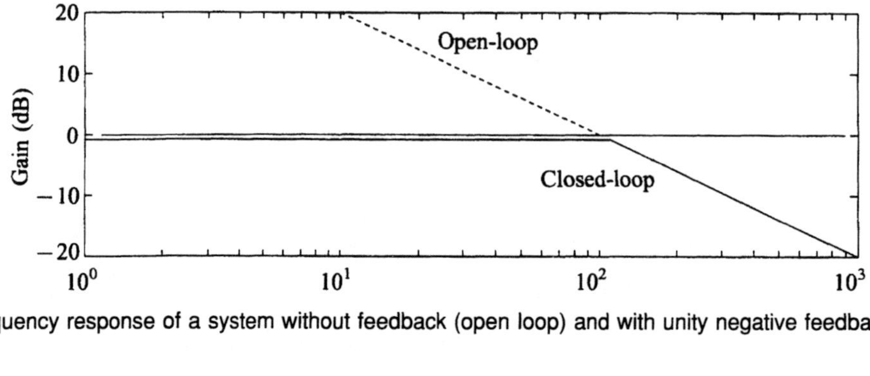

Fig. 6.25 Frequency response of a system without feedback (open loop) and with unity negative feedback.

incremental gain ($\Delta r/\Delta c$) for inputs with magnitude less than 1 is 100, but for inputs with magnitudes greater than 1 the incremental gain drops to 50 or one-half of the former gain. Now make G part of a feedback system as shown in Fig. 6.27. Then for $\epsilon(t) < 1$,

$$\begin{aligned}\Delta r &= 100\Delta\epsilon,\\ \Delta\epsilon &= \Delta c - \Delta r,\\ \Delta r/100 &= \Delta c - \Delta r,\\ \Delta c &= \Delta r(1 + 1/100),\\ \Delta r &\approx \Delta c.\end{aligned} \tag{6.70}$$

For $\epsilon(t) > 1$,

$$\begin{aligned}\Delta r &= 50\Delta\epsilon,\\ \Delta c &= \Delta r(1 + 1/50),\\ \Delta r &\approx \Delta c.\end{aligned} \tag{6.71}$$

Thus, the linear range of operation of the feedback system has been greatly extended, although the system gain is lower.

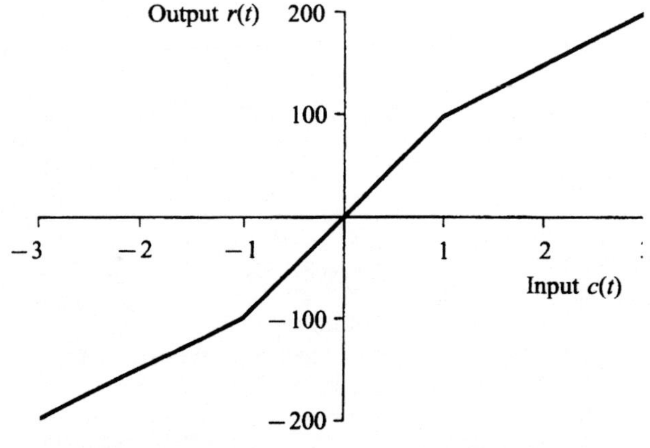

Fig. 6.26 Steady-state input–output repsonse of a system that is only linear over the range of input values from −1 to +1.

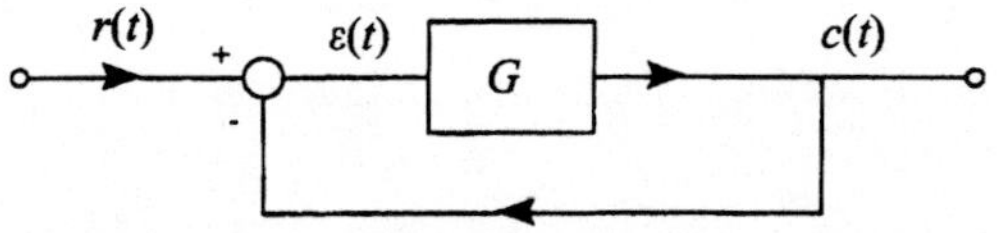

Fig. 6.27 Increasing the linear operating range of a system with negative feedback.

Decreased output variation in response to input noise

Again consider a system G without feedback, where additive noise is getting into the system at the output stage, as shown in Fig. 6.28. Then $R = CG + N$, i.e. the full noise appears in the system's output. But, with feedback,

$$R = C\left(\frac{G}{1+G}\right) + N\left(\frac{1}{1+G}\right). \tag{6.72}$$

If $|G|$ is large, the last term (the noise present in the output) becomes small.

Stability considerations for feedback systems

Feedback systems can show oscillations (instability) in the output that open-loop systems never have. To see the reason for this in a heuristic way, consider the feedback system shown in Fig. 6.29. If the feedback signal, $f(t)$, ever approaches 180° of phase lag at the input to the summing junction, and since 180° more of phase lag is added by the negative feedback, then the output signal adds regeneratively (in phase) with the input and grows each time it travels around the loop (if the loop gain is greater than 1). Notice that two conditions are necessary for growing oscillations or instability: (1) phase lag of

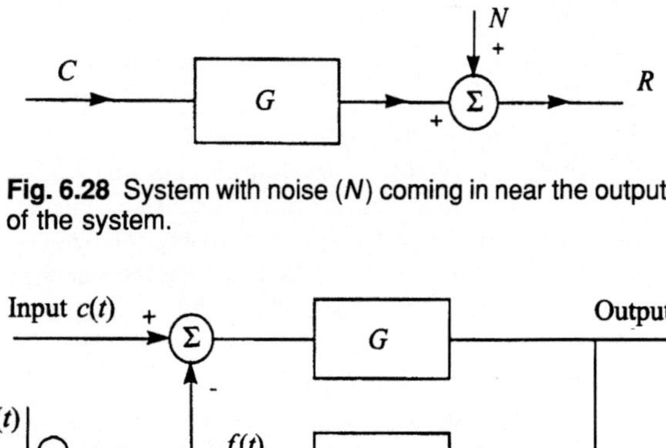

Fig. 6.28 System with noise (N) coming in near the output of the system.

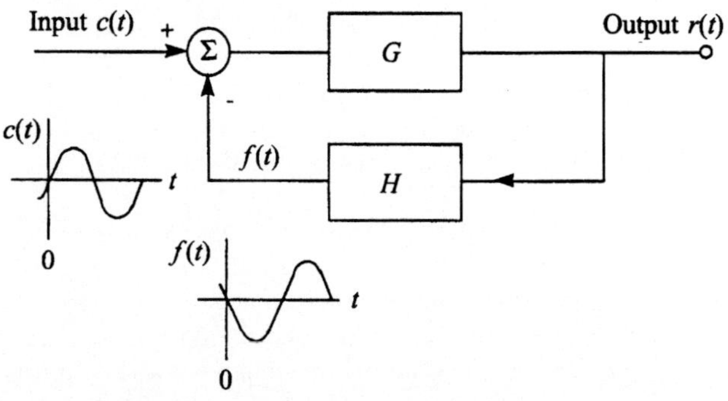

Fig. 6.29 Demonstration of instability in a system caused by feedback.

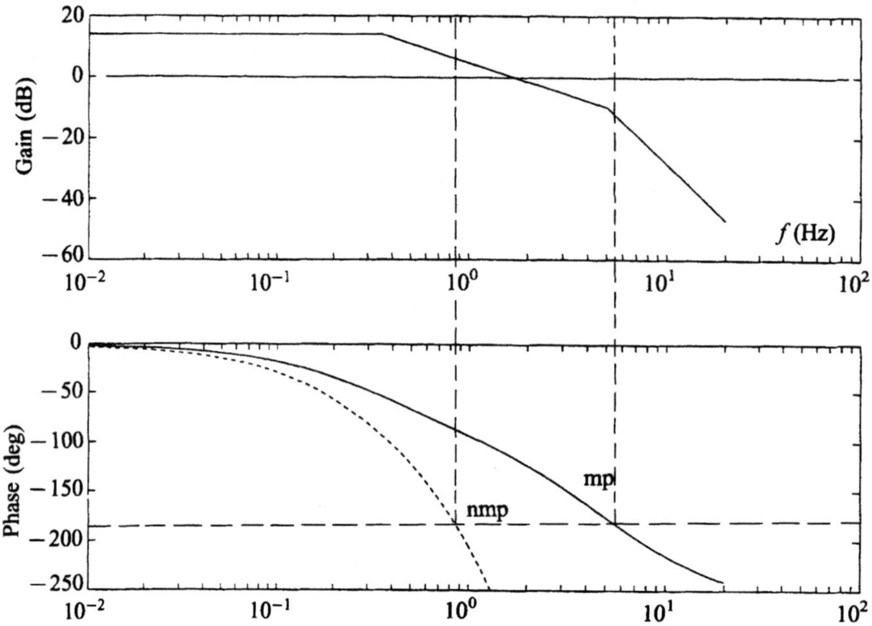

Fig. 6.30 Conditions for stability on the system's Bode diagram. Separate phase curves are given when the system is a minimum-phase system (mp) and when it has non-minimum-phase (nmp) characteristics.

180° through the loop gain, *GH*; and (2) loop gain greater than 1. We will now show these conditions on a sample Bode plot.

Consider the negative feedback system in Fig. 6.29 with open-loop transfer function *GH*(*s*). Suppose its Bode gain plot is given in Fig. 6.30. It certainly has a region of gain greater than unity (zero dB) so that condition (2) is met. Will it be an unstable system when the loop is closed? We have to consider condition (1) first before we can answer this question. Examine the Bode open-loop phase plot in Fig. 6.30 which is labelled mp, for minimum phase. The dashed horizontal line at −180° shows that this system does have a frequency where $\phi = -180°$. Now follow up the vertical dashed line from where the −180° phase line intersects the phase curve. Notice that the gain curve shows that the gain is less than unity at this frequency. Thus both conditions are not met at any frequency, which implies that the closed-loop system is stable.

Now consider the system with the open-loop phase curve labelled nmp, for non-minimum phase. It has the same open-loop gain curve as the previous system. However, the −180° phase line now intersects the phase curve at a much lower frequency. When the vertical dashed line from this lower-frequency intersection is traced up to the gain curve, it is easy to see that the gain is much greater than unity at this frequency. Therefore, both conditions are met in this non-minimum-phase, open-loop transfer function and this system will be unstable in closed-loop operation. It will have oscillations at a frequency of 0.8 Hz (the 180° phase lag point on the phase curve) that grow larger with time. The fact that in this example the system is of the non-minimum-phase type is immaterial to the consideration of stability. One only has to check for the presence of conditions (1) and (2) on the Bode plot. The example does illustrate that delay elements, which are always present in biological systems, lead to large phase lags, and thus are possible problems for stability design.

Open-loop measurements of internal system parameters

It is often desirable to make estimates of internal system parameters based on measurements made at the input and output of a system. When this is done on systems with internal feedback loops, it is usually necessary to devise some way of making these measurements with the loop open. The reason for this is illustrated in the following example.

Example 9. Consider the ocular tracking system as shown in Fig. 6.31. This is a closed-loop control system whose purpose is to move the eyes at the same velocity (V_E) as that of a target (V_T) moving in space. When V_E is nearly the same as V_T, the slip of target image on the retina (ϵ) is minimized and maximum visual acuity is maintained. We wish to estimate the value of the system's internal parameter A in a group of patients to see if this value declines with disease progression. We make V_T be a step function of $10°\ s^{-1}$ and measure the steady-state eye velocity as the patient tracks the target. We then analyze the results using the feedback eqn (6.63):

$$V_E(s) = V_T \cdot G/(1+G)$$
$$= V_T \cdot \frac{A/(1+\tau s)}{1+A/(1+\tau s)}. \quad (6.73a)$$

Rearranging the right-hand side,

$$V_E(s) = V_T \cdot \frac{A/(A+1)}{1+\frac{\tau}{A+1}s}. \quad (6.73b)$$

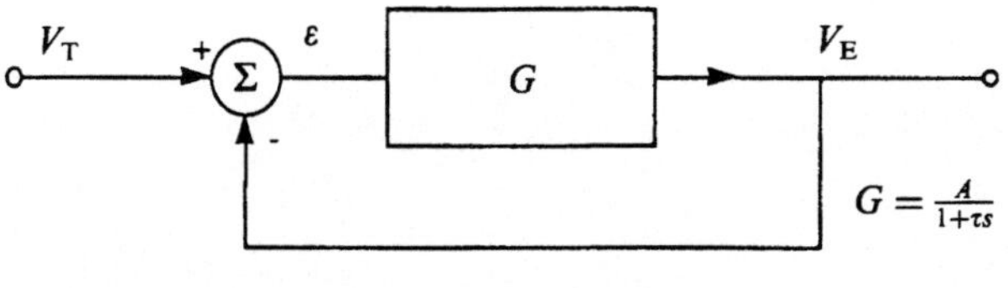

Fig. 6.31 Closed-loop ocular tracking system.

Applying the final value theorem,

$$\lim_{t\to\infty} V_E = \lim_{s\to 0}(10s/s) \cdot \frac{A/(A+1)}{1+\frac{\tau}{A+1}s},$$
$$V_{E,\text{steady state}} = 10 \cdot \frac{A}{A+1}. \quad (6.74)$$

The last equation tells us that the value of A can be computed from the ratio of measured eye velocity to target velocity ($10°\ s^{-1}$). Suppose we measure the steady-state eye velocity to be $9.5°\ s^{-1}$. Then A will be estimated to be

$$A/(A+1) = 0.95$$
$$A = 19 \quad (6.75)$$

But measurements of biological signals like eye velocity are always subject to error. Consider how large the error in the estimation of A would be if we have made a measurement error in V_E of 5 per cent, i.e. steady-state eye velocity is actually about $9.05°\ s^{-1}$. The actual value of A is therefore 9.5, not the 19 estimated from the eye velocity measure. This is an error of 100 per cent in the estimation of A from closed-loop measurements for only a 5 per cent measurement error.

Now consider if we make an estimate of A based on open-loop measurements of this same system. With the feedback loop opened, we have

$$V_E = V_T \cdot A/(1+\tau s). \quad (6.76)$$

In steady-state $V_E/10 = A$. Now a 5 per cent measurement error in V_E will result in only a 5 per cent error in the estimate of A. This example makes very clear why it is desirable to make open-loop measurements when estimating internal system parameters.

These examples have illustrated the use of control systems theory to physiological systems. For a more detailed and rigorous development of the control theory the interested reader should consult a systems theory text, such as Saucedo and Schiring (1968).

Problems

P6.1 Muscles are typically fairly slow devices compared to a motor, for example. Suppose we have a model for a muscle as shown in Fig. 6.32. Note that $n(t)$ is the muscle nerve input signal and $x(t)$ is the muscle length or output signal.

(a) In what units would $n(t)$ be typically given?

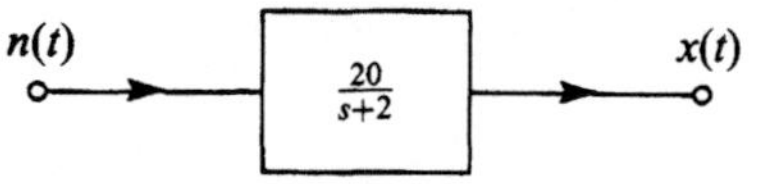

Fig. 6.32 System model of a muscle, where $n(t)$ is the muscle nerve input signal, $x(t)$ is the muscle length, and the muscle transfer function is given in the block diagram.

(b) What is the time constant for this muscle?

Consider putting unity-gain feedback around the muscle to change its properties.

(c) How would this feedback affect the time constant of the muscle and the gain?

(d) What physiological entity might compose this feedback pathway?

P6.2 Consider the system shown in Fig. 6.33 as a model of the human pursuit system (eye tracking system). Note that A is a scalar and the summing junction is the retina.

(a) Describe how you might open the loop of this system to make open-loop gain measurements.

(b) Once you have opened the loop, derive an expression for loop transfer function. Simplify as much as possible.

(c) Will this system be stable? Give a reason for your answer.

(d) If $A = 100$ and the target is moving at $100° \text{ s}^{-1}$, how fast will the eyes be tracking in the steady state?

(e) Realistically we must add a pure delay element to the system. Add an $e^{-\tau s}$ between the s and the A blocks. Make $\tau = 0.2$ s, $T = 0.1$ s, and $A = 100$. Will the system still be stable? If it is unstable, at what frequency (rad s^{-1}) will it oscillate?

(f) Suppose we actually measure how fast the eyes are tracking in (d) above, but make a -5 per cent measurement error. We use this measured value of output to estimate the value of A. How big a percentage error do we make in the estimation of A? If the loop had been open and we made the same -5 per cent measurement error of output, how big would the estimation error in A be?

P6.3 Describe the gain (magnitude of the transfer function) and phase characteristics at low and high frequencies of the following systems. Use briefly worded descriptions.

(a) $1/(s + a)$

(b) $(s + b)/(s + a)$

(c) $s/(1 + Ts)$

(d) $e^{-\tau s}/(1 + Ts)^2$

(e) $1/(s - a)$

P6.4 The following (non-minimum-phase) transfer function is used to approximate a pure time delay:

$$G(s) = \frac{a - s}{a + s}$$

What value of time delay is best simulated? (*Hint:* plot the Bode plot of this function and compare it to the Bode plot of a pure time delay.)

P6.5 (a) Find the output $y(t)$ for a system with transfer function $H(s) = 10/(1 + Ts)$, if the input is $x(t) = u(t)$ (a step function).

(b) Now find the output if the system above is followed (in cascade) by a differentiator. Check your work by differentiation of the answer in (a) above.

(c) Suppose that past experiments have suggested that the transfer function of a muscle stretch receptor may be represented as $H(s) = A(1 + 10s)/(1 + 0.1s)$. Find the steady-state response of these muscle receptors to a sinusoidal input of $10 \sin \omega t$ (in mm). $A \approx 10$ spikes/s per mm, $\omega = 1$ (rad s^{-1}).

P6.6 One type of vestibular receptor organ (the semicircular canals) is often modelled as a second-order system:

$$\frac{k}{(1 + T_1 s)(1 + T_2 s)},$$

where $T_1 = 10$ s and $T_2 = 5$ ms in man, and the input is head acceleration and the output is neural discharge rate, $n(t)$.

(a) What is the bandwidth of this transducer in Hz?

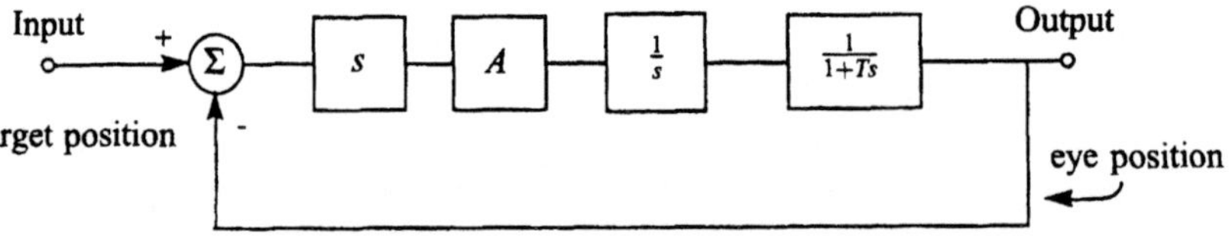

Fig. 6.33 System model of the ocular tracking system, where A is a scalar and the summing junction is the retina.

(b) Since natural head movements are the principal motions this system has evolved to measure, and such movements have no frequency components above 10 Hz, simplify the above expression (transfer function) for head movements as the only input.

(c) If a sinusoidal head acceleration at 0.4 Hz is the receptor's input, is the system's output approximately in phase with head acceleration or velocity? Answer the same question if the input was at 0.001 Hz.

(d) Show why a person rotated at constant velocity in total darkness from $t = 0$ until $t = 100$ s will feel after a while that they have stopped rotating, but will claim that they are being rotated in the opposite direction when stopped at $t = 100$ s. (*Hint:* plot the time form of the response; consider the head acceleration to be an impulse at $t = 0$ and a negative impulse at $t = 100$ s.)

(e) From (d) above it is easy to see why this system is thought to be a transducer of head velocity. The Bode plot for head velocity as the input has a bandpass type gain curve and the output is in phase with head velocity over a mid-range of frequencies. The lower frequency cut-off is at $\omega = 0.1$ rad s^{-1} or about 0.016 Hz. However, when actual measurements are made of neural discharge rate within the central nervous system, the bandpass of the system extends down to 0.001 Hz. A central neural network in series with the canal system with positive feedback, forward loop gain of a (where a is a scalar, $0 < a < 1$) and feedback transfer function of $H = 1/(1 + 10s)$ has been proposed as the way in which the bandpass of the canal system is improved to lower frequencies. Explain how this idea would work.

P6.7 Suppose that the feedback to motoneurons from a muscle spindle was such that the muscle developed a force proportional to the square of the stretch imposed on the spindle.

(a) Draw a block diagram of this system, labelling all variables.

(b) How much energy will the muscle store when it is stretched by an applied force?

(c) How much energy would it store if the feedback loop were opened and the muscle became a simple linear spring?

P6.8 Often in a biological system, the feedback loop can be opened at one point, but measurements of gain have to be made at another. In the system shown in Fig. 6.34, the loop can be opened at point X, but only the input and output can be observed. Your job is to estimate $H(s)$ from the Bode plots shown in the figure.

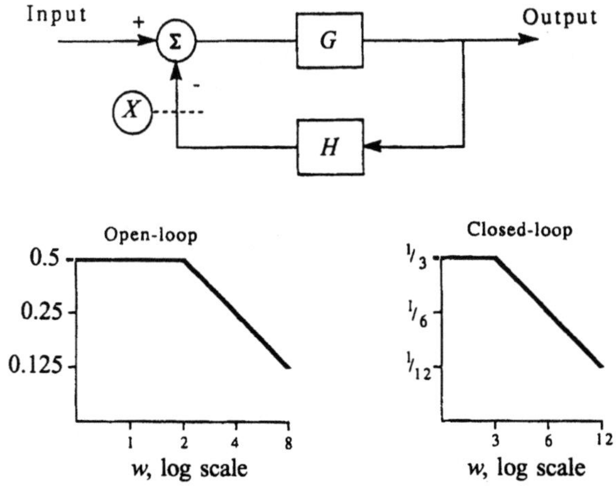

Fig. 6.34 Feedback system with open-loop and closed-loop gain measurements.

P6.9 You propose that a proper model for a sensory receptor that you are studying is $H(s) = 1/(s^2 + b^2)$. You are asked what the final value (t approaches infinity) will be of the output of this system when a unit-step function is applied as an input. Use the final-value theorem to provide an answer. Why doesn't the theorem work for this particular transfer function? (*Hint:* the inverse transform for this particular system is $(1/b)\sin(bt)$. How might you restrict the class of functions to which the final value theorem can be applied to avoid difficulties like this?

P6.10 Frequency response measurements on a muscle spindle yield the Bode plot shown in Fig. 6.35. Estimate the form of the transfer function and all parameters in it. Is this a minimum- or non-minimum-phase system?

P6.11 The discharge frequency of vestibular nerve cells is proportional to and in phase with the velocity of head movements. The neuronal pathways to the eye muscles from these vestibular nerves consist of a direct pathway with transfer function of k (a scalar) and a parallel pathway with transfer function of $1/s$.

(a) What is the complete transfer function of the parallel pathways taken together?

(b) Suppose the transfer function of the eye muscles to eye position is given by $(1/k)/(s + 1/k)$. What is the overall transfer function from vestibular nerves to eye

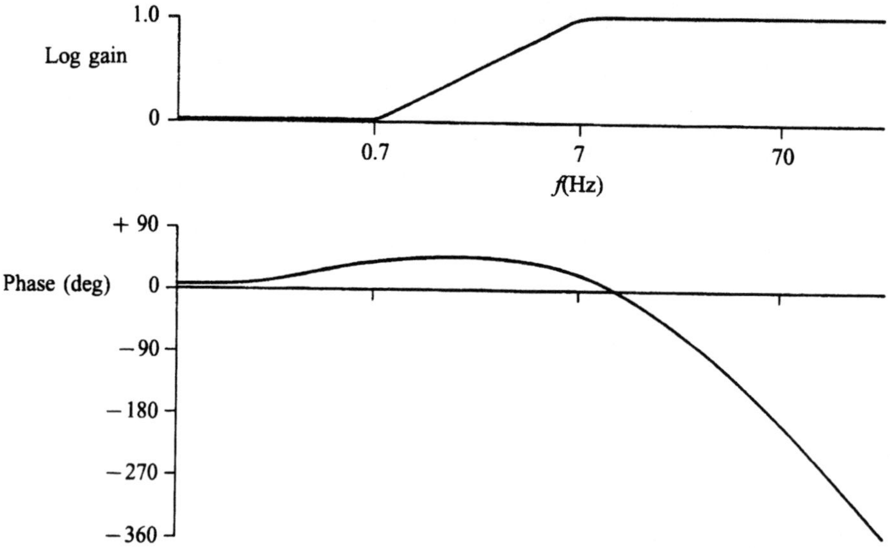

Fig. 6.35 Bode diagram of the response of a muscle spindle.

position?

(c) Based on this model, what phase lag (in degrees) would you expect to see between head velocity and eye position during sinusoidal head movements? Would this phase angle depend on frequency?

P6.12 Suppose you had been able to open the loop on the feedback system controlling convergence eye movements. When you applied a unit-step input signal, you found that the output was zero for 200 ms and then began to increase linearly at a rate of 100 units s^{-1} and continued at this rate for the rest of the observed response.

(a) Show a block diagram with transfer functions for each block that describes this open-loop system.

(b) If the normal system has a feedback pathway with transfer function of -1 between output and input, give an expression for the closed-loop response of the system to a unit-step function input.

(c) Determine the steady-state difference between the input (a unit step) and the output for the closed-loop system.

(d) Using Bode plots, determine if the system should be stable.

(e) If the delay is ignored (set to zero), then what will be the time constant of the response for a unit-step input?

(f) If we put a unit step into the open-loop system, as in part (a) above, and instead of a ramp output we found that, after the delay, the movement started with a high initial slope and the movement output reached a steady-state value of 100 units (with an exponential-like curve shape), what are the open- and closed-loop transfer functions? (Again, assume unit negative feedback.)

P6.13 Many sensory receptors show a power function response of the form $f(t) = c\bar{t}^k$ to sudden impulsive stimulation applied at $t = 0$, and k is usually found to be between 0.2 and 0.8 (Thorson and Biederman-Thorson 1974). The Laplace transform pair for this time function is:

$$L[t^{k-1}] = s^{-k}\,\Gamma(k), \quad 0 < k < 1.0,$$

where $\Gamma(k)$ is the gamma function and $\Gamma(1/2) = \pi^{1/2}$. Plot the frequency response (Bode plot) of the transfer function of the receptor for $k = 0.5$. Plot both gain and phase.

P6.14 Consider the model shown in Fig. 6.36 of the feedback control system for the human pupil in answering the questions below. Note that $G = 0.2/(1 + 0.1s)$; $H = e^{-0.18s}/(1 + 0.1s)^2$; A and B are signal measurement points.

(a) Draw Bode diagrams of the open-loop response (both gain and phase) of this system.

(b) Can this system show spontaneous oscillations?

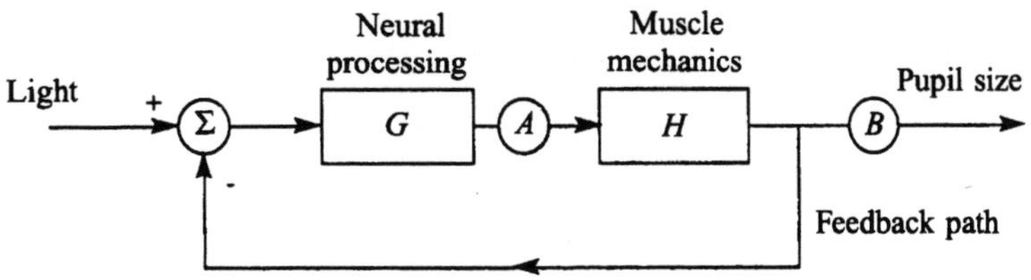

Fig. 6.36 Pupillary control system.

Explain why or why not.

(c) If the open-loop gain of the system is increased artificially by increasing the gain of the feedback pathway to 20, at what frequency (Hz) might you expect the system to oscillate?

(d) If a positive light step of amplitude 1 is applied at $t = 0$, what will be the final steady-state pupil size (in arbitrary units)?

(e) An oscillation of pupil size sometimes occurs and is called *hippus*. The frequencies present in this random oscillation are all greater than 10 Hz. Is it more likely that this oscillation gets into the system at point A or B? Explain. (*Hint:* consider the transfer functions for noise injected at the two points. At which point will the noise be attenuated and not appear in the output?)

References

Bahill, A. T. (1981). *Bioengineering*. Prentice-Hall, Englewood Cliffs, NJ.

Carpenter, R. H. S. (1977). *Movements of the eyes*. Pion, London.

Desoer, C. A. and Kuh, E. S. (1969). *Basic circuit theory*. McGraw-Hill, New York.

Kamath, B. Y. and Keller, E. L. (1974). A neurological integrator for the oculomotor control system. *Mathematical Biosciences*, **30**, 341–52.

Kuffler, S. W., Nicholls, J. G., and Martin, A. R. (1984). *From neuron to brain*. Sinauer Associates, Sunderland, MA.

MacGregor, R. J. and Lewis, E. R. (1977). *Neural modeling*. Plenum, New York.

Marmarelis, P. Z. and Marmarelis, V. Z. (1978). *Analysis of physiological systems*. Plenum, New York.

Milhorn, H. T., Jr (1966). *The applications of control theory to physiological systems*. Saunders, Philadelphia.

Optican, L. M. and Zee, D. S. (1984). A hypothetical explanation of congenital nystagmus. *Biological Cybernetics*, **50**, 119–34.

Robinson, D. A. (1981). The use of control systems analysis in the neurophysiology of eye movements. *Annual Reviews of Neuroscience*, **4**, 463–503.

Saucedo, R. and Schiring, E. E. (1968). *Introduction to continuous and digital control systems*. Macmillan, New York.

Talbot, S. A. and Gessner, U. (1973). *Systems physiology*. Wiley, New York.

Thorson, J. and Biederman-Thorson, M. (1974). Distributed relaxation processes in sensory adaptation. *Science*, **183**, 161–72.

Wurtz, R. H. and Goldberg, M. E. (ed.) (1989). *The neurobiology of saccadic eye movements*. Elsevier, Amsterdam.

Zuber, B. L. (ed.) (1981). *Models of oculomotor behavior and control*. CRC Press, Boca Raton, FL.

7

A BRIEF INTRODUCTION TO NETWORK THEORY

Edwin R. Lewis

Contents

Symbols

a_i	specific (tracer) activity of compartment i
A	area
$\boldsymbol{A}$	matrix of rate constants
B	magnetic flux density
c	compliance or capacity per unit length
c_i	concentration at locale i
C_n	incremental capacity or compliance of branch n
d	distance; diameter
D_n	diffusivity for particles in path n
D_{sr}	denominator of transfer function
E_B	bulk modulus (isothermal)
E_B^*	adiabatic bulk modulus
E_L	elastic modulus for longitudinal waves in solids
E_S	shear modulus
F	force
$\boldsymbol{F}$	faraday constant
F_1, F_2	potentials at ports 1 and 2
F_{ij}	transformed operator in signal-flow graph; effort or potential between nodes i and j
F_n	effort or potential (Gibbs free energy per unit of stuff) across nth branch
g	conductance per unit length
$\boldsymbol{g}$	equivalent acceleration of gravity
g_{ij}	hybrid two-port parameter
G	Gibbs free energy; power gain of two-port element in context
G_A	available gain of two-port element in context
G_n	conductance of branch n
G_T	transducer gain of two-port element in context
h_{ij}	hybrid two-port parameter
$h(t)$, $h_{12}(t)$	impulse response, probability density function
$\boldsymbol{H}$	angular momentum (vector)

H_{12}	generating function
H_{sr}	transfer function
i	inertia per unit length
I	inertia, moment of inertia
I_n	inertia of branch n
$\mathrm{Im}\{Z\}$	imaginary part of complex number Z
j	square root of -1
J_1, J_2	flows (e.g. into ports 1 and 2)
$J_{ij}(n)$	flow from node i to node j through branch n
J_n	flow through nth branch
k_b	Boltzmann constant
k_{ij}	two-port parameter in generalized notation
k_{ji}	rate constant for transfer of stuff from compartment i to compartment j
$\{k_i\}$	set of values $k_1, k_2, \ldots$ taken over all values of i
L	length, distance
L_m	product of operators around mth signal-flow loop
m	mass
M	moment amplitude (mechanical)
$\boldsymbol{M}_o$	moment (vector) about o
$\boldsymbol{M}$	molar mass
M_i	impedance or admittance in generalized two-port equations
M_r	rth moment (statistical)
n	index of discrete time; index of circuit branch; number of moles
N_A	Avogadro constant
N_i	expected number of particles in state i
N_{sr}	numerator of transfer function
p_i	pressure at locale i
p_{ji}	conditional probability that particle will move from state i to state j in δt
P_i	probability that particle is in state i
P_n	power delivered to branch n
P_{DL}	power delivered to load
PE_n	potential energy in branch n
q_i	quantity of tracer in compartment i
Q_i	quantity of conserved stuff in state, locale, or branch i
$\boldsymbol{Q}$	vector with elements Q_i
r	resistance per unit length
r_i	common ratio
r, R	radius
$\boldsymbol{R}$	molar gas constant
R_n	resistance of branch n
$\mathcal{R}_n$	remainder of Taylor's series
$\mathrm{Re}\{Z\}$	real part of complex number Z
s	complex variable of the Laplace transform
S_i	entropy at locale i
t	time
$t_{ij}, \bar{t}_{ij}$	two-port transmission parameters
T_D	diffusion time
T_i	torque about axis i; temperature at locale i
u_i	input flow of tracer to compartment i
U_i	input variable, generalized source variable
$\boldsymbol{U}$	vector with elements U_i
U_{TN}	source variable for Thévenin or Norton equivalent circuit
V	volume
V_i	electrical potential at locale i relative to the reference locale
y	admittance per unit length
y_{ij}	two-port (short-circuit) admittance parameter
Y_{ij}	driving-point admittance between nodes i and j
Y_n	admittance of branch n
z	impedance per unit length; complex variable of the z-transform
z	ionic valence of charged particle
z_{ij}	two-port (open-circuit) impedance parameter
Z_{ij}	driving-point impedance between nodes i and j
Z_n	impedance of branch n
Z_S, Z_L	source impedance, load impedance
α_k	exponential coefficient; eigenfrequency; natural frequency
ϵ	dielectric permittivity
ζ_{ij}	rate at which stuff in compartment i is transferred to compartment j
η	dynamic viscosity
Λ	entropy transport parameter
μ	coefficient of friction; mechanical mobility
Ξ_k	product of operators along kth signal-flow path
ρ	density; electrical resistivity
σ	Stefan–Boltzmann constant
ω	angular or rotational frequency (rad s^{-1})

7.1 Introduction

In Chapter 6 the reader was introduced to the physiological and biophysical applications of linear systems models—employing the concepts of the transfer function, open loop gain, and closed loop gain. In Chapter 7, among other things, we shall explore in considerable detail how one may use linear network models to derive such transfer and gain functions from basic physical laws. Working with the concepts of impedance and admittance, we also shall explore the use of linear network theory in thinking about distributed (wave and diffusion) processes of interest to bioengineers and in the design and interfacing of sensors and actuators (transducers). Along the way, the reader will be introduced to a computational system (SPICE) that allows both linear and nonlinear networks to be analyzed by digital computer.

7.1.1 Physical realms

Network theory is applicable to idealized physical processes in which some sort of entity or stuff flows or moves in some way and has been identified as being conserved. Such processes make up the basic analytic models in many of the traditional subject areas of engineering. For each kind of conserved entity or stuff, one can identify a distinct *physical realm*. Thus one can identify a hydraulic realm, in which water flows from place to place and is conserved as it does so; a pneumatic realm, in which air flows from place to place and is conserved; a particle diffusion realm, in which a particular species of dissolved particles moves from place to place and is conserved; a chemical realm, in which a particular atomic or molecular species moves from chemical state to chemical state and is conserved; and so forth. The notion of distinct physical realms will become especially useful later in this chapter when transducers are discussed.

The choice of the identity of the stuff presumed to be conserved and the measure of the amount of that stuff represented in a network model depend not only on the physical realm being modeled, but also on ease of model manipulation and on tradition. Conservation of mass is a presumption common to models in several realms, yet that conservation traditionally is represented differently in the various realms. For example, for continuum mechanics of solids, for nearly incompressible fluids, and for compressible fluids under negligible variation in density, there is a tradition of using volume as a measure of mass. For chemical reactions and diffusion of particles, there is a tradition of using the number of particles of substance (measured in moles) in the population. In rigid-body mechanics, the shape of the individual rigid mass element is presumed to be conserved, but that presumption traditionally is represented by taking translational and rotational displacements to be conserved as they are transferred from place to place through that element. In the electric realm, charge traditionally is taken to be the conserved stuff. In the thermal and optical realms, thermal or optical energy often is taken to be the conserved stuff. However, when those realms are coupled to others through certain (reversible, passive) transducers, the thermal or optical energy is not conserved, but entropy is; and there is a tradition of treating entropy as the conserved stuff that flows.

Some of the traditional fields of engineering have been defined in terms of the physical realms with which they deal. Electrical engineering specializes in electrical and optical realms. Mechanical engineering traditionally is more general, dealing with rigid-body and continuum mechanical realms, fluid realms, thermal realms, and particle diffusion realms. Chemical engineering also is general, traditionally dealing with fluid, thermal and particle diffusion realms, as well as chemical and (occasionally) electrical realms. Bioengineering is the most general of all; none of the physical realms of engineering is excluded from the bioengineer's repertoire.

7.1.2 Definition of a network model

For the purposes of this chapter, a network model is defined to be a collection of discrete (*lumped*) *locales* or *states* (e.g. chemical states) in which identified kinds of conserved stuff (fluid, charge, chemical reactants, particles, etc) can accumulate, and a set of *lumped paths* connecting neighboring states or locales. Two lumped states or locales are considered to be neighbors if the stuff in question can flow from one of them to the other without passing through a third. The paths provide the routes over which that flow can occur. The stuff is considered to be *conserved* if it is neither created nor destroyed in any of the states or locales or in any of the paths. Furthermore, although the stuff is allowed to accumulate at the various states or locales, it is not allowed to accumulate in the paths. For example, if two large reservoirs for water were connected by a pipe, Fig. 7.1, one might construct a model in which each reservoir

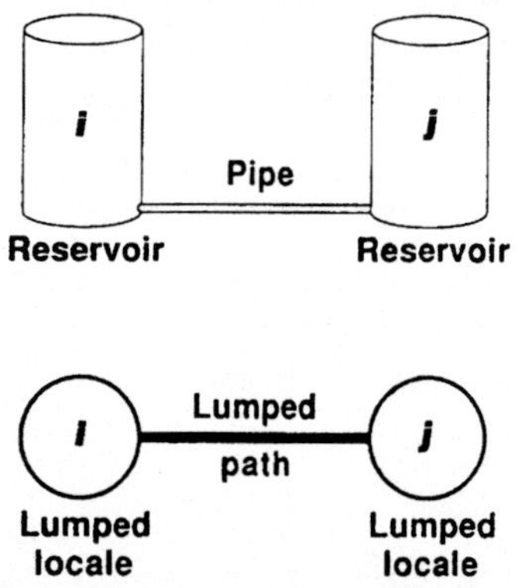

Fig. 7.1 A pair of water reservoirs connected by a pipe, depicted in the upper diagram, is modeled as two lumped locales connected by one lumped path.

is represented as a single lumped locale and the pipe is represented as a single lumped path. In that case, the implied assumption is that all accumulation of water takes place in the reservoirs (the accumulation of water in the pipe is ignored); and all flow of water takes place through the pipe (flow of water in the reservoirs is ignored). Often in real systems, the same physical locale will serve conspicuously as both reservoir and flow path. A volume of matter, for example, could serve both as a reservoir for heat and as a path for heat flow to neighboring volumes. In that case, one might construct a model in which the volume in question is represented as a single lumped locale (for bookkeeping storage of heat) and one or more lumped paths (for bookkeeping heat flow), Fig. 7.2. In network models, the process of accumulation and the process of flow always are depicted separately—even though they are intermingled in the volume being modeled.

Skillful lumping is the quintessence of successful network modeling. There are many instances where lumped models are inappropriate for a particular engineering analysis or design task; and one must apply distributed-parameter models, such as the Navier–Stokes equations in fluid mechanics or Maxwell's equations in electromagnetics or electrical dynamics. On the other hand, one occasionally witnesses laborious applications of these equations to deduce dynamics or design criteria that would have been obvious almost immediately if the

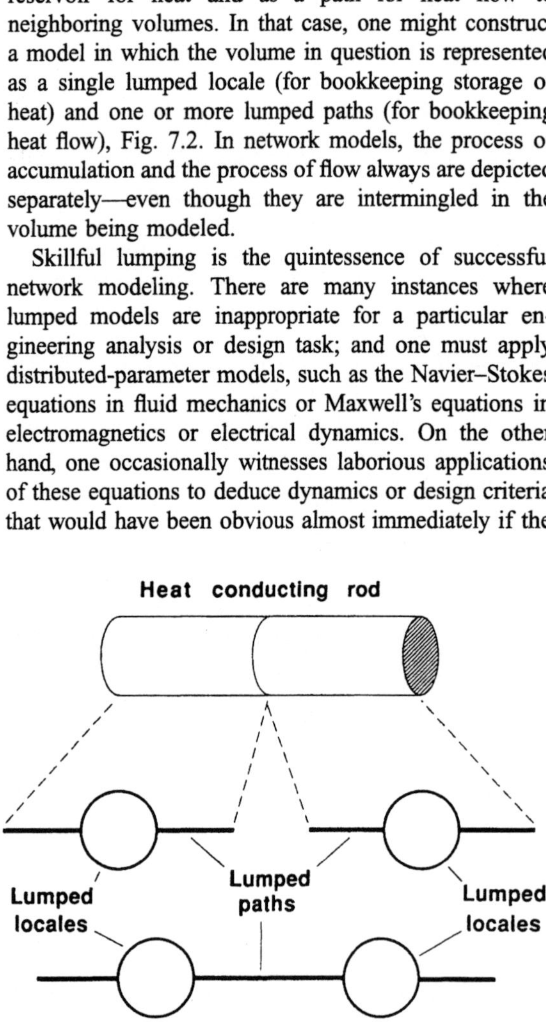

Fig. 7.2 A heat-conducting rod has been divided (arbitrarily) into two segments, each of which is modeled as one lumped locale and two lumped paths. The model segments are connected (lower diagram) to represent the entire rod.

appropriateness of lumping had been recognized. Regardless of the lumped-modeling scheme used, the criteria for lumping always are based on the spatial and temporal resolution required in the deductions or conclusions to be drawn from the model. This issue is discussed in more detail later in the chapter.

7.1.3 Advantages of network theory

Network theory provides the bioengineer a unified set of analytic and synthetic tools that can be applied to all of the realms. It is especially useful when the processes in those realms are highly interactive. The analytical side of network theory, now automated with available computational packages such as SPICE, allows one to deduce the dynamic behavior expected from a set of interacting physical processes whose parameters are known, or from a hypothetical set of interacting processes whose parameters are suspected. For example, one might use network analysis to deduce the effectiveness of a particular cardiac catheter in transmitting pressure from the aortic arch to a transducer outside the body. With available computer power, this side of network theory is effective for both linear and nonlinear processes. On its synthetic side, network theory allows one to establish bounds on the dynamic behavior achievable with various sets of interacting processes (regardless of how the individual processes are connected to one another). In other words, when one is attempting to design a particular device or system with a given set of elements, network theory can be used to determine what is possible and what is not. Presently, this synthetic side of network theory is powerful for dealing with physical processes that are behaving linearly or nearly linearly, and not very powerful for dealing with strongly nonlinear processes. Fortunately, Taylor's theorem (sometimes combined with the second law of thermodynamics) assures us that if processes are continuous they should behave linearly in response to small changes. If one has data regarding the dynamic behavior of an existing device or system, the synthetic side of network theory can be used to translate those data into bounds on the underlying physical processes and their interactions. For example, a scuba diving enthusiast might use network theory to determine which hypotheses about nitrogen gas absorption in tissue are consistent with the experimental data in the US Navy dive tables, and which are not.

A major advantage of network models is the fact that one can display them graphically, which for many people is a great aid to intuition. The graphical representations that conventionally are used for network models include *bond graphs*, *compartmental models*, and *circuit models*. Of these, the last two are discussed in this chapter. Because they are widely used in biochemistry, ecology, and physiology, and because they can be used easily to represent random processes, compartmental models are discussed first.

7.2 Compartmental models

7.2.1 Theoretical and empirical bases

In compartmental models, each locale or state is represented by a 'compartment' (often depicted graphically as a circle), Fig. 7.3. The path between each connected pair of states or locales (compartments) is represented by a pair of oppositely oriented, *directed lines* (lines with arrowheads). Each directed line represents flow in the direction of the arrowhead. Paths thus are represented explicitly as being *bidirectional*. Sometimes the flow represented by a line may be so small that it is taken to be zero, in which case the line may be dropped from the network diagram, and the path appears to be unidirectional. When a compartment represents a locale in the form of a volume in physical space, that volume need not comprise contiguous elementary volumes and it need not be stationary. For example, the interiors of all of the red blood cells of an animal

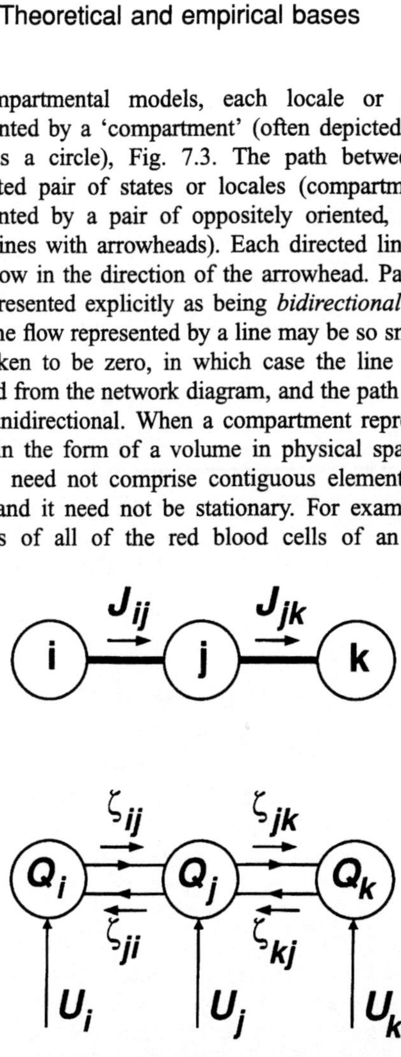

Fig. 7.3 A model comprising three lumped locales connected by lumped paths (upper diagram) is represented as three compartments connected by bidirectional paths, with an input path depicted for each compartment.

sometimes are lumped together as a single compartment, even though the elementary volumes are distributed throughout the circulatory system and are continuously undergoing changes in position, volume, and shape. Similarly, the fatty tissues might be taken to be a single compartment; so might muscle tissues, or the kidneys, or the leaf litter of a tropical rain forest, or the phytoplankton of the oceans.

The rules regarding compartmental dynamics are based in part on a simple *conservation relationship*:

$$\frac{dQ_i}{dt} = \sum_{j=1}^{N} [\zeta_{ji} - \zeta_{ij}] + U_i(t), \qquad (7.1)$$

where Q_i is the amount of conserved stuff accumulated in compartment i, ζ_{ij} is the rate at which stuff currently in compartment i is being transferred to compartment j, and $U_i(t)$ is the rate at which stuff is being introduced into compartment i from outside the system being modeled. The sum is taken over all compartments that represent states or locales that are neighbors of the state or locale represented by compartment i (that is, connected directly to it). The net flow (J_{ij}) from compartment i to compartment j through the path connecting them is:

$$J_{ij} = \zeta_{ij} - \zeta_{ji}. \qquad (7.2)$$

A key assumption here is that the stuff leaving i and going to j does not linger in the path; it arrives at j the instant it leaves i.

The rules for conventional compartmental dynamics are based on a simple *constitutive relationship*:

$$\zeta_{ij} = f(Q_i), \qquad (7.3)$$

where $f(Q_i)$ is a function of the present value of Q_i (i.e. not a function that depends on the history of Q_i as well as its present value). The key assumption here is that the tendency for stuff to flow from compartment i to any other compartment is not affected at all by the history of flow into or out of compartment i. This would be true if the flow in and out of compartment i affects neither the spatial distribution nor the temperature distribution of the stuff within that compartment. For example, if one were dealing with a particle diffusion process, it would be possible that stuff just entering the volume represented by a compartment would be closer to the volume's spatial borders and thus more likely than other stuff in the same volume to turn around and diffuse out again. A key assumption in compartmental modeling is that this is not so, that all of the stuff within the volume is distributed in a steady state fashion at all times, and that whenever stuff flows in or out, the distribution *instantly* relaxes to steady state. This is an approximation. No matter how small the volume represented by a compartment, relaxation to steady-state distribution takes time. The approximation is a good one if relaxation takes place in times that are short in comparison to the unit of temporal resolution (for example, 100 s, 0.1 ps) desired in the deductions from the model. In other words, the validity of the selection of the boundaries of the state or locale represented by a compartment depends on the temporal resolution desired. The time required for relaxation to steady state normally decreases as the size of the volume or hypervolume decreases. As greater temporal resolution is demanded in the modeling deductions (e.g. as one asks questions in terms of microseconds instead of milliseconds), one usually must increase the number of compartments and let each new compartment represent a smaller volume in physical space or a smaller volume or hypervolume in state space. Thus, increased temporal resolution requires increased spatial resolution. As greater spatial resolution is demanded in the modeling deductions, one must increase the number of compartments accordingly, even if temporal resolution already is adequate.

If the conserved stuff comprised discrete particles (for example, atoms or molecules) and an individual particle were able to move from one state or locale (represented by a compartment) to another only if it had sufficient energy, then flow of particles out of the state or locale could remove the more energetic particles and leave the less energetic ones behind, leading to flow-dependent cooling of the stuff left behind and making the tendency to flow dependent on the history of flow. This is not acceptable in compartmental modeling. If flow does tend to remove the more energetic particles, then validity of compartmental modeling requires that heat instantly be added to the ones that are left behind so that cooling does not occur. In other words, thermal equilibrium between the environment and the stuff in the state or locale must be reached in times that are short in comparison with the unit of temporal resolution desired in the deductions from the model. Since heat cannot be transferred instantly, thermal equilibrium is never instantaneous, and the time required for it usually increases with increasing size of the volume over which it is to be achieved. Thus, again, improved temporal resolution in modeling deductions usually requires increased numbers of compartments, each representing smaller volumes or hypervolumes.

The set $\{U_i(t)\}$ provides for the possibility of stuff being injected into the various compartments of the model from sources outside the system being modeled. It often is useful to provide as well for the possibility of stuff irreversibly exiting the system being modeled. This can be accomplished by adding a *sink* compartment for which $f(Q)$ equals zero (i.e., regardless of how much stuff accumulates in the sink, there is no tendency for it to leave).

7.2.2 Linear compartmental models

The Taylor series for a function $f(Q)$ of one variable, Q, takes the following form:

$$f(Q) = K_0 + K_1 Q + K_2 Q^2 + K_3 Q^3 + \dots, \qquad (7.4)$$

where $K_0, K_1, K_2, K_3, \dots$ are constant coefficients. When Q is much less than one unit (that is, $\log[Q] \ll 0$), $f(Q)$ is approximated well by the affine function

$$f(Q) = K_0 + K_1 Q. \qquad (7.5)$$

If $f(Q)$ is the flow of stuff out of a volume, and if the values of Q are constrained to be non-negative, then K_0 must be zero and $f(Q)$ is the linear function

$$f(Q) = K_1 Q. \qquad (7.6)$$

This is one set of reasons for representing ζ_{ij} as a linear function of Q_i. There are many reasons that $f(Q)$ might be a nonlinear function.

When linear functions are used in compartmental models, they usually are written in a form such as

$$\zeta_{ij} = k_{ji} Q_i, \qquad (7.7)$$

where k_{ji} is the *rate constant* for the transfer of stuff presently in compartment i to compartment j (Fig. 7.4). The arrangement of the indices here corresponds to the conventional notation for Q_i being an element of a column vector. Given a linear function of this type for every flow in a compartmental model, one can write a complete set of linear differential equations to describe its behavior. Conventionally, they are written in terms of

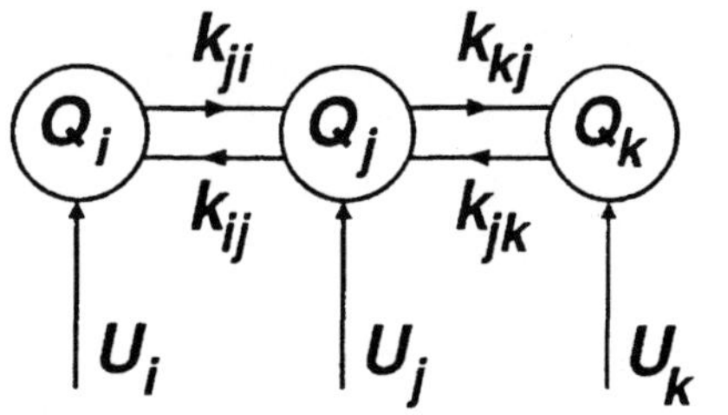

Fig. 7.4 A three-compartment model with conventional compartmental variables and parameters (e.g. see Atkins 1969; Godfrey 1983).

the accumulations (Q's) of stuff in the various compartments:

$$\left\{\frac{dQ_i}{dt} = \sum_{j=1}^{N} \left[k_{ij}Q_j - k_{ji}Q_i\right] + U_i(t)\right\} \tag{7.8}$$

or, in vector form (see Sections 1.1.5 and 1.1.6),

$$\frac{d\boldsymbol{Q}}{dt} = \boldsymbol{A} \cdot \boldsymbol{Q} + \boldsymbol{U}, \tag{7.9}$$

where N is the number of compartments in the model (each representing a state or locale in which the conserved stuff may be accumulated), $\boldsymbol{Q}$ is a column vector comprising the elements $\{Q_i\}$, $\boldsymbol{U}$ is a column vector comprising the elements $\{U_i\}$, and $\boldsymbol{A}$ is a matrix whose off-diagonal elements make up the set of rate constants $\{k_{ij}\}$, i and j being the row and column indices respectively, and whose diagonal elements are given by

$$k_{ii} = -\sum_{j=1}^{N} k_{ji}. \tag{7.10}$$

When it is complete, the set of parameters $\{k_{ij}\}$ (or, equivalently, the matrix $\boldsymbol{A}$) uniquely defines the structure of the conventional compartmental model—that is, the number of separate compartments and which pairs of compartments represent neighboring states or locales. The values of those parameters define the constitutive relationships for all of the flow paths.

7.2.3 Natural frequencies

The differential equation (7.8) or (7.9) for the conventional compartmental model is linear, is nonhomogeneous, and has constant coefficients. The forms of the solutions of such equations and the methods for solving them should be well known to all engineering students. The complete solution for Q_i comprises a solution of the homogeneous version of the equation and a particular solution. The solution of the homogeneous version usually takes the form

$$Q_i(t) = \sum_{k=1}^{n} A_k e^{-\alpha_k t}. \tag{7.11}$$

A_k depends on three things: (1) the structure of the entire model, including paths and rate constants (that is, on the set $\{k_{ij}\}$ of scalar values); (2) the initial values (at $t = 0$) of the accumulations of stuff in each state (that is, on set $\{Q_i(0)\}$ of scalar values); and (3) the input flows to the system (that is, on the set $\{U_i(t)\}$ of functions). The exponential coefficients $\{-\alpha_k\}$, on the other hand, depend only on the structure of the model (i.e., on the set $\{k_{ij}\}$), and often are called the *natural frequencies* of the model or the *eigenfrequencies* of the model (i.e. the model's own frequencies). A *well-connected* model is one in which stuff can flow from every compartment either directly or indirectly (through intervening compartments) to every other compartment. In a well-connected compartmental model, every natural frequency ($-\alpha_k$) depends upon the entire set $\{k_{ij}\}$; no individual natural frequency is attributable to any single rate constant or to any proper subset of rate constants. This easily demonstrated fact has profound implications for reductionist biology. In a well-connected compartmental model, the response of every variable to injected flow anywhere in the system exhibits the same natural frequencies. In other words, the natural frequencies are properties of the entire network model and they are encountered in the deduced dynamics everywhere in the model.

The particular solution for $Q_i(t)$ depends on all three sets, $\{k_{ij}\}$, $\{Q_i(0)\}$, and $\{U_i(t)\}$. Because the model is linear, it can be analyzed in steps. For example, one can derive a solution for $Q_i(t)$ for the given set of values of $\{Q_i(0)\}$, but with all members of $\{U_i(t)\}$ taken to be zero. The result, known as the *zero-input solution*, is part of the homogeneous solution. One also can derive a combined particular and homogeneous solution for $Q_i(t)$ for each member of the set $\{U_i(t)\}$, with the other members of the set taken to be zero and all members of $\{Q_i(t)\}$ also taken to be zero. The result is known as a *zero-state solution*. Linearity allows one to compute these various solutions independently and to sum them to yield the complete solution.

7.2.4 Steady state

It often is useful to compute the values of the members of the set $\{Q_i\}$ under the assumptions that all members of

$\{dQ_i/dt\}$ equal zero (i.e., that the accumulation, Q_i, in each state remains constant) and the members of $\{U_i\}$ are constant. Under those conditions, the model is in *steady state*, and the corresponding values of $\{Q_i\}$ compose the steady-state response of the model. If there are paths to a sink compartment in the model, then the members of $\{Q_i\}$ will remain constant at nonzero values only if at least one member of the set $\{U_i\}$ is not equal to zero. If there is no sink, then steady state occurs only when $\Sigma U_i = 0$. A commonly considered situation is that in which all members of both $\{dQ_i/dt\}$ and $\{U_i(t)\}$ are zero. The values of $\{Q_i\}$ under those conditions compose the *zero-input, steady-state response* of the model; and they depend only on $\{k_{ij}\}$ and $\Sigma Q_i(0)$. Algebraically, the zero-input steady-state response is an eigenvector corresponding to zero for the matrix A of eqn (7.9).

Example 1. Given a three-compartment model with

$$\begin{array}{lll} k_{12} = 1 & k_{23} = 1 & k_{31} = 4 \\ k_{32} = 2 & k_{21} = 1 & k_{13} = 2 \end{array} \tag{7.12}$$

and

$$Q_1(0) = 1.0 \quad Q_2(0) = 0 \quad Q_3(0) = 0, \tag{7.13}$$

find the zero-input steady state.

Under steady-state conditions,

$$\begin{aligned} \frac{dQ_1}{dt} &= -(k_{21} + k_{31})Q_1 + k_{12}Q_2 + k_{13}Q_3 = 0 \\ \frac{dQ_1}{dt} &= -5Q_1 + Q_2 + 2Q_3 = 0 \\ \frac{dQ_2}{dt} &= Q_1 - 3Q_2 + Q_3 = 0 \\ \frac{dQ_3}{dt} &= 4Q_1 + 2Q_2 - 3Q_3 = 0, \end{aligned} \tag{7.14}$$

from which one can deduce that

$$Q_1 = Q_2 = \tfrac{1}{2}Q_3. \tag{7.15}$$

The total quantity of conserved stuff in the system at $t = 0$ was

$$Q_1(0) + Q_2(0) + Q_3(0) = 1.0. \tag{7.16}$$

Because there is no flow of stuff into the system from external sources, nor is there a sink, conservation requires that

$$Q_1(t) + Q_2(t) + Q_3(t) = 1.0 \tag{7.17}$$

be true for all time greater than zero. Therefore, the steady-state values of the accumulations of stuff in the various compartments are

$$Q_1 = Q_2 = \tfrac{1}{4}, \; Q_3 = \tfrac{1}{2}. \tag{7.18}$$

7.2.5 Detailed balance

A special case of the zero-input steady state occurs when

$$J_{ij} = k_{ji}Q_i - k_{ij}Q_j = 0 \tag{7.19}$$

for every pair of compartments. This is known as *detailed balance* or *detailed equilibrium*, and requires that

$$\frac{Q_j}{Q_i} = \frac{k_{ji}}{k_{ij}} \tag{7.20}$$

for every pair of compartments representing neighboring states or locales. Recall that the pair of lines (with arrowheads) connecting two compartments was defined to be a single (bidirectional) flow path. A sequence of such flow paths leading from a compartment and back to the same compartment, without encountering any other compartment more than once, is defined to be a *loop*, Fig. 7.5.

In steady state, the total flow into each compartment must be precisely equal to the total flow out of it. If a compartmental model contains one or more loops, then

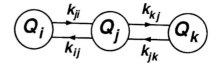

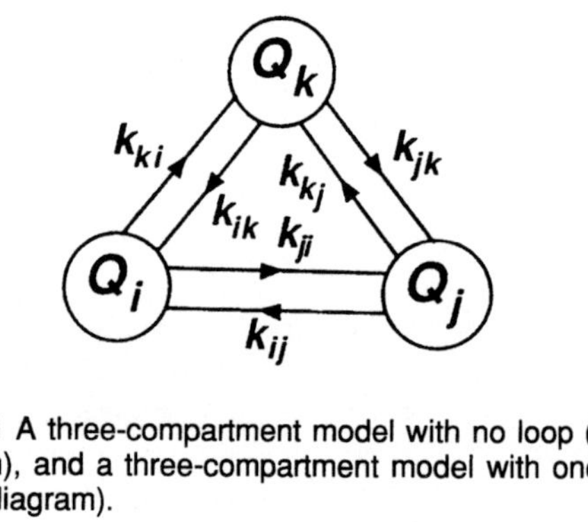

Fig. 7.5 A three-compartment model with no loop (upper diagram), and a three-compartment model with one loop (lower diagram).

this condition can be met with nonzero flow circulating around a loop. Under conditions of detailed balance, however, the circulating flow in every loop is zero. With zero input, in the absence of loops, steady state always is accompanied by detailed balance. Therefore, all compartmental models that have no loops are able to achieve detailed balance under zero-input, steady-state conditions. In compartmental models with loops, however, the ability to achieve detailed balance depends on a precise relationship among the rate constants of the paths composing each loop. Consider a loop with paths from compartment i to compartment j to compartment k to various other compartments, then to compartment y to compartment z and back to compartment i:

$$\frac{Q_i}{Q_z} = \frac{Q_i}{Q_j} \cdot \frac{Q_j}{Q_k} \cdots \frac{Q_y}{Q_z}. \qquad (7.21)$$

Applying eqn (7.20), one obtains

$$\frac{Q_i}{Q_z} = \frac{k_{ij}}{k_{ji}} \cdot \frac{k_{jk}}{k_{kj}} \cdots \frac{k_{yz}}{k_{zy}} \qquad (7.22)$$
$$\frac{Q_z}{Q_i} = \frac{k_{zi}}{k_{iz}};$$

therefore,

$$\frac{k_{ij}}{k_{ji}} \cdot \frac{k_{jk}}{k_{kj}} \cdots \frac{k_{yz}}{k_{zy}} \cdot \frac{k_{zi}}{k_{iz}} = 1.0 \qquad (7.23)$$

or

$$k_{ij}k_{jk} \cdots k_{yz}k_{zi} = k_{iz}k_{zy} \cdots k_{kj}k_{ji}. \qquad (7.24)$$

Example (7.24) is a statement of the *loop constraint*, which must be met for each loop in a compartmental model in order for detailed balance to be possible. Stated in words, the product of the rate constants for all lines whose arrows are directed clockwise around the loop must equal the product of rate constants for all lines whose arrows are directed counterclockwise.

Example 2. Given the three-compartment model of the previous example,

$$k_{12} = 1 \quad k_{23} = 1 \quad k_{31} = 4$$
$$k_{32} = 2 \quad k_{21} = 1 \quad k_{13} = 2 \qquad (7.25)$$
$$Q_1(0) = 1.0 \quad Q_2(0) = 0 \quad Q_3(0) = 0,$$

one can see that there is one loop and that the loop constraint is met for that loop

$$k_{12}k_{23}k_{31} = k_{13}k_{32}k_{21} = 4. \qquad (7.26)$$

Therefore, under steady-state conditions,

$$\frac{Q_1}{Q_2} = \frac{k_{12}}{k_{21}} = 1, \qquad (7.27)$$
$$\frac{Q_1}{Q_3} = \frac{k_{13}}{k_{31}} = \frac{1}{2}.$$

Taking the total accumulation of stuff initially in the model to be conserved, one obtains

$$Q_1 = Q_2 = \tfrac{1}{4}, \quad Q_3 = \tfrac{1}{2}. \qquad (7.28)$$

7.2.6 Tracer dynamics in linear compartmental models

Biological and biomedical scientists often are faced with the task of deriving a compartmental model (that is, a set $\{k_{ij}\}$) for a specified conserved stuff in a specified biological context. Among many other possibilities, the goal might be to gain insight into the organization of a physiological or biophysical process such as the transport and utilization of a given metabolite in a single organism, or to understand the logistics of an ecological process such as the transport and utilization of a key chemical element over an entire ecosystem. Observation of the dynamic behavior of Q even in just one place would allow the investigator to make initial estimates of the structure of the underlying system (for example, estimates of the number of compartments that might be required to construct an adequately predictive model of the process). A common constraint in such circumstances, however, is the inability of the investigator to observe or control the value of Q anywhere directly. Even if one were able to observe Q somewhere, inability to control it would greatly hinder one's ability to see dynamic behavior. In fact it is not uncommon that the system being observed is close to steady state, so that Q is not changing much at all.

Under such circumstances, investigators often turn to *tracers*, materials that can be injected into the system and that are presumed to follow precisely the same dynamics as the stuff being studied. Many different things are used as tracers in various situations. A molecular species labeled with a radioisotope often is used as a tracer to study the kinetics of that same species in a particular situation. The presence of the isotope is presumed not to change to chemical nature of the molecule (until, of course, the isotope undergoes its nuclear reaction). Dye, heat, and labeled particles of various sorts often are used as tracers in fluid-dynamic systems. In one form of simple ecosystem, known as a chemostat, single-celled organisms often are used as tracers. Since tracers can be injected locally (for example, into a locale modeled, putatively, as a single compartment), the investigator can establish a situation that is not steady-state (as far as the tracer itself is concerned). This leads to the opportunity to observe dynamics, and to infer candidate compartment models from those dynamics. The quality of each model can be tested by comparing its dynamics with those observed in the actual system. Thus, a major component of the inference process is analysis of candidate models. For linear processes, the analysis of compartmental models with tracers is elementary.

Let Q_i be the total amount of the conserved stuff (including that which is labeled) in compartment i. Let q_i be the amount of the stuff in compartment i that is labeled (that is, $Q_i - q_i$ is the amount of unlabeled stuff in compartment i). The key assumption in tracer analysis is that the equations governing the dynamics of the members of the set $\{q_i(t)\}$ are the same as those governing the dynamics of $\{Q_i(t)\}$. Thus, in the linear case,

$$\left\{\frac{\mathrm{d}q_i}{\mathrm{d}t} = \sum_{j=1}^{N}\left[k_{ij}q_j - k_{ji}q_i\right] + u_i(t)\right\}, \tag{7.29}$$

where $u_i(t)$ is the rate at which tracer substance is being introduced into compartment i from outside the system being modeled. If one were able to observe a subset of $\{q_i(t)\}$ directly, then this set of equations would be the best complete description of the model (for comparison with data). Typically, however, one is able to take only a small sample from one or more putative compartments, thus obtaining a mixture of labeled and unlabeled stuff. The investigator typically can determine the proportions of labeled and unlabeled stuff in the sample mixture, which can be translated into an estimate of the ratio q_i/Q_i for each sampled locale or state. Therefore, compartmental dynamics for tracers often are described in terms of the set $\{a_i(= q_i/Q_i)\}$, where a_i is the *specific activity* of compartment i. The differential equation for the dynamics of a_i under zero-input conditions is

$$\left\{\frac{\mathrm{d}a_i}{\mathrm{d}t} = \frac{\mathrm{d}}{\mathrm{d}t}\left(\frac{q_i}{Q_i}\right) = \left(\frac{1}{Q_i}\right)\frac{\mathrm{d}q_i}{\mathrm{d}t} - \left(\frac{q_i}{Q_i^2}\right)\frac{\mathrm{d}Q_i}{\mathrm{d}t}\right\}, \tag{7.30}$$

$$\left\{\frac{\mathrm{d}a_i}{\mathrm{d}t} = \sum_{j=1}^{N}\left(\frac{Q_j}{Q_i}\right)k_{ij}(a_j - a_i)\right\}. \tag{7.31}$$

Under the special circumstances that steady state exists for the members of $\{Q_i\}$, Q_j/Q_i is constant. In that case the dynamics of specific activity are described by a linear differential equation with constant coefficients—easily analyzed. In the case of detailed balance, eqn (7.20) applies and eqn (7.31) becomes

$$\left\{\frac{\mathrm{d}a_i}{\mathrm{d}t} = k_{ji}(a_j - a_i)\right\}, \tag{7.32}$$

in which case, the specific activity (a) becomes analogous to an effort or a potential (see Section 7.5.2).

7.3 Markov chains

7.3.1 Markovian ideal

The compartmental models presented so far in this chapter conform to the *deterministic ideal*, which can be stated as follows: Given the present state of the model (for example, the complete set of initial accumulations, $\{Q_i(0)\}$) and all subsequent inputs to the model (for example, the complete set of input functions, $\{U_i(t)\}$), one can deduce the state of the model at all future times (for example, one can deduce all members of the set $\{Q_i(t)\}$). If the conserved stuff is particulate (for example, a gas made up of individual molecules), the deterministic ideal is approximated in a real system when the numbers of particles accumulated in the various locales or states are large. This is assured by the various laws of large numbers from probability theory. When the number of particles is small (for example, a single gated ion channel in the plasma membrane of a nerve cell), the deterministic ideal is not an appropriate basis for modeling. In such cases, an appropriate alternative may be provided by the *Markovian ideal.*

Imagine that each circle in a compartmental model

represents one state or locale that an individual particle might occupy, and that each directed line (line with arrowhead) represents a possible path that the particle might take to change its state or locale. Because there is only one particle being considered, the notion of accumulation, Q_i, of particles in state i is inappropriate. It is replaced by P_i, the probability that the one particle in question is in state i. A conditional probability (p_{ji}), rather than a rate constant, now is associated with each directed line in the network model, Fig. 7.6. *Definition*: p_{ji} is the probability that the particle will be in state j at the end of the next time increment δt, given that it presently is in state i.

If it is possible for the particle to remain in state i, then a directed line from state i to itself must be added to the network diagram. Associated with that line is the conditional probability p_{ii}. The *Markovian ideal* can be stated as follows: the values of the members of the set $\{p_{ij}\}$ (including p_{ii}) are independent of the history of the travels of the particle among the states or locales. In other words, in a network model conforming to the Markovian ideal, the direction that the particle will take next is not influenced at all by the direction it took to reach its present state, nor by how long it has occupied its present state.

7.3.2 Discrete-time models

If the network model includes all of the possible states (or locales) that the particle is presumed to occupy, and if the values of all members of the set $\{p_{ij}\}$ are specified, then all one needs to add to complete this model of the particle's journey is a description of the initial location of the particle (i.e., values of the members of the set $\{P_i(0)\}$). These two sets, $\{p_{ij}\}$ and $\{P_i(0)\}$, define a *Markov chain*; and the network diagram, with circles and directed lines, is a graphical representation of the Markov chain. If time is taken to be a discrete variable, with irreducible increment δt, then the dynamics of the particle are described by the Chapman–Kolmogorov equation, which has the following forms:

$$P_i(n) = \sum_{j=1}^{N} p_{ij} P_j(n-1) \tag{7.33}$$

for state i, or

$$\boldsymbol{P}(n) = \boldsymbol{A} \cdot \boldsymbol{P}(n-1) = \boldsymbol{A}^n \cdot \boldsymbol{P}(0) \tag{7.34}$$

in vector form for all states, where $\boldsymbol{P}(n)$ is a column vector whose elements are the members of the set $\{P_i(n)\}$, $P_i(n)$ being the probability that the particle is in state i at the end of the nth time increment, n being an integer; $\boldsymbol{A}$ is a square matrix whose elements are the members of $\{p_{ij}\}$, i and j being row and column indices, respectively; n is the number of time increments (each being δt in duration) that have passed since $t = 0$; and N is the number of states.

Conservation is imposed on this model by requiring that

$$\sum_{i=1}^{N} P_i = 1.0, \tag{7.35}$$

which implies that the particle certainly will remain in the set of N states included in the model, and

$$\sum_{i=1}^{N} p_{ij} = 1.0, \tag{7.36}$$

which implies that a particle in state j at the end of one time increment will be in one of the states connected to j (possibly including state j itself) at the end of the next time increment.

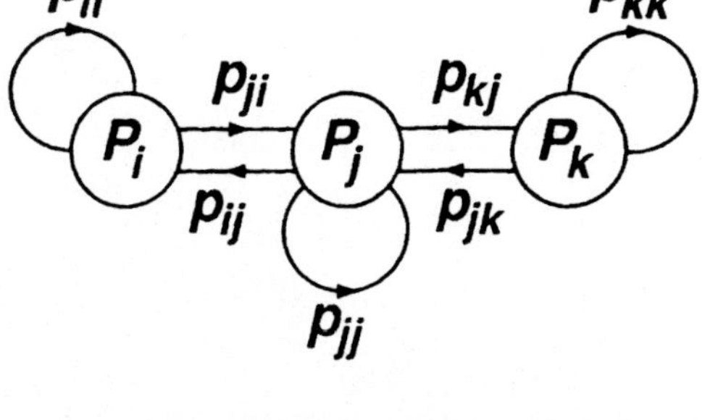

Fig. 7.6 A stochastic three-compartment model, depicting a Markov chain (see, for example, Feller 1968).

7.3.3 Population models

If several particles are moving *independently* among the same set of states or locales, with the same values of $\{p_{ij}\}$ and $\{P_i(0)\}$, then $P_i(n)$ in eqn (7.33) can be interpreted to be the *expected proportion* of particles in state i at the end of the nth time increment. In that case, when P_i is multiplied by the total number of particles in all states, the product is the expected number of particles (N_i) in state i at the end of the nth time increment. The

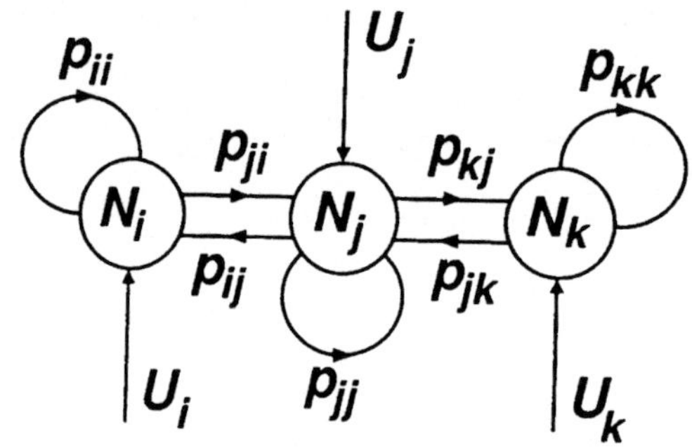

Fig. 7.7 A stochastic three-compartment model converted, by a law of large numbers, into a model of particle population dynamics.

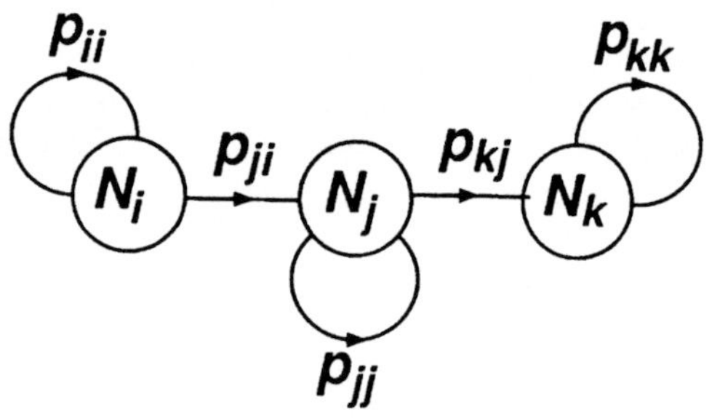

Fig. 7.8 A typical configuration for life-cycle dynamics of plants and animals. The bidirectional paths of Fig. 7.7 have been replaced by unidirectional paths—depicting graduation from one life-cycle stage to the next.

Chapman–Kolmogorov equation becomes

$$N_i(n) = \sum_{j=1}^{N} p_{ij}N_j(n-1) + U_i(n), \quad (7.37)$$

where $U_i(n)$ is the expected number of particles flowing into state i from external sources during the nth time interval. In this case, p_{ij} can be interpreted as the proportion of the particles in state j at the end of the previous time increment (for example, $n-1$) that will be in state i at the end of the present time increment (e.g., n).

Equation (7.37) often is used to describe populations of organisms, where each individual is treated as a particle (see Fig 7.8). In that case, the various states of the model typically represent stages in the life cycle of the organism; and most of the directed lines represent graduation from one stage to the next. When it is associated with graduation, p_{ij} normally is the proportion of individuals in stage j at the end of one time increment that are expected to survive and graduate to stage i by the end of the next time increment. The number of dead organisms usually is not included in the bookkeeping; so no state explicitly accumulating dead individuals is included in the model. Therefore, one expects

$$p_{ij} < 1.0. \quad (7.38)$$

Usually at least one directed line represents reproduction and the coefficient associated with that line is a reproduction rate rather than a graduation probability. The only constraint normally imposed on a reproduction rate is that it be greater than zero. If the life-cycle stages in the model are defined strictly by age classes (Fig. 7.9), a common practice, then all but the oldest age class should be of equal duration, and that duration should be equal to the duration of the time increment. Otherwise the model may depict a logical absurdity, such as a finite probability of a newborn becoming ten years old in one year.

7.3.4 Common ratios

When the members of $\{p_{ij}\}$ are constant and when the members of $\{U_i\}$ in eqn (7.37) are zero, eqns (7.33), (7.34), and (7.37) are homogeneous difference equations with constant coefficients. The solutions to such equations take the form

$$P_i(n) = a_0 + a_1r_1^n + a_2r_2^n + \ldots \quad (7.39)$$

or

$$N_1(n) = a_0 + a_1r_1^n + a_2r_2^n + \ldots, \quad (7.40)$$

where the coefficients $\{a_i\}$ depend on the values of the members of $\{P_i(0)\}$ or $\{N_i(0)\}$, and the values of the members of the set $\{r_i\}$ are determined by those of the set

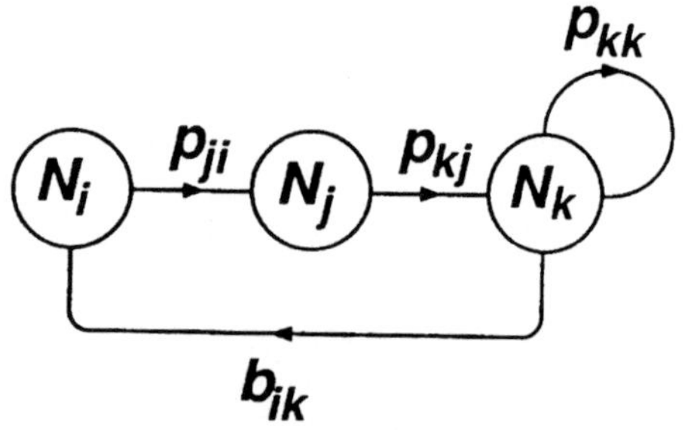

Fig. 7.9 A model of life-cycle dynamics in which the stages are age classes (see, for example, Leslie 1945; Pielou 1969).

$\{p_{ij}\}$. Each r_i is the *common ratio* of a geometric sequence, r_i^n describing part of the dynamics of P_i. Like the natural frequencies of a linear, continuous-time model, the common ratios of linear, discrete-time models are properties of the models themselves and are independent of initial conditions and flow inputs. For discrete-time, Markov-chain models of diffusion processes and chemical kinetics, the common ratios normally are positive real numbers with magnitudes less than 1.0. For discrete-time models of populations of organisms, the common ratios may be real or complex numbers (in conjugate pairs), some with magnitudes greater than 1.0. To understand the implications of various real and complex common ratios, it is instructive to consider what happens to r^n as the value of the integer n increases.

7.4 Signal flow graphs

7.4.1 Introduction

If one can reduce a linear, time-invariant compartmental model or discrete-time network model to a differential or difference equation describing a single-input–single-output relationship, then the process of solution by means of one of the three transforms of the Appendix is straightforward. The reduction to a single-input–single-output equation is a major step in network analysis. For large or complicated network models, one almost certainly will want to employ a digital computer for this task. For small or simple networks, however, one can apply the reduction algorithms on a single sheet of paper or perhaps the back of an envelope. Furthermore, facility with the algorithms adds to an engineer's ability to gain insight about the general properties of network models and the relationships between those properties and network structure.

For compartmental models and discrete-time network models of Markov chains and populations, signal-flow-graph analysis is an especially simple and effective method for reduction to single-input–single-output functions. This reduction is made simple by the fact that each signal flow graph is constructed in the domain of one of the three transforms—Laplace, z, or phasor.

7.4.2 Construction of signal flow graphs

A signal flow graph comprises a set of nodes connected by directed lines. Each node represents the transform of either a state variable (e.g. an accumulation Q_i in a compartmental model, or a probability P_i in a Markov chain, or a population element N_i in a population model) or an input variable; and each directed line represents a dependence of the variable toward which the corresponding arrowhead points on the variable from which it points. Explicitly associated with the arrowhead is the transformed version of the set of mathematical operations involved in that dependence. Thus the combination of each directed line and the node from which it is directed represents a transformed function of time. Two or more directed lines pointing toward the same node represent addition of the corresponding functions. One of the advantages of signal flow graphs constructed in the three transform domains is the fact that complicated operations can be represented simply while they are being manipulated.

In Fig. 7.10, for example, $F_{12}(s)$ could be any transfer function. If it were the ratio of two polynomials in s, $N_{12}(s)/D_{12}(s)$, with coefficients that are real numbers, the relationship between Q_1 and Q_2 would be that of the transformed differential equation $D_{12}(s)Q_2(s) =$

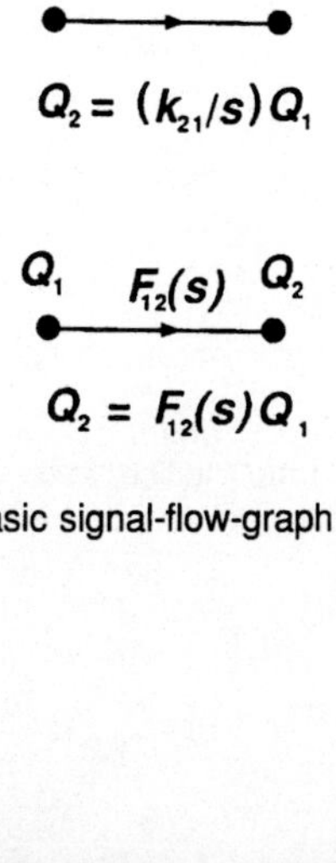

Fig. 7.10 Some basic signal-flow-graph representations.

$N_{12}Q_1(s)$, where the time course of Q_1 is independently specified and the time course of Q_2 is determined by that of Q_1 and by the differential equation. In principle, there are no limits on the order of the differential equation (e.g. on the degrees of the polynomials $D_{12}(s)$ and $N_{12}(s)$).

Given a diagram of a compartmental model or a discrete-time network (with lumped states) representing a Markov chain or a population model, one constructs a signal flow graph such as Fig. 7.11 as follows:

1. Replace every compartment or lumped state in the original model with a single node, keeping the directed lines intact.
2. Replace the rate constant (k_{ij}) associated with each arrowhead in a continuous-time model with k_{ij}/s or $k_{ij}/\mathrm{j}\omega$, depending on whether signal-flow-graph analysis is being carried out with Laplace transforms or phasors; replace the conditional probability or other constant (p_{ij}) associated with each arrowhead in a discrete-time network model with p_{ij}/z.
3. For compartmental models, draw a directed line from each node (i) to itself and label the arrow with k_{ii}/s or $k_{ii}/\mathrm{j}\omega$, where

$$k_{ii} = -\sum_{j=1}^{N} k_{ji}. \tag{7.41}$$

4. Draw a node for the input function, $U_s(s)$, $U_s(\omega)$, or $U_s(z)$, label it accordingly, and draw one directed line connecting the input node to the node representing the state or locale into which the input flow is applied. The arrowhead associated with this line is directed away from the input node and should bear the label $1/s$, $1/\mathrm{j}\omega$ or $1/z$, depending on the transform domain being used. An input node is characterized by the absence of any directed paths leading into it.
5. Select the response variable. If that variable is one of the state variables already represented by nodes, then label the corresponding node with the appropriately transformed function (e.g. if $Q_i(t)$ is selected to be the response variable, label the corresponding node with $Q_r(s)$, $Q_r(\omega)$ or $Q_r(z)$). If the response variable is not already represented by a node (e.g. if it is a flow rather than an accumulation in a compartmental model) then draw a node to represent it and draw and label appropriate directed lines to reflect its dependence on the state variables that already are represented by nodes.

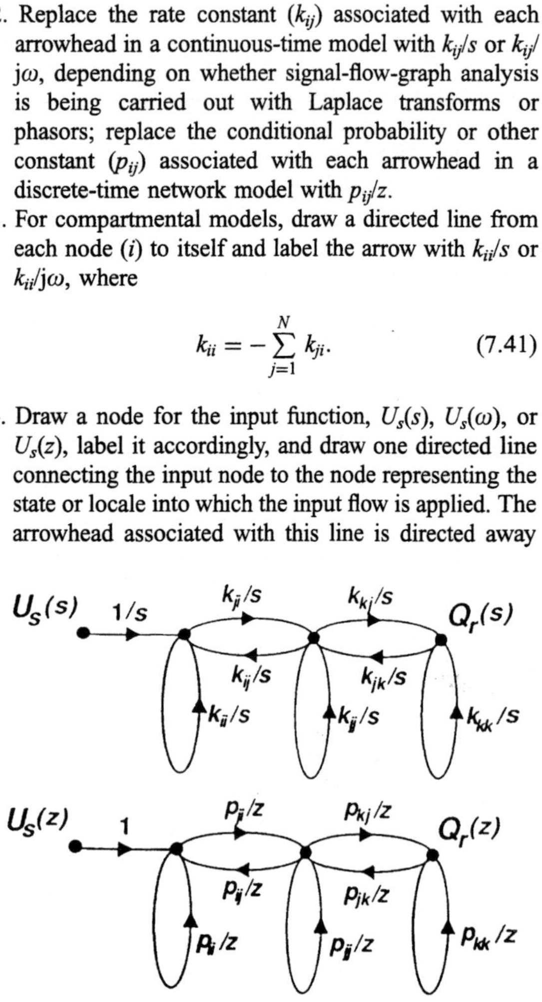

Fig. 7.11 Signal-flow graph for a three-compartment model without a loop. Notice, however, that the signal-flow graph itself has five signal-flow loops.

One can show easily that when the signal flow graph is constructed in this manner for a compartmental model, the differential equation (7.8) describing the dynamics of each accumulation, Q_i, is represented correctly for either zero-state analysis or sinusoidal steady-state analysis. One also can show easily that when the signal flow graph is constructed in this manner for a discrete-time model, the difference equation (7.33) or (7.37) is represented correctly.

7.4.3 Reduction of signal flow graphs

The objective of signal-flow-graph analysis usually is the deduction of one single-input–single-output relationship. Traditionally, the process of deduction was carried out graphically, by removal of one unlabeled node at a time until only the input node and the node representing the response variable were left, along with a directed single line with its arrowhead pointing toward the response node. Associated with that arrowhead would be a single function,

$$H_{\mathrm{sr}} = \frac{N_{\mathrm{sr}}}{D_{\mathrm{sr}}} \tag{7.42}$$

representing the transformed, reduced differential or difference equation

$$D_{\mathrm{sr}}X_{\mathrm{r}} = N_{\mathrm{sr}}U_{\mathrm{s}}, \tag{7.43}$$

where D_{sr} and N_{sr} are polynomials in s, ω or z, X_{r} is the transform of the time function describing the response, and U_{s} is the transformed time function of the input.

The algorithm presented here is a one-step reduction

process first published by S. J. Mason (see Mason and Zimmerman 1960). Readers interested in the traditional multistep process should consult a standard textbook on linear systems theory (e.g. Kuo 1967). To apply Mason's algorithm, one requires the following definitions:

(1) A directed line *begins* at the node from which its arrowhead is pointing and *ends* on the node toward which its arrowhead is pointing.
(2) Two directed lines are *connected* if one ends on the node at which the other begins.
(3) A *signal-flow route* is a series of one or more connected, directed lines.
(4) A signal-flow route *passes through* a node if one of its directed lines ends there *and* another of its direct lines begins there.
(5) a *signal-flow loop* is a signal-flow route that leads from any node back to itself but passes through no node more than once.
(6) A *signal-flow path* is a signal-flow route that begins at the input node, ends at the response node, and contains no signal-flow loops (i.e. does not return to any node).
(7) Two signal-flow loops *touch* if they both pass through the same node.
(8) A signal-flow loop *touches* a signal-flow path if it passes through the input node, the response node, or any of the nodes through which the signal-flow path passes.
(9) Two signal-flow routes (loops or paths) are distinct if either one passes through a node that the other one does not pass through.

Mason's algorithm: Associated with each directed line in a signal-flow route is a transformed operator, F_{ij}. Let Ξ_k be the product of all operators associated with the directed lines composing the kth signal-flow path, and let L_m be the product of all operators associated with the directed lines composing the mth signal-flow loop. The denominator on the right-hand side of eqn (7.42) is computed as follows:

$$D_{\text{sr}} = 1 - \sum L_m + \sum L_m L_n - \sum L_m L_n L_o + \cdots \pm L_m L_n L_o \ldots L_z, \qquad (7.44)$$

where the sum in the second term on the right is taken over all distinct signal-flow loops; the sums in the third and subsequent terms are taken over all products of L's corresponding to distinct signal-flow loops that do not touch each other. Thus, if the mth signal-flow loop touched the qth signal-flow loop, there would be no products on the right-hand side of eqn (7.44) that contained both L_m and L_q. If no two signal-flow loops touched each other, then the number of terms in the last product on the right would equal the number of distinct signal-flow loops in the graph. Another way to represent this computation is

$$D_{\text{sr}} = \prod (1 - L_m)^{**}, \qquad (7.45)$$

where the product is taken over all distinct signal-flow loops, and ** indicates that all terms involving touching signal-flow loops are dropped. Using similar notation, one can express the computation of the numerator in eqn (7.42) as follows:

$$N_{\text{sr}} = \sum \left[\Xi_k \prod (1 - L_m) \right]^{***}, \qquad (7.46)$$

where the sum is taken over all distinct signal-flow paths and the product over all distinct signal-flow loops, and *** indicates that all terms involving touching signal-flow routes (paths or loops) are dropped. The algorithm is completed by application of eqn (7.42). An example of application of Mason's algorithm is depicted in Fig. 7.12.

7.4.4 Transfer and driving point functions

As they have been described so far, $H_{\text{sr}}(s)$, $H_{\text{sr}}(j\omega)$ and $H_{\text{sr}}(z)$ all describe causal relationships between a flow input represented by node U_s and a response variable at

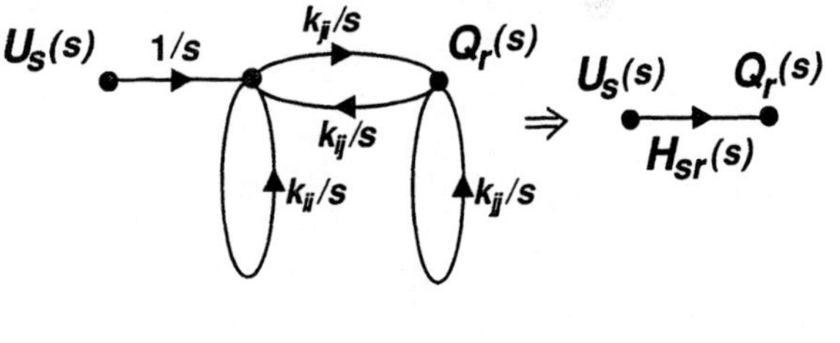

$$L_1 = k_{ii}/s \quad L_2 = (k_{ji}/s)(k_{ij}/s) \quad L_3 = k_{jj}/s \quad \Xi_1 = (1/s)(k_{ji}/s)$$

$$\prod_m (1 - L_m)^{**} = 1 - L_1 - L_2 - L_3 + L_1 L_3 \qquad \sum_k \left[\Xi_k \prod_m (1 - L_m) \right]^{***} = \Xi_1$$

$$k_{ii} = -k_{ji} \quad k_{jj} = -k_{ij} \qquad H_{sr} = k_{ji}/(s)(s + k_{ji} + k_{ij})$$

Fig. 7.12 The objective of signal-flow-graph analysis is reduction of the graph to a single path (representing a single transformed relationship, H_{sr}) connecting a node representing the input or source variable (U_s) to a node representing the response variable (Q_r).

node Q_r (or node J_r if the response variable is a flow). Often, the input node is connected to the rest of the (unreduced) signal flow graph through one node, which represents a single state variable of the model. If that node happens to represent the response variable (that is, node U_s is connected through a single, directed line to node Q_r or J_r), then H_{sr} is a *driving-point function*. If node U_s is connected directly to a node other than Q_r or J_r (and therefore is not connected directly to node Q_r or J_r), H_{sr} is a *transfer function*.

7.5 Circuit models

7.5.1 Nodes and branches

A circuit model can be considered a graph comprising a set of nodes connected by a set of branches (Fig. 7.13). Among other things, nodes provide graphical devices for bookkeeping conservation of flowing stuff. Stuff is not allowed to accumulate at nodes; this leads to the following (node) rule: The sum of the flows into each node at all times is instantly equal to the sum of the flows out of that node. In general, one node (designated as the reference node) is not subject to this rule. This node represents a reference or ground state (or locale), and in circuit models applied to most physical realms it is treated as an inexhaustible source or sink for the conserved stuff. Each of the other lumped states or locales in the model is assigned a unique node and is represented by a branch connected directly from that node to the reference node. A branch with one end connected to the reference node is defined to be in a shunt configuration. Each lumped path in the model is represented by a branch connected directly between the two nodes assigned to the neighboring lumped states considered to be linked by the path. A branch connected between two nodes other than the reference node is defined to be in series configuration. Two branches are defined to be connected in series *with one another* if one end of each of them is connected to the same node, and no other branches are connected to that node. Two branches are said to be in parallel with one another if they both are connected between the same pair of nodes.

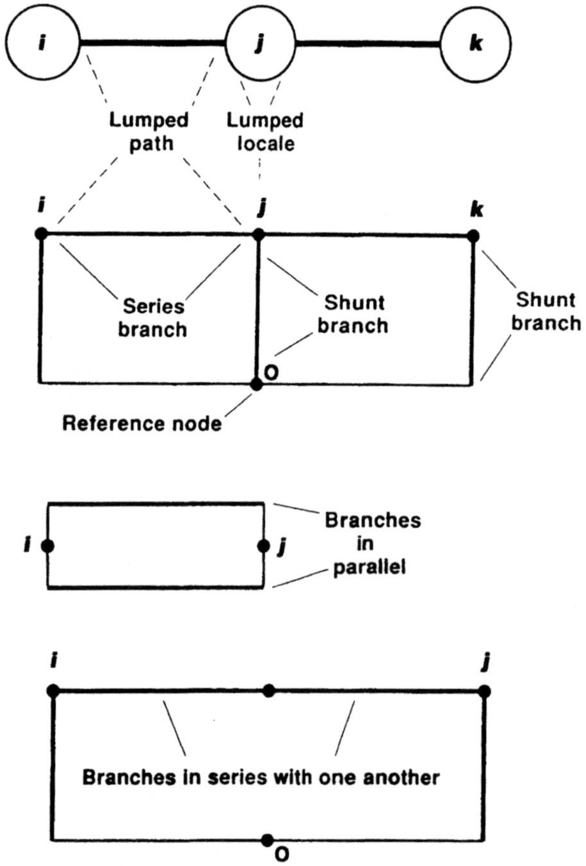

Fig. 7.13 Circuit graphs, comprising nodes and branches.

Circuit theory traditionally involves two sets of variables: a set $\{J_n\}$ of flow variables (where J_n is the model's representation of the flow of conserved stuff through branch n) and a set $\{F_n\}$ of effort differences or potential differences (where F_n is the model's representation of the tendency for stuff to flow spontaneously through branch n). The potential or effort difference for each branch is said to be *conjugate* to the flow for that branch, and vice versa (J_n and F_n are conjugates to one another); variables associated with different branches (e.g. J_n and F_m) are nonconjugates to one another. If branch n connects nodes i and j, then J_n and F_n also could be labeled as $J_{ij}(n)$ and F_{ij}, respectively, where $J_{ij}(n)$ is the model's representation of the flow from node i to node j through branch n, and F_{ij} is the model's representation of the tendency for conserved stuff to flow spontaneously from node i to node j. F_{ij} may be positive (representing tendency for stuff to flow spontaneously from i to j) or negative (representing tendency for stuff to flow in the opposite direction); and $J_{ij}(n)$ may be positive (representing stuff flowing from i to j) or negative (representing stuff flowing in the opposite direction). Defined in this way, F_{ij} and $J_{ij}(n)$ are said to have associated reference directions. In circuit models, F_{ij} and $J_{ij}(n)$ are taken to be unique in the sense that at any instant F_{ij} has the same sign and magnitude for all branches connected between nodes i and j, and at any instant $J_{ij}(n)$ has the same sign and magnitude throughout branch n.

The strict separation of flow (restricted to lumped paths) and accumulation (restricted to lumped states or locales) in network models is represented explicitly in circuit models by this uniqueness of $J_{ij}(n)$. The lumping of a path in a network model implies that the flow into one end of the path is taken instantly to emerge as the flow out of the other end; none is left behind to accumulate within the path. When a modeler considers whether or not to represent a real structure, such as a pipe carrying water between two reservoirs, as a lumped path, the decision ultimately must be based on the extent to which the structure exhibits a unique flow. For candidate paths that do not leak, the decision will depend on two things: (1) the rate at which steady-state flow is established through the real candidate path, and (2) the meaning of 'instantly' in the situation at hand. Item (1) depends on the physics involved in the flow process and on what is connected to the two ends of the candidate lumped path. Item (2) depends on the temporal resolution that one demands in the deductions to be drawn from the network model, and that in turn depends on the purpose for which the model is constructed. Problems arise because any real path that is a candidate for treatment as a lumped path in a network model will be capable of accumulating a finite quantity of conserved stuff, leading to temporary deviations from the unique-flow property whenever the potential or effort difference *across* the path (i.e. from one end of the path to the other) is changed. When steady-state flow is not established in a real path in times that are short in comparison to the temporal resolution desired in the deductions, then the standard remedy is to represent that path as a cascade of lumped locales connected by lumped paths. For a real path through which flow is diffusion-like, temporal resolution in the deductions usually increases as the square of the number of lumped locales in the model of the path.

As it has been described so far, the circuit model has two variables associated with each branch: a flow J_n through it and an effort or potential difference F_n across it. In that case, flow is called the *through* variable and effort or potential the *cross* variable. Circuit models also may be constructed with effort or potential as the through variable and flow as the cross variable. A pair of circuit models that represent precisely the same physical relationships, thus leading to precisely the same dynamic equations, employing these two different bases are said to be *duals* of one another. In the translational and rotational realms of rigid-body mechanics, circuit models traditionally are based on potential (force, torque, or moment) being the through variable and flow (translational or rotational velocity) the cross variable (e.g. see Fig. 6.5). For other physical realms, circuit models usually conform well to physical intuition when they are based on flow being the through variable, effort or potential the cross variable. If one envisions each rigid mass element as a path (with inertia) along which uniform velocity (unique flow) is established instantly, then the same form of circuit model (rather than its dual) will conform to physical intuition in rigid-body mechanics as well (Fig. 7.14).

7.5.2 Potentials and free energy

There are at least two ways to define the variable F_{ij}. The traditional approach has been to invoke the linear empirical laws of nonequilibrium thermodynamics, including the following: Fourier's law, which states that the heat flow along a thermal conduction path is directly

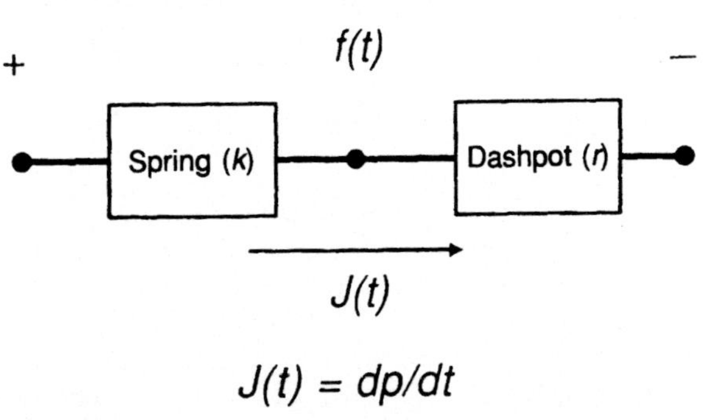

Fig. 7.14 Dual-circuit representation of the model in Fig. 6.5.

proportional to the temperature difference between the two locales connected by the path (see Section 5.2.2); Fick's law, which states that the diffusional flow of particles along a path is directly proportional to the difference in particle concentration at the two locales connected by the path (see Section 4.2.3); Poiseuille's equation, which states that the (fully developed, laminar) flow of a Newtonian fluid through a horizontal pipe is directly proportional to the pressure difference between the two locales connected by the pipe (see Section 2.5.1); and Ohm's law, which states that the charge flow along a path is directly proportional to the voltage difference between the two locales connected by the path. The variables associated with the flows are temperature, concentration, pressure, voltage, and so forth. These are not always linear measures of *potential*, in the thermodynamic sense, and therefore often are categorized as *efforts*.

The alternative approach is derived from the thermodynamic definition of *potential difference*. If stuff tends to flow spontaneously and predictably from locale i to locale j, and if one could harness that flow appropriately, it could be made to do work. The reason for including 'predictably' in this statement is the principle that work cannot be obtained from the random motions associated with thermal energy under conditions of thermal equilibrium. Taking the thermodynamic approach to circuit theory, one defines F_{ij} to be the maximum possible work available to the observer when a unit quantity of stuff moves from state or locale i to state or locale j. F_{ij} specifically excludes the pressure–volume work that must be done against the atmosphere in order for the stuff to move (that component of work is not available to the observer).

With this exclusion, F_{ij} by definition is the change in the *Gibbs free energy* (G) that takes place in the model when a unit quantity of stuff moves from state j to state i:

$$F_{ij} = \frac{\delta G}{\delta Q_i} - \frac{\delta G}{\delta Q_j}, \tag{7.47}$$

where Q_i and Q_j are the total quantities of stuff stored in states i and j respectively. Thus, a positive value of the potential F_{ij} implies free-energy increase when stuff is moved from state j to state i, and the availability of work to the observer when the stuff returns again to state j. In circuit models based on this definition of F_{ij}, it is easy to evaluate the flow of free energy. For that reason, such circuits are especially useful for carrying out analysis and design involving transducers, through which free energy (but not conserved stuff) can flow from one physical realm to another.

Following the tradition of nineteenth century thermodynamics, one can derive the forms of the potentials for the various physical realms by a series of thought experiments in which ideal transducers harness the flow of conserved stuff in each realm and convert it to force-times-distance ($F \times d$) work in the translational, rigid-body mechanical realm. The Carnot engine is an example of such a transducer. In that case the conserved flow of entropy between two thermal reservoirs is harnessed to do $F \times d$ work. If the imagined transducer produces the maximum mechanical work possible for the amount of conserved stuff transferred between two states (e.g., the transduction process is imagined to be carried out so slowly that none of the potential for work is lost through friction), if the potential difference (F_{ij}) between the two states or locales is imagined to remain constant throughout the transduction process, and if any component of the $F \times d$ work that must be done against the atmosphere is excluded, then the remaining $F \times d$ work divided by the total amount of conserved stuff transferred from state or locale i to state or locale j is defined to be F_{ij}.

When the potential difference is a log function (see Table 7.2 later), it represents potential to extract heat from the thermal realm and convert it to work (that is, by virtue of an increase in entropy in the physical realm for which the potential is being derived). For the (ideal) pneumatic realm, the expression for F_{ij} can be deduced by invoking the ideal-gas law and employing an imaginary transducer (for example, involving a piston) that uses pressure difference to generate $F \times d$ work as 1.0 mol of gas particles is transferred from a reservoir at pressure p_i to a reservoir at pressure p_j. For the (ideal) diffusional realm, F_{ij} can be deduced by invoking the Pfeffer–van't Hoff law and employing an imaginary transducer that uses osmotic pressure to generate $F \times d$ work as 1.0 mol of particles is transferred from a reservoir at concentration c_i to a reservoir at concentration c_j.

For (conserved) flow of heat from a reservoir at temperature T_i to one at T_j, imagine that the process is carried out in small temperature steps through a sequence of intervening reservoirs (each with a slightly lower temperature than the previous one), with a separate Carnot engine doing the transduction at each step. Each Carnot engine actually converts a fraction of the flowing heat to work. In the process being imagined here, the flowing heat is taken to be conserved. This can be accomplished by replacing the heat converted to work at each step with heat from the thermal realm (that is, each transduction step will begin with the same amount of heat being drawn from the hotter reservoir). Thus, the conserved, multistep transfer of heat from the hottest reservoir in the sequence (at T_i) to the coolest (at T_j) is taken to produce a second transfer of heat, all of which is converted to work. The logarithmic relationship between converted heat and T_i/T_j arises when one assumes that the temperature steps become infinitesimal and the number of steps infinite.

Example with ideal gas

It is instructive to carry out this sort of thought experiment in the pneumatic realm. Consider an ideal gas in which isothermal expansion or compression is described by the isothermal gas law: $pV = n\mathbf{R}T$, where p is the pressure of the gas, V is its volume, n is the number of gas molecules (given in moles), $\boldsymbol{R}$ is the molar gas constant, and T is the absolute temperature and is taken to be constant. Imagine two very large reservoirs of the gas, one at pressure p_i and the other at pressure p_j, where $p_i > p_j$, with both reservoirs at temperature T. Estimate the $F \times d$ work that can be done by a frictionless transducer that takes 1.0 mol (Avogadro's number) of gas molecules from the reservoir at p_i and transfers it to the reservoir at p_j, assuming that p_i and p_j remain constant during the process (i.e., the transfer of this small amount of gas has negligible effect on the pressures in the very large reservoirs). Imagine that the reservoirs share one wall and that the transducer includes a cylindrical tube connected to that wall and extending into the reservoir at p_j, as depicted in Fig. 7.15. A frictionless piston sits in the tube, and there are two frictionless valves, one in the piston itself and one in an otherwise intact wall between the inside of the tube and the reservoir at p_i.

Beginning with the piston tight against the shared wall between the two reservoirs and with both valves closed, one can imagine work being accomplished in two steps, followed by effortless restoration of the initial state of the piston. In the first step, the valve in the shared wall is opened and the piston is allowed to move toward the reservoir at p_j, carrying out $F \times d$ work on an observer in that reservoir. This process continues until 1.0 mol of gas has been transferred from the reservoir at p_i into the tube (in the volume that now exists between the shared wall and the face of the piston). The valve in the wall is closed, and the gas between the wall and the piston (still at pressure p_i) is allowed to expand isothermally until its pressure reaches p_j, again doing $F \times d$ work on an observer in the reservoir at p_j. Next, the valve in the piston is opened, and the piston is moved slowly and effortlessly back to the shared wall. Finally, the valve in the piston is closed and the transducer is restored to its initial state. One mole of gas has been transferred and an easily calculated amount of work has been done on the observer. The work (W_1) done in the first step is

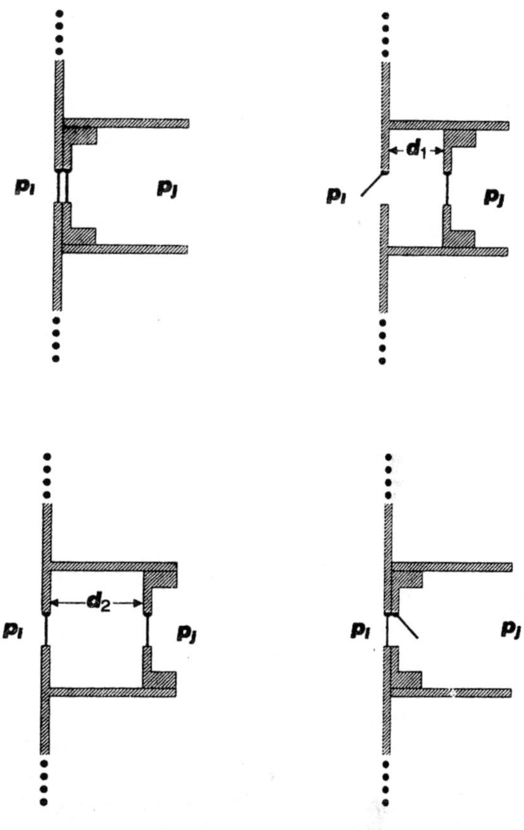

Fig. 7.15 A simple transducer for transfer of energy from a compressible-fluid realm to a translational rigid-body mechanical realm.

$$W_1 = (p_i - p_j)Ad_1 = (p_i - p_j)V_1, \qquad (7.48)$$

where A is the area of the piston, d_1 is the distance the piston is displaced by the 1.0 mol of gas, and $V_1 = Ad_1$ is the volume of that 1.0 mol of the gas at pressure p_i and temperature T. The work (W_2) done in the second step (during which the pressure of the gas being transferred varies according to the gas law) is

$$W_2 = \int_{x=d_1}^{d_2} (p - p_j)A \, dx = \int_{V=V_1}^{V_2} \left(\frac{RT}{V} - \frac{RT}{V_2} \right) dV$$
$$= RT \ln\left[\frac{V_2}{V_1}\right] - RT\left[\frac{(V_2 - V_1)}{V_2}\right]$$
$$= RT \ln\left[\frac{p_i}{p_j}\right] - (p_i - p_j)V_1, \qquad (7.49)$$

where, according to the isothermal gas law for 1.0 mol of gas,

$$p_i V_1 = p_j V_2 = RT. \qquad (7.50)$$

The net work done on the observer is

$$W_{\text{net}} = W_1 + W_2 = RT \ln\left[\frac{p_i}{p_j}\right]$$
$$= RT \ln p_i - RT \ln p_j. \qquad (7.51)$$

It also is instructive to consider the sources of the various components of work in the previous paragraph. Work W_1 was done *by* the gas in the reservoir at p_i as it pushed the piston into the other reservoir. Exactly this same amount of work was done *on* the gas in the reservoir at p_j. It was the remaining work that was available to be captured by the observer. This work must have come from the 1.0 mol of gas trapped between the wall and the piston. The internal (thermal) energy of an ideal gas is a function of the temperature of the gas, but is independent of its pressure and volume. Because the process was assumed to be isothermal, the internal energy of the 1.0 mol of gas did not change as work was done on the observer. Where did the energy for the work arise? The answer to this question is in the thermal realm. As the gas expanded, it extracted heat from the thermal realm and converted it to $F \times d$ work. One can imagine this process taking place as a series of pairs of small steps: (1) The 1.0 mol of gas expands adiabatically, converting a small amount of its internal energy into $F \times d$ work, and thereby cooling very slightly. (2) The piston is held in position while the cooled gas draws heat from the environment and returns to its original temperature (T). As long as the steps remain infinitesimal, the transduction process remains essentially isothermal.

A classical definition of entropy

The ability to draw heat, essentially isothermally, from the thermal realm and convert it into work in the mechanical realm or any other realm was given a quantitative measure by classical thermodynamicists; the name of that measure is *entropy*. Imagine that 1.0 mol of ideal gas has expanded slowly (to avoid frictional generation of heat) and isothermally (at temperature T) from pressure p_i to pressure p_j. In doing so, the gas extracted an amount of heat ΔQ from the environment and converted it to $F \times d$ work. The potential for doing this is inherent in the 1.0 mol of gas itself, and is measured as the difference between its entropy, S_j, at pressure p_j and its entropy, S_i, at pressure p_i. Entropy is defined in such a way that the amount of heat, ΔQ, converted to work is equal to the entropy change in the gas times the temperature of the gas:

$$\Delta Q = (S_j - S_i)T. \qquad (7.52)$$

An increase in the entropy of the gas corresponds to conversion, by the gas, of heat to work; a decrease in the entropy of the gas corresponds to conversion, by the gas, of work to heat. Thus, for example, one can imagine a 'gas spring' constructed from a frictionless piston pushing against an ideal gas in a closed cylinder. If the spring were allowed to operate isothermally (i.e. very slowly), then during compression, $F \times d$ work would be transformed to heat and transferred to the thermal realm; the entropy of the gas would decrease, but its internal energy would remain unchanged. During expansion, heat would be converted (isothermally) to $F \times d$ work; the entropy of the gas would increase but its internal energy again would remain unchanged. One usually thinks of a spring as a device that is capable of storing energy (in the form of potential energy). In this case, the energy itself is stored in the thermal realm; but the ability to recover it (as work) is stored in the entropy decrease of the gas. Many other types of springs operate similarly under isothermal conditions, storing the potential for work in decreased entropy and storing energy in the thermal realm. During adiabatic operation, gas springs, and other kinds of springs, must store thermal energy within their own volumes, rather than passing it to or withdrawing it from the environment. In that case, the springs will heat and cool during their operation.

Additivity of effort or potential and the uniqueness of free energy

An axiom of circuit theory is additivity of effort or potential differences: $F_{ao} - F_{io} = F_{ab} + F_{bc} + \cdots + F_{gh} + F_{hi}$, where this relationship holds for all possible routes between nodes a and i. Experience tells us that this relationship holds for the variables (for example, pressure, concentration, temperature, and voltage) usually employed as measures of effort (for example, it holds for $F_{ij} = T_i - T_j$). If that is true, then it also will hold for any affine transformation of the effort: for example, it will hold for $F_{ij} = (aT_i + b) - (aT_j + b)$. For absolute measures of effort variables (that is, efforts referenced to absolute zero levels), additivity would hold for any potential of the form $F_{ij} = g(X_i) - g(X_j)$, where $g(X)$ is an arbitrary function (for example, log X, as in Table 7.2) and X is the measure of effort (for example, pressure, concentration, temperature). For potentials defined as in Tables 7.1 and 7.2, the additivity axiom translates to a statement of uniqueness of free energy: The change in free energy when any amount of stuff is transferred from one lumped state or locale to another is taken to be independent of the route used in the transfer.

There also is a temporal aspect to the uniqueness of free energy. The terms on the right-hand side of eqn (7.47), and therefore the potential, are well defined only if we impose the following rule: the free-energy change associated with the transfer of stuff from one lumped state to another is independent of the time elapsed since the transfer took place (as long as the stuff remains in the new locale). In other words, when a unit of stuff is represented as being transferred from state or locale j to state or locale i, the free energy in the network model is assumed instantly to change to the (unique) value corresponding to the new distribution of stuff. When finite volumes (or hypervolumes) in state space or physical space are represented as lumped states or locales, then the validity of the additivity axiom is based on the assumption that all of the stuff *within* each volume is distributed (for example, spatially and energetically) in a steady-state fashion at all times, and always in the same steady-state fashion for a given amount of stuff (that is, whenever stuff flows in or out of the volume, the distribution of the total stuff in the volume *instantly* relaxes to a unique steady state). As was the case with compartmental models, the assumption of instantaneous relaxation is a good one if relaxation takes place in times that are short in comparison to the temporal resolution desired in the deductions from the circuit model. The time required for relaxation to within an acceptable proximity of steady state normally decreases as the size of the volume or hypervolume decreases. As greater temporal resolution is demanded in the modeling deductions, one usually must increase the number of lumped states or locales in the circuit model and let each one represent a smaller volume in physical space or a smaller volume or hypervolume in state space.

The operations of many biological structures (such as

Table 7.1 Examples of conjugate potentials and flows in various physical realms

Physical realm	Conserved stuff, SI unit	Flow, SI unit	Traditional effort, SI unit	Potential (F_{io}), joules/unit stuff
Chemical	Number of molecules mol	Reaction rate mol s^{-1}	Chemical potential (μ) J mol^{-1}	$\mu_i - \mu_o$
Continuum mechanical	Volume m^3	Volume flow $m^3\ s^{-1}$	Stress (σ) N m^{-2} (J m^{-3})	$\sigma_i - \sigma_o$
Electric	Charge C	Charge flow A (C s^{-1})	Electric potential (V) V (J C^{-1})	$V_i - V_o$
Hydraulic (horizontal flow)	Volume m^3	Volume flow $m^3\ s^{-1}$	Pressure (p) Pa (J m^{-3})	$p_i - p_o$
Magnetic (total dipoles conserved)	Dipole polarization Wb	Polarization rate Wb s^{-1}	MMF (F_m) A (J Wb^{-1})	$F_{mi} - F_{mo}$
Rigid body (translational motion along one horizontal axis)	Shape m	Velocity m s^{-1}	Force (F) N (J m^{-1})	$F_i - F_o$
Rigid body (rotational motion about one vertical axis)	Shape rad	Rotational velocity rad s^{-1}	Torque (T) or moment (M) N m (J rad^{-1})	$T_i - T_o$
Thermal	Entropy J K^{-1}	Entropy flux J $K^{-1}\ s^{-1}$	Temperature (T) K (J per J K^{-1})	$T_i - T_o$

Table 7.2 Further examples of conjugate potentials and flows in various physical realms

Physical realm	Conserved stuff, SI unit	Flow, SI unit	Traditional effort, SI unit	Potential (F_{io}), joules/unit stuff
Optical (black-body radiator)	Radiant energy J	Energy flux W	Absolute temperature (T) K ($\neq$ J J^{-1})	$\ln T_i - \ln T_o$
Particle diffusion (ideal)	Number of molecules mol	Diffusion rate mol s^{-1}	Absolute concentration (c) mol m^{-3} ($\neq$J mol^{-1})	$\mathbf{R}T(\ln c_i - \ln c_o)$
Particle diffusion (non-ideal)	Number of molecules mol	Diffusion rate mol s^{-1}	Absolute concentration (c) mol m^{-3} ($\neq$J mol^{-1})	$\mathbf{R}T(\ln a_i - \ln a_o)$ (a = activity)
Pneumatic (ideal gas)	Number of molecules mol	Particle flow mol s^{-1}	Absolute pressure (p) Pa ($\neq$J mol^{-1})	$\mathbf{R}T(\ln p_i - \ln p_o)$
Pneumatic (non-ideal gas)	Number of molecules mol	Particle flow mol s^{-1}	Absolute pressure (p) Pa ($\neq$J mol^{-1})	$\mathbf{R}T(\ln f_i - \ln f_o)$ (f = fugacity)
Thermal	Thermal energy J	Heat flow W	Absolute temperature (T) ($\neq$J J^{-1})	$\ln T_i - \ln T_o$

R = gas constant = 8.317 joule/mol K

neurons and muscle cells) as well as the operations of biophysical measuring devices (such as electrodes and electrophoresis equipment) are based on flows of charged particles (ions) in fluids. If the fluid itself is not moving, then the particle will move as a consequence of its random thermal motion (diffusion), as a consequence of electric fields acting on it, as a consequence of chemical reactions in which it is involved, and as a consequence of gravity. For particles (such as inorganic ions) with low mass, the effects of gravity usually are ignored. They become important, however, in some devices used to sort or sense large biological molecules. The effects of gravity also are important in hydraulics when vertical flow in a gravity field is involved, and in many instances of rigid-body and continuum mechanics of solids in a gravity field. Therefore, the bioengineer often must deal with situations in which more than one physical phenomenon contributes to the total potential for some stuff. If the contributed components of potential are defined in such a way that they all refer to the same measure of particle quantity (for example, all given as free energy per mole of the particle species in question), then they can be summed: for example, $F_{ij}(\text{total}) = F_{ij}(\text{diffusion}) + F_{ij}(\text{electric}) + F_{ij}(\text{chemical}) + F_{ij}(\text{gravity})$. With number of particles as the conserved stuff and 1 mol as the unit measure of that stuff, under conditions of uniform gravitational field this expression would be $F_{ij}(\text{total}) = \boldsymbol{RT}\ln(c_i/c_j) + \boldsymbol{zF}(V_i - V_j) + (\mu_i - \mu_j) + \boldsymbol{gM}(h_i - h_j)$, where $\boldsymbol{M}$ is the molar mass of the particle (Avogadro's constant times the mass of the single particle), z is the ionic valence of the particle, $\boldsymbol{F}$ is the Faraday (approximately 96 500 coulombs), $\boldsymbol{g}$ is the equivalent acceleration of gravity (approximately 9.8 m s^{-2} at the earth's surface, depending slightly on latitude), and h is vertical height. In hydraulics, with volume as the measure of stuff and the possibility of vertical flow, $F_{ij}(\text{total}) = p_i - p_j + \rho g(h_i - h_j)$ is the potential, where ρ is the density of the fluid.

Free-energy flow

When potential is defined as Gibbs free energy per unit of conserved stuff, then the product of (conjugate) potential and flow for a given branch in a circuit model is equal to the rate at which free energy is represented as flowing between that branch and the rest of the circuit. When associated reference directions are used and the product is positive, the free-energy flow is represented as being into the branch. Thus, free energy is depicted as flowing from branch to branch, in which case energy conservation is represented by requiring that, at each instant, the sum of the flows of free energy into all (N) branches is zero:

$$\sum_{n=1}^{N} F_{ij}J_{ij}(n) = 0. \tag{7.53}$$

Although it represents the first law of thermodynamics when F_{ij} is free energy per unit of conserved stuff, eqn (7.53) is asserted by Tellegen's theorem to be more general. According to Tellegen's theorem, eqn (7.53) is true for any set of potential or effort values that conform to the additivity rule taken together with any set of flow values (for the same graph) that conform to the node rule (Desoer and Kuh 1969).

7.5.3 Constitutive relationships

Whether or not the product $F_n J_n$ has the dimensions of energy flow (power), each branch in a circuit model represents a specific (constitutive) relationship between the potential or effort, F_n across the branch and the flow, J_n, through the branch. For shunt branches representing lumped states or locales, the essential relationship is between the total accumulation of conserved stuff, Q_n, and F_n. In that case, the shunt branch represents a reservoir of some sort, and the modeler may envision the flow of conserved stuff to take place at one end of the branch only—the end not connected to the reference node. A flow of stuff in and out of that one end conforms to one's physical intuition about accumulation of stuff in reservoirs. The potential or effort at the entrance to the reservoir, however, will be measured relative to the reference state or locale; connection of the other end of the shunt branch to the reference node serves as a graphical representation of that fact. Thus

$$F_n = F_{io}, \tag{7.54}$$

where node i corresponds to the state represented by branch n, and node o is the reference node. To attach a reference direction to the flow, one can label it $J_{io}(n)$. Thus, for such branches, a positive value of $J_{io}(n)$ represents flow into the reservoir and a negative value corresponds to flow out of the reservoir. Conservation requires that

$$J_{io}(n) = \frac{dQ_n}{dt}, \tag{7.55}$$

where Q_n is the total accumulation of stuff in the reservoir. The constitutive relationship between F_{io} and $J_{io}(n)$ for a branch representing a reservoir can be written as follows:

$$\frac{dF_{io}}{dt} = \frac{dF_{io}}{dQ_n} J_{io}(n). \tag{7.56}$$

By convention, this expression usually is translated to

$$J_{io}(n) = C_n \frac{dF_{io}}{dt}, \tag{7.57}$$

where C_n is the *incremental capacity* or *incremental compliance* of branch n:

$$C_n = \frac{dQ_n}{dF_n}. \tag{7.58}$$

Examples of lumped states and locales

For idealized situations, one often can estimate C_n from the physics and geometry of the reservoir.

Example 1. Consider the situation depicted in Fig. 7.16. An (idealized) incompressible fluid of density ρ flows in and out of an entrance at the (flat, horizontal) bottom of a cylindrical reservoir (with its axis oriented vertically and its horizontal cross-sectional area $= A$, a constant). Let the reservoir be open at the top to constant atmospheric pressure, p_{atm} and let the equivalent acceleration of gravity be constant, g, throughout the reservoir. The pressure, p, at the bottom of the reservoir, under these ideal circumstances, would be given by

$$p = \rho g(h - h_0) + p_{\text{atm}} = \frac{\rho g V}{A} + p_{\text{atm}} \tag{7.59}$$

where $h - h_0$ is the height of the fluid surface measured from the bottom of the reservoir, and V is the volume of fluid stored in the reservoir (see Section 2.3). If the reservoir were represented by branch n (connected between node i and the reference node) in a circuit model, with Q_n taken to be equal to V (that is, volume taken to be the measure of conserved stuff) and $F_n = F_{io}$ taken to be $p - p_{\text{atm}}$, then

$$C_n = \frac{dQ_n}{dF_n} = \frac{A}{\rho g} \tag{7.60}$$

Example 2. Imagine a massless ideal gas stored at

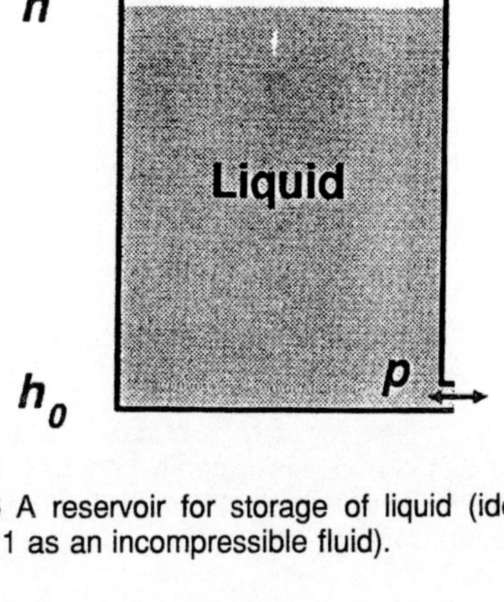

Fig. 7.16 A reservoir for storage of liquid (idealized in Example 1 as an incompressible fluid).

constant temperature, T, in an ideal rigid-walled reservoir with constant volume, V. Then

$$p = \frac{RT}{V}n. \tag{7.61}$$

If this reservoir were represented by branch k (connected between node j and the reference node) in a circuit model, with Q_k taken to be n (i.e. the number of particles is taken to be the measure of conserved stuff) and $F_k = F_{jo}$ taken to be $RT[\ln p_j - \ln p_o]$ (see Table 7.2), then

$$\frac{dF_k}{dt} = \frac{RT}{p_j}\frac{dp_j}{dt} = \frac{RT}{Q_k}J_{jo}(k) \tag{7.62}$$

and

$$C_k = \frac{Q_k}{RT}. \tag{7.63}$$

Example 3 (Fig. 7.17). If the reservoir of the Example 2 were cylindrical, with its axis oriented vertically, its inner vertical wall smooth, and its horizontal cross-sectional area constant (A), and its ceiling were a piston of mass m, making a perfect, frictionless seal against the inner vertical wall and this piston floated on the gas stored within the reservoir, and if the reservoir resided in a uniform gravitational field (equivalent acceleration g), then

$$p = p_{\mathrm{atm}} + \frac{mg}{A}. \tag{7.64}$$

If this reservoir is represented by branch k (connected between node j and the reference node) in a circuit model, with Q_k taken to be n (that is, the number of particles taken to be the measure of conserved stuff) and F_k taken to be $RT[\ln p - \ln p_{\mathrm{atm}}]$, then

$$F_k = F_{jo} = RT \ln\left[\frac{p_{\mathrm{atm}} + (mg/A)}{p_{\mathrm{atm}}}\right], \tag{7.65}$$

where p_{atm} (atmospheric pressure) is taken to be the reference pressure (i.e. the pressure corresponding to the ground state). In that case, $dF_k/dQ_k = 0$ and the incremental capacity is infinite.

Notice that if the horizontal cross-sectional area of the reservoir in the first example had not been constant, C_n would have been a function of the height of the fluid surface and thus a function of Q_n. In the second example, C_k would be nearly constant if Q_k were nearly so (that is, if Q_k were varying by small amounts about some average value, Q_o). On the other hand, if we had selected the $p_j - p_o$ as our measure of F_k in the second example, with n (number of moles) remaining as our measure of Q_k, then C_k would have been constant regardless of how much Q_k varied.

When it has been assigned its constitutive relationship, each branch in a circuit model becomes an element of that model. In the circuit graph, each element of this kind has two terminals, and the elements are connected to one another at the nodes. So far, we have three classes of constitutive relationship, all based on eqn (7.57): (1) C_n is a finite, positive constant; (2) C_n is a function of Q_n; and (3) C_n is infinite (F_n is constant). Only in the first case (C_n a finite constant) does the relationship exhibit the properties (additivity and homogeneity) of linearity (see Appendix). In both the first and second cases, the element is called a *capacitive element*. In the third case, it is called an *independent potential source* or an *independent effort source*. In this chapter, each capacitive element will be depicted graphically as a box surrounding an appropriate label (for example, C_n), with two terminals (represented by lines), and each independent potential or effort source will be depicted as a circle surrounding the label F_x or F_{xx}, with two terminals (Fig. 7.18).

Independent potential or effort sources can represent other situations in which the Gibbs free energy change per unit of stuff placed in a state or locale is independent of the amount of stuff already accumulated there. With respect to the chemical energy available, this would be true for the accumulation of any atomic or molecular species in a particular chemical state (e.g. the accumula-

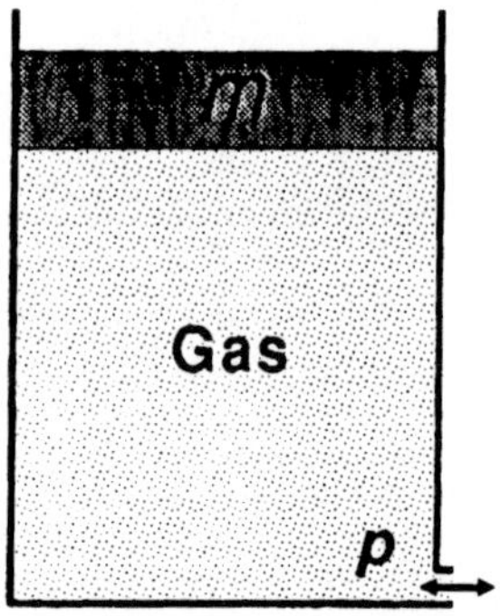

Fig. 7.17 Reservoir for storage of gas. The pressure of the stored gas is maintained nearly constant by a mass floating on the top of the gas and held against the gas by gravity.

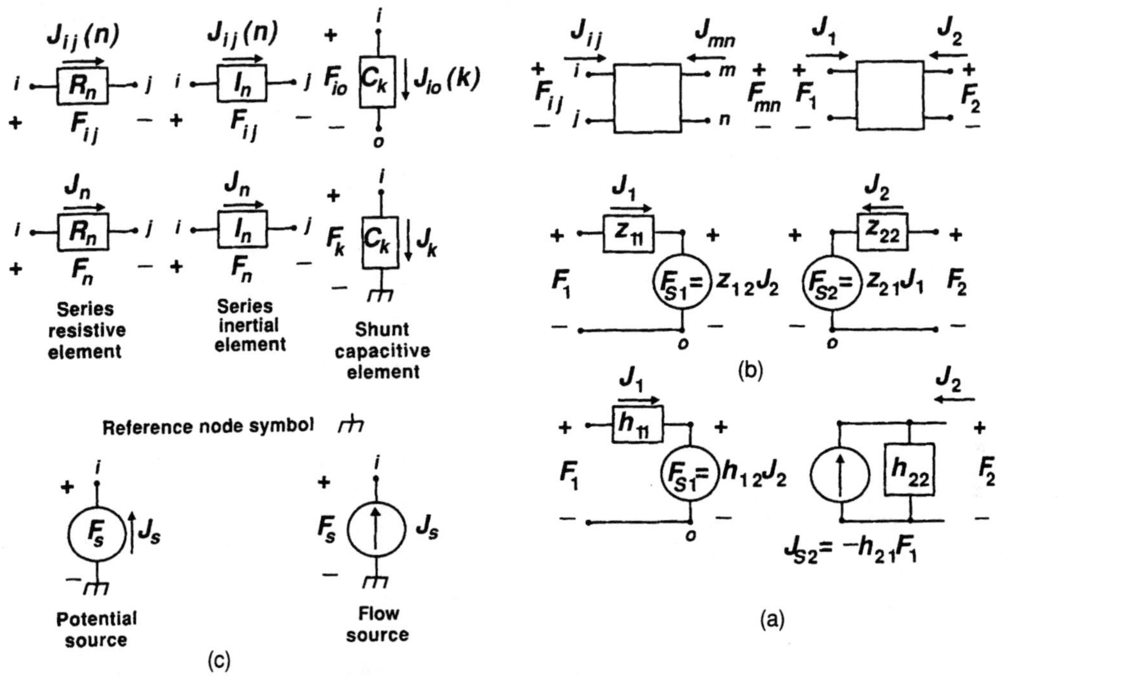

Fig. 7.18 Graphical representations of (a) the five basic two-terminal elements (flow source, potential source, capacitive element, resistive element, inertial element) and (b) the basic two-port element. The $+$ and $-$ signs indicate the reference direction for F_{ij}. When F_{ij} is a positive number, stuff will tend to flow spontaneously from node i to node j. The arrow indicates the reference direction for J_{ij}. When J_{ij} is positive, stuff flows from node i to node j. When the direction of the arrow for a flow is from $+$ to $-$ for the conjugate potential or effort, the reference directions are said to be *associated*. In (c) are two circuit models (constructed from two-terminal elements) that are equivalent to the two-port element.

tion of phosphorus in a pool of adenosine triphosphate molecules). It also is true with respect to energy stored in the displacement of a rigid body against a constant force (for example, a mass being displaced vertically in a uniform gravity field, as in Example 5, or an object with constant charge being moved against a uniform electric field).

When a rigid body is displaced against an elastic element, on the other hand, the force will vary with the accumulation of displacement (that is with the compression or expansion of the elastic member). In circuit models with force, moment, or torque taken to be the cross variable and velocity taken to be the through variable for each branch, elastic members always are represented as shunt branches—depicting locales at which displacement can accumulate (Fig. 7.19). If an elastic member were ideal in the sense of conforming to Hooke's law (see Section 1.3.5), then the parameter C_n representing it would be constant (related to an appro-

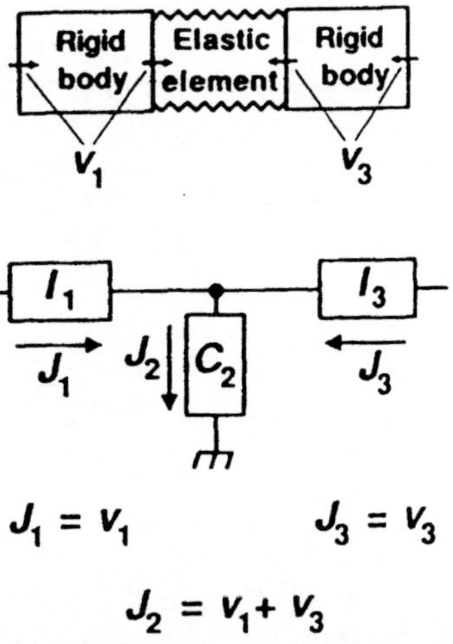

Fig. 7.19 The mass–spring configuration depicted in the upper diagram is represented the lumped circuit model in the lower diagram.

priate elastic modulus).

When volume is taken to be the measure of conserved stuff, as it often is in continuum-mechanical models of solids or liquids, and pressure is taken to be the potential, then accumulation of the conserved stuff corresponds to bulk compression, and the appropriate elastic modulus for computation of C_n is the bulk modulus, E_B (see Section 1.3.5 for an extended discussion of elastic moduli):

$$\frac{dF_n}{dQ_n} = \frac{dp}{dV} = \frac{E_B}{V_o}, \qquad (7.66)$$

$$C_n = \frac{V_o}{E_B}. \qquad (7.67)$$

One can visualize this as a very small increment (dV) of volume of matter compressed into a fixed volume (V_o), of space that is already filled with the material (thus the term on the far right in eqn (7.66) is positive). In circuit models, each appropriately lumped volume of filled space is represented as a shunt branch—depicting a locale into which small additional volumes of material can be squeezed. Imagine, for example, a rigid-walled pipe that is filled with a nearly incompressible fluid and through which that same fluid is flowing (axially). If the fluid in the pipe is undergoing slight compression, then the flow out of the pipe will not be equal to the flow into the pipe. In a circuit model of the pipe, this difference would be represented as flow into one or more shunt branches (Fig. 7.20).

In some instances, the variation of gas density in a pneumatic system is so small that one can equate conservation of mass with conservation of volume. In that case, lumped locales are local volumes of gas-filled space in which the pressure is varying by small amounts about some average pressure, p_o. In a circuit model, each of these locales can be represented by a shunt branch with C_n given by eqn (7.67). For an ideal gas undergoing isothermal (very slow) compression and expansion, the bulk modulus is p_0 (from $pV =$ constant). When local compression and expansion are very fast, as they typically are in acoustics, the adiabatic bulk modulus (γp_o, from $pV^\gamma =$ constant) should be used.

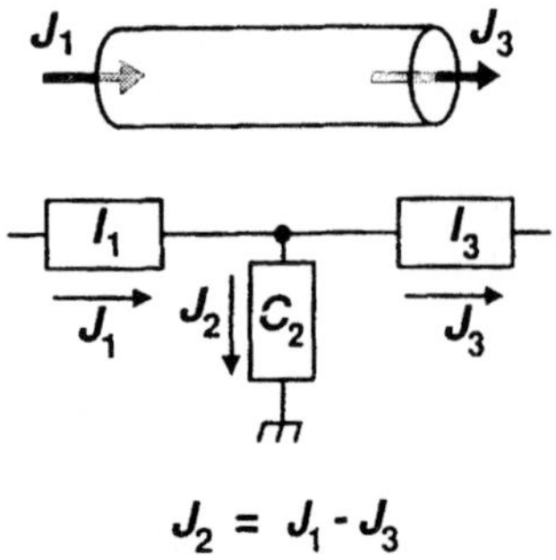

Fig. 7.20 The liquid depicted in the upper diagram as flowing through a pipe is represented by the lumped circuit model in the lower diagram.

For circuit models in the thermal realm, if F_n is taken to be temperature and Q_n is taken to be heat, C_n for a given locale will be that locale's thermal capacity, which (barring state changes in the matter occupying the locale) can remain relatively constant over physiological temperature ranges. For circuit models of particle diffusion, if F_n is taken to be concentration (moles per unit volume) and Q_n the number of particles (moles), then C_n for a given locale will be that locale's volume.

Potential energy

On the other hand, when one wishes to use a circuit model to deduce the flow of energy in the system being modeled, F_n should be the Gibbs free energy per unit of conserved stuff, in which case C_n for the realms represented in Table 7.2 typically will depend on Q_n, and the constitutive relationship represented by the corresponding shunt branch will be nonlinear. This arises from the empirical fact that the relationships between incremental accumulations of conserved stuff and incremental changes in the traditional efforts (temperature, pressure, concentration) typically are linear or nearly linear, and the Gibbs free energy per unit of conserved stuff, in each of those cases, is logarithmically related to the effort. If the effort in question undergoes only small changes (ΔX) about a mean value (X_o), then the logarithmic relationship will be approximately linear:

$$\ln\left[\frac{X_o + \Delta X}{X_o}\right] \approx \frac{\Delta X}{X_o} \qquad (7.68)$$

when

$$\frac{\Delta X}{X_o} \ll 1.0. \qquad (7.69)$$

This is known as *small-signal operation* about the *operating point* X_o. Note also that

$$\ln X_i - \ln X_j \approx 2\frac{X_i - X_j}{X_i + X_j} \qquad (7.70)$$

when

$$X_i - X_j \ll (X_i + X_j)/2. \quad (7.71)$$

It is instructive to check these approximations with a pocket calculator. It also is a simple exercise to find and consider the first-order error terms for each of them. Doing so, one discovers that both of them are robust. When F_n is taken to be Gibbs free energy per unit of stuff and the approximation of eqn (7.68) is applied, then C_n will be approximately constant and the constitutive relationship for the shunt branch will be approximately linear.

For all physical realms, the energy represented as being accumulated in the shunt branches (capacitive elements or independent potential sources) that represent lumped states or locales is called *potential energy, PE*. When F_n is the Gibbs free energy per unit of stuff,

$$\Delta PE_n = F_n \Delta Q_n. \quad (7.72)$$

In a capacitive element for which C_n is absolutely constant (a *linear, time-invariant element*),

$$PE_n = \frac{Q_n^2}{2C_n} = \frac{C_n F_n^2}{2}. \quad (7.73)$$

In an independent potential source for which F_n is absolutely constant,

$$PE_n = F_n Q_n, \quad (7.74)$$

the potential energy is directly proportional to the accumulation of stuff.

Examples of lumped paths

The second law of thermodynamics assures us that whenever stuff flows from one state or locale to another, a finite amount of free energy is dissipated. In circuit models, dissipation of free energy traditionally is represented by flow through a resistive branch. Some forms of flow, such as the flows of liquids and the motions of solids, exhibit conspicuous momentum; once the flow has begun, it tends to continue, and infinite potential or effort would be required to halt the flow instantly. In circuit models, momentum of flow is represented by inertial branches. The energy represented as being accumulated in such branches is the energy of motion—kinetic energy. In the circuit model, dissipation of energy through resistive branches leads to reduction in the sum of potential and kinetic energies in the capacitive and inertial branches.

A lumped path that exhibits both momentum and free-energy dissipation conventionally is represented by a resistive branch connected in series with an inertial branch (representing one flow associated with two additive components of potential or effort). The node shared by the two branches is assigned a potential or effort that is not associated with a state or locale. Instead, it is a computational convenience. For a resistive branch, the basic constitutive relationship is between $J_{ij}(n)$ and F_{ij} ($=F_n$). The second law of thermodynamics implies that $J_{ij}(n)$ will be zero when F_{ij} is zero, in which case the constitutive relationship may be written

$$J_{ij}(n) = G_n F_{ij}, \quad (7.75)$$

where G_n may depend on $J_{ij}(n)$ or F_{ij} or both. As long as G_n is not zero, then the constitutive relationship also may be written

$$F_{ij} = R_n J_{ij}(n), \quad (7.76)$$

where

$$R_n = 1/G_n. \quad (7.77)$$

For an inertial branch, the basic constitutive relationship is between F_{ij} and acceleration of flow, $\mathrm{d}J_{ij}(n)/\mathrm{d}t$; and for all mechanical systems, Newton's first law of motion (Section 1.1.7) states that $\mathrm{d}J_{ij}(n)/\mathrm{d}t$ will be zero when F_{ij} is zero, in which case

$$\frac{\mathrm{d}J_{ij}(n)}{\mathrm{d}t} = K_n F_{ij}, \quad (7.78)$$

where K_n may depend on $J_{ij}(n)$ or F_{ij} or both. If K_n is not zero, then the constitutive relationship also may be written

$$F_{ij} = I_n \frac{\mathrm{d}J_{ij}(n)}{\mathrm{d}t}, \quad (7.79)$$

where

$$I_n = 1/K_n. \quad (7.80)$$

Resistive elements and inertial elements will be depicted graphically as boxes surrounding appropriate labels (for example, R_n and I_n, respectively, as in Fig. 7.18).

For idealized situations, one often can estimate R_n and I_n from the physics and geometry of the path. Consider, for example, an incompressible Newtonian fluid (constant density ρ) undergoing *undeveloped* laminar flow through a cylindrical pipe with rigid walls, length L, and cross-sectional area A. Let $p_i - p_j$ be the pressure difference from one end of the pipe to the other, and let $J_{ij}(n)$ be the fluid flow through the pipe (given in volume of fluid per second). From Newton's second law of motion, one can easily derive the following relationship between $p_i - p_j$ and the acceleration of the flow, $\mathrm{d}J_{ij}(n)/\mathrm{d}t$ (see Section 3.2.2):

$$p_i - p_j = \frac{\rho L}{A}\frac{\mathrm{d}J_{ij}(n)}{\mathrm{d}t}. \qquad (7.81)$$

For the same fluid undergoing *fully developed* laminar flow in the same pipe, Poiseuille's equation (based on Newton's law of viscosity) provides the following relationship between $p_i - p_j$ and $J_{ij}(n)$ (Section 2.5.1):

$$p_i - p_j = \frac{8\pi\eta L}{A^2} J_{ij}(n), \qquad (7.82)$$

where η is the dynamic viscosity of the fluid. If F_{ij} ($=F_n$) is taken to be $p_i - p_j$, then from eqn (7.81),

$$I_n = \frac{\rho L}{A} \qquad (7.83)$$

for undeveloped flow, and from eqn (7.82),

$$R_n = \frac{8\pi\eta L}{A^2} \qquad (7.84)$$

for fully developed flow.

Next, consider a rigid object (Fig. 7.21) of mass m being pushed in a straight line along a horizontal surface with coefficient μ for the friction between the object and the surface (see Section 1.1.12) and constant equivalent acceleration of gravity, g. Taking $J_{ij}(n)$ to be the velocity of the object, the force difference ($F_{ij} = F_i - F_j$) being applied to push the object will be

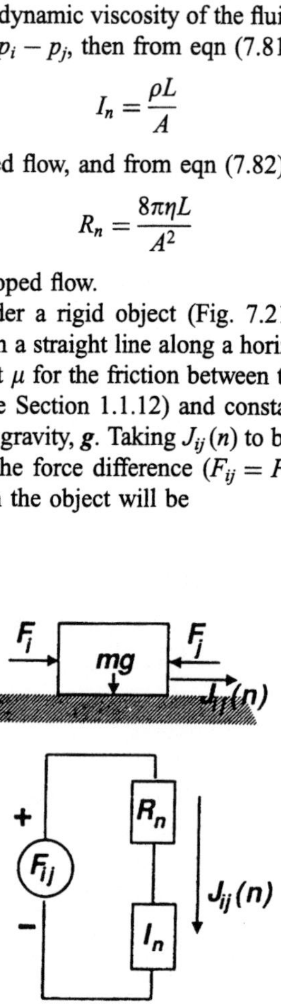

Fig. 7.21 The rigid mass depicted in the upper diagram as being pushed along a horizontal surface with friction is represented by the lumped circuit model in the lower diagram.

$$\begin{aligned} F_{ij} &= \mu m g + m\frac{\mathrm{d}J_{ij}(n)}{\mathrm{d}t}, \quad J_{ij}(n) > 0, \\ F_{ij} &= -\mu m g + m\frac{\mathrm{d}J_{ij}(n)}{\mathrm{d}t}, \quad J_{ij}(n) < 0. \end{aligned} \qquad (7.85)$$

In a circuit model, the first term on the right can be represented by a nonlinear resistive element with

$$R_n = \frac{\mu m g}{J_{ij}(n)}\mathrm{sgn}[J_{ij}(n)], \qquad (7.86)$$

where

$$\begin{aligned} \mathrm{sgn}[x] &= 1.0, \quad x > 1, \\ \mathrm{sgn}[x] &= -1.0, \quad x < 1. \end{aligned} \qquad (7.87)$$

The second term on the right in eqn (7.85) would be represented by a linear inertial element with

$$I_n = m. \qquad (7.88)$$

Next, consider a spherical object (diameter d) moving at a constant velocity, $J_{ij}(n)$, with no turbulence in its wake in a Newtonian fluid with viscosity η. According to Stokes's law, which applies at low velocities (see Section 2.5.4), the force difference (F_{ij}) required to sustain the velocity is given by

$$F_{ij} = 3\pi\eta d\, J_{ij}(n). \qquad (7.89)$$

In a circuit model, this relationship would be represented by a linear resistive element for which $R_n = 3\pi\eta d$. Linear resistive elements also are used commonly in the electrical, thermal, and particle-diffusion realms, where linear empirical laws often prevail in the relationships between conventional measures of flow and conventional measures of effort. Among those three realms, however, only in the electrical realm does the measure of effort equal the Gibbs free energy per unit of stuff that is flowing. In the other two realms, the empirical laws lead to nonlinear resistive elements when F_{ij} is taken to be Gibbs free energy per unit of stuff. According to the linear empirical laws, when flow of stuff occurs through a path of uniform material with constant cross-sectional area A_n and length L_n, the resistance of the path is

$$R_n = k_n L_n / A_n, \qquad (7.90)$$

where k_n is a constant associated with the material of the path and the flowing stuff. In Ohm's law, the stuff is charge, F_{ij} ($=F_n$) is the voltage difference from one end of the path to the other, and k_n is the electrical resistivity (reciprocal of electrical conductivity) of the material. In Fourier's law (Section 5.2.2), the stuff is heat, F_{ij} is the temperature difference from one end of the path to the other, and k_n is the thermal resistivity (reciprocal of thermal conductivity) of the material. In Fick's law (Section 9.2.3), the stuff is the population of particles, F_{ij} is the particle concentration difference from one end of the path to the other, and k_n is the reciprocal of the diffusivity for the (diffusing) particle species in the material (through which it is diffusing). When circuit models are to be used to deduce the flow of energy in a thermal or particle-diffusion system, or in a pneumatic system in which gas is flowing in tubes and Poiseuille's equation is applicable, then the approximation of eqn (7.68) could be used to achieve linear approximations of the resistive elements in the model. Thus, for example, if $c_i = c_o + \Delta c_i$ and $c_j = c_o + \Delta c_j$, where c_i and c_j are particle concentrations at the two ends of a diffusional path, and Δc_i and Δc_j both are small in comparison to c_o, if F_{ij} is taken to be $(\boldsymbol{RT}/c_o)(\Delta c_i - \Delta c_j)$, and if number of particles (in moles) is taken to be the measure of conserved stuff, then eqn (7.90) would translate to

$$R_n = \frac{\boldsymbol{RT}L_n}{D_n c_o A_n}, \tag{7.91}$$

where D_n is the diffusivity for the particle species in the material of the path.

In at least three physical realms—electrical, hydraulic, and pneumatic—there are structures that can be modeled as capacitive branches in series configuration. The parallel-plate electrical capacitor derives its large capacity for charge storage (i.e. its ability to accumulate large amounts of charge without large, concomitant changes in electric potential) by storing charge in the form of compact dipole pairs: every positive charge stored on one plate is accompanied by a nearby negative charge on the other plate. A terminal is connected to each plate, and when charge flows onto one plate, the same amount of charge is displaced from the other plate, leaving the appropriate countercharge behind. The overall capacitor is left with no net charge. Charge has effectively been transferred, however, from one plate to the other; and the electric potential difference between the plates has been changed. To the extent that the displacement of charge occurs instantly, the parallel-plate capacitor exhibits a unique flow (the flow of charge into one terminal is instantly equal to the flow of charge out of the other terminal) and thus behaves in the same manner as a lumped path. Therefore, the (capacitive) branch representing the capacitor in a circuit model may be placed in series configuration. An elastic, deformable barrier placed in the path of fluid flow is an analogous device. Fluid flowing into one side of the device deforms the elastic barrier, displacing the same amount of fluid from the other side and creating a change in the pressure difference from one side to the other. The cupula of the semicircular canal in a vertebrate inner ear is such a device, and it often is represented as a series capacitive branch in circuit models of the canal (Fig. 7.22).

7.5.4 Modeling philosophy

Descriptive, structural, and synthetic models

The physical laws (for example, Newton's laws of motion, Newton's law of viscosity, Fourier's law, Ohm's law, Fick's law) upon which we have based our estimates of R_n and I_n can be considered to be descriptions (that is, *descriptive models*) of natural phenomena, derived from repeated observation. In each case, we combined the descriptive model with a set of structural constraints (i.e. a *structural model*) to derive the estimate of R_n or I_n. Thus, for example, constraining the flow of fluid to be fully developed and entirely in the axial direction on the

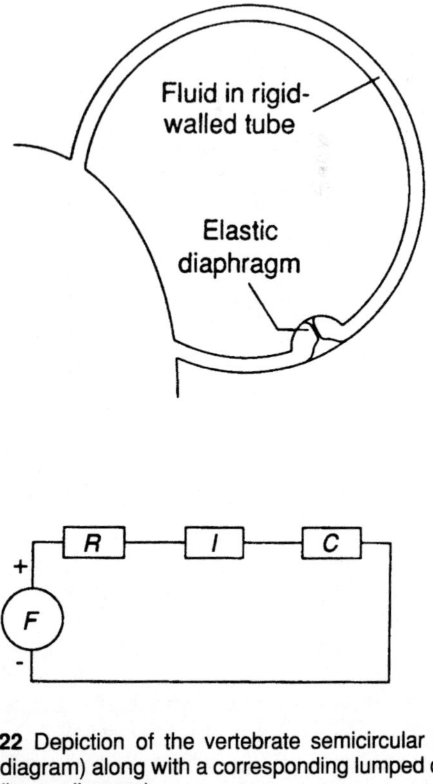

Fig. 7.22 Depiction of the vertebrate semicircular canal (upper diagram) along with a corresponding lumped circuit model (lower diagram).

inside of a rigid-walled, cylindrical pipe allows one to translate Newton's law of viscosity into a parabolic velocity profile and into Poiseuille's equation relating volume velocity to pressure difference. The combination of a structural model and one or more descriptive models is called a *synthetic model.* One uses a synthetic model to deduce the consequences of the descriptive models in particular situations (for example, the consequences of Newton's law of viscosity in fully developed laminar flow in a rigid-walled cylindrical pipe). The deductions are *emergent properties* or *emergent behavior* of the synthetic model, and may or may not be good descriptions of the properties or behavior of the real physical thing being modeled. The synthetic model always represents a set of hypotheses about the structural constraints and physical laws underlying the properties of a physical thing, and ideally one would want to test the synthetic model by comparing its emergent properties with an accurate description of the actual properties of the thing being modeled (i.e. a good descriptive model of the thing).

Scientific and engineering advances often involve an interplay between synthetic and descriptive models. The first evidence for the inverse-square law of gravitational force is an example. Kepler's laws compose a descriptive model of the properties of planetary orbits. Newton combined a structural model (comprising two point mass elements along with their initial positions and motions) with descriptive models (the laws of motion) and a conjectural model (concerning the force of gravity) to form a synthetic model of an orbiting object. Newton tested his synthetic model by comparing its emergent behavior with the behavior of Kepler's descriptive model, and he sculpted his conjectural model of gravity to make the synthetic model fit the descriptive model (only the inverse-square law would work).

Another example is provided by the Hodgkin–Huxley model of electrical impulse (spike) generation by a nerve fiber. An extremely good descriptive model of spike generation had been published independently by Rashevsky (1933), Monnier (1934), and Hill (1936). Hodgkin and Huxley (1952) concluded that the elements underlying spike production were charge flows carried by sodium ions and potassium ions, interacting through accumulation of charge dipoles across a dielectric membrane. Through extensive observations, they constructed descriptive models of the ion currents and the dipole accumulation process. Then they combined these descriptive models with a structural model (placing the ion-flow paths in parallel with each other and with a capacitive branch representing the locale of dipole accumulation) to form a synthetic model. In testing this synthetic model, Hodgkin and Huxley compared its emergent behavior with that of the descriptive model of Rashevsky, Monnier, and Hill. The emergent properties of the Hodgkin–Huxley model are remarkably close to those of the real nerve fiber; but with respect to the behavior (accommodation of threshold) that the Rashevsky, Monnier, and Hill model was intended to describe, that model remains a better descriptor of real nerve fibers. Therein, I believe, lies an important message for reductionists. Synthetic models are excellent tools for advancing science, but their emergent properties always should be taken to be hypothetical until they are tested against observations on the object being modeled.

Circuit theory provides an excellent, widely applicable basis for synthetic modeling. When constructing a circuit model, one may estimate the constitutive relationships of its elements by combining the various laws (descriptive models) of physics with presumed structural constraints, as we did earlier in this section, in which case the elements of the circuit model are themselves synthetic models. Alternatively, one may estimate the constitutive relationships of the elements by direct observation on the object being modeled, in which case the elements of the circuit model are themselves descriptive models. The graph of the circuit model (the relative positions of nodes and branches) is the structural model.

Axiomatic science and natural science

All hypotheses that people propose to account for natural phenomena are human inventions or designs. Therefore it seems reasonable for a biological scientist, such as a bioengineer, interested in a particular biological phenomenon or function to seek inspiration for his or her hypotheses among the inventions or designs that already exist or are available. The engineer has a special advantage in this regard because he or she is trained to be able to put known components or processes together in a systematic way to realize the designs that are available with current knowledge.

Among human inventions are the axiomatic sciences (mathematics and statistics), which provide structures and functions that natural scientists can use in the inductive phases of their studies—when they formulate descriptive models of various phenomena or processes. Thus, for example, we commonly use the real numbers and their properties (e.g. the ordering property) to quantify our observations, and we commonly use linear or nonlinear mathematical functions to describe the causal relationships that we observe. The axiomatic sciences also provide deductive tools that natural scientists can apply when they combine descriptive models of several elementary processes (guided by a structural model) to form a synthetic model of a more complex process. The descriptive models in that case are elevated to axioms and are combined with axiomatic statements about interactions of the elementary processes. The consequences of the system of axioms (that is, the emergent properties of the synthetic model) are revealed by mathematical tools, such as the conventional methods for solving differential equations or difference equations. Circuit theory provides a convenient axiomatic framework for this process.

7.5.5 Further energy considerations

Reference directions

So far in this chapter, the reference directions for potentials or efforts as well as those for flows have been given by the order of indices (e.g. a positive value of the variable J_{ij} taken to mean flow of that magnitude from node i to node j, a negative value of J_{ij} to mean flow in the opposite direction). Another convention, which is followed in the figures of this chapter, is to use an arrow to define the reference direction for each flow and appropriately placed + and − signs to define the reference direction for each potential or effort. Thus, for F_{ij}, the + sign would be placed near node i, the − sign near node j. For J_{ij}, the arrow would be directed from node i toward node j. In all circuit graphs, the positions of the + and − signs and the directions of the arrows remain fixed; and a change in the polarity of a potential or effort or in the direction of a flow is indicated by a change in the sign of the value of the corresponding variable. By convention, for all elements other than sources, the reference directions are associated—that is, the arrow for each branch is directed from the + sign toward the − sign for that branch.

Recall that when associated reference directions are used, and when F_n is taken to be the Gibbs free energy per unit of stuff, then the product F_nJ_n is the rate of free energy flow into the branch. For a two-terminal resistive element, this is the rate at which free energy is represented as being dissipated in the branch. Often it means that the energy is being transferred irreversibly from the physical realm being modeled to some other realm. In the hydraulic realm, dissipation of energy typically is equated with its irreversible transfer to the thermal or acoustic realms (hydraulic energy is converted either to heat or to sound energy); in the electrical realm, dissipation typically corresponds to irreversible transfer of energy to the thermal or optical realms (electrical energy is converted to heat or to low-energy photons). When heat is conserved in the thermal realm, however, the dissipative flow of heat in that realm leads to reduction in free energy without transfer of heat to another realm.

For a two-terminal capacitive element, the product F_nJ_n is the rate at which free energy is accumulating in the branch (as potential energy). For a two-terminal inertial element, F_nJ_n is the rate at which free energy is accumulating in the branch (as kinetic energy). In each case, a negative value of the product means that energy is flowing out of the branch, back into the rest of the circuit.

Sources

The independent potential or effort source already has been mentioned, in the discussion concerning branches representing states. That is one of four types of sources important to circuit modeling. The others are the independent flow source, the dependent flow source, and the dependent potential or effort source. A source is defined as a branch for which one of the two variables (called the *source variable*) is independent of its conjugate. Thus, if J_n were independent of F_n, then branch n would be a flow source; and if F_n were independent of J_n, branch n would be a potential source. For an independent source, the source variable is independent of all other branch variables in the circuit model. For a dependent source, the source variable depends on one or more potentials or flows other than its own conjugate flow or potential. Independent sources commonly are used to represent application of external stimuli or signals to the system being modeled. Dependent sources (e.g. branch n, connecting nodes i and j, with $J_{ij}(n)$ being independent of F_{ij} but dependent on F_{ab}) are especially useful for engineering design of systems involving transducers and for modeling biological processes in which free energy is transferred from one physical realm to another. Dependence on nonconjugate variables can be extended to resistive branches, capacitive branches, and inertial branches. In resistive branch n (connecting nodes i and j), for example, $J_{ij}(n)$ might be dependent not only on F_{ij}, but also on F_{ab} or on other potentials (or flows). Resistive branches of this sort are especially convenient in SPICE circuit models for representing nonlinear biophysical interactions at the cellular level (for example, those involving enzyme kinetics and those involving ion-channel gating).

In a circuit diagram, each flow source is depicted graphically as a circle surrounding an arrow that shows the reference direction for the flow (Fig. 7.18). A positive value of the flow variable (e.g. J_s) associated with the source is taken to mean that the flow is in the direction of the arrow; a negative value is taken to mean that the flow is in the opposite direction. The reference direction for the potential or effort (e.g. F_s) of a flow source traditionally is depicted by a + sign near the node toward which the arrow is directed and a − sign near the other node. For each potential or effort source, the reference direction for the potential or effort is indicated with appropriately placed + and − signs, and the reference direction for the flow is given by an arrow directed from the − sign toward the + sign. A positive value of the product $F_s J_s$ therefore means that free energy is being delivered from the source to the rest of the circuit. If F_s is taken to be the Gibbs free energy per unit of conserved stuff, then the magnitude of the product will be the magnitude of the free energy flow.

Two-port and multiport elements

A two-port element serves at once as two branches that share either no common nodes or at most one common node. In fact the two branches of a two-port element may represent different physical realms, in which case the stuff represented as flowing through one branch will be different from that represented as flowing through the other branch. The principal function of the two-port element is to represent the transfer of free energy directly from one branch to the other. When this transfer is between different physical realms, the two-port element represents a *transducer*. Multiport elements serve the same function, but for more than two branches and, possibly, more than two physical realms.

In contrast to those for two-terminal elements, two-port and multiport constitutive relationships are based on explicitly stated directions of causality. Each of the element's branches is called a port and has the usual pair of variables (a potential and a flow), making four variables for the two-port and $2n$ variables for the n-port. For each set of constitutive relationships, half of the variables are selected as dependent variables, and half as independent variables. Thus, for the two-port there are six sets of constitutive relationships. If the constitutive relationships are linear, however, the modeler can derive all the other five sets from any one. To simplify the notation, each port traditionally is assigned a single, unique index, and each branch variable is assigned the single index of its port. For example, let port 1 be branch h, connected between nodes i and j, and let port 2 be branch k, connected between nodes m and n:

$$F_{ij} = F_1 \quad F_{mn} = F_2$$
$$J_{ij}(h) = J_1 \quad J_{mn}(k) = J_2. \tag{7.92}$$

Passive and active elements

A circuit model usually comprises a set of connected elements that represent local accumulation or dissipation of free energy that already is present in the system being modeled, plus one or more elements that represent processes by which free energy may enter the system from elsewhere. The former are said to be *passive elements*, the latter are *active elements*. When F is taken to be the Gibbs free energy per unit of conserved stuff, then the role played by a given element can be ascertained from its constitutive relationships. For a passive two-terminal element (that is, one that represents only the accumulation or dissipation of free energy) the constitutive relationship must be such that the inequality

$$\int_{\tau_0}^{\tau} F_n(t) J_n(t)\, \mathrm{d}t + E_0; \geqslant 0 \tag{7.93}$$

(F_n and J_n having associated reference directions) will be true for all initial time τ_0, for all time $\tau \geqslant \tau_0$, and for all possible functions $J_n(t)$ or $F_n(t)$, whichever is taken to be the input, E_0 being the free energy stored in the element at time $t = \tau_0$ (see Desoer and Kuh 1969, p. 802). If this condition is not met, the element is active. The corresponding inequality for a passive two-port element is

$$\int_{\tau_0}^{\tau} F_1(t) J_1(t)\, \mathrm{d}t + \int_{\tau_0}^{\tau} F_2(t) J_2(t)\, \mathrm{d}t + E_0 \geqslant 0. \tag{7.94}$$

Viewed separately as a two-terminal element, an individual port of a two-port need not meet the inequality of eqn (7.93) in order for the two-port element, as a whole, to be passive. This leads to ambiguity with respect to circuit models of transducers. To a circuit modeler considering two physical realms at once, a transducer that simply passed free energy between the two realms would translate into a passive two-port element. To a circuit modeler considering only one of those realms, the same transducer would translate into a source, which is an active two-terminal element. For example, an electromagnetic motor/generator might be modeled as a (two-terminal) torque source by a mechanical engineer, a (two-terminal) voltage source by an electrical engineer, or a passive two-port element by a transducer engineer. Thus, the definitions of passive and active are based on the boundary assigned to the system being modeled. The travels of free energy outside the boundary of the system being modeled are not represented in the circuit model. A circuit element is active only when it represents the capability of transferring free energy across the boundary, *into* the system being modeled. That capability never is represented in a passive element. Either kind of element may be used to represent storage of free energy that already has entered the system or transfer of free energy across the boundary in the other direction (*out of* the system being modeled).

Signal energy

When a person conducts experiments on a biological system, it often is convenient to apply one or more carefully controlled stimuli (beginning at some designated starting time) and to observe variables in the system that appear to be causally related to the stimuli. Usually in such situations one wishes to consider only the effects of the designated stimuli. Unfortunately, the energy associated with every observed variable will include the energy of random thermal motion (which must be present in any system at temperatures above 0 K). It also may include residues of energy already in the system (at the designated starting time) as a result of previous stimuli, and energy entering the system (after the starting time) from stimuli not controlled by the experimenter. Mixed with all of this other energy will be the energy that the experimenter wants to observe—the energy of the response itself (i.e. the energy associated with accumulations and flows that are caused by the stimuli). This energy sometimes is called *signal energy*. Hoping that the properties of the system are stationary, the experimenter may use various statistical signal-processing techniques to estimate the time course of signal energy; and in constructing a circuit model of the system, he or she may wish to consider only the signal energy. In that case, the definition of the boundary of the system being modeled would include energy: signal energy always resides within the boundaries, all other energy resides outside the boundaries. In that case, conversion of any form of nonsignal energy to signal energy can be represented only by an active element; and a circuit element will be active only if it represents the capability of such conversion. Otherwise it will be passive.

For example, when an acoustic stimulus source is used to excite the auditory sensors of the ear, the signal energy evidently takes predominantly the following forms as it moves inward:

(1) pressure times volume displacement of the air in the external ear canal;

(2) force times translational displacement of the tympanum (ear drum);

(3) force times translational displacement and moment times angular displacement of the malleus and incus;

(4) force times translational displacement of the stapes;

(5) force times translational displacement of the oval window;

(6) pressure times volume displacement of the inner-ear fluids;

(7) force times translational displacement and moment times angular displacement of the micromechanical structures of the organ of Corti, up to the hair bundle of the sensory cell (hair cell);

(8) moment times angular displacement of the hair bundle;

(9) chemical potential times state displacement (from closed to open state) of the strain-gated channel molecules of the hair bundles;

(10) voltage times charge displacement of the (mostly potassium) ions passing through the strain-gated channels;

(11) voltage times charge displacement of various ion species in and about the cell body of the hair cell;

(12) chemical potential times state displacement (from closed to open state) of the voltage-gated calcium-channel molecules of the hair-cell body;

(13) voltage times charge displacement and diffusion potential times particle displacement of the calcium ions passing through the channels;

(14) chemical potential times state displacement of calcium-activated molecules associated with synaptic vesicles;

(15) diffusion potential times particle displacement of synaptic transmitter molecules;

(16) chemical potential times state displacement (from closed to open state) of transmitter-gated channel molecules of nerve fibers (auditory afferent axons) at the afferent synapses of the hair cells;

(17) voltage times charge displacement and diffusion potential times particle displacement of the ions passing through the transmitter-gated channel molecules;
(18) voltage times charge displacement of ions in and about the initial segments of the auditory afferent axons;
(19) chemical potential times state displacement of sodium-ion channels associated with the axonal spike triggers;
(20) voltage times charge displacement and diffusion potential times particle displacement of the sodium ions passing through the sodium-ion channels;
(21) electrochemical energy associated with nerve impulses (spikes) being propagated to the auditory brainstem.

In steps 1 through 9, the signal energy from the stimulus source evidently is simply converted from one form to another, ultimately being converted to the opening of ion channels against a chemical potential. The energy involved in steps 10 through 12, on the other hand, arises from an electrical source (often called the endolymphatic potential) in the inner ear. Much as a gardener might use the energy in his body to open the valve of a water faucet, allowing the hydraulic energy of the water source to distribute the water to the garden, the energy from the sound source opens electric valves, allowing the electric energy of the endolymphatic system to distribute charge over the cell body of the sensory cell. As it flows from the valve into the system, this gated energy becomes *signal energy.* Its conversion from non-signal energy (e.g. stored hydraulic or electric energy) was caused by the signal energy from the stimulus source (the gardener or the acoustic source), but it contains little if any of the energy provided by the original source. In a circuit model, the gardener's valve and the gated ion-channels of step 10 both would be modeled as active elements. All of the processes up to step 10 would be represented by passive elements. Beyond step 10 are several more gated ion-channels, all of which would be represented by active elements.

Circuit analysis that excludes the effects of residual energy in the modeled system is called *zero-sate analysis* (see Appendix A). In the zero state, a circuit is presumed to represent a system that initially contains no stored energy; there is no potential energy represented as being in any of its capacitive branches and no kinetic energy represented as being in any of its inertial branches. In zero-state analysis, the modeler designates one or more independent sources to represent the stimuli to the system being modeled; and, at the designated starting time, the circuit is assumed to be in the zero state. In constructing a circuit model of a biological system, the bioengineer has moved from the domain of the natural sciences (biology, chemistry, physics) to the domain of the axiomatic sciences (mathematics, statistics). The observations on the system have been translated to a set of axioms (constitutive relationships, conservation relationships, etc.), and noise-free, interference-free revelation of the deduced responses to the stimulus requires only the solutions of a set of differential or difference equations.

7.5.6 Linear circuit models

A linear circuit model is one in which the constitutive relationship of every branch is a linear function. Recall that a linear function or transformation is one that exhibits the properties of *additivity* and *homogeneity.* Thus, for example, a time derivative of any order provides a linear transformation of any analytic function of time:

$$h[f(t)] = \frac{d^n f(t)}{dt^n},$$
$$h[af_1(t) + bf_2(t)] = ah[f_1(t)] + bh[f_2(t)]. \tag{7.95}$$

The common linear two-terminal constitutive relationships are those of the resistive, capacitive and inertial branches: $F_{ij} = R_n J_{ij}(n)$, $J_{ij}(n) = C_n\, dF_{ij}/dt$, and $F_{ij} = I_n\, dJ_{ij}(n)/dt$, when R_n, C_n and I_n are constants.

Note that the only independent sources that meet the requirements for linearity are the trivial ones, $F = 0$ and $J = 0$. Nontrivial independent sources are nonlinear circuit elements and thus are excluded from linear circuit models. The linear two-terminal constitutive relationships imply no direction of causality: neither F_{ij} nor $J_{ij}(n)$ is taken to be cause or effect, or to be dependent variable or independent variable.

Transformed constitutive relationships: admittances and impedances

The concepts of impedance and admittance have been extremely effective tools for dealing with linear circuits. They arise from the use of Laplace transforms in zero-state analysis and from the use of phasor transforms in sinusoidal steady-state analysis. Applying the transform pairs of Tables A.2 and A.8 to the linear two-terminal constitutive relationships, one obtains the entries in Table 7.3. All of these transformed constitutive relationships can be represented by multiplicative factors that are generalized functions of s or ω:

$$\begin{aligned} F_n(s) &= Z_n(s)J_n(s), \quad F_n(\omega) = Z_n(\omega)J_n(\omega) \\ J_n(s) &= Y_n(s)F_n(s), \quad J_n(\omega) = Y_n(\omega)F_n(\omega) \\ Z_n(s) &= \frac{1}{Y_n(s)} \qquad Z_n(\omega) = \frac{1}{Y_n(\omega)}. \end{aligned} \tag{7.96}$$

$Z_n(.)$ is the *impedance* of branch n, and its reciprocal $Y_n(.)$ is the *admittance* of branch n.

In single-input–single-output analysis of a circuit model, the goal often is to find a general relationship between a designated independent source variable and a designated response variable. In zero-state or sinusoidal steady-state analysis of linear networks, that relationship can be stated in the form of an impedance, an admittance, or a dimensionless transfer ratio. Suppose, for example, that the independent source is connected from node i to node j and that the designated source variable is the flow J_s from the source into node i. If the designated response is the resulting potential, $F_s = F_{ij}$, between nodes i and j, one can describe the relationship as follows:

$$F_s = Z_{ij}J_s, \tag{7.97}$$

where Z_{ij} is a *driving-point impedance*. If the designated source variable had been the potential, $F_s = F_{ij}$ applied from node i to node j, and the designated response had been the flow, J_s, from the source (into node i), then the relationship might have been described as

$$J_s = Y_{ij}F_s, \tag{7.98}$$

where Y_{ij} is a *driving-point admittance*. Because they were obtained for the same pair of nodes, the driving-point impedance and driving-point admittance in this case are reciprocal. Thus, a driving-point impedance or admittance bears no implication regarding causality—it relates potential to flow but does not imply that one is cause, one effect.

Transfer relationships, on the other hand, always imply a direction of causality. If the designated source is $F_{ij} = F_s$ and the designated response is $F_{mn} = F_r$, the generalized relationship would be:

$$F_{mn} = H_{sr}F_{ij}, \tag{7.99}$$

where H_{sr}, a function of s or ω, is a dimensionless transfer ratio. If the situation had been reversed, with F_{mn} as designated source and F_{ij} as designated response, the transfer ratio would not be $1/H_{sr}$ (it also would not, generally, be H_{sr}). The computation of H_{sr} was based on F_{ij} being cause, F_{mn} being effect; and that is the only situation to which H_{sr} can be expected to apply. The same thing is true when H_{sr} is a transfer admittance

$$J_{mn} = H_{sr}F_{ij} \tag{7.100}$$

or a transfer impedance

$$F_{mn} = H_{sr}J_{ij}. \tag{7.101}$$

In general, the reciprocal of H_{sr} has no meaning. As a convention in this chapter, the variable designated as cause always will appear immediately to the right of a transfer function. Furthermore, as was done in Chapter 6 to simplify notation, the arguments of the various transformed functions often will be omitted. When the Laplace transform is being used, Y, Z, F, J, and H will imply the functions $Y(s)$, $Z(s)$, $F(s)$, $J(s)$, and $H(s)$; and when the phasor transform is being used, Y, Z, F, J, and H will imply the functions $Y(\omega)$, $Z(\omega)$, $F(\omega)$, $J(\omega)$, and $H(\omega)$.

Driving-point and transfer functions are illustrated in Fig. 7.23.

Table 7.3 Linear transforms of two-terminal constitutive relationships[a]

Element	For zero-state analysis	For sinusoidal steady-state analysis
Resistive	$F_n(s) = R_nJ_n(s)$	$F_n(\omega) = R_nJ_n(\omega)$
Capacitive	$J_n(s) = sC_nF_n(s)$	$J_n(\omega) = j\omega C_nF_n(\omega)$
Inertial	$F_n(s) = sI_nJ_n(s)$	$F_n(\omega) = j\omega I_nJ_n(\omega)$

[a] With R_n, C_n, and I_n constants

Examples

Refer to Fig. 7.24.

Definitions:

$$Z_{12} = R_{12} \tag{7.102a}$$

$$Y_1 = sC_1 \text{ or } j\omega C_1 \tag{7.102b}$$

$$Y_2 = sC_2 \text{ or } j\omega C_2 \tag{7.102c}$$

Additivity of F:

(Fig. 7.24b, d) $\quad F_S = F_1 = F_2 + F_{12}$ (7.102d)

(Fig. 7.24c,e) $\quad F_S + F_1 = F_2 + F_{12}$ (7.102e)

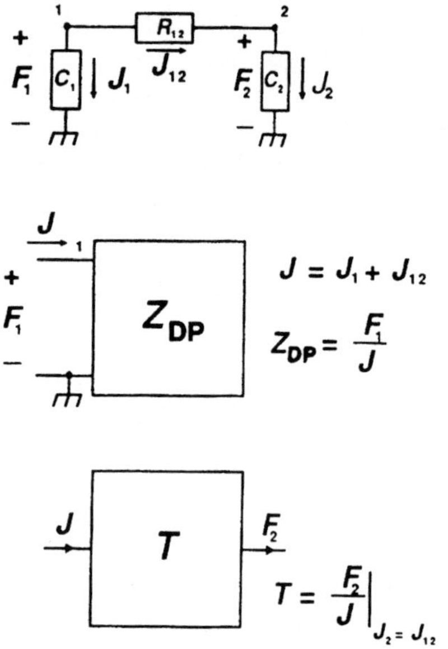

Fig. 7.23 Driving-point and transfer functions. When appropriate, an entire circuit can be reduced to a single element: a two-terminal circuit element—representing a driving-point impedance or admittance—or a two-terminal systems element (black box)—representing a context-independent driving-point or transfer relationship.

Conservation:

$$\text{(Fig. 7.24b, c,d,e)} \quad J_{12} = J_2 \tag{7.102f}$$

$$\text{(Fig. 7.24b,d)} \quad J_S = J_1 + J_{12} \tag{7.102g}$$

$$\text{(Fig. 7.24c,e)} \quad J_S = -J_1 = J_{12} \tag{7.102h}$$

Constitutive relations:

$$J_1 = Y_1 F_1 \tag{7.102i}$$

$$F_{12} = Z_{12} J_{12} \tag{7.102j}$$

$$J_2 = Y_2 F_2 \tag{7.102k}$$

Zero-state and sinusoidal steady-state analyses:

$$\text{(Fig. 7.24b,d)} \quad J_{12} = J_2 = Y_2 F_2 \tag{7.102l}$$

$$F_{12} = Z_{12} J_{12} = Z_{12} Y_2 F_2 \tag{7.102m}$$

$$F_1 = F_{12} + F_2$$

$$= (1 + Z_{12} Y_2) F_2 \tag{7.102n}$$

$$J_1 = Y_1 F_1 = Y_1 (1 + Z_{12} Y_2) F_2 \tag{7.102o}$$

$$J_S = J_1 + J_{12}$$

$$= (Y_1 + Y_2 + Y_1 Z_{12} Y_2) F_2 \tag{7.102p}$$

$$F_S = F_1 = (1 + Z_{12} Y_2) F_2 \tag{7.102q}$$

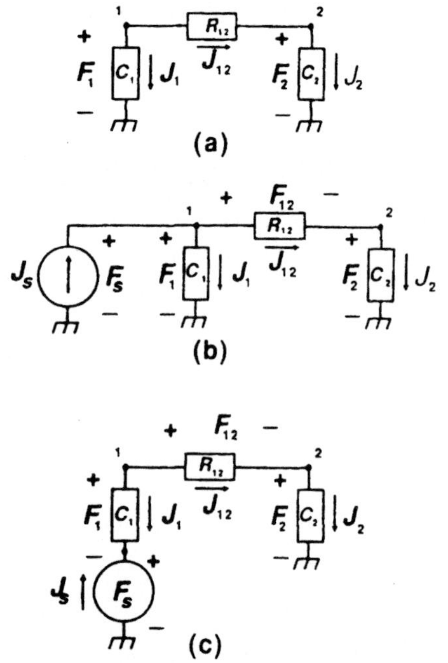

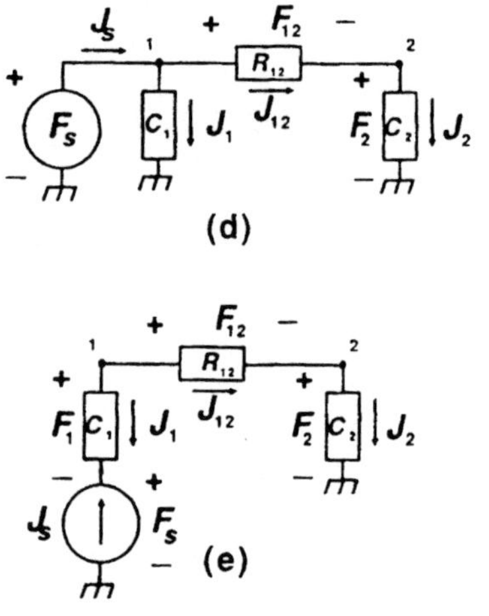

Fig. 7.24 Circuit models for the examples in Section 7.5.6. In (b) and (c) the sources are connected in ways that do not alter the natural frequencies of the original model in (a). In (d) and (e) the sources are connected in ways that alter the natural frequencies.

(Fig. 7.24b) $H_{sr}(1) = F_2/J_S$
$$= 1/(Y_1 + Y_2 + Y_1 Z_{12} Y_2) \quad (7.102r)$$

$$Z_{DP} = F_S/J_S = (1 + Z_{12}Y_2)/(Y_1 + Y_2 + Y_1 Z_{12} Y_2) \quad (7.102s)$$

$$H_{sr}(2) = J_2/J_S = Y_2/(Y_1 + Y_2 + Y_1 Z_{12} Y_2) \quad (7.102t)$$

(Fig. 7.24d) $H_{sr}(1) = J_2/F_S$
$$= Y_2/(1 + Z_{12}Y_2) \quad (7.102u)$$

$$Y_{DP} = J_S/F_S = (Y_1 + Y_2 + Y_1 Z_{12} Y_2)/(1 + Z_{12}Y_2) \quad (7.102v)$$

$$H_{sr}(2) = F_2/F_S = 1/(1 + Z_{12}Y_2) \quad (7.102w)$$

(Fig. 7.24c,e) $F_S = F_{12} + F_2 - F_1$ (7.102x)
$$J_S = -J_1 = J_{12} = J_2 \quad (7.102y)$$
$$F_S = Z_{12}J_{12} + J_1/Y_1 + J_2/Y_2 = (Z_{12} + 1/Y_1 + 1/Y_2)J_S \quad (7.102z)$$
$$J_S = Y_1 Y_2 F_S/(Y_1 + Y_2 + Y_1 Z_{12} Y_2) \quad (7.102aa)$$
$$F_2 = J_2/Y_2 = J_S/Y_2 = Y_1 F_S/(Y_1 + Y_2 + Y_1 Z_{12} Y_2) \quad (7.102bb)$$

(Fig. 7.24c) $H_{sr}(1) = J_2/F_S = J_S/F_S$
$$= Y_1 Y_2/(Y_1 + Y_2 + Y_1 Z_{12} Y_2) \quad (7.102cc)$$
$$Y_{DP} = J_S/F_S = Y_1 Y_2/(Y_1 + Y_2 + Y_1 Z_{12} Y_2) \quad (7.102dd)$$
$$H_{sr}(2) = F_2/F_S = Y_1/(Y_1 + Y_2 + Y_1 Z_{12} Y_2) \quad (7.102ee)$$

(Fig. 7.24e) $H_{sr}(1) = F_2/J_S = 1/Y_2$ (7.102ff)
$$Z_{DP} = F_S/J_S = (Y_1 + Y_2 + Y_1 Z_{12} Y_2)/Y_1 Y_2 \quad (7.102gg)$$
$$H_{sr}(2) = J_2/J_S = 1. \quad (7.102hh)$$

In the circuit models of Fig. 7.24(b) and (e), the designated independent source variable is J_S; in the models of Fig. 7.24(c) and (d) it is F_S. Notice that all of the driving point and transfer functions for the models of Fig. 7.24(b) and (c) have the same denominator, and notice the symmetry in that denominator. When zero-state analysis is carried out in Laplace transforms, the zeros of the denominator polynomial (in s) are the natural frequencies of the circuit model. When an independent flow source is connected to existing nodes in a circuit model (in parallel with any elements already connected between those nodes), the natural frequencies of the original circuit model are not altered. The same thing is true when an independent potential source is connected in series with any element in the circuit (i.e. by addition of one node to the model). As long as independent sources are connected in these ways, the denominators of all driving-point and transfer functions will be identical, regardless of where the source is connected or which circuit variable is designated to be the response. When an independent flow source is connected in series with any element, or an independent potential source is connected between any pair of existing nodes, the natural frequencies will not be the same as those of the original circuit, and the denominators of driving-point and transfer functions will be altered accordingly. To gain physical intuition about the effects of various source connections on natural frequencies, imagine the limiting situations in which $F_S = 0$ and $J_S = 0$. When $J_S = 0$, the independent flow source becomes an open circuit; if it is connected in series with any element, that element is effectively removed from the circuit model. When $F_S = 0$, the independent potential source becomes a short circuit; if it is connected between any pair of existing nodes in the model, those two nodes are reduced to one. In either case, the basic structure of the circuit is altered, and its natural frequencies also will be altered.

Linear two-port elements

For a linear two-port element, whether ports 1 and 2 correspond to the same physical realm or to different physical realms, one can obtain driving-point and transfer impedances. With an independent flow source, $J_1 = J_s$, connected to port 1 and nothing connected to port 2 (that is, $J_2 = 0$), one obtains

$$F_1 = z_{11} J_1 \quad (7.103)$$

and

$$F_2 = z_{21}J_1. \tag{7.104}$$

With an independent flow source, J_2, connected to port 2 and nothing connected to port 1 ($J_1 = 0$), one obtains

$$F_1 = z_{12}J_2 \tag{7.105}$$

and

$$F_2 = z_{22}J_2. \tag{7.106}$$

When the two-port element is linear, one can apply the property of additivity, to obtain the complete constitutive relationships

$$\begin{aligned} F_1 &= z_{11}J_1 + z_{12}J_2 \\ F_2 &= z_{21}J_1 + z_{22}J_2. \end{aligned} \tag{7.107}$$

Although in this derivation, J_1 and J_2 are taken to be cause, F_1 and F_2 to be effect, the complete constitutive relationship transcends the direction of causality, allowing the modeler to deduce the behavior of the two-port element regardless of the nature of the independent source and regardless of the port to which it connected. This will be demonstrated in the section on transducers.

As an example of the general approach for fitting a two-port element to an idealized physical situation, consider the flow of radiant energy from two black-body radiators facing one another through a medium in which radiation travels—a situation of potential interest to bioengineers concerned with radiation thermography (see also Section 5.2.2). If we lump each black body into a single locale for energy accumulation, with a single temperature, then we can write

$$\begin{aligned} J_1 &= f_{11}(T_1) - f_{12}(T_2) \\ J_2 &= f_{21}(T_1) - f_{22}(T_2), \end{aligned} \tag{7.108}$$

where J_i is the net radiant energy flow from black body i, and T_i is the temperature of black body i. For idealized situations, the functions on the right-hand sides of eqns (7.108) can be estimated from synthetic models. Imagine, for example, that each of the two bodies is a spherical, isotropic radiator and absorber, with emissivity equal to 1.0, and that the medium is one in which radiation travels without loss of energy. Let the radii of the two spheres be r_1 and r_2, respectively, and the distance separating the centers of the spheres be R, where R is very much greater than r_1 and r_2. According to the Stefan–Boltzmann law [eqn (5.3)], the total radiant energy flow, J_i, from black body i is

$$J_i = 4\pi r_i^2 \sigma T_i^4 \tag{7.109}$$

where σ is the Stefan–Boltzmann constant. The proportion of the radiant energy emitted by black body i that will be captured by black body j (at distance R) is the ratio of the cross-sectional area of black body j to the surface area of a sphere of radius R, which is $(r_j/2R)^2$. Therefore,

$$\begin{aligned} J_1 &= 4\pi r_1^2 \sigma T_1^4 - \frac{\pi r_1^2 r_2^2}{R^2}\sigma T_2^4 \\ J_2 &= -\frac{\pi r_1^2 r_2^2}{R^2}\sigma T_1^4 + 4\pi r_2^2 \sigma T_2^4. \end{aligned} \tag{7.110}$$

For small variations about a large reference temperature, T_o, the relationships between flow variation and temperature variation become nearly linear. Let

$$\begin{aligned} T_i &= T_o + \delta T_i, \\ J_i &= J_{io} + \delta J_i. \end{aligned} \tag{7.111}$$

If

$$\frac{\delta T_i}{T_o} \ll 1.0, \tag{7.112}$$

then

$$T_i^4 \approx T_o^4\left[1 + 4\frac{\delta T_i}{T_o}\right] \tag{7.113}$$

and

$$\begin{aligned} J_{1o} &= 4\pi r_1^2 \sigma T_o^4 - \frac{\pi r_1^2 r_2^2}{R^2}\sigma T_o^4 \\ J_{2o} &= -\frac{\pi r_1^2 r_2^2}{R^2}\sigma T_o^4 + 4\pi r_2^2 \sigma T_o^4 \\ \delta J_1 &= 16\pi r_1^2 \sigma T_o^4\left(\frac{\delta T_1}{T_o}\right) - \frac{4\pi r_1^2 r_2^2}{R^2}\sigma T_o^4\left(\frac{\delta T_2}{T_o}\right) \\ \delta J_2 &= -\frac{4\pi r_1^2 r_2^2}{R^2}\sigma T_o^4\left(\frac{\delta T_1}{T_o}\right) + 16\pi r_2^2 \sigma T_o^4\left(\frac{\delta T_2}{T_o}\right). \end{aligned} \tag{7.114}$$

If we identify radiant energy as the conserved stuff and $\ln\{T_i/T_o\}$ as the thermodynamic potential (Gibbs free energy per unit of radiant energy) of the black body at temperature T_i, then we can use eqns (7.68) and (7.69) to obtain

$$F_1 = \ln\left[\frac{T_1}{T_o}\right] \approx \frac{\delta T_1}{T_o}, \quad F_2 = \ln\left[\frac{T_2}{T_o}\right] \approx \frac{\delta T_2}{T_o} \tag{7.115}$$

from which we can identify the linear two-port y parameters for the radiation path:

$$\delta J_1 \approx y_{11}F_1 + y_{12}F_2$$
$$\delta J_2 \approx y_{21}F_1 + y_{22}F_2$$
$$y_{11} = 16\pi r_1^2 \sigma T_o^4, \quad y_{22} = 16\pi r_2^2 \sigma T_o^4 \qquad (7.116)$$
$$y_{12} = y_{21} = -4\pi\sigma T_o^4 \left[\frac{r_1 r_2}{R}\right]^2.$$

The equality of y_{12} and y_{21} is called *reciprocity.* According to the Onsager reciprocity theorem, all processes that display microscopic reversibility—including the transfer of radiant energy between black bodies—will also exhibit reciprocity (Onsager 1931).

7.5.7 Distributed-parameter models

One-dimensional flow paths

Often it is simplest to model a flow path as though it comprised an infinite number of infinitesimal segments, each with its own incremental resistance, ΔR, or inertia, ΔI. If, distributed along its length, the flow path has the capability to store the conserved stuff, then each infinitesimal segment also may include an incremental storage element, ΔC. In fact, each segment may be a complicated network model comprising several incremental elements and representing complicated dynamics that are distributed along the path. With an infinite number of infinitesimal segments, such a construction becomes a *distributed-parameter model.*

Resistive path segment for particle flow

Consider massless ions of a given species flowing through a resistive path segment of length ΔL, with no chemical potential difference. Let the measure of conserved stuff be the number of ions given in moles. In that case,

$$\Delta F(\text{electrical}) = z\boldsymbol{F}(\Delta V) \qquad (7.117)$$

and

$$\Delta F(\text{diffusional}) = \boldsymbol{R}T \ln[(c + \Delta c)/c], \qquad (7.118)$$

where ΔF is the thermodynamic potential difference from one end of the segment to the other, ΔV is the voltage difference from one end of the segment to the other, Δc is the ion concentration difference from one end of the segment to the other, z is the ionic valence, and $\boldsymbol{F}$ is the Faraday. If c is not zero and the concentration gradient is not infinite, then we always can choose a segment length that is sufficiently short to make $\Delta c/c \ll 1.0$, in which case

$$\Delta F(\text{diffusional}) = \boldsymbol{R}T\ \Delta c/c \qquad (7.119)$$

and

$$\Delta F(\text{total}) = z\boldsymbol{F}\ \Delta V + \boldsymbol{R}T\ \Delta c/c. \qquad (7.120)$$

The flow through the segment should be given by the following equation:

$$J(\text{total}) = J(\text{electrical}) + J(\text{diffusional})$$
$$= \Delta F(\text{total})/\Delta R, \qquad (7.121)$$

where ΔR is a general resistance to particle flow. ΔR should be independent of the potentials driving that flow (that is, independent of whether the flow is a consequence of concentration gradients or a consequence of voltage gradients, or both). In principle, the same resistance should apply to flow resulting from gravity as well, if the particle is not massless. According to eqn (7.91), which was derived from Fick's law for a uniform segment of length ΔL and cross-sectional area A,

$$\Delta R = \boldsymbol{R}T\ \Delta L/DcA. \qquad (7.122)$$

Combining eqns (7.120), (7.121), and (7.122), one obtains

$$J(\text{total}) = z\boldsymbol{F}\ \Delta V\ DcA/\boldsymbol{R}T\Delta L + DA\ \Delta c/\Delta L. \quad (7.123)$$

The first term on the right is a statement of Ohm's law, which often is written

$$J(\text{electrical}) = z\boldsymbol{F}\mu cA\ \Delta V/N_A\Delta L, \qquad (7.124a)$$

$$\mu = N_A D/\boldsymbol{R}T = D/k_b T, \qquad (7.124b)$$

where μ is the mechanical mobility of an individual particle (for example, a sodium ion) in the medium in which it is moving (for example, water), D is the diffusivity of the same solute particle in the same medium; $\boldsymbol{R}$ is the molar gas constant; N_A is the Avogadro constant, T is absolute temperature, and k_b is the Boltzmann constant (the molar gas constant divided by the Avogadro constant). Stokes's law predicts that μ will be inversely proportional to the dynamic viscosity (η) of the medium; for particles dissolved in water, that prediction is consistent with the observation that the temperature dependence of μ is the same as the temperature dependence of $1/\eta$. Equation (7.124b) is called the Einstein relation.

In deriving eqn (7.123), we used the conventions of lumped-element circuit theory—including the use of associated reference directions for J and F. Therefore, the reference directions for J and ΔV and for J and Δc also are associated. If we consider ΔL to be a measure of distance along a translational coordinate, x, and if the reference direction for J is taken to be in the direction of increasing positive values of x, then according to standard convention,

$$\Delta c = -\frac{\mathrm{d}c}{\mathrm{d}x}\Delta L$$
$$\Delta V = -\frac{\mathrm{d}V}{\mathrm{d}x}\Delta L \quad (7.125)$$

and, as ΔL becomes infinitesimal, the combination of Ohm's law and Fick's law becomes

$$J(x) = -[zFDc(x)A/RT]\frac{\mathrm{d}V}{\mathrm{d}x} - DA\frac{\mathrm{d}c}{\mathrm{d}x}, \quad (7.126)$$

Dividing both sides of eqn (7.126) by A, one obtains the *Nernst–Planck equation* for *flow density*, $j(x)$,

$$j(x) = \frac{J(x)}{A} = -[zFDc(x)/RT]\frac{\mathrm{d}V}{\mathrm{d}x} - D\frac{\mathrm{d}c}{\mathrm{d}x}. \quad (7.127)$$

Equation (7.127) relates the particle flow density (for example, in mol s^{-1} m^{-2}) to the gradients of voltage and concentration. In it, the total flow density is decomposed into two components: a *drift flow* owing to the electrical voltage gradient, and a *diffusion flow* owing to the concentration gradient. The negative sign on the drift-flow term means that positively charged particles (z a positive integer) drift *down* the voltage gradient; negatively charged particles drift in the opposite direction. Both types of particles diffuse *down* the concentration gradient.

The uniform, linear, one-dimensional path

Equation (7.127) is one example of a general class of equations of the form

$$\Delta F = -z(x, s)\Delta x\, J(x, s), \quad (7.128)$$

where ΔF is the potential difference across an incremental segment of a path, Δx is the length of the segment, $J(x)$ is the flow through the path, and $z(x, s)\Delta x$ is the impedance to that flow. Thus $z(x, s)$ is the series impedance per unit length of path in the immediate vicinity of point x. The impedance is given in the form of a Laplace transform to indicate that it may include dynamic operations. By convention for distributed-parameter models (eqn 7.125), the reference directions for ΔF and J are opposite to the associated reference directions.

As the flow passes through the segment, it may be decreased (through leakage or local storage) by an amount $-\Delta J$. This process is represented as a shunt admittance, $y(x, s)\Delta x$, with one end connected to the path at point x and the other connected to the reference node:

$$\Delta J = -y(x, s)\Delta x\, F(x), \quad (7.129)$$

where $F(x, s)$ is the potential in the immediate vicinity of point x in the path, relative to the reference state, and the reference directions for ΔJ and F are opposite to the associated reference directions. The function $y(x, s)$ is the shunt admittance per unit length of path in the immediate vicinity of point x. Dividing both sides of eqns (7.128) and (7.129) by Δx and taking the limits of both of equations as Δx approaches zero, one obtains the following equations:

$$\frac{\mathrm{d}F(x, s)}{\mathrm{d}x} = -z(x, s)J(x, s), \quad (7.130)$$

$$\frac{\mathrm{d}J(x, s)}{\mathrm{d}x} = -y(x, s)F(x, s), \quad (7.131)$$

where x is distance along the path being modeled.

If z and y are independent of x, the path is *uniform*. Uniform, linear paths are approximated by many situations of potential interest to bioengineers. When z and y are dependent on x, but exhibit gradual rather than sudden changes, the path dynamics often are estimated by solving the equations for the uniform case, then incorporating the x-dependence in the solutions. This approach is part of a process known as the WKB approximation. For more precise solutions in the presence of x-dependence of z and y, finite-element analysis with a digital computer is the choice.

When z and y are independent of x, the two equations can be combined to yield

$$\frac{\mathrm{d}^2F(x, s)}{\mathrm{d}x} = -z(s)\frac{\mathrm{d}J(x, s)}{\mathrm{d}x} = z(s)y(s)F(x, s)$$
$$\frac{\mathrm{d}^2J(x, s)}{\mathrm{d}x} = -y(s)\frac{\mathrm{d}F(x, s)}{\mathrm{d}x} = z(s)y(s)J(x, s). \quad (7.132)$$

The solutions to these equations usually are presented in one of two standard forms

$$F(x, s) = F_f(x, s) + F_r(x, s)$$
$$F_f(x, s) = F_f(0, s)e^{-\sqrt{zy}x}, \quad F_r(x, s) = F_r(0, s)e^{\sqrt{zy}x}$$
$$J(x, s) = J_f(x, s) - J_r(x, s)$$
$$J_f(x, s) = J_f(0, s)e^{-\sqrt{zy}x}, \quad J_r(x, s) = J_r(0, s)e^{\sqrt{zy}x}, \quad (7.133)$$

where

$$F_f(x,s)=\sqrt{\frac{z}{y}}J_f(x,s),\quad F_r(x,s)=\sqrt{\frac{z}{y}}J_r(x,s),\quad (7.134)$$

or

$$F(x,s)=F(0,s)\cosh(\sqrt{zy}\,x)-\sqrt{\frac{z}{y}}J(0,s)\sinh(\sqrt{zy}\,x)$$

$$J(x,s)=J(0,s)\cosh(\sqrt{zy}\,x)-\sqrt{\frac{y}{z}}F(0,s)\sinh(\sqrt{zy}\,x),\quad (7.135)$$

where

$$\cosh q=\frac{e^{q}+e^{-q}}{2}$$
$$\sinh q=\frac{e^{q}-e^{-q}}{2}.\quad (7.136)$$

In the first set of equations, F_f and J_f normally are interpreted as being the potential and flow components of a wave traveling in the *forward* direction (positive x-direction) along the path being modeled; F_r and J_r are interpreted as the components of a wave traveling in the *reverse* direction. The product $F_f(x, t)J_f(x, t)$ is the rate at which energy flow past the point x in the forward direction at time t; $F_r(x, t)J_r(x, t)$ is the rate at which energy flows past point x at time t in the reverse direction. The term

$$\sqrt{\frac{z}{y}}=\sqrt{\frac{z(s)}{y(s)}}\quad (7.137)$$

is the *characteristic impedance* of the path, often depicted as Z_o; for example,

$$F_f(x,s)=Z_oJ_f(x,s),$$
$$F_r(x,s)=Z_oJ_r(x,s).\quad (7.138)$$

A uniform, linear one-dimensional path of length L can be modeled as a two-port element, the terminals of port 1 being the input end of the path and the reference node, and the terminals of port 2 being the other end of the path and the reference node. The two-port element has the following z-parameters:

$$z_{11}=z_{22}=Z_o\frac{\cosh(\sqrt{zy}\,L)}{\sinh(\sqrt{zy}\,L)}$$
$$z_{12}=z_{21}=Z_o\frac{1}{\sinh(\sqrt{zy}\,L)}.\quad (7.139)$$

When an impedance, Z_L, is connected across port 2, the driving-point impedance at port 1 is

$$Z_{DP}=Z_o\frac{Z_L\cosh(\sqrt{zy}\,L)+Z_o\sinh(\sqrt{zy}\,L)}{Z_o\cosh(\sqrt{zy}\,L)+Z_L\sinh(\sqrt{zy}\,L)}.\quad (7.140)$$

See eqn (7.179) below.

When a wave is propagated along the path from a source at port 1 port toward the impedance Z_L, the ratio of the forward wave to the reflected wave in the path immediately adjacent to port 2 can be deduced as follows:

$$J_L=J_f(L)-J_r(L),\quad F_L=Z_LJ_L=F_f(L)+F_r(L)$$
$$F_f(L)=Z_oJ_f(L),\quad F_r(L)=Z_oJ_r(L)\quad (7.141)$$
$$\frac{F_r}{F_f}=\frac{Z_L-Z_o}{Z_L+Z_o},\quad \frac{F_L}{F_f}=\frac{2Z_L}{Z_L+Z_o},$$

where F_L is the potential across the impedance Z_L, and J_L is the flow through it. When Z_L is equal to Z_o, there is no reflected wave; all of the energy of the forward wave is transferred to Z_L.

One-dimensional, uniform diffusion path

For diffusion of particles or heat along a uniform path,

$$z=r,\quad y=sc,\quad (7.142)$$

where r is the diffusional resistance per unit length of path and c is the local storage capacity per unit length of path. Thus

$$\sqrt{zy}=\sqrt{rcs},\quad (7.143)$$

$$Z_o=\sqrt{\frac{r}{sc}}.\quad (7.144)$$

The key Laplace-transform pairs in this case are presented in Table 7.4. Other useful pairs can be derived

Table 7.4 Laplace transforms for diffusion processes

$F(s)=\frac{1}{\sqrt{s}}$	$f(t)=\frac{1}{\sqrt{\pi t}}$
$F(s)=e^{-\sqrt{s}}$	$f(t)=\frac{e^{-\frac{1}{4t}}}{2\sqrt{\pi t^3}}$
$F(s)=\frac{e^{-\sqrt{s}}}{s}$	$f(t)=\mathrm{erfc}\left(\frac{1}{2\sqrt{t}}\right)$
$F(s)=\sqrt{s}\,e^{-\sqrt{s}}$	$f(t)=(1-2t)\frac{e^{-\frac{1}{4t}}}{4\sqrt{\pi t^5}}$

by manipulating these with the relationships in Table A.2, p. 492, and by applying the generating function

$$\frac{1}{1-x} = \sum_{n=0}^{\infty} x^n, \quad |x| < 1.0 \qquad (7.145)$$

to transform $1/\cosh x$ and $1/\sinh x$ into exponential series.

For example, the forward wave in a uniform, one-dimensional diffusion path is described by

$$F(x,s) = F(0,s)\mathrm{e}^{-\sqrt{rcx^2 s}}. \qquad (7.146)$$

Applying the relationship

$$F(as) = L\left[\frac{1}{a} f\left(\frac{t}{a}\right)\right], \qquad (7.147)$$

one concludes that the corresponding time function is

$$f(x,t) = \left[\sqrt{\frac{rcx^2}{r\pi t^3}}\,\mathrm{e}^{-\frac{rcx^2}{4t}}\right] * f(o,t) \qquad (7.148)$$

(see Section A.2.1, p. 490), which has an interesting kernel. Comparing this result with the conventional one-dimensional diffusion equation, one concludes that the diffusivity (D) and diffusion time (T_{D}) for the path of length x are

$$D = 1/rc, \quad T_{\mathrm{D}} = rcx^2. \qquad (7.149)$$

The complete description of $F(x,t)$ will take the relatively simple form of eqn (7.148) only if the path either is infinitely long or is terminated with an impedance equal to Z_{o}. In either case, F_{r} will be zero. Otherwise, $F(x,t)$ will comprise a sum of terms each of which is similar to the kernel in eqn (7.148). Inspection of such a solution will show that it can be interpreted as the sum of the effects of a single diffusional wave that has been reflected back and forth between the source (usually taken to be at $x = 0$) and the termination at $x = L$.

It is instructive to consider the phasor transform of the characteristic impedance of a uniform, one-dimensional diffusion path:

$$Z_{\mathrm{o}} = \sqrt{\frac{r}{\mathrm{j}\omega c}} = \sqrt{\frac{r}{2\omega c}} - \mathrm{j}\sqrt{\frac{r}{2\omega c}}. \qquad (7.150)$$

Notice that the amplitude declines as the reciprocal of the square root of ω, and the phase angle is constant at $-\pi/4$ (−45°). This sort of steady-state frequency response has been observed in the responses of some sensory cells (see Problem 13 in Chapter 6), and it is commonly seen in the electrical impedances of bare-metal electrodes in electrolytic solutions. Wherever one observes it, this behavior should suggest involvement of a diffusional path. In a bare-metal electrode, for example, the conventional explanation (Warburg model) for it is the presence of a concentration gradient in the solution immediately adjacent to the electrode surface, through which the charge must flow by diffusion. According to the Warburg model, the impedance of the electrode is dominated by the driving-point impedance of that diffusion path.

Leaky diffusion paths

The diffusion path of the previous paragraphs can be modified to include distributed leakage from the path to the ground state:

$$z = r, \quad y = g + sc, \qquad (7.151)$$

where g is the leakage conductance per unit length of the path. The resulting model often is used to represent charge flow along biological structures (such as neuronal dendrites) that are bounded by membranes. To create Laplace-transform pairs useful for this model, one can use the following relationship (from Table A.2) to modify the transform pairs used for diffusion without leakage:

$$F(s+\alpha) = L\left[\mathrm{e}^{-\alpha t} f(t)\right]. \qquad (7.152)$$

When the potential and flow along the path do not vary with time, the dynamic aspects of diffusion are eliminated and

$$F_{\mathrm{f}}(x) = F_{\mathrm{f}}(0)\mathrm{e}^{-x/\lambda}, \quad \lambda = 1/\sqrt{rg}. \qquad (7.153)$$

Thus, in such a model, the potential or flow associated with a time-invariant signal will diminish by a factor of 1/e for every increment λ added to the distance of the signal from the source. The parameter λ is the *length constant* of the path. This model is employed commonly in cellular neurophysiology and in leaky heat-flow paths.

Wave paths

For propagation of mechanical or electromagnetic waves,

$$z = si, \quad y = sc, \qquad (7.154)$$

where i is the series inertia per unit length of path and c is the shunt capacity or compliance per unit length of path.

Under these circumstances,

$$\begin{aligned} Z_o &= \sqrt{i/c} \\ \sqrt{zy} &= \sqrt{ic}\, s \\ F_f(x,s) &= F_f(0,s)e^{-\sqrt{ic}\,xs} \\ F_f(x,t) &= F_f(0,t-\tau),\ t>\tau \\ \tau &= \sqrt{ic}\, x. \end{aligned} \tag{7.155}$$

This represents a wave that does not change shape as it travels. Its velocity, v, is

$$v = 1/\sqrt{ic}. \tag{7.156}$$

The distributed series elements in this case are inertial, and the distributed shunt elements are capacitive. In the case of the forward wave, one may envision energy flow as a continual transfer from potential energy (accumulated in the local capacitive element) to kinetic energy (accumulated in the subsequent local inertia) and then back to potential energy (accumulated in the following local capacitive element). Thus the energy moves in the forward direction, from one element to the next. Different kinds (or modes) of waves typically are associated with different ways to accumulate potential energy locally—leading to different values of c. In liquids and gasses, for example, potential energy can be accumulated in local bulk compression, but not in local shear or torsion. Thus, both media are capable of propagating compressional waves. At the surface of a liquid, potential energy can be stored in vertical displacement in a gravity field—leading to another, familiar mode of wave propagation. Potential energy also can be stored in small displacements against surface tension—leading to still another mode of wave propagation. Solids are capable of storing potential energy locally in bulk compression, shear, or torsion—leading to three different modes of propagation. For longitudinal compressional waves in fluids, with negligible changes in density, and with volume taken to be the conserved stuff,

$$\begin{aligned} i &= \rho/A, \\ c &= A/E_B^*, \end{aligned} \tag{7.157}$$

where A is the cross-sectional area of the wave path (normal to the direction of wave propagation), E_B^* is the adiabatic bulk modulus of the material in the path, and ρ is its density (e.g., see Section 3.2.2). For the analogous wave in solids, again with volume taken to be the conserved stuff,

$$\begin{aligned} c &= A/E_L, \\ E_L &= E_B + \frac{4}{3}E_S, \end{aligned} \tag{7.158}$$

where E_B is the bulk modulus and E_S is the shear modulus. E_L reflects the concomitance of shear with compression, as well as the fact that solids can sustain elastic shear stress (see Section 1.3.5).

Lumping revisited

For a uniform diffusion path of length L, the total resistance R and capacity C are

$$R = rL,\ C = cL, \tag{7.159}$$

and the diffusion time, T_D, is

$$T_D = rcL^2 = RC. \tag{7.160}$$

Following a sudden change in the flow into one end of the path or the potential at one end of the path, the transient responses within the path (corresponding to the redistribution of diffusing stuff) will continue to be relatively large for times of the order of a few to many diffusion times, depending on what is connected to the ends of the path. If one represents the path as a lumped resistive flow path or as a lumped capacitive storage element, the deductions from the resulting circuit model will not include the effects of the transient responses within the path. This is justified if those responses become negligible in times that are short in comparison to the temporal resolution desired in the deductions. Thus, in general, the diffusion time for a path should be very much shorter than the desired unit of temporal resolution in order to justify lumping the path. If that is not the case, then the path can be divided into segments, with each segment represented by a series lumped resistive element and a shunt lumped capacitive element. For modeling of driving-point properties (viewed at one end of the path), the lengths of the segments should be graded (e.g. $\Delta L, 2\Delta L, 4\Delta L, \ldots, 2^{N-1}\Delta L$, with the shortest segment at the end being viewed). In that case, the temporal resolution of the model will improve approximately as 2^{2N}, where N is the number of segments. For modeling of transfer properties, segments of equal lengths should be used. In that case, temporal resolution of the model will improve as the square of the number (N) of segments (i.e., T_D for each segment will decrease as $1/N^2$). For a radial diffusion in a spherical volume (often assumed in simplified models of mass transport in cells), the segments can be as the layers of an onion—leading to the label *onion-skin model.*

For a uniform wave path of length L, the propagation time (T_w) for the wave is

$$T_w = \frac{L}{v} = \sqrt{ic}\,L = \sqrt{IC}. \tag{7.161}$$

As long as T_w is much shorter than the desired unit of temporal resolution in the deductions from the model, one would be justified in representing the path as a lumped inertial flow path or as a lumped capacitive storage element. If T_w is not sufficiently short, the path may be divided into segments of equal length, and each segment can be modeled as a series inertia and shunt capacity. In that case, the temporal resolution of the model will improve in direct proportion to the number (N) of segments (rather than as N^2). Alternatively, one may incorporate the distributed-parameter model of the path into the circuit model.

Flow paths as random time delays

Consider a realm in which the conserved stuff comprises a population of identical particles (e.g. molecules) that move independently of one another. In that realm, imagine a path through which the particles can flow without being lost, so that all of the particles entering one end of the path eventually exit at the other end. The relationship between the flow (J_1) into the entrance and that (J_2) out of the exit will be

$$\begin{aligned} J_2(t) &= h_{12}(t) * J_1(t) \\ J_2(s) &= H_{12}(s)J_1(s). \end{aligned} \tag{7.162}$$

If $J_1(t)$ is an impulse (for example, one mole of particles delivered to the entrance at t), then $h_{12}(\tau)\delta\tau$ is the proportion of those particles expected to emerge from the exit during an increment of time $\delta\tau$ long, centered about time $t+\tau$. Because the particles move independently, $h_{12}(\tau)\delta\tau$ also must be the probability that a given particle will emerge during $\delta\tau$. Therefore, $h_{12}(\tau)$ is the *probability density function* for the travel time (τ) of the individual particle through the path, and τ itself is a random variable. Because the particle is certain to emerge eventually,

$$\left.\int_{\tau=0}^{T} h_{12}(\tau)\,d\tau\right|_{T\to\infty} = 1.0. \tag{7.163}$$

One can translate this to a condition on $H_{12}(s)$ as follows. Let

$$f(t) = \int_{\tau=0}^{t} h_{12}(\tau)\,d\tau \tag{7.164}$$

and

$$L[h_{12}(t)] = H_{12}(s). \tag{7.165}$$

It follows that

$$L[f(t)] = L\left[\int_{\tau=0}^{t} h_{12}(\tau)\,d\tau\right] = H_{12}(s)/s. \tag{7.166}$$

The desired function is

$$\lim f(t)_{t\to\infty}. \tag{7.167}$$

Applying the final-value theorem (eqn 6.21), one obtains

$$\lim f(t)|_{t\to\infty} = \lim(s\,L[f(t)])|_{s\to 0} = s\frac{H(s)}{s}|_{s\to 0}, \tag{7.168}$$

from which,

$$H_{12}(0) = 1.0. \tag{7.169}$$

In the theory of probability, $H_{12}(s)$ is known as a *generating function*; from it one can obtain the various moments of the random variable τ. The definition of the rth moment (M_r) is

$$M_r = \int_{\tau=0}^{\infty} \tau^r h_{12}(\tau)\,d\tau. \tag{7.170}$$

Noting in Table A.2 (see Appendix, p. 492) that multiplication by τ transforms to the operation $-d/ds$, one can demonstrate rather easily that

$$M_r = \left[\left(-\frac{d}{ds}\right)^r H_{12}(s)\right]_{s\to 0}. \tag{7.171}$$

If τ is a discrete variable, then the appropriate transform of h_{12} is $H_{12}(z)$. In that case,

$$M_r = \left[\left(-z\frac{d}{dz}\right)^r H_{12}(z)\right]_{z\to 1} \tau_u, \tag{7.172}$$

where τ_u is the discrete time increment (unit of discrete time). The certainty of ultimate emergence of the particle requires that $M_0 = 1$. One can think of the other members of the set $\{M_r\}$ as representing the shape of the graph of $h_{12}(\tau)$. Usually, the members of $\{M_r\}$ are combined to form normalized descriptors of shape, such as the coefficient of variation and the coefficient of skewness:

coefficient of variation:

$$\frac{\sqrt{M_2 - M_1^2}}{M_1}; \qquad (7.173)$$

coefficient of skewness:

$$\frac{M_3 - 3M_1(M_2 - M_1^2) - M_1^3}{\left(\sqrt{M_2 - M_1^2}\right)^3}. \qquad (7.174)$$

M_1 is the expected value of τ, and $M_2 - M_1^2$ is its variance. If $M_0 < 1.0$, implying that the particle has a finite probability of disappearing before it can emerge at the exit, one may create a new probability density function

$$h_{12}(\tau') = \frac{h_{12}(\tau)}{M_0} \qquad (7.175)$$

that corresponds to the random variable τ' for those particles that do emerge eventually. The normalized descriptors of shape may be applied to τ'.

After estimating $h_{12}(\tau)$ experimentally (for example, by repeated observations of passages of single particles through the path), one can infer those classes of network models that can yield the same transfer or driving-point relationship (the exit and entrance could be one and the same) and those classes that cannot (Section 7.5.8). By examining those classes, one can infer in turn some of the attributes of the physical processes in the path. This approach is commonly taken in studies of channel kinetics in cell membranes. One also could consider the effects of random time delays on the dynamics of the systems of which they are part. This has been done in population modeling, where the effects of random time to sexual maturity, random gestation periods, and the like, have been estimated through modeling studies. Shape descriptors have found especially widespread use in cellular neurobiology, where they have been used to infer aspects of dendritic and synaptic organization.

7.5.8 Attributes of passive, linear, time-invariant circuit models

A passive, linear circuit model with elements whose parameters are fixed (time-invariant) is known as a passive LTI circuit model. In addition to the properties of additivity and homogeneity, which combine to yield linearity, a passive LTI circuit model exhibits general properties, some of which depend on the kinds and numbers of elements it contains as well as the way in which those elements are connected, and some of which are independent of the elements. These properties should be familiar to any engineer attempting to use such models. Some of them are presented below, without proofs.

Dynamic order

The dynamic order of a circuit model is the number of *independent* energy storage elements (capacitive elements and inertial elements) that it contains. Two capacitive elements (C_i and C_j) are independent if their potentials (F_i and F_j) can vary independently. Two inertial elements (I_{ij} and I_{jk}) are independent if their flows (J_{ij} and J_{jk}) can vary independently. In the amplitude Bode diagram of the phasor transform of any transfer function (i.e. log–log plot of the amplitude of $H_{sr}(\omega)$ versus ω; see Example 1 in Section 6.4) of a given circuit, the difference between the asymptotic slope at low frequencies and that at high frequencies is less than or equal to the dynamic order of the circuit. The total variation in the phase shift of the transfer function, divided by 1/4 cycle ($\pi/2$ radians) as ω varies from zero to infinity is less than or equal to the dynamic order of the circuit. By making sinusoidal steady-state measurements of a physiological system and displaying the data in Bode diagrams, investigators can estimate the dynamic order required in circuit models of the system. This has been done, for example, for circuit models in vision research and hearing research.

Natural frequencies

The number of natural frequencies exhibited by a circuit model is equal to the dynamic order. The nature of those frequencies depends on the kinds of elements used in the model. The passive nature of the model requires that all natural frequencies will have nonpositive real parts. Any circuit model that conforms to the second law of thermodynamics will have only natural frequencies with negative real parts (nonzero as well as nonpositive real parts). In models comprising only resistive elements and capacitive elements, or only resistive elements and inertial elements, the natural frequencies all will be negative real numbers (representing decaying exponential waveforms), and no two of those numbers will be the same. In models comprising inertial elements and capacitive elements, but no resistive elements (and therefore not conforming to the second law of thermodynamics), the natural frequencies all will be purely imaginary numbers (in complex conjugate pairs), and no two of those numbers will be the same. In networks comprising all three types of elements, the natural frequencies will be negative real numbers, complex numbers (in conjugate pairs) with the real parts being negative, or purely imaginary numbers; no two purely imaginary natural frequencies will be the same, but complex and negative real natural frequencies can be repeated (two or more natural frequencies may be the same).

Zeros

When a driving-point or transfer impedance or admittance is given as the ratio of two polynomials in s (the independent variable of the Laplace transform), the zeros of the numerator polynomial are denoted as the *zeros* of that impedance or admittance, and the zeros of the denominator polynomial are denoted as its *poles* (see Section 6.3.2 and Example 2 in Section 6.4). If a flow source is connected between any two existing nodes in a circuit model (i.e. not inserted into a branch, thereby creating a new node), then the poles of the corresponding driving-point and transfer impedances are natural frequencies of the intact circuit model and must conform to the properties described in the previous paragraph. The poles of the corresponding driving-point admittance (the reciprocal of the driving-point impedance) are natural frequencies of a modified version of the circuit (that is, that formed by replacing the flow source by a potential source). If a potential source is inserted into any branch (i.e. not simply connected from one existing node to another), then the poles of the corresponding driving-point admittance also are natural frequencies of the intact circuit. The poles of the corresponding driving-point impedance (the reciprocal of the driving-point admittance) are natural frequencies of another modified version of the circuit model (i.e. the circuit formed when the potential source at the driving point is replaced by a flow source). It follows that the poles and zeros of all driving-point impedances and admittances of passive LTI circuits are natural frequencies of a passive LTI circuit. Therefore, all such poles and zeros must conform to the properties described in the previous paragraph. In driving-point impedances and admittances of passive LTI circuit models comprising only two of the three kinds of two-terminal elements, poles and zeros alternate in order of increasing magnitude.

In circuits comprising resistive elements and capacitive elements or inertial elements or both, the zeros of transfer functions will be positive or negative real numbers or complex numbers with positive, negative, or zero real parts. Complex zeros will occur in conjugate pairs. A *minimum-phase function* is one in which all of the zeros have nonpositive real parts (see Example 3 in Section 6.4). Driving-point impedances and admittances of passive LTI circuit models are minimum-phase functions; transfer impedances, admittances, and ratios need not be.

Reciprocity

Any passive LTI circuit model representing a single physical realm can be treated as a two-port element, with any pair of nodes serving as port 1, and any pair serving as port 2. Regardless of which nodes are selected for the ports, the two-port element will exhibit reciprocity:

$$z_{12} = z_{21}, \quad y_{12} = y_{21}. \tag{7.176}$$

A proof of this statement can be found in standard circuit-theory texts, where it is based on Tellegen's theorem. According to the Onsager reciprocity theorem (Onsager 1931), any passive LTI two-port element representing transduction between two physical realms will exhibit reciprocity if the transduction process does not involve rotation. When a transducer involving rotating mass or charge is modeled as a passive LTI two-port element, on the other hand, that element will exhibit antireciprocity ($y_{12} = -y_{21}$).

7.6 Elementary transducer theory

7.6.1 General considerations

Transduction is the process of passing free energy from one physical realm to another. Engineers design and use transducers for three purposes:

(1) to convert free energy from one physical realm into a generally available (nonsignal) form in another physical realm (for example, conversion of optical

energy into electrical energy that is stored in a battery);

(2) to convert signal energy into purposeful action;

(3) to convert signal energy into a form in which the information it contains can be processed easily.

Transducers used for purpose (1) sometimes are called *generators*, those used for purpose (2) are *actuators*, and those used for purpose (3) are *sensors*. In this section, we focus on actuators and sensors. Sensor and actuators, of course, were essential elements in the feedback control systems of Chapter 6.

As implied in Section 7.5.5 under 'Two-port and multiport elements', actuators and sensors usually can be modeled as two-port elements, with one port connected to each of two physical realms. Many sensors and actuators transfer signal energy bidirectionally between physical realms, operating equally well in both directions, without converting any nonsignal energy to signal energy. When bidirectional transduction of this sort is carried out in steps that are small enough to make the process linear (and reversible in the thermodynamic sense), then the process also is either reciprocal or antireciprocal (see Section 7.5.8 under 'Reciprocity', and Onsager 1931). In that case, the sensor or actuator can be represented as a passive linear two-port element. When a sensor or actuator uses signal energy in one physical realm to control a conversion of nonsignal energy to signal energy, with some or all of the latter being in a second physical realm, then the transduction process is not reciprocal or antireciprocal and does not operate equally well in both directions (in fact it typically does not operate at all in the reverse direction). In that case, the sensor or actuator must be represented as an active two-port element.

When one is designing systems involving transducers, a consideration that often arises, especially with actuators, is how to maximize the power transferred through the device. This is accomplished by impedance matching, which in turn often is accomplished through use of another kind of device, known as a *transformer*. Transformers transfer signal energy from one place to another within a single physical realm, but do not convert nonsignal energy to signal energy. A transformer can be modeled as a passive two-port element, but with both ports connected to the same physical realm. In many cases, transformers comprise two bidirectional transducers connected back-to-back. Linear transformers of this sort can be modeled as two passive linear (reciprocal or antireciprocal) two-port elements connected in cascade.

In modern instrumentation, most signal processing is done in the electrical realm; and for that reason, many sensors and actuators are constructed to operate between that realm and some other realm. In many situations, however, the coupling between the two realms is accomplished in two or more steps, often involving other realms and intermediate transduction. A capacity microphone, for example, couples the electrical realm and the pneumatic realm; but an intermediate step is coupling between the pneumatic realm (pressure and volume velocity) and the realm of translational rigid-body mechanics (force and translational velocity)—the intermediate transducer being a piston or elastic diaphragm. Coupling to and from the electric realm ultimately must involve coupling between a conjugate pair of potential and flow variables (electric potential and charge flow) in that realm and a conjugate pair of potential and flow variables in another realm. There is a large class of passive transducers in which coupling is nearly direct and involves some of the more interesting empirical laws of classical physics—such as Coulomb's law, which relates charge displacement to force, the Seebeck–Peltier effect, which relates charge flow to heat flow; the Henry–Faraday law, which relates charge flow to translational motion in a magnetic field; and Fick's Law, which relates charge flow to ion concentration differences. There also is a large class of active transducers, including many strain gauges, based on modulation of the constitutive relationship of an electrical two-terminal element (resistor, inductor, or capacitor) or an electrical two-port element (e.g. an electric transformer).

Prior to the emergence of the modern electric era, sensors (e.g. barometers, thermometers, voltmeters, tachometers, psychrometers, etc.) often involved coupling between the translational-motion realm or the rotational-motion realm and other realms; and the sensed information was carried in the displacement of a meter or of a column of liquid—where it could be captured by visual observation. Such devices, of course, are used widely today. In automated instrumentation systems, however, the human visual system is bypassed and the signals are taken to the electrical realm. In some cases, this is done by adding one more transducer, such as a strain gauge, to the classical device.

While bioengineers surely must be interested in sensors and actuators employed in instrumentation systems, they also must be concerned with the transducers and transformers of physiological or biophysical systems, such as those described in Chapter 6 and those (related to auditory signals) described briefly in Section 7.5.5 under 'Signal energy'. Such devices will not necessarily involve the electrical realm.

7.6.2 Two-port parameters and equations

Basic definitions

The basic notion of a linear two-port element was introduced in Section 7.5.6, as were the z and y parameter sets. Each port serves as a two-terminal element, with a potential across it and a flow through it. The criteria for lumping are the same as those described previously for two-terminal elements. The constitutive relationships involve both pairs of conjugate potential and flow simultaneously. Thus, the potential across one port is related not only to its conjugate flow, but also to the potential or flow at the other port. Two-port elements conventionally are represented by parameter sets derived by taking two of the port variables to be *independent* and the other two to be *dependent*. The following equations define all six sets of linear two-port parameters that can be generated this way:

$$\begin{bmatrix}F_1\\F_2\end{bmatrix}=\begin{bmatrix}z_{11}&z_{12}\\z_{21}&z_{22}\end{bmatrix}\cdot\begin{bmatrix}J_1\\J_2\end{bmatrix}\qquad\begin{bmatrix}J_1\\J_2\end{bmatrix}=\begin{bmatrix}y_{11}&y_{12}\\y_{21}&y_{22}\end{bmatrix}\cdot\begin{bmatrix}F_1\\F_2\end{bmatrix}$$
$$\begin{bmatrix}F_1\\J_2\end{bmatrix}=\begin{bmatrix}h_{11}&h_{12}\\h_{21}&h_{22}\end{bmatrix}\cdot\begin{bmatrix}J_1\\F_2\end{bmatrix}\qquad\begin{bmatrix}J_1\\F_2\end{bmatrix}=\begin{bmatrix}g_{11}&g_{12}\\g_{21}&g_{22}\end{bmatrix}\cdot\begin{bmatrix}F_1\\J_2\end{bmatrix}$$
$$\begin{bmatrix}F_1\\J_1\end{bmatrix}=\begin{bmatrix}t_{11}&t_{12}\\t_{21}&t_{22}\end{bmatrix}\cdot\begin{bmatrix}F_1\\-J_2\end{bmatrix}\quad\begin{bmatrix}F_2\\-J_2\end{bmatrix}=\begin{bmatrix}\bar{t}_{11}&\bar{t}_{12}\\\bar{t}_{21}&\bar{t}_{22}\end{bmatrix}\cdot\begin{bmatrix}F_1\\J_1\end{bmatrix}. \tag{7.177}$$

Each parameter set is represented as a matrix that maps a vector comprising two *independent variables* into a vector comprising two *dependent variables*. Each parameter is an admittance, an impedance, or a transfer ratio. As such, it need not simply be a real constant. Thus, z_{11} is the driving-point impedance at port 1 when no flow is allowed to occur at port 2, and z_{12} is the transfer impedance relating the potential at port 1 to the flow into port 2 when no flow is allowed to occur at port 1. The following relationships demonstrate the approach that can be used to estimate each parameter from a physical model of a device or to measure the parameter directly in the actual device:

$$F_1 = h_{11}J_1 + h_{12}F_2$$
$$h_{11} = \left.\frac{F_1}{J_1}\right|_{F_2=0} \tag{7.178}$$
$$h_{12} = \left.\frac{F_1}{F_2}\right|_{J_1=0}.$$

Input and output impedances

In spite of the explicit representation of causal direction, unlike the conventional black-box system element (Section 6.2), in which input and output are completely isolated and the transfer relationship is taken to be independent of context, the two-port model embodies context dependence. This is its great advantage; it allows the modeler to incorporate context dependence (for example, the impacts of source impedance and load impedance) in design and analysis. For example, when a two-terminal load impedance, Z_L, is connected across port 2, the driving-point impedance at port 1 is easily found to be

$$Z_{DP} = z_{11} - \frac{z_{12}z_{21}}{z_{22}+Z_L} = h_{11} - \frac{h_{12}h_{21}}{h_{22}+1/Z_L}. \tag{7.179}$$

Impedance is an important consideration in the design of systems involving transducers, especially actuators—where the object often is to maximize the delivery of power. Working with sinusoidal steady state and phasor transforms (Appendix, Section A.7), consider the configuration at the top of Fig. 7.25, in which a nonideal source (represented by an ideal potential source, with potential F_s, in series with impedance Z_s) is connected to a load with impedance Z_L, where

$$Z_s = R_s + jX_s, \quad Z_L = R_L + jX_L. \tag{7.180}$$

Averaged over a full cycle of the sinusoidal signal, the power (P_n) delivered to any branch n is given by

$$P_n = \tfrac{1}{2}\mathrm{Re}\{F_n\bar{J}_n\},$$
$$\text{where } \mathrm{Re}\{\bar{J}_n\} \triangleq \mathrm{Re}\{J_n\}, \quad \mathrm{Im}\{\bar{J}_n\} \triangleq -\mathrm{Im}\{J_n\}. \tag{7.181}$$

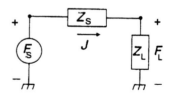

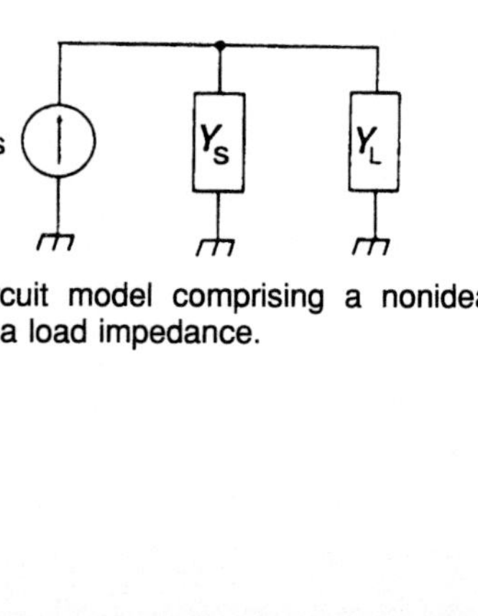

Fig. 7.25 Circuit model comprising a nonideal source connected to a load impedance.

Therefore, the power (P_{DL}) delivered to the load can be deduced as follows:

$$J = \frac{F_s}{Z_s + Z_L}, \quad F_L = Z_L J = \frac{Z_L}{Z_s + Z_L} F_s$$
$$P_{DL} = \tfrac{1}{2}\mathrm{Re}\{F_L \bar{J}\} = \frac{1}{2}\frac{|F_s|^2}{|Z_s + Z_L|^2} R_L. \quad (7.182)$$

If Z_s is determined by the nature of the nonideal source, then it is easy to show that P_{DL} is maximum when Z_L is adjusted to be

$$Z_L = \bar{Z}_s, \quad (7.183)$$

i.e.,

$$R_L = R_s, \quad X_L = -X_s. \quad (7.184)$$

By similar analysis, one can deduce that P_{DL} is maximum for the network at the bottom of Fig. 7.25 when

$$Y_L = \bar{Y}_s, \quad (7.185)$$

i.e. when

$$\frac{1}{R_L} = \frac{1}{R_s}, \quad \frac{1}{X_L} = -\frac{1}{X_s}. \quad (7.186)$$

Of greater concern in this section are the situations depicted in Fig. 7.26, in which the power from the nonideal source is delivered to the load through a transducer or a transformer. In order to maximize the power delivered to the load, one would like to maximize the power delivered to the transducer or transformer through port 1 and the power delivered to the load from port 2. Design constraints, usually based on the properties of available components, often limit this maximization process. For example, if one is given a nonideal source (and its properties) and a transducer (and its properties), but is allowed to adjust Z_L, then the design that maximizes the power delivered to the transducer may not be the one that maximizes the power delivered to the load.

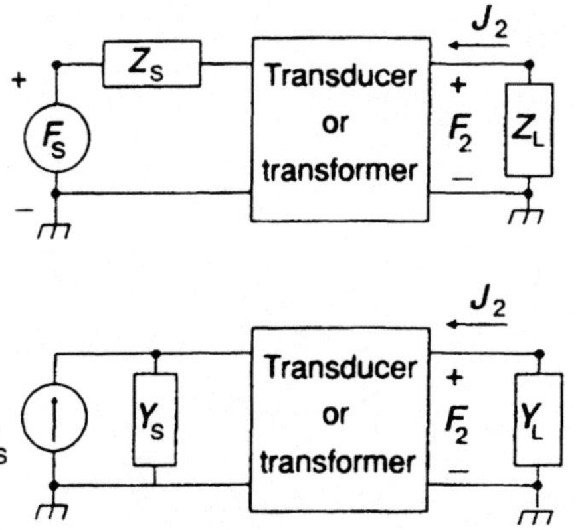

Fig. 7.26 Circuit model of Fig. 7.25 modified by insertion of a two-port element between the nonideal source and the load impedance.

A useful step in the design (or analysis) process is the construction of a Thévenin or Norton equivalent for the entire circuit to the left of port 2. The Thévenin equivalent circuit has a configuration identical to that for the nonideal source in Figs 7.25 (top) and 7.26 (top), with a potential source in series with an impedance; the Norton equivalent circuit has a configuration identical to that for the nonideal source in figs 7.25 (bottom) and 7.26 (bottom), with a flow source in parallel with a source admittance. The values of the elements in the Thévenin and Norton equivalent circuits for the combinations of two-port elements and nonideal sources in Fig. 7.26 are computed as follows:

$$F_T = F_2|_{J_2=0}, \quad J_N = -J_2|_{Z_L=0}$$
$$Z_T = \frac{1}{Y_N} = \frac{F_T}{J_N}, \quad (7.187)$$

where F_T is the source potential and Z_T the source impedance for the Thévenin equivalent circuit, and J_N is the source flow and Y_N the source admittance for the Norton equivalent circuit. The power delivered to the load will be maximum when

$$Z_L = \bar{Z}_T \quad (7.188)$$

or

$$Y_L = \bar{Y}_N. \quad (7.189)$$

To maximize the power delivered from the source to port 1 of the two-port in Fig. 7.26, one must adjust the source, the load, or the parameters of the two-port circuit so that

$$Z_{DP1} = \bar{Z}_s,$$
$$Y_{DP1} = \bar{Y}_s, \quad (7.190)$$

where Z_{DP1} and Y_{DP1} are the driving-point impedance and admittance, respectively, into port 1 when Z_L is connected across port 2. Note that for a given source

$$Z_{DP1} = 1/Y_{DP1},$$
$$Z_S = 1/Y_S. \quad (7.191)$$

The general equations for the driving-point impedance (or admittance) at port 1 (M_1) and the Thévenin impedance (or Norton admittance) at port 2 (M_2) are

$$M_1 = \frac{\det[k] + k_{11}M_L}{k_{22} + M_L} = k_{11} - \frac{k_{12}k_{21}}{k_{22} + M_L},$$
$$M_2 = \frac{\det[k] + k_{22}M_s}{k_{11} + M_s} = k_{22} - \frac{k_{12}k_{21}}{k_{11} + M_s}, \quad (7.192)$$
$$\det[k] = \det\begin{bmatrix} k_{11} & k_{12} \\ k_{21} & k_{22} \end{bmatrix} = k_{11}k_{22} - k_{12}k_{21},$$

where k_{ij} is z_{ij}, y_{ij}, h_{ij}, or g_{ij} (depending on the two-port parameter set being used), M_1 is the driving-point impedance or admittance looking into port 1 when impedance or admittance M_L is connected across port 2, and M_2 (the Thévenin impedance or Norton admittance) is the driving-point impedance or admittance looking into port 2 when impedance or admittance M_s is connected across port 1. M_1 and M_s are admittances when k_{11} is an admittance, and impedances when k_{11} is an impedance; M_2 and M_L are admittances when k_{22} is an admittance, and impedances when k_{22} is an impedance. Thus, when k_{11} is an impedance, the nonideal source is represented by its Thévenin equivalent circuit; when k_{11} is an admittance, the nonideal source is represented by its Norton equivalent circuit. The expressions for M_1 and M_2 are easily derived from the two-port equations and the constraints added by the presence of M_s and M_L.

Two-port amplification

Extending the general notation of the previous paragraph (for Fig. 7.26), let the independent source variable (J_s or F_s) be U_s, and let it be of the same type (flow or potential) as the dependent variable of port 1 in the two-port parameter set being used; let the dependent variable (J_2 or F_2) of port 2 be U_2; and let the source variable (J_N or F_T) of the Thévenin or Norton equivalent circuit looking into port 2 be U_{TN}, and let it be of the same type as U_2. The ratio of U_2 to U_s (with the load, M_L, connected) is easily calculated:

$$\frac{U_2}{U_s} = \frac{k_{21}M_L}{(k_{22} + M_L)(k_{11} + M_s) - k_{12}k_{21}}. \quad (7.193)$$

For the z parameter set, for example, U_s and U_2 are potentials, making this a ratio of potentials (possibly in different physical realms). For the y parameter set, it is a ratio of flows; for the h parameter set, it is the ratio of a flow to potential; and for the g parameter set, it is the ratio of a potential to a flow. Regardless of its dimensions, the ratio corresponds to a general sort of amplification provided by the two-port element in the context of Fig. 7.26.

U_{TN} can be found by taking the limit of the ratio as M_L goes to infinity and then multiplying by U_s:

$$U_{TN} = \left.\frac{U_2}{U_s}\right|_{M_L \to \infty} = \frac{k_{21}}{k_{11} + M_s} U_s. \quad (7.194)$$

Keep in mind that U_s is a potential when k_{11} is an impedance, and a flow when k_{11} is an admittance; U_2 and U_{TN} are potentials when k_{22} is an impedance, and flows when k_{22} is an admittance.

Power gains

The power delivered to the load (P_{DL}) from port 2 when both M_L and M_s are present can be computed as follows:

$$P_{DL} = \frac{1}{2}\mathrm{Re}\{J_L\bar{F}_L\} = \frac{1}{2}\mathrm{Re}\{M_L\}\frac{|U_{TN}|^2}{|M_2 + M_L|^2}. \quad (7.195)$$

The power delivered to port 1 (P_{D1}) when both M_L and M_s are present is

$$P_{D1} = \frac{1}{2}\mathrm{Re}\{J_1\bar{F}_1\} = \frac{1}{2}\mathrm{Re}\{M_1\}\frac{|U_s|^2}{|M_s + M_1|^2}. \quad (7.196)$$

Combining this result with eqns (7.190) and (7.191), one can show that the maximum power available from the source (P_{AVS}) is

$$P_{AVS} = \frac{1}{8}\frac{|U_s|^2}{\mathrm{Re}\{M_s\}}. \quad (7.197)$$

By similar argument, one can show that the maximum power available from port 2 (P_{AV2}) is

$$P_{AV2} = \frac{1}{8}\frac{|U_{TN}|^2}{\mathrm{Re}\{M_2\}}. \quad (7.198)$$

It is convenient to consider three measures of power (see Linvill and Gibbons 1961): the power gain, G, the transducer gain, G_T, and the available gain, G_A:

$$G = P_{DL}/P_{D1}$$
$$G_T = P_{DL}/P_{AVS} \quad (7.199)$$
$$G_A = P_{AV2}/P_{AVS}.$$

G_T provides a figure of merit for a transducer, giving the ratio of the power actually delivered by the transducer to the power that could be delivered by an ideal passive device with appropriate impedance matching at both ports. For a passive LTI model of a transducer or transformer, G must be less than or equal to 1.0, and G_T must be less than or equal to G. For a given source coupled through a given transducer, G_A gives the power gain that would be achieved if $R_L = \mathrm{Re}\{M_2\}$ and $X_L = -\mathrm{Im}\{M_2\}$.

Constraints on passive LTI two-port elements

When phasor notation is employed (for sinusoidal steady-state conditions), passivity requires that the real part of any impedance not be negative. In eqn (7.179), for Z_L with nonnegative real part, Z_{DP} also must have a nonnegative real part. Therefore, conditions that must be met by an LTI two-port element in order for it to be passive include

$$\mathrm{Re}\left\{k_{11} - \frac{k_{12}k_{21}}{k_{22}}\right\} \geq 0. \tag{7.200}$$

Passive LTI two-port elements also are either reciprocal or antireciprocal. The conditions for reciprocity in the various parameter sets are

$$\begin{aligned} z_{12} &= z_{21} \quad y_{12} = y_{21} \\ g_{12} &= -g_{21} \quad h_{12} = -h_{21} \\ \det[t] &= 1 \quad \det[t'] = 1. \end{aligned} \tag{7.201}$$

The conditions for antireciprocity are found by multiplying the right-hand side of each of these equations by -1.

Conversion table

$$y_{11} = \frac{z_{22}}{\det[z]} \quad y_{12} = \frac{-z_{12}}{\det[z]} \quad y_{21} = \frac{-z_{21}}{\det[z]} \quad y_{22} = \frac{z_{11}}{\det[z]}$$

$$h_{11} = \frac{\det[z]}{z_{22}} \quad h_{12} = \frac{z_{12}}{z_{22}} \quad h_{21} = \frac{-z_{21}}{z_{22}} \quad h_{22} = \frac{1}{z_{22}}$$

$$g_{11} = \frac{1}{z_{11}} \quad g_{12} = \frac{-z_{12}}{z_{11}} \quad g_{21} = \frac{z_{21}}{z_{11}} \quad g_{22} = \frac{\det[z]}{z_{11}}$$

$$t_{11} = \frac{z_{11}}{z_{21}} \quad t_{12} = \frac{\det[z]}{z_{21}} \quad t_{21} = \frac{1}{z_{21}} \quad t_{22} = \frac{z_{22}}{z_{21}}$$

$$\bar{t}_{11} = \frac{z_{22}}{z_{12}} \quad \bar{t}_{12} = \frac{-\det[z]}{z_{12}} \quad \bar{t}_{21} = \frac{-1}{z_{12}} \quad \bar{t}_{22} = \frac{z_{11}}{z_{12}}$$

$$t_{11} = \frac{-\det[h]}{h_{21}} \quad t_{12} = \frac{-h_{11}}{h_{21}} \quad t_{21} = \frac{-h_{22}}{h_{21}} \quad t_{22} = \frac{-1}{h_{21}}$$

$$h_{11} = \frac{t_{12}}{t_{22}} \quad h_{12} = \frac{\det[t]}{t_{22}} \quad h_{21} = \frac{-1}{t_{22}} \quad h_{22} = \frac{t_{21}}{t_{22}}$$

$$z_{11} = \frac{y_{22}}{\det[y]} = \frac{\det[h]}{h_{22}} = \frac{1}{g_{11}} = \frac{t_{11}}{t_{21}} = \frac{\bar{t}_{22}}{\bar{t}_{21}}$$

$$z_{12} = \frac{-y_{12}}{\det[y]} = \frac{h_{12}}{h_{22}} = \frac{-g_{12}}{g_{11}} = \frac{\det[t]}{t_{21}} = \frac{1}{\bar{t}_{21}}$$

$$z_{21} = \frac{-y_{21}}{\det[y]} = \frac{-h_{21}}{h_{22}} = \frac{g_{21}}{g_{11}} = \frac{1}{t_{21}} = \frac{\det[\bar{t}]}{\bar{t}_{21}}$$

$$z_{22} = \frac{y_{11}}{\det[y]} = \frac{1}{h_{22}} = \frac{\det[g]}{g_{11}} = \frac{t_{22}}{t_{21}} = \frac{\bar{t}_{11}}{\bar{t}_{21}}.$$

7.6.3 Passive reciprocal or antireciprocal transducers

Passive transducers pass signal energy back and forth between two physical realms, without augmenting it by conversion of nonsignal energy to signal energy. In such a device, all of the energy transferred to either of the realms is derived from a corresponding reduction in free energy in the other realm. In some instances, a passive transducer couples two nonthermal realms, but exchanges heat bidirectionally with the thermal realm during the transduction process. Thus, for example, a transducer that derives $F \times d$ work from isothermal transfer of gas from a higher-pressure reservoir to a lower-pressure reservoir is deriving all of the actual energy for that work from heat, transferred isothermally from the thermal realm. The transformation of that heat to work is accomplished at the expense of free energy reduction in the pneumatic realm. Perhaps the most celebrated passive transducer is the Carnot engine, which transfers energy back and forth between the thermal realm and a rigid-body mechanical realm, using an ideal gas as an intermediary.

Occasionally, sensors and actuators are idealized as being able to transfer free energy between two physical realms, but not being able (by themselves) to store or dissipate free energy. In terms of the z and h parameters, the constitutive relationships of the two-port models for linear versions of such ideal transducers are

$$\begin{bmatrix} F_1 \\ F_2 \end{bmatrix} = \begin{bmatrix} 0 & z_{12} \\ z_{21} & 0 \end{bmatrix} \cdot \begin{bmatrix} J_1 \\ J_2 \end{bmatrix} \quad \begin{bmatrix} F_1 \\ J_2 \end{bmatrix} = \begin{bmatrix} 0 & h_{12} \\ h_{21} & 0 \end{bmatrix} \cdot \begin{bmatrix} J_1 \\ F_2 \end{bmatrix} \tag{7.202}$$

According to the Onsager reciprocity theorem for passive linear transducers involving microscopic reversibility

(Onsager 1931),

$$z_{12} = z_{21}, \quad h_{12} = -h_{21} \tag{7.203}$$

(see Section 7.5.8 under reciprocity). Combining this with the constraint on the driving point impedance of passive LTI two-port elements, eqn (7.200), one has the following constraints on the transfer impedance (z_{12}) and transfer ratio (h_{12}) of the ideal linear, passive, reciprocal transducer of eqn (7.202):

$$\mathrm{Re}\{z_{12}^2\} \leqslant 0, \quad \mathrm{Re}\{h_{12}^2\} \geqslant 0. \tag{7.204}$$

Transducers that can be modeled well as passive, reciprocal LTI two-ports include pistons and diaphragms (between hydraulic or pneumatic realms and translational rigid-body mechanical realm), pulleys, semilevers and rack-and-pinion gears (between rotational rigid-body mechanical realm and translational rigid-body mechanical realm), and Seebeck–Peltier devices (between electrical and thermal realms). A two-port model of an ideal (e.g. rigid, massless, frictionless) version of such a device can take the following form:

$$\begin{aligned} h_{11} &= 0, \quad h_{12} = \lambda_t, \\ h_{21} &= -\lambda_t, \quad h_{22} = 0. \end{aligned} \tag{7.205}$$

Wire coils (transducers between electric and magnetic realms) and electromagnetic velocity transducers (between electric and mechanical realms) are passive but antireciprocal (see, for example, Merhaut 1981). In fact, for any transducer in which potential in one realm is directly proportional to flow in the other realm (that is, in which phasor-transform versions of z_{12} and z_{21} are real numbers) and in which the self-impedances (z_{11} and z_{22}) can be made arbitrarily small (which is true, for example, of geophones, electromagnetic loudspeakers, electromagnetic motors/generators, and other electromagnetic velocity transducers), eqn (7.200) implies that reciprocity is inconsistent with passivity. Based on the fact that they involve rotational motion (for example, rotating charge), Onsager (1931) excluded such devices from his reciprocity theorem. A two-port model of ideal versions of these antireciprocal transducers takes the following form:

$$\begin{aligned} z_{11} &= 0, \quad z_{12} = r_t, \\ z_{21} &= -r_t, \quad z_{22} = 0. \end{aligned} \tag{7.206}$$

Both two-port models, eqns (7.205) and (7.206), translate passive resistors connected across port 2 into positive-real driving-point impedances at port 1 (see eqn 7.179). Flow through port 1 thus is dissipative; free energy is represented as being lost from the circuit connected to that port. With these ideal models, however, all of that free energy is represented as being transferred to the resistive element connected across port 2. With the reciprocal two-port model of eqn (7.205), a capacitive element connected across port 2 translates to a capacitive driving-point impedance at port 1, and an inertial element connected across port 2 translates to an inertial driving-point impedance at port 1. The antireciprocal two-port of eqn (7.206) does the opposite: it translates a capacitive element across port 2 to an inertial driving-point impedance at port 1, and vice versa. The passive, antireciprocal two-port element represented by eqn (7.206) is known as a *gyrator.* (Another device with antireciprocal properties is a rotating top or gyroscope serving as a transducer between orthogonal translational rigid-body mechanical realms; see the next subsection.) Thus a passive device (capacitive element) that accumulates potential energy is converted, by a gyrator, into a passive device that appears to possess inertia and to accumulate kinetic energy (see the subsection 'Between electrical and magnetic realms' below).

For physical realms in which inertia is negligible or nonexistent (e.g. chemical, thermal, diffusional), which includes most of the physical realms associated with cellular biophysics, the ability to emulate inertia and kinetic energy would open the door to a range of dynamic behavior (e.g. resonance) not available with linear, passive systems employing potential energy alone—just as the inductor did for electrical circuit designers (see Guillemin 1957, Table 1). The range of dynamic behavior of linear systems without inertia also can be extended by inclusion of active elements. This may be the basis, for example, of electrical resonances (believed by some to be involved in tuning) in various sensory cells (Zakon 1986; Crawford and Fettiplace 1981). On the other hand, the possibility of reciprocal transduction between mechanical structures outside the cell (for example, via motile cilia; see Weiss 1982) and the cell's electrical realm opens the door to the possibility of translation of mechanical inertia to electrical inductance and incorporation of that inductance into an electrical resonance.

Just as we did for two-terminal elements, one can use synthetic models to estimate the parameters of two-port elements. Alternatively, of course, for transducers taken from the shelf, one could estimate the parameters by direct observation (for example, applying an approach analogous to that implied by eqn (7.178)). In the following paragraphs, some traditional bidirectional transducers are presented and, through synthetic modeling, reduced to a common representation—the passive LTI two-port element.

Between orthogonal translational realms: the rotating top

Consider a spinning top (as depicted in Fig. 7.27), in an environment without gravity. Let I be the top's moment of inertia about the axis of spin (see Section 1.1.13) and ω be the rotational velocity of its spin; and let the tip of the top be fixed in place. If, at the point on the spin axis at the other end of the top, distance L from the tip, one applies a translational force, magnitude F_1, normal to the spin axis, then the point will move incrementally in a direction that is perpendicular to both the spin axis and the direction of applied force. The direction of the incremental motion, J_2, is given by the right-hand rule (applied to the angular momentum, $\boldsymbol{H}$, of the spin and to the moment, $\boldsymbol{M}_o$, about the tip of the top, created by the applied translational force):

$$\frac{d\boldsymbol{H}}{dt} = \boldsymbol{M}_o$$

$$J_2 = \frac{L}{|\boldsymbol{H}|}\left|\frac{d\boldsymbol{H}}{dt}\right|$$

$$|\boldsymbol{H}| = I\omega \qquad (7.207)$$

$$\left|\frac{d\boldsymbol{H}}{dt}\right| = |\boldsymbol{M}_o| = LF_1$$

$$J_2 = \frac{L^2}{I\omega}F_1.$$

Applying the same physical laws to relate F_2 and J_1, one finds

$$J_1 = -\frac{L^2}{I\omega}F_2. \qquad (7.208)$$

The two-port model for this idealized device, serving as a microscopic or incremental transducer between orthogonal translational realms, would have the follow-

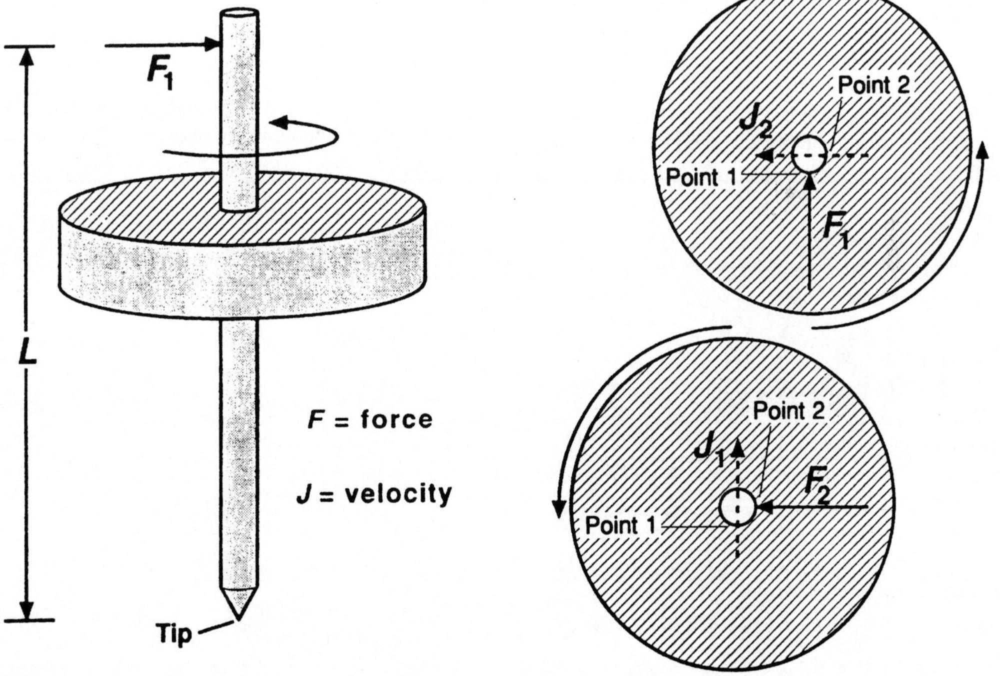

Fig. 7.27 Depiction of a rotating top, serving as a transducer (for microscopic displacements) between orthogonal translational mechanical realms. A force applied into point 1 leads to velocity out of point 2 (direction opposite that of the arrow through point 2); a force applied into point 2 leads to velocity into point 1 (same direction as the arrow through 1). This device could be represented in a lumped circuit model by an antireciprocal two-port element. Imagine what would happen to the microscopic (translational) motion as a consequence of a force applied into point 1 if point 2 were connected to a spring.

ing transfer parameters:

$$y_{12} = -L^2/I\omega, \quad y_{21} = L^2/I\omega. \tag{7.209}$$

Thus the two-port element representing this rotational system exhibits antireciprocity (that is, the element is a gyrator). The reader interested in finding a biophysical example of this type of device should explore the literature on the haltere organs of the flies and their relatives (dipterans).

Between pneumatic and rigid-body mechanical realms: the piston

Consider a piston (Fig. 7.28) that is massless and absolutely rigid within itself, but can move freely (with no resistance) in the surrounding cylinder, with area A, serving as the transducer between a pneumatic realm with negligible density variation (F_1 = pressure difference across piston, J_1 = volume velocity) and a translational rigid-body mechanical realm (F_2 = force, J_2 = velocity). If one takes the volume velocity to be directed into the piston on one side and the mechanical velocity to be directed into it on the other, then

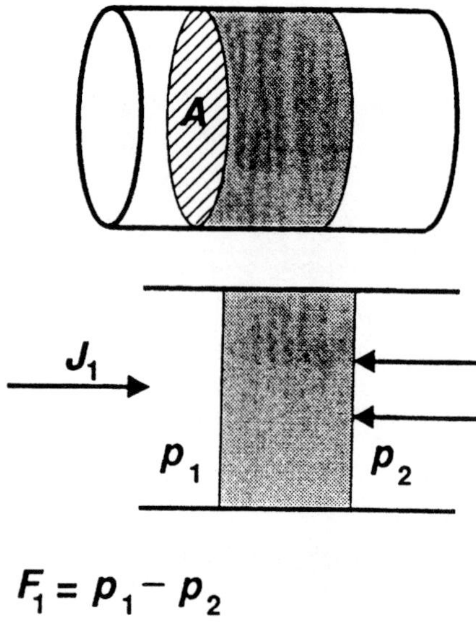

Fig. 7.28 Depiction of a massless, frictionless piston serving as a transducer.

$$\begin{aligned} F_2|_{J_2=0} &= AF_1 \quad J_1|_{F_1=0} = -AJ_2 \\ g_{11} &= 0 \quad g_{12} = -A \\ g_{21} &= A \quad g_{22} = 0 \\ h_{11} &= 0 \quad h_{12} = \frac{1}{A} \\ h_{21} &= -\frac{1}{A} \quad h_{22} = 0. \end{aligned} \tag{7.210}$$

The reference directions for potentials and flows are associated, so that when the product F_2J_2 is positive, energy is being transferred from the mechanical realm to the pneumatic realm (i.e. the F_1J_1 product will be negative) and vice versa. Note that the two-port element representing this transducer is reciprocal. It is instructive (and left to the reader) to consider the changes that would occur in the parameters of that element if one relaxed the assumptions about the piston being massless, frictionless and absolutely rigid.

Between rotational and translational mechanical realms: wheel and semilever

Consider a massless, rigid wheel (radius L) operating on a frictionless axle, as depicted in Fig. 7.29 (top), serving as a transducer between a rotational realm (F_1 = torque on axle, J_1 = rotational velocity conjugate to F_1) and a translational realm (F_2 = tangential force at the rim of the wheel, J_2 = tangential velocity of rim). The two-port representation is

$$\begin{aligned} F_2|_{J_2=0} &= \frac{F_1}{L} \quad J_1|_{F_1=0} = -\frac{J_2}{L} \\ g_{11} &= 0 \quad g_{12} = -\frac{1}{L} \\ g_{21} &= \frac{1}{L} \quad g_{22} = 0 \\ h_{11} &= 0 \quad h_{12} = L \\ h_{21} &= -L \quad h_{22} = 0. \end{aligned} \tag{7.211}$$

The same representation applies for a massless, rigid, frictionless semilever of length L (Fig. 7.29, bottom), serving as a transducer between a rotational realm (F_1 = moment about fulcrum, J_1 = rotational velocity conjugate to F_1) and a translational realm (F_2 = force normal to the end of semilever opposite the fulcrum, J_2 = velocity conjugate to F_2). Again, in both cases the reference directions for potentials and flows are associated, so that when the product $F_2 J_2$ is positive, energy is being transferred from a translational mechanical realm

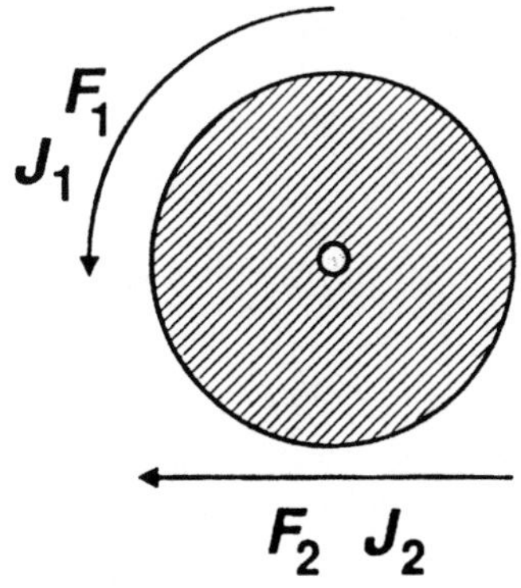

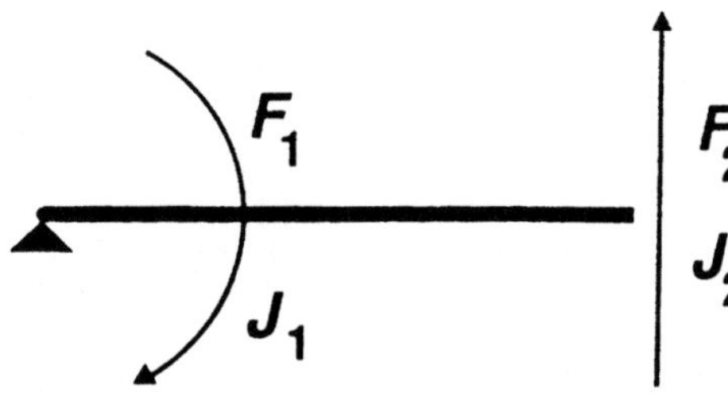

Fig. 7.29 Depiction of a massless wheel and a frictionless axle (upper diagram) and a semilever (lower diagram) serving as transducers.

to a rotational mechanical realm ($F_1 J_1$ will be negative), and vice versa. Again, the two-port element is reciprocal.

Between electrical and translational mechanical realms

Fixed-charge capacitive strain gauge

Consider two massless parallel plates (area $= A$) separated by distance D ($D^2 \ll A$), each plate being a perfect electrical conductor. One plate has a positive electric charge, $+Q$, the other has an equal countercharge, $-Q$. This device will serve as a transducer between a translational rigid-body mechanical realm ($F_1 =$ force, $J_1 =$ velocity) and an electrical realm ($F_2 =$ voltage, $J_2 =$ electric current). According to Coulomb's law, a massless, very small positive charge δq placed between the plates would experience a force F, given by

$$F = \frac{Q\,\delta q}{A\epsilon}, \tag{7.212}$$

where ϵ is the dielectric permittivity of the medium separating the plates, and F is directed toward the negatively charged plate. The change in Gibbs free energy that occurs when such a charge is transferred from the negative plate to the positive plate would be equal to the work required to carry out the transfer

$$\Delta G = \int_{x=0}^{D} F\,\mathrm{d}x = \frac{Q\,\delta q}{A\epsilon}D. \tag{7.213}$$

Therefore

$$F_2 = \frac{\Delta G}{\delta q} = \frac{QD}{A\epsilon}. \tag{7.214}$$

The change in Gibbs free energy that occurred when charge $+Q$ was transferred from one plate to the other, leaving $-Q$ behind, would have been

$$\Delta G = \int_{q=0}^{Q} \frac{D}{A\epsilon} q\,\mathrm{d}q = \frac{1}{2}\frac{D}{A\epsilon}Q^2 \tag{7.215}$$

An amount of work equal to this ΔG could be recovered in the mechanical realm by allowing the positively charge plate to move (under coulombic force) toward the negatively charge plate, and capturing the $F \times d$ work:

$$\Delta G = \int_{x=D}^{0} F\,\mathrm{d}x = \frac{1}{2}\frac{D}{A\epsilon}Q^2. \tag{7.216}$$

Thus, evidently,

$$F_1 = \frac{1}{2}\frac{Q^2}{A\epsilon}. \tag{7.217}$$

Notice that this mechanical potential (force) is independent of displacement or velocity. The velocity conjugate to F_1 and the electric current conjugate to F_2 are

$$J_1 = \frac{\mathrm{d}D}{\mathrm{d}t}, \quad J_2 = \frac{\mathrm{d}Q}{\mathrm{d}t}. \tag{7.218}$$

The corresponding transducer equations are

$$\frac{\mathrm{d}F_1}{\mathrm{d}t} = \frac{Q}{A\epsilon}J_2, \quad \frac{\mathrm{d}F_2}{\mathrm{d}t} = \frac{Q}{A\epsilon}J_1 + \frac{D}{A\epsilon}J_2, \tag{7.219}$$

which are nonlinear. Taking

$$\begin{gathered} Q = Q_o + \delta q, \quad D = D_o + \delta D, \\ F_1 = F_{10} + \delta F_1, \quad F_2 = F_{20} + \delta F_2, \\ F_{10} = \frac{1}{2}\frac{Q_o^2}{A\epsilon}, \quad F_{20} = \frac{Q_o D_o}{A\epsilon}, \\ \text{with } \delta q \ll Q_o, \quad \delta D \ll D_o, \end{gathered} \tag{7.220}$$

where Q_o and F_{20} are the charge and voltage when the device is at rest, D_o is the resting distance between the plates, and F_{10} is the force tending to pull the plates together at rest, and adding a spring with compliance C to the system so that the resting force F_{10} is balanced by the force on the spring, then the two-port model of the transducer becomes

$$\delta F_1 = z_{11}J_1 + z_{12}J_2, \quad \delta F_2 = z_{21}J_1 + z_{22}J_2,$$
$$z_{11} = \frac{1}{j\omega C}, \quad z_{22} = \frac{D_o}{j\omega A\epsilon}, \quad z_{12} = z_{21} = \frac{Q_o}{j\omega A\epsilon}. \tag{7.221}$$

Electromagnetic velocity transducer.

Consider a perfect electrical conductor of length L moving normal to the direction of magnetic flux of uniform density B, and serving as the transducer between a translational rigid-body mechanical realm (F_1 = force, J_1 = velocity) and an electrical realm (F_2 = voltage, J_2 = electric current). Taking F_1 to be the force on the wire in the direction of the velocity and F_2 to be the voltage from one end of the conductor to the other, and invoking Faraday's law of induction to estimate F_2 and the definition of the derived SI unit (1.0 Wb m^{-2}) of magnetic flux density (and the original observation by Oersted) to estimate F_1,

$$F_1 = -BLJ_2, \quad F_2 = BLJ_1,$$
$$z_{11} = 0, \quad z_{12} = -BL, \quad z_{21} = BL, \quad z_{22} = 0. \tag{7.222}$$

Again, the reference directions for potentials and flows are associated, so that when the product F_1J_1 is positive, energy is being transferred from the mechanical realm to the electric realm (i.e. the F_2J_2 product will be negative). Thus, for example, if two-terminal resistor (electrical resistance R) is connected across port 2,

$$F_2 = -RJ_2 = BLJ_1, \quad F_1 = -BLJ_2 = \frac{(BL)^2}{R}J_1,$$
$$F_1J_1 = \frac{(BLJ_1)^2}{R} = -F_2J_2. \tag{7.223}$$

Note that the two-port model of this transducer is antireciprocal.

Between electrical and thermal realms

Consider the wire depicted in Fig. 7.30, comprising a single species of metallic conductor (M_I) with voltage difference F_1 ($=V_a - V_b$) and temperature difference F_2

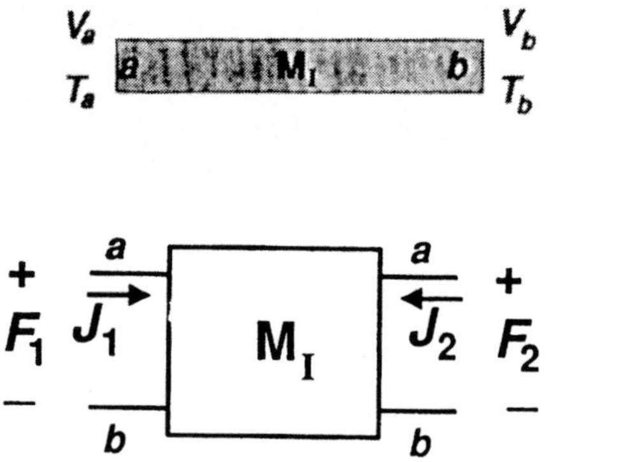

Fig. 7.30 Depiction of a metal wire serving as a transducer between electrical and thermal realms (upper diagram), and a two-port model of the transducer.

($=T_a - T_b$) from end a to end b, and charge flow J_1 and entropy flow J_2 from a to b. Represent the wire as a two-port element, with end points, a and b, corresponding to the nodes for both an electric port and a thermal port. For linear operation, the h parameters are

$$h_{11} = R_{\mathrm{EI}}, \quad h_{12} = -\Lambda_{\mathrm{I}},$$
$$h_{21} = \Lambda_{\mathrm{I}}, \quad h_{22} = G_{\mathrm{TI}}, \tag{7.224}$$

where Λ_{I} is the entropy transport parameter for this species of metal, R_{EI} is the electrical resistance of the wire, and G_{TI} is its thermal conductance. In this model we envision entropy being carried along with the flow of charge (i.e. the flow of electrons, in the case of a metallic conductor). When the temperature of the wire is uniform, T_o, the entropy flow translates directly into heat flow:

$$J_{\mathrm{heat}}|_{F_2=0} = T_0J_2. \tag{7.225}$$

The heat that is flowing is the random kinetic energy of the electrons. One may interpret Λ_{I} as a measure of the heat capacity of the conduction electrons.

When F_1 and F_2 both are nonzero, electrons flow as a consequence of the thermal potential difference from one end of the wire to the other (diffusing from the hotter end, where they have faster thermal motion, to the cooler end, where their thermal motion is slower) and as a consequence of the electric potential difference. When the effects of the two potentials are just balanced, the charge flow is zero ($J_1 = 0$), and

$$F_1|_{J_1=0} = -\Lambda_{\mathrm{I}}F_2 \tag{7.226}$$

F_1 in this case is the Seebeck potential. Because Λ_1 depends slightly on temperature, eqn (7.226) is a linear approximation of a more general expression:

$$F_{\mathrm{I}}|_{J_1=0} = \int_{T=T_a}^{T_b} -\Lambda_{\mathrm{I}}\,\mathrm{d}T. \quad (7.227)$$

If the electrical resistance (R_{EI}) of the conductor is finite, in addition to carrying heat with it, the flow of electrons also generates heat (known as Joule heat). Furthermore, if the temperature is not uniform along the length of the wire, Λ_{I} will not be constant. Owing to the change in their thermal capacity, electrons will release heat as they flow from a region of higher Λ_{I} to a region of lower Λ_{I}, and they will absorb heat as they flow in the opposite direction. This released or absorbed heat is known as Thomson heat. Joule heat and Thomson heat both accumulate in the wire itself. The rate of accumulation, J_{acc}, is

$$\begin{aligned} J_{\mathrm{acc}} &= R_{\mathrm{EI}}J_1^2 + [\Lambda_{\mathrm{I}}(T_a) - \Lambda_{\mathrm{I}}(T_b)]J_1 \\ &\approx R_{\mathrm{EI}}J_1^2 + \frac{\mathrm{d}\Lambda_{\mathrm{I}}}{\mathrm{d}T}J_1F_2. \end{aligned} \quad (7.228)$$

In a simple *thermocouple* (Fig. 7.31), two wires (composed of different metallic conductors, M_{I} and M_{II}) are connected in series electrically (a single charge flow passes through both wires) and in parallel thermally (for each wire, the same temperature difference occurs from one end to the other). Therefore, for linear operation, the thermocouple can be represented by two two-port elements (each representing one metallic conductor) connected as shown in Fig. 7.31 (bottom), with ports 1 in series and ports 2 in parallel. By returning to the original two-port equations for h parameters, it is easy to show that this connection of two-port elements can be reduced to a single, resultant two-port element with the following h parameters:

$$\begin{aligned} h_{11} &= R_{\mathrm{EI}} + R_{\mathrm{EII}}, \quad h_{12} = -(\Lambda_{\mathrm{I}} - \Lambda_{\mathrm{II}}), \\ h_{21} &= (\Lambda_{\mathrm{I}} - \Lambda_{\mathrm{II}}), \quad h_{22} = G_{\mathrm{TI}} + G_{\mathrm{TII}}. \end{aligned} \quad (7.229)$$

A *thermopile*, as depicted in Fig. 7.32, can be modeled as a combination of n identical two-port elements, each representing a simple thermocouple, with electrical ports all in series and thermal ports all in parallel. It is easy to show that the single, resultant two-port element has the following parameters:

$$\begin{aligned} h_{11} &= n(R_{\mathrm{EI}} + R_{\mathrm{EII}}), \quad h_{12} = -n(\Lambda_{\mathrm{I}} - \Lambda_{\mathrm{II}}), \\ h_{21} &= n(\Lambda_{\mathrm{I}} - \Lambda_{\mathrm{II}}), \quad h_{22} = n(G_{\mathrm{TI}} + G_{\mathrm{TII}}). \end{aligned} \quad (7.230)$$

The thermocouple and thermopile are instructive examples of the ease with which two-port elements can be combined in circuit models to represent combinations of physical elements. The various combinations of pairs of two-port elements are shown in Fig. 7.33.

Between electrical and magnetic realms

Although the electrical and magnetic realms may be inextricably linked, if one treats them as being separate then the electrical inductor can be represented, in a synthetic model, as a capacitive element in the magnetic realm coupled by an antireciprocal two-port element (gyrator) to the electric realm. A common inductor comprises a wire coil wound around a core with high magnetic permeability. It is this core that can be

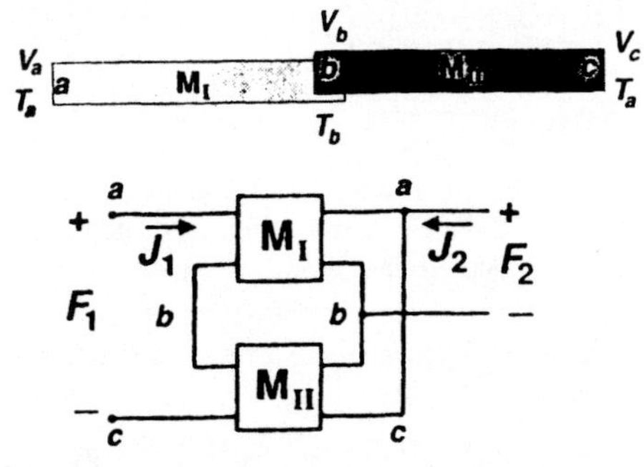

Fig. 7.31 Depiction of two metal wires (composed of different metals, M_{I} and M_{II}) connected to form a thermocouple (upper diagram), and a circuit model of the thermocouple (lower diagram).

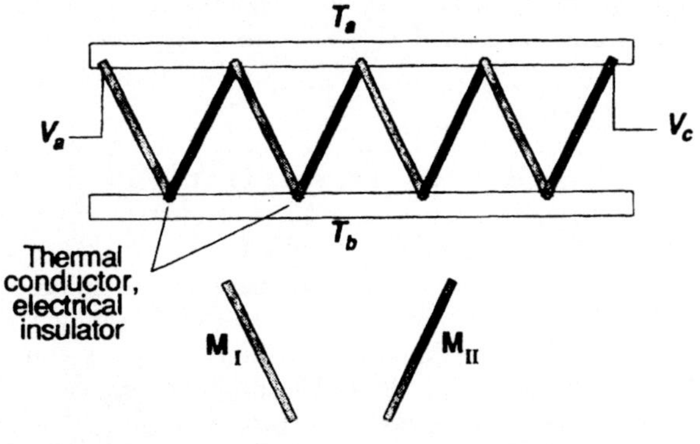

Fig. 7.32 Depiction of metal wires connected to form a thermopile.

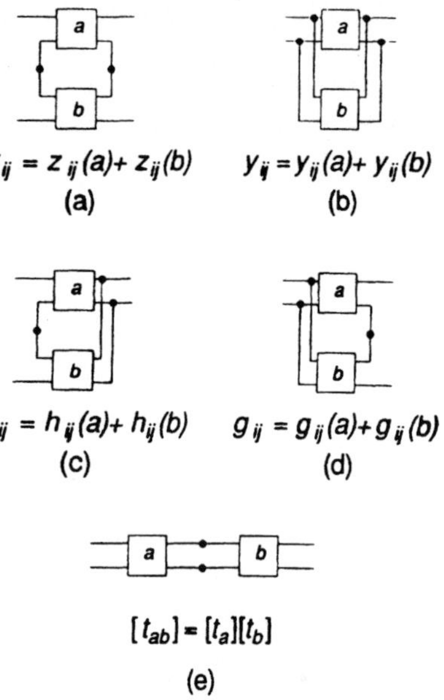

Fig. 7.33 Reduction of combinations of two-port elements to single equivalent two-port elements. In series configuration (a) the z parameters are added. In parallel configuration (b) the y parameters are added. In series–parallel configuration (c) the h parameters are added. In parallel–series configuration (d) the g parameters are added. In cascade configuration (e) the t matrices are multiplied.

represented as a capacitive element. In it, storage of potential energy is accomplished through alignment of magnetic dipoles with respect to the applied magnetomotive force vector. The total number of (atomic-level) magnetic dipoles in the core is taken to be constant (just as the total charge in an electrical capacitor is). Therefore, accumulation of aligned dipoles is accomplished by reorientation of those already present in the core (just as accumulation of charge dipoles in the parallel-plate electrical capacitor is accomplished by redistribution of the charge that already is present). Letting F_1 be electric potential (measured in J C^{-1}), F_2 be magnetomotive force (MMF, measured in J Wb^{-1}), J_1 be charge flow (measured in C s^{-1}), and J_2 be magnetic flow (rate of reorientation of magnetic dipoles, measured in Wb s^{-1}), and invoking Faraday's law of induction for z_{12} and Ampere's law for z_{21}, one obtains the following two-port model for an ideal wire coil with N turns: $z_{11} = 0$, $z_{12} = -N$, $z_{21} = N$, $z_{22} = 0$. The core serves as a magnetic capacitor, accumulating alignment (measured in Wb) of its (fixed number of) magnetic dipoles, leading to increase of potential energy (and MMF). The value of the magnetic capacity (C_m, measured in Wb per unit MMF) is the reciprocal of the magnetic reluctance. The driving-point impedance into port 1 (the electrical port) can be derived from eqn (7.179):

$$\begin{aligned} Z_{DP}(1) &= z_{11} - \frac{z_{12}z_{21}}{z_{22} + Z_L}, \\ Z_L &= 1/j\omega C_m, \\ Z_{DP}(1) &= j\omega N^2 C_m. \end{aligned} \tag{7.231}$$

Thus a capacitive element and a gyrator provide emulation of inertial impedance ($j\omega N^2 C_m$) and momentum ($N^2 C_m J_1$) in the electric realm.

7.6.4 Transformers

From passive transducers

In principle, any pair of passive reciprocal transducers that operate between the same two physical realms can be coupled in one realm to yield a transformer in the other. Consider, for example, the ideal (massless, frictionless) piston, wheel, or semilever, or the ideal thermopile (zero electrical resistance, zero thermal conductance), all of which have analogous sets of h parameters:

$$\begin{bmatrix} F_1 \\ J_2 \end{bmatrix} = \begin{bmatrix} h_{11} & h_{12} \\ h_{21} & h_{22} \end{bmatrix} \cdot \begin{bmatrix} J_1 \\ F_2 \end{bmatrix} = \begin{bmatrix} 0 & \lambda_t \\ -\lambda_t & 0 \end{bmatrix} \cdot \begin{bmatrix} J_1 \\ F_2 \end{bmatrix}. \tag{7.232}$$

If the two-port element were reversed, so that port 1 became port 2 and vice versa, then

$$\begin{bmatrix} F_1 \\ J_2 \end{bmatrix} = \begin{bmatrix} h_{11} & h_{12} \\ h_{21} & h_{22} \end{bmatrix} \cdot \begin{bmatrix} J_1 \\ F_2 \end{bmatrix} = \begin{bmatrix} 0 & \dfrac{1}{\lambda_t} \\ -\dfrac{1}{\lambda_t} & 0 \end{bmatrix} \cdot \begin{bmatrix} J_1 \\ F_2 \end{bmatrix}. \tag{7.233}$$

If two such devices were connected back-to-back in cascade, as depicted (by a and b) in Fig. 7.34, then the parameters of the resulting two-port element (ab) could be computed as follows (see Fig. 7.33 and the conversion table in Section 7.6.2):

$$\begin{bmatrix} F_1 \\ J_1 \end{bmatrix} = \begin{bmatrix} F_1(a) \\ J_1(a) \end{bmatrix} \qquad \begin{bmatrix} F_2(a) \\ -J_2(a) \end{bmatrix} = \begin{bmatrix} F_1(b) \\ J_1(b) \end{bmatrix}$$

$$\begin{bmatrix} F_2(b) \\ -J_2(b) \end{bmatrix} = \begin{bmatrix} F_2 \\ -J_2 \end{bmatrix}$$

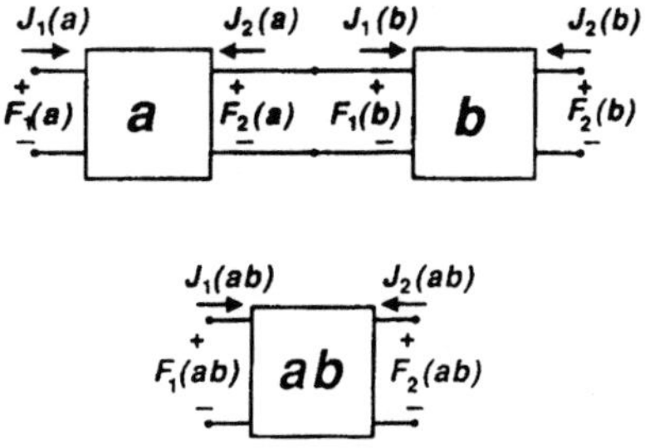

Fig. 7.34 Lumped circuit model of two transducers coupled rigidly in one realm to yield a transformer in the other realm.

$$\begin{bmatrix} F_1 \\ J_1 \end{bmatrix} = \begin{bmatrix} t_{11} & t_{12} \\ t_{21} & t_{22} \end{bmatrix} \cdot \begin{bmatrix} F_2(a) \\ -J_2(a) \end{bmatrix}$$

$$= \begin{bmatrix} t_{11}(a) & t_{12}(a) \\ t_{21}(a) & t_{22}(b) \end{bmatrix} \begin{bmatrix} t_{11}(b) & t_{12}(b) \\ t_{21}(b) & t_{22}(b) \end{bmatrix} \cdot \begin{bmatrix} F_2 \\ -J_2 \end{bmatrix}$$

$$t_{11}(ab) = t_{11}(a)t_{11}(b) + t_{12}(a)t_{21}(b)$$

$$t_{12}(ab) = t_{11}(a)t_{12}(b) + t_{12}(a)t_{22}(b)$$

$$t_{21}(ab) = t_{21}(a)t_{11}(b) + t_{22}(a)t_{21}(b)$$

$$t_{22}(ab) = t_{21}(a)t_{12}(b) + t_{22}(a)t_{22}(b)$$

$$t_{11}(a) = -\lambda_t(a) \quad t_{12}(a) = 0 \quad t_{21}(a) = 0 \quad t_{22}(a) = \frac{1}{\lambda_t(a)}$$

$$t_{11}(b) = -\frac{1}{\lambda_t(b)} \quad t_{12}(b) = 0 \quad t_{21}(b) = 0 \quad t_{22}(b) = \lambda_t(b).$$

$$\begin{aligned} t_{11}(ab) &= \frac{\lambda_t(a)}{\lambda_t(b)} \quad t_{12}(ab) = 0 \\ t_{21}(ab) &= 0 \quad t_{22}(ab) = \frac{\lambda_t(b)}{\lambda_t(a)} \\ h_{11}(ab) &= 0 \quad h_{12}(ab) = \frac{\lambda_t(a)}{\lambda_t(b)} \\ h_{21}(ab) &= -\frac{\lambda_t(a)}{\lambda_t(b)} \quad h_{22}(ab) = 0. \end{aligned} \tag{7.234}$$

The two-port parameters of eqn (7.234) are those of an ideal transformer, with transformer ratio equal to $\lambda_t(a)/\lambda_t(b)$.

Ideal gyrating transducers, such as the electromagnetic velocity transducer and the wire coil of an electric inductor, have analogous z-parameter sets:

$$\begin{aligned} z_{11} &= 0, \quad z_{12} = r_t, \\ z_{21} &= -r_t, \quad z_{22} = 0. \end{aligned} \tag{7.235}$$

Coupled back-to-back in one realm, as in Fig. 7.34, a pair of such devices (a and b) also would yield an ideal transformer (in the other realm):

$$z_{11}(a) = 0 \quad z_{12}(a) = r_t(a) \quad z_{21}(a) = -r_t(a) \quad z_{22}(a) = 0$$

$$z_{11}(b) = 0 \quad z_{12}(b) = -r_t(b) \quad z_{21}(b) = r_t(b) \quad z_{22}(b) = 0$$

$$t_{11}(a) = 0 \quad t_{12}(a) = -r_t(a) \quad t_{21}(a) = -\frac{1}{r_t(a)} \quad t_{22}(a) = 0$$

$$t_{11}(b) = 0 \quad t_{12}(b) = r_t(b) \quad t_{21}(b) = \frac{1}{r_t(b)} \quad t_{22}(b) = 0.$$

$$\begin{aligned} t_{11}(ab) &= -\frac{r_t(a)}{r_t(b)} \quad t_{12}(ab) = 0 \\ t_{21}(ab) &= 0 \quad t_{22}(ab) = \frac{r_t(b)}{r_t(a)} \\ h_{11}(ab) &= 0 \quad h_{12}(ab) = -\frac{r_t(a)}{r_t(b)} \\ h_{21}(ab) &= \frac{r_t(a)}{r_t(b)} \quad h_{22}(ab) = 0. \end{aligned} \tag{7.236}$$

For back-to-back, non-ideal passive transducers, the effects of nonzero z_{11} and z_{22} or h_{11} and h_{22} (e.g. nonzero viscosity or mass) can be estimated by similar analysis.

Quarter-wave path

Under sinusoidal steady state, for a given frequency, a uniform, lossless wave path whose length is one-quarter of a wavelength at that frequency can serve as a transformer. The two-port z parameters of such a structure can be found by making the following substitution in eqn (7.139):

$$\sqrt{zy}L = \mathrm{j}\frac{\pi}{2}, \tag{7.237}$$

which leads to

$$z_{11} = 0, \quad z_{12} = -\mathrm{j}Z_0, \quad z_{21} = -\mathrm{j}Z_0, \quad z_{22} = 0, \tag{7.238}$$

where Z_0 is the characteristic impedance of the path.

Impedance transformation

In the examples of the previous paragraphs there are two kinds of transformers: one with two-port h parameters of the form

$$h_{11} = 0, \quad h_{12} = \pm\frac{k_a}{k_b}, \quad h_{21} = \mp\frac{k_a}{k_b}, \quad h_{22} = 0, \tag{7.239}$$

and one with two-port z parameters of the form

$$z_{11} = 0, \quad z_{12} = -jZ_0, \quad z_{21} = -jZ_0, \quad z_{22} = 0. \tag{7.240}$$

With the former, when a load with impedance Z_L is connected to port 2, the driving-point impedance at port 1 is

$$Z_{DP}(1) = h_{11} - \frac{h_{12}h_{21}}{h_{22} + 1/Z_L} = \left(\frac{k_a}{k_b}\right)^2 Z_L. \tag{7.241}$$

With the latter, under the same circumstances,

$$Z_{DP}(1) = z_{11} - \frac{z_{12}z_{21}}{z_{22} + Z_L} = \frac{Z_0^2}{Z_L}. \tag{7.242}$$

Transformers described by eqn (7.241) include the simple machines, such as the lever (two semilevers, rigidly connected in the rotational mechanical realm), the hydraulic press (two pistons, coupled rigidly in the hydraulic realm), and the gear system (wheels, rigidly connected in the translation mechanical realm). They also include the conventional electromagnetic transformer. The usual application of such devices is the transformation of a load impedance to better match it to the capabilities of the available source of power (such as a human worker) and thus maximize the rate of energy transfer from the source to the load. Biophysical and physiological systems have many transformer-like devices—including compound transformers such as the combination of tympanum, ossicular chain, and oval window of the mammalian ear, which evidently serves to match the acoustic impedance of the air (and external ear canal) to the driving-point impedance of the cochlea. The middle-ear transformers in some lower vertebrates are simpler, approximating the inverse of the hydraulic press (by having two back-to-back pistons more or less rigidly coupled in the translational mechanical realm).

Quarter-wave paths (eqn 7.242) usually are used to match the characteristic impedances of two media, so that wave energy (e.g. acoustic energy or optical energy) can pass back and forth between them without reflection. Quarter-wave paths are commonly used in ultrasonic imaging (see Chapter 12).

7.6.5 Sensors based on modulation of electrical circuit elements

If a nonelectric potential or flow can be made to alter the properties of an electrical circuit element, then information about that potential or flow can be translated directly into an electrical signal. For example, if the impedance (Z_{ij}) of a two-terminal electrical element is a single-valued function of a nonelectric potential or flow (X), then one should be able to infer the value of X from the relationship between F_{ij} and J_{ij}.

Bridge circuits

A common configuration for such sensors is the four-element bridge, depicted in the circuit model of Fig. 7.35. Here, V_o is an AC or DC voltage of constant amplitude (derived from an independent voltage source), and V_x is the voltage used to infer the nonelectric potential or flow, X:

$$V_x = \left[\frac{Z(X) - Z_o}{Z(X) + Z_o}\right] V_o. \tag{7.243}$$

This relationship can be expanded as follows. Let X_o be the resting value of X, and define

$$x = X - X_o. \tag{7.244}$$

If $Z(x)$ is an analytic function, then it will have a Maclaurin series

$$Z(x) = Z(0) + \frac{\delta Z(0)}{\delta x}\frac{x}{1!} + \frac{\delta^2 Z(0)}{\delta x^2}\frac{x^2}{2!} + \frac{\delta^3 Z(0)}{\delta x^3}\frac{x^3}{3!} + \cdots + \frac{\delta^{(n-1)} Z(0)}{\delta x^{(n-1)}}\frac{x^{(n-1)}}{(n-1)!} + \mathcal{R}_n \tag{7.245}$$

where $\mathcal{R}_n$ is a remainder. Regardless of the form of the

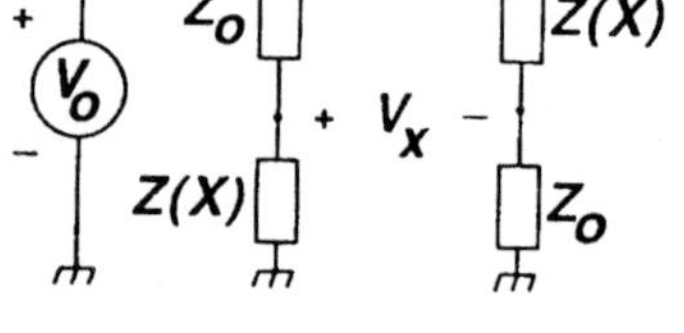

Fig. 7.35 A simple bridge circuit.

function $Z(x)$, the responsiveness of the bridge will be greatest when $dV_x/dZ(x)$ is maximum. Usually, one wishes to make the sensor system maximally responsive at the resting position ($x = 0$). To do so, one must set Z_o equal to $Z(0)$, which leads to:

$$V_x = \left[\frac{\frac{Z'(0)}{1!}x + \frac{Z''(0)}{2!}x^2 + \frac{Z'''(0)}{3!}x^3 + \cdots}{2Z(0) + \frac{Z'(0)}{1!}x + \frac{Z''(0)}{2!}x^2 + \frac{Z'''(0)}{3!}x^3 + \cdots}\right] V_o, \tag{7.246}$$

where

$$Z^{n\prime} = \frac{d^n Z}{dx^n} \tag{7.247}$$

A linear relationship between V_x and x would allow the engineer to characterize the sensor system with just one parameter and thus would be especially convenient. To evaluate the linearity of V_x for very small values of x, use the generating function

$$\frac{1}{1+\Delta} = 1 - \Delta + \Delta^2 - \Delta^3 + \cdots, \tag{7.248a}$$

which is valid when $|\Delta| < 1.0$ and leads to

$$V_x = \frac{1}{2Z(0)}\left[\frac{Z'(0)}{1!}x + \frac{Z''(0)}{2!}x^2 + \frac{Z'''(0)}{3!}x^3 + \cdots\right] \times \left[1 - \frac{Z'(0)}{1!2Z(0)}x - \frac{Z''(0)}{2!2Z(0)}x^2 - \cdots + \left(\frac{Z'(0)}{1!2Z(0)}\right)^2 x^2 + \cdots\right] V_o. \tag{7.248b}$$

When the value of x is exceedingly small, the zeroth-order term, $Z(0)$, should dominate the Maclaurin series for $Z(x)$. As the magnitude of x increases, one expects the first-order term to become important first, followed by the second-order term, and so forth. Thus, as the first-order term becomes important, $Z(x)$ will approximate an affine function. The bridge configuration, with Z_o equal to $Z(0)$, has removed the corresponding constant term in $V_x(x)$ and thus has made an important step toward linearization.

Occasionally, modulated impedances can be arranged in a *push–pull* configuration, as depicted in Fig. 7.36. With identical sensing elements in each of the four arms of the bridges,

$$V_x = \left[\frac{Z(x) - Z(-x)}{Z(x) + Z(-x)}\right] V_0. \tag{7.249}$$

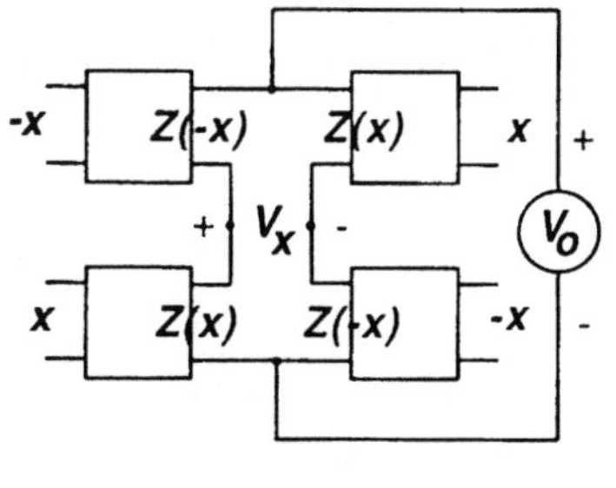

Fig. 7.36 A push–pull bridge circuit. The transducer elements are connected so that increased inputs to two of them are concomitant with decreased inputs to the other two (see Fig. 7.37).

By creating an odd function in the numerator and an even function in the denominator, the push–pull bridge configuration has removed all of the even-order terms from the function V_x:

$$V_x = \frac{1}{2Z(0)}\left[\frac{Z'(0)}{1!}x + \frac{Z'''(0)}{3!}x^3 + \cdots\right] \times \left[1 - \frac{Z''(0)}{2!2Z(0)}x^2 - \cdots + \left(\frac{Z'(0)}{1!2Z(0)}\right)^2 x^2 + \cdots\right] V_0 \tag{7.250}$$

Elimination of both the zeroth-order term and the second-order term has extended the range of values of x over which V_x is expected to be approximately linear.

Normally, the impedance Z is sensitive not only to the potential or flow for which it was designed, but also to several other potentials or flows. The presence of those others will interfere with the inferences concerning x. Let y be the difference between the actual value (Y) of an interfering variable and the value (Y_o) for which the sensing system has been calibrated. The effect of the resulting interference can be estimated from the Maclaurin's series for $Z(x, y)$:

$$Z(x, y) = Z(0, 0) + \sum_{i+j=1}^{n}\left[\frac{\delta^{i+j}Z(0, 0)}{\delta x^i \delta y^j}\frac{x^i y^j}{i!\, j!}\right] + \mathcal{R}_n, \tag{7.251}$$

where the sum is taken over all possible combinations of positive integral values of i and j that sum to values equal to or greater than 1.0 and less than or equal to n. If, in the bridge of Fig. 7.36, the potential or flow corresponding to x is applied in a push–pull fashion and that corresponding to y is not (which is the typical situation), then

$$V_x = \left[\frac{Z(x,y) - Z(-x,y)}{Z(x,y) + Z(-x,y)}\right] V_o. \qquad (7.252)$$

The numerator has only odd-order terms in x alone and even-order terms in y alone, while the denominator has only even-order terms in x alone and both even and odd terms in y alone. Thus, in addition to improving the linearity of $V_x(x, y)$ with respect to x by removing the zeroth- and second-order terms in x alone, the push–pull bridge has reduced the impact of y on $V_x(x, y)$ by removing the first-order term in y alone. The lowest-order term remaining in y is the second-order cross-modulation contribution from the numerator,

$$\frac{\delta^2 Z(0,0)}{\delta x \delta y} xy. \qquad (7.253)$$

Resistive sensors

If the electrical resistivity (ρ) of the material in a charge-flow path is uniform, and the cross-sectional area (A) of the path also is uniform, then

$$R = \rho \frac{L}{A}, \qquad (7.254)$$

where L is the length of the path. For such a path, there evidently are two ways that a nonelectrical flow or potential could alter R: (1) by altering the geometry of the path or (2) by altering ρ.

Alteration of L is accomplished in *potentiometric sensors*, for example, by placing a movable electric contact (a *wiper*) in the middle of the path, creating the resistance $\rho(L + 2x)/2A$ between the wiper and one end of the path, and the resistance $\rho(L - 2x)/2A$ between the wiper and the other end of the path (x being the distance of the wiper from the precise center of the path). The wiper often can be coupled to the translational or rotational motion of a mechanical element in such a way that x is a linear function of the corresponding displacement. Such a sensor is ideally suited for a push–pull bridge configuration.

Alteration of L is accomplished in some *strain gauges* simply by fastening the two ends of the path to different points in a mechanical network, so that changes in the distance between those two points produce concomitant changes in L. In order for L to contract when the distance between the two points decreases, the material of the strain gauge must be elastic and already under tension. By using two outlying points and one central point, one can fasten four matched strain gauges in a push–pull bridge configuration in which V_x tracks the relative position of the central point. When the length of a resistive path is stretched, there are concomitant changes in ρ and A. In wire strain gauges, these changes normally enhance the responsiveness of R to changes in L. For appropriately designed semiconductors, in response to a change (δL) in length the relative change in ρ can be much greater than the relative change in L itself (that is, $\delta\rho/\rho \gg \delta L/L$). Other nonelectric stimuli that can alter ρ include magnetic field, temperature, photon flux, humidity, and electrolyte concentration. Figure 7.37(a) shows a push–pull arrangement of two strain gauges.

Capacitive sensors

When two conductors are separated by an electrically insulating region, an electrical capacitor is formed. The magnitude, C, of the capacity generally increases as the mean dielectric permittivity of the insulating region increases, and generally decreases as the distance between the conductors increases. For the parallel conducting plates separated by a very thin, uniform insulating region,

$$C = \epsilon A/d, \qquad (7.255)$$

where A is the area of each plate, ϵ is the dielectric permittivity of the insulating region, and d is the thickness of that region (that is, the distance between the plates). The common ways to alter C are to vary the distance between the conductors and to vary ϵ.

Capacitive strain gauges, of course, are based on variation of d, and it is easy to connect two capacitive strain gauges in a push–pull configuration (Fig. 7.37b). The capacitive probe, a variation on the capacitive strain gauge, is a single electrical conductor that is brought close to a vibrating surface that is electrically conductive (Fig. 7.37c). The vibratory motion is translated into variations in the capacity of the electrical capacitor formed by the probe and surface. One merely needs to make electrical connection from the probe and the surface to a bridge circuit or some other circuit to translate the capacity variations into voltage variations. Capacitive sensors also are used widely to sense proximity of objects (Fig. 7.37d). If the two conductors of an electric capacitor are relatively far apart, then a conducting or insulating object coming close to the gap between the conductors will alter the electric field configuration and thus alter the capacity. The capacity of a capacitor formed by a wire strung parallel to the

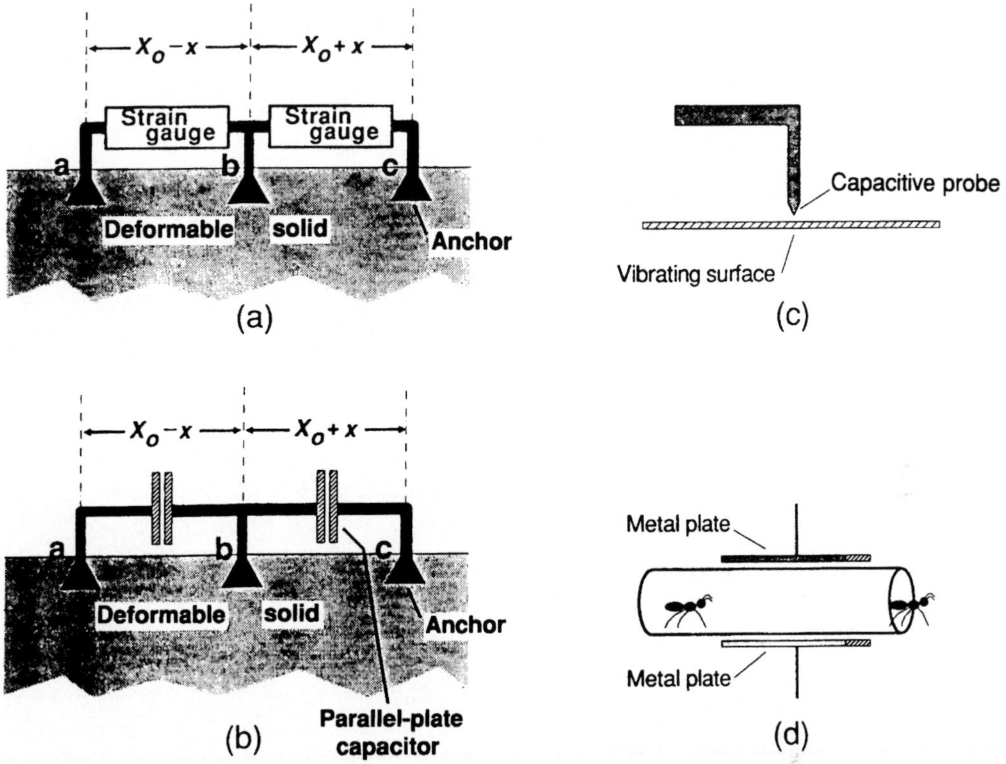

Fig. 7.37 (a) Depiction of push–pull arrangement of two strain gauges. Displacement (x) of point b relative to points a and c leads to a push on one sensor, a pull on the other. (b) Depiction of parallel-plate capacitor strain gauges in push–pull configuration. (c) Depiction of a capacitive probe. (d) Diagram of a parallel-plate capacitive sensor for counting ants.

ground and the ground itself, for example, will be altered by the proximity of an animal.

Inductive sensors

The paths along which dipoles in the core tend to be aligned when electric current flows through the coil of an inductor are closed loops perpendicular to the loops of the coil. If the core forms such a loop, then these paths of magnetization will be largely confined to the core itself, and nearby objects will not have much effect on C_m in eqn (7.231). If the core does not form a loop through the coil, then the magnetization paths will not be confined to the core, and objects (especially ferromagnetic objects) coming into those paths will alter C_m and thus alter $Z_{DP}(1)$ in eqn (7.231). This is the basis of magnetic proximity sensors.

An electromagnetic transformer comprises two coils of wire effectively sharing part or all of the same core. It can be represented in a circuit model with two two-port elements connected through an intervening magnetic circuit (Fig. 7.38). The transfer ratio for electric potentials, F_4/F_1, can be deduced by circuit analysis. Assume $Z_L \to \infty$, then

$$Z_{DP}(3) = z_{11} - \frac{z_{12}z_{21}}{z_{22} + Z_L} = 0$$

$$F_3 = 0 \quad F_4 = N_2 J_3 = N_2 F_2 j\omega C_m \quad F_2 = \frac{1}{N_2 j\omega C_m} F_4$$

$$F_2 = N_1 J_1 \quad J_1 = \frac{1}{N_1 N_2 j\omega C_m} F_4$$

$$F_1 = J_1 Z_{DP}(1) = \frac{Z_{DP}(1)}{N_1 N_2 j\omega C_m} F_4$$

$$Z_{DP}(1) = z_{11}z_{22} - \frac{z_{12}z_{21}}{z_{22} + Z_L(\text{port2})} = \frac{N_1^2}{Z_L(\text{port2})}$$

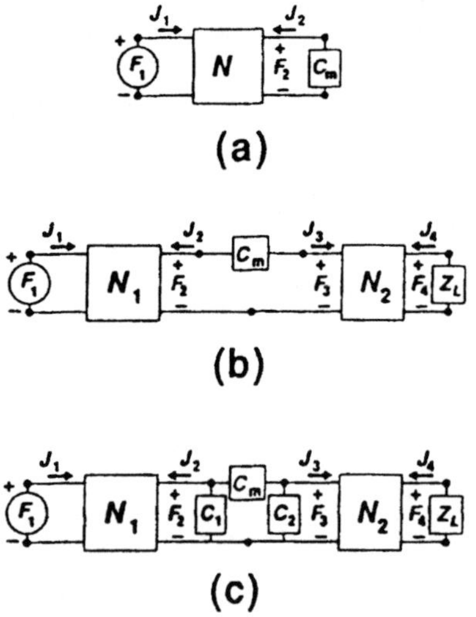

Fig. 7.38 Lumped circuit models of electromagnetic elements. (a) Electrical inductor comprising a coil with N turns wrapped around a core with magnetic reluctance $1/C_M$. In the ideal situation, the z parameters of the two port are $z_{11} = z_{22} = 0$, $z_{21} = -z_{12} = N$. (b) Ideal electrical transformer comprising two electromagnetic transducers (coils) rigidly coupled in the magnetic realm, i.e. all of the magnetic flow (rate of change of magnetic flux) generated by one coil passes through the other, via the series element C_M). (c) Nonideal electrical transformer.

$$Z_L(\text{port2}) = \frac{F_2}{-J_2}$$

$$-J_2 = F_2(j\omega C_1 + j\omega C_M) \quad Z_L(\text{port2}) = \frac{1}{j\omega(C_1 + C_m)};$$

$$F_1 = \frac{N_1}{N_2}\frac{C_1 + C_m}{C_m}F_4, \quad T_{14} = \frac{F_4}{F_1} = \frac{N_2}{N_1}\frac{C_m}{C_1 + C_m}. \tag{7.256}$$

Numerous rotational and translational displacement sensors are based on variation of C_1 and C_m resulting from displacement of one coil relative to the other (as in numerous servo devices) or from displacement of a core shared by the two coils (as in a linear variable differential transformer).

Another lumped-circuit model of a transformer is shown in Fig. 7.39. The transformer is represented by two equivalent circuits comprising two-terminal elements (as at the bottom of Fig. 7.18), connected back to back but through an intervening structure (represented by the impedances z_{c1} and z_{c2} and the admittance y_c) that is not perfectly rigid. The constitutive relationships for the dependent sources are $F_{s1} = \lambda_1 F_{A2}$, $J_{s1} = \lambda_1 J_{A1}$, $F_{s2} = \lambda_2 F_{B2}$, and $J_{s2} = \lambda_2 J_{B1}$. If this circuit represented the middle ear of a frog (comprising two pistons connected by a nearly-rigid rod), for example, then impedance Z_s could represent the acoustic impedance of the air, z_1 could represent the combined elastic, inertial and resistive impedances of the external (left-hand) piston (the tympanum) when it is allowed to move freely,

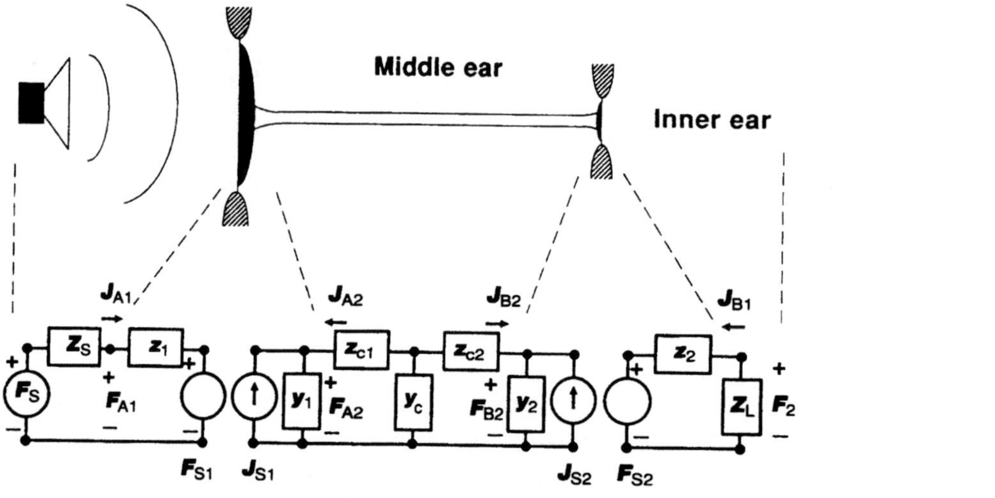

Fig. 7.39 Another lumped circuit model of a transformer, along with a depiction of the middle ear of a frog.

admittance y_1 the compressibility of that piston when it is held rigidly in place, admittance y_c the compliance of the rod connecting the two pistons, impedances z_{c1} and z_{c2} the inertial impedance of that rod, admittance y_2 the compressibility of the internal piston (at the oval window of the inner ear) when it is held rigidly in place, impedance z_2 the combined elastic, inertial and resistive impedances of that piston when it is allowed to move freely (unimpeded by the rod or the inner ear), and impedance Z_L the driving-point impedance of the inner ear as seen from the oval window. In an ideal transformer, $z_1, z_2, y_1, y_2, y_c, z_{c1}$, and z_{c2} will be negligibly small. For consistency in the model, however, the two dependent flow sources should be shunted by a finite admittance, a finite impedance should be connected in series with the independent potential source F_s and the dependent potential source F_{s1}, and dependent potential source F_{s2} should have finite impedance connected across it. As z_1, $z_2, y_1, y_2, y_c, z_{c1}$, and z_{c2} become negligible, the driving-point impedance at port 1 (F_{A1}/J_{A1}) will approach $(\lambda_1/\lambda_2)^2 Z_L$. The acoustic energy flow into the middle ear will be maximum when that driving-point impedance equals the complex conjugate of Z_s.

7.6.6 Driving-point impedance sensors

In the previous section, sensors based on electric impedances (e.g. temperature sensors based on the dependence of electrical resistance on temperature) were described. This design strategy occasionally is extended to impedances in other physical realms. For example, the hot-wire anemometer and the thermostromuhr both are used to sense fluid flow rate on the basis of a thermal driving-point impedance. Recently, sensors based on mechanical driving-point impedances and acoustic driving-point impedances have been proposed for specific large biological molecules.

7.6.7 Equilibrium sensors

Whenever two or more potential differences occur across a single path, acting upon a single flow variable, it is possible to achieve a nontrivial equilibrium—in which nonzero potential differences just balance one another and bring the flow through the path to zero. If all but one of those potentials can be sensed or predetermined, the remaining one can be inferred. This is the basic mode of operation of a wide variety of common sensors.

Equilibria involving gravity

For example, the total potential along a nonhorizontal path for a nearly incompressible fluid in which volume is taken to be the measure of conserved stuff is

$$F_{ij} = (P_i - P_j) + \rho g(h_i - h_j). \qquad (7.257)$$

In its normal operation, a *manometer* is such a flow path, with one end connected to a location at which the height (h_i) is known but the pressure (P_i) is not known. In an *open manometer* the pressure (P_j) at the other end is the present value of atmospheric pressure, and the height (h_j) at that end is allowed to vary until equilibrium is achieved (that is, $J_{ij} = 0$). At equilibrium, F_{ij} must equal zero, and P_i can be inferred from knowledge of h_i and P_j and observation of h_j.

For a point mass (m) in a gravity field (with equivalent acceleration $\mathbf{g}$), the free energy contribution from gravity is

$$\Delta G_{\text{gravity}} = m\mathbf{g}\Delta h. \qquad (7.258)$$

If, as is commonly the case, displacement is taken to be the conserved stuff and velocity the flow variable, then the potential (for downward vertical flow) corresponding to $\Delta G_{\text{gravity}}$ is

$$F_{\text{gravity}} = \frac{\Delta G_{\text{gravity}}}{\Delta h}. \qquad (7.259)$$

If an upward force, F_{upward}, also acts upon the point mass, then it also contributes to the potential for downward flow:

$$F_{\text{mechanical}} = -F_{\text{upward}}. \qquad (7.260)$$

Under equilibrium conditions, the downward flow is zero and

$$F = F_{\text{mechanical}} + F_{\text{gravity}} = -F_{\text{upward}} + m\mathbf{g} = 0. \qquad (7.261)$$

With a weighing scale, F_{upward} often is provided by a mechanical spring calibrated for displacement vs. force, $\mathbf{g}$ is known, and m is inferred from the displacement of the spring at equilibrium.

In an old-fashioned electrometer, two pith balls are tethered to the same point and hang side by side in the gravity field. When electric charge $+Q$ is placed on each of them, the resulting electric field tends to drive them apart horizontally, the gravity field tends to drive them downward, and, at equilibrium, the tethers hold them in place against both. If each tether is deflected by an angle

θ from the vertical, the potential for horizontal motion is

$$F_{\text{hrz}} = F_{\text{hrz mech}} + F_{\text{hrz electr}} = -F_{\text{teth}} \sin\theta + \epsilon \frac{Q^2}{4d_{\text{hrz}}^2}, \tag{7.262}$$

where F_{teth} is the magnitude of the total force applied by the tether on the pith ball, ϵ is the dielectric permittivity of the air surrounding the pith balls, and d_{hrz} is the horizontal displacement of the center of the pith ball from the rest state (tether hanging vertically). Coulomb's law has allowed us to express the electric potential as *free energy per unit displacement*, allowing the units of mechanical and electrical potentials to be the same. The potential for vertical motion is

$$F_{\text{vrt}} = F_{\text{mech vrt}} + F_{\text{grav}} = -F_{\text{teth}} \cos\theta + \text{m}\boldsymbol{g}. \tag{7.263}$$

The tether allows the gravity potential and electric potential to interact, in spite of being conjugate to orthogonal motions. At equilibrium, F_{hrz} and F_{vrt} both will be zero, and Q can be inferred from knowledge of m and observation of θ.

Equilibria involving diffusion potentials

Imagine two well-stirred aqueous solutions of strong electrolytes, separated by a membrane that is permeable to a subset of the ion species present. Each solution can be considered a lumped locale for accumulation of ions, and the membrane a flow path between the two lumped locales. If ion flow through the membrane is driven only by concentration gradient and electric field, according to the discussion in Section 7.5.2 under 'Additivity of effort...', the potential for the flow of the ith *permeating* ion species (with ionic valence z_i) from solution 1 to solution 2 is

$$\begin{aligned} F_{12}(i) &= F_{12}(\text{electr}) + F_{12}(\text{diff}) \\ &= z_i \boldsymbol{F} V_{12} + \boldsymbol{R}T \ln\left[\frac{c_i(1)}{c_i(2)}\right], \end{aligned} \tag{7.264a}$$

where $c_i(1)$ and $c_i(2)$ are the concentrations of the ith species in the two solutions. An alternative, for solutions in which the ion concentrations are not very low (see Table 7.2) is

$$F_{12}(i) = z_i \boldsymbol{F} V_{12} + \boldsymbol{R}T \ln\left[\frac{a_i(1)}{a_i(2)}\right], \tag{7.264b}$$

where $a_i(1)$ and $a_i(2)$ are the activities of the ith species in the two solutions. If equilibrium prevails for the ith permeating ion species ($J_{12}(i) = 0$ and therefore $F_{12}(i) = 0$), and if the concentration (or activity) of that species in one solution is known, the temperature is known, and the voltage between the two solutions is observed, then the concentration (or activity) of that ion species in the other solution can be inferred.

If a chemical reaction at the interface between a solid electrode (lumped locale 1) and an aqueous solution (lumped locale 2) leads to the production of an ion species in solution, then

$$\begin{aligned} F_{12}(i) &= F_{12}(\text{chem}) + F_{12}(\text{electr}) + F_{12}(\text{diff}) \\ &= \mu_{12}(i) + z_i \boldsymbol{F} V_{12} + \boldsymbol{R}T \ln\left[\frac{a_i(2)}{1.0}\right], \end{aligned} \tag{7.265}$$

where $a_i(2)$ is the activity of the ion species in the solution and $\mu_{12}(i)$ is the free energy decrease per mole of ions produced by the reaction at a given temperature when $a_i(2) = 1.0$. At equilibrium at that same temperature ($F_{12} = 0$), the present value of $a_i(2)$ can be inferred from the difference between observed value of V_{12} and that when $a_i(2) = 1.0$.

The equilibria described in the previous two paragraphs are the bases of a wide variety of ion-sensing systems, such as pH electrodes:

$$\text{pH} = -\log_{10}\left[\frac{a_{H^+}}{1.0}\right]. \tag{7.266}$$

At 25°C,

$$\frac{\boldsymbol{R}T}{\boldsymbol{F}} \approx 25.6\,\text{mV}. \tag{7.267}$$

Noting that

$$\ln[x] \approx 2.3 \log_{10}[x], \tag{7.268}$$

one concludes that at 25°C the equilibrium voltage of a pH electrode will change by 59.1 mV per unit change of pH. In ion-sensing electrodes, such as the pH electrode, equilibrium (or near-equilibrium) is imposed by allowing ions of the observed species (e.g. hydrogen ions for the pH electrode) to cross the path unaccompanied by counterions, leading to charge accumulation on each side, and to establishment of an electric potential that just balances the other potentials in the system. In most electrodes, the number of ions that must cross in order to establish equilibrium is an exceedingly small fraction of the total number of such ions in the locales at each end of the path.

Other equilibria

As a final example of an equilibrium sensor, consider the wet-bulb thermometer or sling psychrometer, both of which are used to sense the partial pressure of water vapor in the ambient air. Consider water to be either in lumped state 1, liquid, or lumped state 2, vapor. Evaporation and condensation provide the flow path between the lumped states. The free-energy change that takes place in water when one mole of it evaporates is

$$F_{12} = -\boldsymbol{R}T \ln\left[\frac{p_{\text{wat}}}{p_0}\right] - \Delta H_{\text{vap}}, \qquad (7.269)$$

where ΔH_{vap} is the heat of vaporization (for one mole of water at the present temperature and pressure) and p_{wat} is the partial pressure of water vapor over the surface of the water. Under equilibrium conditions ($J_{12} = 0$, rate of evaporation equals rate of condensation), F_{12} is zero. For a particular value of p_{wat}, equilibrium will occur only at a specific temperature, the dew point. One expects the dew point normally to be equal to or less than the ambient temperature. If the dew point is less than the ambient temperature, then the rate of evaporation exceeds that of condensation (there is a net flow from state 1 to state 2). With a wet-bulb thermometer or a sling psychrometer, p_{wat} is that of the ambient air and J_{12} is forced to occur nearly adiabatically, reducing the temperature of the water–water-vapor interface until it is close to the dew point and J_{12} is nearly zero. By measuring that temperature, one can infer p_{wat}. In principle, the same sensing scheme could be applied to any type of equilibrium.

7.6.8 Tracking sensors

Occasionally, proportional-error feedback-control systems (see Chapter 6) are employed as sensors; and in such cases, an amplified version of the error is used as the measure of the thing being sensed. This is the basis, for example, of some seismometers, some optical position sensors, and various scanning microscopes (tunneling, atomic force, capacitive probe).

7.7 SPICE

In Chapters 3 and 5, electrical analogs of fluid and thermal systems were presented. In this section, we invoke the analogy that arises from the fundamental equivalence of electric variables and general nonequilibrium thermodynamic variables (charge is conserved stuff, current is flow of charge, voltage is free energy per unit charge). SPICE was designed to analyze electric circuit models, but it works perfectly well for all of the network models in this chapter. The translation from the notation of this chapter to the notation of SPICE is simple: V in SPICE represents potential or effort (F); I in SPICE represents flow (J); L in SPICE represents inertia (I); R in SPICE is resistance (R); and C in SPICE is capacity or compliance (C).

7.7.1 Steps for constructing a circuit model with SPICE

1. Identify physical realm.
2. Identify flowing stuff that is conserved.
3. Identify the corresponding effort or potential variable.
4. Imagine the realm divided *appropriately* into lumped states or locales (the choice of divisions being governed, in part, by the temporal resolution desired in the deductions from the model).
5. Assign a node on a circuit graph to each lumped state or locale and designate one state or locale as the ground (reference) state or locale.
6. Assign the label '0' to the ground state or locale and to the node representing it; assign a unique nonzero integer to each of the other states or locales and to the node representing it.
7. To represent the constitutive relationship between the accumulation of stuff in state or locale i and the effort or potential F_i ($=F_{i0}$), connect an appropriate capacitive element between node i and node 0 on the circuit graph. In a SPICE program this is accomplished with a one-line instruction. For a linear

constitutive relationship, use

```
Ci i 0 K IC=M
```

where `Ci` is the literal label of the capacitive element, K is the magnitude of the capacitance (Ci = K), and M is the initial value of F_i. Thus the line

```
C1 1 0 20 IC=0
```

represents a capacitive element C_1 of constant magnitude 20 (unit appropriate to realm represented in model), connected from node 1 to node 0, with $F_i = 0$ when $t < 0$. For a nonlinear constitutive relationship, use

```
Ci i 0 POLY A0 A1 A2 A3 ... IC=M
```

where A_k is the parameter of the kth term of a Taylor's series describing the relationship between the magnitude of the capacitance and F_i,

$$C_i = A_0 + A_1(F_i) + A_2(F_i)^2 + A_3(F_i)^2 + \cdots$$

8. Repeat step 7 for every state or locale with a nonzero label.
9. To represent the flow path between each pair of neighboring states or locales, connect a lumped resistive element or a lumped inertial element (or a combination of the two) between the corresponding nodes on the circuit graph. In a SPICE program, this also is accomplished with a one-line instruction.

   For a linear resistive flow path between state i and state j, use

```
Rij i j K
```

where R_{ij} is the literal label of the element, and K is the magnitude of the resistance. Thus the line

```
R12 1 2 100
```

represents resistive element R_{12} of constant magnitude 100 (unit appropriate to realm represented in model), connected from node 1 to node 2.

For a nonlinear resistive flow path, use

```
Gij i j POLY i j P1 P2 P3 P4 ... IC=M
```

where P_k is the parameter of the kth term of a Taylor series describing the resistive constitutive relationship between J_{ij} and F_{ij},

$$J_{ij} = P_1(F_{ij}) + P_2(F_{ij})^2 + P_3(F_{ij})^3 + \cdots$$

and M is the initial value of (F_{ij}).

For a linear inertial flow path, use

```
Lij i j K IC=M
```

where K is the magnitude of the inertia, and M is the initial value of J_{ij}. For a nonlinear inertial flow path, use

```
Lij i j POLY P1 P2 P3 ... IC=M
```

where P_k is the parameter of the kth term of a Taylor series describing the dependence of inertia on flow,

$$L_{ij} = P_0 + P_1(J_{ij}) + P_2(J_{ij})^2 + \cdots$$

and M is the initial value of J_{ij}.

When a flow path is to have both resistance and inertia, place a new (intermediate) node on the graph between the nodes representing the pair of neighboring states or locales. This new node does not represent a state or locale, it is merely a topographical convenience. Therefore, it will not have a capacitive element connected to it. Label the node with an integer not already used in the model and use it to connect a resistive element in series with an inertial element. For example, if the integer 100 is not already used in the model, then in a SPICE program the two lines of instruction

```
Rij i 100 K1
```

and

```
Lij 100 j K2 IC=0
```

would represent a path from state or locale i to state or locale j, with a new node (node 100) used as a connecting point. This would represent the following constitutive relationship between J_{ij} and F_{ij}:

$$F_{ij} = K_1 J_{ij} + K_2 \frac{dJ_{ij}}{dt}.$$

10. To represent an experimentally controlled flow of conserved stuff between any two states or locales (for example, from the ground state to any other state), connect an independent flow source between the nodes representing the two states. In a SPICE program, each independent flow source can be represented by a one- or two-line instruction, such as the following:

```
Iij i j DC K
```

or

```
Iij i j AC K
.AC DEC ND FSTART FSTOP
```

or

```
Iij i j PULSE(K1 K2 TD TR TF TW)

.TRANT STEP TSTOP TSTART UIC
```

where node i corresponds to the state or locale from which the stuff is taken and node j to the state or locale to which it flows; DC K indicates a constant flow of magnitude K, AC K indicates a sinusoidally varying flow with peak amplitude K, and the .AC line specifies that a whole series of sinusoidal steady-state analyses will be carried out, each at a different frequency—beginning at the lower frequency FSTART (Hz) and ending at the higher frequency FSTOP (Hz), with the frequency being varied logarithmically, and ND frequency values being used per decade. PULSE(K_1 K_2 TD TR TF TW) indicates a single pulse of flow, with amplitude K_2, superimposed on a constant flow of amplitude K_1; the pulse begins at time TD seconds, rises to the full amplitude ($K_1 + K_2$) in TR seconds, maintains the full amplitude for a period of TW seconds, and then falls back to the constant flow amplitude (K_1) in TF seconds. The .TRAN line specifies that the results of the analysis will be plotted with a time increment of TSTEP seconds, beginning at time TSTART and ending at time TSTOP; regardless of the value specified for TSTART, the actual analysis always begins at time zero (that is, the time for which the initial values of potential or efforts on capacitive elements and flows through inertial flow paths all are specified); UIC specifies that those initial values will be used.

11. To represent a path through which the flow is controlled in part by efforts in some other realm or efforts not conjugate to the flow (such as control of flow through ion channels by various efforts or potentials acting on the channel molecules themselves rather than on the ions flowing through them), connect a dependent flow source between the nodes representing the two states connected by the path. In a SPICE program, a dependent flow source is given the same designation as a nonlinear resistive flow path, but with different control nodes. Here is an example in which only one nonconjugate potential or effort (F_k) is involved:

```
Gij i j POLY(2) i j k 0 P0 P1 P2 P3 P4 ... IC=M1,M2
```

This represents flow from state i to state j, through the path being modeled. The amplitude of the flow depends on F_k (F_{k0}) as well as on F_{ij}, and P_n is the parameter of the nth term of a two-dimensional Taylor series describing the relationship between the flow through the path and the two potentials:

$$J_{ij} = P_0 + P_1(F_{ij}) + P_2(F_k) + P_3(F_{ij})^2 + P_4(F_{ij})(F_k) + P_5(F_k)^2 + P_6(F_{ij})^3 + P_7(F_{ij})^2(F_k) + P_8(F_{ij})(F_k)^2 + P_9(F_k)^3 + \cdots$$

M1 and M2 are the initial values of F_{ij} and F_k, respectively.

12. To represent an effort- or potential-source in series with a resistive flow path, follow the procedure for combined inertia and resistance in a single flow path, but replace the inertial element with an independent effort source. In a SPICE program, a constant effort source is represented by a single line of instruction, with the same format as that for the independent flow source, but with the first letter of the element label being V instead of I. For example, if the integer 200 is not already used as a node label in the graph, then the two lines of instruction

```
Ri0 i 200 K1

Vi0 200 0 DC K2
```

represent a path between state or locale i and state or locale 0, with

$$F_{i0} = K_2 + K_1 J_{i0},$$

where J_{i0} is the flow through the path (from state or local i).

13. Begin each SPICE program with a title line, using capital letters; for example,

```
TWO-STATE CHEMICAL REACTION
```

14. Include a line that describes the analysis to be carried out:

```
.TRAN TSTEP TSTOP TSTART UIC
```

specifies conventional time-domain analysis (in which the dynamics of specified variables are computed as functions of time);

```
.AC DEC ND FSTART FSTOP
```

specifies sinusoidal steady-state analysis (in which the amplitude and phase of specified variables are computed as functions of frequency).

15. Include a line that tells which variables are to be plotted or displayed:

```
.PLOT TRAN V(i) V(j)
```

specifies plotting F_i and F_j as functions of time;

```
.PRINT AC V(1)
```

specifies printing of the values of amplitude and phase of the sinusoidal steady-state response of F_i to one or more (sinusoidal) flows.

16. End the program with the line

```
.END
```

17. SPICE requires that there be at least one non-capacitive route from every node to the ground node. To be sure that requirement is met, insert a line for a huge linear resistance between any node (for example, node 1) and node 0:

```
RHUGE 1 0 1.0E9
```

where 1.0E9 is interpreted by SPICE as 1.0×10^9. Be sure that this resistance is at least 1000 times as big as the largest of the other resistances connected to the same non-reference node (node 1 in this case).

7.7.2 Example programs

Two-state chemical reaction

```
TWO-STATE CHEMICAL REACTION
C1 1 0 1 IC=1
C2 2 0 2 IC=0
R12 1 2 1
RHUGE 1 0 1.0E9
.TRAN 0.1 4.0 0 UIC
.PLOT TRAN V(1) V(2)
.END
```

Here, RHUGE is 100 000 000; the forward rate constant is 1 s^{-1} ($1/R_{12}C_1$); the reverse rate constant is 0.5 s^{-1} ($1/R_{12}C_2$); the initial condition is $Q_1 = 1$ mol, $Q_2 = 0$; and $Q_1/1$ and $Q_2/2$ will be plotted as functions of time, in 0.1 s steps, beginning at time 0 and ending at 4 s.

Simplified membrane patch model under current clamp

```
MEMBRANE PATCH MODEL
RNA 1 100 5.0E4
VNA 100 0 DC 5.0E-2
RK1 200 4.0E3
VK 200 0 DC -8.0E-2
RCL 1 300 5.0E4
VCL300 0 DC -5.0E-2
ICLAMP 1 0 PULSE(0 1.0E-60 0.2E-3 0.2E-
    3 10E-3)
.TRAN 0.2E-3 10E-30
.PLOT TRAN V(1)
.END
```

Here RNA (sodium ion resistance), RK (potassium ion resistance) and RCL (chloride ion resistance) are 5×10^4, 4×10^3 and $5 \times 10^4\ \Omega$, respectively; VNA (reversal potential for sodium ions), VK (reversal potential for potassium ions), and VCL (reversal potential for chloride ions) are 50×10^{-3}, -80×10^{-3}, and -50×10^{-3} V, respectively; and CM (membrane capacitance) is 1×10^{-6} F. The amplitude of the current clamp (ICLAMP) is 10×10^{-6} A; it begins at $t = 0$, rises in 0.2×10^{-3} s and remains on for 10×10^{-3} s. The transient response will be plotted at 0.2×10^{-3} s intervals, from $t = 0$ to $t = 10 \times 10^{-3}$ s. The modeled membrane potential, V(1), will be computed and plotted against time.

Problems

Once a network model has been established, with appropriate care in lumping, one can apply a wide variety of computational tools to carry out the desired deductive process. For this purpose, there are numerous software packages available. From time to time, however, the engineer will find it useful to be able to carry out simple network computations without a computer (for example, on the back of an envelope). The problems presented in this section are designed to exercise basic skills in network computations and, in some instances, to illustrate further properties of LTI networks.

P7.1 When it is driven only by the chemical potential difference between two states, a chemical reaction will continue until the higher-potential state is empty. In most situations, however, the accumulation of particles in each of the two states corresponds to finite concentration or partial pressure—leading to a second potential acting on the same flow path. Taking the total potential, F_{ij}, in that case to be

$$F_{ij} = \mu_i - \mu_j + \boldsymbol{RT} \ln\left[\frac{a_i}{a_j}\right],$$

where μ_i is the chemical potential of state i, relative to the ground state; and a_i is the activity of the particles in state i, carry out the following tasks:

(a) Find the conditions for equilibrium from one end of the path to the other ($J_{ij} = 0$).
(b) For small deviations (ΔF_{ij}) of F_{ij} from its equilibrium value, derive a linear relationship between ΔF_{ij} and $\Delta a_i - \Delta a_j$ (where $\Delta a_i = a_i - a_i(0)$),
(c) Assuming that the activity coefficients, γ_i and γ_j (defined by the relationship $a_i = \gamma_i c_i$, where c_i is the concentration of particles in state i) are constant and that the volume confining the reaction is constant, and taking $\Delta F_{ij} = \Delta F_i - \Delta F_j$, find a linear relationship between ΔF_i and ΔQ_i (where Q_i is the accumulation of particles in state i).
(d) Using a linear capacitance to represent each state and a linear resistance to represent the reaction flow path between the states, construct a circuit model of the system. Use conservation, the uniqueness of free energy, and the constitutive relationships of *R*s and *C*s to derive differential equations describing the dynamic behavior of $\Delta F_i(t)$ and $\Delta F_j(t)$.
(e) Assuming that the flow J_{ij} through the reaction path is linearly related to the accumulations (Q_i and Q_j) of particles in states i and j, construct a compartmental model representing the two states and the reaction path. Write the differential equations describing the dynamics of Q_i and Q_j.
(f) Translate the compartmental model of part (e) into a circuit model; and relate the values of the circuit elements to the parameters already defined in parts (a) through (e).
(g) Imagine that the system modeled in part (f) has been taken slightly out of equilibrium, so that $Q_i = Q_i(0) + \Delta Q$ and $Q_j = Q_j(0) - \Delta Q$ [$\Delta Q \ll Q_i(0)$; $\Delta Q \ll Q_j(0)$]. If the flow from state i to state j during the return to equilibrium could be harnessed to do work, how much (W) could it do?
(h) An impulse of flow (J_s) is applied from the ground state to state i in the circuit model of part (f), placing ΔQ particles in C_i at $t = 0$ (where $\Delta Q \ll Q_i(0)$). Use Laplace transforms and ladder analysis (beginning with the circuit branch farthest from the source and working toward the source, branch by branch) to deduce $\Delta F_j(t)$ in response to this stimulus.
(i) Transform the compartmental model of part (e) into a signal flow graph based on the Laplace transform, include a source appropriate for part (h), and use signal-flow-graph analysis to find $\Delta Q_j(t) = Q_j(t) - Q_j(0)$.
(j) A single molecule is placed in state i at $t = 0$, its time (τ) of *first* arrival at state j is a continuous random variable. Use a signal-flow-graph model to deduce the density function and the generating function for that variable. From the generating function, find the expectation (mean) and coefficient of variation of τ.
(k) Find the phasor transform of the flow J_{ij} in part (j), then sketch a piecewise-linear approximation of its amplitude Bode diagram. What is the total range of phase shift in $J_{ij}(\omega)$?
(l) Show that for every LTI RC circuit model there exists an *equivalent* linear compartmental model (i.e., a model whose dynamic behavior is described by precisely the same set of differential equations).
(m) Show that the loop constraints are met in every compartmental model that is equivalent to an LTI RC circuit model.

P7.2 The *ladder circuit* in Fig. 7.40 can be used to model a wide variety of systems of interest to bioengineers. The circuit branches are numbered sequentially, beginning next to the (flow or potential) source (U_s), and alternate between shunt admittances and series impedances. Let F_i be the potential difference across branch i and J_i be the flow through branch i; and let Y_i or Z_i be the Laplace or phasor transform of the corresponding constitutive relation (see Dutta Roy 1964 and Lewis 1974 for further discussion of ladder circuits).

(a) Find a sequential algorithm for analyzing the network.
(b) Find appropriate functions for Z_i and Y_j for a model of the laminar flow of water through a cylindrical pipe with rigid walls and constant cross-sectional area A.

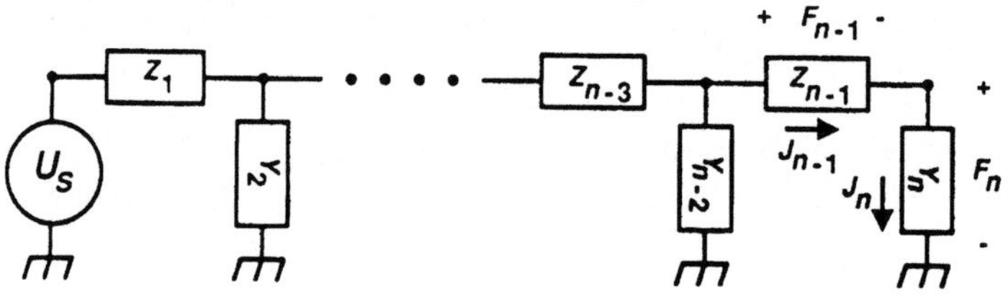

Fig. 7.40 Ladder circuit for Problem P7.2.

Assume that each ladder segment represents a pipe segment of length L.

(c) For sinusoidal flow in the pipe of part (b), the degree of flow development will depend on frequency. Find a frequency that approximately marks the boundary between undeveloped laminar flow and fully-developed laminar flow in a pipe of length L, area A. *Note*: this frequency often is taken to be an important parameter in various canal organs (that sense rotational motion) in vertebrate and invertebrate animals (for example, see Bolis and Keynes 1984).

(d) A mechanical vibration isolation filter is constructed with layers of soft viscoelastic elements (for example, rubber inner tubes, inflated to low pressures) separated by boards and layers of bricks. The intention of the designer is to isolate the top layer from vibrations in the floor. Find appropriate expressions for the shunt and series elements when flow is taken to be velocity normal to the floor and potential to be the conjugate force.

(e) For $n = 8$ in the model of part (d), and with the stiffnesses, viscous resistances, and masses of the various layers being equal, find an expression for $J_n(\omega)/J_s(\omega)$. What is the relationship between the magnitude of this ratio and ω at high frequencies? What are the advantages of a multi-layer filter of this sort over a filter with one mass layer and one viscoelastic layer.

(f) If the ladder circuit is intended to represent diffusion of particle species k along a path with uniform temperature and uniform cross-sectional area A, what would the appropriate forms of Z_i and Y_j be if the unit measure of particle population were 1.0 mol?

(g) Show that when the ladder network includes Cs or Is or both, $F_n(s)/X_s(s)$ has the form $1/P(s)$, while $F_{n-j}(s)$ has the form $Q(s)/P(s)$, where $Q(s)$ and $P(s)$ are polynomials in s.

(h) A neuronal dendrite often is represented by a ladder network in which

$$Y_j = G_j + sC_j, \qquad Z_i = R_i,$$

where G_j and C_j are the electrical conductance and capacitance of the membrane bounding the dendrite surface, and R_i is the resistance of the core of the dendrite to axial flow. Assuming that the dendrite is cylindrical, with diameter D, and that the length represented by each segment of the ladder network is L, find expressions for G_j, C_j, and R_i in terms of the electrical properties of the material of the cell and the dendrite dimensions.

(i) Converting the ladder circuit to a distributed-parameter model, find the characteristic impedance of a dendrite with constant diameter, uniform membrane thickness, conductivity and permittivity, and uniform core resistivity. If the dendrite is connected at one end to several infinitely long branches that have the same resistivity and membrane thickness, conductivity and permittivity, but smaller diameters, what relationship would be required between the diameters of the branches and that of the parent dendrite in order to provide maximum transfer of signal power from the parent to the branches?

(j) The mammalian cochlea sometimes is modeled (a *long-wave, one-dimensional* model) as a distributed-parameter path with $z = j\omega i$, $y = 1/(j\omega m + 1/j\omega c)$. This particular model sometimes is called the *critical-layer resonance* model because this form of shunt admittance (y) corresponds to a resonance at (angular) frequency $\omega_0 = 1/[mc]^{1/2}$. Explore the behavior of this model as (angular) frequency varies from very low values to values close to ω_0. In the model, the path actually is tapered so that ω_0 is largest at the input end of the path and decreases steadily (for example, over eight octaves or more) with increasing distance from the input end. Explore the behavior of this model for an input whose angular frequency is constant, at a value less than ω_0 at the input end but greater than ω_0 at the other end. *Hint*: use the WKB approach for the second part.

P7.3 A geophone, depicted by the circuit model of

Fig. 7.41, typically is constructed as a slightly underdamped second-order translational-mechanical circuit, with the mass element rigidly linked to the translational-mechanical port of an electromagnetic velocity transducer.

(a) All of the elements on the translational mechanical side of the transducer, including port 1 of the transducer itself, are connected in series in the model of Fig. 7.41. What does this imply about the mechanical elements?

(b) Taking the transducer to be ideal, with

$$z_{11} = 0, \quad z_{12} = -k,$$
$$z_{21} = k, \quad z_{22} = 0,$$

the electrical port to be open-circuited (that is, Z_L = infinity), and the mass (m), viscous resistance (R), and mechanical capacitance (or compliance, C) to be lumped linear elements, find the driving-point admittance J_1/F_s, the ratio of velocity directed into port 1 to the force applied by the source, under sinusoidal steady-state conditions.

(c) For an open-circuit electric port and fixed values of m and C, find the value of R for which the translational mechanical circuit is critically damped.

(d) Assume that this ideal geophone is sold with the mechanical circuit slightly underdamped and sealed in a can, leaving only the electric port (a pair of electric terminals) exposed. Assume further that the device is sold with the electric port open circuited. For the given mass (m) and spring compliance (C), in order to yield the fastest response without ringing the mechanical circuit must be critically damped. How would you achieve critical damping without opening the sealed can?

(e) With m and C fixed and the transducer critically damped, find the transfer ratio F_2/F_s (the electric potential difference at port 2 per unit force at the source).

(f) The force, F_s, is developed by acceleration of the mass, m, which acts as the *sensing mass* (sometimes called the *seismic mass*).

$$F_m = j\omega mv + j\omega mJ_1,$$
$$= F_s + j\omega mJ_1,$$
$$F_s = j\omega mv,$$

where v is the velocity of the sealed can enclosing the mechanical circuit, J_1 is the velocity of the mass relative to the can, and $v + J_1$ is the total velocity of the mass. Find an expression for the frequency range over which the critically-damped device will produce an electric potential at port 2 that is directly proportional to the velocity of the can.

Note: the reader interested in biological examples of motion sensors based on sensing masses should explore the literature of otolithic or otoconial sensory organs, or statolithic sensory organs in various invertebrate and vertebrate animals (for example, see Bolis and Keynes 1984).

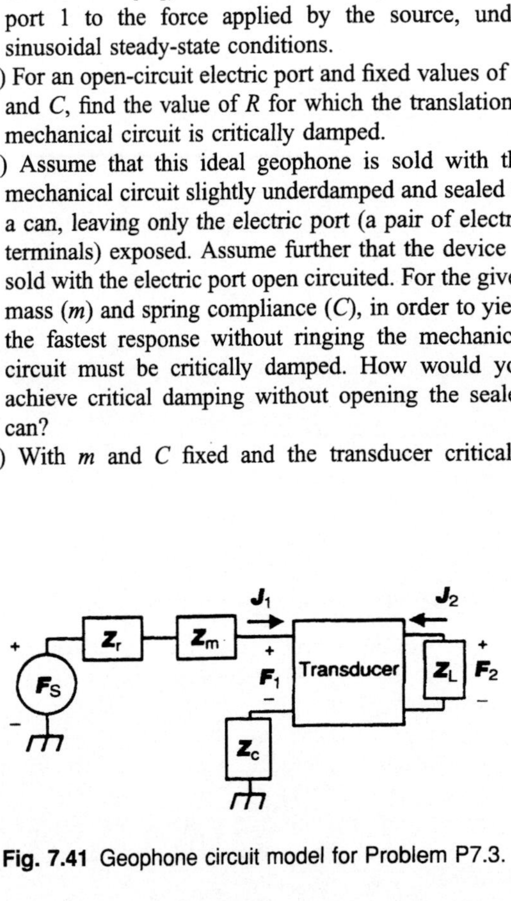

Fig. 7.41 Geophone circuit model for Problem P7.3.

P7.4 If a convenient, simple transducer is not available for direct coupling between two physical realms, engineers may use multiple transducers—routing the energy flow between the two realms through one or more other realms. The capacitance microphone is such a system. It typically comprises a piston, which passes energy between the pneumatic realm and the translational mechanical realm, in cascade with a fixed-charge capacity strain gauge, which passes energy between the translational mechanical realm and the electrical realm. The overall system is a passive transducer that passes energy between the pneumatic realm and the electrical realm.

(a) Assuming a frictionless piston, with mass m and area A, rigidly coupled to one plate of a linearly-operating strain gauge with resting distance D_o (between plates), resting charge Q_o, area A, and dielectric permittivity ε for the volume between the plates, and assuming that the strain gauge itself is massless and frictionless, find a two-port model (for sinusoidal steady-state analysis) for the transducer system.

(b) Notice that when the electrical port is an open circuit ($J_2 = 0$), the driving-point impedance (z_{11}) at the pneumatic port is an undamped resonance. How might one introduce damping into the system?

P7.5 Compartmental modeling and related problems.

(a) When tissue is prepared for microscopic observation (as in pathology labs), the water often must be replaced by another solvent, such as ethanol. The exchange of solvents must be carried out gradually or the tissue will be damaged. One method is continuous dilution, as in the following example: A small piece of tissue is placed in a 500 mL container full of water (plus appropriate salts). Pure ethyl alcohol then is allowed to flow into the container at 100 mL per minute; the solution is constantly and thoroughly stirred, and the excess mixture is allowed to flow out (i.e., the total volume remains constant at 500 mL).

Assuming that the tissue itself is very small and remains essentially at equilibrium with the solution in the container, construct a compartmental model to represent the container and the water in it (that is, let $Q(t)$ be the residual volume of water in the container at time t). What is the rate constant for the removal of water from the container? How much ethanol must be used in order to reduce the water content of the container by $1/e^2$?

(b) A drug has been injected intravenously so that 100 mg of it is well distributed in the circulating blood. As the blood passes through the brain, the concentration of drug in the blood equilibrates with the concentration of unbound drug in the brain cells. The drug in the brain exists in two states, bound and unbound, with a reversible transition between them. The rate constant for the transition from bound to unbound is 0.01 min^{-1}, while the rate constant for the transition from unbound to bound is 1.0 min^{-1}. No other tissues take up the drug. Assuming that the total blood volume is 6 liters, that the total volume of the brain is 1.5 liters, and that the blood flow through the brain is 750 mL min^{-1}, construct a compartmental model that will allow you to deduce the dynamics of drug uptake in the brain. Provide numerical values for all of the rate constants.

(c) Water is flowing from a river through a well-mixed reservoir that contains 10^7m^3 of water. The flow of the river is 10^3 m^3 min^{-1}. A crop duster accidentally dumped a load of pesticide into the reservoir. Construct a compartmental model to represent the dynamics of pesticide removal by the flushing action of the river. If the initial pesticide concentration in the reservoir is 100 parts per billion, how long would it take the river to reduce the concentration to 1 part per billion?

(d) The brain (including the blood circulating through it) has a mass of approximately 1.5 kg, with a specific heat very close to that of water. It is perfused by 750 mL min^{-1} of blood, which comes to thermal equilibrium with the brain tissue as it passes through the capillaries of the brain. The mass of the rest of the body is approximately 65 kg, again with specific heat close to that of water. Construct a compartmental model to represent the flow of heat between the brain and the rest of the body, specifying the rate constants.

(e) Solomon and Gold allowed fresh human red blood cells to equilibrate with radioactive potassium ions ($^{42}K^+$), then transferred the cells to plasma and measured the rate of release of the tracer into the plasma (see Atkins 1969). On the basis of their data, they concluded that K^+ exists in two states in the red blood cells, and that the equilibrium K^+ concentrations were distributed as follows:

plasma, 2.64; RBC state 1, 0.34;
RBC state 2, 44.1;

and they inferred the following rate constants:

RBS state 2 to RBC state 1: 1.77 h^{-1}
RBC state 1 to plasma: 0.65 h^{-1}
Plasma to RBC state 2: 0.34 h^{-1}.

Assuming that the transport of potassium ions is passive, allowing thermodynamic equilibrium to be achieved, apply the principle of detailed balance to determine the unknown rate constants.

(f) For the three-state model of the sodium channel gate (see Goldman 1964; Hille 1992), derive a single differential equation for one of the variables. To make things easy, assume that all of the rate constants are equal to 500 s^{-1}.

The three-state gate model

$$Q_1,\ Q_2,\ Q_3$$

$$k_{12},;k_{21},\ k_{23},\ k_{32},\ k_{31},\ k_{13}$$

Detailed balance holds.

Channel open when all gate molecules are in state 2

(g) (See Part (b)) A drug has been injected intravenously so that 100 mg of it is well distributed in the circulating blood. As the blood passes through the brain, the concentration of drug in the blood equilibrates with the concentration of unbound drug in the brain cells. The drug in the brain exists in two states, bound and unbound, with a reversible transition between them. The rate constant for the transition from bound to unbound is 0.01 min^{-1}, while the rate constant for the transition from unbound to bound is 1.0 min^{-1}. Each time the blood passes through the kidneys, 10 per cent of the drug is removed. No other tissues take up the drug. Assuming that the total blood volume is 6 L, that the total volume of the brain is 1.5. L, and that the blood flow through the brain is 750 mL min^{-1}, and the flow through both kidneys (together) is 1.2 L/min^{-1}, construct a compartmental model that will allow you to deduce the dynamics of drug uptake in the brain and excretion by the kidneys. Provide numerical values for all of the rate constants. What steady-state concentration of drug do you expect for the brain?

(h) Water is circulated at a constant rate between two well-mixed 50-gallon aquariums. To estimate the circulating flow rate between the two aquariums, an investigator dumps a small amount (q) of dye into one of the aquariums, then takes small (for example, less than 1 mL) samples of water from each aquarium. One set of samples, taken just after the dye has been placed in the system (and has been well mixed in its aquarium), shows a concentration of 10 ppm of dye in the first aquarium and no measurable dye concentration at all in the second aquarium. The second set of samples, taken after 10 minutes, shows 0.5 ppm in the second aquarium. What is the magnitude of the circulating flow?

(i) The model shown in Fig. 7.42 appeared in *Science* in January 1978. It represents global flow and storage of CO_2. Your task is to deduce from it the mean (expected) time required for a molecule of CO_2 in the deep-sea compartment to reach the atmosphere.

(j) Construct a compartmental model to represent the dynamics of a population of protein molecules, each of which is undergoing thermally-induced transitions back and forth between two states, 1 and 2, where $k_{12} = 1.0\ s^{-1}$ and $k^{21} = 2.0\ s^{-1}$. Convert your model to a signal flow graph and to an RC network. Show numerical values of the model parameters in each case. Find the driving-point impedance looking into state 1. Interpret the impedance (that is, say to what variables it relates).

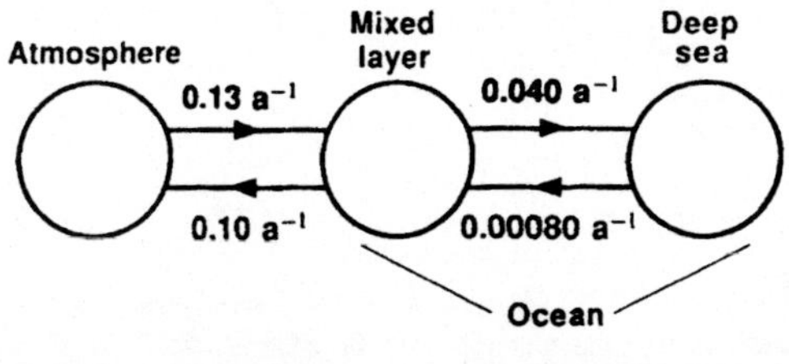

Fig. 7.42 Model of global CO_2 circulation for Problem P7.5(i).

(k) Use nonlinear dependent flow sources (and other elements) to construct a SPICE circuit model of the Michaelis–Menten reaction: substrate (concentration S) plus free enzyme (concentration E) goes by second-order mass action (rate constant k_1) to an enzyme–substrate complex (concentration ES); enzyme–substrate complex goes by first-order mass action (rate constant k_2) to free enzyme plus unaltered substrate; enzyme–substrate complex also goes by first-order mass action (rate constant k_3) to free enzyme plus product (concentration P). For a fixed, total number of enzyme molecules ($E + ES =$ constant), design and carry out an experiment with your SPICE model to deduce the time-course of product formation in response to a very brief (impulse) injection of substrate (S)—and how that time-course changes as the amount of injected substrate is changed.

(l) Construct a SPICE circuit version of the following model of the potassium-channel-gating kinetics (see Hille 1992):

A five-state gate model

$$Q_1,\ Q_2,\ Q_3,\ Q_4,\ Q_5$$

$$k_{12} = \beta,\ k_{21} = 4\alpha,\ k_{23} = 2\beta,\ k_{32} = 3\alpha,$$
$$k_{34} = 3\beta,\ k_{43} = 2\alpha,\ k_{45} = 4\beta,\ k_{54} = \alpha.$$

For α equal to twice β, use your SPICE model to deduce the time-course of Q_4 (the number of channels in state 4), given that they all started in state 1 at the same time. For the same values of α and β, design and carry out an experiment with your SPICE model that will allow you to deduce the probability density function for the random variable T_r, the time that a single channel will be away from state 4, given that it just left that state.

(m) The female members of a local population of herring gulls (*Larus argentatus*) exhibit the following life cycle. Nesting occurs on the fourth month of every year. The newly-fledged female faces a probability of 0.6 of not surviving to age 36 months. At that time, she is mature and capable of nesting. The probability that she will do so is 0.35. Once she has participated in nesting, she will continue to do so once each year for the rest of her life. If she did not participate at age 36 months, the probability that she will do so at age 48 months is 0.6; failing this, she is certain to participate the following year (at age 60 months). All adult birds face probability 0.06 of not surviving from one nesting season to the next. The expected number of female fledglings per nesting pair per year is 0.5. Lumping the nesting adult females into a single class and using time resolution of 1 year, construct a discrete-time compartmental model of this population. Using z transforms, convert the compartmental model to a signal flow graph. Find the common ratios for the model (1.0886; $0.227 \pm j0.147$; $-0.301 \pm j0.238$). Show that the population is expected to double in approximately 8 to 9 years. Show that the population will exhibit damped oscillations in response to abrupt perturbations.

P7.6 Constructing a circuit-modeling system. For non-relativistic, translational rigid-body mechanics in one-dimension (for example, velocity parallel to the x-axis), take momentum to be the conserved entity, and take the unit of momentum to be 1.0 kg m s^{-1}. Momentum accumulates in rigid bodies with mass and flows from one body to another through resistive or elastic coupling.

(a) Show that $v_1 - v_2$ is the thermodynamic potential difference between a body with mass m_1 and velocity v_1 and a body with mass m_2 and velocity v_2.

(b) In a circuit model based on this scheme, each rigid body will be represented by a shunt capacitive element. Show that the capacitive admittance for a body of mass m_1 is $j\omega m_1$.

(c) If the two bodies are coupled through viscous resistance, such that the force (f_{12}) between them is given by $f_{12} = \Psi_{12}(v_1 - v_2)$, then that resistance is a path through which momentum will flow between the bodies. Show that the flow of momentum, J_{12}, is equal to f_{12}, and that the path is resistive and has resistance equal to $1/\Psi_{12}$.

(d) The direction of momentum flow may be parallel to the velocities of the rigid bodies, or it may be perpendicular to them. An example of the latter is the flow of momentum between lamina in a fluid undergoing transition from free laminar flow to fully-developed laminar flow. Construct a lumped circuit model of such a fluid, flowing in a horizontal cylindrical pipe with rigid walls. Using parameters for a specific situation (with specified fluid viscosity and density and specified pipe diameter), use SPICE to compute the time-course of flow development in response to a stepwise change in pressure from one end of the pipe to the other. *Hint:* Use laminae with equal volumes (equal cross-sectional areas) and assume that the model represents a cross section through the pipe far from its entrance.

(e) If two rigid bodies are coupled through an elastic element, such that $f_{12} = (x_1 - x_2)/c_{12}$, where x_1 and x_2 are the positions of the centers of the bodies, then that elastic element is a path through which momentum will flow between the bodies. Show that the path is inertial, with impedance $j\omega c_{12}$.

(f) What are the properties that one would assign to the reference body (the body depicted by the ground or reference node) in this modeling system?

References

Atkins, G. L. (1969). *Multicompartmental models for biological systems.* Methuen, London.

Bolis, L. and Keynes, R. F. (ed.) (1984). *Comparative physiology of sensory systems.* Cambridge University Press.

Borsellino, A. I. and Fuortes, M. G. F. (1964). Changes in time scale and sensitivity in the ommatidia of Limulus. *Journal of Physiology (London)*, **172**, 239–63.

Crawford, A. C. and Fettiplace, R. (1981). An electrical tuning mechanism in turtle cochlear hair cells. *Journal of Physiology (London)* **364**, 359–79.

Desoer, C. A. and Kuh, E. S. (1969). *Basic circuit theory.* McGraw-Hill, New York.

Dutta Roy, S. C. (1964). Formulas for the terminal impedances and transfer functions of general multimesh ladder networks. *Proceedings of the Institute of Electrical and Electronic Engineers*, **52**, 738–9.

Feller, W. (1968). *An introduction to probability theory and its applications.* Wiley, New York.

Godfrey, K. (1983). *Compartmental models and their applications.* Academic Press, London.

Goldman, D. E. (1964). A molecular structural basis for the excitation properties of axons. *Biophysical Journal*, **4**, 167–88.

Guillemin, E. A. (1957). *Synthesis of passive networks.* Wiley, New York.

Hill, A. V. (1936). Excitation and accommodation in nerve. *Proceedings of the Royal Society (London)*, **119**, 305–55.

Hille, B. (1992). *Ionic channels of excitable membranes*, (2nd edn). Sinauer, Sunderland, MA.

Hodgkin, A. L. and Huxley, A. F. (1952). A quantitative description of membrane current and its application to conduction and excitation in nerve. *Journal of Physiology (London)*, **117**, 500–44.

Kuo, T. (1967). *Linear networks and systems*. McGraw-Hill, New York.

Leslie, P. H. (1945). On the use of matrices in certain population mathematics. *Biometrika*, **33**, 183–212.

Lewis, E. R. (1974). A note on transfer and driving-point functions of iterated ladder networks. *IEEE Transactions on Circuits and Systems*, **21**, 334–8.

Lewis, G. N. and Randall, M. (1951). *Thermodynamics and the free energy of chemical substances*. McGraw-Hill, New York.

Linvill, J. G. and Gibbons, J. F. (1961). *Transistors and active circuits*. McGraw-Hill, New York.

Mason, S. J. and Zimmerman, H. J. (1960). *Electronic circuits, signals and systems*. Wiley, New York.

Merhaut, J. (1981). *Theory of electroacoustics*. McGraw-Hill, New York.

Monnier, A. (1934). *L'excitation électrique des tissus*. Hermann, Paris.

Onsager, L. (1931). Reciprocal relations in irreversible processes. I, II. *Physical Review*, **37**, 405–26; **38**, 2265–79.

Pielou, E. C. (1969). *An introduction to mathematical ecology*. Wiley, New York.

Rashevsky, N. (1933). Outline of a physico-mathematical theory of excitation and inhibition. *Protoplasma*, **20**, 42–56.

Weiss, T. F. (1982). Bidirectional transduction in vertebrate hair cells: a mechanism for coupling mechanical and electrical processes. *Hearing Research*, **7**, 353–60.

Zakon, H. H. (1986). The electroreceptive periphery. In *Electroreception* (ed. T. H. Bullock and W. Heiligenberg), pp. 103–56. Wiley, New York.

Further reading

Allocca, J. A., and Allen, S. (1984). *Transducers, theory and application*. Reston, Reston, VA.

Beckwith, T. G., Buck, N. L., and Marangoni, R. D. (1982). *Mechanical measurements*. Addison-Wesley, Reading, MA.

Cadzow, J. A. (1973). *Discrete-time systems*. Prentice Hall, Englewood Cliffs, NJ.

Chua, L. O., Desoer, C. A., and Kuh, E. S. (1987). *Linear and nonlinear circuits*. McGraw-Hill, New York.

Cobbold, R. S.C. (1974). *Transducers for biomedical measurements*. Wiley, New York.

Daniels, F. and Alberty, R. A. (1955). *Physical chemistry*. Wiley, New York.

Derenzo, S. E. (1990). *Interfacing*. Prentice Hall, Englewood Cliffs, NJ.

De Silva, C. W. (1988). *Control sensors and actuators*. Prentice Hall, Englewood Cliffs, NJ.

Doebelin, E. O. (1983). *Measurement systems*. McGraw-Hill, New York.

Fox, R. W. and McDonald, A. T. (1978). *Introduction to fluid mechanics*. Wiley, New York.

Gardner, M. F. and Barnes, J. L. (1942). *Transients in linear systems*. Wiley, New York.

Goldstein, R. J. (ed.) (1983). *Fluid mechanics measurements*. Hemisphere, New York.

Hudspeth, A. J. and Lewis, R. S. (1988). A model for electrical resonance and frequency tuning in saccular hair cells of the bullfrog, *Rana catesbeiana*. *Journal of Physiology (London)*, **400**, 275–97.

Jack, J. J. B., Noble, D., and Tsien, R. W. (1975). *Electric current flow in excitable cells*. Clarendon, Oxford.

Karnop, D. C. and Rosenberg, R. C. (1975). *System dynamics: a unified approach*. Wiley, New York.

Kittel, C. (1958). *Elements of statistical physics*. Wiley, New York.

Kittel, C. (1976). *Introduction to solid state physics*. Wiley, New York.

Mason, S. J. (1956). Feedback theory—further properties of signal flow graphs. *Proceedings of the Institute of Electrical and Electronic Engineers*, **44**, 920–6.

Normann, R. A. (1988). *Principles of bioinstrumentation*. Wiley, New York.

Norton, H. N. (1989). *Handbook of transducers*. Prentice Hall, Englewood Cliffs, NJ.

Oster, G. F., Perelson, A. S., and Katchalsky, A. (1973). Network thermodynamics: dynamic modelling of biophysical systems. *Quarterly Reviews of Biophysics*, **6**, 1–134.

Plonsey, R. and Fleming, D. G. (1969). *Bioelectric phenomena*. McGraw-Hill, New York.

Quarles, T. (1989). *SPICE 3 version 3C1 user's guide*. University of California/Electronics Research Laboratory, Berkeley.

Robinson, D. A. (1968). The electrical properties of metal microelectrodes. *Proceedings of the Institute of Electrical and Electronic Engineers*, **56**, 1065–71.

Seippel, R. G. (1983). *Transducers, sensors, and detectors*. Reston, Reston, VA.

Skilling, H. H. (1957). *Electrical engineering circuits*. Wiley, New York.

Skilling, H. H. (1951). *Electric transmission lines*. McGraw-Hill, New York.

Thoma, J. U. (1990). *Simulation by bondgraphs*. Springer-Verlag, Berlin.

Timoshenko, S. and Young, D. H. (1951). *Engineering mechanics*. McGraw-Hill, New York.

Van Putten, A. F. P. (1988). *Electronic measurement systems*. Prentice Hall, New York.

Yourgrau, W., van der Merwe, A., and Raw, G. (1982). *Treatise on irreversible and statistical thermophysics*. Dover, New York.

Zador, A. and Koch, C. (1994). Linearized models of calcium dynamics: formal equivalence to the cable equation. *Journal of Neuroscience*, **14**, 4705–15.

BIOMATERIALS

R. B. Martin

8

Contents

Symbols

E	elastic modulus	t	time	ε	strain
M	molecular weight	T	absolute temperature	ρ	mass density
R	universal gas constant	V	volume fraction	σ_Y	yield stress

8.1 Introduction

The term *biomaterials* is frequently applied to man-made materials used to construct prosthetic or other medical devices for implantation in a human being. For the purposes of this discussion, a broader definition shall be adopted, including also those biological materials which form the bodies of not only humans, but other living organisms, from insects to whales. The reason for doing this is that it is necessary to understand the properties of the original materials in various biological organs before one can appreciate fully the problems inherent in trying to replace such organs with a prosthesis.

In this context, before going any further, one should cast off any notion that modern technology has the ability to replace any part of a living organism with an artificial organ which will be superior to the original structure. While it is possible to imagine situations in which this might be true in some limited sense (for example, to replace a bone with a similar structure of titanium alloy having greater strength), one always finds that the organism as a whole will never work better than when the original organ was in place. Rather than proving superior, the prosthesis is likely to fail, or provoke an adverse response from the animal's other organs, or otherwise detract from the animal's overall function. (In the case of replacing a bone with another material, failures stem primarily from awkwardness in connecting the prosthesis to muscles and other related tissues, and from the inability of the metal to repair fatigue damage the way bone can.) The best one can achieve is a kind of neutral equilibrium, as, for example, when a screw is used to repair a fractured bone (as described in Section 9.2.3), and remains in place after the bone has healed and remodeled itself to eliminate the stress concentration[1] caused by the screw hole. Another example would be the situation when a heart pacemaker allows a patient to function within limits much broader than he could without the device, but certainly restricted relative to those of a healthy heart.

The reason for this barrier to improvement is, of course, that all living organisms are the result of evolution—a process of 'cut and try' engineering involving trials in millions of individuals over millions of years, thus exceeding by orders of magnitude anything which human engineers can manage. Furthermore, to this point at least, theoretical engineering and computer simulations have not been able to manage any better solutions than evolution has contrived. Thus, evolution has such a tremendous head start that it is doubtful if biomedical engineers will ever catch up using the methods currently in practice. The ultimate solution to most problems involving internal prostheses (Section 9.2.4) will probably come when we are able to control cell function well enough so that organs can be replaced biologically. With the rapid development of molecular biology and genetic engineering in the 1980s, this has become a plausible goal.

In the meantime, however, despite rapid advances in cell biology, millions of people are going to need prostheses to improve their health, limited though these devices may be in their function. It seems clear that the materials presently available can be significantly improved, and the goal of this chapter is to frame the boundaries of this problem: the biological 'supermaterials' on one side, and the current substitutes on the other. In the interest of fitting this subject into a general survey of bioengineering, the emphasis will be on the mechanical or structural aspects of these materials.

[1] Stress concentration (briefly described (in Section 1.4.4)) refers to the local elevation of stress found near a hole or other irregularity in the geometry of a loaded structure. Typically, a hole drilled through one side of a bone reduces its torsional energy-to-failure by 50%.

8.2 Biological materials

It is useful to begin this discussion of the mechanical properties of biological materials by distinguishing between rubbery and crystalline elasticity.

8.2.1 Crystalline elasticity

If we take conventional engineering materials as examples, metals have crystalline elastic behavior, as shown in the engineering stress–strain curves on the left in Fig. 8.1. Their atoms are locked into highly ordered geometric arrangements by strong interatomic forces, and the elastic modulus is a direct function of these forces. Deformation involves an increase in the internal energy of the material. The stress–strain curve is linear in the region where the atomic structure is nondestructively stretched, and only becomes nonlinear (that is, *yields*, producing a concave downward shape) when some of the atoms are permanently displaced relative to one another. The elastic modulus is of the order of 100 GPa, and the strain at failure is a few tenths of one per cent.

8.2.2 Rubbery elasticity

Rubber, of course, is a good example of a material having rubbery elasticity. (Although natural rubber is a biological material, it has long been used by conventional engineers.) Its tensile stress–strain curve, shown on the right in Fig. 8.1, exhibits strain-hardening (that is, upward concavity) as opposed to the yielding behavior of metals. The reason for this is that the internal structure consists of innumerable long, chain-like molecules which lack a symmetric arrangement. Therefore, they cannot be

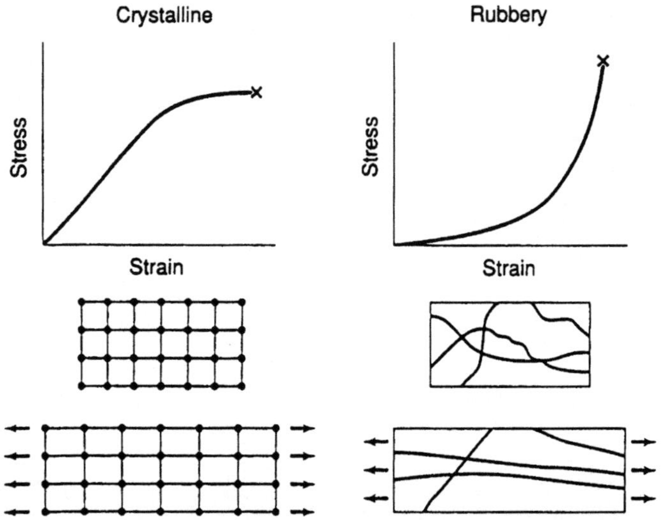

Fig. 8.1 Comparison of crystalline and rubbery elasticity. Crystalline elasticity is characterized by an initially linear stress–strain curve which becomes more horizontal as molecular or atomic forces are irreversibly overcome. Rubbery elasticity is characterized by a nonlinear stress–strain curve which has increasing slope as chain molecules of disparate directionality are aligned by deformation of the materia.

parallel to the direction of the load, and the initial displacement of the structure tends to reduce the load–molecule angles rather than to stretch the molecules. Later, when the molecules have been brought into greater alignment with the load direction, intramolecular stretching occurs, and greater forces are experienced. Therefore, the material becomes stiffer as stretching proceeds, and deformation alters the internal entropy as well as the internal energy of the material. The initial elastic modulus of rubbery materials is much less than that of crystalline materials, of the order of 1 MPa, and the strain at failure may be several hundred per cent.

This behavior is characteristic of materials called *high polymers*, whose long molecules are connected by cross-links to variable degrees. Examples include natural and synthetic rubbers, plastics, proteins (e.g. collagen), and polysaccharides (for example, chitin). It is intuitively clear that cross-linking would increase the stiffness of such materials, and this idea has been used to derive a mathematical relationship between the elastic modulus, E (in pascals), and the mean molecular weight (a dimensionless number) of the chain segments between cross-links, M, in an ideal rubbery material:

$$E = \frac{3000\rho RT}{M}, \tag{8.1}$$

where ρ is the density of the material (kg m^{-3}), T is its absolute temperature (K), and R is the universal gas constant (8.314 J K^{-1} mol^{-1}). (See Problem 8.1 at the end of this chapter for an example of how this equation might be applied.) This equation implies that the elastic modulus of a rubbery material is inversely proportional to the distance between cross-links, and that if more and more cross-links are added, the stress–strain behavior should become more like that of a crystalline material. That is, in fact, what happens, so that some plastics (e.g. poly(methyl methacrylate)) have linear, brittle stress–strain curves.

8.2.3 Strain crystallization

As a polymer is stretched and its chain molecules are brought into greater alignment, more cross-links may spontaneously form because the sites at which they attach are brought to greater proximity. This phenomenon is known as *strain crystallization*. It substantially alters the stiffness and strength of the polymer; for example, it can triple the tensile strength of cellophane film, and increase the elastic modulus of nylon twenty times. A biological example occurs in the strands of silk pulled from the spinneret of a silkworm or spider. As the monomer passes through the spinneret, it polymerizes due to shearing, and as the silk is stretched by the animal's motion, it crystallizes further, acquiring the greater stiffness which is needed within the cocoon or web. (One problem for these animals is keeping the monomer from polymerizing within the silk gland, because its protein concentrations are apparently kept at a nearly critical level. Dead spiders have been found with polymerized silk glands; whether their death caused or resulted from the polymerization is unknown.)

8.2.4 Some important biological constituent materials

The laws of chemistry are such that cells cannot produce metal structures. They are therefore forced to do all their structural engineering with the other two primary building materials of human engineers: nonmetallic minerals and polymers. Nine of the more prominent of these constituent materials will be briefly described, and then some of the *tissues* formed by these materials will be characterized. Table 8.1 summarizes some of the physical properties of the constituent materials to be discussed.

Collagen is the most important structural substance in the bodies of vertebrates. It constitutes 20 per cent of the protein in a mouse. There are more than a dozen different types of collagen, distinguished by variations in the sequence of amino acids in their polymer chains. All, however, contain three tropocollagen molecules with left-handed spiral structures which are in turn wound into a right-handed superhelix. The most abundant type of collagen is type I, consisting of two identical and one different amino acid chains. It is found in bone, tendon, ligaments, skin, and elsewhere. Each end of the collagen polypeptide molecule has an extension of 15 amino acids containing lysine and hydroxylysine residues, which can be enzymatically attached to aldehydes to form the beginnings of a cross-link. Cross-links can be *intramolecular*, between two of the three polypeptides in a single molecule, or *intermolecular*. In the latter case, exact fiber alignment is necessary for cross-linking to occur, and such periodicity is a cardinal sign of collagen in electronmicrograph images (see Fig. 8.2). In bone, tendon, and ligaments, cross-linking is abundant, so that the stress–strain behavior is fairly crystalline (Fig. 8.3). The elastic modulus of collagen (as measured in tendons) is about 1 GPa, and its maximum strain is 10–20 per cent. Cross-linking increases with age, which increases strength but reduces the ability of skin, cornea, and other tissues to stretch in an elastic manner.

Table 8.1. Mechanical properties of various biological materials and tissues[a]

Materials	Tensile strength (MPa)	Elastic modulus (MPa)	Strain at failure (%)
Abductin		3	
Apatite		114 000	
Bone	150	20 000	1.5
Cellulose	1000	80 000	2.4
Cartilage (costal)	1	14	8.0
Chitin	50	45 000	1.3
Collagen (tendon)	75	1300	9.0
Cuticle	50	20 000	
Elastin		1	
Enamel		95 000	
Keratin[b]	50	5000	2.0
Resilin		2	300.0
Silk (cocoon)	400	2000	35.0
Wood (hard)	100	10 000	
Rubber			1

[a] Values in this and subsequent tables are approximations compiled from a variety of sources listed at the end of this chapter. Given the variability of biological tissues, and variations in the manufacturing processes of the man-made materials discussed here, precise values cannot be given for mechanical properties.
[b] For the alpha-form region of tensile tests on wool.

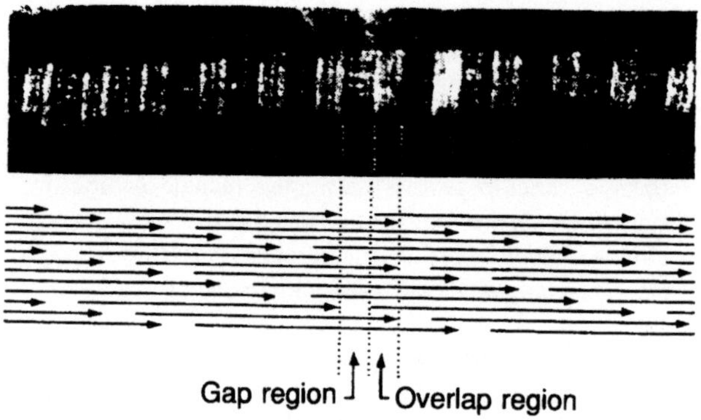

Fig. 8.2 Electron microscope image of collagen showing its characteristic 6.4 nm banding, and a schematic conception of how this banding is produced by gaps and overlaps in the molecular structure. There is still debate about how accurate the representation shown here may be.

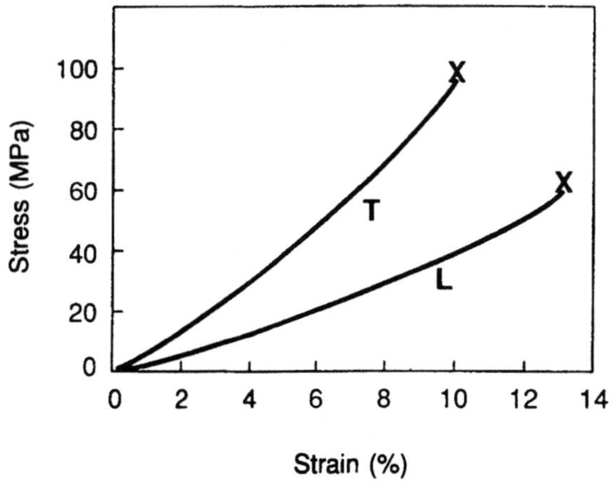

Fig. 8.3 Typical stress–strain curves for tendon (T) and ligament (L). Tendons are composed almost entirely of well-aligned collagen fibers.

All solids store strain energy when they are deformed. However, the next three biological materials seem to be specialized for this function rather than strength. They are cross-linked, but their elasticity remains much more rubbery than that of collagen. The first two of these materials, elastin and resilin, are named after their ability to store strain energy. The word resilient is a synonym for elastic. Physicists usually define an elastic deformation as one in which no energy is lost; engineers are inclined to think of it as one in which there is no permanent deformation. These are two sides of the same coin, but one may think of elastin as a constituent material which helps tissues return to their original dimensions when loading is reduced (the engineer's definition), while resilin helps tissues store energy (the physicist's definition). In the long run, of course, these concepts are inextricably connected (see Sections 1.2.2 and 1.2.3).

Elastin is a protein which is not so abundant as collagen in vertebrates, but it is important because of its elasticity at high strains. It is found mixed to varying degrees with collagen in several tissues, including skin, and in the walls of blood vessels, where it acts as a spring to accommodate pressure changes in the blood. It is also found in the ligamentum nuchae, a ligament at the back of the neck which is small in humans but large in horses and other grazing animals. It stores energy when these animals put their head down to graze, so that less muscle energy is required to raise it up again. Unlike collagen, elastin has no discernable periodicity in its molecular structure, but it is known to have cross-links. In the electron microscope, elastic fibrils appear to be composed of a central, amorphous region of elastin surrounded by *microfibrillar protein*, which may be glycoprotein. Both collagen and elastin are manufactured by cells called fibroblasts, and while elastin is similar in some respects to collagen, it has different sequences of amino acids and its molecules are composed of four rather than three polypeptides wound together. Both collagen and elastin are viscoelastic, but elastin is less so. Also, the elastic modulus of elastic fibers is less than 0.1 per cent of that of collagen, and because it is so much more 'rubbery', its cross-linking must be quite different from that of collagen. However, like collagen, its cross-linking increases with age, reducing the elasticity of such tissues as blood vessels and skin.

Resilin is a protein somewhat similar in function to elastin, but found in arthropods. It is found, for example, in the thorax of insects, where it stores and returns energy during wing-flapping. Also, fleas jump onto the backs of dogs by storing energy in resilin. There is an important difference between the functioning of organs containing resilin and elastin, however. Elastin is used in organs where the rates of energy input and output are similar and relatively slow (e.g. the horse raising and lowering its head). In the case of resilin, on the other hand, energy transfer is very fast on the output side, and either slow (the jumping flea) or fast (insect wing-flapping) on the input side. In order to jump high or fly, insects must generate much more power than can be developed by a contracting muscle. For example, the power output when a flea jumps has been estimated by Alexander (1968) at 3 kW kg^{-1} of muscle, about 10 times greater than that which a frog can produce in a single contraction of its leg muscles. (R. McNeill Alexander's book, *Animal mechanics*, is highly recommended for its readable discussion of this and many related problems.) The solution which nature has devised for this problem is to store energy in a spring, but unlike the elastin spring, this one must be capable of releasing its energy very quickly. Therefore, unlike most biological materials, resilin has almost no viscoelastic creep or relaxation, as quantified by eqns (1.129) and (1.130), so that it returns to its original shape instantaneously. It does this despite consisting of 50–60 per cent water. (The lack of relaxation behavior also allows the flea to slowly 'cock' the springs in its hind legs with a muscle and wait an indeterminate time for its next meal to come along.) When loaded and released, resilin returns 97 per cent of its strain energy, compared to about 91 per cent for similar man-made polymers, but this is merely a refinement compared to its lack of viscous behavior. Its elastic modulus is about twice that of elastin, and its ultimate strain is more than 300 per cent. In terms of the amount of energy that can be stored per unit volume, resilin is more efficient than a steel spring.

Abductin is another energy-storing material, found in scallops. It serves as an antagonist for the adductor muscle in this animal, since any muscle placed on the outside of the shell would be in jeopardy. Its composition is quite different from either resilin or elastin. Its elastic modulus is similar to that of resilin, but its resiliency is only about 91 per cent. Its mechanical properties are made more interesting by the fact that scallops, unlike other bivalves, swim by opening and closing their shells at about 3 Hz. While the resiliency of scallop abductin is not remarkably high, it is substantially higher than the 80 per cent value of nonswimming mollusks. It would be of considerable interest to conduct a consistent experimental study of the viscoelastic properties of abductin, resilin, and elastin, in order to learn more about the ways in which these materials are adapted to the special conditions under which they function.

Keratin is a protein found in such diverse tissues as hair, horn, hooves, birds' beaks and feathers, and whale baleen. It is crystalline, not rubbery, and is much more rigid in compression than collagen, although it usually shares collagen's basic structure: three alpha-helices are twisted into a protofibril, which combines with others to form 8 nm diameter microfibrils. (Feathers and reptilian scales have a pleated molecular structure, known as the beta-form.) These chains are heavily cross-linked, both internally and to a surrounding polymer matrix, by disulfide bonds. The resulting stiffness of keratin in compression is, ironically, what gives two very soft materials, down and fur, their superior thermal insulating qualities. The structure of these materials can be minute to trap large quantities of air, yet stiff enough not to collapse under their own weight. Much of the information about the mechanical properties of keratin comes from studies related to wool in textiles (see Wainwright *et al.* 1976, p. 190). From these studies, it is known that keratin will transform from the alpha- to the beta-form under tensile load, exhibiting a great deal of plastic deformation in the process. Keratin has about 25 per cent of the stiffness of bone, and its properties are very susceptible to variations in its moisture content.

Cellulose is by far the most abundant biological structural material. It is a cross-linked, crystalline polymer whose basic building block appears to be a fibril 3.5 nm in diameter with a 4.0 nm periodicity. In the manufacturing of textiles, long, aligned fibers of cellulose can be formed. Much is known about the mechanical properties of cellulose because it is used so commonly in textiles, paper, and rope. Its tensile elastic modulus and extensibility are in the neighborhood of those of cortical bone. The long hairs on the seeds of the cotton plant (the cotton fiber used in textiles) are pure cellulose. Wood is a composite of cellulose, lignin, and other substances, and has less tensile stiffness than cellulose; it is frequently transversely isotropic (see Section 1.3.4).

Chitin (pronounced kite-in, and named after *Chiton*, a genus of molluscs having a chitinous dorsal plate) is a polysaccharide found in many invertebrates, but primarily in the insect exoskeleton (cuticle). Next to collagen, it is the most common component of connective tissues in animals. Connective tissue is that which serves to transmit loads or hold structures in place by virtue of an abundance of intercellular substance or interlacing processes. The opposite of connective tissue would be tissue which performs nonmechanical functions, such as a neuron. The chemical structure of chitin is similar to cellulose, and it has a lamellar structure similar to wood or bone. It is usually found mixed with protein to adjust its mechanical properties to a particular application. Its elastic modulus is about the same as that of hardwoods or bone. Its energy-storing properties are apparently not known, but are of interest because the thorax wall of insects like dragonflies contains chitin, which stores and releases strain energy during wing-flapping. Flexible chitin is found in the tendons of insects, and a mineralized form of chitin is found in the 'cuttlebone' of the cuttlefish (a chiton), where it forms a porous structure used for buoyancy. The mineral is not the hydroxyapatite of bone, but an orthorhombic form of $CaCO_3$ called aragonite, the same substance used to mineralize stony corals.

Apatite, or hydroxyapatite, is the mineral found in bone and calcified cartilage. It is a calcium phosphate, $Ca_{10}(PO_4)_6(OH)_2$. In the past it has been common for biologists to consider the mammalian skeleton as a large calcium reservoir. However, it should be appreciated that the skeletons of vertebrates are mineralized for mechanical reasons; the amount of skeletal calcium is far more than is needed as a physiological reservoir. The elastic modulus of apatite is about 100 times greater than that of collagen, and its density is about 3000 kg m^{-3}, about three times greater than that of most other biological materials. There is, therefore, a significant metabolic cost in using this material, since it substantially increases body weight.

Silk, like cellulose and keratin, is a biological material which has been usurped by man as a commercial material, so that much is known about the particular form used in industry. Silk fabric is made from the silk produced by the larva of a moth, *Bombyx mori*, for the construction of its cocoon, but many other arthropods produce similar materials. Silk in general is a strain-crystallized composite polymer consisting of two proteins: two strands of tough *fibroin* are embedded in a matrix of gummy *sericin*. Fibroin is the mechanically important component, with the sericin apparently serving to stick fibers together. The elastic modulus of larval silk is similar to bone and hardwoods, but its extensibility is much greater, about 20 per cent. Spiders produce two types of silk. Dragline silk is used in the radial fibers of a web, and when the animal 'rappels', as in visiting Little Miss Muffett. This material exhibits a yielding phenomenon at about 2 per cent strain, followed by a linear elastic region after 6 per cent strain. Consequently, this silk is several times stronger than cocoon silk, yet almost as extensible. Figure 8.4 compares this behavior to that of the transient yielding seen in wool, which was caused by a change in the molecular structure of the keratin. The other kind of silk produced by spiders is used for the spiral fibers of the web. It is very sticky (for trapping prey) and extremely rubbery, with maximum strains reaching 1600 per cent. It has been suggested that this behavior is achieved by adding proline to the monomer, which distorts the polymer structure so that cross-linking is prevented.

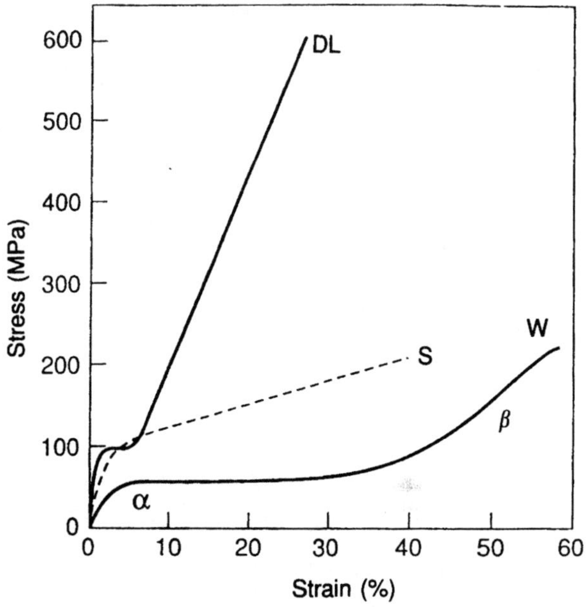

Fig. 8.4 Stress–strain curves of dragline (DL) and cocoon or spiral (S) silk are compared to that of the keratin fibers of wool (W). When wool is strained, its molecular structure changes from the alpha to the beta form. Cocoon silk yields at a strain of about 4–5 per cent, but dragline silk seems to experience a state change at a similar strain which gives it added strength and stiffness. In this respect, wool and dragline silk may be similar, but the effect occurs within a narrower strain window in the latter.

8.2.5 Biological tissues

Tissues are composed of mixtures of biological materials and the cells which produce or maintain them. The following are examples of structurally important, or *connective* tissues.

Bone is a composite of hydroxyapatite (about 43 per cent by weight) and type I collagen (about 36 per cent by weight). It also contains about 14 per cent water and a small amount of mucopolysaccharides, or ground substance. The primary determinant of its mechanical properties is its porosity. Cortical or compact bone is about 90 per cent solid bone tissue; spongy or trabecular bone is 80–90 per cent cavities filled with marrow. The elastic modulus of compact bone is about 20 GPa, and its tensile strength is about 200 MPa. Its failure strain is about 1.5 per cent, which is 750 per cent greater than the maximum strains measured in vigorously exercising animals. Bone is demonstrably viscoelastic, but it can usually be modeled as a brittle solid (Wood 1971). (Bone failure using linear fracture mechanics analysis is briefly mentioned in Section 1.4.4). Bone usually has a lamellar structure which is anisotropic (see Table 8.2).

Bone is manufactured by *osteoblasts*, which produce the organic components, with water in place of the mineral phase. This material, called *osteoid*, is mineralized over a period of weeks. Cells called *osteoclasts*

Table 8.2. Anisotropy of compact bone in tension

Direction	Elastic modulus (GPa)	Strength (MPa)
Longitudinal	23.1 ± 3.2	150 ± 11
Transverse	10.4 ± 1.6	49 ± 7

(Data from D. T. Reilly *et al.* 1974.)

resorb bone. The skeleton is constantly being removed and replaced in small packets in order to repair fatigue damage and reshape the structure to minimize stress. This process is known as *remodeling*, and the axiom that remodeling is governed by mechanical as well as metabolic factors is known as *Wolff's law.* Julius Wolff popularized this concept at the end of the nineteenth century, suggesting that bone structure was mechanically determined with mathematical precision, but the explication of this theory has still not been accomplished.

Cartilage is a tissue which comes in various forms for various purposes. Articular cartilage forms the bearing surfaces in joints like the knee. It contains more water than most soft connective tissues—about 75 per cent. Water aside, its primary constituents are proteoglycan (10–15 per cent) and type II collagen (60–65 per cent). Type II collagen contains three identical amino acid chains. Proteoglycans are extremely important in determining the mechanical properties of cartilage because they are hydrophilic. The collagen and proteoglycan molecules form a complex network of long molecules which hold water, making it possible for the tissue to support large loads. Figure 8.5 shows the structure of proteoglycan. Its core is hyaluronic acid, which is the key ingredient of synovial fluid. During loading, a filtrate of synovial fluid is forced through the cartilage; this is important for both the lubrication of the joint and the nutrition of *chondrocytes*, the cells which produce cartilage and lie distributed through it. The elastic modulus of cartilage is about 30 MPa. It is highly viscoelastic.

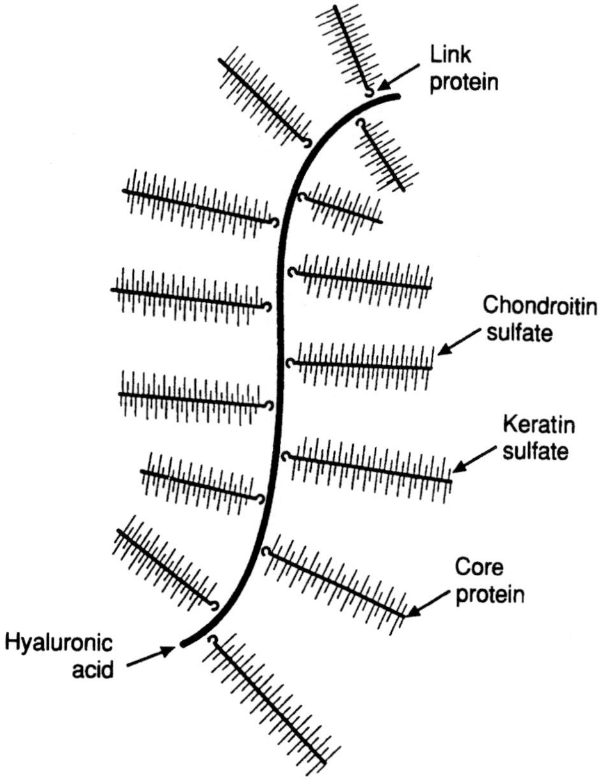

Fig. 8.5 The molecular structure of proteoglycan is sketched. It contains a central core of hyaluronic acid, to which brush-like glycosaminoglycans (GAGs) are attached by link proteins. Each GAG consists of a core protein and bristles of keratin sulfate or chondroitin 4- or 6-sulfate. The length of the proteoglycan molecules varies from 4 to 400 μm.

Tendons serve to connect muscles to bone. They are composed primarily of water (60–70 per cent) and type I collagen. For most tendons, elastin forms less than 3 per cent of the dry weight, and proteoglycans about 1 per cent. Their collagen fibers are well aligned, and they exhibit crystalline elasticity, as was shown in Fig. 8.3. This is important because it means that the forces exerted by muscles are not modified by creep or relaxation of the tendons. The mechanical properties of tendon are essentially those of collagen. Like bone, these properties are modified by cells (*tenocytes*) to match the mechanical conditions. In this case, however, the primary controlling variable is probably the cross-linking of the collagen. During locomotion, much energy can be wasted in slowing body parts and then accelerating them again. Some tendons, such as the Achilles and patellar tendons in humans, or the digital flexors in a horse, improve the efficiency of locomotion by storing large amounts of strain energy during one part of the gait cycle, and returning it later, in order to reduce muscle work. The ultimate example of this is the Achilles tendon of the kangaroo, which stores 70 per cent of the gait cycle energy during the animal's hopping locomotion.

Ligaments have compositions similar to tendons, but with slightly more elastin and proteoglycans. They run between bones to stabilize joints; important examples are the longitudinal ligaments which stabilize the spine, and the four major ligaments which stabilize the knee. Of these, the anterior cruciate ligament is often damaged in athletes. Ligaments also have aligned type I collagen fibers, and their mechanical properties are similar to collagen, except in the case of specialized organs like the ligamentum nuchae, in which abundant elastin reduces the elastic modulus and increases the extensibility. It is very difficult to measure the mechanical properties of tendons and ligaments because gripping them creates stress concentrations which perturb the results. The tensile strength of ligaments and tendons is so high that at low rates of loading they usually avulse their bony attachments rather than break. At high rates of loading, however, the viscoelasticity of bone enters the picture. Since bone is stronger when quickly loaded, the ligament or tendon may then break at mid-substance, and at a higher load.

Arteries like the aorta have about the same water content as tendons, but the dry weight is 40–50 per cent elastin, and only about 30 per cent collagen. The collagen is primarily type III, a still different combination of amino acid chains. This mixture causes a very rubbery elasticity and relatively low modulus of elasticity (about 0.3 MPa).

Skin has slightly less water and collagen than tendon, but 5–10 per cent of the dry weight is elastin. The collagen is type I. The primary determinant of the mechanical properties, however, is the random orientation of the collagen fibers. This makes the elastic modulus exceedingly low until strains of more than 50 per cent are reached, after which the elastic modulus is similar to that of articular cartilage. Figure 8.6 compares the stress–strain behavior of skin with that of tendon, ligament, cartilage, aorta, and collagenous tissue from the mitral heart valve.

Dentin forms the inner mass of a tooth. It is much more mineralized than bone—about 80 per cent apatite and 20 per cent organic matrix. Of the latter, about 90 per cent is collagen. It is, of course, a relatively brittle, crystalline material.

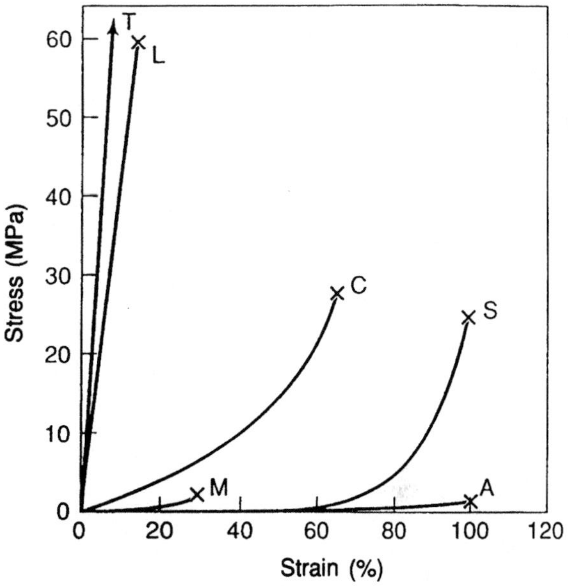

Fig. 8.6 Typical stress–strain behaviors of tendon (T), ligament (L), cartilage (C), aorta (A), skin (S), and mitral valve tissue (M) are compared. The X's at the ends of the curves indicate failure; tendon failure is beyond the boundary of the graph.

Enamel forms the outer layer of the crown of a tooth, and its content of hydroxyapatite is even greater than that of dentin—about 95 per cent.

Cuticle forms the exoskeleton of insects. Using chitin and protein, but without benefit of mineral, it attains about the same elastic modulus as bone. It also has a lamellar architecture like that of bone.

It should be emphasized that as biological materials are organized into organs, additional factors come into play. In general, biological molecules do not occur in pure form, but are organized into composite materials. Thus, we find collagen mixed with apatite in bone, cellulose with lignin in wood, collagen with proteoglycans in cartilage, and so forth. In addition, most biological materials are anisotropic, exhibiting very specific orientations of their fibrous components. For example, an artery is a composite of elastin and collagen in which alternating sheets of material have fibers helically wound around the axis of the vessel in opposing directions. (A rare example of a pure biomaterial is found in the cornea of the eye, the majority of which is made entirely of type I collagen. In addition, the corneal fibers must be of uniform diameter in order for the cornea to be transparent, and their orientation determines the shape of the cornea, and hence its optical properties.) Because of their composite nature and anisotropy, biological materials can exhibit very specific mechanical properties which are usually extremely difficult to duplicate in a man-made material.

8.2.6 Summary of biological materials

From the brief overview presented above, one obtains some idea of the enormous variability in the structural materials found in living organisms. Actually, however, this specialization of materials goes far beyond what we have just seen. First, each of the above materials has been further specialized by evolution of the various phyla and species, to suit the particular functional needs of an ecological niche. Secondly, in many instances the connective tissues of individual organisms are able to modify themselves within a small fraction of the lifetime of that individual, so as to cope with changing conditions. This is known to be true for bone, cartilage, and wood, and is probably true for other tissues as well, in both animals and plants. Thus, in order to duplicate the engineering feats of evolution, the bioengineer would not only have to create a virtually infinite variety of biologically compatible structural materials, but would also have to give these materials the ability to repair and alter themselves in an adaptive way. In all likelihood, it will prove far easier to learn the secrets of cell biology well enough to repair diseased tissues by controlling cells. In the meantime, bioengineers should approach their jobs with realistic expectations, and attempt to imitate and learn from nature's solutions whenever it is practical to do so.

8.3 Man-made biomaterials

Man-made biomaterials may be broadly categorized into metals, polymers, ceramics, carbons, and composites. One might also include here a few natural materials, in the sense of natural rubber, or porcine heart valves, which are treated and implanted in humans as a material which is not quite the same as a xenograft. (A graft is biological tissue transplanted from one place to another. An autograft is tissue transplanted within the same individual, e.g. abdominal skin moved to cover a hand wound. An allograft is transplanted within the same species; for example, a human-to-human heart transplant. A xenograft is transplanted from one species to another.) Each of these categories may contain materials in several forms, such as solids, membranes, fibers, or coatings. These materials may serve many purposes, including replacing structural organs or organs which carry out chemical exchange, housing electronic devices, repairing damaged or congenitally defective cardiovascular or connective tissue, or delivering drugs.

8.3.1 History

One may break the history of man-made biomaterials into three eras. Prior to 1850, nonmetallic materials, such as wood and ivory, were used, along with common metals like iron, gold, silver, and copper, to fabricate various simple prosthetic devices, such as teeth and noses, and in crude attempts to fix fractured bones (that is, to hold them together while they healed). In 1829 Levert experimented with lead, gold, silver, and platinum wire in dogs, but these metals clearly did not have the desired mechanical attributes (Williams and Roaf 1973). Furthermore, without anesthesia, human patients could not endure long surgeries in order to implant meaningful prostheses or fixation devices.

The second era of biomaterials was associated with the rapid development of surgery as something other than an emergency procedure; this occurred between 1850 and 1925. The advent of anesthesia just prior to the middle of the nineteenth century precipitated this development by making surgery infinitely more tolerable to the patient, and allowing the surgeon time to do precise work. Also, x-rays were discovered by Röntgen and found immediate application in orthopedics in the late 1800s, revealing for the first time the true nature of many skeletal problems which had previously been misunderstood. Finally, the acceptance of the aseptic surgical procedures propounded by Lister (Major 1954), which all too gradually occurred in this period, reduced the rate of post-surgical infections to a more reasonable level. (Surgical masks were not standard garb until after 1925, and even in the 1940s, many surgeons still only wore them over their mouths, leaving their noses free.)

Finally, the period from 1925 to the present may be viewed as the third era, in which the primary advances in the various surgical specialties have been due to three important developments. The first was the development of cobalt chrome and stainless steel alloys, in the 1930s and 1940s respectively. The second was the development of polymer chemistry and plastics in the 1940s and 1950s. (These alloys and polymers were developed primarily for industrial rather than medical applications.) Finally, the third advance was the discovery of ways to produce useful quantities of penicillin and other antibiotics. The ability to further reduce surgical infection rates, and to fabricate many devices which were compatible with biological tissues, significantly advanced the ability of surgeons to treat a great variety of problems. It may be noted that most of the implant materials commonly in use today were developed more than 25 years ago, and the intervening years have been ones of gradual refinement.

8.3.2 Problems associated with surgical implants

At this point, it is useful to present an overview of the problems which implant designs must ideally overcome.

Anatomic compatibility. The implant must be geometrically compatible with the body. From a materials point of view, this means that the material must be formable into the desired shape using economical methods and without compromising its mechanical properties.

Corrosion. The material chosen for the implant must not corrode in the presence of body fluids. This includes the avoidance of crevice and fretting corrosion.

Toxicity and infection. The material must not poison the patient; therefore, it must either be free of toxic substances, or they must be adequately locked into the material's structure. Also, it must be easy to completely sterilize the implant (using steam under pressure, radiation, or ethylene oxide gas) without damaging its material properties.

Strength. The implant must not break, either due to an occasional acute overload or due to fatigue from repeated functional loads. The strength and fatigue properties must combine with the shape of the implant to keep stresses within safe limits, particularly where stress concentrations cannot be avoided (see Section 1.4.4).

Fastening. The implant must be fastened to the anatomy in such a way that it can function. In the case of a cardiac pacemaker, this may be simply a matter of maintaining an approximate position of the electronic package. In the case of a bone implant, some sort of rigid, load-bearing fixation is usually required. This frequently complicates the materials problems—for example, crevice corrosion can occur under screw heads.

Long-term physiological compatibility. Usually, the design must work not simply for a year or two, but for many years. Changes which may occur in the implant must not be physiologically harmful. Toxic ions which may be gradually released must not accumulate or lead to a long-term immunological response. A connective tissue implant must not perturb the stresses in adjacent tissues into a state which prohibits normal remodeling to repair fatigue damage.

Revision surgery. The implant must be designed so that it can be removed and replaced if it fails. Since most prostheses involve the destruction of the original organ, and removal of the prosthesis may further destroy important tissues, one must have contingency plans for replacing the prosthesis which take these factors into account. This is particularly important in the case of orthopedic surgery, since if a prosthesis fails even though rigid fixation of a portion of it to the bone was successful, it may be necessary to destroy part of the remaining bone to detach the implant.

It should be emphasized that these ideals are only approximately achieved by most implants. This will be seen in the description of problems associated with total hip replacement described below.

8.3.3 Evolution and biocompatibility

In considering the problem of choosing a man-made biomaterial capable of successfully functioning inside the body, it is useful to begin by reviewing some fundamental concepts about biology and, specifically, about evolution. Living organisms have evolved with one overriding principle which motivates their actions—survival. One threat to survival is to be invaded by another organism. Therefore, all organisms seek to prevent foreign matter from gaining entry to them, and if they fail in this, they work very hard either to destroy the invading object if its molecules look destructible, or to encapsulate it if it looks impregnable. However, this rather simple principle becomes more complicated in the light of an evolutionary theory which holds that cells evolved from bacteria which specialized and combined to function as a single organism, and that higher organisms are actually giant conglomerates of cells and bacteria which have achieved a capacity to work together for their common survival. If animals are biological 'melting pots', then the problem of detecting foreign material becomes even more complex, and the problem of defeating an immune system which successfully copes with this task looms very large indeed.

Today, people pay surgeons to put foreign objects into them, but their bodies still stick to the old principle of encapsulate or destroy. In some cases, encapsulation does not have an adverse effect on an implant, but as a general rule, biomaterials should be able to avoid the body's natural inclination to either encapsulate them or break them down. Since doing this by destroying the body's defenses (that is, the immune system) is a poor tactic, the best way to proceed is to make the implant invisible to the host's chemical sensors. This is difficult to do, since most materials will be easily recognized as 'outsiders', and will have a difficult time avoiding the consequences. One principle to keep in mind while playing this game is that materials with molecules which look like biological molecules will be more readily attacked by destructive cells. For example, nylon and polyethylene both have a core of carbon atoms with hydrogens attached along their lengths. The main difference between them is that polyethylene has a CH_3 terminal group, while nylon has an NH_2 terminal group. Since proteins have the latter kind of terminus, nylon tends to be more susceptible than polyethylene to degradation by the immune system.

On the other hand, impregnable (i.e. inert) materials like polyethylene are usually distasteful to cells to the extent that it is very difficult to induce the body to attach tissue to them. In fact, no polymers have been found which are both immune to degradation and amenable to tissue adhesion. Other materials are very attractive to connective tissue cells; for example, hydroxyapatite is nearly identical to bone mineral, and will quickly become integrated with bone when implanted in the skeleton. Unfortunately, hydroxyapatite is not nearly as strong as bone. The commonly used materials which are strong—metals—cannot be destroyed by the immune system, so the body tends to encapsulate them in fibrous tissue. Obviously, this frustrates attempts to rigidly attach them to bone. One begins to see the nature of the problem.

8.3.4 Orthopedic implant biomaterials

The emphasis in this section will be on orthopedic implant materials, because these applications use primarily the same materials as other types of implants, they usually are extremely demanding from a structural point of view, and in terms of sheer mass of material (if not numbers) they constitute a very large percentage of non-dental materials implanted each year. In the mid-1980s there was, for example, an enormous increase in the annual number of total hip replacements (or *arthroplasties*) (see Section 9.1). This growth stems from the inability of medicine to treat various forms of arthritis pharmacologically, and the remarkable success of hip arthroplasty techniques perfected by Charnley (1973) and others in the 1960s, and used with increasing frequency in the 1970s. After these devices had been in place for 10–15 years, it became clear that many of them were mechanically failing. Ironically, however, the full realization of the magnitude of the problems associated with Charnley's methods was concurrent with the increased implantation of total hip prostheses in the 1980s. The reason for this is that Charnley's technique is reasonably dependable for the expected lifetime of the typical patient (about 15 years). Also, patients and physicians have no other options with as much promise of relief from pain and loss of motion. Therefore, total hip arthroplasties continue to be done at an increasing rate, with alternative techniques sometimes tested in patients before animal tests are done, because of the exigencies of the risk-to-benefit ratios in this area of medicine. With this background as context, this procedure may be used as an important example for the discussion of the problems associated with the selection and use of implant materials.

8.3.5 Total hip arthroplasty

The fundamental problems in total hip arthroplasty were wear of the ball and socket surfaces and attachment of the prosthesis to the bone. The first of these was solved when it was found that acceptable wear was obtained when a stainless steel ball was used in a cup lined with ultra-high-molecular-weight polyethylene (UHMWPE). To solve the second problem, Charnley borrowed the polymer poly(methyl methacrylate) (PMMA) from dental practitioners to serve as a grouting material between the implant and bone. (In orthopedics, PMMA is frequently called 'bone cement', but it is not truly a cement.) Figure 8.7 shows the end result of the procedure, which proceeds as follows. The neck of the femur is removed and a hole is reamed through spongy bone down into the medullary canal. Then a batch of PMMA is prepared by mixing the liquid monomer with a powder of PMMA polymer; this forms a dough which is workable for several minutes. Charnley stuffed this dough into the medullary canal and pushed the femoral component in on top of it, extruding the PMMA into the interstices of the

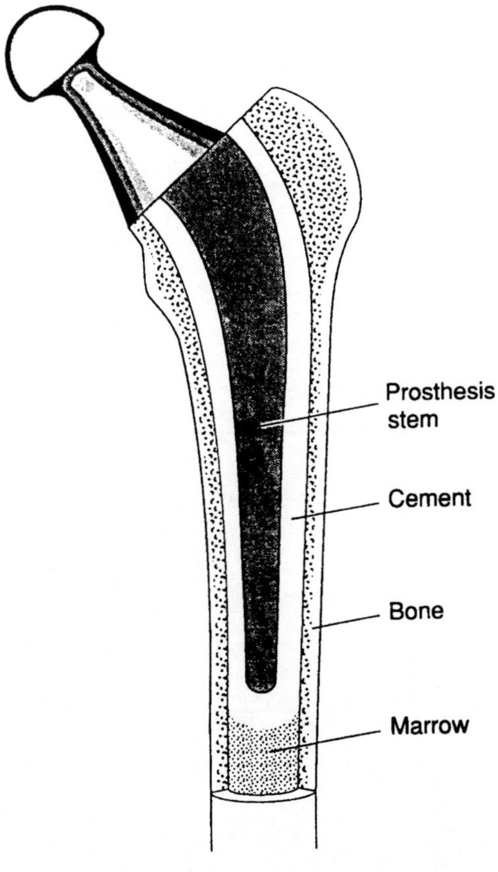

Fig. 8.7 Sketch of the femoral component of a total hip prosthesis in place. Following removal of marrow, the stem of the prosthesis is grouted into the medullary canal of the femur by a layer of PMMA (bone cement). The ball at the top of the prosthesis fits into a cup of UHMWPE, which is fastened to a metal hemisphere. This half of the prosthesis is screwed or 'cemented' (with PMMA) into the reamed-out acetabular cup of the pelvis. The assembly is held together by the tension of the surrounding membranes and muscles.

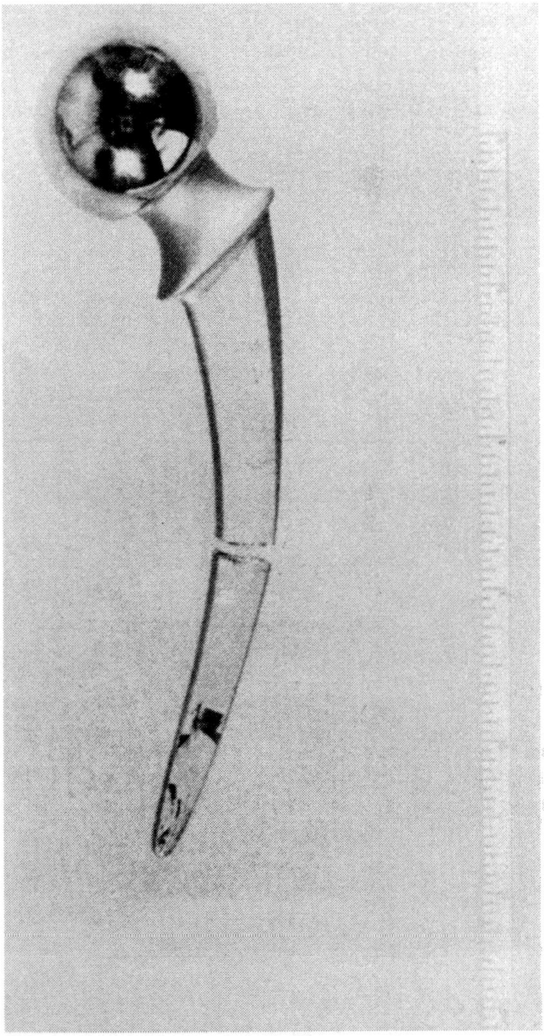

Fig. 8.8 Photograph of a failed hip prosthesis, showing the typical appearance of a fatigue fracture caused by proximal loosening.

spongy medullary bone and forming a congruent layer of polymer around the stem of the prosthesis. A similar procedure was followed on the acetabular (or socket) side of the joint.

The procedure soon became very common. It appears to be very forgiving of mistakes on the part of the surgeon because the PMMA fills gaps and allows the position of the implant to be adjusted during the procedure. Subsequently, however, it has been learned that its success depends greatly on consequences of variations in the details of the technique and the patient's anatomy—e.g. air bubbles or blood in the cement, uniformity of the cement mantle around the stem, and the quantity and quality of the adjacent bone. Eventually, it became clear that the Charnley technique was very successful for about 10 years, but failures were frequent after 15 years or so. Typically, the bone around the proximal stem deteriorated, the proximal PMMA fractured, and the prosthesis was only gripped by the distal 2–3 cm. Consequently, a fatigue fracture would break off the tip of the stem, and the remainder of the femoral component would become very loose, as depicted in Fig. 8.8.

The strategy most investigators are taking to solve these problems is to get rid of the bone cement. The alternative most widely proposed, in one form or another, is to coat the stem of the prosthesis (and the back of the acetabular component) with a porous material which bone can grow into, creating a bond, as indicated in Fig. 8.9.

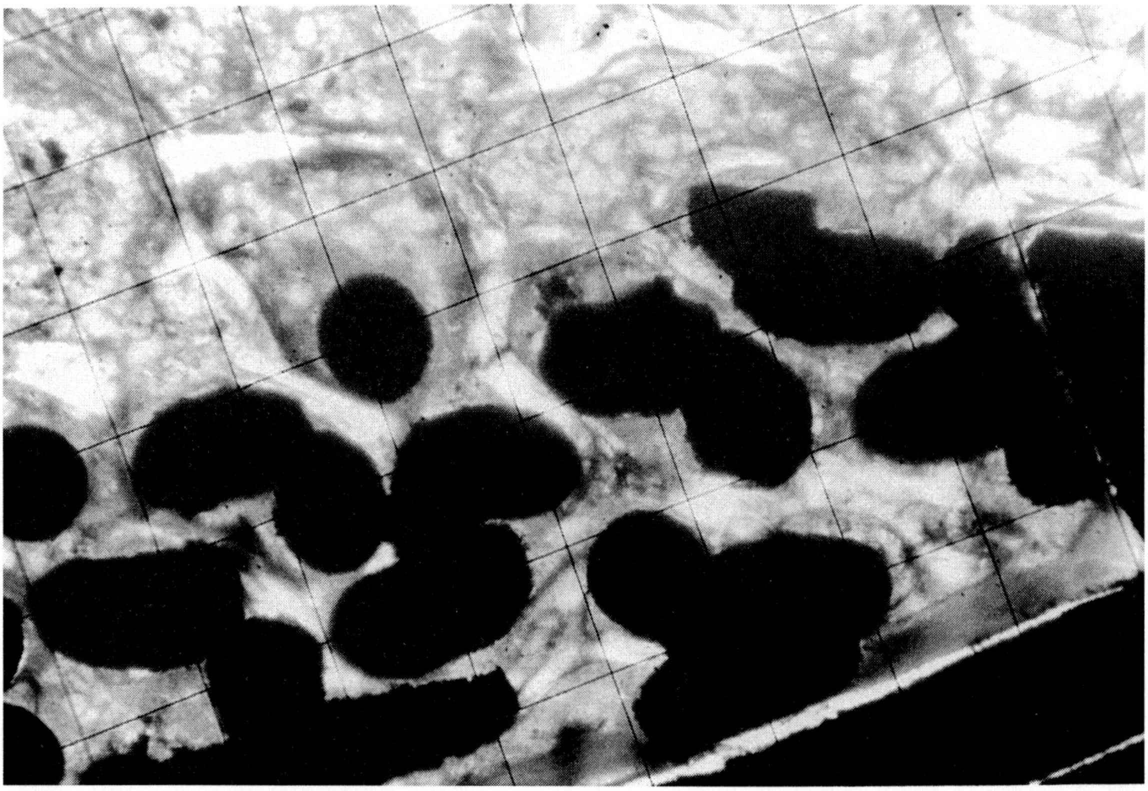

Fig. 8.9 Photomicrograph of bone growth into a porous-surfaced bone implant. The black material is titanium alloy wire woven and folded to form a porous surface about 1 mm thick. The photograph was made with polarized light, so that the bony struts appear as bright regions invading the pores. A layer of fibrous tissue may be seen as a dark gray band at the base of the porous region. In human patients with porous-surfaced joint prosthesis, fibrous tissue ingrowth frequently exceeds bony ingrowth. The superimposed grid was used for analysis of the ingrown tissues; the grid lines are 0.23 mm apart.

8.3.6 Other orthopedic implants

Other common orthopedic implant devices include replacement knee and finger joints, and a variety of internal fracture fixation devices: screws driven across the fracture, plates screwed to the side of the broken bone, rods placed down the medullary canal, wires wrapped around a comminuted fracture, and rods fastened to the spine to stabilize a fracture, correct scoliosis, or fuse two or more vertebrae. There are also various external fracture fixation devices. These consist of pins driven into each half of a broken bone and protruding out through the skin to attach to a framework designed to hold the bone securely together while it heals. Generally speaking, fracture fixation devices and screw-fixed systems work well because the loading on them can be controlled. The major problems are associated with total joint replacement, where functional loads must be transmitted across a bone–metal interface for a long period of time. So far, no one has found a way to do this with the materials at hand, which are as follows.

8.3.7 Commonly used implant materials in orthopedics

Almost all orthopedic implants involve some combination of the following three metals and two polymers.

Cobalt–chromium alloys were the first corrosion-resistant alloys to be developed, and have proven very effective in surgical implants beginning in 1938 when Venable reported the use of such an alloy in orthopedics

Table 8.3 Approximate compositions (wt%) of metal alloys commonly used in biomedical implants

Element	Stainless steel	Cobalt–chrome alloy	Titanium alloy	MP 35N alloy
Al			6	
Co		62		33
Cr	18	28		20
Fe	64			1
Mn	2	1		
Ni	13	3		35
Ti			90	
V			4	

Table 8.4 Mechanical properties of metals and ceramics commonly used in biomedical implants

Material	Elastic modulus (GPa)	Strain at failure (%)	Tensile strength (MPa)
Stainless steel	193	10	1000
Cast Co–Cr alloy	235	8	670
Wrought Co–Cr alloy	235	12	1170
Ti–6Al–4V alloy	117	10	900
Pure titanium	100	15	550
Alumina	380	0	50[a]
Hydroxyapatite	62	0	690
C–Si composite	21	0	690

[a] Compressive strength = 4 GPa.

(Omer 1987). They are usually regarded as the metal of choice in skeletal prostheses. They have historically been available in cast and wrought forms; as with stainless steel, the wrought material is substantially stronger. The compositions and mechanical properties of each of the alloys discussed in this section are shown in Tables 8.3 and 8.4 respectively.

A modification of Co–Cr alloy was introduced in 1972 by a European manufacturer of orthopedic implants. This alloy contains 35 per cent nickel, and is thus known MP 35N. It can be forged and heat-treated to obtain tensile strengths significantly above those of stainless steel and Co–Cr alloy—as high as 1800 MPa.

Stainless steel (usually 316L) is a workhorse industrial alloy which has been very successful as a surgical implant material. Both stainless steel and cobalt–chrome alloys owe their corrosion resistance to the formation of a ceramic-like CrO_2 coating on the surface, and it is important that this coating not be scratched during implantation. The ductility of these alloys can be increased by heat treatment (for example, heat to 1050°C for 1 h and cool in air), and their strength can be increased by cold working (e.g. rolling a thick bar into a thin one). Cast stainless steels are unsuitable for orthopedic applications because of their large grain sizes and low fatigue strengths. Type 316LVM (low carbon, vacuum melt) material is preferred.

Titanium alloys are primarily Ti–6Al–4V (that is, 6 per cent aluminum, 4 per cent vanadium, and 90 per cent titanium). (Pure titanium is also used, primarily in dental applications.) This metal is becoming increasingly popular because its strength as is good as that of the two previous choices, but it is only half as stiff (Table 8.4). This is potentially important because a large elastic modulus mismatch between implant and bone causes stress concentrations in some places and tends to unload the bone in others. However, the modulus of Ti–6Al–4V is still several times greater than bone, and marked changes in the behavior of the adjacent bone have not been seen when this alloy is used.

Poly(methyl methacrylate) (PMMA) is an extremely common acrylic plastic otherwise known as Lucite. Its monomer has the form

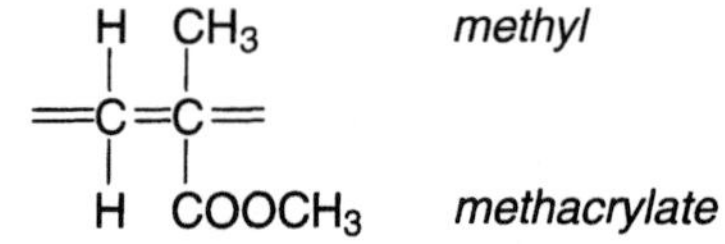

These units are connected to form the polymer. Frequently, the polymer will have an additive: barium to increase its radiographic visualization, or an antibiotic to prevent infection following surgery. The mechanical properties of PMMA are shown in Table 8.5; barium and antibiotic additions do not substantially affect these properties.

PMMA polymerizes with an exothermic reaction which causes the mass of dough used in total hip surgery to reach temperatures in the vicinity of 90°C, and it is thought that this may kill bone cells which would otherwise survive, but this problem may be insignificant compared to the vascular interruptions and other trauma

Table 8.5. Mechanical properties of polymers commonly used in biomedical implants

Polymer	Elastic modulus (MPa)	Strain at failure (%)	Tensile strength (MPa)
Silicone rubber[a]	2.4	700	
Polyether urethane		700	41
Biopolymer[a]	1.5	350	13
UHMWPE	500.0	350	35
PMMA	2000.0	2	30[c]
TCF[b]	20,000.0		250

[a] Modulus at 100% strain.
[b] Carbon-fiber-reinforced triazone resin.
[c] Compressive strength = 90 MPa.

produced by the surgery. Similarly, while there is something about the PMMA which causes blood pressure to drop momentarily when it is implanted, this problem is not serious compared to another cardiovascular possibility associated with the surgery: embolism caused by fat released from the femur's medullary canal.

PMMA is also used in the manufacture of hard contact lenses, and of intraocular lenses for cataract patients. For this application, polymerization is initiated with heat (rather than heat being the result of polymerization, as in bone cement) to produce an extremely clear material with good optical properties. Soft contact lenses are usually hydrogels made of homo- or copolymers of hydroxyethyl methacrylate; other methacrylates and silicones have also been used, however.

Ultra-high-molecular-weight polyethylene (UHMWPE) has a very simple chemical structure:

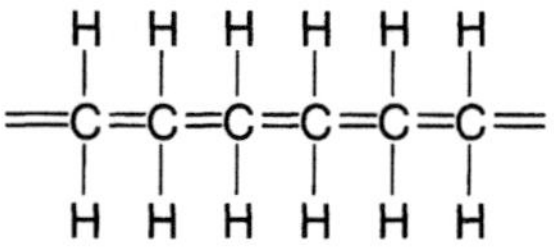

In the 'ultra-high-molecular-weight' form, these chains achieve a molecular weight of 1–4 million. When this material is used as a bearing surface against one of the above metal alloys, ordinary body fluids are usually sufficient to lubricate the artificial joint and prevent wear problems. While a variety of other polymers have been tried, UHMWPE remains the best one for use in this application because of its relatively low wear against metal. Polyethylene has also been used to replace the ossicles of the inner ear.

The great majority of orthopedic implants are made from these five materials. (A notable exception is a type of small joint replacement made from silicone rubber, as described below.) While PMMA appears to be on its way out as a staple in the orthopedist's cupboard, there are no indications that any other materials are about to be placed in the cupboard. It seems realistic to accept the remaining four materials as having reasonably good mechanical and corrosion properties, although their ability to form intimate bonds with bone is greatly limited. If one accepts such an assessment, then almost all failures in fixation devices occur for reasons other than inadequate materials; for example, the surgeon uses poor technique or judgment, or is forced to use a device improperly by the nature of the fracture. On the other hand, while some failures of total joint prostheses may be the result of poor technique or judgment, most are probably due to a combination of poor design and the inability to form a good tissue–implant interface. As a result, the bone adjacent to the prosthesis remodels in a pathological way, experiences fatigue damage, and deteriorates, overloading the implants until it fails. Thus, the primary problem is one of biomaterials only to the extent that a dependable and gradual load transfer between implant and bone depends on (a) forming intimate bone–implant bonds and (b) matching the elastic modulus of the implant to that of bone. Otherwise, the problem is one of designing the implant so that it produces physiologically acceptable stresses in the bone, and stresses within itself which result in a fatigue life greater than the lifetime of the patient.

It is clear, therefore, that the problems in this sphere involve bone physiology at least as much as materials science. However, in concluding this chapter by surveying additional biomaterials used in other areas of medicine, several substances will be mentioned which are being experimentally tested with the above orthopedic problems in mind.

8.3.8 Other man-made biomaterials

Table 8.6 lists many other areas of surgery and medicine where biomaterials are important. This chapter will conclude with a brief survey of materials that are commonly used in these applications, as well as some materials still being studied experimentally.

Table 8.6. Some biomedical implants used in various branches of medicine

Specialty	Implants
Cardiovascular surgery	Heart valves
	Blood vessels
	Assist devices
	Pacemakers
	Total heart prostheses
	Blood oxygenators[a]
Dentistry	Prosthetic teeth
Medicine	Kidney dialysis machines[a]
	Implanted drug release devices
Neurosurgery	Hydrocephalus shunt
	Bone plates
Obstetrics and gynecology	Implanted birth control devices (e.g. IUDs, cervical caps)
Ophthalmology	Contact lenses
	Intraocular lenses
	Artificial eyes
	Detached retina sponges
Orthopedic surgery	Joint prostheses
	Fracture fixation devices
	Bone, tendon, ligament substitutes
Otolaryngology	Ossicle prostheses
	Cochlear prostheses
	Drainage tubes
Plastic/oral surgery	Bone, cartilage, skin substitutes
	Breast implants
Urology	Incontinence devices
	Impotence devices

[a] While these are not implanted, they come in direct contact with blood, and hence require biocompatible materials.

Polymers

Man-made polymers may be classified as *elastomers*, which have fewer cross-links and are thus more rubbery, and *plastics*, which are more crystalline. PMMA and UHMWPE are examples of plastics; silicone rubber is the premier example of an elastomeric biomaterial.

Poly(dimethyl silixane) (silicone rubber) is a widely used polymer (made from sand!) which was first applied biomedically in a hydrocephalus shunt in 1955. It has a long history of biological compatibility and clinical testing in a great variety of applications, and a 'medical grade' (Silastic) has been developed with superior biocompatibility and mechanical properties. It can be sterilized by steam, radiation, or ethylene oxide.

This polymer can be manufactured with various degrees of cross-linking to adapt its mechanical properties for other purposes as well. The cross-linking typically takes the following form:

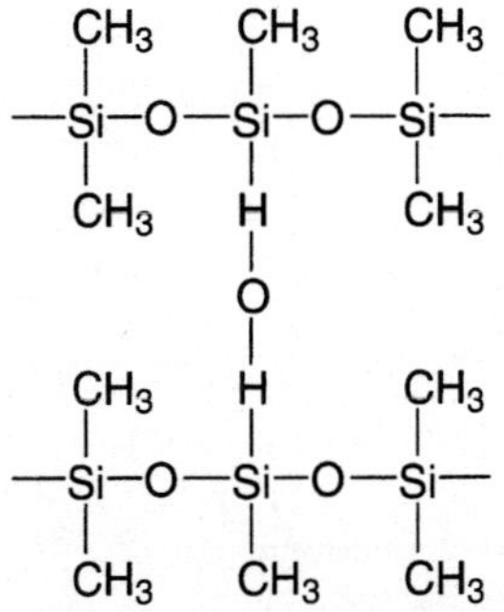

Silicone rubber is commonly used for catheters. A gel silicone has often been used for breast prostheses or augmentation, and rubbery sheets have been used in hernia repair. Harder versions have served as balls in ball-and-cage heart valves. Silicone rubber is widely used in pacemakers, and in biomedical research, to form seals against body fluids for electrical leads.

Silicone rubber is used as the functional membrane in both kidney dialysis and extracorporeal blood oxygenator machines. It has also been used in drug delivery implants, where it serves as a membrane through which the drug slowly passes. Silicone also serves as a structural material in prosthetic heart valves, and to replace the auricle or the ossicles of the ear.

In orthopedics, silicone rubber is formed into soft, flexible 'strap hinges' for the replacement of arthritic finger, wrist, or toe joints. Typically, the metaphyseal portion of each bone is cut off, and the prosthesis has a pair of stems to go into the two medullary canals. There is no bearing surface; the rubbery connection between the stems simply bends to accommodate the motion. However, its fatigue resistance is not as good as it should be for this application; failure rates after 2–3 years have been reported to be as high as 38 per cent.

Dacron has been used for blood vessel prostheses since 1951. Dacron is thrombogenic, so the pores in the fabric soon become filled with coagulated blood, which is then replaced by a tissue called neointima, which serves as a biological wall between the Dacron and the blood. However, if the vessel diameter is smaller than 6 mm, the neointima will occlude the tube. Dacron has also been used in prosthetic heart valves.

Polytetrafluoroethylene (PTFE, Teflon, or Gor-Tex) has been used experimentally to try to produce blood vessel prostheses smaller than 6 mm in diameter, and the neointima appears to be thinner with this material. However, it too is unsatisfactory for most human blood vessels, which are smaller than 3 mm. PTFE is also used to make prosthetic heart valves, ligaments, and artificial ossicles for the ear. It was tried by Charnley (1973) as a bearing surface in total hip prostheses, but rejected due to its poor wear behavior against metal.

Polyether urethane has been used for years in blood bags and tubing for kidney dialysis machines. It has also been tried as a material for blood vessel replacement. It is frequently found in intra-aortic balloon pumps and in artificial hearts, where it lines the chambers and forms the pumping diaphragm. The primary requirement in the latter application is fatigue resistance; a prosthesis which is to last 10 years must flex about 360 million times. Materials which have reportedly been tried for this purpose and found wanting include several other polymers used in tubing for kidney dialysis machines: poly(vinyl chloride), silicone rubber, and natural and synthetic rubbers.

Polyalkylsulfones are also used in blood oxygenator membranes.

Hexsyn and a variation, *Bion*, are commercial names of a recently developed elastomer which apparently has good biological compatibility and an extraordinarily high fatigue life (more than 300 million cycles to failure compared to 0.6 million for silicone rubber in an ASTM D430 flexure test). It has been tested in human finger joint prostheses.

Epoxy resins have been used to encapsulate electronic implants.

Polydepsipeptides and *polylactic acid* polymers have been tested with some success as biodegradable implants. Implant materials which disintegrate gradually in body fluids are useful as sutures, and potentially useful as bone plates, to fill defects in bone or other tissues, and for the delivery of embedded drugs.

Ceramics

Ceramics generally have hydrophilic surfaces and are amenable to intimate bonding with tissues. They are very biocompatible but brittle relative to biological materials, including bone. Their primary application may ultimately be as a coating for metals to promote attachment to bone.

Hydroxyapatite and *tricalcium phosphate* (TCP, $Ca_3(PO_4)_2$) in various forms have proven to be very biocompatible, and form intimate bonds with bone. Unfortunately, their strength is not great. They are primarily viewed as materials which may be made in porous or granular forms and used to fill bone defects in lieu of a bone graft. It is possible that when their pores are fully ingrown with bone, and the region is buttressed by peripheral new bone, the resulting structure will be strong enough to carry functional loads. There is also some indication that TCP or similar materials (for example, calcium sulfate, calcium carbonate) may

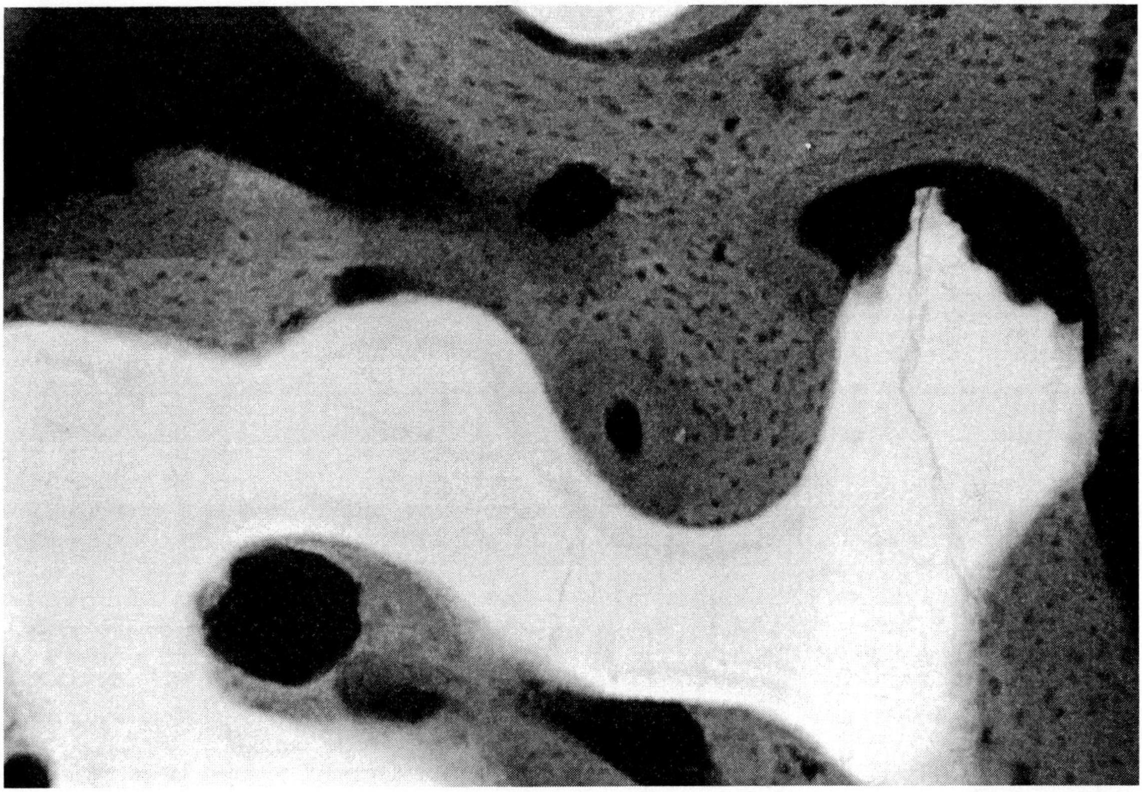

Fig. 8.10 Microradiograph of a porous hydroxyapatite implant (white material) ingrown with bone (gray material). The black areas are vascular and marrow spaces. Note the intimate contact between bone and hydroxyapatite. Note also the evidence of resorption of the implant in the upper right corner. Original magnification 31×.

eventually be resorbed and replaced entirely by bone. Figure 8.10 shows bone growth into a porous hydroxyapatite implant.

Alumina is a widely used industrial ceramic; for example, it is used to make spark plug insulators. Alumina is a polycrystalline form of Al_2O_3. (Rubies and sapphires are single crystals of Al_2O_3.) It has relatively good strength characteristics, but is very stiff and brittle compared to bone. It is sintered from alumina powder under pressure at 1600°C to form structures with relatively smooth surfaces. Currently, alumina is being used for the heads of total hip prostheses, particularly in Europe. In this application, it has proven to be wear- and corrosion-resistant, and highly biocompatible.

Zirconia (polycrystalline ZrO_2) is another ceramic which is being used for the heads of total hip prostheses in Europe. Its wear resistance is reported to be inferior to that of alumina, however.

Bioglass (Na_2O–CaO–P_2O_5–SiO_2) is a glass with soluble additives designed to form a silica gel at its surface and thereby aid chemical bonding to bone. To date, however, this material has not been widely accepted as a good orthopedic implant candidate.

Carbon materials

Since carbon is the basis for organic chemistry, one might think that this element would serve also as the basis for biocompatible materials. This is generally true. Graphite has a weak, anisotropic crystalline structure, but turbostratic carbon is isotropic and relatively strong. The two most useful forms of turbostratic carbon in biomedical applications are vitreous (or glassy) carbon, and pyrolytic low-temperature isotropic (LTI) carbon. These are both isotropic materials, but the latter is more wear-resistant and stronger. Turbostratic carbons, carbon-fiber–PMMA, and carbon–silicon composites have excellent biocompatibility and can have elastic moduli similar to that of bone. However, it typically turns out that in order to reduce the stiffness of these materials to that of bone, the strength must be reduced to less than is available in metals; see Table 8.5. In the last decade a number of these materials have been tested for various biomedical applications, but few have been widely used. Carbon–silicon and carbon-coated graphite have, however, been used to some extent in heart valves, and pyrolytic carbon coating has been shown to improve bone growth into porous metal surfaces.

8.4 Current avenues of biomaterials research

Many new ideas are being explored for the improvement of biomaterials. With the continued rapid growth of molecular biology, it is becoming possible to construct artificial proteins of uniform structure. It may eventually be possible to construct polymers which the body cannot recognize as 'foreign' and which have mechanical or other properties that are optimized for a particular application. In the meantime, one step in this direction is the development of 'star polymers' and dendrimers. These polymers have 'spokes' radiating from a central core to which biomolecules can be attached, 'hiding' the central (i.e. the functional) molecule from the immune system.

Polymers for soft tissue (for example, tendon) replacement are being developed which avoid problems associated with the leaching of biologically problematic, unreacted compounds by using benign processing methods. Poly(vinyl alcohol) is an example of such a material. Poly(acrylic acid) and carboxymethyl cellulose are being examined as alternatives to medical adhesives such as cyanoacrylate ('superglue') which degrade unpredictably. Eventually, it may be possible to mimic the effectiveness of natural adhesives, like those used by barnacles, for example.

Resorbable orthopedic biomaterials, which would serve a mechanical role until the replaced tissues reconstitute themselves and then disappear, have been difficult to achieve because the correct combination of biocompatibility, resorbability, and strength has not been found. Polydioxanone is a recent candidate in this realm which has received FDA approval in the USA, but so far only for sutures. Another idea which has not yet been turned into a clinical success is the development of polymers which would change phase under physiologically compatible conditions.

There are many possibilities for new biomaterials, but the history of this subject suggests that success is slow and infrequent. The infinite variety of tissues which nature creates from a myriad of 'natural biomaterials' (some of which were reviewed at the beginning of this chapter) stands as a vast ocean in comparison to the few man-made biomaterials which have been successful enough to achieve common use. Certainly, there is every reason to expect greater success in the future, and the problems are extremely interesting, but the student of biomaterials needs to have a realistic understanding of the challenge.

8.5 Further information

The primary sources of the material used in this chapter are listed below. Several of these are very enjoyable to read and are highly recommended to those with a deeper interest in this subject. For example, Alexander's *Animal Mechanics* (1968), and *Mechanical design in organisms* by Wainwright *et al.* (1976), contain a wealth of information about nature's biomaterials, and many of the key concepts presented here come from these volumes. Two other books which contain excellent reviews of both biological and man-made biomaterials are *Natural and living biomaterials*, edited by Hastings and Ducheyne (1984), and *The biomechanics of trauma*, edited by Nahum and Melvin (1985). The concept that polymers having molecular structures like those of proteins may be more immune to degradation in the body is discussed by Kawahara, writing in *Perspectives on biomaterials*, edited by Lin and Chao (1985). Each of the other works in the bibliography is also recommended for further reading in those areas suggested by their titles.

Problems

P8.1 Chemical analyses suggest that the molecular weight between cross-links in resilin is about $M = 3200$ kg mol^{-1}. Compare this value to that estimated mechanically using eqn (8.1). In using this equation, the density should be given as kilograms of dry protein per cubic meter; assume this is 500 kg m^{-3} for resilin. Assume that the temperature is 20°C and that the elastic modulus of resilin is 1.8 MPa. See Wainwright *et al.* (1976) and Alexander (1968) for further discussion of the result.

P8.2 Resilin is able to store 2×10^6 J of energy per cubic meter when loaded to near the failure point. (a) Show that this is much more than the energy per unit volume than can be stored by compressing or stretching a steel column having a yield stress $\sigma_Y = 425$ MPa and an elastic modulus $E = 200$ GPa. Assume the steel is loaded to the yield point, and is linearly elastic to this point. (b) What if the steel spring is in the form of a beam loaded as a cantilever with a point load at its end? Assume the beam width and length are 2 and 16 times its thickness, respectively. (c) If a flea weighing 0.45 mg has dual resilin springs containing a total of 3×10^{-13} m^3 resilin, how high could it jump?

P8.3 Suppose that the stem of a hip replacement prosthesis is a stainless steel cylinder 12 mm in diameter. Assume that the top of the femur is cut off perpendicular to the shaft and the stem is inserted into the medullary canal, which has been reamed to a diameter of 12 mm. The outer diameter of the femur is 22 mm. Assume also that the prosthesis has a collar which distributes the applied load such that the bone and the metal stem experience equal strains in simple compression (see Fig. 8.11). (a) Show that the stress in the bone depends on the ratio of elastic modulus of the bone relative to that of the metal. (b) How do the relative cross-sectional areas of bone and metal affect the bone stress?

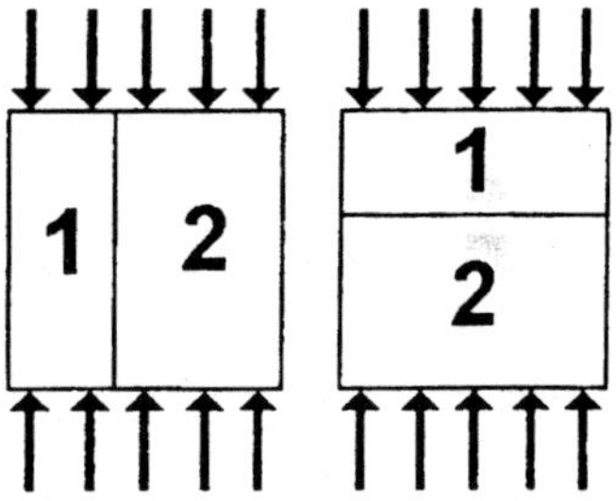

Fig. 8.11 Sketch of Voigt (left) and Reuss (right) models. Each model contains two materials, of volume fractions V_1 and V_2 and moduli E_1 and E_2, respectively. In the Voigt model the two materials experience the same strain but different stresses; in the Reuss model, they experience the same stress but different strains. The effective modulus of each combination is to be derived in Problem 8.4.

P8.4 Many biological materials are composed of mixtures; for example, bone may be considered a mixture of collagen and mineral. Consider the elastic modulus of a mixture of two materials whose individual moduli are E_1 and E_2. The volume fractions of materials 1 and 2 are V_1 and V_2, respectively. It is known that the elastic modulus of the mixture has upper and lower limits given by mathematical models known as the Voigt and Reuss models, respectively. Both models assume that the two materials are lumped into a solid mass as shown in Fig. 8.11. The Voigt model assumes strain is equal in the two materials, the Reuss model assumes stress is the same in the two materials. Derive the following results for the modulus of the mixture:

Voigt model, upper bound $E = V_1E_1 + V_2E_2$

Reuss model, lower bound $E = E_1E_2/(V_1E_2 + V_2E_1)$.

P8.5 For bone, assume that the volume fraction of apatite (with a modulus of 114 GPa) in the calcified matrix is about 45 per cent. Consequently, the volume fraction of collagen is about 55 per cent (with a modulus of 1.3 GPa). Use the results given in Problem 8.4 to calculate upper and lower bounds for the elastic modulus of bone and compare the results with the values given in Tables 8.1 and 8.2. Discuss the results in relation to the assumptions used in each model.

8.6 A frequently cited empirical relationship between the compressive elastic modulus of bone and its density is that of Carter and Hayes (1977),

$$E = 3790(d\varepsilon/dt)^{0.06}\rho^3,$$

where E is in MPa, $d\varepsilon/dt$ is strain rate in s^{-1}, and ρ is the 'apparent' density of the bone, in g cm^{-3} (the usual units of specification in the literature). Apparent density means the mass of a specimen, including the void spaces that are filled with soft tissues, divided by its total volume. Sketch a plot of this equation for strain rates of 1000, 1, and 0.001 s^{-1}. Assume the greatest density of cortical bone is 2 g cm^{-3}. What does this equation predict for the modulus of trabecular and compact bone having apparent densities of 0.3 and 2.0, respectively? Assume a strain rate of 1 s^{-1}. If the trabecular bone suffered a 30 per cent reduction in apparent density due to osteoporosis, what percent change in modulus would be predicted? (Osteoporosis is a condition in which bones become thinner and/or more porous, but retain their normal mineral content. It is extremely common among elderly women. Typically, 30–50 per cent of bone mass must be lost before the condition is apparent on a radiograph.)

References and further reading

Alexander, R. McN. (1968). *Animal mechanics*. University of Washington Press, Seattle.

Carter, D. R. and Hayes, W. C. (1977). The compressive behavior of bone as a two phase porous structure. *Journal of Bone and Joint Surgery*, **59A**, 954–62.

Charnley, J. (1973). The classic. Arthroplasty of the hip: a new operation. *Clinical Orthopaedics and Related Research*, **95**, 4–25.

Fung, Y. C. (1990). *Biomechanics: motion, flow, stress, and growth*. Springer-Verlag, New York.

Gebelein, C. B. (1984). *Polymeric materials and artificial organs*. American Chemical Society, Washington, DC.

Ham, A. W. and McCormack, D. H. (1979). *Histology*. Lippincott, Philadelphia.

Hastings, G. W. and Ducheyne, P. (1984). *Natural and living biomaterials*. CRC Press, Boca Raton, FL.

Hench, L. L. and Ethridge, E. C. (1982). *Biomaterials: an interfacial approach*. Academic Press, New York.

Kossowsky, R. and Kossovsky, N. (1986). *Materials sciences and implant orthopaedic surgery*. Martinus Nijhoff, Boston.

Lin, O. C. C. and Chao, E. Y. S. (ed.) (1985). *Perspectives on biomaterials*, Proceedings of the 1985 International Symposium on Biomaterials. Elsevier, Amsterdam.

Major, R. A. (1954). *A history of medicine*. C. C. Thomas, Springfield, IL.

Nahum, A. M. and Melvin, J. (ed.) (1985). *The biomechanics of trauma*. Appleton-Century-Crofts, Norwalk, CT.

National Research Council (1972). *Internal structural prostheses*, Report of a Workshop on Fundamental Studies for Internal Structural Prostheses. National Academy of Sciences, Washington, DC.

Omer, G. E. (1987) Orthopaedics in 1921–1953: advances and current problems. *Journal of Bone and Joint Surgery*, **69A**, 1262–4.

Oxlund, H., Manschot, J., and Viidik, A. (1988). The role of elastin in the mechanical properties of skin. *Journal of Biomechanics*, **21**, 213–18.

Peppas, N. A. and Langer, R. (1994). New challenges in biomaterials. *Science*, **263**, 1715–20.

Reilly, D. T., Burstein, A. H., and Frankel, V. H. (1974). The elastic modulus for bone. *Journal of Biomechanics*, **7**, 271–5.

Wainwright, S. A., Biggs, W. D., Currey, J. D., and Gosline, J. M. (1976). *Mechanical design in organisms*. Princeton University Press, Princeton, NJ.

Williams, D. F. and Roaf, R. (1973). *Implants in surgery*. W. B. Saunders, Philadelphia.

Woo, S. L.-Y., Gomez, M. A., and Akeson, W. H. (1985). Mechanical behaviors of soft tissues: measurements, modifications, injuries, and treatment. In *The biomechanics of trauma* (ed. A. M. Nahum and J. Melvin), pp. 109–133. Appleton-Century-Crofts, Norwalk, CT.

Wood, J. L. (1971). Mechanical properties of human cranial bone. *Journal of Biomechanics*, **4**, 1–12.

Yamada, H. (1973). *Strength of biological materials*. R. E. Krieger, Huntington, NY.

9

THE INTERACTION OF BIOMATERIALS AND BIOMECHANICS

Harry B. Skinner

Contents

Symbols

a	major diameter of ellipse; distance from cantilever support to load
A	area, cross section
b	width; minor diameter of ellipse
c	one-half plate thickness; proportionality constant
d	distance from neutral axis of cross section to neutral axis of construct
E	Young's modulus
F	load
G	shear modulus
h	plate thickness
I'	area moment of inertia
J	polar moment of inertia; joint reaction force
K	stress amplification factor
L	total beam length; load
M	moment
p	hardness of surface
P	load
r	radius
t	thickness
T	torque
V	volume fraction; wear debris volume
x	distance from plate; distance of sliding
δ	beam deflection
ϵ	strain
θ	angle
$\mu\epsilon$	microstrain
ν	Poisson ratio
σ	stress
τ	shear stress

Subscripts

act	actual
ave	average
b	bone
c	contact
calc	calculated
cort bone	cortical bone
f	fiber
ID	inside diameter
m	matrix

max	maximum	*Acronyms*		PEEK	poly(ether ether ketone)
OD	outside diameter	ACL	anterior cruciate ligament	SCFE	slipped capital femoral epiphysis
parallel	composite parallel	AML	anatomic medullary locking	UHMWPE	ultra-high-molecular-weight polyethylene
perp	perpendicular	BW	body weight		
pl	plate	ID	inner diameter		
pr	prosthesis	MCL	medial collateral ligament		
xx	In x-direction acting on area perpendicular to x	OD	outer diameter		

9.1 Introduction

Biomaterials play a major role in the management of many musculoskeletal problems. Without them, treatment of disorders such as fractures or arthritis would be rudimentary, with a major increase in cost in terms of quality of life, lost productivity, and health care expense. Many patients with fractures and serious arthritis are treated with surgery using implant materials to return patients to early, painfree function. In the United States, an estimated 11 million people have implants of one type or another, with about 51 per cent of these resulting from the treatment of a musculoskeletal disorder (Praemer *et al.* 1992). Many fracture fixation devices are removed after they have served their purpose. Thus, the number *in situ* is an estimate of the numbers intended to remain permanently, combined with the difference of those being implanted and removed. The number of fractures each year is 2.9/100 for men and 2.3/100 for women, and many of these are being treated with fixation devices. The direct (hospital, doctor etc.) and indirect (work loss, etc.) cost of fractures each year is more than 20 billion dollars. Joint replacement is a biomechanical solution to destruction of a joint from arthritis or trauma. The hip and knee are the most common joints replaced, but the shoulder, the elbow, the finger joints, and the wrist are also replaced, although with lesser frequency. Approximately 180 000 hips and 95 000 knees are replaced each year but the total cost for the medical and surgical treatment of all aspects of arthritis is approximately 8.6 billion dollars.

The interaction of biomechanics and biomaterials is a complicated process because of two factors that seldom combine in physical systems. These are the cyclic nature of loading experienced by biological organisms and the ability of these organisms to change with time due to adaptation and healing processes. Many of the applications of biomaterials are accomplished through the implantation of devices. Hence, a healing process to some steady state occurs. However, the presence of the foreign device is well known to stimulate an adaptation by the organism's tissue that can change the loading applied to the implanted device. Many failures of implantable devices have resulted from a failure to account for, or adequately account for, these two factors.

The body has been found to be very efficient in adapting to new stress conditions. It has long been realized that bone, tendon, and ligament adjust to accommodate new stress levels by remodeling either to take away that which is not necessary or to add additional load-bearing capacity where needed. This concept, when applied to bone remodelling, has been called Wolff's law (Roesler 1981) and has achieved the level of understanding whereby strain levels of 800 to 1000 $\mu\epsilon$ are identified as signals to remodel (Rubin and Lanyan 1987). For example, bone strains greater than -1000 $\mu\epsilon$ (compressive) at higher than 100 load cycles per day were associated with bone apposition, while strains of -500 $\mu\epsilon$ and lower were associated with bone resorption. Thus, it has become apparent that stable implant–bone constructs must maintain adequate bone loading or face resorption of bone with potential adverse effects that may occur over the long term. Similar considerations apply to other tissues such as ligaments or tendons.

9.1.1 Rigidity

Rigidity is an important concept because healing processes frequently require close apposition of surfaces. A low rigidity fracture fixation device may adequately appose two fracture fragments but allow motion in excess (believed to be > 50 μm) of that needed for fracture healing without callous formation. Rigidity is interpreted in different ways: axial, flexural, torsional. Since these terms are a function of both material properties and geometry, both quantities are necessary to define them. Thus, axial, flexural, and torsional rigidity are defined by the product of a geometric term and the elastic modulus: AE, IE, and JG, where E is Young's modulus and G is the shear modulus, A is the area, I is the area moment of inertia, and J is the polar moment of inertia. One approach under consideration to control both rigidity and strength is composite technology. Variation in the orientation of a strong fiber in an appropriate matrix of a composite material may permit low rigidity to be combined with high strength.

9.1.2 Uniaxial failure modes

Failure of biological–prosthetic constructs can occur in a single-load mode or with a fatigue mechanism (see Section 1.4.4). Single loads can lead to immediate failure after surgery or at a later time. Only very unstable constructs would fail immediately after surgery, because loading is markedly reduced due to the inhibition of muscle activity in response to surgical pain. Later, after acute surgical pain has resolved, failure due to physiological loads is a potential problem. The major problem in predicting these situations is mainly in defining the conditions correctly that lead to failure. This problem is solved by developing and then applying a thorough knowledge of the characteristics of the biological–prosthetic system. Frequently, exact loads are not available, so appropriate worst-case estimates must be made, with improvement of the estimates as more data become available.

Several static (or quasi-static) failure modes will be examined to illustrate the interaction of materials with biological systems. These examples illustrate the necessity of defining failure mechanisms and having a thorough knowledge of response characteristics of biological–prosthetic systems.

Physical failure must be defined in the context of each particular application. Failure can occur because of fracture, plastic deformation, or lack of rigidity. Permanent or plastic deformation is the failure mode found in cast stainless steel hip prostheses used in early total hip arthroplasty (Fig. 9.1). Any plastic deformation results in unacceptable angulation of the hip and impending failure. Obviously, catastrophic (ultimate) failure is not acceptable. Lack of rigidity in fracture fixation devices may permit excessive motion and failure of the healing process.

Fatigue of a material (failure under cyclic stress below the static yield stress) can quickly occur in the normally functioning human. For example, the typical human heart beats 70 pulses min^{-1}, or about 100 000 times per day, or more than 36 million times per year. Thus, the cyclic loading on a heart valve is tremendous. Further, the average sedentary human walks several kilometers per day at about 90 m min^{-1} (115 steps min^{-1}, step length 780 mm). Thus, 5 km d^{-1} results in about 8000 cycles d^{-1} or 2.9 million cycles per year. Over a period

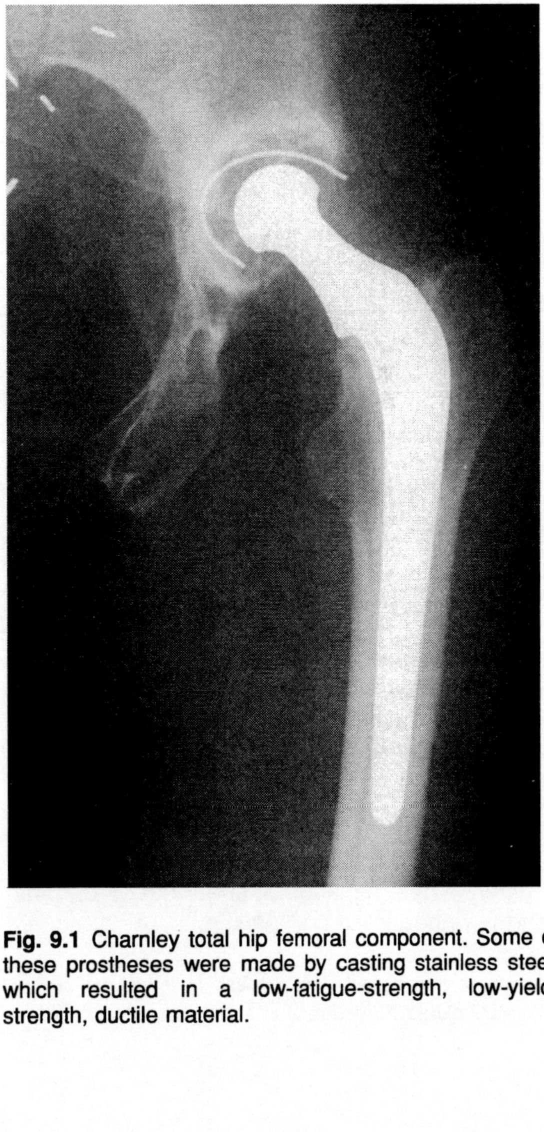

Fig. 9.1 Charnley total hip femoral component. Some of these prostheses were made by casting stainless steel, which resulted in a low-fatigue-strength, low-yield-strength, ductile material.

of 4 to 10 years, the stress endurance limit at 10^7 cycles could be exceeded for many materials.

Cyclic stress can become especially important in design of prostheses and other devices such as bone plates because incorrect design can result in stress concentration sites. These increase the local stress and predispose an implant to failure. Design of implants requires an accurate understanding of the loads that implants must be able to resist. This in turn requires an appreciation for the internal loads on joints, which first of all are frequently dynamic; however, if considered to be static, they are indeterminate because of redundant muscle action. Dynamic loads can result in much higher stresses than expected from static approximations. Thus, design of implants is highly dependent on an understanding of anatomy and physiology that determines loading conditions as well as the long-term response of biological materials to loading and the behavior of implant materials to cyclic loading. With the basis of the information of previous chapters, we will build on this understanding using examples of clinical problems.

9.1.3 Composite beams

Composites offer the opportunity to combine the best characteristics of two or more materials to produce a better material. One type of composite in use today is a biological–biomaterial composite characterized by the total hip replacement (bone–bone-cement–prosthesis) and another is the bone plate for fracture fixation. These types of composites have generally worked well, but problems have occurred with fixation due to low fatigue endurance of bone cement and the stress-shielding effects of the prosthesis or plate on bone. Another type of composite is under consideration for use in hip replacement and, if developed, will probably be composed of carbon fiber with matrix materials such as poly(ether ether ketone) (PEEK) or polysulfone.

9.1.4 Wear

Wear phenomena take on importance in implant surgery with 'fretting' or micromotion between screws and bone plates and with the movement that is fundamental in joint replacement. Serious sequelae from wear between screws and plates after fracture fixation can be resolved by removal of the fixation after fracture healing. Replacement of a joint utilizes a metallic or ceramic surface that produces wear of a plastic surface. While complete destruction of a joint replacement due to wear can occur, failure is more commonly caused by the products of wear. Wear debris unfortunately has an adverse biological effect in that it stimulates the resorption (removal) of bone.

9.2 Static load considerations

9.2.1 Introduction

Many of the problems to be addressed in this section consider dynamic loading but are treated as quasi-static problems. Perhaps, more accurately, they are cyclic loading problems, but instead are analyzed as single-load failures.

9.2.2 Subcapital femoral epiphysis/femoral neck fracture—determination of shear stress on pins, and bearing stress on bone

A frequent problem that occurs in adolescence is a disorder called slipped capital femoral epiphysis (SCFE). This hip problem arises at the proximal cartilaginous interface surface at which lengthwise growth of a long bone occurs (termed a growth plate or physis). In a small percentage of individuals in the 11–15 year age range, the interface is susceptible to shear failure (Fig. 9.2a). The portion of bone proximal to the growth plate (physis) is called the epiphysis. A mechanically similar disorder occurs in elderly adults due to failure of the bone of the femoral neck. These two processes are somewhat analogous due to anatomical location of the problem. In the child, the problem results from slippage on the growth plate that traverses the approximate equatorial surface of the femoral head. This growth plate is made of cartilage and has a somewhat roughened surface, but has a relatively low shear resistance because of the properties of the cartilage. The analogous problem, frequent in the elderly adult, is the femoral neck fracture which occurs across the midcervical or subcapital area of the femoral neck in osteoporotic bone. While the cancellous bone has a much higher intrinsic resistance to shear compared with the cartilage growth plate, its strength is decreased by a loss of bone mineral due to age. Thus, it is possible for the trabecular bone to fail in bending or shear, resulting in transverse or oblique failure of the neck. Both these disorders can be treated with pins placed from the lateral aspect of the femur through the femoral neck into the femoral head (Fig. 9.2b). These pins made of 316L stainless steel are removed in young people with SCFE (Fig. 9.2c). Patients with femoral neck fractures frequently have thee pins left in place permanently.

The question that arises is: how many pins should be placed across the growth plate or across the fracture in order either to prevent the fracture from displacing, or to prevent further slip of the femoral head on the neck of the femur? If excessive slip occurs in either case, the femoral head will suffer from disruption of the blood supply, which primarily comes from the femoral neck. In the SCFE, a malformed femoral head–neck angle will result from posterior and inferior slippage of the femoral head in relation to the neck and shaft, which would increase the risk of early osteoarthrosis of the hip. An associated problem is: how many pins would be necessary to prevent compressive failure of the cancellous bone of the femoral head which would permit slipping of the head on the neck at the fracture or the physeal plate?

The joint reaction force, J (Fig. 9.3), in either case can approach three times body weight at the stance phase of gait. The shear force is equal to $J \cos \theta$ (θ is the angle between the force J and the fracture line) and the shear stress on the pins is equal to $J \cos \theta / A$, where A is the cross-sectional area of the pins. The assumption is made that little resistance to shear is generated by either the physeal plate or the fracture. This is a reasonably good estimate in SCFE because the shear stress at failure for the physeal plate has been measured at about 1 to 1.2 MPa (Chung *et al.* 1976). For a 60 kg mass 13-year-old who weighs 270 N, the joint force will be approximately 800 N. The shear force is 275 N with an angle of 70° between the joint load and the plate. Fixation pins are 3.2 mm in diameter (cross-sectional area of 8×10^{-6} m^2) so that the shear stress on the pin is about 3 MPa (about 9 MPa for the 80 kg adult). The shear yield stress of steel is about 165 MPa; thus, one pin would be adequate to prevent slippage in either of these cases except in rotation around the pin (Aronson and Carlson 1992; Ward *et al.* 1992). Thus, two pins are preferable to prevent potential rotation. Two-pin fixation would be more than adequate to prevent further slippage of either the fracture or epiphyseal fragment of the femoral head.

The ultimate compressive strength of dense cancellous bone can be taken to be about 27 MPa (Anderson *et al.* 1992; Carter and Hayes 1976). If the pin crosses the physeal plate or the fracture by 20 mm, the bearing surface of that pin would be about 64 mm^2, suggesting that the shear force that this bone could tolerate before experiencing compressive failure would be approximately 1700 N. This is significantly less than the 800 N predicted for a 13-year-old child, and this is under the condition of having only one pin. This would be doubled to 3400 N that would be tolerated by the cancellous bone if two pins were present; this would be necessary in the case of the adult because the shear force in the adult could conceivably be 800–900 N. Thus, the safety factor would still be adequate with just two pins if the cancellous bone was half as strong in an older patient.

9.2.3 Variation in bending stiffness of bone plates with thickness

One of the potential solutions to a clinical problem such as fracture of a bone plate is to increase the thickness of the plate. This would obviously make the plate stronger. The maximum stress (ignoring holes) in the plate, $\tau_{xx,\max}$, subjected to bending moment M can be calculated from the flexure formula (eqn 1.146)

$$\tau_{xx_{\max}} = \frac{Mc}{I'} \equiv \sigma_{\max}, \qquad (9.1)$$

where c is one-half of the thickness of the plate, and the area moment of inertia (I') is $bh^3/12$, where h is the thickness of the plate and b is the width of the plate. The stress would be

$$\sigma_{\max} = \frac{6M}{bh^2}. \qquad (9.2)$$

Doubling the thickness decreases the stress to

$$\sigma_{\max} = \frac{3M}{2bh^2}. \qquad (9.3)$$

Thus the stress is reduced to one-quarter of the previous stress by doubling the plate and, therefore, the plate is four times stronger in bending. Another way of looking

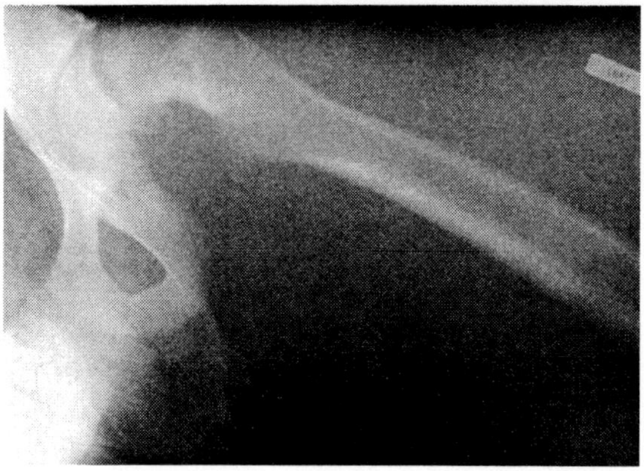

(*a*)

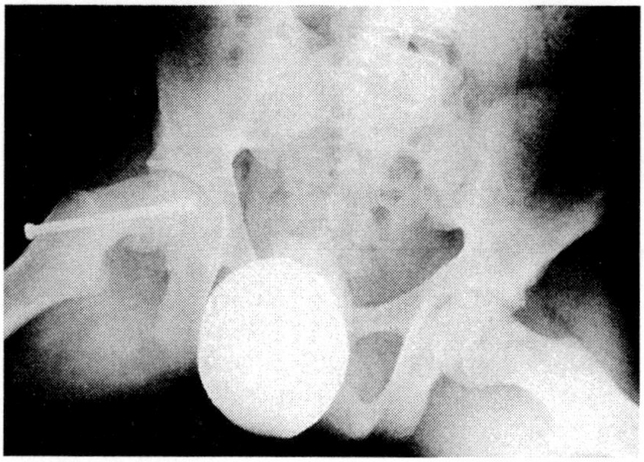

(*b*)

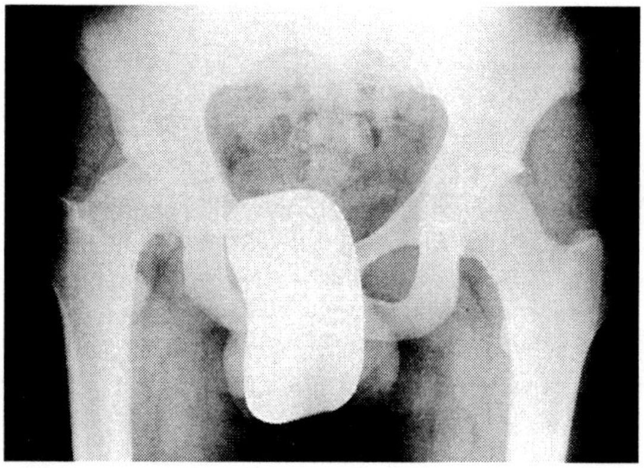

(*c*)

Fig. 9.2 Radiographs of (a) a left hip, showing the area of acute slip, (b) a pelvis with a screw in place, (c) the pelvis after screw removal, showing the residual deformity of the right hip (left hip is nearly normal).

at this is that the load would have to be four times higher before the failure stress would be reached. Similarly, since the endurance limit (10^7 cycles) of steel (Section 1.4.4) is approximately one-half the ultimate strength, four times higher cyclic loads can be tolerated without fatigue failure.

The bending stiffness of the bone plate is given by the flexural rigidity, EI'. Assuming a rectangular cross section, the flexural rigidity is given by:

$$EI' = \frac{Ebh^3}{12}. \qquad (9.4)$$

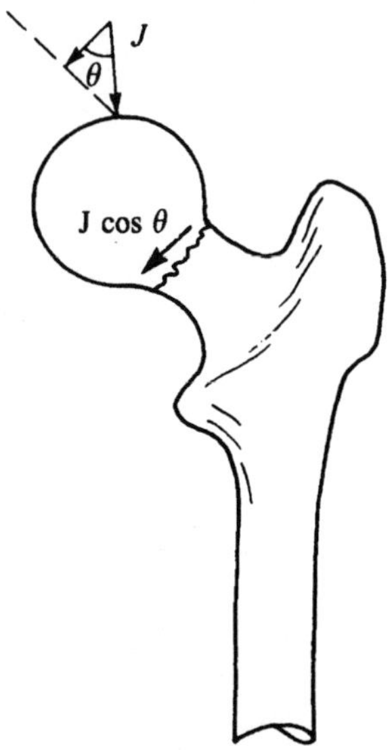

Fig. 9.3 Free-body diagram of the loads on the proximal femur after fracture or 'slip'.

Thus if the bone plate thickness is doubled, the 'stiffness' (flexural rigidity) becomes:

$$EI' = \frac{Eb(2h)^3}{12} = \frac{2Ebh^3}{3}, \tag{9.5}$$

or eight times stiffer. This is comparable to a strength increase of only fourfold. Failure of the bone plate is possible through either mechanism: insufficient stiffness or insufficient strength. The former may result in delayed fracture healing due to too much motion, and the latter may result in plastic deformation or catastrophic fatigue failure, resulting in displacement of the fracture.

It would seem that the most desirable bone plate would be the thickest bone plate. However, constraints are placed on the size of the bone plate by the need to cover it with soft tissue, to be able to shape it to conform to the contour of the bone, and by the necessity for the bone to experience strain.

9.2.4 Axial load sharing by a bone plate on a cylindrical bone

The level of strain in a bone is very important in determining its long-term morphology. After fracture fixation of long bones, healing of the fracture results in a composite beam sharing the load between the fracture fixation device and the tubular bone. In the case of a bone plate, remodeling removes most of the stress concentration effects of the screw holes. For simplicity, we consider axial loading only, although the stress state (and strain state) of bone is a result of flexural and other loading that results in higher strains than simple axial loading. A tubular bone and a firmly attached plate share the axial load from weight-bearing and muscle force (Fig. 9.4). The axial compressive strain in the midshaft of the bone prior to the fracture is given by:

$$\epsilon_{xx,\mathrm{b}} = \frac{\sigma}{E} = \frac{F_\mathrm{b}}{A_\mathrm{b}E_\mathrm{b}}. \tag{9.6}$$

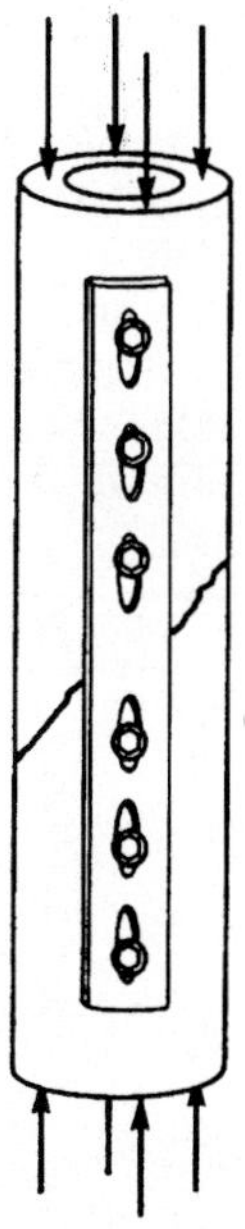

Fig. 9.4 Diagram of bone with axial load after bone-plating for fracture.

For a load (F) of 1500 N, an area of 4.5×10^{-4} m^2 (OD = 30 mm, ID = 18 mm) and a modulus of 20 GPa, a strain of 166 $\mu\epsilon$ is calculated. The strain in a bone plate will be given by

$$\epsilon_{xx,\mathrm{pl}} = \frac{\sigma_{\mathrm{pl}}}{E_{\mathrm{pl}}} = \frac{F_{\mathrm{pl}}}{A_{\mathrm{pl}}E_{\mathrm{pl}}}. \quad (9.7)$$

Because the plate and bone are held firmly together by screws preventing shear, the strain in both must be equal over the length of the plate:

$$\epsilon_{\mathrm{b}} = \epsilon_{\mathrm{pl}}, \quad \epsilon_{xx,\mathrm{b}} = \epsilon_{xx,\mathrm{pl}}, \quad (9.8)$$

$$\frac{F_{\mathrm{pl}}}{A_{\mathrm{pl}}E_{\mathrm{pl}}} = \frac{F_{\mathrm{b}}}{A_{\mathrm{b}}E_{\mathrm{b}}}, \quad (9.9)$$

$$\frac{F_{\mathrm{pl}}}{F_{\mathrm{b}}} = \frac{A_{\mathrm{pl}}E_{\mathrm{pl}}}{A_{\mathrm{b}}E_{\mathrm{b}}} = 2.99. \quad (9.10)$$

Furthermore, the sum of the force on the plate and bone is equal to the total force:

$$F = F_{\mathrm{pl}} + F_{\mathrm{b}}. \quad (9.11)$$

For a plate with width 0.02 m, height of 0.007 m, and modulus of 193 GPa, it can be seen that the plate carries 74.9 per cent of the load and the strain in the bone under the plate is reduced to 41.7 $\mu\epsilon$ (assuming uniform strain across the bone cross section). Thus, the bone has been 'shielded' from axial stress by the bone plate. Other modes of loading are also prevented from straining the bone by the plate. Over a period of time the bone will respond to this new stress environment by remodelling. The remodelling process will remove bone until a new equilibrium stress state is achieved, probably with average strain in the bone in the physiological 500–1000 $\mu\epsilon$ region, with final strength approximately equivalent to pre-fracture strength (Woo *et al.* 1976; Rubin and Lanyan 1987).

9.2.5 Axial, flexural, torsional load sharing by total hip prosthesis and femur: effect on bone remodelling

The material properties of bone are different for cortical (compact) bone and cancellous bone because of density (Reilly and Burstein 1975). The cortical bone density is much higher but is probably continuous with cancellous bone. Mechanical properties of both are highly dependent on density to the second or third power. Alterations in density result from changes in the stress state. The remodelling process deposits or resorbs bone (Rubin and Lanyan 1987). This occurs in the cortex after uncemented hip arthroplasty and has a long-term effect on the properties of the bone and eventually the stress on the bone and prosthesis. This process of remodelling (resorption proximal to the tip of the prosthesis) has been observed in all uncemented prostheses but is most easily considered for the AML (anatomic medullary locking) prosthesis (DePuy, Inc., Warsaw, IN) because it is fully porous-coated and is longer than many prostheses (165 mm vs. 120 mm). Thus, the assumption of perfect transmission of axial, torsional, and bending loads at the interface between bone and prosthesis is reasonably valid.

The most important loading mode is uncertain. High bending and axial loads are applied while standing. High torsional loads are produced by stair-climbing and arising from a chair. The change in stress (strain) for all modes must be considered. The femur is 30 mm OD and 15 mm ID with a 15 mm diameter prosthesis firmly implanted with complete contact to bone (Fig. 9.5). A load of 3 body weight (BW) = 2100 N is applied at 30 mm from the axis of the prosthesis while standing at a plane 75 mm below the collar. A 4 BW (2800 N) load is applied perpendicular to the plane of the prosthesis at the center of the femoral head at 0.04 m from the axis while

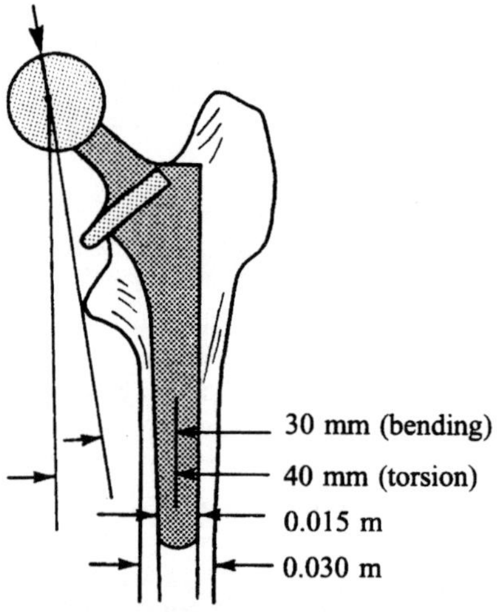

Fig. 9.5 Loading conditions for an uncemented hip replacement. $A_{\mathrm{b}} = 5.30 \times 10^{-4}\mathrm{m}^2$, $A_{\mathrm{pr}} = 1.77 \times 10^{-4}\mathrm{m}^2$; $I_{\mathrm{b}} = 3.73 \times 10^{-8}\mathrm{m}^4$, $I_{\mathrm{pr}} = 2.49 \times 10^{-9}\mathrm{m}^4$; $J_{\mathrm{b}} = 7.46 \times 10^{-8}\mathrm{m}^4$, $H_{\mathrm{pr}} = 4.97 \times 10^{-8}\mathrm{m}^4$; $E_{\mathrm{b}} = 20$ GPa; $E_{\mathrm{pr}} = 200$ GPa; $G_{\mathrm{b}} = 7.46$ GPa.

arising from a chair. For axial stress, the analysis is similar to that given in the previous section. The strain in the bone prior to implantation from an axial load of 3 BW is equal to 198 $\mu\epsilon$. After implantation, this same load reduces the strain in the bone to approximately 47 $\mu\epsilon$.

Flexural loading results from the offset of the femoral head from the shaft. The bending moment while standing is 63 N m. From the flexure formula, eqn (9.2), the maximum stress (strain) in the bone prior to implantation is calculated to be 25.3 MPa (1270 $\mu\epsilon$). After implantation, correction for the difference in moduli of the prosthesis and bone has to be made. The flexure formula is given by

$$\tau = \frac{Mc}{I'_{\mathrm{b}} + I'_{\mathrm{pr}}}, \tag{9.12}$$

where M is the bending moment, c is the distance from the neutral axis, and I'_{b} and I'_{pb} are the area moments of inertia of bone and prosthesis respectively. Correction of the modulus of the prosthesis is done by changing the effective area of the prosthesis from a circle to an ellipse with the major (neutral axis) diameter equal to the ratio of $E_{\mathrm{pr}}/E_{\mathrm{b}}$. This ratio is 200 GPa/20 GPa, or 10, and thus:

$$\sigma_{\max} = \frac{Mc}{\frac{\pi}{4}(r_{\mathrm{OD}}^4 - r_{\mathrm{ID}}^4) + \frac{\pi ab^3}{4}} \tag{9.13}$$

where a is the corrected dimension (major diameter) of the ellipse on the neutral axis ($10r_{\mathrm{pr}}$) and b is the dimension in the bending direction (minor diameter) equal to r_{pr}. The maximum bone stress is then calculated to be 15.3 MPa with a strain of 765 $\mu\epsilon$, a reduction of peak strain of 40 per cent.

Torsional loading also results from the offset of the femoral head. The torque is estimated to be 112 N m during the act of rising from a chair. The shear stress in the cortical bone prior to implantation is given by

$$\tau = \frac{Tc}{J}, \tag{9.14}$$

where T is the torque, c is the distance from the center, and J is the polar moment of inertia. The maximum shear stress (strain) in the cortical bone, the shear modulus of which is $G = 7.4$ GPa, is calculated to be 22.5 MPa (3040 $\mu\epsilon$). After implantation, the polar moment of inertia of the construct is calculated to be 1.243×10^{-7} m^4. Use of the shear modulus for Co–Cr alloy ($v = 0.342$, $G = 74$ GPa) allows calculation of the shear stress to be reduced to 13.5 MPa. The shear strain with the prosthesis in place is then calculated to be 1800 $\mu\epsilon$.

While the actual strains may be different from those that are calculated here, the trends that occur with implantation will be the same. Thus, implantation is likely to cause a resorption response due to the lowered strain in the bone as suggested in the previous section. This resorption has been observed in patients with this prosthesis (Kilgus *et al.* 1993) and in animal studies (Bobyn *et al.* 1985; Husby *et al.* 1989).

9.2.6 Calculation of failure of a bone plate placed on tensile and compressive surfaces

Because of the shape and muscle loading of long bones, bending in certain planes is more likely than others. For example, most of the muscle mass of the leg is posteriolateral to the tibia, implying that the anteromedial surface is in tension. Similarly, the 'bow' of the femur and the posterior location of hamstring musculature tend to place the anterior surface of the femur in tension. It has been found that this loading can have significant impact on the failure of bone plates used to reduce fractures to anatomic alignment.

Placement of the bone plate on the 'tension' surface with posterior contact of the bone moves the neutral axis out of the bone plate and closer to the neutral axis of the bone, because the fracture site can transmit compressive stress. The anterior surface is the tension surface because the femur is a bowed bone and the hamstring muscles therefore have an advantage over the quadriceps in moment arm (as well as in bulk). Further, the quadriceps is medial and lateral as much as anterior. The neutral-axis calculation is somewhat complicated because the application of higher loads to the construct will cause more contact and therefore a larger area transmitting compressive stress. For simplicity, assume that the fracture is transverse (perpendicular to the shaft) and that one-half of the posterior cortex will transmit compressive stress. Figure 9.6 demonstrates the problem. It is necessary to correct for having a composite beam consisting of steel and bone and to use the parallel-axis theorem (Section 1.1.13). The area moment of inertia for the metal plate and the bone (half cylinder) are:

$$I'_{\mathrm{pl}} = \frac{bh^3}{12}, \tag{9.15}$$

Table 9.1 Area moment of inertia of common cross sections

Dimensions		Area	Axis	I'
Rectangle	height h, width b	bh	$\frac{h}{2}$	$\frac{bh^3}{12}$
Triangle	height h, base b	$\frac{bh}{2}$	$\frac{h}{3}$	$\frac{bh^3}{36}$
Circle	radius r, $d = 2r$	πr^2	r	$\frac{\pi r^4}{4}$ or $\frac{\pi d^4}{64}$
Thin tube	d_{ave} = mean diameter, r_{ave} = mean radius, t = thickness	$2\pi r_{ave} t$	r_{ave}	$\pi r^3_{ave} t$
Thick tube	$2r_{ID} = d_{ID}$ = inside diam., $2r_{OD} = d_{OD}$ = outside diam.	$\frac{\pi}{4}(r_{OD}^2 - r_{ID}^2)$		$\frac{\pi}{4}(r_{OD}^4 - r_{ID}^4)$

$$I'_b = 0.095\pi r^3_{ave} t, \tag{9.16}$$

where b and h are width and height of the plate, respectively, and r_{ave} and t are average radius and thickness of the bone, respectively (Popov 1968), as indicated in Fig. 9.6(b) and Table 9.1. The distance from the centroid of the bone to the centroid of the plate is 0.02299 m. Thus, the neutral axis is a distance x from the plate

$$xI_{pl} = I_b(0.02299 - x), \tag{9.17}$$

$$x = 0.00825 \text{ m},$$

$$0.02299 - x = 0.0147 \text{ m}.$$

Each area moment of inertia must be moved to obtain the I' for the structure:

$$I' = I_{pl} + A_{pl}d_{pl}^2 + I_b + A_b d_b^2, \tag{9.18}$$

where d is the distance from the neutral axis of the cross section to the neutral axis of the construct. Assume that the muscle force arises from the hamstrings 0.03 m from the plate. The maximum stress in the plate σ_{max} is then given by eqn (9.1)

$$\sigma_{max} = \frac{Mc}{I'}, \tag{9.19}$$

where $M = 700 \text{ N} \times 0.03 \text{ m} = 21 \text{ N m}$, $c = 0.0086$ m, and $I' = 4.33 \times 10^{-6} \text{ m}^4$. The stress in the plate is calculated to be 0.042 MPa (tensile).

Placement of the plate on or near the posterior cortex gives the hamstrings the opportunity to distract the fracture site, subjecting the bone plate to all of the loading. The bone carries none of the load. The maximum stress in the bone plate is given by eqn (9.1), where M is 21 Nm, c is 0.0035 m (one-half the thickness of the bone plate), and I'_{pl} is $5.72 \times 10^{-10} \text{ m}^4$. The extreme fiber stress in the plate (assuming no hole at the fracture site) is 128 MPa. Depending on the width of the plate, a screw hole may more than double the actual stress.

The placement of the plate on the lateral surface yields stress in the plate intermediate between these extremes (tensile on the anterior edge, compressive on the posterior). The importance of contact of the cortical bone in compression and proper placement of the plate to achieve compression of the bone is obvious. Without this compression, stress in the bone plate can reach levels that will result in fatigue failure. The fatigue endurance limit for stainless steel in a bone plate is approximately 300 MPa.

9.2.7 Calculation of the rupture of MCL/ACL with valgus stress to knee: relation to prosthetic ligament Gore-Tex/Dacron. Graft relation to shear failure, or compressive failure of cancellous bone

The mechanical properties of bone and ligament (Anderson *et al.* 1992; Carter and Hayes 1976; Reilly and Burstein 1975; Woo *et al.* 1983) and the materials to replace these must be understood in the context in which they function. A frequent injury incurred during football

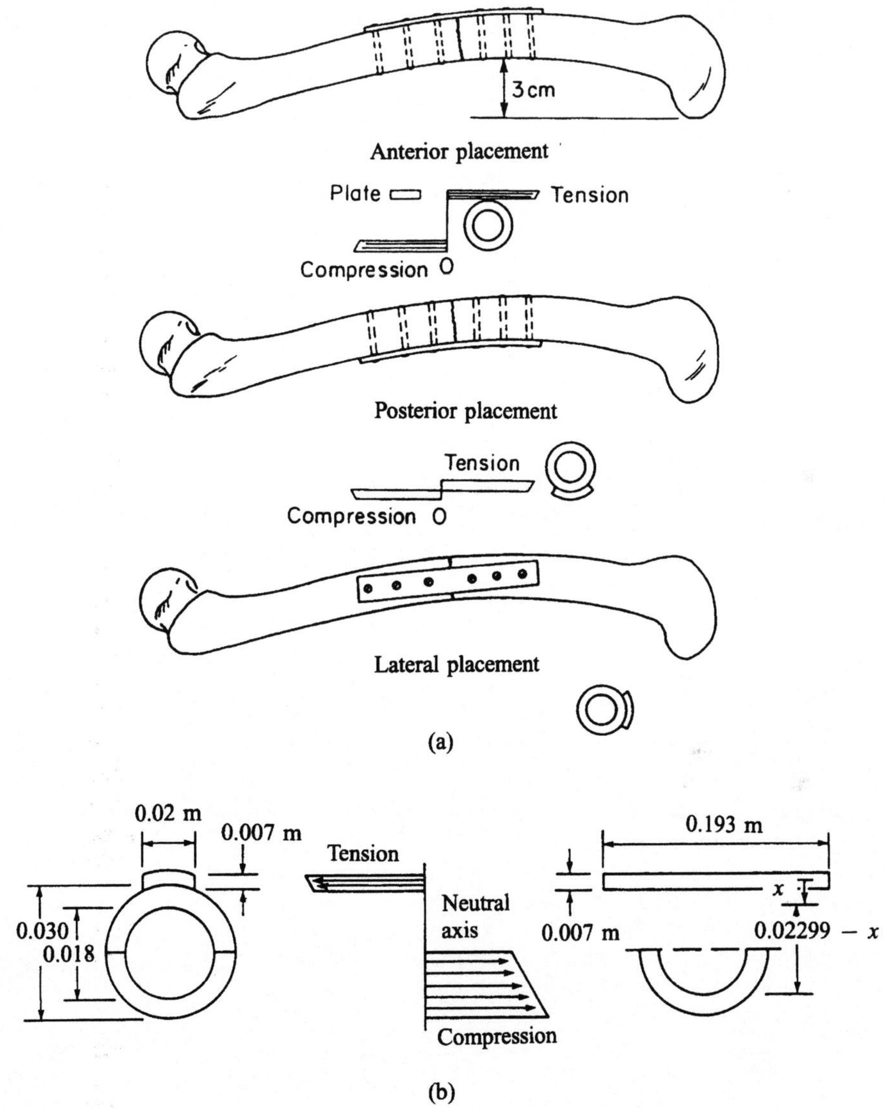

Fig. 9.6 Effect of placement of a bone plate on different surfaces of a bone and how it can lead to material failure: (a) physical system; (b) stress analysis.

	Bone	*Metal*
I (m^4)	$0.095\pi r^3 t = 3.09 \times 10^{-9}$	$\frac{1}{12}bh^3 = 5.52 \times 10^{-9}$
	$[b = \frac{193}{20}(0.2) = 0.193]$	
A (m^3)	$\pi r_{ave} t = 2.26 \times 10^{-4}$	$bh(\frac{193}{20}) = 1.35 \times 10^{-3}$
d (m)	0.01974	0.00825
I_{total} (m^4)	3.33×10^{-6}	1.00×10^{-6}

is rupture of the medial collateral ligament, rupture of the anterior cruciate ligament, and, less often, fracture of the lateral tibial plateau of the tibia. While rupture of the ligaments occurs more frequently in sports such as football or basketball, tibial plateau fractures are more frequently suffered from bumper injuries to the lateral aspect of the leg. Which of these injuries occurs is probably a direct function of the strength of the two structures. The indirect factors that have a bearing on which fails probably include the strain rate of the injury, the degree of mineralization of the bone, and therefore the age and activity of the injured person, and muscle loading at the time of injury.

Some concept of what is happening during such an injury can be inferred from considering the loads in a typical football injury (Fig. 9.7). Assume an individual has his foot planted on the ground and has a 90 kg (200 lb) opponent fall against him at the level of the knee joint on the lateral aspect. Assume this produces a force on the knee of 882 N with a moment arm of approximately 0.4 m (ground to knee area), resulting in a bending moment of 353 N m. Further, assume no muscle contraction in the soon-to-be-injured knee. This moment is resisted by the lateral condyle of the femur pressing on the lateral plateau of the tibia and by the medial collateral ligament, which is in tension (until rupture, where the anterior cruciate takes over the major stress). The distance between these two structures is approximately 50 mm, suggesting that the load applied to the lateral plateau is 7060 N; the medial collateral ligament cross-sectional area is approximately 5×15 mm or 7.5×10^{-5} m^2. This results in a stress on the medial collateral ligament of 7060 N/7.5×10^{-5} m^2 = 94.1 MPa. This compares to a failure strength in tension of 80 MPa. After rupture of the medial collateral ligament, this moment is resisted by the anterior cruciate ligament, which is approximately 20 mm from the lateral tibial plateau point of contact, resulting in a tensile load of 17 650 N acting on the ligament of about 125 mm^2 cross-sectional area, producing a stress on the anterior cruciate of 141 MPa. This compares to the ACL failure strength in tension of 37.8 MPa (Noyes and Grood 1976).

The medial collateral ligament is frequently treated without surgery because it is outside the joint and has a good blood supply. Healing results in near-normal strength of the ligament. The ACL, however, is inside the joint, surrounded by synovial fluid and when ruptured, frequently loses all continuity. Thus, natural biological repair is usually unsuccessful and reconstruction is necessary. Biological materials, such as a portion of the patellar tendon, are candidates for replacement as well as nonbiological materials such as Gore-Tex, an expanded Teflon material. The strength of these materials can be compared to the loads that were just calculated and to the strength of the ACL. The middle one-third of the patellar tendon has a strength of 58.3 MPa (Noyes *et al.* 1984) and the Gore-Tex ligament fails at 4830 N (Bolton and Bruckman 1985).

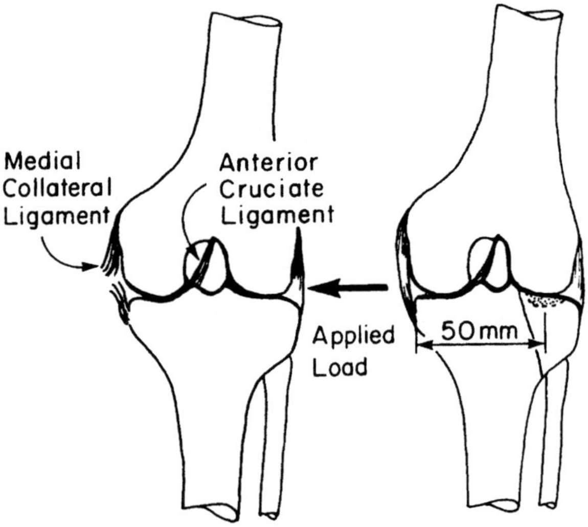

Fig. 9.7 Loading the knee with a planted foot by an individual blocking from the side.

The alternative to failure of the medial collateral ligament is either compressive failure of the cancellous bone in the lateral plateau (and/or femoral condyle) or shear failure of the whole lateral plateau. Both of these processes occur; failure of the femoral condyle is much less frequent. Application of 7060 N to an area of 15 mm diameter results in a compressive stress of about 40 MPa. This stress is likely to be normal to the surface because of the low coefficient of friction of the articular cartilage. Frequently, a 10–30 mm diameter depression is observed in the lateral plateau from such an injury. If the diameter is 0.03 m, the area is 7.07×10^{-4} m^2 and a load of 7060 N produces an *average* compressive stress of 9.99 MPa. For comparison, cancellous bone strength (0.310 g/cm^3) is approximately 5 MPa in compression (Anderson *et al.* 1992). Alternatively, failure can occur by the whole plateau shearing off from the rest of the tibia. While there is some cortex involved (Fig. 9.8), it is quite thin, and the primary bone failure is in the cancellous bone; the total area at that location is approximately 2000 mm^2, from which an average shear at failure of 3.5 MPa is calculated. Shear strength of cancellous bone is reported to be about 5 MPa (Saha and Gorman 1981) and is dependent on density. Thus, any of the failure modes are possible and the operative mode probably depends on loading rate, loading direction, muscle contraction, cancellous bone density, and age of the individual. Strength of the ACL is sensitive to the direction of loading and age of the patient. Cancellous bone density (and therefore strength) is very sensitive to age for women above age 45 and less sensitive for men.

The assumption of no muscle action is probably not correct. Certainly muscle contraction could be minimal in an individual who is unaware that he is soon to be a casualty. The medial muscles, the sartorius, gracilis, and semitendinosis, have a line of action that is not optimal for applying medial compression force unless the knee is nearly fully extended. Contraction of medial and lateral hamstrings would tend to increase both medial and lateral compressive forces and would probably shift the failure mode away from medial collateral ligament rupture and towards lateral tibial plateau fracture failure.

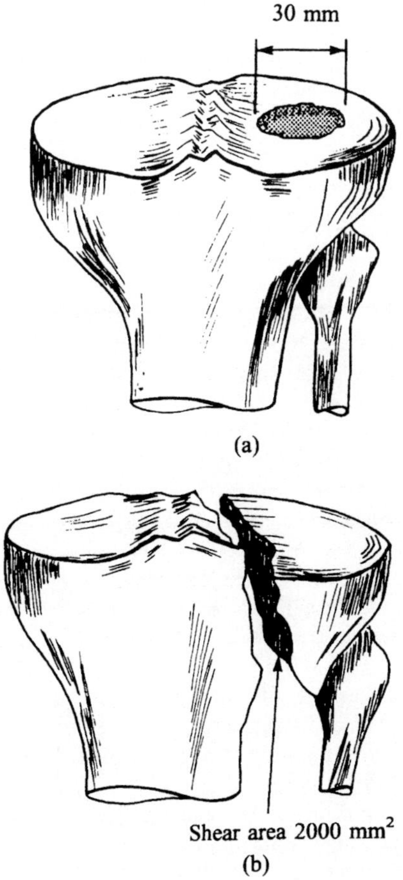

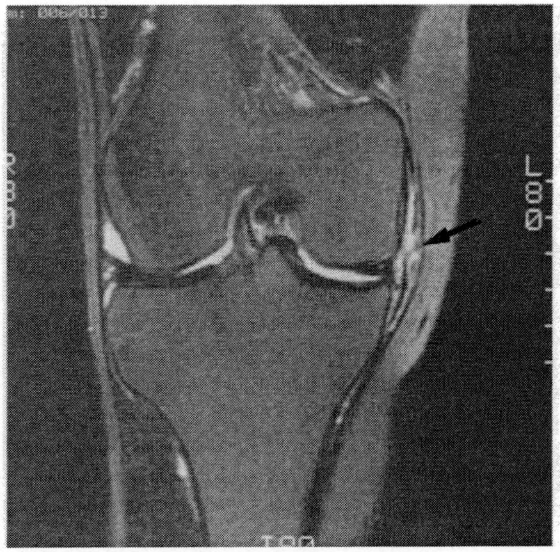

Fig. 9.8 Diagram of a proximal tibia demonstrating (a) compression failure, (b) shear failure of the lateral plateau. (c) Magnetic resonance image of the knee (coronal section) showing the anatomic structures: MCL (torn, see arrow), lateral femoral condyle, medial femoral condyle. (Courtesy of Charles Peterfy, MD.)

9.3 Cyclic loading considerations in biomaterial/biomechanical problems

9.3.1 Introduction

As previously described, materials in the human body undergo significant cyclic stress that can place them at risk of failure in a catastrophic manner. Failure of materials under alternating stress is known as fatigue. Materials other than metals suffer fatigue, and bone is a particular example that will be described later. Materials almost always fail under tensile cyclic stress, but that cyclic stress could be applied from tension to compression to tension, or from one level of tension to a higher level of tension and back to the original level of tension, or any combination of stress. Fatigue failure begins with the initiation of a microscopic crack with gradual propagation of that crack through the material with each additional sufficient tensile cycle (see Section 1.4.4). As the crack grows, the cross-sectional area of the load-bearing member decreases, increasing the overall average stress, causing the local stress to increase, and eventually catastrophic failure occurs. The higher the applied stress, the more likely is crack growth to occur, and the fewer the number of cycles that would be necessary to achieve failure. For metals, geometric stress concentration sites increase local stress (Section 1.4.4), and can accentuate the effect of microscopic flaws such as voids or inclusions in the material. Design considerations and surface finish are critically important in lowering fatigue failure. Geometric stress concentrations (as opposed to material stress concentrations such as voids) are a result of design. These design features have variable effect depending on the type of loading to which a structure is subjected. The usual equations for bending, torsion, or axial loading are used to determine the stress, σ. The actual stress (Popov 1968) is higher than the calculated stress by a factor, K:

$$\sigma_{\text{act}} = K\sigma_{\text{calc}}. \tag{9.19}$$

The value of K can be obtained for various geometries from graphs such as Fig. 9.9 (Peterson 1976).

As indicated in Section 1.4.4, the maximum cyclic stress that causes failure decreases as the number of cycles increases until a level of stress is reached, for most steels and some other metals, at which no further failure occurs despite cycling to 10^8 or more cycles. In such cases, the stress below which failure does not occur is called the endurance limit and is usually defined at 10^7 cycles. For steels, the endurance ratio (fatigue limit divided by tensile strength) is approximately 0.5. Many nonferrous alloys do not demonstrate an endurance limit. Thus, the fatigue strength of a material may be specified as the failure strength at a defined number of cycles. Cobalt–chromium alloys demonstrate a continuous decrease in strength with increasing number of cycles as shown in Fig. 9.10 (Georgette and Davidson 1986).

9.3.2 Stress concentration in titanium alloys. Fatigue failure of titanium hip

Fatigue is a matter of concern in stainless steel, titanium alloy (Ti–6Al–4V) and cast cobalt–chromium alloys. Titanium alloy demonstrates a property not specifically noted in other implant alloys in that it is 'notch'-sensitive. Cast cobalt–chromium alloys have many voids, inclusions, and other imperfections that act as stress concentration sites for the initiation of fatigue failure. Titanium alloy is relatively free of these problems, and thus demonstrates a relatively high fatigue strength of about 600 MPa. However, surface finish and design stress concentration sites can drastically change the fatigue endurance limit, producing a behavior termed *notch sensitivity*. Such a design change would be the application of a porous coating through the process of sintering. Through this heat treatment process, 100–500 μm titanium beads are fused to the surface of the titanium alloy prosthesis to permit attachment to the musculoskeletal system through bone ingrowth. While these bead attachment sites cause a drastic change in the fatigue endurance limit of titanium alloy (Cook *et al.* 1984), little effect is noted in the cast cobalt–chromium alloys because of the pre-existing high concentration of imperfections from casting (Fig. 9.11). Part of the problem with titanium alloys is the phase change that occurs with heating to temperatures necessary to achieve sintering of porous coatings. While techniques have been

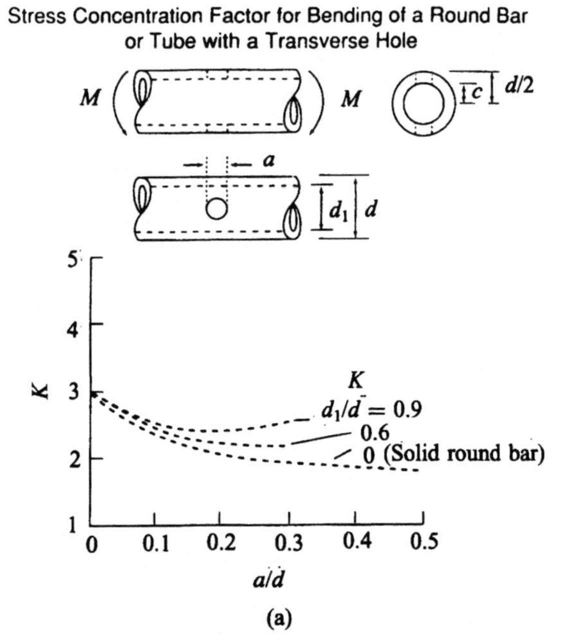

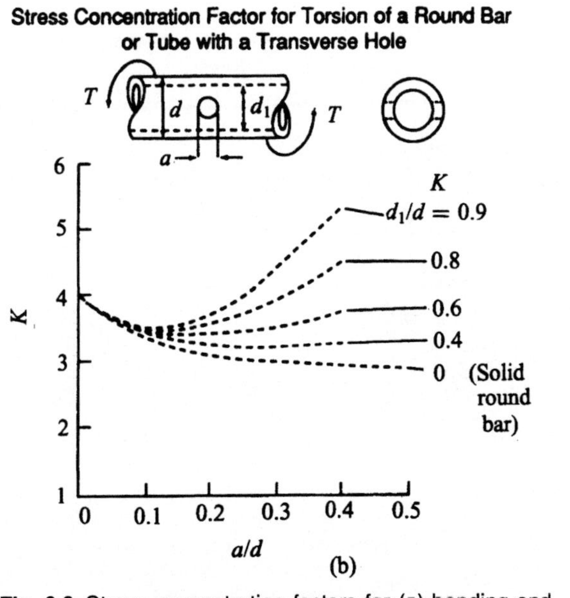

Fig. 9.9 Stress concentration factors for (a) bending and (b) torsion of a tubular structure with a hole. (Adapted from Peterson 1976, and reproduced with permission.)

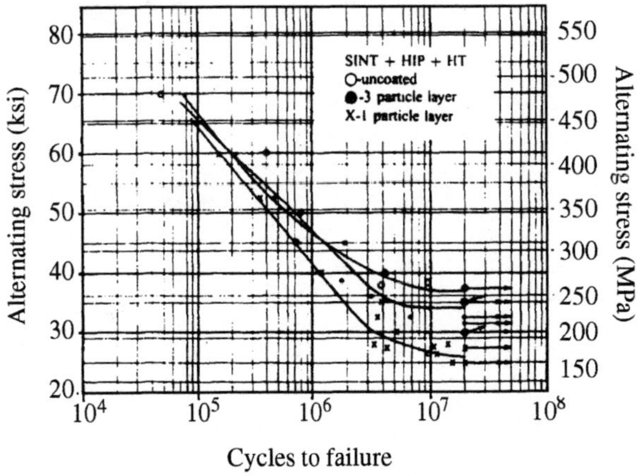

Fig. 9.10 Fatigue endurance curves for Co–Cr alloy. (From Georgette and Davidson 1986.)

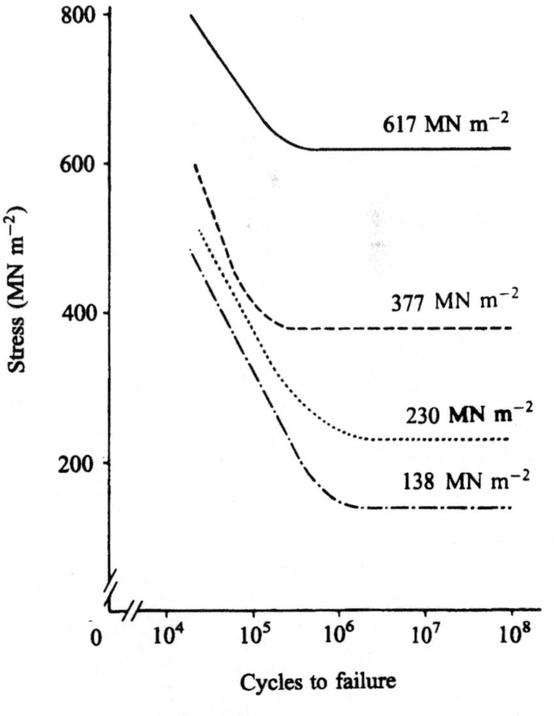

Fig. 9.11 Fatigue endurance curves for titanium alloy. (From Cook *et al.* 1984.)

developed to obviate this problem to some extent, it still remains a serious consideration in the design of prosthetic implants.

Many current designs of femoral component implants obtain initial fixation to the bone to permit ingrowth by press-fitting the prosthesis into the bone. This requires reaming the inside of the bone to a cylindrical shape of the exact diameter of the prosthesis, providing very good distal fixation of the femoral component that allows bone growth into a porous coating on the proximal surface (Fig. 9.12). Firm fixation of the distal stem without proximal fixation implies a significant bending moment applied at the porous coating junction with a smooth stem. The stance phase bending moment on the prosthesis is given by 3 BW (2100 N) times approximately 0.03 m. The cross-sectional area of a typical stem with a diameter of 12 mm is 1.13×10^{-4} m^2. The area moment of inertia of that cross section is $I' = \pi r^4/4 = 1.02 \times 10^{-9}$ m^4. The stress in the extreme fiber of the prosthesis assuming the prosthesis acts as a cylindrical beam is then given by 370 MPa. This is to be compared to the fatigue curves from Cook *et al.* (1984) shown in Fig. 9.11. It can be seen that the stress of walking without ingrowth is sufficient to cause failure of a notched specimen at 10^5 cycles (less than one year of normal walking). This has resulted in careful design of titanium alloy prostheses to avoid stress concentration sites. Similar concerns arise in acetabular components (the socket side of hip replacement) because holes are utilized for screw attachment and to allow the surgeon to check depth of placement. These holes constitute stress concentration sites and have been shown to result in stresses near fatigue endurance limits (Burton and Skinner 1989).

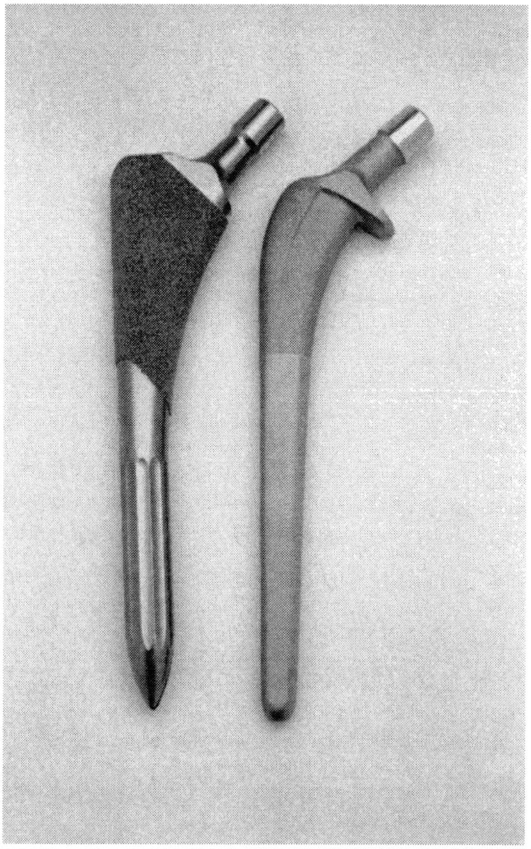

Fig. 9.12 The Golden Gait hip prosthesis (left) for use without cement and the Ultimate-C hip prosthesis (right). (Courtesy DePuy, Inc.)

9.3.3 Fatigue failure of dynamic hip screw with and without medial support

Evaluation of the forces involved with screw–plate fixation devices is simplified by resolving the joint loading force into a force that is parallel with the screw device and a force that is perpendicular to it. Two types of loading conditions should be considered (Fig. 9.13). The first of these, (a), is the case in which there is no

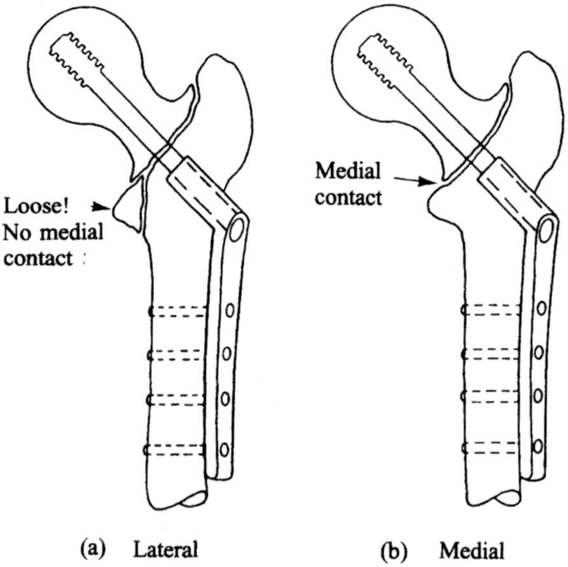

Fig. 9.13 Diagram of an intertrochanteric fracture with a screw plate device showing (a) no medial bone contact, (b) medial bone contact.

bony contact on the medial aspect and the entire joint reaction force is borne by the screw device. For a 135° screw–plate device, the parallel force is 30° from the joint reaction force, and thus the parallel force is 0.866 times the resultant force. Similarly, the perpendicular force is 60° from the resultant force and is therefore equal to cos 60° times the joint reaction force or 0.5 times the reaction force.

Reasonably accurate estimates for certain variables permit the calculation of the stress in the screw during loading. For example, a 70 kg man might be expected to load the screw with approximately three times body weight (2000 N) during the stance phase of gait. The screw device has a diameter of 10 mm, and an area moment of inertia of 4.91×10^{-10} m^4 is calculated. The moment arm would be approximately 60 mm, assuming that the cancellous bone in the greater trochanteric region supported the screw device. The moment arm would, of course, be longer if this were not the case. The moment then becomes 1000 N × 0.06 m, or 60 N m. Use of eqn (9.1) with this moment, a value of c of 0.005 m, and an I' of 4.91×10^{-10} m^4 yields a stress at the surface in tension or compression of 611 MPa. This is to be compared with the yield strength of stainless steel, which is approximately 790 MPa. Thus, stresses significantly over three times body weight will put the screw plate device close to the yield range. Similarly, increases in the moment arm of 20 mm will also produce adequate stress to cause yielding in the screw. The effect of compression from the parallel force has to be accounted for, but this constitutes only a small percentage of the stress caused by the bending force.

It should also be noted that the forces across the hip joint will be reduced in cases of fracture of the greater trochanter because much of this force is a result of the abductor muscle tension. On the other hand, because of the lack of damping of some of the muscles around the hip after fracture, dynamic forces from impact may exceed three times body weight during walking.

Contact of the two fracture fragments of the bone medially in the calcar region, Fig. 9.13(b), does not significantly change the moment arm of the calculation of the stress, but does significantly change the area moment of inertia of the construct. Medial contact changes the neutral axis of bending from the center of the screw device to somewhere medial and parallel to it, and thus the screw device is withstanding only tension while the bone is withstanding the compression forces. This is called a 'tension band' effect. The area moment of inertia now includes the portion of medial bone that is in contact in compression and the screw–plate device laterally. If the contribution of the bone is included and the neutral axis is moved 15 mm medially, the area moment of inertia contribution of the screw–plate device is increased by a factor of approximately 35, reducing the stress in the screw–plate device by approximately a factor of 27. Thus medial bony contact is extremely important to the longevity of the plate. This longevity becomes a factor with a delayed union or nonunion of the fracture. Fracture healing may take a year or more although the usual time to union is three months.

Bending moments applied to the screw–plate device cause significant frictional forces between the screw and the barrel of the device. These frictional forces can be minimized by using nail–plate fixation devices with a higher angle. Similarly, engaging more of the screw within the barrel device decreases the bending and permits distal migration of the femoral head and compression of the fracture.

9.3.4 Fatigue failure of bone

Fatigue failure of bone is a relatively common occurrence. Military training subjects untrained individuals to intensive, repetitive stress which results in fatigue fractures of bones in the foot, leg, thigh, and femoral neck. These fractures are termed *stress fractures* by the medical profession for some obtuse reason. These fractures also occur in individuals who train for distance running or other physical activities such as basketball without adequate preparation. Bone will accommodate to new stress states but must have adequate time to do it. Bone responds to increased stress by remodeling, but begins this process by removing old bone prior to replacing it with new, stronger bone. Thus, bone gets weaker prior to getting stronger. Continued physical activity after initiating the remodelling process can result in pain and, in extreme cases, catastrophic failure (Skinner and Cook 1982). In the report by Skinner and Cook, a patient ran about 161 km (100 miles) for an estimated 10^5 cycles over a two-month period. The patient suffered an undisplaced fracture of the femoral neck with an estimated failure stress of 22 MPa, much below the single load failure of bone. The remodeling process in humans takes weeks, so increases in physical activity should be arithmetic rather than geometric, and increments should occur at 4–6 week intervals. This would allow adequate time for remodeling.

6

Stress concentration can also occur in bone. The usual mode of failure is with a single application of load, frequently in torsion although stress concentration can be a factor in fatigue. Single load failures occur after biopsies of bone for tumors or after placing screws or screw holes. Failure strength in these areas is reduced by about one-half (Burstein *et al.* 1972).

9.3.5 Fatigue behavior of C–SiC alloys: comparison to heart valves

One biomaterial application in the human body that must withstand significant cyclic loading is the heart valve. Designs that utilize a structural ring to support the valve components must undergo approximately 36 million cycles per year. Obviously, replacement of heart valves every 2–3 years would not be desirable. A material that tolerates high cyclic stress is carbon–silicon carbide alloy. This material is produced by gas-phase decomposition of propane and methyltrichlorosilane (CH_3SiCl_3) in a fluidized bed (Bokros 1969). The amount of silicon in the C–SiC alloy can be controlled by the amount of methyltrichlorosilane in the gas mixture. These materials are termed pyrolytic carbons and are deposited on a graphite substrate. The mechanical properties are quite good and vary with density and fraction of silicon (Kaae and Gulden 1974). Densities range from 00000 for pure pyrolytic carbon to 00000 for SiC alloys while Young's modulus ranges from 19 GPa to 105 GPa. The interesting property is the endurance limit, which is essentially the same as the single-cycle failure stress. This property significantly simplifies design of devices undergoing cyclic loading (Shim 1974).

Consider one leaflet of a heart valve (Fig. 9.14a) with radius r supported by pegs A and B with radius r_1. This valve is designed to be supported by the two pegs when the leaflet is closed (Fig. 9.14b). The loading is at four points on the valve leaflet. There is minimal loading of the valve when open (Fig. 9.14c); since the blood is flowing through the valve, loading occurs with every closing. The highest pressure differential occurs when the aortic valve is closed (left ventricle to aorta) because the ventricle is filling and the aortic valve is maintaining diastolic pressure in the vascular system. Typical blood pressure is 120/80 mmHg (systolic/diastolic); 80 mmHg is equivalent to a pressure of 0.01 MPa. The area of the orifice of the largest St Jude's valve is 5.31×10^{-4} m^2 ($r = 13$ mm). The size of each peg is approximately 1 mm × 2.5 mm. The load on each peg is given by

$$\text{(fraction carried by each peg)} \times \text{(pressure differential)} \times \text{(area of valve)},$$

or

$$1/4 \times (0.01\ \text{MPa}) \times (5.31 \times 10^{-4}\ \text{m}^2) = 1.3\ \text{N up and } 1.3\ \text{N down.}$$

Thus, each peg experiences 2.6 N average load with a stress of approximately 1 MPa with every closure of the heart valve. Bending stress between the two supports of each peg may reach an order of magnitude higher. This is a static approximation and seems to have a generous safety factor, since the failure stress of pyrolytic carbon with 20 per cent SiC is approximately 500 MPa. However, all the actual stresses must be considered for the entire load cycle to determine the safety factor. This complicated problem is probably best solved using a finite-element analysis.

9.3.6 Fatigue of tibial plateau prostheses

Extension of the preceding analyses such as in Section 9.2.4 to the knee joint reveal that the same problems exist. Designs of total knee prostheses for the tibial component have frequently employed some sort of a stem that extends into the bone of the tibia for a variable length, but typically 50–70 mm. The same problems that occur with total hip replacement and bone plates also occur for tibial plateau prostheses, in that the stem is quite rigid in relation to the bone that it is adjacent to, and stress from the plateau is transmitted along the stem to its tip and from there to the bone of the tibia (Fig. 9.15). Thus, the cancellous bone in the proximal tibia bears less stress than it would otherwise under normal conditions. The stemmed metal components came about because of the perceived need to support the ultra-high molecular-weight polyethylene insert and reduce loosening of the cement–bone interface. When loosening occurs, the interface that develops is a fibrous material with a very low modulus (1–50 MPa) and a high fluid content. If this membrane forms, it obviously provides much less support to the tibial plateau surface than cancellous bone (modulus 500 MPa), and the prosthesis may be considered to be a column with two cantilever beams. There have been reports of failure of these metal plateaux, and an analysis can explain this phenomenon. For simplicity, the prosthesis is assumed to be fixed, with

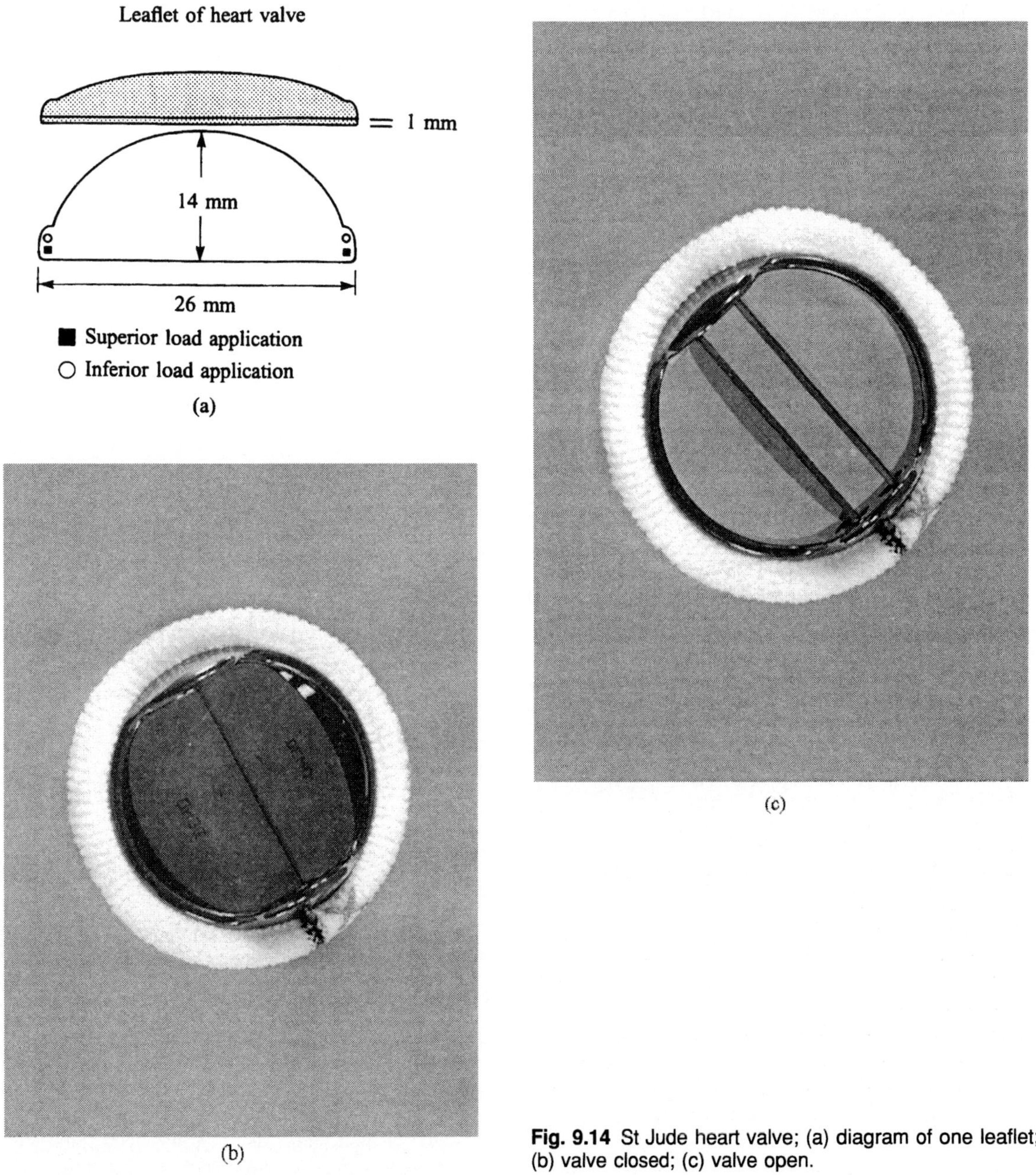

Fig. 9.14 St Jude heart valve; (a) diagram of one leaflet; (b) valve closed; (c) valve open.

the plateau cantilevered with no support. A load of 1.5 body weight (1200 N) is applied to each plateau 24 mm from the center. The plateau is assumed to be rectangular (75 mm medial–lateral dimension by 37.5 mm anterior–posterior dimension) for simplicity, and 3 mm thick made of cobalt–chromium alloy. The area moment of inertia, then, is 9×10^{-11} m^4, and EI' is equal to 18.0 N m^2. Referring to a standard text for beam

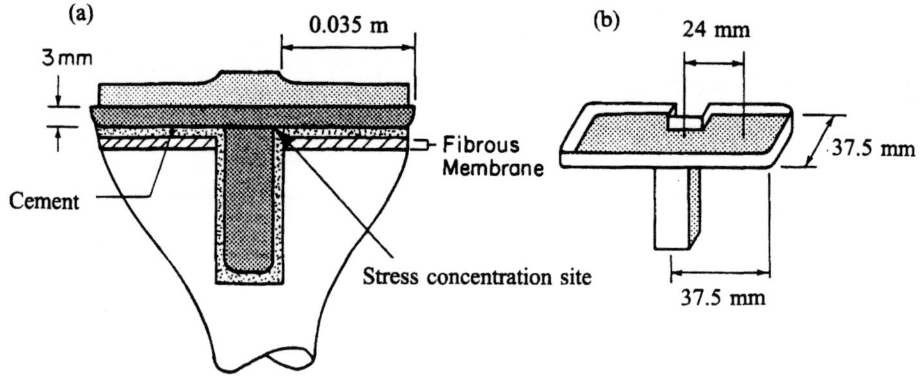

Fig. 9.15 Diagram showing (a) coronal view of tibia and tibial plateau prosthesis and (b) oblique view of the tibial plateau prosthesis, with loading conditions.

deflection formulae, we find that the maximum deflection $y_{\max}$ is given by the equation

$$y_{\max} = \frac{Fa^2}{6EI'}(3L - a), \qquad (9.20)$$

where F is the load, a is the distance from the cantilever support to the load, and L is the total length of the beam. The deflection is estimated to be very large at 5.66×10^{-4} m ($\sim$0.6 mm). The flexure formula can be used to calculate the stress in the metal plateau at the cantilever and yields a relatively high stress of about 53 MPa. Stress concentrations from geometric irregularities or material defects can significantly increase the stress, while plateau support from cancellous bone or cement can reduce the bending moment and stress (Paganelli *et al.* 1988; Skinner *et al.* 1987).

9.4 Composite materials—the impetus for more 'flexible' prostheses

Composite materials are a combination of two or more materials such that the combination results in significantly better mechanical properties than either material alone (Skinner 1988). Composite materials have gained attention in the orthopedic surgery area because of the potential to vary the mechanical properties, particularly the elastic modulus, to tailor a prosthesis to meet certain specified criteria. Initial experience with composite materials in orthopedic surgery began with carbon-fiber-reinforced polyethylene for bearing surface applications in total joint replacements. These, however, did not work out well because of problems of attachment of the matrix to the carbon fiber material. Pull-out of fibers caused joint debris and increased wear. The increased modulus of the composite results in higher contact stress at the bearing surface and thus higher wear.

The perceived need for composites in orthopedic surgery results from a recognition of the stress shielding of bone by high-modulus (flexural, axial, or torsional rigidity) prostheses presently made of metallic alloys. Although exact design criteria for femoral component replacement have not been defined (Skinner 1991; Skinner and Curlin 1990), it is felt that a lower-modulus (rigidity) prosthesis will result in a more biologically stable prosthetic construct. Unfortunately, the goal in such a design is high strength and low modulus, which is difficult to achieve even using composite materials.

A typical composite consists of a matrix material and a fiber material. Both of these materials must be biocompatible and stable in a physiological environment. The matrix materials that are being investigated at the present time include polysulfone, poly(ether ether ketone), and

Table 9.2. Properties of composite matrix materials

	Polysulfone (UDEL)	Poly(ether ether ketone) (Victrex PEEK)
Density (g cm^{-3})	1.25	1.32
Tensile strength, yield (MPa)	70.3	92
Compressive strength, yield (MPa)	96	118
Tensile modulus (GPa)	2.48	3.6
Elongation (%)	50–100	50
Flexural strength (MPa)	106	170
Poisson ratio	0.37	0.42
Fatigue endurance limit (MPa)	6.9	70

(Adapted from Skinner 1988.)

variations on these materials. The primary fiber under consideration is produced from carbon, although Kevlar (polyamide) is also under consideration. The carbon-based fibers typically have a longitudinal Young's modulus ranging from 250 to 390 GPa. Tensile strength of the fiber is in the range of 2.2 to 2.7 GPa, with the diameter about 8 μm. Properties of the composite matrix materials are given in Table 9.2.

Structure and properties are directly related for composites; thus a rule of mixtures applies for composites with the fibers aligned in the direction of stress or transverse to that plane. When the stress is in line with the fibers, the strain in the matrix and the strain in the fibers are the same, and the load is distributed in relation to the elastic modulus and a cross-sectional area of each component. The elastic modulus for the total composite, then, is given by

$$E_{\text{par}} = \frac{E_f A_f}{A} + \frac{E_m A_m}{A}, \tag{9.21}$$

where E_{par} is the modulus of the composite parallel to the fibers, A is the total cross-sectional area, and the other terms are the equivalent parameters for each component. This equation reduces to

$$E_{\text{par}} = E_f V_f + E_m V_m, \tag{9.22}$$

where V_f and V_m are the volume fractions of the fiber and matrix.

Stress applied transverse to the fibers results in equivalent stress in both components and a relationship for the elastic modulus perpendicular to the fibers given by

$$E_{\text{perp}} = \frac{E_f E_m}{E_f V_m + E_m V_f}. \tag{9.23}$$

While the relationship for elastic modulus parallel to the fiber is quite accurate in predicting modulus, the elastic modulus calculated from eqn (9.23) is not as reliable for the perpendicular stress application. Chopped fiber composites result in a significant improvement in the strength of the composites but involve a much more complicated modulus relationship because of the possibilities with various lengths and orientations of the fibers.

Strength is not so easily determined, although obviously anisotropic behavior is observed. At typical volume fractions of fibers (0.4 to 0.7), the strength of the fiber and the direction of the fiber are the determining variables. Transverse to the fibers, the strength is highly dependent on the matrix properties. A composite with multiple layers and multiple orientations has its strength determined by a summation of the individual layers. Layers are frequently included in a laminate in the primary stress-bearing direction, termed longitudinal, and at ±45° and 30° from that direction and even 90° if transverse strength is required. It is important to remember that, while the composite industry typically uses composites for applications with clearly defined loading conditions, such as pressure vessels or turbine blades, the orthopedic industry is considering these materials not as a thin-walled pressure vessel but as a structural element that will have to undergo significant axial, bending, and torsional loading.

The improvement in flexural rigidity can be estimated by calculating the ratio of bone to composite for a typical press-fit total hip replacement of a cylindrical design. The required data for composites reveals that a poly(ether ether ketone) (PEEK) carbon fiber composite in a 45–45 lay-up has a modulus of about 20 GPa. Laid up in a 0° alignment, the modulus is about 121 GPa. When laid up at 90° it is about 9 GPa, and an isotropic PEEK 30 per cent carbon fiber has a 13 GPa flexural modulus. The modulus of bone is taken as 20 GPa; the flexural rigidity, EI', for the femoral shaft is given by

$$(EI')_b = \frac{E\pi}{4}(r_{OD}^4 - r_{ID}^4) \tag{9.24}$$

The flexural rigidity, EI', of the prosthesis with a press-fit would be given by

$$(EI')_{pr} = \frac{E\pi r_{ID}^4}{4}. \tag{9.25}$$

Therefore, the EI' ratio of bone to prosthesis is

$$\frac{(EI)'_b}{(EI')_{pr}} = \frac{E_b}{E_{pr} r_{ID}^4}(r_{OD}^4 - r_{ID}^4). \tag{9.26}$$

For a prosthesis that is 0.015 m diameter inside a 0.030 m bone, this relation reduces to

$$\frac{(EI')_b}{(EI')_{pr}} = 15\frac{E_b}{E_{ppr}}, \tag{9.27}$$

where $E_b = 20$ GPa.

The rigidity ratios for various composites are given in Table 9.3. It is obvious that a prosthesis with a modulus as low as 9 GPa is going to be much more rigid than the material replaced by the prosthesis. That material will be fat and at most cancellous bone with a modulus of approximately 500 MPa. The question that must be answered then is: how close to ~500 MPa does the prosthesis have to be in order to cause acceptable response in the bone. This question has yet to be answered.

Table 9.3. Ratio of bending stiffness of cortical bone (femur) to that of a cylindrical prosthesis filling the inside of the bone[a]

Material	EI_b/EI_{pr}
CF–PEEK (45°/ – 45°)	15
CF–PEEK (0°)	2.5
CF-PEEK (90°)	33.3
Isotropic 30% CF–PEEK	23.1
Co–Cr	1.4

[a] Assuming no shear transmission.

9.5 Wear problems with UHMWPE (ultra-high-molecular-weight polyethylene)

9.5.1 Introduction

Wear is the removal of one material by another material due to sliding contact. The primary wear process in orthopedics is in artificial joints, which must have sliding to accomplish their function. Another source of wear debris results from the interaction between bone plates and screws. Because of micromotion, the protective oxide layer and some underlying metal are rubbed off the screw or plate, producing particulate material in a process called 'fretting'. This process tends to be an insignificant problem compared to artificial joint wear debris. The wear debris problem is particle-size-dependent much more than chemistry-dependent. Obviously, particulate debris has a large surface area and would be susceptible to chemical degradation with resultant toxicity if harmful ions could be produced.

9.5.2 Effect of particulate material wear debris

The main problem that arises from wear debris is the tissue reaction that is mounted by the body to remove the debris (Skinner and Mabey 1987). The process involves cells that try to engulf the debris (typically in the micrometer size range) and degrade it with enzymes. The material is polymeric (ultra-high-molecular-weight polyethylene) and not soluble. Because degradation is not possible, cell necrosis (death) results from an excess of this particulate debris. Cells then agglomerate to form 'giant cells' to try to remove this necrotic material and debris. Incidental to this process, bone is frequently removed, which leads to loosening of the joint prosthesis and failure of the implant. Minimization of the wear process is essential to the long-term success of joint replacement.

Wear processes vary depending on the sliding materials. The materials in common usage include UHMWPE as one surface and either Co–Cr alloy or one of two ceramic materials, ZrO_2 or Al_2O_3, as the other. Cobalt–chromium alloy vs. cobalt–chromium alloy was used as a bearing in the recent past, but produces significant debris despite a low wear rate and has raised concerns about neoplasm from the absorption of ions by the body. Although this wear couple has been mentioned recently to avoid the UHMWPE wear problem, the potential neoplastic problem raises safety concerns regarding reintroduction.

9.5.3 Wear mechanisms

Wear occurs through adhesive or abrasive modalities. Obviously, minimizing abrasive wear of the UHMWPE is dependent on the maintenance of a smooth, hard mating surface. Other variables include surface conformation and UHMWPE thickness, both of which affect contact stress. Increased contact stress increases the chance of adhesive and abrasive wear because contact is more intimate. UHMWPE can also be transferred to its mating surface through adhesion, bringing surface wettability to importance as a variable. The introduction of a hard third body can significantly increase the abrasive wear process.

Independent of mechanism, the total volume and particle size of the wear debris are the important variables. One cubic centimeter of wear volume converts to 2.4×10^{16} particles of 1 μm diameter, and cell dimensions are of the order of 10 μm. The body seems to tolerate relatively small loads, but the threshold for an adverse response is unknown.

The relation for adhesive wear is given by

$$V = \frac{cLx}{p}, \tag{9.28}$$

where V = volume of wear debris, c is a proportionality constant, L is the load on the surfaces, x is the distance of sliding, and p is the hardness of the surface suffering wear. Abrasive wear is described by a similar relation, dependent on the shape model assumed for the asperity of the hard material plowing through the soft material. Basically, abrasive wear increases with increasing wear distance and load, and decreases if the wearing material hardness increases. Wear data seem to be quite dependent on the method of measurement.

Minimizing adhesion of UHMWPE can be realized by utilizing harder bearing materials with ionic character that will therefore have a low contact angle for water solutions (i.e. are quite wettable). Thus, a fluid film intervening between the two bearing surfaces will tend to prevent adhesion. Alumina (Al_2O_3) and zirconia (ZrO_2) are ionic and therefore have lower contact angles than Co–Cr alloys. Co–Cr alloys have a passive oxide film that prevents corrosion and reduces the contact angle. This film, however, can be removed by shear, resulting in adhesion of the UHMWPE.

Abrasion can be minimized by maintaining a hard, smooth-bearing surface against the UHMWPE. Achieving a smooth surface for the ceramic materials or Co–Cr is relatively easily accomplished, but even against a soft material such as UHMWPE, burnishing of the Co–Cr surface occurs. Thus, the ceramic materials, in addition to being harder, will maintain the smooth surface longer, thereby decreasing adhesive wear.

Third-body wear is a confounding factor in these wear couples. Bone chips, with small crystals of hydroxyapatite, can introduce a new form of wear of either bearing surface. Further corrosion or shear of the Co–Cr surface can result in fine (<1 μm) debris that can accelerate the deterioration of the polished Co–Cr surface.

The contact angle of a liquid on a solid is the resultant of the surface energies of the liquid–vapor, liquid–solid, and solid–vapor interfaces. Low contact angles occur when the liquid has an affinity for the solid because of similar chemical character. The inverse produces high, nonwetting contact angles (Fig. 9.16).

The contact stress (σ_c) is important in UHMWPE wear processes; it is equal to load divided by contact area, $\sigma_c = L/A_c$. Thus, higher contact stress implies higher local loads from decreased area despite constant general loads. High contact stress from nonconforming surfaces results in higher wear.

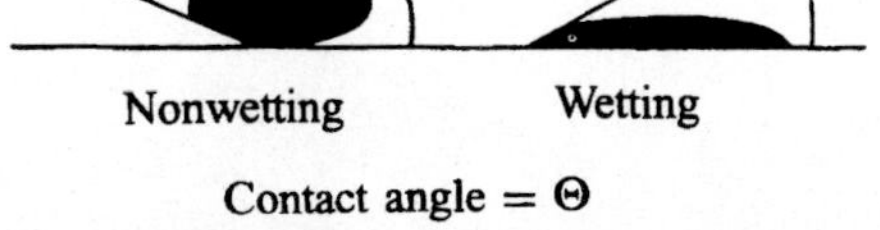

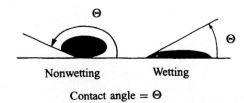

Fig. 9.16 Contact angle, defined by the tangent to the interface between liquid and solid.

Problems

P9.1 With regard to the lateral placement of a bone plate on a bone (as shown in Figure 9.6), if the muscle load is 4 cm from the center of the plate, and if you assume that no load is borne by the bone, what is the maximum compressive stress in the plate and the maximum tensile stress? Assume there are no screw holes in the area of the fracture (no stress concentration).

P9.2 A.B. is a healthy 14-year-old male (1.95 m, femur length 0.4 m, 67 kg) who presented with chronic left hip pain over a ten-month period. He was noted to be overweight at the onset of weight training, but gradually decreased his body weight. The pain as insidious in onset, intermittent in duration, and described as 'achy' in nature. The patient denied any trauma or injury. Three weeks before presentation, A.B. complained of an acute increase in left hip pain while performing resisted knee extensions on a weight machine. His knee extensions did not lift his trunk off the seat and were done with 667 N weights applied at 0.3 m from the knee joint axis. A.B. had been limping since that incident. On physical examination, the left hip had abduction 10°, adduction of 30°, internal rotation of $-10°$, and external fixation of 60°. Full extension without a flexion contracture was noted. Leg lengths were equal. Radiographs by his pediatrician revealed an acute and chronic grade-III SCFE of the left hip (see Fig. 9.2a). The right hip was unremarkable both clinically and radiographically.

Upon admittance, the patient was placed in Russell skin traction with an internal rotation strap for two days. Because no improvement in the femoral head position was obtained after traction, no attempt at closed reduction was performed. *In situ* pinning with a single centrally placed cannulated screw (6.5 mm cancellous, Synthes, Paoli, PA) under fluoroscopic assistance was performed through lateral approach (see Fig. 9.2b). Pin placement was measured to be within 5 mm of the articular surface.

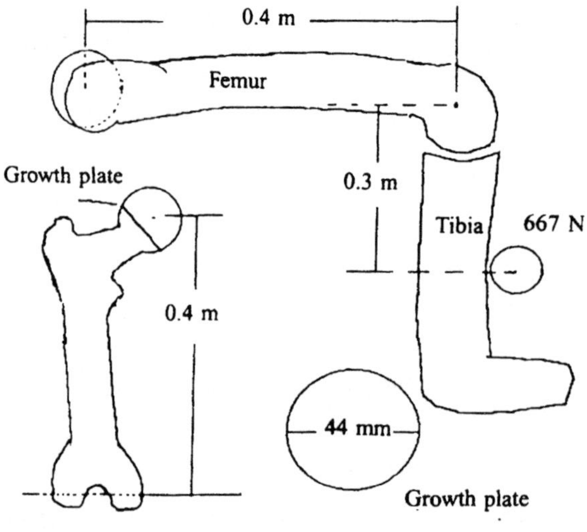

Fig. 9.17

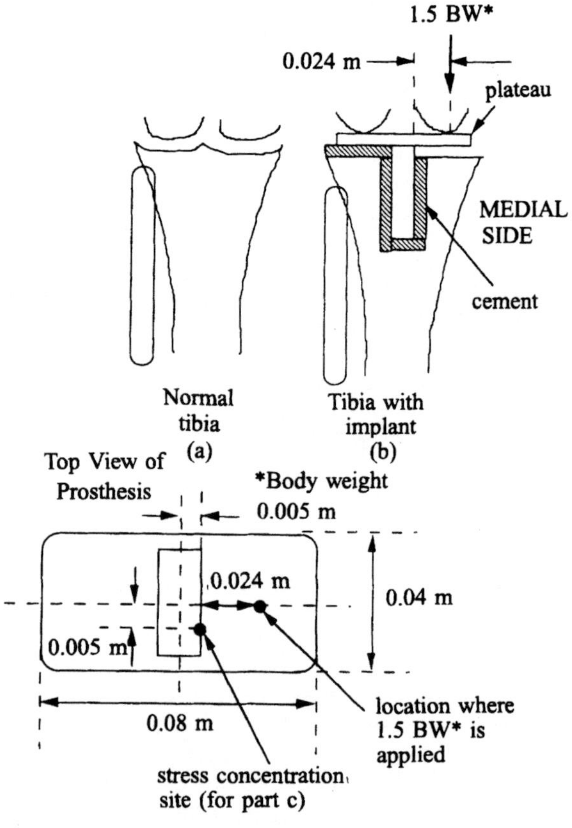

Fig. 9.18

Assume (see Fig. 9.17) that

(i) the mean diameter of the growth plate is 44 mm,

(ii) the plane of the growth plate is vertical,

(iii) the mass of the trunk is applied at the center of the femoral head.

(a) Calculate the shear stress on the physis (growth plate) (hint: statics problem).

(b) Compare that to 1 ± 0.5 MPa (reported shear stress at failure of the growth plate).

References: Chung *et al.* (1976); Pritchet and Perdue (1988).

P9.3 Assume that an orthopedic surgeon made a technical error in implanting a tibial plateau prosthesis (see Fig. 9.18). The medial side of the prosthesis did not get any cement. The body weight is 800 N, the prosthesis is 4 mm thick and rectangular (0.08 m × 0.04 m), and $E_{CoCr} = 200$ GPa.

(a) Calculate the deflection of the plateau.

(b) Calculate the maximum stress in the plateau.

(c) Assuming that there is a stress concentration site ($k = 3.2$), calculate the stress in plateau.

(d) Discuss the effect of the stress and the deflection, if the plateau (without stress concentration site) is supported by cancellous bone, $E = 0.5$ GPa, or by bone cement, $E = 2$ GPa.

References

Anderson, M. J., Keyak, J. H., and Skinner, H. B. (1992). Compressive mechanical properties of human cancellous bone after gamma irradiation. *Journal of Bone and Joint Surgery*, **74A**, 747–52.

Aronson, D. D. and Carlson, W. E. (1992). Slipped capital femoral epiphysis: a prospective study of fixation with one screw. *Journal of Bone and Joint Surgery*, **74A**, 810–19.

Bobyn, J. D., Dilliar, R. M., Binnington, A. G., and Szivek, J. A. (1985). The effect of partially and fully coated porous coated canine hip stem design on bone remodelling. *Journal of Orthopaedic Research*, **5**, 393–408.

Bokros, J. C. (1969). Deposition, structure and properties of pyrolytic carbon. In *Chemistry and physics of carbon*, vol. 5, (ed. P. L. Walker). Dekker, New York.

Bolton, C. W. and Bruchman, W. C. (1985). The Gore-Tex expanded polytetrafluoroethylene prosthetic ligament. *Clinical Orthopaedics and Related Research*, **186**, 202–13.

Burstein, A. H., Currey, J., Frankel, V. H., Heiple, K. G., Lunseth, P., and Vessely, J. L. (1972). Bone strength: the effect of screw holes. *Journal of Bone and Joint Surgery*, **54A**, 1143–56.

Burton, D. and Skinner, H. B. (1989). Stress analysis of a total hip acetabular component: an FEM study. *Biomaterials, Artificial Cells and Artificial Organs*, **17**, 371–83.

Carter, D. R. and Hayes, W. C. (1976). Bone compressive strength: the influence of density and strain rate. *Science*, **194**, 1174–6.

Chung, S. M. K., Batterman, S. C., and Brighton, C. T. (1976). Shear strength of the human femoral capital epiphyseal plate. *Journal of Bone and Joint Surgery*, **58A**, 94–103.

Cook, S. D., Georgette, F. S., Skinner, H. B., and Haddad, R. J., Jr (1984). Fatigue properties of carbon- and porous-coated Ti–6Al–4V alloy. *Journal of Biomedical Materials Research*, **18**, 497–512.

Georgette, F. S. and Davidson, J. A. (1986). The effect of HIPing on the fatigue and tensile strength of a cast, porous coated Co–Cr–Mo alloy. *Journal of Biomedical Materials Research*, **20**, 1229–48.

Husby, O. S., Gjerdet, N. R., Ericksen, E. S., Rykkje, J. M., and Molster, A. O. (1989). Porosity of rat femora following intramedullary reaming and nailing. *Clinical Orthopaedics and Related Research*, **246**, 305–12.

Kaae, J. L. and Gulden, T. D. (1971). Structure and mechanical properties of codeposited pyrolytic C–SiC Alloys. *Journal of the American Ceramic Society*, **54**, 605–9.

Kilgus, D. J., Shimaoka, E. E., Tipton, J. S., and Eberle, R. W. (1993). Dual energy X-ray absorptiometry of bone mineral density around cementless femoral implants. *Journal of Bone and Joint Surgery*, **75B**, 279–87.

Leichter, H. L. and Robinson, E. (1970). Fatigue behavior of a high-density graphite and general design correlation. *Journal of the American Ceramic Society*, **53**, 197–204.

Noyes, F. R., Butter, D. L., Grood, E. S., Zernicke, R. F., and Hefzy, M. S. (1984). Biomechanical analysis of human ligament grafts used in knee ligament repairs and reconstructions. *Journal of Bone and Joint Surgery*, **66A**, 344–52.

Noyes, F. R. and Grood, E. S. (1976). The strength of the anterior cruciate ligament in humans and Rhesus monkeys. *Journal of Bone and Joint Surgery*, ***58A***, 1074–82.

Paganelli, J. V., Skinner, H. B., and Mote, C. D., Jr (1988). Prediction of fatigue failure of a total knee replacement tibial plateau using finite element analysis *Orthopedics*, **11**, 1161–8.

Peterson, R. E. (1976). *Stress concentration factors*. Wiley, New York.

Popov, E. P. (1968). *Introduction to mechanics of solids*, p. 555. Prentice Hall, Englewood Cliffs, NJ.

Praemer, A., Furner, S., and Rice, D. P. (1992). *Musculoskeletal conditions in the United States*. American Academy of Orthopedic Surgeons. Park Ridge, IL.

Pritchett, J. W. and Perdue, K. D. (1988). Mechanical factors in slipped femoral epiphysis. *Journal of Pediatric Orthopedics*, **8**, 385–8.

Reilly, D. T. and Burstein, A. H. (1975). The elastic and ultimate properties of compact bone tissue. *Journal of Biomechanics*, **8**, 393–405.

Roesler, H. (1981). Some historical remarks on the theory of cancellous bone structure (Wolff's Law). In *Mechanical properties of bone*, (ed. S. C. Cowin), AMD 45, pp. 27–42 ASME, New York.

Rubin, C. T. and Lanyan, L. E. (1987). Osteoregulatory nature of mechanical stimuli: function as a determinant for adaptive remodelling of bone. *Journal of Orthopaedic Research*, **5**, 300–10.

Saha, S. and Gorman, P. H. (1981). Strength of human cancellous bone in shear and its relationship to bone mineral content. *Transactions of the Orthopaedic Research Society*, **6**, 217.

Shim, H. S. (1974). The behavior of isotropic pyrolytic carbons under cyclic loading. *Biomaterials, Medical Devices, Artificial Organs*, **2**, 55–65.

Skinner, H. B. (1988). Composite technology for total hip arthroplasty. *Clinical Orthopaedics and Related Research*, **235**, 224–36.

Skinner, H. B. (1991). Isoelasticity and total hip arthroplasty. *Orthopedics*, **14**, 323–8.

Skinner, H. B. and Cook, S. D. (1982). Fatigue failure stress of the femoral neck: a case report. *American Journal of Sports Medicine*, **10**, 245–7.

Skinner, H. B. and Curlin, F. J. (1990). Decreased pain with lower flexural rigidity of uncemented femoral prostheses. *Orthopedics*, **13**, 1223–8.

Skinner, H. B. and Mabey, M. F. (1987). Soft tissue response to total hip surface replacement. *Journal of Biomedical Materials Research*, **21**, 569–84.

Skinner, H. B., Mabey, M. F., Paganelli, J. V., and Meagher, J. M. (1987). Failure analysis of PCA revision total knee replacement tibial component: a preliminary study using the finite element method. *Orthopedics*, **10**, 581–4.

Ward, W. T., Stefko, J., Wood, D. B., and Stanitski, C. L. (1992). Fixation with a single screw for slipped capital femoral epiphysis. *Journal of Bone and Joint Surgery*, **74A**, 799–809.

Woo, S.L-Y., Akeson, W. H., Coutts, R. D., Rutherford, L., Doty, D., Jemmott, G. F., and Amiel, D. (1976). A comparison of cortical bone atrophy secondary to fixation with plates with large differences in bending stiffness. *Journal of Bone and Joint Surgery*, **58A**, 190–201.

Woo, S.L-Y., Gomez, M. A., Seguchi, Y., Endo, C. M., and Akeson, W. H. (1983). Measurement of mechanical properties of ligament substance from a bone–ligament–bone preparation. *Journal of Orthopaedic Research*, **1**, 22–29.

Further reading

Cochran, G. V. B. (1982). *A primer of orthopaedic biomechanics*. Churchill Livingstone, New York.

Fu, F. H., Harner, C. D., Johnson, D. L., Miller, M. D., and Woo, S.L-Y. (1993). Biomechanics of knee ligaments—basic concepts and clinical application. *Journal of Bone and Joint Surgery*, **75A**, 1716–27.

Hirokawa, S. (1993). Biomechanics of the knee joint—a critical review. *Critical Reviews in Biomedical Engineering*, **21**(2), 79–135.

Holmes, C. A., Edwards, W. T., Myers, E. R., Lewallen, D. G., White, A. A., 3rd, and Hayes, W. C. (1993). Biomechanics of pin and screw fixation of femoral neck fractures. *Journal of Orthopaedic Trauma*, **7**, 242–7.

Keaveny, T. M. and Hayes, E. C. (1993). A 20-year perspective on the mechanical properties of trabecular bone. *Journal of Biomechanical Engineering*, **115**, 534–42.

Mow, V. C. and Hayes, W. C. (ed.) (1991). *Basic orthopaedic biomechanics*. Raven Press, New York.

Nordin, M., Frankel, V. H., Forssen, K., and Nachamie, H. (1989). *Basic biomechanics of the musculoskeletal system*, (2nd edn). Lea & Febiger, Philadelphia.

Radin, E. L., Simon, S. R., Rose, R. M., and Paul, I. L. (1992). *Practical biomechanics for the orthopedic surgeon*, (2nd edn). Churchill Livingstone, New York.

Swanson, S. A. V. and Freeman, M. A. R. (1977). *The scientific basis of joint replacement*, pp. 49–51. Wiley, New York.

Tencer, A. F., Johnson, K. D., Kyle, R. F., and Fu, F. H. (1993). Biomechanics of fractures and fracture fixation. *Instructional Course Lectures*, **42**, 19–55.

Woo, S.L-Y. (1982). Mechanical properties of tendons and ligaments. I. Quasi-static and non-linear viscoelastic properties. *Biorheology*, **19**, 385–96.

Woo, S.L-Y., Gomez, M. A., Seguchi, Y., Endo, C. M., and Akeson, W. H. (1983). Measurement of mechanical properties of ligament substance from a bone–ligament–bone preparation. *Journal of Orthopaedic Research*, **1**, 22–29.

Yamada, H. (1970). *Strength of biological materials*, (ed. F. G. Evans). Williams & Wilkins, Baltimore.

LOCOMOTION AND MUSCLE BIOMECHANICS

10

Steven L. Lehman, Rodger Kram, and Claire T. Farley[1]

Contents

10.1 Introduction

This chapter will introduce the large part of biomechanics dealing with animal movement, and will focus on locomotion. Most of the necessary formalism was developed in Chapter 1. This chapter uses that formalism to investigate an interesting set of biological problems—those involved in animal, and especially human, locomotion.

Biomechanists also study the movement of individual joints or limbs and their control (an area called motor control), the physiology and mechanics of human work (ergonomics), the optimization of human performance, and the man–machine interface (human factors). They ask questions at many levels, from the whole-body to the atomic scale. Their motivation may be to advance basic science or to solve a clinical problem. The methods and technologies introduced in this chapter can be applied to any of these areas. The recommended readings at the end of this chapter should give an idea of the span of this active area of bioengineering.

This chapter will view animal locomotion through a sequence of increasingly complicated models. In Section 10.2, the animal is approximated as a mass. A surprising amount can be learned about energetics of movement by measuring external forces and tracking the center of mass! The model in Section 10.3 recognizes that the animal has parts that might move independently. The body is approximated as a linkage made of rigid bodies, and data on the kinematics of the body segments and external forces allow the deduction of joint moment and power. Section 10.4 shows how muscle forces are measured or inferred from other data, and why such an inference is often not possible. This section also introduces the anatomy and mechanical properties of muscle, and points out the many open problems regarding this most common and intriguing biological actuator.

The primary aim of this chapter is to show how physics and anatomy can elucidate physiology.

[1] The authors would like to thank D. R. Wilkie for inspiration and Paul de Vita for the data in Figs 10.11 and 10.12.

Experimental methods, instrumentation, and methods of data analysis are sketched with the intention to give an idea of which data are available and how they are gathered and manipulated, rather than to produce a laboratory manual. The secondary aim is to list open research problems explicitly. Finally, this chapter should serve as a proper introduction, giving enough of the flavor of the endeavor to excite interest, and enough of a pointer to the literature to allow that interest to be followed.

10.2 Ground reaction force

The definition of ground reaction force follows from Newton's third law (Section 1.1.7). It states that any force that one body exerts on second body is counteracted by an equal and opposite force that the second body exerts on the first body. A ground reaction force is the force that the ground exerts on a body, and it is equal and opposite to the force that the body exerts on the ground. For example, when a human runner exerts a force on the ground with his foot, the ground exerts an equal and opposite force on the runner's foot.

Biomechanists measure the ground reaction force to gain insight into the force and mechanical work required to perform physical activities such as walking, running, jumping, or lifting. The ground reaction force is the largest force that acts on the body during many such activities. Measurements of the ground reaction force therefore give information about the magnitude of the force experienced by an organism's musculoskeletal system during these activities. This information gives insight into a central issue in functional biomechanics: the relationship between the mechanical loading and the structure of the musculoskeletal system. A second reason why biomechanists make measurements of the ground reaction force during physical activity is that they can be used to calculate the amount of mechanical work required for moving the body's center of mass during a physical activity. Finally, ground reaction force measurements have clinical applications. They can be used as a tool in diagnosis and rehabilitation of injuries and disorders of the musculoskeletal and nervous systems.

10.2.1 Measurement of ground reaction force

Special force-sensing devices, called 'force platforms', are used to measure ground reaction forces. Force platforms are much like sophisticated scales that can be used to measure three components of the ground reaction force: vertical (up and down), horizontal (forward–backward or anterior–posterior relative to a human subject) and lateral (side-to-side or medial–lateral relative to a human subject). Force platforms are usually set into the ground so that their top surface is level with the floor surface. With this set-up, people or other animals can walk or run over them while using normal movements. Force platforms are also sometimes built into treadmills to allow measurements of the vertical component of the ground force reaction during many strides of steady-speed walking or running (Kram and Powell 1989). This set-up is ideal for studying prolonged steady-speed locomotion and for clinical examination of gait. However, it does not allow the measurement of the horizontal component of the ground reaction force, because the treadmill belt slides over the force platform, applying a horizontal frictional force.

Force platforms come in all shapes and sizes. The most common type is designed to study human biomechanics. It is as big as a meter and a half long and half a meter wide, and is used for studying physical activity in humans as well as other large mammals. Miniature force platforms that are more sensitive have been designed for the measurement of the ground reaction force during locomotion in animals as small as cockroaches, small lizards and frogs (Biewener and Full 1993).

The mechanism for force measurement in force platforms involves either piezoelectric or resistive force transducers. Piezoelectric materials change surface charge when loaded, due to deformation of the crystal lattice structure. Typically, very small deformations (of the order of 1 μm) produce measureable charge differences, so that piezoelectric materials make very good force transducers. A charge amplifier converts the surface charge into a proportional output voltage. Resistive force transducers (also called strain gauges) also measure force by transducing small strains in elastic materials. Resistive strain gauges are fashioned from metals or semiconductors. When a metal is stretched, its cross-sectional area is reduced, increasing the resistance. The structure of either a metal or a semiconductor is altered by strain, also increasing resistance. This strain-sensitive resistance is included in a Wheatstone bridge to produce a change in voltage due to strain. In many types of force platforms, beams are built into the platform that have strain gauges mounted on them. Typically the beams are very stiff, so displacements are small (around 20 μm), and the platform is nearly isometric. Force transduction is more fully described in Cobbold (1974), and force platform technology in Biewener and Full (1993).

To maximize the accuracy of ground reaction force measurements, force platforms are designed to have a high natural frequency relative to frequency of the force of interest (for human walking and running, greater than 200 Hz), a linear response over large range of forces, a uniform response regardless of position of force application to the platform, and an ability to resolve the point of force application to the platform.

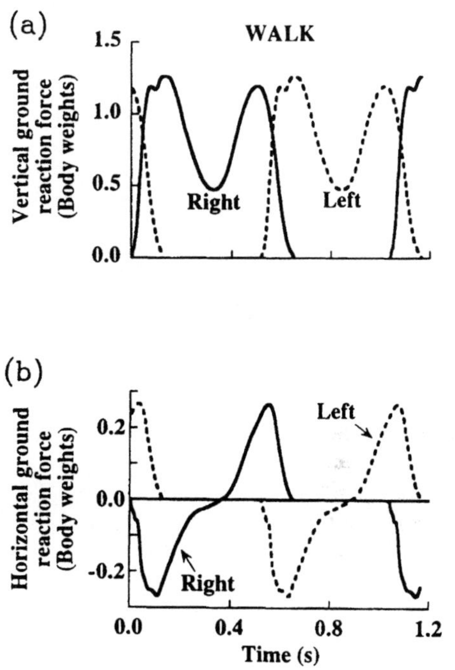

Fig. 10.1 Representative ground reaction force as a function of time for walking in a human. The solid line represents the stance phase of the right foot, and the dashed line represents the stance phase of the left foot. (a) Vertical component. (b) Horizontal component. This graph was created by repeating the ground reaction force for a single footfall at the appropriate interval.

10.2.2 Ground reaction forces in locomotion

Ground reaction forces in human walking

Figure 10.1 shows the vertical and horizontal ground reaction force for a human walking at a moderate speed (1.25 m s^{-1}). The lateral component is not shown, because its magnitude is generally less than 5 per cent of the magnitude of the vertical component in normal walking.

In Figure 10.1(a), the vertical component of the ground reaction force begins to rise as the right foot hits the ground. This event marks the beginning of the 'stance' phase for the right foot, the phase when the right foot is in contact with the ground. The vertical ground reaction force reaches a maximum equivalent to about 1.25 times body weight after about a quarter of the stance phase. The vertical ground reaction force then decreases to a minimum that is equivalent to about half of the body weight at the middle of the stance phase. Finally, the vertical ground reaction force increases to a second maximum that is equivalent to about 1.25 times the body weight before decreasing to zero as the foot leaves the ground. Before the right foot leaves the ground, the left foot hits the ground. The stage when both feet are on the ground is called the 'double support phase', and it constitutes about a quarter of a stride in walking. Because the subject is walking on level ground and the motion is uniform, the average vertical ground reaction force for a stride is equivalent to one body weight.

Figure 10.1(b) shows the horizontal component of the ground reaction force for human walking. It is immediately obvious that the peak magnitude of the horizontal ground reaction force is much smaller than the peak magnitude of the vertical ground reaction force. In this example for moderate-speed walking, Fig. 10.1(b), the peak magnitude of the horizontal ground reaction force is 20 per cent of the peak magnitude of the vertical ground reaction force. The horizontal ground reaction force is negative in the first half of the stance phase and is positive in the second half of the stance phase. The negative horizontal ground reaction force in the first half of the stance phase indicates that the ground is applying a backward force to the subject's foot. The positive horizontal ground reaction force in the second half of the stance phase indicates that the ground is applying a forward force to the subject's foot. Because the subject is walking at a steady speed over level ground, the average horizontal ground reaction force over a stride is zero.

Ground reaction forces in human running

Figure 10.2 shows the ground reaction force as function of time for two footfalls of running at a moderate speed. Notice that in running, unlike walking, there is no time in a stride when both the left and right feet are in contact with the ground. In fact, in running, there is a time when the ground reaction force is zero while neither foot is in contact with the ground. This part of the stride is commonly called the 'aerial phase'.

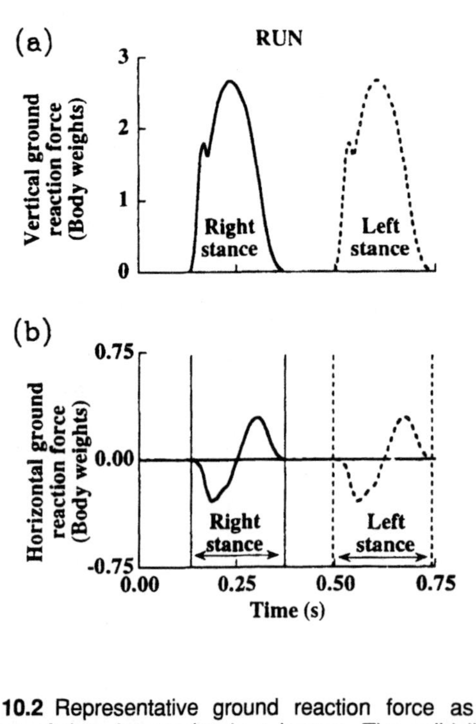

Fig. 10.2 Representative ground reaction force as a function of time for running in a human. The solid line represents the stance phase of the right foot, and the dashed line represents the stance phase of the left foot. (a) Vertical component. (b) Horizontal component. This graph was created by repeating the ground reaction force record for a single footfall at the appropriate interval.

The pattern of ground reaction force during a stance phase in running has some fundamental similarities to the pattern in walking. First, in both gaits, the magnitude of the vertical component of the ground reaction force is much greater than the magnitude of the horizontal component (compare Figs 10.1 and 10.2). The ratio of the peak magnitude of the vertical component to the peak magnitude of the horizontal component is about five in walking and about nine in running. The magnitude of the lateral component of the ground reaction force is less than 5 per cent of the magnitude of the vertical component in both gaits, and as a result, the lateral component of the ground reaction force is not shown for either gait. The general pattern of horizontal ground reaction force as a function of time is similar in both gaits (compare Figs 10.1b and 10.2b). In both walking and running, there is a backward force on the foot in the first half of the stance phase, and a forward force on the foot in the second half of the stance phase.

There are also important differences between the pattern of ground reaction force between walking and running. First, the peak magnitude of the vertical ground reaction force is generally 2–3 times greater for running than for walking (Figs 10.1a and 10.2a). Second, in running, there is an impact peak in the first 50 ms of the stance phase (Fig. 10.2a). This impact peak is thought to be associated with the deceleration of the lower leg shortly after the foot hits the ground, and it is only present when the heel is the first part of the foot to touch the ground. Finally, the time course of the stance phase is 2–3 times longer for walking than for running (Figs 10.1 and 10.2).

10.2.3 Mechanical energy of the center of mass

Components of mechanical energy

As defined in Chapter 1, external forces are those exerted on a system by agents not belonging to the system, while internal forces are those exerted by the particles of a system on one another. Energy can likewise be divided into external and internal parts.

It was also shown in Chapter 1 that the vector sum of the internal forces, over all particles in a system, is zero, and therefore the center of mass of the system acts as if the sum of external forces acts on it as a point mass. The mechanical energy of the center of mass can therefore be called the 'external mechanical energy'. Force platform data can be used to calculate the mechanical energy of the center of mass during movement.

The center of mass of the body moves as if the entire mass of the body and all of the external forces were concentrated at that point. The location of the center of mass of the body changes when the limbs or other body segments are moved relative to the rest of the body. In a human who is standing erect with arms hanging normally, the center of mass is located at about the level of the navel in the middle of the trunk. It is useful to know how the mechanical energy of the center of mass fluctuates during a physical activity because this gives insight into how much mechanical work must be produced by the musculoskeletal system to lift and accelerate the center of mass. The mechanical work required to increase the kinetic energy or the gravitational potential energy of the center of mass is called 'external mechanical work'.

Internal mechanical energy is that corresponding to internal forces, and cannot be measured from force platform data. The energy of body segments relative to the center of mass is an important component of internal energy in the study of human and animal movement. This energy can be computed from the kinematics, masses, and moments of inertia of the segments (Winter 1990; this chapter, Section 10.3).

Method of calculation of the mechanical energy of the center of mass of the body from force platform measurements

To calculate the external mechanical energy from ground reaction force data, one must first determine the kinematics of the center of mass: acceleration, velocity, and displacement. The velocity of the center of mass is then used to calculate kinetic energy. The displacement of the center of mass in the vertical direction is used to calculate its gravitational potential energy. Because the magnitudes of the lateral component of the ground reaction force, acceleration, velocity, and displacement are small compared to the vertical and horizontal components, this component is often ignored by locomotion biomechanists and will not be considered in the subsequent calculations in this section.

The first step in these calculations is to determine the acceleration of the center of mass by dividing the ground reaction force by the mass of the animal. Gravitational acceleration must be added to calculate the net vertical component of acceleration of the center of mass. In human walking, because more than one foot is in contact with the ground at a time (Fig. 10.1), the acceleration of the center of mass reflects the sum of both ground reaction forces at a given time. The second step in this procedure is to calculate the velocity of the center of mass by integrating the acceleration of the center of mass with respect to time. Finally, the displacement of the center of mass in the vertical direction is determined by integrating the velocity of the center of mass with respect to time. For each integration, boundary conditions are used to determine integration constants.

The kinetic energy of the center of mass, E_{kin}, is calculated from the mass of the animal, M, and the vertical and horizontal components of the velocity of the center of mass, v_z and v_y:

$$E_{\text{kin}}(t) = 0.5Mv_z^2 + 0.5Mv_y^2. \tag{10.1}$$

The gravitational potential energy of the center of mass, E_{grav}, is calculated from the mass of the animal, the gravitational acceleration, g, and the vertical component of the displacement of the center of mass, d_z:

$$E_{\text{grav}}(t) = Mgd_z(t). \tag{10.2}$$

The total mechanical energy of the center of mass, E_{tot}, is calculated by taking the sum of E_{grav} and E_{kin} at each time. The mechanical work associated with increasing the external mechanical energy is calculated from all the positive increments in the total mechanical energy of the center of mass during a stride.

It is important to notice that these calculations only consider the effects of the ground reaction force on the motions of the center of mass. They do not consider other forces such as air resistance. However, this is a reasonable approach for analysis of walking and moderate-speed running because the mechanical work required to overcome air resistance is much smaller (< 5 per cent) than the total mechanical work associated with locomotion (Shanebrook and Jaszczak 1976). At the highest sprinting speeds, the mechanical work required to overcome air resistance becomes more substantial.

Velocity and displacement of the center of mass during human walking

Figure 10.3 shows the vertical velocity, vertical displacement and horizontal velocity of the center of mass during walking. The vertical velocity of the center of mass cyclically oscillates around an average velocity of zero (Fig. 10.3a). The vertical velocity is equal to zero near

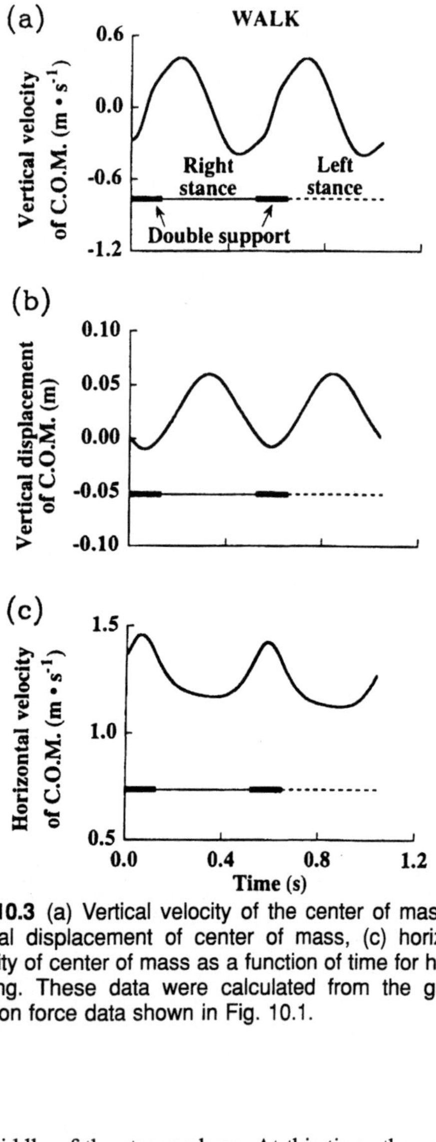

Fig. 10.3 (a) Vertical velocity of the center of mass, (b) vertical displacement of center of mass, (c) horizontal velocity of center of mass as a function of time for human walking. These data were calculated from the ground reaction force data shown in Fig. 10.1.

the middle of the stance phase. At this time, the vertical displacement of the center of mass reaches its maximum value (Fig. 10.3b). The integration constants are defined so that the vertical displacement of the center of mass is zero at the beginning of the stance phase of the right foot. This convention is convenient because the calculation of the mechanical work associated with increasing the vertical displacement of the center of mass requires only knowledge of the changes in the vertical displacement over a stride, and not knowledge of the absolute magnitude of the vertical displacement.

The horizontal velocity of the center of mass oscillates around its average value of 1.25 m s^{-1} (Fig. 10.3c), the walking speed of the subject. Each time a foot hits the ground, the ground applies a backward force to the foot and the horizontal velocity of the subject's center of mass decreases. Then, during the second half of the stance phase, the ground applies a forward force to the foot and the horizontal velocity increases to its value at the beginning of the stance phase.

Velocity and displacement of the center of mass during human running

The vertical velocity, vertical displacement, and horizontal velocity of the center of mass during running are shown in Fig. 10.4. During the aerial phase, the center of mass moves like a projectile, as it is primarily under the influence of gravitational acceleration. Thus, during the first half of the aerial phase, the vertical velocity of the center of mass decreases toward zero, and the vertical displacement increases toward its maximum. At approximately the middle of the aerial phase, the vertical velocity of the center of mass is zero, and the vertical displacement is maximal. During the second half of the aerial phase, the center of mass falls toward the ground (the vertical velocity is negative). It should be noted that, in human running, the center of mass is at a lower vertical displacement at the beginning of the stance phase than at the end of the stance phase (Fig. 10.4b). This asymmetrical pattern occurs because the knee is more bent as the foot hits the ground at the beginning of the stance phase than it is as the foot leaves the ground at the end of the stance phase.

In the stance phase, the vertical velocity is negative during the first half, as the center of mass moves downward (Fig. 10.4a and b). At approximately the middle of the stance phase, the vertical velocity is zero as the vertical displacement reaches its minimum. The vertical velocity then increases for the second half of the stance phase as the center of mass moves upward. The pattern of change in the vertical displacement is quite different from the pattern for walking. In walking, the vertical displacement is maximized at the middle of the stance phase, compared to running in which the vertical displacement of center of mass reaches its lowest point at the middle of the stance phase. This difference in the pattern of vertical displacement reflects a difference in potential energy storage between walking, in which the stance leg carries the body like an inverted pendulum,

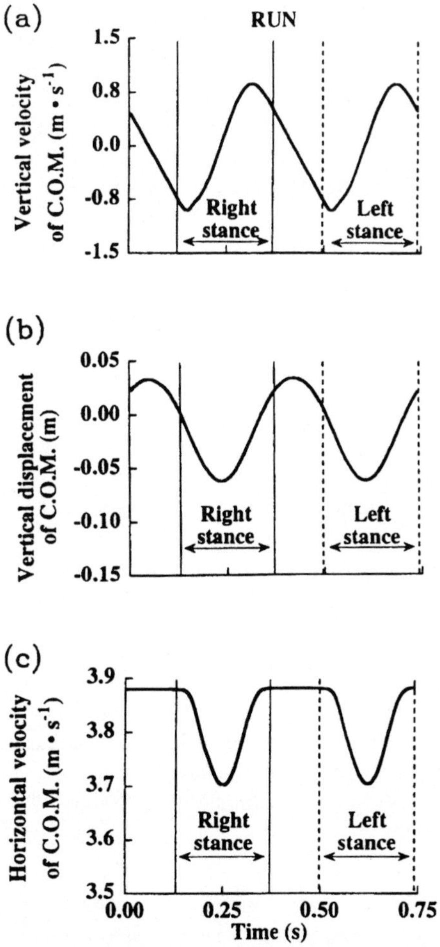

Fig. 10.4 (a) Vertical velocity of the center of mass, (b) vertical displacement of center of mass, (c) horizontal velocity of center of mass as a function of time for human running. These data were calculated from the ground reaction force data shown in Fig. 10.2.

and running, in which the stance leg stores energy elastically.

The horizontal velocity of the center of mass first decreases and then increases during the stance phase (Fig. 10.4c). The horizontal velocity fluctuates about the average forward velocity of the runner, 3.8 m s^{-1}. It is interesting to notice that the pattern of fluctuation of the horizontal velocity during the stance phase is quite similar for running and walking (Figs 10.4 and 10.3).

Kinetic energy and gravitational potential energy of the center of mass during human walking

The kinetic energy, gravitational potential energy and total mechanical energy of the center of mass during a walking stride are shown in Fig. 10.5. Figure 10.5(a) shows that in walking, the kinetic energy decreases during the first half of the stance phase, reaches its minimum at the middle of the stance phase, and increases during the second half of the stance phase. The gravitational potential energy is completely out of phase

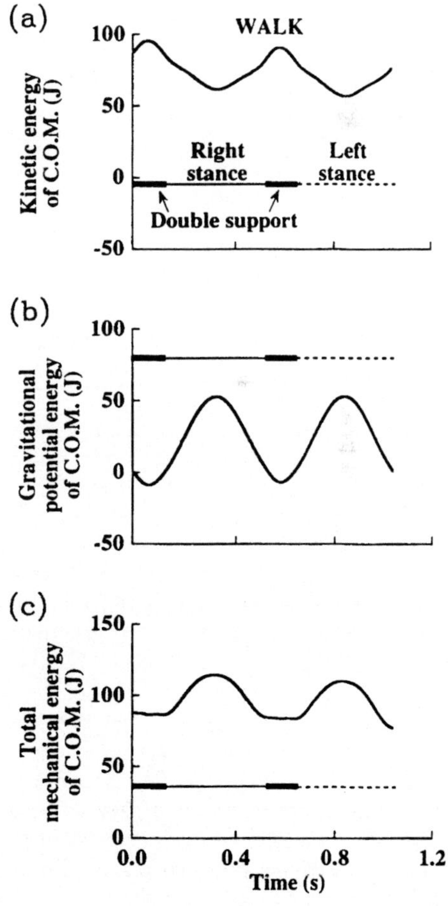

Fig. 10.5 (a) Kinetic energy of center of mass, (b) gravitational potential energy of center of mass, (c) total mechanical energy of center of mass as a function of time for human walking. These data were calculated from the data presented in Fig. 10.3.

with the kinetic energy. It increases during the first half of the stance phase, reaching its maximum at the middle of the stance phase, then decreases during the second half (Fig. 10.5b).

From the positive increments in the mechanical energy of the center of mass, it is possible to calculate the mechanical work associated with increasing its kinetic energy and gravitational potential energy. Over the stride shown in Fig. 10.5, the sum of the positive increments in kinetic energy is about 60 J, and the sum of the positive increments in gravitational potential energy is about 120 J. Thus, you might think that the subject's muscles have to perform 180 J of mechanical work to lift and accelerate the center of mass over the course of a stride. However, because of the 'inverted pendulum mechanism', much less mechanical work is required than this calculation suggests.

The inverted pendulum mechanism allows exchange between the kinetic energy and the gravitational potential energy of the center of mass. A comparison of Fig. 10.5(a) and (b) shows that the kinetic energy and gravitation potential energy are almost exactly out of phase. When the kinetic energy is minimized at the middle of the stance phase, the gravitational potential energy is maximized. As a result, when the total mechanical energy of the center of mass is calculated by summing the kinetic energy and the gravitational potential energy at each instant in the stride, the positive increments in the total mechanical energy are much smaller than the sum of the positive increments of the kinetic energy and gravitational potential energy (Fig. 10.5c). More specifically, the sum of the positive increments in total mechanical energy (60 J) is one-third of the sum of the positive increments of the kinetic and gravitational potential energy (180 J). The positive increments in the total mechanical energy of the center of mass are small because there is exchange between the kinetic energy and gravitational potential energy of the center of mass. During the first half of the stance phase, the walker loses kinetic energy and gains gravitational kinetic energy. During this phase, kinetic energy is converted to gravitational potential energy. During the second half of the stance phase, the center of mass loses gravitational potential energy, and it is converted to kinetic energy. This mechanism of energy exchange is often called the 'inverted pendulum mechanism'. In the example shown in Fig. 10.5, there is a saving of about 67 per cent of the mechanical work that the muscles would have had to perform without the exchange of energy by this mechanism.

In the example shown in Fig. 10.5, the subject's speed is about 1.25 m s^{-1}. At this speed, a human walker can save more energy by the inverted pendulum mechanism than at higher or lower speeds (Cavagna *et al.* 1976). The primary reason for the change in energy exchange with walking speed is that the kinetic energy fluctuations of the center of mass increase as walking speed increases. At speeds above 1.5 m s^{-1}, the exchange of energy by the inverted pendulum mechanism is reduced because the magnitude of the fluctuations in kinetic energy is substantially greater than the magnitude of the fluctuations in gravitational potential energy. At speeds below 1 m s^{-1}, the exchange is reduced because the magnitude of the fluctuations in kinetic energy is substantially less than the magnitude of the fluctuations in gravitational potential energy. It is interesting to note that people normally prefer to walk at speeds between 1 and 1.5 m s^{-1}. It is not surprising that the metabolic energy consumed per unit distance traveled is lower in this speed range than at other walking speeds (Rose *et al.* 1994).

Many types of legged animals use the inverted pendulum mechanism of energy exchange for walking. These animals include humans, horses, dogs, lizards, and ghost crabs (Cavagna *et al.* 1976, 1977; Blickhan and Full 1987; Ko and Farley 1992). However, none of the other animals saves as much energy as humans by the inverted pendulum mechanism. Figure 10.6 shows a simple inverted pendulum model of a single stance phase of human walking.

Kinetic energy and gravitational potential energy of the center of mass during human running

Figure 10.7 shows the kinetic energy, gravitational potential energy, and total mechanical energy of the center of mass during a stride of running. In the aerial phase of running, the center of mass moves much like a projectile, with the total mechanical energy of the center of mass remaining constant (Fig. 10.7c). After the foot leaves the ground and the aerial phase begins, the kinetic energy of the center of mass begins to decrease, and it reaches zero near the middle of the aerial phase (Fig. 10.7a). This decrease in kinetic energy is entirely due to the decrease in the vertical component of the velocity of the center of mass. At the time that the kinetic energy reaches its minimum, the gravitation potential energy of the center of mass attains its maximum (Fig. 10.7b). Then, as the center of mass begins to fall, its gravitational potential energy decreases and its kinetic energy increases (Fig. 10.7a and b).

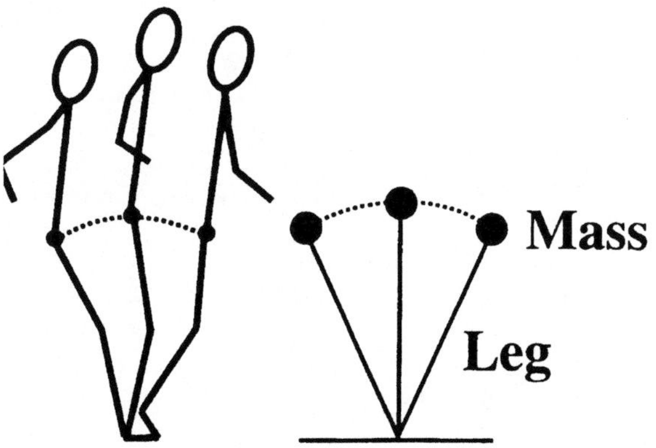

Fig. 10.6 An inverted pendulum model and a stick figure representation of a single stance phase of human walking. The model consists of a mass and a stiff strut that connects the foot and the center of the human. This figure depicts for a human moving from left to right the beginning of the stance phase (leftmost position), the middle of the stance phase (center position), and the end of the stance phase (rightmost position).

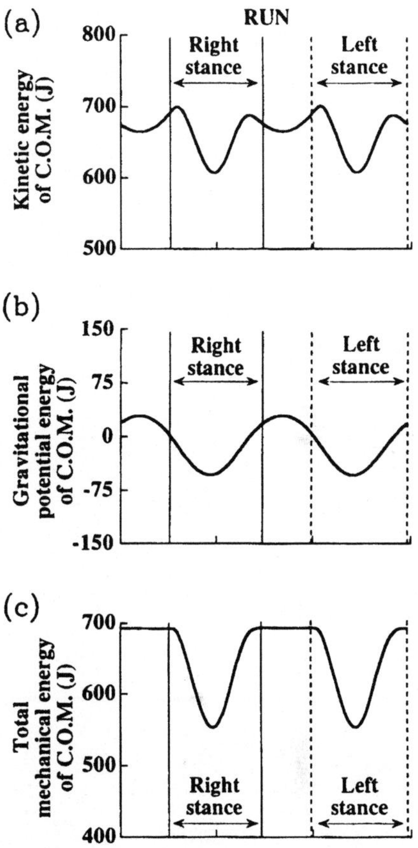

Fig. 10.7 (a) Kinetic energy of center of mass, (b) gravitational potential energy of center of mass, (c) total mechanical energy of center of mass as a function of time for human running. These data were calculated from the data presented in Fig. 10.4.

In the stance phase, the kinetic energy of the center of mass decreases during the first half, reaches a minimum at approximately the middle, and increases during the second half (Fig. 10.7a). This pattern is similar to that for a stance phase of walking. The major difference is that the magnitude of the kinetic energy fluctuations is greater in running, due to the higher forward speed.

By contrast, the pattern of change in the gravitational potential energy of the center of mass during the stance phase is completely different for running than it is for walking. In the stance phase of running, the gravitational potential energy decreases for approximately the first half, reaches its minimum at about the middle, and increases in the second half (Fig. 10.7b). Thus, for running, the gravitational potential energy of the center of mass is at its minimum at the middle of the stance phase, and for walking, it is at its maximum at the middle of the stance phase.

An important difference between running and walking is that the kinetic energy and gravitational potential energy of the center of mass are almost exactly in phase during the stance phase of running (Fig. 10.7a and b). Thus, in running, the kinetic energy and the gravitational potential energy both reach their minimum values at the same time. This pattern is completely different from the pattern during walking. In walking, the kinetic energy and gravitational potential energy are almost completely out of phase during the stance phase (Fig. 10.5a and b). Because the kinetic energy and gravitational potential energy of the center of mass are in phase in running, the inverted pendulum mechanism for energy exchange does not function.

A simple spring-based model has been developed for running. It consists of single linear spring called the 'leg spring' and a point mass equivalent to the mass of the runner (Fig. 10.8) (Alexander and Vernon 1975; Blickhan 1989; McMahon and Cheng 1990). In this model, as the leg spring hits the ground after the aerial phase, the spring compresses and the mass moves

Fig. 10.8 A spring–mass model and a stick figure representation of a single stance phase of human running. The model consists of a mass and a single leg spring that connects the foot and the center of mass of the animal. This figure depicts the model at the beginning of the stance phase (leftmost position), at the middle of the stance phase (leg spring is oriented vertically) and at the end of the stance phase (rightmost position).

downward. The mass reaches its lowest point at the middle of the stance phase just as occurs in human running (Fig. 10.4b). During the first half of the stance phase, the leg spring compresses and stores elastic energy. During the second half of the stance phase, the elastic energy is released by the leg spring as it recoils. This model has been shown to describe and predict the mechanics of running gaits remarkably well (Blickhan and Full 1993b; Farley *et al.* 1991, 1993; He *et al.* 1991). In addition, it applies equally well to the running gaits of a variety of mammals including humans, dogs, horses, kangaroos, cockroaches, and ghost crabs (Blickhan and Full 1993a; Farley *et al.* 1991, 1993; He *et al.* 1991).

There is a physiological justification for the use of a spring-based model for running. Muscles, tendons and ligaments all have spring-like properties and are capable of storing and returning elastic energy (Alexander 1988). In running, mechanical energy is conserved by the storage and return of elastic energy by muscles, tendons, and ligaments. These tissues store elastic energy when they are stretched and return much of the elastic energy when they recoil. In human running at high speed, the Achilles tendon alone stores and returns 35 per cent of the mechanical energy required for a stride (Alexander 1988).

10.3 Joint moments and joint power

10.3.1 Joint moments

Often biomechanists want to understand the musculoskeletal system at the next level of organization, the joints. Data on joint kinematics and moments are used to characterize human gait and to design leg prostheses. In ergonomics, joint kinematics and moments are used to determine the best way to perform a task, e.g. the safest posture for squat lifting (Holmes *et al.* 1992). Finally, joint kinematics and moments are often easily available. Joint kinematics can be measured noninvasively, and joint moments can be derived from kinematics and measurements of external forces.

A schematic of a human ankle joint for a person hopping on his toes is shown in Fig. 10.9. The most obvious forces that exert a moment about the ankle joint are the ground reaction force and the force from the ankle extensor muscles. Figure 10.9(a) indicates the resultant ground reaction force, $\boldsymbol{F}_g$, and its moment arm about the ankle joint, $\boldsymbol{R}$. The moment arm is the perpendicular distance between the joint center and the line of action of the force. Figure 10.9(a) also shows the force exerted by an ankle extensor muscle, $\mathbf{F}_{m1}$, and its moment arm, $\mathbf{r}_{m1}$. The static moment balance equation for this situation is:

$$\boldsymbol{r}_{m1} \times \boldsymbol{F}_{m1} = \boldsymbol{R} \times \boldsymbol{F}_g. \qquad (10.3)$$

The muscle force $\boldsymbol{F}_{m1}$ at each instant can be calculated when $\boldsymbol{r}_{m1}$, $\boldsymbol{R}$ and $\boldsymbol{F}_g$ are known.

In addition to the ground reaction force and the muscle force, the gravitational force acts on the center of mass of the foot (Fig. 10.9b). This gravitational force, $\boldsymbol{F}_w$, exerts a moment about the ankle equal to $\boldsymbol{r}_w \times \boldsymbol{F}_w$, where $\boldsymbol{r}_w$ is

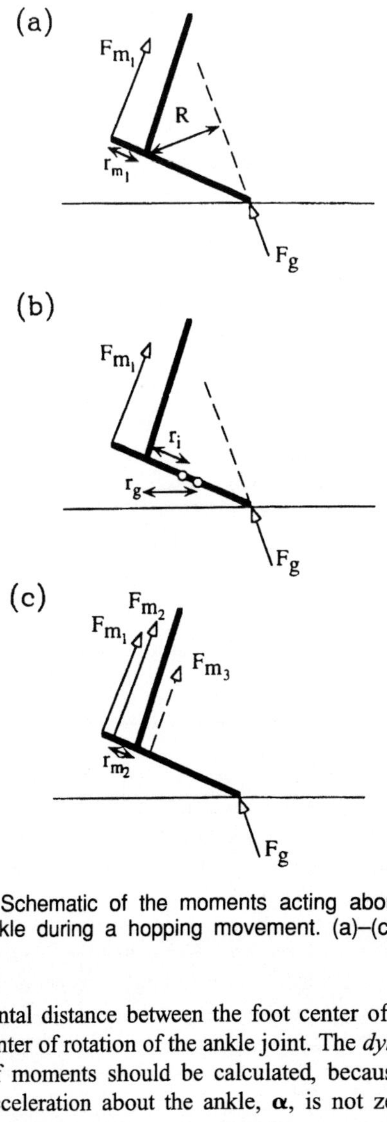

Fig. 10.9 Schematic of the moments acting about the human ankle during a hopping movement. (a)–(c) See text.

the horizontal distance between the foot center of mass and the center of rotation of the ankle joint. The *dynamic* balance of moments should be calculated, because the angular acceleration about the ankle, $\boldsymbol{\alpha}$, is not zero in hopping:

$$\boldsymbol{r}_{\mathrm{m1}} \times \boldsymbol{F}_{\mathrm{m1}} - \boldsymbol{R} \times \boldsymbol{F}_{\mathrm{g}} - \boldsymbol{r}_{\mathrm{w}} \times \boldsymbol{F}_{\mathrm{w}} = I\boldsymbol{\alpha}. \qquad (10.4)$$

Practically, the gravitational force exerts a rather small moment about the ankle in humans, and the product of moment of inertia I and angular acceleration (sometimes called the 'inertia moment') is small compared to the other terms. To be complete, joint frictional forces would also need to be considered, but these forces are generally negligible. These terms are therefore often neglected. However, at the human knee and hip these terms can be substantial (Wells 1981), and in other joints and animals, the gravitational force needs to be considered. Of course, when there is zero reaction force acting on the object (e.g. during the swing phase of gait) the gravitational and muscle moments are the only ones that need to be considered.

So far this example has included only one extensor muscle. In fact, most joints have multiple extensor muscles which exert force. A second extensor muscle is shown in Figure 10.9(c). If we ignore the gravitational joint moments, the static moment balance equation in this situation is

$$\boldsymbol{r}_{\mathrm{m1}} \times \boldsymbol{F}_{\mathrm{m1}} + \boldsymbol{r}_{\mathrm{m2}} \times \boldsymbol{F}_{\mathrm{m2}} = \boldsymbol{R} \times \boldsymbol{F}_{\mathrm{g}}. \qquad (10.5)$$

This equation is indeterminate without invasive measurements. Various schemes have been proposed for proportioning the force between muscles which act in concert (agonists), like m_1 and m_2, or in opposition (antagonists), like m_3. Some of these methods are described in Sections 10.4.2 and 10.4.3.

In this example, it was quite reasonable to utilize a planar analysis for estimating joint moments. Often a three-dimensional analysis is required. Three-dimensional analysis requires resolving the resultant force into its three components, and locating the joint and other anatomical landmarks in three dimensions.

The indirect method outlined above allows only estimates of *net* joint moment. Nevertheless, net joint moments are an important step towards understanding muscle function.

10.3.2 Joint power

Net joint power can be calculated from the product of net joint moment and joint angular velocity (see Chapter 1, p. 32):

$$\text{net joint power} = \boldsymbol{M} \times \boldsymbol{\omega} \qquad (10.6)$$

Figure 10.10 is a diagram of a human elbow when a person is holding a weight. This illustrates the concepts of net power production and absorption. If the elbow flexor muscles are activated enough to shorten, both the moment and angular velocity are positive and thus the net joint power is positive, i.e. the muscles produce net mechanical power. If the elbow flexors are activated slightly less, the weight can be lowered in a controlled manner. The net joint moment during the lowering would still be in the same (anticlockwise) direction but the angular velocity of the forearm relative to the upper arm would be negative as the elbow extends. Thus, the net

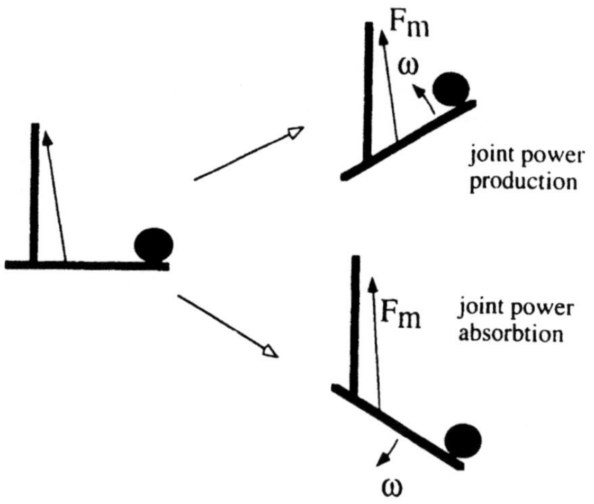

Fig. 10.10 Illustration of the concept of joint power production and absorption about the elbow.

joint power would be negative, indicating that the muscles absorb mechanical power. The sign of net joint power may be obvious in this simple example, but is often less intuitively available in the analysis of more complicated movements, such as locomotion.

10.3.3 Measurement of kinematics

Motion analysis

Position, velocity, and acceleration data are needed for the calculation of net joint moments and powers. Position data can be obtained from video images, cinematic film images, or various optoelectronic methods. In the past, positions of visible markers placed on a subject were measured manually on each frame of a moving picture frame. Videorecording has replaced motion picture photography in recent years. Automated video analysis has largely replaced manual marker discrimination. Optoelectronic methods, such as Vicon and Selspot, which use special markers and sensors, have reduced the complexity of motion analysis by sensing only the marker, and not the entire image.

All motion analysis methods require calibration of the field of interest with an object of known length, and transformation from camera coordinates to laboratory coordinates. For two-dimensional planar analysis, a single camera is adequate. Depending on the principle used in each optoelectronic system, a different number of 'cameras' may be needed. It is possible and sometimes important to characterize motion in three dimensions. This requires at least two cameras and a three-dimensional calibration object. By a technique called 'direct linear transformation', two or more two-dimensional views can be combined to determine three-dimensional position.

Markers

For determination of joint kinematics, one seeks the position of joints and other bony landmarks. Investigators have used pins screwed into the bone which protrude from the skin (e.g. Lafortune *et al.* 1992), but less invasive markers are more common. Common methods of marking include small dots of ink or reflective tape, special three-dimensional reflective markers (Vicon), and markers that emit light (such as light-emitting diodes) or sound. Usually, markers are placed on the skin over the joint or other landmark of interest. During movements, the skin does not move in exactly the same way as the bones, and inaccuracy is unavoidable, especially in subjects with a large amount of subcutaneous fat.

Data conditioning

Even if the markers were perfectly placed, uncertainty in marker position, quantization noise, and human error would still produce small errors of position. Such errors are not systematic mistakes but errors which tend to be randomly distributed. Small errors in position can result in very large errors in velocity because velocity is the first derivative of position. Thus, position data are usually smoothed digitally using low-pass filters or spline curve fitting.

Sampling rate

As with all measurements of time-varying signals, it is important to sample the signal at high-enough frequency to capture the true movement. Too infrequent sampling can miss fast movements or allow aliasing of high-frequency events into lower frequencies (Hamming 1983). An example of aliasing due to sampling is the apparent backward motion of stagecoach wheels in Western movies, during rapid forward motion of the stagecoach. The motion of a wheel is sampled at each frame of the movie, a slow sampling rate compared to the rate of rotation. Therefore a spoke in one frame is captured in the next frame after nearly a full cycle of rotation. Thus the rapid forward rotation of the wheel is aliased to a much slower, backward apparent rotation. According to the Nyquist theorem, aliasing can be avoided by sampling at least twice the highest frequency in the sampled signal. Sampling should also be at high enough a rate to allow for smoothing of random digitizer error without distortion.

Optoelectronic systems have a wide variety of sampling rates. Standard video in the USA and Japan is recorded at 60 Hz, and 50 Hz in Europe and many other parts of the world. High-speed video is available at up to 6000 images per second, a rate that may be required for extremely rapid movements, such as insect wing beats.

Moment arms

The moment arm of the ground reaction or other force is determined from a combination of the kinematic analysis described above and the force platform technique described in Section 10.2. Muscle moment arms are often estimated with a ruler or calipers and external palpation of the tendon and joint. Medical imaging techniques, such as x-radiography (see, for example, Smidt 1973), computed tomography (e.g. Nemeth and Ohlsen 1985) and magnetic resonance imaging (e.g. Rugg *et al.* 1990) have been used to refine this measurement. These techniques can also aid in locating the joint center of rotation.

10.3.4 Examples of net joint moments and power analyses

Joint moments during locomotion

Figures 10.11 and 10.12 present joint moments at the ankle, knee, and hip as functions of time, for normal gait of a healthy young man. During walking (Fig. 10.11), the greatest joint moment occurs at the ankle, about 150 N m near the end of the stance phase. The positive sign of this moment is by convention called an extensor moment (also called plantarflexion). The ankle extensor muscles (located in the back of the calf) produce most of the joint moment. During the swing phase, when the foot is off the ground, the gravitational moment is nearly zero, so no net joint torque is needed.

In walking, a peak moment about 75 N m is exerted at the knee, early in the support phase. A smaller extensor moment peak also occurs toward the end of the stance

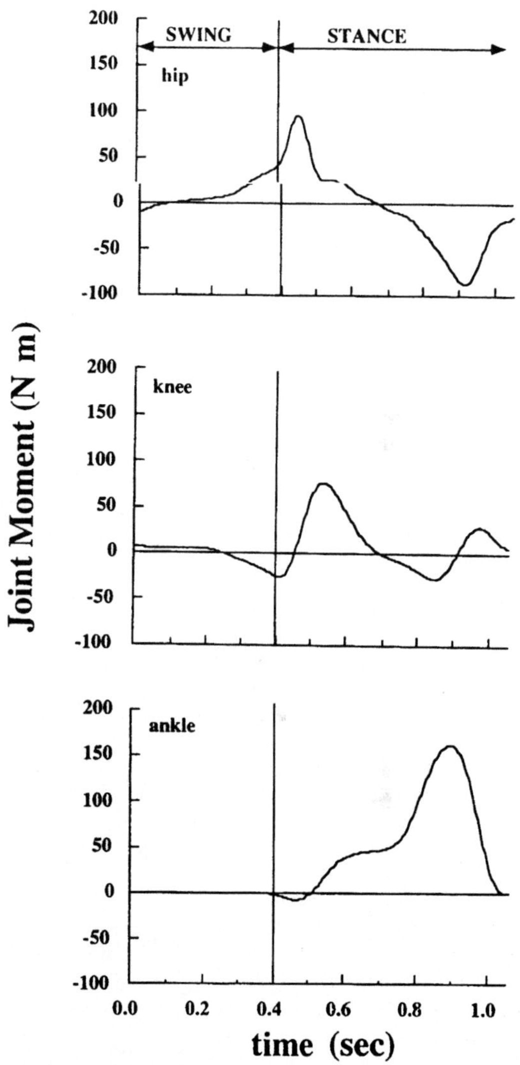

Fig. 10.11 Net joint moments for the hip, knee, and ankle joints during walking by an average-sized male subject. Positive joint moments act to extend the joint. (Data were kindly provided by Paul DeVita of Southern Illinois University. Details of how all joint moment and joint power data were collected are available in DeVita (1994).)

phase. The positive sign here indicates a moment in the direction of knee extension (i.e. straightening the knee). The quadriceps muscles, in the front of the thigh, dominate this moment. At the hip, during the first part of the stance phase the net moment is in the extension

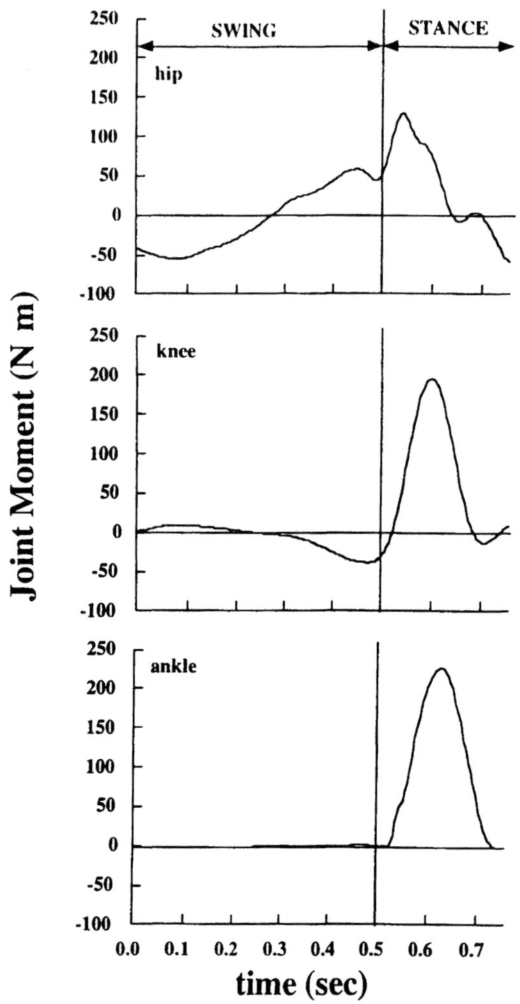

Fig. 10.12 Net joint moments for the hip, knee, and ankle during running. Positive joint moments act to extend the joint. Moments are substantially higher for running than for walking.

direction, caused by the hamstring muscles in the back of the thigh and the gluteal muscles. During the latter half of the stance phase the sign reverses, so thigh flexor muscles (such as rectus femoris) must be active.

During running, the ankle, knee, and hip all show net joint extensor moments during the stance phase (Fig. 10.12). The magnitude of the moment at the three joints is similar. As one would expect, the net joint moments are substantially higher than in walking. During the swing phase, there is little moment required at the ankle or knee, but the hip musculature must exert a sizable flexor moment during the swing phase to reverse the direction of the leg.

Joint power during locomotion

Although the ankle and knee show net extensor moments for the contact phase, it is essential to combine the kinematics of the movements with the joint moment data to calculate the joint power. The moment data from Fig. 10.12 have been combined with kinematic data, and the joint power values for a person running are shown in Fig. 10.13. At the ankle, during the early stage of stance phase, the ankle is being dorsiflexed, but the net joint moment is extensor or plantar flexion. Thus, the calf musculature is actively resisting lengthening and is thus absorbing power. During the second half of the stance phase, the net joint moment is positive (extension direction) and the ankle joint is being extended. Thus, net joint power is being produced at the ankle. A nearly identical pattern is seen at the knee: the knee flexes, then extends, and net joint power is first absorbed, then produced. Although the peak net moment about the knee is similar to that about the ankle, a great deal more power is produced and absorbed at the ankle. At the hip there is much less power absorption and production, but the pattern is similar. This pattern of power absorption and production during the stance phase of running is consistent with the earlier discussion (Section 10.2) of how a running animal behaves as a spring–mass system.

10.3.5 Open issues

One of the most active areas of research in musculoskeletal biomechanics relates to problems of multiple muscle systems (Winters and Woo 1990). The action of muscles that cross two joints is particularly vexing. The simple examples in this chapter have dealt with muscles that cross only one joint. However, many muscles produce moments about two joints. For example, the gastrocnemius muscle in the calf, which acts to extend (plantar flex) the ankle, is also a weak knee flexor. The rectus femoris muscle in the thigh acts as both a knee extensor and a hip flexor. The angles of both joints must be considered when analyzing the functions of such multi-joint muscles.

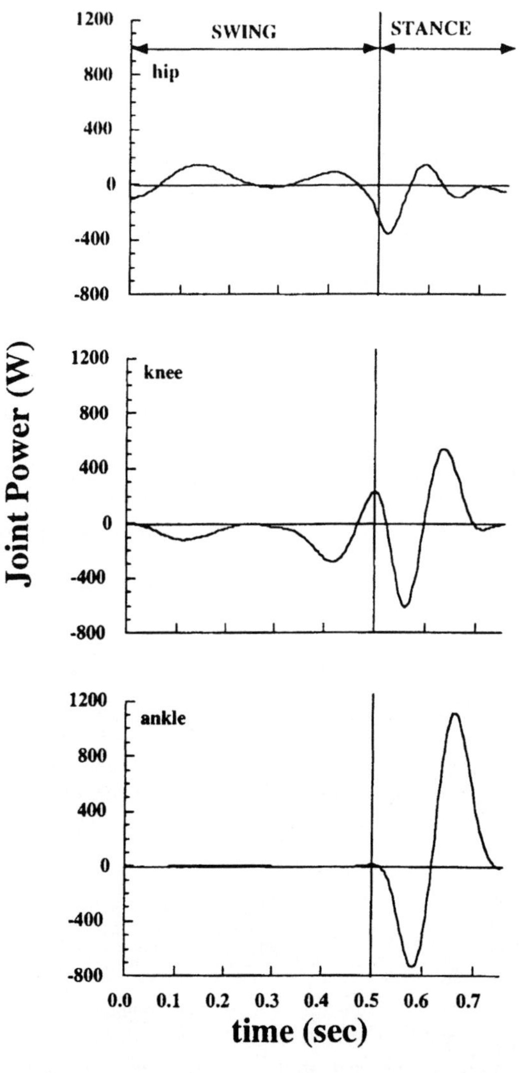

Fig. 10.13 Net joint power for the hip, knee, and ankle during running. Positive power indicates production by the muscles and negative power indicates absorption. Note the characteristic absorb–produce pattern in all the three joints during the stance phase. The ankle produces almost three times as much power as either the knee or hip.

It is generally believed that movement patterns emerge which are economical in terms of metabolic energy. Various schemes for minimizing co-contractions or minimizing net joint moments have also been proposed (Section 10.4.3). Some of these models do not take into account the physiological processes that determine the metabolic cost of muscular force production. Models which do incorporate muscle physiology must rely on *in vitro* muscle data.

Understanding the mechanical basis of metabolic energy consumption during locomotion and other activities remains a major challenge. We know that muscle consumes energy when it is active and isometric (at constant length), as well as when it actively shortens or lengthens. Many investigators have tried to relate the mechanical work rate to metabolic energy consumption, with limited success. Others have tried to relate the time integral of joint moments (total angular impulse) to the rate of metabolic energy. A third approach is to consider the time course of force development as well as the muscular impulse. Each of these methods explains some of the observations of metabolic energy consumption. However, no single method has accounted for all cases. There is a wide-open opportunity for someone to synthesize these different approaches, or improve on them. Two recent articles lay out these problems in more detail (van Ingen Schenau and Cavanagh 1990; Martin *et al.* 1994).

In all of these areas, technical problems remain to be solved. Medical imaging tools may play an important role as better methods of measuring the anatomical sites, such as centers of joint rotation. The field would be greatly advanced by techniques that measure forces from both resting and active muscle, and tendon dimensions and forces.

10.4 Muscle force

The following notice was posted for a lecture to the Institution of Electrical Engineers in London, February 11, 1969:

Available now. LINEAR MOTOR. Rugged and dependable: design optimized by worldwide field testing over an extended period. All models offer the economy of "fuel-cell" type energy conversion and will run on a wide range of commonly available fuels. Low stand-by power, but can be switched within msecs to as much as 1 kW mech/kg (peak, dry). Modular construction, and wide range of available subunits, permit tailor-made solutions to otherwise intractable mechanical problems.

Choice of two control systems:

(1) Externally triggered mode. Versatile, general-purpose units. Digitally controlled by picojoule pulses. Despite low input energy level, very high signal-to-noise ratio. Energy amplification 10^6 approx. Mechanical characteristics: (1 cm modules) max. speed: optional between 0.1 and 100 mm/sec. Stress generated: 2 to 5×10^{-5} N m^{-2}.

(2) Autonomous mode with integral oscillators. Especially suitable for pumping applications. Modules available with frequency and mechanical impedance appropriate for

(a) Solids and slurries (0.01–1.0 Hz).

(b) Liquids (0.5–5 Hz;): lifetime 2.6×10^9 operations (typ.) 3.6×10^9 (max.)—independent of frequency.

(c) Gases (50–1000 Hz).

Many optional extras e.g. built-in servo (length and velocity) where fine control is required. Direct piping of oxygen. Thermal generation. etc.

Good to eat.

The lecture is by Professor D.R. Wilkie. The subject is muscle.

The subject of this section is muscle. Previous sections show that many of the important unresolved questions regarding human locomotion particularly, and animal movement more generally, come down to questions regarding the mechanics and energetics of muscle. The physics of the body's center of mass can reveal how mechanical energy changes during locomotion, but cannot reveal where energy is stored. Joint moments reveal the loading of each joint when a person lifts a heavy object, but cannot show which structures are stressed, or which muscles might be strengthened to prevent injury or lessen fatigue. Joint kinematics and moments have revealed some important regularities in motor control, but the next level, individual control of each actuator, remains enigmatic.

This section deals with the ways in which muscle forces are measured, or inferred from less direct data. Direct measurement of muscle force is possible in some special cases, but is too invasive to be generally useful. The multiplicity of muscles and the intuitive appeal of some biological optimization have led many to attempt to infer muscle forces from joint moments and external forces, assuming the optimization of some criterion. This section sketches those attempts, and gives examples of their lack of success. A different approach is to deduce muscle force from data closer to the muscle level: kinematics and activations of muscles themselves. The core of this section is a celebration of the mechanical behavior of isolated muscle: the fact that force is determined by the length, velocity, and activation of each fiber. The celebration is damped somewhat by the fact that some of these determining factors are hard to estimate, and that force is also history-dependent in some circumstances. These complications in muscle mechanics underlie some of the most important open questions in biomechanics, which are the subject of the final portion of this section.

10.4.1 Direct measurement of muscle force

Sometimes, muscle force can be measured directly. Muscles usually attach to bones through tendons. If a tendon is in series with the whole muscle, then the tension in the tendon is necessarily equal to the tension in the muscle. Tendons are usually thinner and less changeable in shape than muscles, so it is sometimes possible to attach transducers to the tendon.

Strain gauge transducers are designed so that the tendon threads through them, like a belt through a buckle. Tension in the tendon produces a small elastic strain in the transducer, and the strain is measured by standard strain gauges (Section 10.2).

Tendon transducers have been implanted in animals and, occasionally, in humans. They have aided in diverse studies, including locomotion of cats (Gregor *et al.* 1991) and humans (Komi *et al.* 1992), ergonomics of the hand

(Schuind *et al.* 1992), and control of fast eye movements by monkeys (Miller and Robins 1992). The use of *in vivo* force measurement by implanted tendon transducers is fully reviewed by Komi (1990).

Despite their obvious value, such studies are rare. The implantation is invasive, requiring a surgery. It is difficult to guarantee that a movement is not changed by the introduction of the transducer. Often, many muscles are capable of producing a given moment about a joint, so many transducers would be necessary to infer all the forces. Finally, the architecture of many muscles does not allow implantation of a transducer: the muscle may attach to bone directly, or through a broad aponeurosis.

10.4.2 Muscle forces from kinematics and external forces

Biomechanists usually cannot measure muscle forces directly. If estimates of muscle force are needed, they must be inferred from other data. Often, the most easily obtainable data are joint kinematics and external forces.

These data are generally not sufficient to solve the problem of assigning a unique set of muscle forces to a given joint moment. The number of muscle forces (and other forces due to joint structures and ligaments) usually exceeds number of equilibrium equations that can be written summing forces and moments: the muscle force assignment problem is usually statically indeterminate (Section 1.1.9), there being many more unknowns than equations relating them.

For example, seven muscles cross the human elbow. Three of them can produce flexion moments. Even if it were known that the four others produced no force, and if forces from ligaments and bone–bone contacts were neglected, the number of variables (3) in the static moment balance equation still exceeds the number of equilibrium equations (1). The problem in assigning muscle forces to a given moment is even clearer at the human ankle, where either the soleus or the gastrocnemius muscles can exert plantarflexion moment. They even have the same moment arm!

It is difficult to find a physiological case in which a moment can only be produced by force from a single muscle. Determinate cases can sometimes be arranged by special positioning, or another measurement can be found to make the assignment of muscle forces determinate. For example, de Luca and Forrest (1973) found a shoulder position in which a moment could be produced by only one muscle, then showed that electromyograms (EMGs, see Section 10.4.4 under 'Estimates of fiber . . .activation') from all the possible moment producers obeyed a fixed relationship.

Sometimes it can be shown that all the muscles capable of producing a given moment act as a unit, activating and deactivating at the same time. Rapid flexion of the wrist uses three muscles, for example, but they are all turned on and off simultaneously (Litvenstev and Seropyan 1977), so they act as a single 'muscle equivalent' (Bouisset 1973). However, even muscles that are anatomically arranged so that their moments are about almost orthogonal axes (e.g. human extraocular muscles) have pairs of muscles acting as antagonists, so that a given moment is still generated by the combination of two (unmeasured) forces. In the more typical example of the elbow, any of the seven muscles may be active during elbow flexion.

10.4.3 Optimization approaches to the redundancy problem

The only general technique used to attack the indeterminacy problem in deducing forces from moments has been constrained optimization. The approach is to produce some criterion that is assumed to be optimized in the movement. For example, an investigator of rapid movements might assume that the movement time is minimized (for example, see Lehman and Stark 1979). The investigator then searches for the pattern of forces (each a function of time) that would optimize the criterion functional (minimize movement time), subject to the constraint that the sum of moments (forces times moment arms) adds up to the moment of inertia times the angular acceleration. Additional constraints on the forces available from muscles (positive, bounded) often aid in finding a global optimum.

For human walking, for example, energy has had appeal as a criterion to be minimized (Nubar and Contini 1961; Chow and Jacobson 1971). Even a linear criterion function (Seirig and Arvikar 1975) produced fair correspondences between predicted muscle forces and measured electromyograms, and a physiologically based nonlinear criterion function that included the fatigue characteristics of muscle (Crowninshield and Brand 1981) produced a better correspondence, but neither was good enough to have predictive value.

The results of using optimization schemes to resolve muscle forces have seldom been tested rigorously, and, when tested, have usually not been very good. In a study of optimization to predict muscle force in human locomotion, Pedotti *et al.* (1978) compared four criteria; the sum of forces from all eleven leg muscles, the sum of the squares of these forces, the sum of forces normalized to maximum force for each muscle, and the sum of squares of normalized forces. Although the sum of squares of normalized forces produced the best correspondence between predicted forces and timing and measured electromyograms, most of the criteria produced similar force pattern predictions. Pedotti *et al.* concluded that some muscles may have a unique function in some part of the gait cycle, so must produce force, regardless of cost. It could also be argued that walking has evolved to be close to optimal for several criteria.

Recently, Buchanan and coworkers have tested the ability of various optimization criteria to predict muscle activation (measured by EMG) for static contractions in different directions about the elbow (Buchanan *et al.* 1989) and the wrist (Buchanan *et al.* 1993). Shreve and Buchanan (1993) predicted muscle forces from moment arms and estimated maximum forces, using several optimization criteria to determine distribution of forces. They then measured force in several directions and the EMG of each muscle crossing the joint. They found no criterion that produced good predictions.

10.4.4 Muscle force from muscle kinematics and activation

Eighty years of study of mechanics of isolated muscle and muscle fibers (see, for example, Hill 1970) have established that the force produced by a muscle can be predicted from very few variables. For shortening muscle, it has long been known (Hill 1938) that the force is a function of just three variables: the activation, the length of the muscle fibers, and the velocity of contraction. There is therefore hope that muscle forces *in vivo* can be deduced from these variables.

Muscle is a heterogeneous, nonisotropic, structurally complex material. The mechanical properties of whole muscle are often attributable to its parts, and modern scientific reductionism has led to the characterization of the mechanics of progressively smaller subunits. A brief anatomy of skeletal muscle will be useful. A more complete anatomy of skeletal muscle, and beautiful illustrations, are available in many muscle physiology texts (e.g. Woledge *et al.* 1985).

Functional anatomy of skeletal muscle

Gross dissection of any skeletal muscle will reveal its orderly arrangement. Most muscles have small bundles of parallel fibers, a few millimeters in diameter and millimeters to centimeters long. These bundles, called fascicles, consist of hundreds of muscle fibers, surrounded by connective tissue.

The arrangement of fascicles and fibers within a muscle is called fiber architecture. Muscles that undergo large length changes or high velocities usually have long fibers, running lengthwise through the muscle, and in line with the muscle's tendon (Fig. 10.14a). For example, extraocular muscles, which move the eye, have fibers running the length of the muscle (about 40 mm). Muscles that undergo only small changes in length, but are required to produce large forces or stiffnesses, such as postural muscles in the human calf, have fibers arranged like the pinna of a feather, at an angle to a tendon, so that many fibers act in parallel (Fig. 10.14b). For example, the human soleus muscle has fibers about 40 mm long, attached to the tendon at an angle of 0.35 radians (Enoka 1988, p. 177).

Each fiber is a single muscle cell, 10 to 200 μm in diameter and typically 2–100 mm long. Each cell is enveloped in a plasma membrane, and has organelles, such as nuclei and mitochondria, similar to those of any other eucaryotic cell. Muscle cells are highly specialized for force production, so some structures common to all cells are greatly elaborated. Each muscle cell has many nuclei. Mitochondria are extensive, as muscles use much metabolic energy. The internal membrane system, which carries electrical excitation from the plasma membrane to

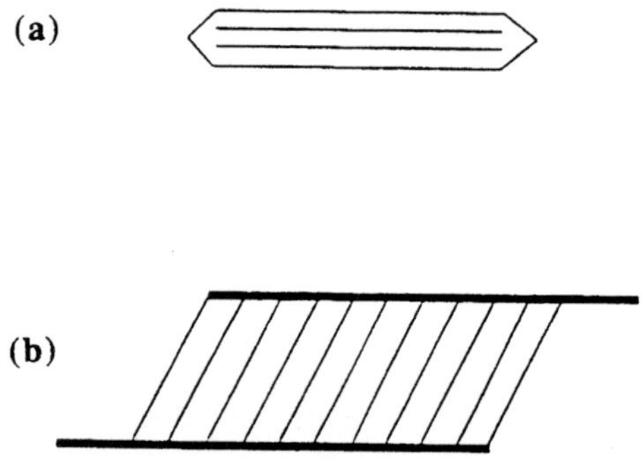

Fig. 10.14 Two common arrangements of fibers within a muscle. In the fusiform arrangement (a), fibers run the length of the muscle. The pennate structure (b) has fibers attached to tendon (bold) at an angle, like the pinna of a feather to its shaft.

the inside of these large cells, and stores calcium for muscle activation, is very elaborate. These elaborations support the greatest specialization in muscle cells, the huge array of force-producing motor proteins.

In skeleton and cardiac muscle, the motor proteins are organized into long fiber arrays, called myofibrils, within each fiber. Myofibrils are typically 1–2 μm in diameter, and run the length of the fiber. In skeletal muscle, these myofibrils appear striped, or striated transversely, at approximately 2 μm intervals, when viewed under a high-power microscope. For this reason, skeletal muscle is often called 'striated muscle'. The striations are due to a regular array of protein filaments within the myofibril: dark stripes, containing thick filaments primarily consisting of the protein myosin, alternate with lighter stripes, containing thin filaments primarily consisting of the protein actin.

Skeletal muscle is capable of large changes in length, sometimes shortening to around half the rest length, and lengthening to about double rest length. These large length changes are due to the relative movements of the thick and thin filaments, which slide past each other without themselves changing length. This 'sliding filament' theory of muscle contraction (A. F. Huxley and Niedergerke 1954; H. E. Huxley and Hanson 1954) has been well established, based on both microscopy and x-ray diffraction.

The mechanism of force production during relative sliding of thick and thin filaments is generally believed to be through the action of individual 'crossbridges', links that stick out from the thick filament to interact with the thin filament. Investigation of the mechanics of interactions between single crossbridges and single actin monomers is an exciting new field of motility research (Scholey 1993).

Muscle force is a function of velocity

The force produced by a muscle depends on the velocity with which it shortens (Fig. 10.15). The faster a muscle shortens, the less force it can produce. For humans lifting a weight by elbow flexion, the relationship between load and velocity is always hyperbolic (Wilkie 1950). This force–velocity relationship is a property of muscle—isolated muscles have the same characteristic, which is well fit with a hyperbola:

$$(F + a)(v + b) = (F_0 + a)b, \qquad (10.7)$$

where F is the magnitude of force, v the magnitude of velocity, and F_0, a and b are constants. It is likely that the

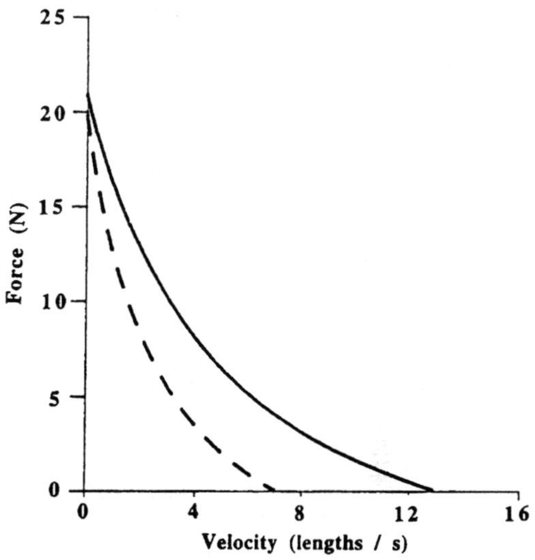

Fig. 10.15 The relationship between force and steady state shortening velocity for a fast-twitch muscle, the rate EDL (solid curve), and for a slow-twitch muscle, the rat soleus (dashed curve), both at 35°C.

hyperbolic relationship is a property of the contractile proteins, as single muscle fibers and isolated actin and myosin have hyperbolic force–velocity characteristics.

The parameter F_0 is the force produced by the muscle at zero velocity (called isometric contraction). To make results more generalizable, stress values, traditionally denoted P_0 (P for pressure), rather than forces are conventionally reported. The "physiological cross sectional area" is usually used for computing stress. This value is obtained by dividing a muscle's volume by its length. The range of P_0 is generally between 100 and 300 kPa (1 Pa = 1 N m^{-2}) (Woledge *et al.* 1985 is an excellent resource for all these parameters), with mammalian muscles in the lower two-thirds of this range. The isometric stress P_0 does not depend strongly on the metabolic fiber type of muscles: a slow oxidative muscle, e.g. rat soleus, and a fast glycolytic muscle, e.g. the rat extensor digitorum longus muscle, both produce around 200 kPa.

The parameter a/P_0 determines the curvature of the force–velocity characteristic. Its value ranges from 0.11 to 0.49, with typical values of about 0.25. The maximum shortening velocity, traditionally denoted v_{max}, is the velocity at which the muscle produces no force. It can be derived from eqn (10.6) that a/p_0 is equal to b/v_{max}. The value of v_{max} is usually estimated by extrapolating the Hill hyperbola to zero force. Maximum shortening velocities range from 1 to 20 muscle lengths per second, and correlate with metabolic fiber type. For example, rat soleus has v_{max} of 7 lengths per second, but EDL contracts at up to 13 lengths per second. v_{max} is very temperature-sensitive: the maximum shortening velocities are halved if the temperature is lowered by 10°C.

Force–velocity relationships are measured in the laboratory by keeping the activation constant (usually maximum), and imposing either constant loads or constant velocities on the muscle. The measurement is a steady-state one—either steady shortening velocity at each load, or steady-state force at each velocity. *In vivo*, the muscle is generally not in the steady state—the load, velocity, and activation usually change during a movement. Luckily, for shortening muscle the transients in response to mechanical changes are very fast. If load or velocity is stepped from one value to another, a new steady state is reached in milliseconds or tens of milliseconds. Human movements occur on time scales of tens to hundreds of milliseconds. For shortening muscle, it is often possible to neglect the mechanical transients, and to treat muscle as if it always obeys the steady-state force–velocity relationship. (Transient responses to changes in activation are much slower, as will be discussed below.)

The mechanical behavior of lengthening muscle has not been characterized nearly as well. For isovelocity lengthening, there is not a unique value of force. Rather, force during isovelocity stretch rises rapidly to a yielding point, after which it may rise at a lower rate, level off, or fall (reviewed in Josephson 1993). The mechanical behavior of lengthening muscle is an important area of current investigation.

Recently, a number of investigators have extended the classical isovelocity or isoload tests of force production. To find muscle work and power in situations closer to those found *in vivo*, they subject isolated muscles to sinusoidal length changes. A plot of muscle force versus length forms a loop, the area of which is work performed in a cycle. Using this 'work loop' technique, biomechanists are beginning to test muscles under more physiological conditions (reviewed in Josephson 1993).

Muscle force is a function of fiber length

The force produced by a muscle also depends on its length. A passive (unactivated) muscle is quite simple mechanically—to a good approximation, the force it produces depends only on its length (Fig. 10.16a, solid line). Below its slack length, the muscle produces no force. Beyond this length, the characteristic relating force

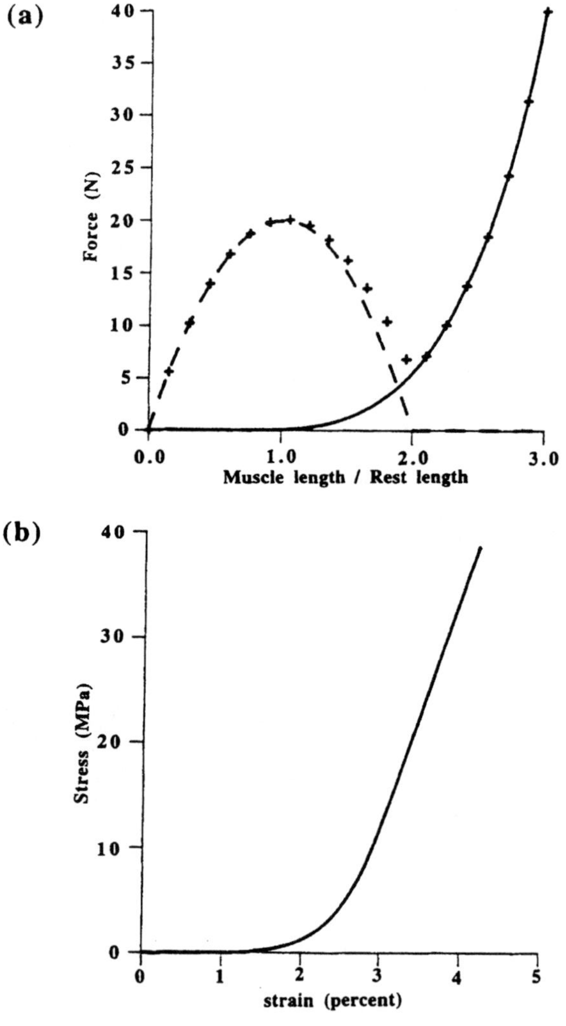

Fig. 10.16 (a) The relationship between length and force for a passive muscle (solid curve), and for an active muscle (+). At each length, the tension developed by active muscle is thought to be the sum of that produced by the passive muscle and the tension produced by the interaction of the sliding filaments (dashed curve). (b) Stress–strain characteristic for a tendon. The characteristic is similar to that for passive muscle, but the tendon is much stiffer.

to length is typically nonlinear. Isolated muscle fibers produce very little force, so most of the force produced by a passive whole muscle is from connective tissue.

The relationship between force and length for a fully activated muscle is much more complicated. If a relaxed muscle is held a fixed length, then tetanized, it produces a maximum force that depends on the length of its fibers, shown as crosses in Fig. 10.16(a). Subtracting the force that a passive muscle would produce, one obtains the dashed curve in Figure 10.16(a), which is thought to come from differing degrees of overlap between the sliding filaments (Gordon *et al.* 1966).

It is important to realize that Figure 10.16(a) does not represent the instantaneous force developed as the muscle changes length. Although the figure depicts a relationship between length and force, the force is measured at discrete lengths, with the muscle relaxed between measurements. The smooth curve is defined by connecting points representing the maximum force available at each length, and is not the characteristic of a nonlinear spring.

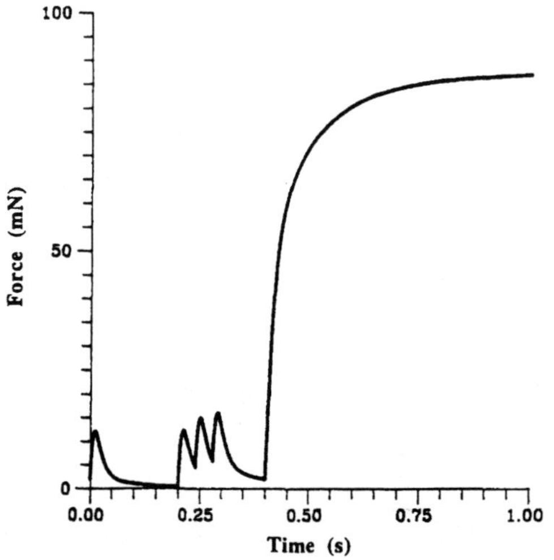

Fig. 10.17 A twitch is the force pattern produced by a motor unit in response to a single stimulus (action potential). As the interval between stimuli decreases, twitches begin to fuse, so that larger forces are developed. Above the 'fusion frequency' the motor unit produces its maximum ('tetanic') force.

Muscle force is a function of activation

The third variable that determines force developed by a muscle is activation. In the body, muscles are activated by neural signals. Each nerve axon is connected to a set of muscle fibers distributed through the muscle, called a motor unit. Each motor unit is independently activated, by impulses (action potentials) in its nerve. the force produced in response to a single impulse is called a twitch.

The force a muscle produces can be modulated by changing the frequency of twitches in each of its motor units (a process called rate coding), or by changing the number of motor units being activated (called recruitment). *In vivo*, muscles are activated to different degrees, from passive to fully activated, by a combination of rate coding and recruitment.

The summation of twitches by rate coding is depicted in Fig. 10.17. At low frequencies, twitches are separate, and each response is like any other. As the stimulus frequency increases, twitches begin to sum, relaxation from one stimulus not being complete before onset of the next. Above a rate termed 'fusion frequency', a smooth maximal contraction called 'tetanus' is obtained. Fusion frequency varies from about 10 Hz for slow-twitch muscles, up to approximately 100 Hz for the fastest fast-twitch. Forces *in vivo* are usually smooth, due to asynchrony in stimuli to motor units, and not to attainment of fusion frequency.

Assessment of the degree of activation of a muscle has always been difficult. A. V. Hill (1938) made a major contribution to muscle mechanics when he conceptually separated the effect of activation from the effect of velocity on muscle force. He made the first attempt to define activation *operationally* when he defined 'active state' as the force that a muscle would produce if its 'contractile element' (the part of the muscle that produces force) were isometric. The active state is thus defined as a force, and the effect of activation is conceptually separated from the effect of velocity.

Discoveries since Hill have illuminated a difficulty with the 'active state' concept. The difficulty is that the 'contractile element' is not separable from the rest of the muscle. That is, the part of the muscle that produces force and has the force–velocity relationship as a characteristic is not separable from other mechanical parts of the muscle, most importantly elastic elements. Therefore, it is not possible to guarantee that the 'contractile element' is isometric. Attempts to deduce the 'active state' from muscle force and kinematics have produced surprising results; for example, a negative active state during a twitch (reviewed in McMahon 1988). Failure of Hill's fundamental separability assumption has led to physically impossible conclusions regarding the 'active state'.

A better measure of activation at the molecular scale is probably the number of potentially cycling crossbridges, i.e. the number of actin binding sites that are open to binding by myosin. However, it is not possible at this time to measure this fraction, so the active-state concept still has utility.

Tendons and the 'series elasticity'

Tendon is an elastic material. A characteristic stress–strain relationship is shown in Fig. 10.16(b). After a toe region extending to 2–3 per cent strain, stress is linear in strain up to between 4 and 5 per cent, with a Young's modulus of about 1 GPa (Wainright *et al.* 1975). Above the linear region, the Young's modulus becomes smaller and there is irreversible damage, up to 8–10 per cent strain, at which point tendon breaks.

Muscles generally have tendons at their ends, so that movement of the muscle may not correspond to movement of the muscle fibers. The discrepancy between muscle length and fiber length has long been recognized (see, for example, Gasser and Hill 1924), and has important mechanical consequences. The most striking example is probably the 'quick release' experiment, in which a muscle is tetanized isometrically, then allowed to lift a load (Fig. 10.18). The step reduction in load causes a corresponding step change in length, followed by a much slower length change as the load is lifted. A. V. Hill (1938) interpreted the near-instantaneous shortening of the muscle as evidence of an elastic element in series with the 'contractile element'. He called this elastic element the 'series elasticity'.

Later investigators have found difficulties in the concept of 'series elasticity'. In addition to the problems involving 'active state', there are two important pieces of evidence indicating that the series spring is not entirely separate from the 'contractile component'. First, the stiffness of the muscle, which should be the stiffness of the 'series elasticity', depends on the degree of activation. Second, much of the muscle's compliance is in the crossbridges themselves, therefore in the 'contractile element'.

The Hill model

It is useful to put the muscle's activation, length, and velocity dependencies together in a single model. Despite the inherent difficulties in the Hill model (Fig. 10.19), it is still the most common representation, and is widely used in biomechanics to predict muscle force.

The 'active state' in Fig. 10.19 is a pure force generator. The dashpot in parallel with the force generator is a mechanism for producing the force–

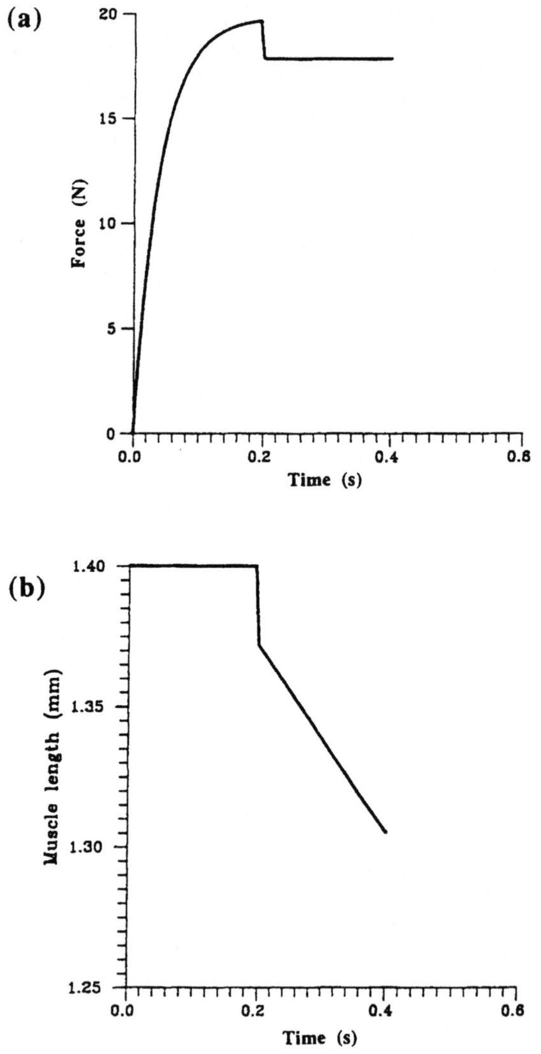

Fig. 10.18 Force (a) and length (b) of a musle subjected to a 'quick release' experiment. The muscle was held at a fixed length and tetanized (rising force). It was then allowed to shorten, lifting the load smaller than F_0. The initial fast length change has been attributed to a 'series elasticity', and the slower length change after it to contraction of the 'contractile element'.

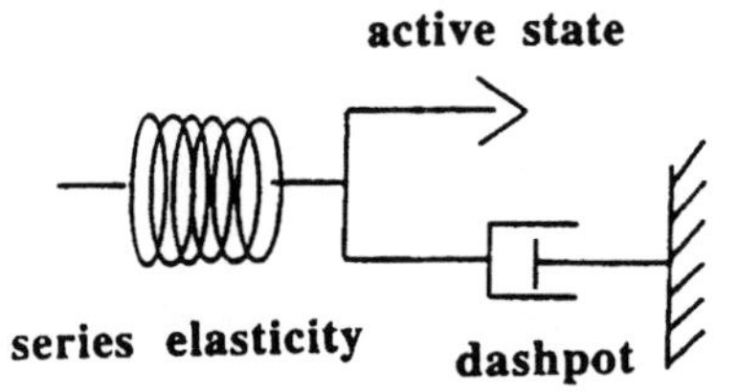

Fig. 10.19 A. V. Hill's model (1938) of muscle mechanics. The 'active state' is a force, which is reduced by the action of the nonlinear dashpot. Together, these form the 'contractile component', which accounts for force production and the force velocity relationship. The 'series elasticity' is necessary to account for the behavior of muscles during quick length or tension changes.

velocity characteristic. The series spring represents the 'series elasticity'. The force–length characteristic is not explicit, but can be included as a length dependence in the active state.

Together, the force generator and dashpot make up the 'contractile element'. The separation of this element from the 'series elasticity' explains the behavior of muscle in a quick-release experiment (Fig. 10.18). The internal stretching of the series elasticity and the force–velocity dashpot can produce the slow rise of tension when a muscle is tetanized (Figs 10.17, 10.18). A rapid rise in active state, followed by a slow decline, produces a force trajectory that matches a twitch (Fig. 10.17) well.

Estimates of fiber velocity, length and activation

It is possible to measure the kinematics of muscle (that is, fiber length and velocity), directly, using implanted markers read by x-ray or fluoroscopy, or by using length transducers (for example, crystals to transmit and receive ultrasound). Such measurements are becoming common in cardiac muscle research, but are quite uncommon in studies of skeletal muscle, because they suffer the same disadvantages as direct force measurements: invasiveness, expense, and sampling problems.

Fiber length and velocity are more often estimated from dissection data. Dissection gives a whole muscle length, fiber pennation angle, tendon length and cross-sectional area, and origin and insertion points. Fiber length is inferred from these geometrical data, usually with the additional assumption that the tendon does not stretch.

Activation is most often estimated from the electromyogram (EMG). The electromyogram is an electrical signal generated by action potentials on the membranes of each muscle fiber. Neural action potentials are translated into muscle action potentials at the neuromuscular junction. Because muscle fibers are much larger than nerve axons, and because each axon generally connects to many muscle fibers in a motor unit, the motor unit action potential is much larger than the neural action potential, and is therefore measurable, even on the surface of the skin. The voltage depends on the size of the motor unit, the amount of dispersion in the signal due to uneven distribution of neuromuscular junctions, and the material between the motor unit and the recording electrodes. Voltage differences of the order of a millivolt are generally measured.

It is rare for one motor unit to fire by itself, so the EMG is usually the sum of several motor unit action potentials, unsynchronized with each other and at various distances from the recording electrodes. Because they are the sum of many overlapping signals, EMGs are very difficult to interpret.

A typical signal from two electrodes placed on the skin over the belly of a muscle is shown in Fig. 10.20. The signal looks like amplitude-modulated noise, and is most often processed as such: the raw signal is rectified, then smoothed, yielding the so-called smoothed, rectified EMG, also known as 'integrated EMG'. It is the low-frequency modulation of a high-frequency 'noisy' carrier. Remarkably, the integrated EMG often correlates well with isometric force or moment (Fig. 10.21 and Problem P10.19).

10.4.5 Open questions

Due to the difficulty in measuring force, important questions that were first posed long ago are still unanswered. Some fundamental biomechanical questions are open: how much mechanical energy is stored in muscle and tendon during locomotion, and what is the role of muscle in energy dissipation? How are muscle mechanics and metabolism related?

Higher-level questions, about the control of movements, have partial answers, but depend on muscle force measurements for completion. Are there motor control primitives (elementary control patterns that can be combined to make more complicated movements)? If so, how are they modulated and connected to each other? What strategies does the brain use to deal with changes in instruction, load? How are multiple joints coordinated

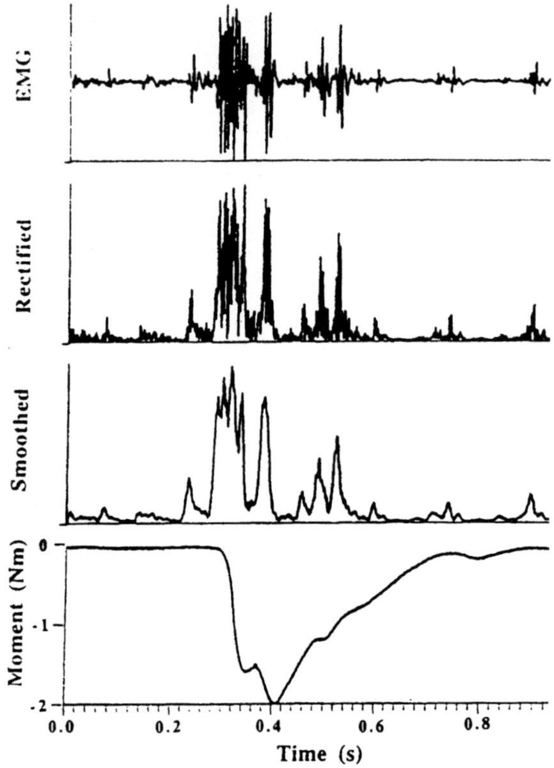

Fig. 10.20 The raw electromyogram is often processed by rectification and smoothing, and is often assumed to vary proportionally with the muscle's 'active state'. In 'phasic contractions', the smoothed rectified EMG does not correlate well with force.

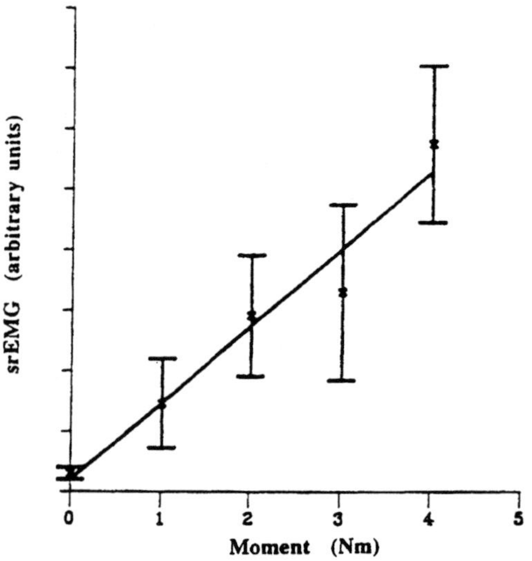

Fig. 10.21 For isometric contractions, integrated smoothed rectified EMG can often be correlated with force.

with each other? Each of these questions has been asked, and answered, at some level. Generally, the questions have been posed about neural strategies, but the answers have been given in terms of integrated EMGs. Direct measurement of neural signals is possible in animals, but most investigators at this level neglect muscle mechanics, and assume that force is proportional to firing frequency of those neurons sampled. If motor control is to be understood, it is vital that the mechanical plant be accounted for. Finally, some of the most interesting questions involve learning. Originally, investigators assumed that motor learning is simpler than any other kind. However, the complicated behavior of the motor plant—the musculoskeletal system—has obscured results in this important field. Progress depends on characterizing the motor plant.

Because direct measurement of force is unusual, and inference of force from other measurements is difficult, the best means of determining force is itself an important open question. Are there better direct methods for force measurement? A smaller and more easily implanted transducer? Likewise, better transducers for measuring fiber length changes during activity would be a great benefit.

Estimation of force from EMG and kinematics seems a hopeful route, but is so far almost entirely untested. Open questions include: How much information is in the EMG, and how should the signal be processed to retrieve that information? Is there fine structure in the EMG 'noise'?

Finally, important fundamental questions regarding mechanics remain unanswered. What are the mechanics of lengthening activated muscle? How are the mechanics and metabolism of muscle related *in vivo*? How do muscle properties change with fatigue?

10.4.6 Conclusion

It is important to estimate muscle force. Many of the open questions listed in this chapter could be answered directly if force were known. Better direct methods of force measurement would help, but direct methods will always be invasive, and therefore of limited applicability.

Moments are relatively easy to deduce, using convenient and noninvasive technology. The problem of deducing muscle forces from moments is, however, almost always indeterminate. Optimization as a general approach has many champions, but tests of the approach are few. The few do not give cause for optimism. Criteria that seem reasonable a priori do not predict muscle activations (EMGs) well, and multiple optimization criteria predict very similar force patterns.

Theoretically, muscle force can be determined from velocity, length, and activation of each muscle's fibers. Estimation of each of these is a technological and theoretical challenge, but preliminary results give hope. A great deal of background data exists on the physiology of isolated shortening muscles. A short-term challenge is to describe the mechanical behavior of lengthening muscle as well.

Finally, deduction of muscle force from fiber kinematics and activation is a synthesis problem—data from many laboratories will be needed. The biomechanics of muscle may serve as a test case for the largest, but least mentioned, assumption behind scientific reductionism: that the many pieces can be synthesized into a meaningful whole.

Problems

P10.1 How does maximum walking speed depend on leg length? To address this question, draw a free-body diagram of a human at mid-stance phase of walking. How does walking speed affect the forces acting on the person? How does this affect the maximum speed at which a person can walk without having an aerial phase?

P10.2 A young boy is walking to the train station with his father. Because they are late for the train, the father begins walking quite fast (2.5 m s^{-1}). The boy has to break into a run to keep pace with his father. What mechanical constraint associated with the inverted pendulum mechanism forces the boy to run at this speed while his father can still comfortably walk? For the purposes of calculations, the father weighs 80 kg and has a leg length of 1.0 m, and the boy weighs 15 kg and has a leg length of 0.4 m.

P10.3 A walker's optimal speed for energy exchange by the inverted pendulum mechanism is substantially higher with stilts than without stilts. Why? *Hint:* draw a diagram of the motions of the center of mass during a stance phase for a person without stilts (leg length of 1 m) and for a person with stilts (total leg length including stilts of 2 m). Assume that the leg sweeps a similar angle during the stance phase in both cases.

P10.4 A person is doing a standing vertical leap. His vertical ground reaction force, $\boldsymbol{F}_z$, as a function of time, t, can be described by: $\boldsymbol{F}_z = 4200 \sin(10.5t)$. The person's body mass is 70 kg and the time that his feet are on the ground is 0.3 s. Assume that he moves downward during the first half of the time that his feet are on the ground (0–0.15 s) and moves upward for the second half (0.15–0.3 s). Calculate the person's vertical excursion while he is in the air.

P10.5 Predict the maximum walking speed for a person on the moon. For your calculations, assume that the moon has one-sixth of the Earth's gravity, the person's leg length is 1 m, and the person's mass is 70 kg.

P10.6 Consider the pros and cons of using various length crank arms on a bicycle with regard to joint moments and the power that can be directed into propelling the bicycle forward. Measure your own leg segment lengths, assume that the pedal reaction force is perpendicular to the crank arm when the crank is at the 3 o'clock position, and sketch out the joint moments. Be sure to consider all three joints. After solving Problem P10.13, regarding the power–velocity curve for muscle, return to this problem and integrate muscle physiology into your reasoning about optimal crank arm length.

P10.7 Inertial and gravitational moments are often unimportant during locomotion, but they may be very important at other joints. List some other joints and movements where these moments are also negligible, as well as movements where they might actually dominate the dynamics.

P10.8 Kangaroo rats are excellent jumping rodents that live in the deserts of the southwest United States. A major predator is rattlesnakes, which kangaroo rats avoid with impressive vertical leaps. Kangaroos, on the other hand, have relatively poor leaping ability, although they travel long distances with remarkable economy in terms of metabolic energy. Given this information, predict the relative magnitudes of $\boldsymbol{R}$ and $\boldsymbol{r}_{\mathrm{m}}$ in kangaroo rats and kangaroos.

P10.9 What is the relationship between the net joint powers shown in Fig. 10.13 and the fluctuations in total mechanical energy of the body center of mass shown in Fig. 10.7?

P10.10 Many people find it surprising that during steady-speed running, the ankle absorbs and produces more power than either the knee or hip. This pattern is unexpected since the hip and knee extensor muscles are so much more massive than the calf muscles. As shown in Section 10.2, during steady-speed running, the peak vertical ground reaction force is much greater than the horizontal force. How would you predict the ground reaction force patterns, joint moments and joint powers would change during accelerations from a standing start?

P10.11 The following data (adapted from Enoka 1988, Table 6.1) show fiber lengths and physiological cross-sectional areas (areas adjusted for pennation angle) for the plantarflexors and dorsiflexors of the ankle:

	fiber length (mm)	*cross-sectional area* (mm^2)
plantarflexors	15.2	13 900
dorsiflexors	29.3	1700

(a) Based on these data, which muscle group is more likely to undergo large changes in length? How much larger?
(b) Which muscle group is stiffer? How much stiffer?

P10.12 The soleus and gastrocnemius muscles act as plantarflexors of the ankle. The physiological cross-sectional area of the soleus is 6610 mm^2, and that of the medial gastrocnemius is 4160 mm^2 (Enoka 1988).
(a) How much force would each muscle produce, isometrically, and fully activated, if the maximum stress is assumed to be 200 kPa?
(b) If the muscles are attached 100 mm from the axis of the ankle, what is the maximum isometric moment?
(c) Estimate the distance from the ankle joint to the ball of the foot, and deduce the maximum force available in plantarflexion of the ankle.
(d) How high could a 70 kg person jump using only these ankle plantarflexors?
(e) Estimate the relative contribution of each muscle to ankle plantarflexion, assuming

(i) that each is to receive equal stress,
(ii) that the sum of normalized forces is to be minimized,
(iii) that the sum of squared normalized forces is to be minimized,
(iv) that endurance is to be maximized.

P10.13 (a) Give a geometric interpretation for the parameters a and b in the hyperbolic fit to the force–velocity data, eqn (10.7).
(b) v_{max}, the maximum shortening velocity, is a physiologically useful parameter. Recast the Hill equation so that v_{max} is a parameter.
(c) v_{max} is usually estimated by extrapolating the hyperbolic fit. How sensitive is the estimate to errors in the curve fit?

P10.14 Power is the product of force and velocity. Plot power as a function of velocity for a muscle.
(a) There is a maximum in the power–velocity relationship. Find it, and find the velocity, relative to v_{max}, at which it occurs.
(b) Exercise increases F_0, but not v_{max}. What is the effect of exercise on the maximum power a muscle can produce? On the velocity at which the maximum occurs?

P10.15 Tendons and the rest of the 'series elasticity' of muscles produce a compliant interface with the world. Is such a compliant interface an advantage or disadvantage if one wishes to produce:
(a) constant position of a limb?
(b) constant force against an object?

P10.16 Muscles *in vivo* generally operate on the rising portion of their length–tension curve: increasing the length increases the available force.
(a) Why would it be bad design to operate a muscle on the descending portion of the curve?

(b) Why can you produce more moment at the elbow with the elbow at 90° than at 130°?

P10.17 Based on his model of muscle, A. V. Hill (1938) supposed that the active state rises very rapidly in a muscle twitch—much more rapidly than the tension does. Use a Hill model to explain why the tension rises so slowly. (Modern measurements show that the time course of tension rise in a twitch is indeed caused mechanically.)

P10.18 Surprisingly often, the electromyogram correlates well with isometric force. Would you expect the EMG to be a better measure of force in:

(a) a muscle with many motor units, or a muscle with a few?

(b) a muscle with faster fibers, or slower?

(c) a muscle with faster motor unit action potentials, or slower?

P10.19 To test whether power is optimized in a frog's jump, Lutz and Rome (1994) filmed frogs as they jumped, and inferred muscle length and velocity from the positions of the frog's legs and anatomical data on the attachments (origin and insertion) of a muscle used in jumping (the semitendinosus). Electromyography indicated that the muscle was maximally activated during the entire jump.

(a) From the films, Lutz and Rome found that the muscle shortened at nearly constant velocity of 3.4 muscle lengths per second during the jump, a period of approximately 50 ms. In a force–velocity test of a single semitendinosus muscle, they found that v_{max} was about 9.5 lengths per second, p_0 was 225 kPa, and a/p_0 was 0.25. What is the maximum power of this muscle? What fraction of its maximum power did the muscle produce during the jump?

(b) Would you expect the muscles to produce constant force during a jump? Explain.

References

Alexander, R. McN. (1988). *Elastic mechanisms in animal movement*. Cambridge University Press.

Alexander, R. McN. and Vernon, A. (1975). Mechanics of hopping by kangaroos (*Macropodidae*). *Journal of Zoology, London*, **177**, 265–303.

Biewener, A. A. and Full, R. J. (1993). Force platform and kinematic analysis. In *Biomechanics: structures and systems*, (ed. A. A. Biewener), pp. 75–96. Oxford University Press.

Blickhan, R. (1989). The spring–mass model for running and hopping. *Journal of Biomechanics*, **22**, 1217–27.

Blickhan, R. and Full, R. J. (1987). Locomotion energetics of the ghost crab. II. Mechanics of the centre of mass during walking and running. *Journal of Experimental Biology*, **130**, 155–74.

Blickhan, R. and Full, R. J. (1993a). Similarity in multilegged locomotion: bouncing like a monopode. *Journal of Comparative Physiology*, **A173**, 509–17.

Blickhan, R. and Full, R. J. (1993b). Mechanical work in terrestrial locomotion. In *Biomechanics: structures and systems*, (ed. A.A. Biewener), pp. 75–96. Oxford University Press.

Bouisset, S. (1973). EMG and muscle force in normal motor activities. In *New directions in electromyography and clinical neurophysiology*, (ed. J. E. Desmedt), pp. 547–83. Karger, Basel.

Buchanan, T. S., Rovai, G. P., and Rymer, W.Z. (1989). Strategies for muscle activation during isometric torque generation at the human elbow. *Journal of Neurophysiology*, **62**, 1201–12.

Buchanan, T. S., Moniz, M.J., Dewald, J. P. A., and Rymer, W. Z. (1993). Estimation of muscle forces about the wrist joint during isometric tasks using an EMG coefficient method. *Journal of Biomechanics*, **26**, 547–60.

Cavagna, G. A., Thys, H., and Zamboni, A. (1976). The sources of external work in level walking and running. *Journal of Physiology*, **262**, 639–57.

Cavagna, G. A., Heglund, N. C., and Taylor, C. R. (1977). Mechanical work in terrestrial locomotion: two basic mechanisms for minimizing energy expenditure. *American Journal of Physiology*, **233**, R243–61.

Chow, C. K. and Jacobson, D. H. (1971). Studies of human locomotion via optimal programming. *Mathematical Biosciences*, **10**, 239–306.

Cobbold, R. S. C. (1974). *Transducers for biomedical measurements: principles and applications*. Wiley, New York.

Crowninshield, R .D. and Brand, R. A. (1981). A physiologically based criterion of muscle force prediction in locomotion. *Journal of Biomechanics*, **14**, 793–801.

DeLuca, C. J. and Forrest, W. J. (1973). Force analysis of individual muscles acting simultaneously in the shoulder joint during isometric aduction, *Journal of Biomechanics*, **6**, 385–93.

DeVita, P. (1994). The selection of a standard convention for analyzing gait data based on the analysis of relevant biomechanical factors. *Journal of Biomechanics*, **27**, 501–8.

Enoka, R. M. (1988). *Neuromechanical basis of kinesiology*. Human Kinetics Books, Champaign, IL.

Farley, C. T., Blickhan, R., Saito, J., and Taylor, C. R. (1991). Hopping frequency in humans: a test of how springs set stride frequency in bouncing gaits. *Journal of Applied Physiology*, **716**, 2127–32.

Farley, C. T., Glasheen, J., and McMahon, T. A. (1993). Running springs: speed and animal size. *Journal of Experimental Biology*, **185**, 71–86.

Gasser, H. S. and Hill, A. V. (1924). The dynamics of muscular contraction. *Proceedings of the Royal Society, London, Series B*, **96**, 398–437.

Gordon, A. M., Huxley, A. F., and Julian, F. J. (1966). The variation in isometric tension with sarcomere length in vertebrate muscle fibers. *Journal of Physiology, London*, **184**, 170–92.

Gregor, R. J., Komi, P. V., Browning, R. C., and Jarvinen, M. (1991). A comparison of the triceps surae and residual muscle moments at the ankle during cycling, *Journal of Biomechanics*, **24**, 287–97.

Hamming, R. W. (1983). *Digital filters*. Prentice Hall, Englewood Cliffs, NJ.

He, J., Kram, R., and McMahon, T. A. (1991). Mechanics of running under simulated reduced gravity. *Journal of Applied Physiology*, **71**, 863–70.

Hill, A. V. (1938). The heat of shortening and the dynamic constants of muscle. *Proceedings of the Royal Society, London, Series B*, **126**, 136–95.

Hill, A. V. (1970). *First and last experiments in muscle mechanics*. Cambridge University Press.

Holmes, J. A., Damaser, M. S., and Lehman, S. L. (1992). Erector spinae activity and movement dynamics about the lumbar spine in lordotic and kyphotic squat lifting. *Spine*, **17**, 327–34.

Huxley, A. F. and Niedergerke, R. (1954). Structural changes in muscle during contraction. Interference microscopy of living muscle fibres. *Nature*, **173**, 971–3.

Huxley, H. E. and Hanson, J. (1954). Changes in the cross-striations of muscle during contraction and stretch and their structural interpretation. *Nature*, **173**, 973–6.

Josephson, R. K. (1993). Contraction dynamics and power output of skeletal muscle. *Annual Review of Physiology*, **55**, 527–46.

Ko, T. C. K. and Farley, C. T. (1992). The rolling egg mechanism in walking lizards. *American Zoology*, **32**(5), 38A.

Komi, P. V. (1990). Relevance of in vivo force measurements to human biomechanics. *Journal of Biomechanics*, **23**, (Supplement 1), 23–34.

Komi, P. V., Fukeshiro, S., and Jarvinen, M. (1992). Biomechanical loading of the Achilles tendon during normal locomotion. *Clinics in Sports Medicine*, **11**(3), 521–31.

Kram, R. and Powell, A. J. (1989). A treadmill-mounted force platform. *Journal of Applied Physiology*, **67**, 1692–8.

Lafortune, M. A., Cavanagh, P. R., Summer, H. J., and Kalenak, A. (1992). Three-dimensional kinematics of the human knee during walking. *Journal of Biomechanics*, **25**, 347–57.

Lehman, S. L. and Stark, L. (1979). Simulations of linear and nonlinear eye movement models. *Journal of Cybernetics and Information Science*, **4**, 21–43.

Litvintsev, A. and Seropyan, V. (1977). Muscular control of movements with one degree of freedom. *Avtomatika i Telekomunika*, **5**, 88–102.

Lutz, G. J. and Rome, L. C. (1994). Built for jumping—the design of the frog neuromuscular system. *Science*, **263**, 370–2.

McMahon, T. A. (1984). *Muscles, reflexes and locomotion*. Princeton University Press.

McMahon, T. A. and Cheng, G. C. (1990). The mechanics of running: how does stiffness couple with speed? *Journal of Biomechanics*, **23**, (Supplement 1), 65–78.

Martin, P. E., Heise, G. D., and Morgan, D. W. (1994). Interrelationships between mechanical power, energy transfers, and walking and running economy. *Medicine and Science in Sports and Exercise*, **25**, 508–15.

Miller, J. M. and Robins, D. (1992). Extraocular muscle forces in alert monkey. *Vision Research*, **32**, 1099–113.

Nemeth, G. and Ohlsen, H. (1985). In vivo moment arm lengths for hip extensor muscles at different angles of hip flexion. *Journal of Biomechanics*, **18**, 129–40.

Nubar, Y. and Contini, R. (1961). A minimal principle in biomechanics. *Bulletin of Mathematical Biophysics*, **23**, 377.

Pedotti, A., Krishnan, V. V., and Stark, L. (1978). Optimization of muscle-force sequencing in human locomotion. *Mathematical Biosciences*, **38**, 57–76.

Rose, J., Ralston, H. J., and Gamble, J. G. (1994). Energetics of walking. In *Human walking*, (2nd edn), (ed. J. Rose and J. G. Gamble), pp. 45–72. Williams and Wilkins, Baltimore.

Rugg, S. G., Gregor, R. J., Mabelbaum, B. R., and Chiu, L. (1990). In vivo moment arm calculation at the ankle using magnetic resonance imaging. *Journal of Biomechanics*, **23**, 495–501.

Scholey, J.M. (ed.) (1993). *Motility assays for motor proteins*, Academic Press, San Diego.

Schuind, F., Garcia-Elias, M., Cooney, W. P., and An, K-N. (1992). Flexor tendon forces: in vivo measurements. *Journal of Hand Surgery*, **17A**, 291–8.

Seirig, A. A. and Avrikar, R. S. (1975). The prediction of muscular load sharing and joint forces in the lower extremity during walking. *Journal of Biomechanics*, **8**, 89–102.

Shanebrook, J. R. and Jaszczak, R. D. (1976). Aerodynamic drag analysis of runners. *Medicine and Science in Sports*, **8**, 43–5.

Shreve, D. A. and Buchanan, T. S. (1993). An evaluation of optimization techniques for estimation of muscle forces based on EMGs during static isometric tasks, *Proceedings of the Institute of Electrical and Electronics Engineers, Engineering in Medicine and Biology Society*, **15**, 1186–7.

Smidt, G. L. (1973). Biomechanical analysis of knee flexion and extension. *Journal of Biomechanics*, **6**, 79–92.

Van Ingen Schenau, G.J. and Cavanagh, P.R. (1990). Power equations in endurance sports. *Journal of Biomechanics*, **23**, 865–82.

Wainright, S. A., Biggs, W.D., Currey, J. D., and Gosline, J. M. (1976). *Mechanical design in organisms*. Princeton University Press.

Wells, R. P. (1981). The projection of ground reaction force as a predictor of internal joint moments. *Bulletin of Prosthetics Research*, **18**, 15–19.

Wilkie, D. R. (1950). The relationship between force and velocity in human muscle. *Journal of Physiology, London*, **110**, 249–80.

Winter, D. A. (1990). *Biomechanics and motor control of human movement*. Wiley, New York.

Winters, J. M. and Woo, S. L-Y. (ed.) (1990). *Multiple muscle systems*. Springer-Verlag, New York.

Woledge, R. C., Curtin, N. A., and Homsher, E. (1985). *Energetic aspects of muscle contraction*. Academic Press, London.

Further reading

In addition to the works by Alexander (1988), Enoka (1988), Winter (1990), Winters and Woo (1990), and Woledge *et al.* (1985), listed under References, the following are recommended for further reading.

Alexander, R. McN. (1983). *Animal mechanics*, (2nd edn). Blackwell, Oxford.

Alexander, R. McN. (1984). Walking and running. *American Scientist*, **72**, 348–54.

Alexander, R. McN. (1992). *The human machine*. Columbia University Press, New York.

Biewener, A. A. (ed.) (1992). *Biomechanics—structures and systems*. Oxford University Press.

Carlson, F. D. and Wilkie, D. R. (1974). *Muscle physiology*, Prentice-Hall, New York.

Cavanagh, P. R. (ed.) (1990). *Biomechanics of distance running*. Human Kinetics Books, Champaign, IL.

Cavanagh, P. R. and Lafortune, M. R. (1980). Ground reaction forces in distance running. *Journal of Biomechanics*, **13**, 397–406.

Fung, Y. C. (1990). *Biomechanics: motion, flow, stress and growth*. Springer-Verlag, New York.

Fung, Y. C. (1993). *Biomechanics: mechanical properties of living tissue*. Springer-Verlag, New York.

Margaria, R. (1976). *Biomechanics and energetics of muscular exercise*. Oxford University Press.

Meglan, D. and Todd, F. (1994). Kinetics of human locomotion. In *Human walking*, (2nd edn), (ed. J. Rose and J.G. Gamble), pp. 23–44. Williams and Wilkins, Baltimore.

Miller, D. I. (1990). Ground reaction forces in distance running. In *Biomechanics of distance running*, (ed. P.R. Cavanagh), pp. 203–224. Human Kinetics Books, Champaign, IL.

Rose, J. R. and Gamble, J. G. (ed.) (1994). *Human walking* (2nd edn). Williams and Wilkins, Baltimore.

PRINCIPLES OF ELECTRO-PHORETIC SEPARATIONS

11

Paul D. Grossman and David S. Soane

Contents

Symbols

a	average pore size (m)
C	concentration (kg m^{-3})
D	diffusion coefficient (m^{2}s^{-1})
D_p	particle diameter (m)
D_{rep}	diffusion coefficient for reptation (m^{2}s^{-1})
D_{tube}	diffusion coefficient for intra-tube migration (m^{2}s^{-1})
E	electric field strength (V m^{-1})
f	translational friction coefficient (N s m^{-1})
F_{drag}	hydrodynamic drag force (N)
F_{elec}	electric force (N)
f_{rep}	translational frictional coefficient for reptation (N s m^{-1})
f_{tube}	translational frictional coefficient for intra-tube migration (N s m^{-1})
g	acceleration of gravity (m s^{-2})
h	heat transfer coefficient between outside surface of electrophoresis tube and the surrounding air (W m^{-2}K^{-1})
I	current density (A m^{-2})
K	proportionality constant in Ogston model (m kg^{-1})
k	Boltzmann constant (J K^{-1})
k_b, k_g	thermal conductivity of fluid and glass wall respectively (W m^{-1}K^{-1})
k_e	electrical conductivity (Ω^{-1}m^{-1})
K_r	retardation coefficient (m^{3}kg^{-1})
l_{rep}	characteristic length scale for reptation (m)
l_{tube}	characteristic length scale for intra-tube migration (m)
m	mass of particle (kg)
MW	molecular weight (kg mol^{-1})
N	number of units in polymer chain
n	number of polymer strands per unit volume in Ogston model (m^{-3})
p	pore size (m)
q	charge (C)
R	resolution
r	particle radius (m)
Ra	Rayleigh number
Re	Reynolds number
R_g	radius of gyration (m)
S	cross-sectional area (m^2)
Se	power generation per volume (W m^{-3})
T	absolute temperature (K)
t	time (s)
v	velocity (m s^{-1})
v_0	initial velocity (m s^{-1})
v_{ss}	steady-state velocity (m s^{-1})
$W_{i,j}$	width of sample zone containing species i or j respectively (m)
x	spatial coordinate (m)
$X_{i,j}$	position of center of mass of

	migrating zone containing species i or j respectively (m)	η	viscosity ($kg\ m^{-1}s^{-1}$)	ρ	fluid density ($kg\ m^{-3}$)
		μ	electrophoretic mobility ($m^2V^{-1}s^{-1}$)	ρ_p	particle density ($kg\ m^{-3}$)
		μ_0	electrophoretic mobility in free solution ($m^2V^{-1}s^{-1}$)	τ_{rep}	characteristic time for reptation (s)
α	thermal diffusivity (m^2s^{-1})				

11.1 Introduction

Biological macromolecules such as proteins and DNA occur naturally in complex mixtures of closely related species. However, the study and utilization of these molecules generally require very pure samples. Hence, separation of these biologically important macromolecules is a key step in biological science and technology.

Electrophoresis has been and continues to be a dominant separation technique which is particularly well suited for biopolymers, typically of high molecular weight, water-soluble, labile and highly charged. Historically, electrophoresis was the first separation method able to fractionate a protein mixture. Now, electrophoresis has become one of the most important and widely used analytical techniques in modern biochemistry and molecular biology. A glance through any issue of a popular science journal such as *Science* or *Nature* will show many examples of the practical utility of electrophoretic separations, highlighting the importance and range of application for this technique.

Modern applications of gel electrophoresis range from the molecular weight determination of proteins (Hames and Rickwood 1984) to DNA sequencing (Trainor 1990) and to the diagnosis of genetic disease (Borst and Miller 1990). Many of the new tools of molecular biology, such as DNA sequencing, require an electrophoretic separation step in order to extract the available information. It is the combination of extremely selective enzymes and electrophoretic separations which makes the available information accessible.

This chapter gives a brief discussion of what electrophoresis is and the physical principles involved in electrophoretic separations. It is not intended to be a thorough review of this large and diverse field, but rather a first introduction to the underlying principles behind this important but often misunderstood technique. We have attempted to highlight the key principles which underlie electrophoretic separations. We have explored the mobility of solutes in free solution and through polymer networks, the effect of solute diffusion on the resolution of species having different mobilities, and the limiting influence of Joule heating on the speed and scale of electrophoretic separations. While our treatment has only skimmed the surface of these complicated phenomena, we hope that an intuitive feel for the process and a springboard for further study have been provided.

At this time the field of electrophoresis is rapidly evolving, both from an experimental and a theoretical standpoint. As ongoing research further develops such emerging opportunities as the effect of higher field strengths, new materials for polymer networks, pulsed fields, and numerical simulation, the capability and range of application of electrophoresis are certain to increase.

11.2 Electrophoretic mobility

11.2.1 Definition

As the name implies, electrophoresis involves the motion of charged particles in solution under the influence of an electric field. The separation of solutes is accomplished as a result of the different rates of migration of the solutes in the mixture in the presence of an electric field. In this section we will discuss the factors which cause differential migration rates among the component molecules of a mixture.

When a charged particle is placed in an electric field, E, it experiences a force, F_{elec}, which is equal to the product of its net charge, q, and the electrical field strength, E:

$$F_{\text{elec}} = q * E. \tag{11.1}$$

In addition to the electrical force, the particle experiences a drag force in the direction opposite to its direction of motion. This drag force, F_{drag}, is proportional to the particle velocity,

$$F_{\text{drag}} = f * v, \tag{11.2}$$

where the proportionality constant f is called the translational friction coefficient. This direct proportionality between drag force and velocity only holds under conditions of creeping flow, i.e. $Re \ll 1$, where Re is the Reynolds number. We will confirm this assumption shortly. For example, for a spherical particle undergoing creeping flow, f is given by Stokes' law (Atkins 1994) as

$$f = 6\pi\eta r, \tag{11.3}$$

where η is the viscosity of the surrounding medium and r is the apparent hydrodynamic radius of the particle. A schematic diagram of this force balance is given in Fig. 11.1. Thus, the equation describing the translational motion of a particle under the influence of an electric field is

$$m\left(\frac{dv}{dt}\right) = F_{\text{elec}} - fv, \tag{11.4}$$

where m is the mass of the particle.

Except for a brief transient when the electrical force is first applied, the electrostatic force is exactly counterbalanced by the drag force, and the particle reaches a steady state velocity, v_{ss}. Using eqns (11.1) and (11.2), the resulting steady-state velocity can be related to the charge and frictional properties of the particle by the simple relationship

$$v_{\text{ss}} = \frac{qE}{f}. \tag{11.5}$$

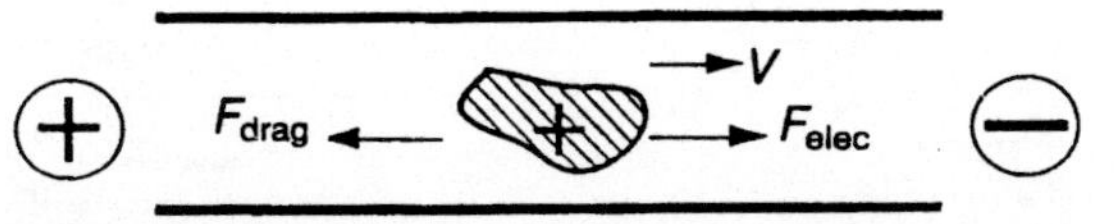

Fig. 11.1 Schematic diagram of a negatively charged particle undergoing electrophoresis.

Furthermore, the electrophoretic mobility, μ, of a particle is defined as the steady-state velocity per unit field strength, or, from eqn (11.5),

$$\mu = \frac{q}{f}. \tag{11.6}$$

Clearly, differences in the electrophoretic mobility of molecules can arise as a result of differences in frictional properties (that is, size or shape), or as a result of differences in the net charge on the molecule. It is these differences in the properties of molecules which form the basis for all electrophoretic separations.

A number of comments should be made regarding the above derivation with respect to the net charge, the assumption of creeping flow, and the assumption that the velocity reaches its steady-state value rapidly.

The net charge of a molecule differs from the actual charge on the molecule because of the influence of counter-ions present in the surrounding solution. For example, if the molecule is negatively charged, an 'atmosphere' rich in positive counter-ions will surround the molecule. This 'atmosphere' is known as the electrical double layer (Hiemenz 1986). The effect of the double layer is to reduce the apparent net charge on the molecule. The extent to which the counter-ion atmosphere shields the charge on the particle depends on the concentration and charge of the counter-ions in solution. As the concentration and charge of the counter-ions increase, the net charge on the particle decreases, resulting in a lower electrophoretic mobility.

Next, we must confirm that the assumption of creeping flow in the derivation of eqn (11.6) is valid. Creeping flow is defined as flow for which the Reynolds number, Re, is very much less than 1. For the case of flow around a spherical object, Re is defined (Denn 1980) as

$$Re = \frac{D_p v \rho}{\eta}, \tag{11.7}$$

where D_p is the diameter of the particle, v is the velocity of the particle, ρ is the density of the solvent, and η is the viscosity of the solvent. Thus, given a spherical molecule with a diameter of 50 Å (5 nm), the dimensions of a typical globular protein, in order for the Reynolds number to exceed a value of 1, the electrophoretic velocity would have to be of the order of 200 m s^{-1}! This is 6 to 7 orders of magnitude greater than would be found in any electrophoresis experiment. Thus, the assumption of creeping flow is justified.

Finally, we must confirm that the molecule achieves a steady-state velocity rapidly relative to the period of an electrophoresis experiment. Solving eqn (11.4) for $v(t)$,

$$v(t) = \frac{F_{\text{elec}}}{f} + \left(v_0 - \frac{F_{\text{elec}}}{f}\right)\exp(-ft/m), \tag{11.8}$$

where v_0 is the initial velocity prior to application of the electric field, m is the mass of the molecule, and t is the time elapsed since application of the electric field. If we assume a spherical protein as before with a radius of 50 Å, a molecular weight of 75 000 u, a net charge of 10 times the charge of an electron, and a field strength of 10 000 V m^{-1}, we can calculate the time required for the molecule to reach 99 per cent of its steady-state velocity. Using eqns (11.1) and (11.3) to find the values for F_{elec} and f, the time required to reach $0.99v_{\text{ss}}$ is of the order of 10^{-11} s. Thus, the assumption of an instantaneous attainment of the steady-state velocity is justified.

11.2.2 Electrophoretic mobility in free solution

Whether in free solution or in a polymer network, the dependence of the electrophoretic mobility on the net charge of the solute is clear. However, the way in which the frictional coefficient, f, is related to the size of the molecule depends on what model is used to describe the conformation of the molecule and on the nature of the medium through which the solute is migrating. In this section we will discuss the case of a solute migrating through free solution, and in the following two sections we will discuss the migration of a solute through a polymer network.

If, in free solution, the migrating molecule is modeled as a solid sphere, the relationship between the translational frictional coefficient and molecular size would be straightforward. From eqn (11.3),

$$f = 6\pi\eta r. \tag{11.9}$$

Furthermore, for a solid sphere, the radius r and the mass m are related by

$$m = \rho_{\text{p}}\left(\tfrac{4}{3}\pi r^3\right), \tag{11.10}$$

where ρ_{p} is the density of the particle. Thus,

$$r \propto m^{1/3} \tag{11.11}$$

and

$$f \propto m^{1/3}. \tag{11.12}$$

If, rather than being a solid sphere, the solute behaves as a loose random coil or a rod, we can no longer use eqn (11.10) to relate the molecular mass to an apparent hydrodynamic radius and thus to a value for f. However, the relationship between mass and the translational friction coefficient has been established for a number of physically important molecular conformations. Some of these are given in Table 11.1, which clearly shows that the way in which electrophoretic mobility in free solution is related to molecular mass is strongly dependent on the model chosen to represent the conformation of the solute.

As it turns out, for many practically important applications, DNA and SDS protein applications in particular, separations based solely on differences in free-solution electrophoretic mobilities are not possible. (SDS proteins are proteins which have been treated with the detergent SDS in order to eliminate any variations in molecular conformation or charge-to-mass ratio between different proteins. This is done to allow one to perform separations based only on molecular size.) The reason for this can be demonstrated through a simple argument using the above relationship for μ with DNA as an example.

The structure of the DNA molecule is such that the total charge on the molecule is directly proportional to its size; i.e. approximately two charges per base pair. Thus,

$$q \propto N, \tag{11.13}$$

where N is the number of units (base pairs) in the DNA chain. In addition, as can be seen from Table 11.1, if the DNA molecule is modeled as a free-draining coil,

$$f \propto N. \tag{11.14}$$

(A free-draining coil is one in which each of the units of the chain contributes equally to the overall drag of the chain.) Finally, by combining eqns (11.13) and (11.14) with the definition for electrophoretic mobility, eqn (11.6), it can be seen that μ is no longer a function of

Table 11.1. Proportionality relationship between frictional coefficient and molecular weight for various molecular models

Molecular model	Proportionality
Solid sphere	$f \propto (MW)^{0.33}$
Random coil—unperturbed chain	$f \propto (MW)^{0.5}$
Long rod	$f \propto (MW)^{0.8}$
Free-draining coil	$f \propto (MW)^{1.0}$

(Source: Cantor and Schimmel 1980a.)

molecular size; that is,

$$\mu = \frac{q}{f} \propto \frac{N}{N} = N^0, \qquad (11.15)$$

where N^0 indicates that μ is constant with respect to changes in N. This argument also applies to the case of SDS proteins where charge is also proportional to N. Therefore, in order to perform a separation of DNA fragments, or any molecule with a constant ratio of charge to frictional coefficient, one must exploit an alternative separation mechanism.

If, instead of allowing the DNA molecule to migrate in free solution, one forces it to travel through a porous polymer network, one can impart a size dependence to the electrophoretic mobility. Two main theories exist to describe the migration of a solute through a polymer gel network. These are the Ogston sieving model and the reptation model. The applicability of each of these models depends on the size of the migrating molecule relative to the mesh size of the polymer network. Both of these models will be discussed in the following two sections.

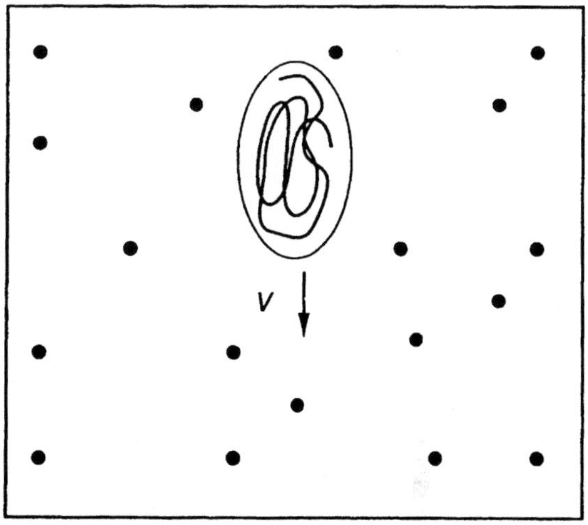

Fig. 11.2 Schematic diagram of a solute migrating through a polymer network by the Ogston mechanism. In this case, because the mesh size of the polymer network is larger than the migrating coil, the coil percolates through the mesh as if it were a rigid particle.

11.2.3 Electrophoretic mobility in a polymer network—Ogston theory

The Ogston model treats the polymer gel network as a molecular sieve. It assumes that the gel consists of a random network of interconnected pores having an *average* pore size, a, and that the migrating solute behaves as an undeformable particle of radius R_g. According to this model, smaller molecules migrate faster because they have access to a larger fraction of the available pores. The mathematical treatment of this problem was first presented by Ogston (1958). Figure 11.2 is a schematic diagram showing a solute migrating through a polymer network by the Ogston mechanism.

In the Ogston theory, the electrophoretic mobility of the migrating solute through the porous structure is assumed to be equal to its free solution mobility, μ_0, multiplied by thc probability that the solute will meet a pore large enough to allow its passage. Thus

$$\mu = \mu_0 P(p \geqslant R_g), \qquad (11.16)$$

where p is the radius of the pore in which the coil resides, and $P(p \geqslant R_g)$ is the probability that a given pore has a radius greater than or equal to the radius of the migrating particle. Thus, now the problem simply becomes one of determining the form of $P(p \geqslant R_g)$.

The Ogston model of the pore size distribution predicts that, in a random network of linear polymers, the fraction of pores large enough to accommodate a spherical particle having a radius of gyration R_g is

$$P(p \geqslant R_g) = \exp - \pi n l (r + R_g)^2, \qquad (11.17)$$

where n is the average number of polymer strands per unit volume, l is the average length of the polymer strands, and r is the thickness of the strands. Furthermore, this model assumes that the product $n * l$ is proportional to the concentration of the gel-forming polymer, C. Thus,

$$P(p \geqslant R_g) = \exp[-KC(r + R_g)^2], \qquad (11.18)$$

where K is a constant of proportionality. The term $K(r + R_g)^2$ is known as the retardation coefficient, which is a number characteristic of a given molecular species in a particular polymer system.

Combination of eqns (11.16) and (11.18) gives the final expression describing the migration of a solute through a polymer network:

$$\mu = \mu_0 \exp[-KC(r + R_g)^2]. \qquad (11.19)$$

If $r \ll R_g$, one would expect that a plot of $\log(\mu/\mu_0)$ vs. C would give a straight line with a slope proportional to

R_g^2. Such plots are known as Ferguson plots (Ferguson 1964). An example of a Ferguson plot is shown in Fig. 11.3. Here R_f is a normalized electrophoretic mobility, defined as μ/μ_s, where μ_s is the electrophoretic mobility of a reference molecule (Cantor and Schimmel 1980b).

As stated before, the Ogston model assumes that the migrating solute behaves as an undeformable spherical particle. It does not take into account the fact that the migrating molecule could deform in order to 'squeeze' through a pore. Therefore, when $R_g > a$, the Ogston model predicts that the electrophoretic mobility of the migrating solute will rapidly approach zero. However, large flexible chain molecules such as DNA still migrate when $R_g \gg a$. This is explained by the second model for migration, the reptation model. The reptation model assumes that instead of migrating as an undeformable particle with radius R_g, the migrating molecule moves 'head first' through the porous network. The reptation model is the subject of the following section.

11.2.4 Electrophoretic mobility in a polymer network—reptation

The basis of the reptation model is the realization that when a long flexible molecule travels through a polymer network having a mesh size smaller than R_g, it does not necessarily travel as an undeformed particle, but rather 'snakes' through the polymer network 'head first'. The migrating solute is assumed to move through 'tubes' which are formed by the gel matrix. The term reptation comes from the reptile-like nature of this motion. Figure 11.4 illustrates this concept further. What follows is a derivation of the size dependence of the translational friction coefficient, f_{rep}, and thus μ, for a chain undergoing reptation. These arguments are scaling arguments rather than complete functional derivations. That is, we are looking for functional dependences rather than exact equations. The first description of the reptation mechanism was presented by de Gennes (1971, 1979) and Doi and Edwards (1978), while the first application of reptation theory to the electrophoresis of biopolymers was presented by Lerman and Frisch (1982).

As a starting point, it is recognized that f_{rep} can be related to an apparent diffusion coefficient, D_{rep}, through

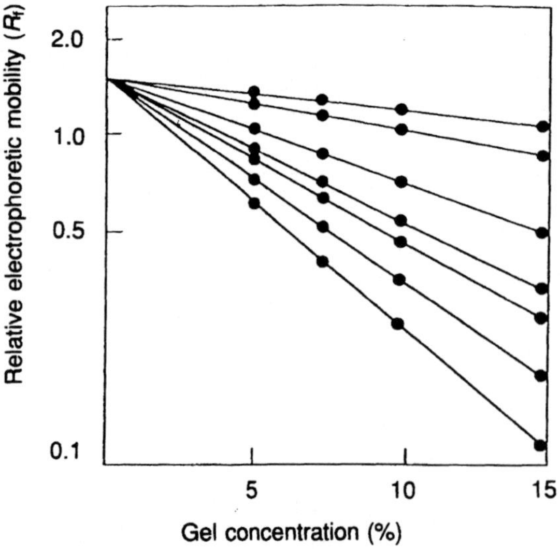

Fig. 11.3 Ferguson plot for seven different proteins ranging in size from 60 000 μ (lowermost curve) to 14 000 μ (uppermost curve). (Ferguson 1964.)

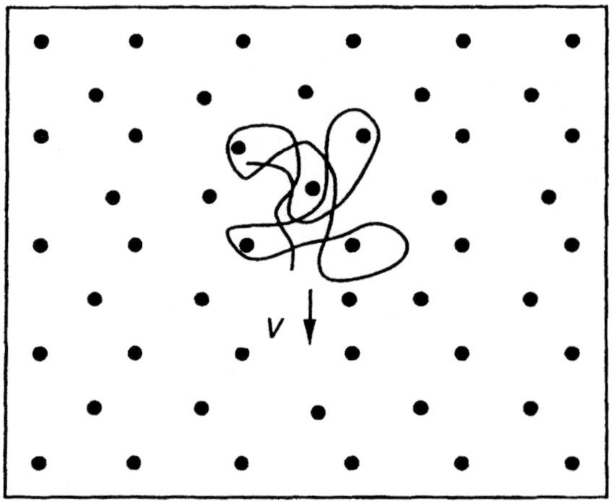

Fig. 11.4 Schematic diagram of a solute migrating by the reptation mechanism. In this case, because the mesh size of the network is much smaller than the radius of gyration of the migrating coil, the coil is forced to travel 'head-first' through the network.

the Stokes–Einstein equation (Atkins 1994):

$$D_{rep} = \frac{kT}{f_{rep}}, \tag{11.20}$$

where k is the Boltzmann constant and T is the absolute temperature. Next, in order to derive an expression for D_{rep}, we will consider the statistical nature of the diffusion process. If diffusion is considered to be simply the random motion of molecules in a concentration gradient (neglecting any nonidealities of the solution), the diffusion coefficient, D, can be defined as

$$D \equiv \frac{l_{rep}^2}{\tau_{rep}}, \tag{11.21}$$

where l_{rep} is a characteristic step length for diffusion and τ_{rep} is the characteristic step time for diffusion. (For a clear description of the diffusion process from the statistical point of view, as well as the derivation of eqn (11.21), see Berg 1983.) In order to use eqn (11.21) to arrive at an expression for D_{rep}, we must determine a characteristic step time, τ_{rep}, and a characteristic step length, l_{rep}, for the reptation process.

First, we consider the characteristic step length, l_{rep}. Implicit in the statistical description of the diffusion process is the assumption that each step is statistically independent of any previous step. For this to be true, in each step the molecule must travel a distance at least as large as its own characteristic dimension. If we assume that the DNA exists in a random coil conformation, this characteristic dimension would be the radius of gyration of the coil. This assumption is only valid for DNA chains larger than about 300 base pairs, because as the fragments become smaller, they begin to behave less like coils and more like rods. For a random coil polymer, the radius of gyration, R_g (Rodriguez 1982), is given by

$$R_g^2 = \frac{Nl^2}{6}, \tag{11.22}$$

where N is the number of independent segments in the polymer and l is the length of each segment. Therefore,

$$l_{rep} = R_g \propto N^{1/2}. \tag{11.23}$$

Thus, the characteristic length for reptation is proportional to the square root of the molecular size, assuming that the DNA behaves as a random coil. Therefore, when the macroscopic displacement of the reptating DNA is l_{rep}, its new location is statistically independent of the previous one.

Next, we must consider the characteristic step time in eqn (11.21). In order to determine a characteristic time for the reptation process, we will define an *intra-tube* diffusion coefficient, D_{tube},

$$D_{tube} = \frac{kT}{f_{tube}}, \tag{11.24}$$

where f_{tube} is the frictional coefficient of the DNA *within* an individual tube. It is important at this point to note the difference between D_{rep} and D_{tube}. Whereas D_{rep} is an apparent 'macroscopic' net translational diffusion coefficient, D_{tube} applies to the motion of the DNA along the tube axis. Because within the pores the DNA is migrating as if in free solution, it is reasonable to assume that f_{tube} is proportional to N; thus,

$$D_{tube} \propto \frac{1}{N}. \tag{11.25}$$

Furthermore, when considering the motion within an individual tube, the characteristic dimension of the DNA now becomes the total extended length of the molecule rather than the radius of gyration. Since the extended length is directly proportional to the molecular length, we obtain

$$l_{tube} \propto N. \tag{11.26}$$

It is this important conceptual difference between the length scales for macroscopic diffusion and microscopic tube diffusion which is the key to providing the molecular size dependence of μ for molecules undergoing reptation. This difference in length scale between macroscopic displacement, l_{rep}, and motion within a tube, l_{tube}, is caused by the need for the molecule to move a longer distance within the pore than the overall macroscopic displacement would indicate, because of the tortuous nature of the tube.

Thus, combining eqns (11.21), (11.25), and (11.26), we can define a characteristic time for diffusion within the tube as

$$\tau_{tube} = \frac{(l_{tube})^2}{D_{tube}} \propto \frac{N^2}{1/N} \tag{11.27}$$

or

$$\tau_{tube} \propto N^3. \tag{11.28}$$

At this point we can determine the dependence of D_{rep}, and thus f_{rep}, on molecular size for molecules undergoing reptation. From eqns (11.21), (11.23), and (11.28) we have

$$D_{rep} \propto \frac{(N^{1/2})^2}{N^3} = \frac{1}{N^2}. \tag{11.29}$$

Thus, from eqn (11.20),

$$f_{\text{rep}} \propto N^2. \tag{11.30}$$

This dependence of f_{rep} on N is in contrast to the free-solution case where f is proportional to N, eqn (11.14).

Finally, we are in a position to determine the relationship between electrophoretic mobility and molecular size for a molecule migrating by reptation. From eqns (11.6), (11.13), and (11.30) we obtain

$$\mu \propto \frac{N}{N^2} = \frac{1}{N}. \tag{11.31}$$

Equation (11.31) is the key result that we have been looking for. It states that for a chain-like molecule undergoing reptile motion under the influence of an electric field, the electrophoretic mobility is inversely proportional to the molecular length. Again, it is important to contrast this result with the free-solution case where μ is independent of N, if the total charge on the molecule is proportional to its length.

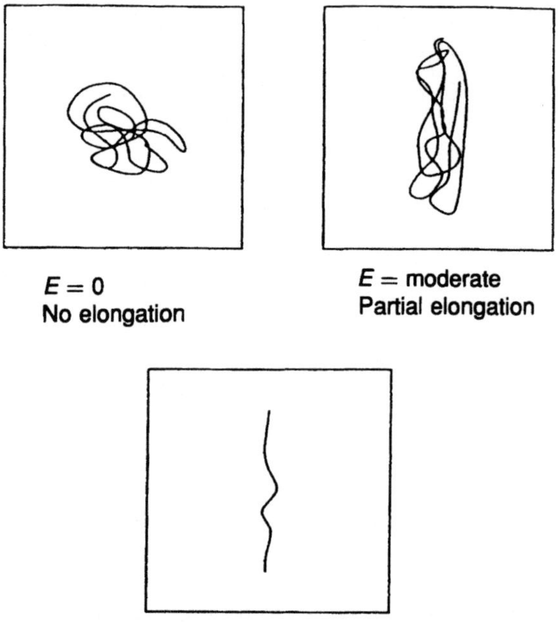

Fig. 11.5 Schematic diagrams showing the elongational influence of the electric field on the radius of gyration of a molecule migrating by the reptation mechanism. When $E = 0$, $R_g \propto N^{0.5}$, whereas for large E, $R_g \propto N$.

A refinement on the reptation model which takes into account the influence of large electric fields is the *biased-reptation* model. When the electric field becomes large, the assumption that the migrating molecule exists as an unperturbed coil, eqn (11.22), is no longer valid. Because of the induced orientation of the leading segment, as the field strength is increased, the coil becomes more elongated. This is shown schematically in Fig. 11.5. In the limiting case, the coil becomes a rod. As can be seen from eqns (11.23) and (11.29), if the migrating molecule becomes a rod, l_{rep} is now proportional to N instead of $N^{1/2}$. If this is substituted into eqn (11.29), $f_{\text{rep}} \propto N$ and $\mu \propto N^0$. Thus, as the coil becomes more elongated, the $1/N$ size dependence of μ disappears. This effect was first described by Lumpkin *et al.* (1985), who arrived at an expression of the form

$$\mu \propto \left(\frac{1}{N} + bE^2\right), \tag{11.32}$$

where b is a function of the mesh size of the polymer network. Note that the first term of eqn (11.32) depends on the size of the migrating molecule but not on the electrical field strength, while the second term does not depend on molecular size but does depend on electrical field strength. Therefore, as the electrical field or the molecular size increases, the dependence of mobility on molecular size decreases. This is the key prediction of the biased-reptation model. The predicted behavior has been observed experimentally (Hervet and Bean 1987). Because of this effect, the maximum size of DNA which can be separated using traditional electrophoretic techniques is approximately 20 000 base pairs.

In order to extend the size range of DNA which can be separated by electrophoresis, recently a new method has been developed to circumvent the problem of size-independent mobilities as N becomes large. This technique is known as pulsed-field electrophoresis (PFE) (Schwartz *et al.* 1982). In PFE, the magnitude and/or the direction of the electric field is changed as a function of time. This forces the migrating molecule to reorient itself periodically, which introduces a size dependence on μ, based on the rate of reorientation. Using PFE, the size range of DNA which can be separated has been increased to 2 500 000 base pairs.

11.3 Resolution—influence of diffusion on separation performance

Thus far we have discussed electrophoretic separations only in terms of how the structural features of the separation medium and the solute affect the average electrophoretic mobility of the solute. Implicitly we have assumed that each molecular species migrates as an infinitely thin zone travelling at a velocity of $\mu * E$. This picture is a vast oversimplification. In order to give a more realistic picture of the electrophoretic separation process, we must take into account the influence of diffusion. As we might expect, diffusion works against the separation process, serving to remix the separated bands.

In order to specify the quality of a separation we need a quantitative measure of the 'separatedness' between two solutes. This measure must take into account both the effects of differential solute mobilities and the dispersing influence of diffusion. Such a quantitative measure is the resolution. The resolution between two sample zones is defined (Littlewood 1970) as

$$R(t) \equiv \frac{X_i(t) - X_j(t)}{\frac{1}{2}[W_i(t) + W_j(t)]}, \qquad (11.33)$$

where X_i and X_j are the locations of bands i and j respectively, and W_i and W_j are the widths of bands i and j measured at half of the peak height respectively. Thus, the resolution is defined as the difference between the average value of the peak positions divided by the average peak widths at a given time. Figure 11.6 shows two such sample zones. Thus far in our discussion, we have only addressed how X_i changes with time; that is,

$$X_i(t) = v_i t = \mu_i E t, \qquad (11.34)$$

where v_i is the velocity of the ith sample zone. Therefore, in order to arrive at an analytical expression for the resolution between two species, we must also obtain an expression for $W_i(t)$.

In order to find $W_i(t)$ we shall solve the convective diffusion equation in one dimension. The convective diffusion equation is

$$\frac{\partial C_i(x,t)}{\partial t} = \frac{D_i \partial^2 C_i(x,t)}{\partial x^2} - v_i \frac{\partial C_i(x,t)}{\partial x}, \qquad (11.35)$$

where D_i is the diffusion coefficient and v_i is the drift velocity of species i. In the case of electrophoresis, $v_i = \mu_i * E$. Note that eqn (11.35) is completely general and independent of the model chosen for electrophoretic mobility. It is simply derived from the conservation of mass and Fick's law. The solution of eqn (11.35), assuming that all the solute molecules are initially confined in an infinitely sharp zone at $x = 0$, is

$$C_i(x,t) = \frac{M}{S\sqrt{4\pi D_i t}} \exp\left(-\frac{(x - \mu_i E t)^2}{4 D_i t}\right), \qquad (11.36)$$

where M is the mass of solute initially in the sample zone at $t = 0$, and S the cross-sectional area of the zone. Equation (11.36) suggests that as the electrophoretic migration process proceeds, the initially thin zone becomes a Gaussian peak, having a standard deviation, σ, of $(2D_i t)^{1/2}$ and a width of $4(2D_i t)^{1/2}$, assuming that $W_i = 4\sigma_i$. Figure 11.6 is a plot of eqn (11.36) for two solutes having the same diffusion coefficient but differing

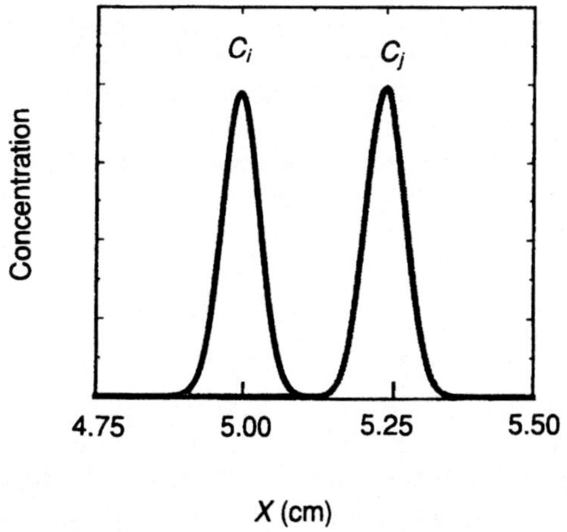

Fig. 11.6 Plot of eqn (11.36) for two species, i and j, where $\mu_i = 1 \times 10^{-8}$ m^2 V^{-1} s^{-1} $\mu_j = 1.05 \times 10^{-8}$m^{2}V $-$ 1s^{-1}, $D_i = D_j = 1 \times 10^{-11}$m^{2}s^{-1}, $E = 1000$ Vm^{-1}, $t = 5000$ s.

in their electrophoretic mobility. Furthermore, eqn (11.36) shows that the average location of peak i at a given time, $X_i(t)$, is $\mu_i Et$, as we would expect. Thus, from eqn (11.36) we have an expression for the width of the sample zone as a function of time,

$$W_i(t) = 4\sqrt{2D_i t}. \tag{11.37}$$

We can now obtain an analytical expression for the resolution expected between two solutes as a function of mobility differences, diffusion coefficients, field strength, and time. The resulting expression is

$$R = \frac{\Delta\mu_{ij} E\sqrt{t}}{2\sqrt{2}(\sqrt{D_i} + \sqrt{D_j})}, \tag{11.38}$$

where $\Delta\mu_{ij}$ is the difference between the mobilities of species i and j.

In the preceding discussion, the only dispersion mechanism that was considered was simple diffusion. In the next section we will examine the effects of thermally induced convection.

11.4 Heating effects

A major limitation on both the speed and feasible scale of electrophoretic separations is the ability of the medium to dissipate the Joule heat which is generated as a result of the electric current. This Joule heating and the resulting temperature gradient can negatively impact the separation process in two ways. First, if the temperature difference is large enough, density gradients in the separation medium can be induced which can cause natural convection. Any such convection would serve to remix separated sample zones. Secondly, even if the temperature gradients are not large enough to cause natural convection, because μ is a strong function of temperature (approximately 2% K^{-1} in aqueous solutions), separation performance can be compromised by introducing a spatial dependence on μ. This spatial dependence of mobility can cause a deformation in the migrating zones, leading to a poorer separation.

A parameter which can be used to correlate the effects of natural convection is the dimensionless Rayleigh number, Ra, defined as

$$Ra = \frac{r^4 g}{\eta\alpha}\left(\frac{\Delta\rho}{\Delta r}\right), \tag{11.39}$$

where r is the radius of the electrophoresis chamber (in this case a tube), g is the acceleration of gravity, η is the viscosity of the medium, α is the thermal diffusivity of the medium, and $\Delta\rho/\Delta r$ is the density change per unit length caused by heating. If $Ra \ll 1$, then the convective flow will be negligibly small, and if $Ra > 1$, natural convection will be significant.

Three main approaches have been used to reduce the value of Ra in electrophoretic processes. The first approach is to perform the electrophoresis in a rigid gel rather than in free solution. This makes the medium viscosity very high, thus reducing the value of Ra. Typical gel matrices include agarose, a glucose polymer, and polyacrylamide. It was this anti-convective property of gels rather than their sieving properties which first prompted their use in electrophoresis. Although gel electrophoresis eliminates the problem of convection, the radial temperature gradient remains. Thus, only low electrical fields are used (typically 1–10 V cm^{-1}). Gel electrophoresis is by far the most common solution to the problem of natural convection. A drawback of using gels is that it makes the scale-up of such process very difficult and it requires that a gel matrix be made before every electrophoresis experiment.

A second approach which has been attempted is to perform the separations in outer space. By performing the separation in a micro-gravity environment, one reduces the value of g in eqn (11.39). This approach is thought to be especially promising for large-scale preparative separations because it eliminates the need for a gel. However, for obvious practical and economic reasons this approach is not in common use.

A relatively new solution to the problem of Joule heating is to perform the electrophoresis in micro-capillary tubes (i.d. ~25–100 μm). This technique is known as capillary electrophoresis (Grossman and Colburn 1992). By operating in a micro-capillary, one makes r small in eqn (11.39), thus reducing the value of Ra. As seen from eqn (11.39), this is a particularly powerful approach because of the fourth-power dependence of Ra on r. With capillary electrophoresis one is able to use electric fields 1 to 2 orders of magnitude higher than those found in traditional electrophoresis, without the need for a stabilizing gel. However, because of the very small sample volumes involved (typically of the order of nanoliters), capillary electrophoresis is practical only as an analytical technique.

To quantitatively determine the temperature gradients caused by Joule heating, one must solve the energy balance equation. In cylindrical coordinates, this is

$$\frac{1}{r}\frac{d}{dr}(rq_r) = Se, \tag{11.40}$$

where q_r is the energy flux at the radial position r and Se is the rate of power generation within the conducting fluid per unit volume. Se is given by

$$Se = \frac{I^2}{k_e}, \tag{11.41}$$

where I is the current density, k_e is the electrical conductivity of the buffer, and q_r is related to T using Fourier's law of heat conduction. This problem is the same as that of an electrically conductive wire (Bird *et al.* 1960). The resulting relationship describing the temperature difference across the separation chamber is

$$T_0 - T_1 = \frac{Se\,R_1^2}{4k_b}, \tag{11.42}$$

where T_0 is the temperature at the center of the tube, T_1 is the temperature at the inside wall, R_1 is the internal radius of the tube, and k_b is the thermal conductivity of the medium. Equation (11.42) assumes that the value of k_b is constant across the tube diameter. Note that eqn (11.42) gives a value for the temperature difference across the chamber, not the total temperature rise. In order to calculate the total temperature rise, one must solve eqn (11.40) inside the tube wall and in the surrounding air. The result is

$$T_0 - T_a = \frac{Se\,R_1^2}{2}\left[\frac{1}{2k_b} + \frac{1}{k_g}\ln\left(\frac{R_2}{R_1}\right) + \frac{1}{R_3 h}\right], \tag{11.43}$$

where k_g is the thermal conductivity of the tube material, R_2 is the outside radius of the tube, and h is the heat transfer coefficient between the outer surface of the tube and the surrounding air. Note that because the value of k_e in eqn (11.41) is a function of temperature, in order to use eqn (11.43) one must perform iterative calculations of $T_0 - T_a$.

Problems

P11.1 Derive eqn (11.8).

P11.2 Assume a cylindrical electrophoresis cell having an inside diameter of 8 mm and a length of 20 cm. Furthermore, assume a buffer which has an electrical conductivity of $2.57 \times 10^{-3}\,\Omega^{-1}\text{cm}^{-1}$ and a thermal conductivity of 5.94×10^{-3} W cm^{-1}K^{-1}.

(a) If one applies a voltage 100 V across this cell, what would be the temperature difference between the center of the chamber and the inside wall (assume a constant value of k_e)?

(b) Calculate the Rayleigh number for this situation at 25°C (assume α and η are constant with respect to temperature). Would you expect any natural convection to occur?

(c) In early electrophoresis experiments, people performed their separations at 4°C. Can you think of a reason why? Would this approach help in the case of the above example?

(d) What diameter tube would one need to use in order to eliminate natural convection (i.e. $Ra = 1.0$) in the above example? How does this compare to the 50 μm diameter tubes typically used in capillary electrophoresis?

P11.3 In this exercise we will see if it is possible to separate two single-stranded oligonucleotides which differ in size by only a single base. This is the separation which is at the heart of DNA sequencing.

Assume that the free-solution mobility of the fragments is 3.0×10^{-8} m^2 V^{-1} s^{-1} and that μ/μ_0 is 0.5 for a 50 base-pair fragment. Furthermore, assume that the diffusion coefficient of both the 50-mer and a 51-mer is 10^{-11} m^2 s^{-1}. Typically, such an experiment would use a 40 cm long gel and a voltage of 1000 V and the separation would be allowed to proceed for 2 hours. We will assume that the conformation of the oligonucleotides is that of a random coil with a segment length of 6 Å per base. Assume that the strand radius of the gel is negligible compared to the radius of the oligonucleotides. Would you expect to be able to separate the 50-base fragment from the 51-base fragment? (A resolution value of 1.0 or greater indicates complete separation.)

References

Atkins, P. W. (1994). *Physical chemistry*, (5th edn), p. 795. Oxford University Press.

Berg, H. C. (1983). *Random walks in biology*, Ch. 1. Princeton University Press.

Bird, R. B., Stewart, W. E., and Lightfoot, E. N. (1960). *Transport Phenomena*, Ch. 9. Wiley, New York.

Borst, M. and Miller, D. M. (1990). DNA isolation and Southern analysis: a clinician's view. *American Journal of the Medical Sciences*, **299**, 356–60.

Cantor, C. R. and Schimmel, P. R. (1980a). *Biophysical chemistry*, Ch. 19. Freeman, New York.

Cantor, C. R. and Schimmel, P. R. (1980b). *Biophysical chemistry*, Ch. 12. Freeman, New York.

De Gennes, P.-G. (1971). Reptation of a polymer chain in the presence of fixed obstacles. *Journal of Chemical Physics*, **55**, 572–8.

De Gennes, P.-G. (1979). *Scaling concepts in polymer physics*. Ch. 8. Cornell University Press, Ithaca, NY.

Denn, M. M. (1980). *Process fluid mechanics*, Ch. 3. Prentice-Hall, Englewood Cliffs, NJ.

Doi, M. and Edwards, S. F. (1978) Dynamics of concentrated polymer systems: Part 1, Brownian motion in the equilibrium state. *Journal of the Chemical Society Faraday Transactions II*, **74**, 1789–1818.

Ferguson, K. A. (1964). Starch-gel electrophoresis—application to the classification of pituitary proteins and polypeptides. *Metabolism*, **13**, 985–1002.

Grossman, P. D. and Colburn (ed.) (1992). *Capillary electrophoresis: theory and practice*. Academic Press, San Diego, CA.

Hames, B. D. and Rickwood, D. (ed.) (1984). *Gel electrophoresis of proteins*. IRL Press, Washington, DC.

Hervet, H. and Bean, C. P. (1987). Electrophoretic mobility of λ phage HINDIII and HAEIII DNA fragments in agarose cells: a detailed study. *Biopolymers*, **26**, 727–42.

Hiemenz, P. C. (1986). *Principles of colloid and surface chemistry*, Ch. 11. Dekker, New York.

Lerman, L. S. and Frisch, H. L. (1982). Why does the electrophoretic mobility of DNA in gels vary with the length of the molecule? *Biopolymers*, **21**, 995–7.

Littlewood, A. B. (1970). *Gas chromatography*, Ch. 5. Academic Press, New York.

Lumpkin, O. J., DeJardin, P., and Zimm, B. H. (1985). Theory of gel electrophoresis of DNA. *Biopolymers*, **24**, 1573–93.

Ogston, A. G. (1958). The spaces in a uniform random suspension of fibres. *Transactions of the Faraday Society*, **54**, 1754–7.

Rodriguez, F. (1982). *Principles of polymer systems*, Ch. 7. McGraw-Hill, New York.

Schwartz, D. C., Saffran, W., Welsh, J., Haas, R., Goldenberg, M., and Cantor, C. R. (1982). New techniques for purifying large DNAs and studying their properties and packaging. In *Cold Spring Harbor Symposia on Quantum Biology*, Vol. 47, pp. 189–96. Cold Spring Harbor, New York.

Trainor, G. L. (1990). DNA sequencing, automation, and the human genome. *Analytical Chemistry*, **62**, 418–26.

12

MEDICAL IMAGING

Thomas F. Budinger

Contents

12.1 Introduction

The four major methods of X-ray transmission, radionuclide emission, nuclear magnetic resonance, and ultrasound comprise contemporary medical imaging (Fig. 12.1). Other methods such as biomagnetic imaging, neutron and charged-particle transmission, electrical current source, electrical conductivity, and light photon absorption imaging are presently under research. This chapter explains the four major methods, including the techniques of reconstruction tomography employed by each method. In addition, the general techniques of image processing are presented.

Each of the four major methods depicts different categories of information reflective of anatomical or physiological processes (Table 12.1). For example, x-ray

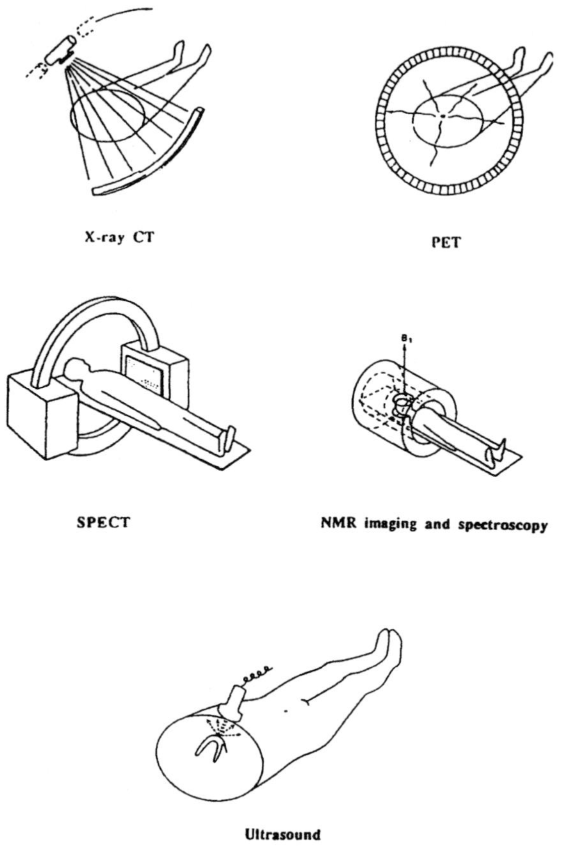

Fig. 12.1 The major modes of medical imaging: x-ray computed tomography (x-ray CT), positron emission tomography (PET), single-photon emission computed tomography (SPECT), nuclear magnetic resonance imaging (MRI) and spectroscopy, and ultrasound.

transmission reveals the attenuation of photons, which is related to tissue density. Thus x-ray imaging gives anatomical information. Emission imaging involves detection of radioactive isotope (radionuclide) distributions, which reflect physiological and biochemical activity (Figs 12.1 and 12.2). Nuclear magnetic resonance imaging MRI gives information on the concentration and relaxation of nuclei such as the hydrogen of water, carbon-13 isotope (1.3 per cent natural abundance) and phosphorus-31 (100 per cent natural abundance) in phosphorus-containing compounds. The contemporary magnetic resonance image reflects the concentration of water and the physical and chemical environment of the water molecules (Figs. 12.1 and 12.3) as well as information on flow comparable to x-ray angiography (that is, blood vessel flow images). Ultrasound imaging displays the echoes returning from reflecting surfaces of tissues with varying acoustic impedance. Further, ultrasound is used in determining blood flow by measuring the frequency shifts (Doppler shifts) in the returning waves.

Table 12.1 Comparison of medical imaging techniques

Method	Parameters measured	Medical applications
Transmission computed tomography	Density and average atomic number	Anatomy, mineral content, flow and permeability from movement of contrast material
Emission computed tomography (positron and single photon)	Concentrations of radionuclides	Metabolism, receptor site concentration, flow
Nuclear magnetic resonance (imaging and spectroscopy)	Concentrations, relaxation parameters T_1 and T_2, and frequency shifts due to chemical form	Anatomy, edema, flow, and chemical composition
Ultrasound	Acoustic impedance mismatches, sound velocity, attenuation, frequency shifts due to motion	Anatomy, tissue structural characteristics, flow

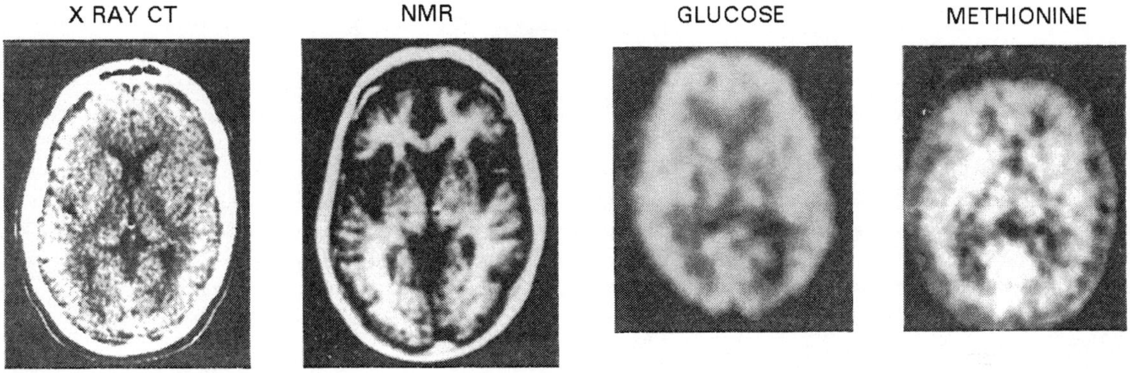

Fig. 12.2 The techniques of x-ray CT, MRI and emission tomography measure different physiological parameters. Brain glucose metabolism and amino acid metabolism are shown by emission tomography, whereas x-ray CT and MRI proton imaging depict anatomy.

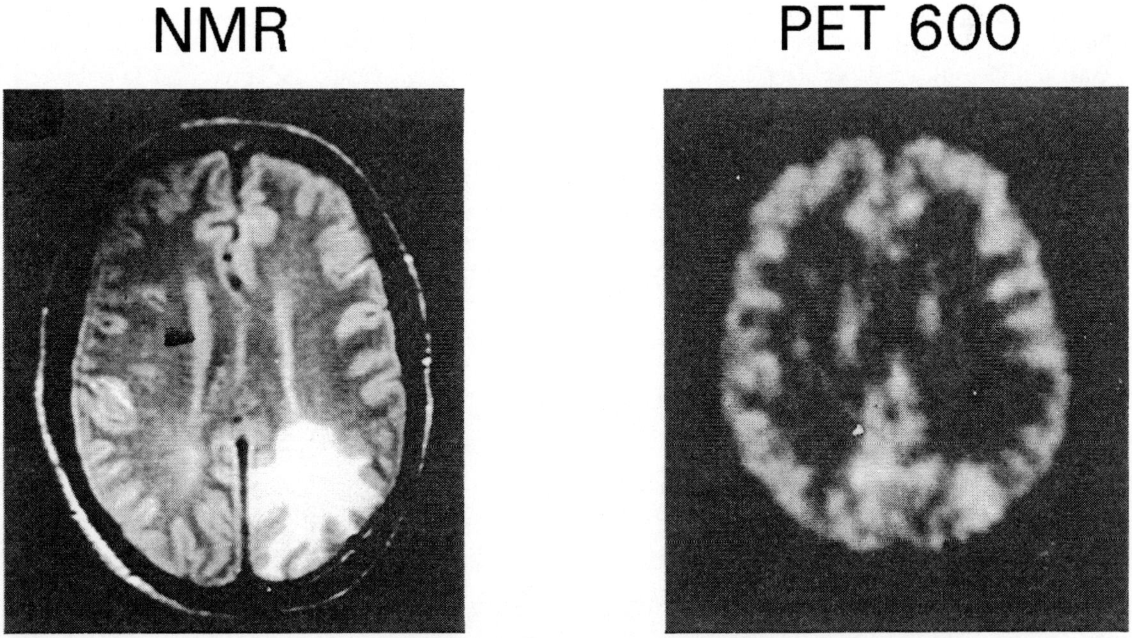

Fig. 12.3 Brain tumor information from MRI (a) shows edema surrounding the tissues being invaded by the tumor. PET (b) shows the increased glucose metabolism of the tumor tissue.

12.2 X-ray transmission imaging

Presently the most widely used imaging modality is the x-ray transmission intensity projection or radiograph (Fig. 12.4), most commonly known as 'the x-ray'. The difference between 'the x-ray', a simple two-dimensional projection image, and the x-ray computed tomograph, 'the CT', is that the latter is the result of mathematically reconstructing an image of a slice through the body from multiple x-ray transmission projection images taken at multiple angles (usually equal) around the body. The x-rays are generated by the interaction of accelerated electrons with a target material such as tungsten. The electrons are produced by a heated cathode and accelerated by applying a voltage of about 100 kV between the cathode and a tungsten anode. The K-shell x-rays from tungsten are about 70 keV, and these as well as other x-rays or photons are emitted from the x-ray tube placed one or more meters from the patient (Fig. 12.4). The x-radiograph is usually a film, not unlike a negative, which is darkened due to the interaction of the photons with the silver halide granules of the film. The image reflects the number or intensity of photons reaching the film but, as is the case for a conventional negative, the greater the intensity, the darker the radiograph.

The intensity of photons transmitted through the body is modulated by the processes of Compton scattering and photoelectric absorption in tissue, which are dependent on electron density and tissue elemental composition, respectively. These modulation processes are lumped into a simple attenuation coefficient. The intensity (or number) of photons arriving at a particular position, (x, y), on the x-ray film is given as

$$I(x, y) = I_o e^{-\mu z}, \qquad (12.1)$$

where $I(x, y)$ is the photon intensity at position (x, y), I_0 is the intensity from the x-ray tube, μ is the attenuation coefficient (units of reciprocal length), and z is the path length through the patient. This equation applies to a situation of constant attenuation along z. The attenuation coefficients of lung, water, tissue and bone differ. Thus the intensity arriving at the film is more generally

$$I(x, y) = I_o \exp\left(-\sum_i \mu(x, y, z_i)\Delta z_i\right), \qquad (12.2)$$

where we have divided the path z into intervals Δz_i. An example of the application of eqn (12.2) is instructive. Suppose we wish to calculate the contrast or intensity difference between a lung tumor region and the surrounding normal tissue as recorded on a conventional projection 'x-ray'. The photons pass through about 35 cm of the chest. The photons first encounter about

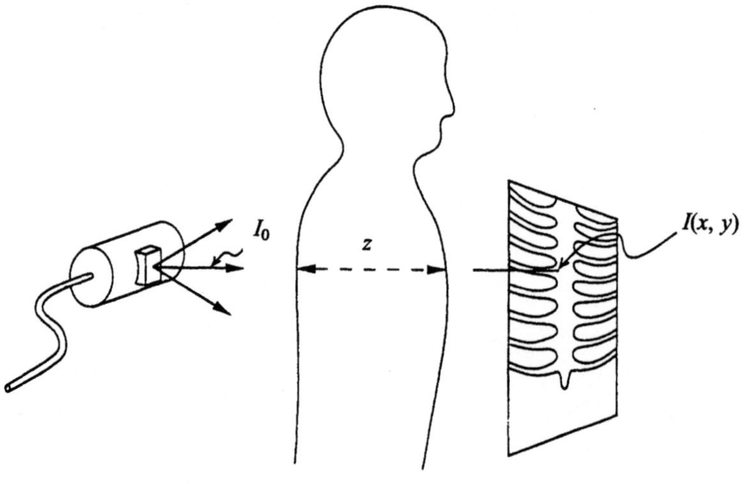

Fig. 12.4 The geometry of x-ray transmission imaging.

3 cm of chest wall tissue, then 29 cm of lung air and tissue, then 3 cm of chest wall tissue if they pass between the ribs. The total attenuation is the following:

$$\begin{aligned} I(x=3) &= I_o e^{-\mu_1 3}, \\ I(x=3+29) &= (I_o e^{-\mu_1 3}) e^{-\mu_2 29}, \\ I(x=3+29+3) &= I_o e^{-\mu_1 3} e^{-\mu_2 29} e^{-\mu_1 3} \\ &= I_o e^{-(\mu_1 6+\mu_2 29)}. \end{aligned} \tag{12.3}$$

The attenuation coefficients for chest wall tissue and lung are $\mu_1 = 0.14\ \text{cm}^{-1}$ and $\mu_2 = 0.05\ \text{cm}^{-1}$. Thus the fraction of incident intensity arriving at the x-ray film is

$$\frac{I}{I_o} = e^{-(0.14\times6+0.05\times29)} = 0.10. \tag{12.4}$$

Now suppose a 3 cm round tumor with attenuation coefficient of 0.14 is in the lung. The added attenuation, $e^{-(0.14\times3)}$, will result in a decrease in intensity by

$$\frac{I}{I_o} = e^{-(0.14\times6+0.05\times26+0.14\times3)} = 0.077. \tag{12.5}$$

This is a change of 23% in contrast. If the photons pass through two 1.5 cm thick ribs with a bone attenuation coefficient of $0.4\ \text{cm}^{-1}$, we observe

$$\frac{I}{I_o} = e^{-(0.14\times6+0.4\times3+0.0\times29)} = 0.046. \tag{12.6}$$

Recall the fact that as the number of photons decreases, the smaller the exposure of the x-ray film and the 'whiter' the image in that region. It is useful to examine the logarithm of the ratio of incoming intensity to the exit intensity:

$$P(x,y) = \ln\frac{I_o}{I(x,y)} = \sum_i \mu(x,y,z_i)\Delta z_i. \tag{12.7}$$

Note that the logarithm of the intensity ratio, which we designate a projection $P(x, y)$, is simply the line integral of attenuation coefficients if the Δz's approach zero:

$$P_{(x,y)} = \int_{\text{Source}}^{\text{Detector}} \mu(x,y,z)\, dz. \tag{12.8}$$

We call this summation a ray sum. The usefulness of this manipulation is in the fact that we can work with projections in a linear fashion to reconstruct the three-dimensional distribution of linear attenuation coefficients and thus create an x-ray computed tomograph using linear superposition principles discussed below.

X-ray imaging is mainly an anatomical procedure. The absorption of x-rays due to differences in elemental composition has an important effect for imaging bone or calcium deposits because this absorption process is proportional to the atomic number. Indeed the reason we can differentiate gray from white matter in the brain in CT images is that the H, C, O, P contents for gray and white matter and cerebral spinal fluid differ slightly. Very little contrast is observable between blood and muscle, as the density (about $1.05\ \text{g cm}^{-3}$) and the elemental compositions are very similar. To provide image contrast between the blood vasculature and surrounding tissue, a dense fluid with elements of high atomic number (e.g. iodine, barium) can be injected or swallowed during the x-ray exposures. The movement through the body vasculature of a 'contrast agent' such as an iodinated compound can be visualized by acquiring a sequence of x-rays. The iodine or barium agent absorbs photons more than blood and tissue because the density is higher and the elements iodine and barium have a high atomic number giving rise to more photoelectric absorption.

12.3 X-ray computed tomography (CT)

Computed tomography became generally available in the mid-1970s and is considered one of the major technological advances of medical science. The general concept of mapping a three-dimensional object from multiple views is to find consistency between actual points or landmarks on or in an object and the corresponding projections imaged from two or more angles. The embodiments of this concept is shown by the physical

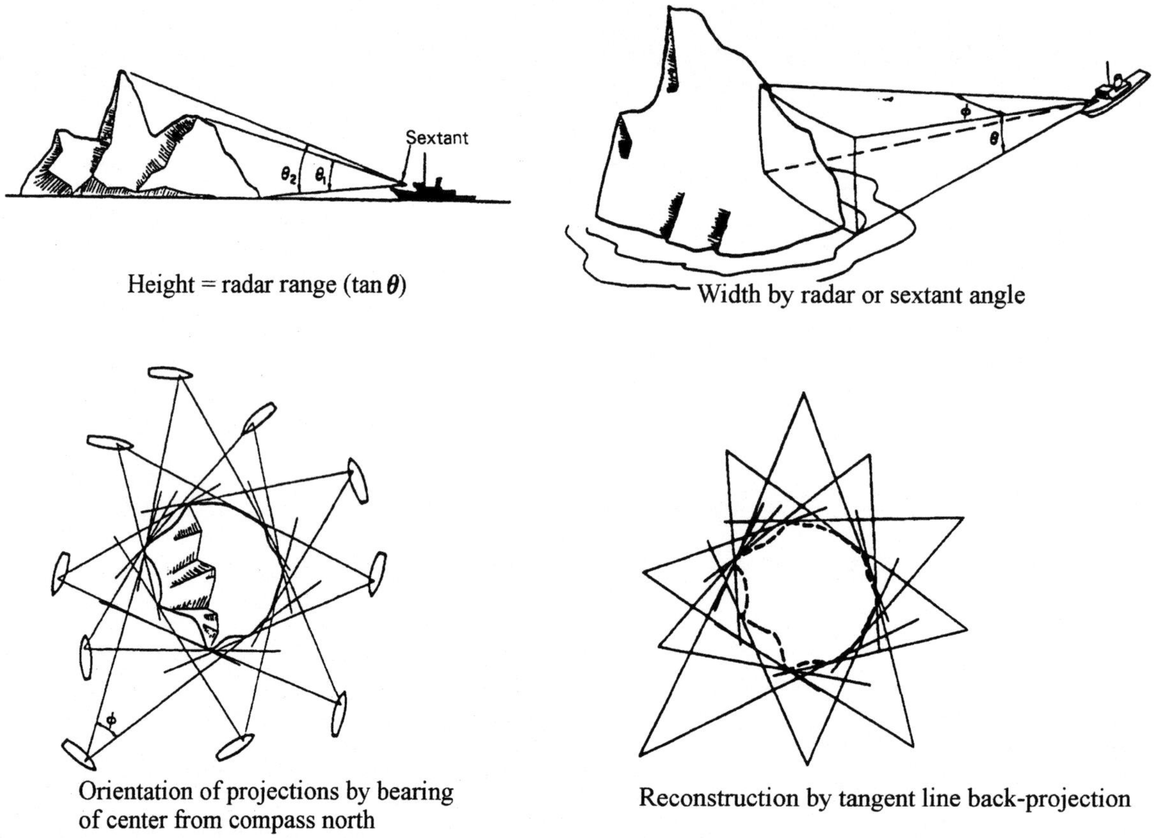

Sextant
Height = radar range (tan θ)
Width by radar or sextant angle
Orientation of projections by bearing of center from compass north
Reconstruction by tangent line back-projection

Fig. 12.5 General concept of mapping or imaging of two- or three-dimensional object from data taken at different angles around the object. The sextant is a hand-held device for measuring angles; the distance from the iceberg is measured by radar; the relative angle is measured by true or magnetic compass bearing between the observer (ship) and the iceberg. The reconstruction is performed on polar coordinate paper.

mapping of an iceberg shape using observations of the tangent angles corresponding to similar edges (Fig. 12.5). Using these angles, radar distances, and relative compass directions at the time of observation, the three-dimensional shape of an object can be estimated, assuming the object is generally convex. This technique is called tangent line reconstruction and was done without computers in 1956.

A second graphic example illustrating the concept of 3-D reconstructure is the back-projection of multiple x-ray projections (Fig. 12.6). We just simplify the 3-D problem by doing 2-D reconstructions from 1-D projections. We take a slice or 1-dimensional projection corresponding to the same axial position from each image. Each of 24 projections in this case can be arranged around a vertical object volume and the intensity at each point projected back into the vertical object. The result of this back-projection and superposition operation is a 2-D transaxial section back-projection. The process leaves us with a blurred or smoothed version of the actual section; and we will describe below how this blurring is removed by high-pass filtering operations before or after back projection. The x-ray CT machines use an x-ray tube, but instead of being detected on film, the intensity is detected from an array of solid-state devices or scintillation crystal and photoelectron multiplier tube systems discussed below in Section 12.4.

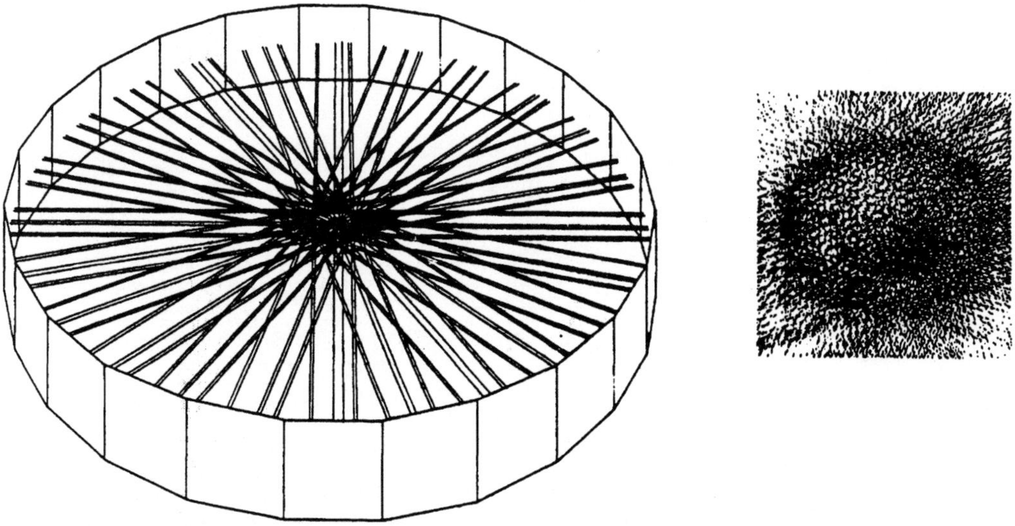

Fig. 12.6 Schematic of performing a back-projection from multiple projections.

12.4 Nuclear medicine projection imaging

In emission imaging we detect the internally emitted photons which arise from the physical decay of radioactive isotopes commonly called radionuclides. The radionuclides are injected into patients as ions or as atoms incorporated into naturally occurring compounds or pharmaceutical molecules, hence the term radiopharmaceuticals. Usually the injected elements or compounds are of specific biological importance; however, some radionuclides have no important biological function but when chemically combined with the molecules of biological importance these 'tags' allow evaluation of flow and physiological information (e.g. brain blood flow, kidney function, bone tissue information).

The radionuclides are divided into two categories: those which emit single photons, and those which decay by positron emission and consequently, due to the annihilation of a positron with an electron, emit two photons at 180° to one another. The physical decay processes which lead to single-photon emission are beta emission, electron caputre, and isometric transition. The energies of the photons of interest in emission CT range from 80 to 511 keV (the x-ray photon energies for x-ray transmission imaging are 70–120 keV). The emitted photons are detected after passing through the patient's body and a lens system which accepts only those photons whose direction will give an image. This lens system is known as a collimator (Fig. 12.7). A collimator consists of a metal plate with channels through it. The plate is usually made of lead and the collimator holes are usually parallel. A single hole could be used to give the equivalent of a pinhole camera, but the sensitivity of such a system is low. The photons then interact with a scintillator crystal by two means: (1) photoelectric absorption, where the energy of a photon is tranferred to a recoil electron; or (2) by Compton scattering, where a fraction of the energy is given to a recoil electron and the photon is scattered at a random angle. Light photons in the visible and near ultraviolet are given off as a result of these interactions. These light photons produce photoelectrons in a multiplier tube (PMT) (Fig. 12.8). The result is an electric current pulse whose magnitude is proportional to the energy of the incoming emission photon. In emission imaging, each photon is counted.

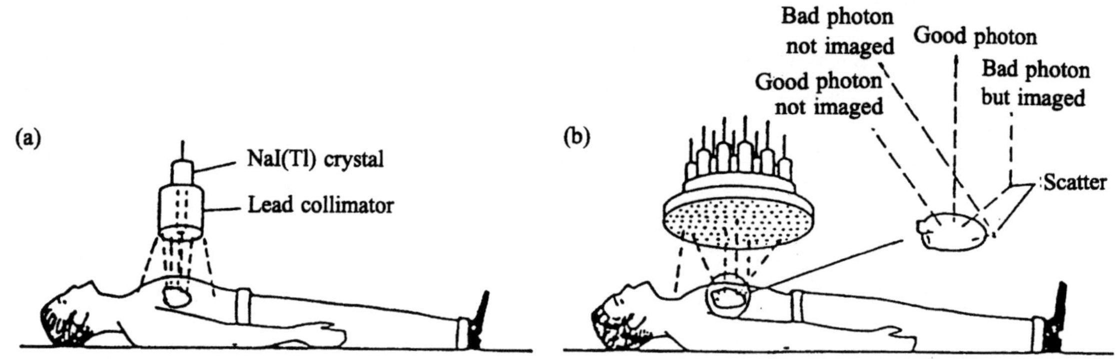

Fig. 12.7 Method of detecting gamma photon emissions from a tracer in the body: (a) single-channel, (b) multi-channel collimator.

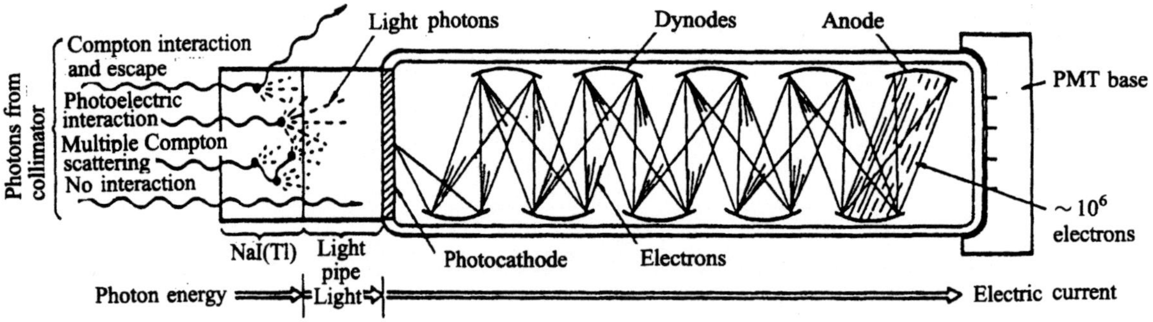

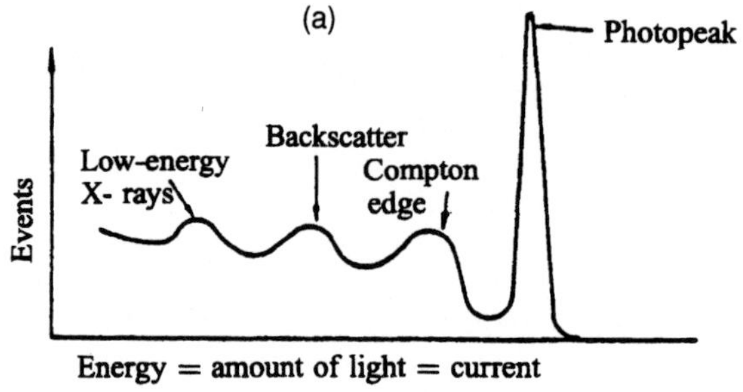

Fig. 12.8 (a) Schematic of the detection of gamma photons using a NaI (Tl) scintillation crystal and a photoelectron multiplier tube. (b) Pulse height spectrum.

Scattered photons enter the collimator holes and some scattering occurs in the collimator. These scattered photons appear at an energy less than the photopeak energy (Fig. 12.8). The current pulses from these photons are rejected by a pulse-height analyzer, as they would blur the image.

An image is obtained by using an array of crystals and PMTs or more practically by a large scintillation crystal backed by a number of PMTs as is shown in Fig. 12.9. The center of gravity of the photopeak interaction is determined electronically by the amount of light which arrives at adjacent phototubes. Thus, with 30–80 phototubes, it is possible to acquire an image corresponding to an array or grid containing as many as 1024 (64 × 64) or even more resolution elements, usually called pixels (picture elements).

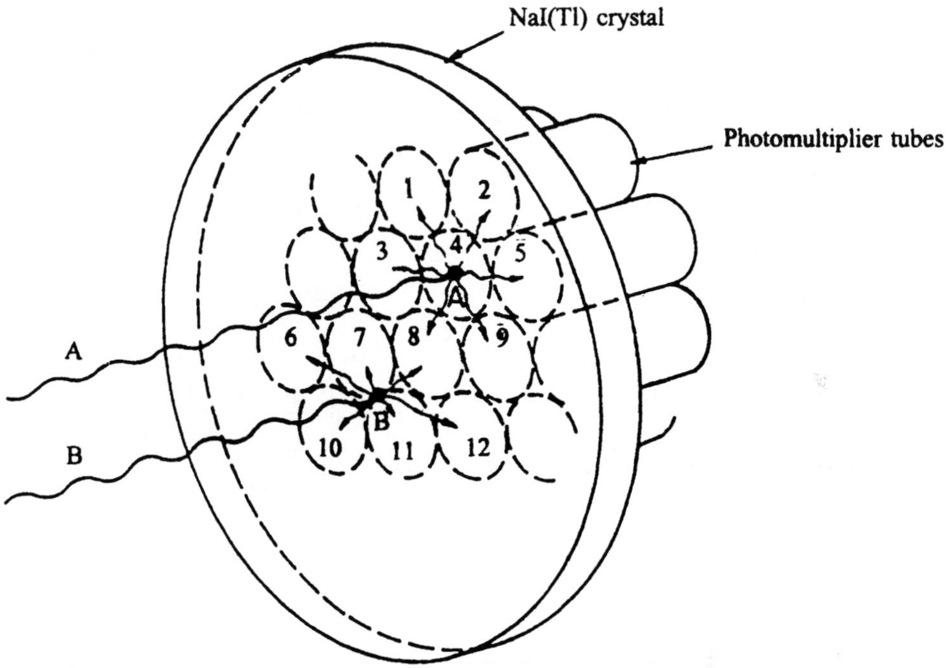

Fig. 12.9 The gamma camera (Anger camera) detects the center of gravity of light. In event A, tube 4 has the greatest response, and tubes 1, 2, 3, 5, 8, and 9 equal responses. In event B, tubes 7 and 11 have the greatest response, while tube 6 = tube 12, and tube 8 = tube 10 in responses.

12.5 Emission computed tomography (SPECT and PET)

Emission computed tomography (ECT) involves detection of photons from radionuclide distributions in the body. There are two general classes of emission tomography. Single-photon computed tomography (SPECT) involves rotation of a photon detector array around the body to acquire data from multiple angles (Fig. 12.10).

Positron emission tomography (PET) utilizes the fact that the annihilation photons of 511 keV arising from the positron–electron interaction are emitted 180° one from the other, thus the electronic detection of scintillations in two detectors of an array gives the line through the object along which the radionuclide must lie (Fig. 12.11). By observing the coincident pairs for thousands of events, we can locate the points of radionuclide concentration. If there are multiple sources, then the operation of back-projecting the multiple lines of coincidence will give a result analogous to that of Fig. 12.6. The general approach is similar to that of x-ray CT; however there are important physical and biological differences between

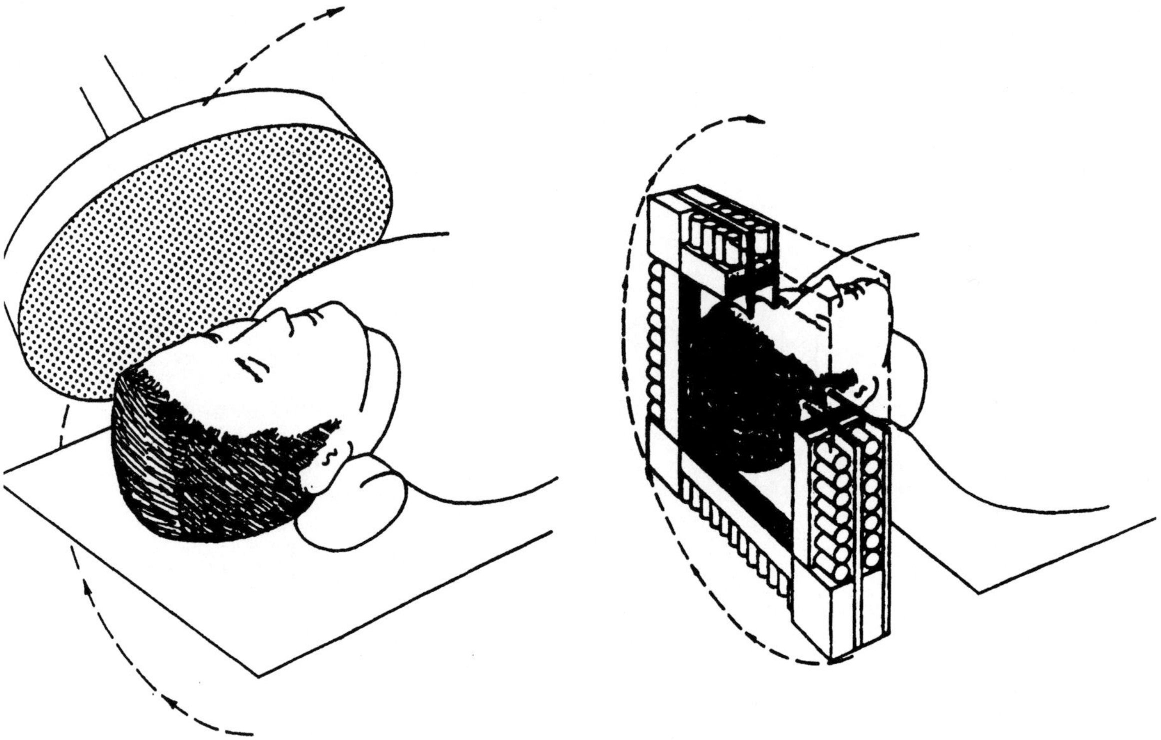

Fig. 12.10 Two methods of collecting tomographic information from single-photon-emitting isotopes.

x-ray CT and ECT (Fig. 12.12). The basic physical difference is that in ECT the information sought is the source and intensity of the gamma radiation emitted by the isotope; in x-ray CT the distribution of the tissue densities or attenuation coefficients is sought. A consequence of this difference is the need in emission tomography to compensate for the effect of attenuation (Fig. 12.13). This is achieved by the incorporation of attenuation coefficients in the image reconstruction methods.

Another important difference is that of available statistics. ECT involves far less (approximately 10 000 times less) available data per transverse image than x-ray CT. The expected error in much of SPECT and PET is 15% or greater. Thus, there are serious statistical problems embedded in the reconstruction strategy. This is one of the reasons why the iterative class of reconstruction (see Section 12.12.4) is used in order to incorporate weighting based on the expected errors. ECT statistics are low because each photon must be detected and analyzed, whereas in x-ray CT the current from a phototube due to many photons is measured. The count-rate capability of detector crystals and associated electronic circuits dictates a practical minimum event processing time of 1 μs. Secondly, radiation dose and practical patient imaging duration limit the number of events which can be collected. Thirdly, imaging time must be short because either the isotope decays by one half-life or more during the 10-20 min study, or the radiopharmaceutical distribution changes within a few minutes. In spite of these limitations, we pursue emission tomography because it is a unique approach to the determination of flow and biochemical kinetics in the body.

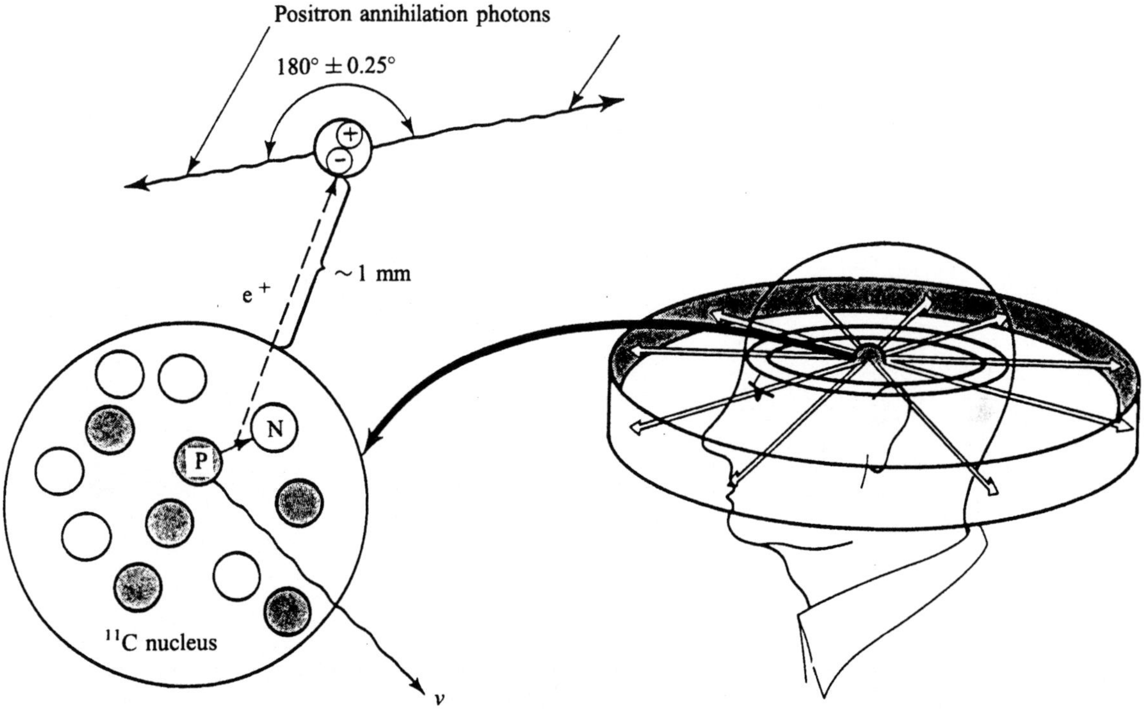

Fig. 12.11 The concept of positron emission tomography.

12.6 Nuclear magnetic resonance imaging (MRI)

Nuclear magnetic resonance (NMR) when used in imaging is denoted magnetic resonance imaging (MRI) and when used for chemical composition is called magnetic resonance spectroscopy (MRS). Most elements have at least one reasonably abundant isotope whose nucleus is magnetic. The proton spin of the hydrogen nucleus imparts the magnetic properties to the water protons of the body. The magnetic nuclei or nuclear spins of high abundance in biological material are those of ^{1}H, ^{13}C, ^{23}Na, ^{31}P, and ^{39}K. The hydrogen nucleus (proton) is abundant in the body because of the high water content of non-bony tissues. When the body is immersed in a static magnetic field, the spins of slightly more protons become aligned with the magnetic field than against it. At 0.25 T (2500 G) and 25°C the difference between these aligned populations of about one proton in a million produces a net magnetization. A rapidly alternating magnetic field at an appropriate resonant frequency in the radiofrequency (RF) range, applied by a coil near the subject or specimen in the static magnetic field, changes the orientation of the nuclear spins relative to the direction of the static magnetic field (Fig. 12.14). These changes are accompanied by the absorption of energy (from the alternating magnetic field) by nuclei which undergo the transition from a lower energy state to a higher one. When the alternating field is turned off, the nuclei return to the equilibrium state with the emission of energy at the same frequency as that of the stimulating

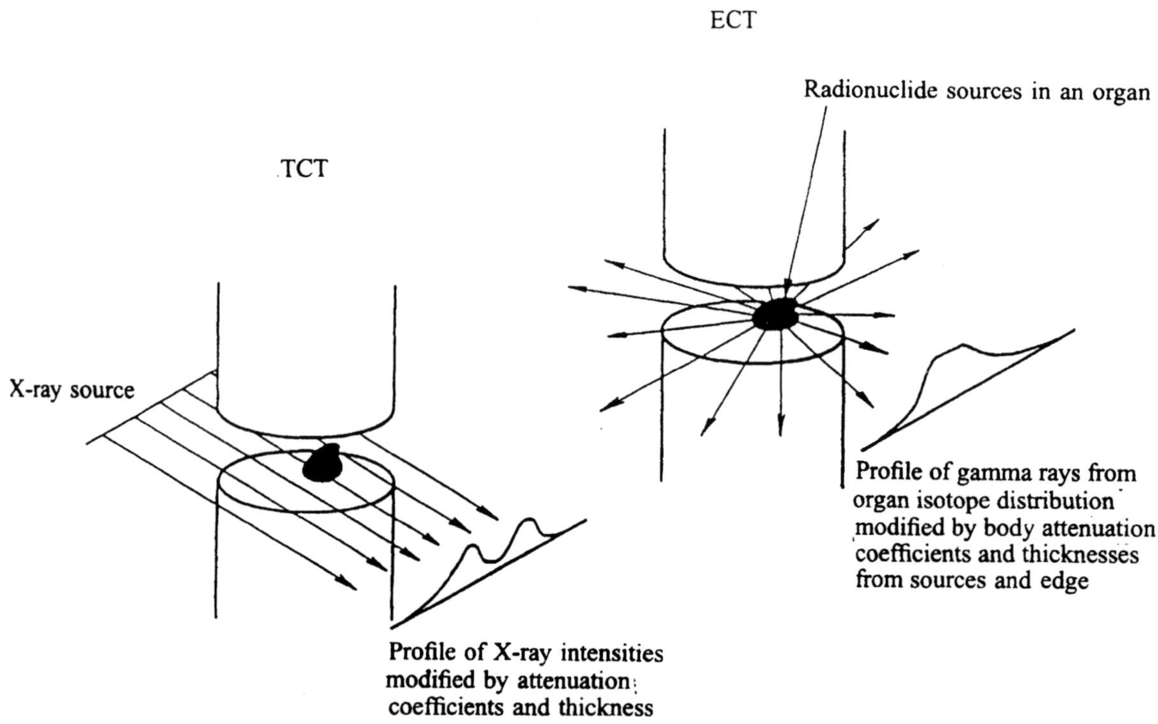

Fig. 12.12 Differences between transmission computed tomography and emission tomography.

alternating field (RF). That frequency is the resonance or Larmor frequency, given by

$$\omega = \gamma B \quad \text{or} \quad \upsilon = \frac{\gamma}{2\pi} B, \qquad (12.9)$$

where γ is the characteristic gyromagnetic ratio of the nucleus and B is the static magnetic field strength. The nuclei of different elements, and even of different isotopes of the same element, have very different resonance frequencies. For a field of 0.1 T (1000 G), the resonance frequency of protons is 4.2 MHz and that of phosphorus is 1.7 MHz. Thus, the magnetic nuclei in the body, when placed in a static magnetic field, become tuned receivers and transmitters of RF energy.

12.6.1 Relaxation parameters

The NMR signal used in MRI and MRS is an oscillating voltage whose amplitude and electronic rate of decay depend on the physicochemical properties of the tissue and the magnetic environment of the nuclei. This variation is a consequence of the interaction of the nuclear spins with a fluctuating magnetic field produced by nearby magnetic moments, including other similar and dissimilar nuclei as well as paramagnetic ions. The imposed RF energy is designed to perturb the thermal equilibrium of the magnetized nuclear spins, and the time dependence of the received signal is determined by the manner in which this system of spins returns to its equilibrium magnetization. The return is characterized by two parameters: T_1, the longitudinal or spin-lattice relaxation time, describes the behavior of the component of the magnetization vector parallel to the applied static field B_o; and T_2, the transverse or spin–spin relaxation time, describes the behavior of the component of the magnetization vector transverse to B_o (Fig. 12.15). Each magnetization component can be envisioned as relaxing to its equilibrium value; the first is precisely the equilibrium magnetization and the second is zero at the equilibrium. T_2 is a measure of the time for the components of

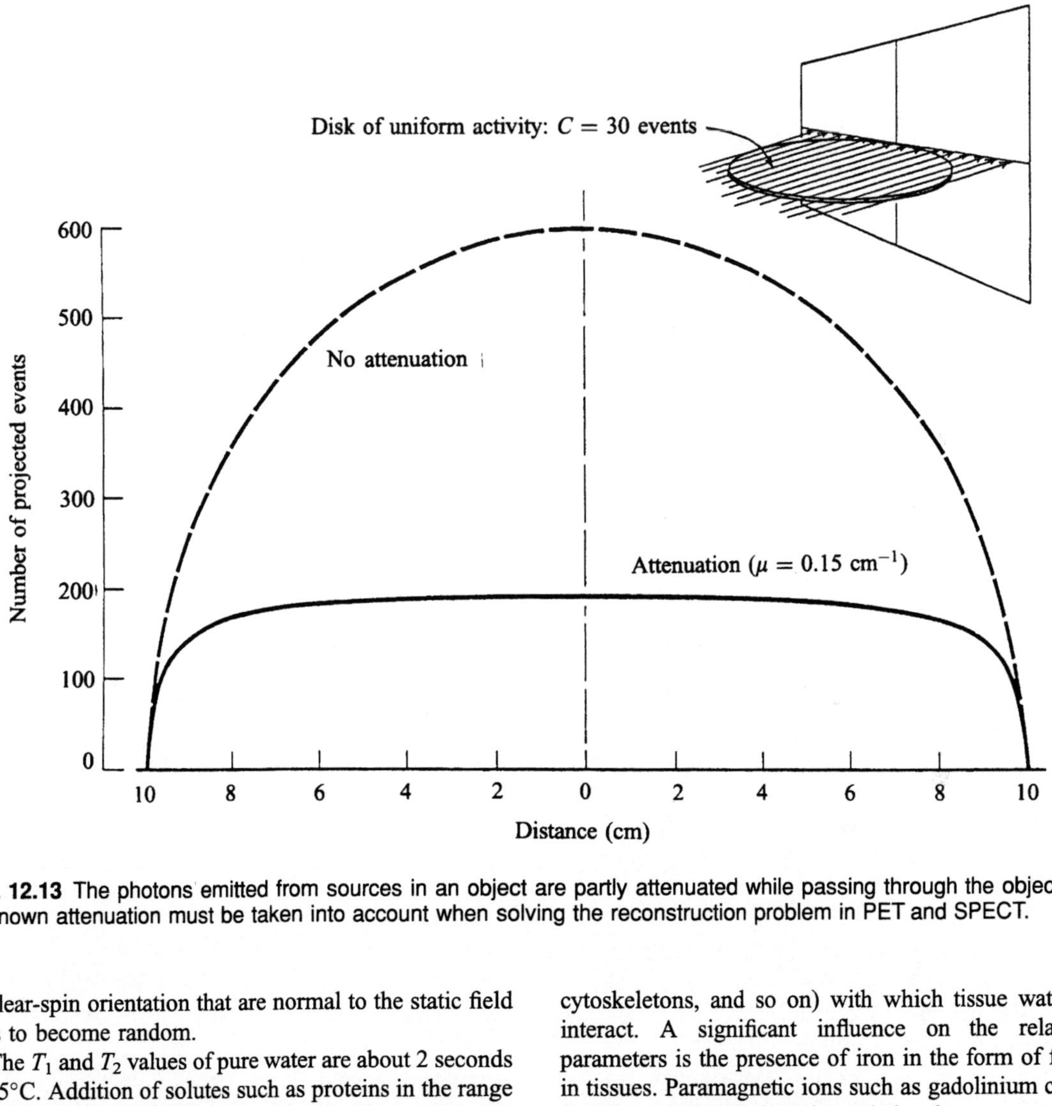

Fig. 12.13 The photons emitted from sources in an object are partly attenuated while passing through the object. This unknown attenuation must be taken into account when solving the reconstruction problem in PET and SPECT.

nuclear-spin orientation that are normal to the static field axis to become random.

The T_1 and T_2 values of pure water are about 2 seconds at 25°C. Addition of solutes such as proteins in the range of physiological concentrations shortens T_1 by a factor of nearly 5 and T_2 by a factor of about 100. The times increase with the field strength. To the extent that T_1 and T_2 are tissue-specific, these differences can be exploited to delinate different tissues in MRI. The reasons for the differences in relaxation behavior of different tissues are not yet known; but it is known that the variation in water content, though it can affect relaxation, is not sufficient to explain all the differences observed. Among the fundamental factors influencing the differences will be the distribution of protein size in any tissue as well as the presence of fixed surfaces and interfaces (membranes, cytoskeletons, and so on) with which tissue water can interact. A significant influence on the relaxation parameters is the presence of iron in the form of ferritin in tissues. Paramagnetic ions such as gadolinium cause a decrease in T_1 and these ions chelated to an appropriate carrier are infused into the body to obtain MRI contrast-enhanced images, useful in tumor detection for example.

Relaxation times can be measured at each point of an image. Because differences in tissue relaxation times determine image contrast, data on the relaxation behavior of different tissues as a function of static field strength B_o are important in the selection of the optimal field strength and RF pulse for acquisition of MR images (for instance, by the spin echo or inversion recovery methods). In the past few years, new MR imaging sequences that sensitize signal intensity to Brownian motion of water molecules

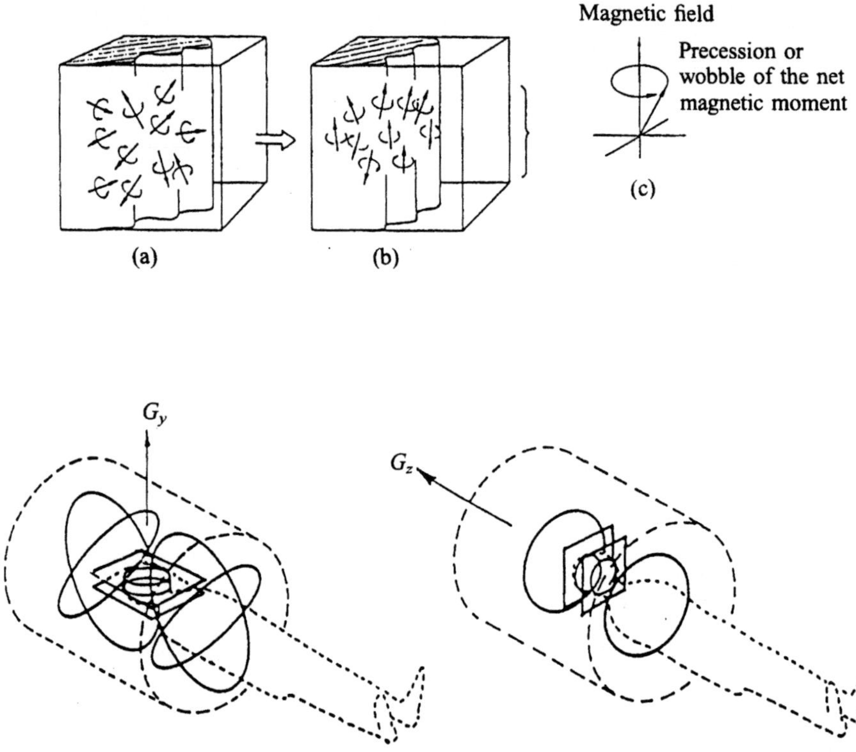

Fig. 12.14 The concept of NMR. (a), (b) Effect of static magnetic field on tissue nuclei: (a) shows the natural state, with randomly spinning nuclear moments; in (b), the two-spin state after application of a magnetic field, hydrogen nuclei (mostly in H_2O are organized into lower and upper energy states. (c) Diagram of net magnetic moment (difference between lower and upper energy states). Lower diagrams show possible configurations for applying field gradients used for imaging.

have become available and thereby have permitted the *in vivo* measurements of the apparent diffusion coefficient (ADC) of tissue water. Diffusion-weighted MR studies of experimental cerebral ischemia models have demonstrated an elevated signal intensity (reduced ADC) in early stroke in regions of absent blood supply.

For spin-echo imaging, a commonly used imaging technique, the signal amplitude is given by

$$S \approx Nf(v)(e^{-TE/T_2})(1 - e^{-TR/T_1}), \tag{12.10}$$

where N is the concentration of nuclei (local spin density), $f(v)$ is the signal modulation due to moving nuclei (blood flow), TE is the elapsed time between the 90° RF pulse and the reception of a spin-echo signal, and TR is the repetition time between successive spin-echo sequences. A decrease in T_1 or an increase in T_2 results in an increase in the NMR signal. Fat has a shorter T_1 and longer T_2 than other tissues; thus the NMR signals from fat produce the strongest signals in proton imaging in accord with eqn (12.10).

Signal intensities from normal tissues will change significantly as the TE and TR timing parameters of the spin echo sequences are changed, but in general the ranking of intensities of signals from normal tissues is that shown in Table 12.2. Tumors and abscesses with T_1s and T_2s longer than those of the normal tissues can be demonstrated with good contrast by choosing a pulse sequence which takes the best advantages of the differences between the relaxation parameters of the diseased tissue and those of the normal tissue (cf. Fig. 12.3). A long TE is imposed on the image by extending the time between the initial RF pulse and the second pulse known as the 180° pulse. T_2-weighted images are those which emphasize T_2 differences by long echo pulse

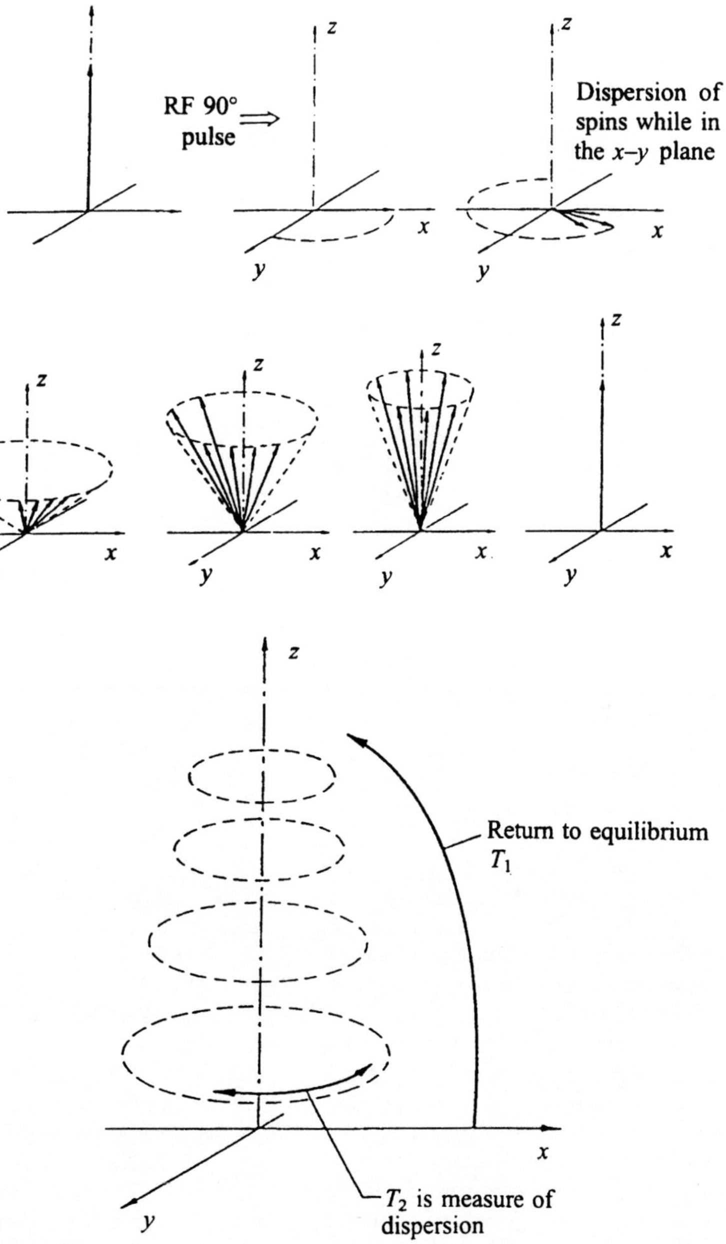

Fig. 12.15 An explanation of the relaxation times associated with NMR. Image contrast is provided by differences in T_1 between tissues if short *TR*/*TE* pulse segments are used, and in T_2 for long *TR*/*TE* pulses.

Table 12.2 Typical NMR signal intensities in T_2-weighted imaging, decreasing top to bottom*

Fat
Marrow and cancellous bone
Brain and spinal cord
Liver, spleen, pancreas
Muscle, kidney
Ligaments and tendons
Blood vessels with rapid flow
Compact bone
Air

*Pulse sequence and state of hydration can change ranking.

intervals and long *TR*. To emphasize T_1 differences such as those differences which distinguish gray from white matter, short *TR* or repetition sequences are used along with short echo times. Multiple-echo sequences are also used, with late echoes used to observe signals with long T_2 at high intensity relative to those with shorter T_2 values.

12.6.2 Imaging strategies

The frequency of the NMR signal is proportional to the magnetic field strength in accord with eqn (12.9). Thus, if the static field varies across an object, the signal frequency will depend on the position of the transmitting nuclei in the object. The strength of the signal at each RF frequency can be interpreted as the nuclear magnetization in a plane within the object where the static magnetic field strength corresponds to a Larmor frequency equal to that RF frequency (Fig. 12.16).

Combinations of gradients in different directions can be used to select points, lines, or planes within the body. Selection of planes or 'slices' is most commonly used in imaging strategies. These can be oriented in any direction, in contrast to the usual transaxial slice orientation of x-ray CT. By application of a gradient during the application of a narrow RF band, the protons are excited in a plane whose position and thickness depend on the bandwidth of the RF pulse.

Once the plane has been selected, for example by a z-gradient, then x- and y-gradients are employed to determine the location within the selected x–y plane of the spin density and relaxation parameters. To obtain projections from different directions, the NMR signal is encoded by using static magnetic field gradients applied in different directions, so that the NMR spectrum corresponds directly to spatial projections. If broadband pulses are used to excite the system in the presence of a linear gradient, the spatial Fourier transform of the received signal is equal to the projection of nuclear spin densities of the object planes orthogonal to the gradient direction. Two- and three-dimensional images are created using algorithms somewhat similar to those used for x-ray CT. Since all of the nuclei in the tissue volume (or plane, in the two-dimensional case) being imaged are excited during each pulse, this technique is capable of achieving the highest signal-to-noise ratio possible for a given resolution per unit acquisition time. Because time must be allowed for the spins to relax after each pulse, there are limits to the minimum performance time required to produce an image.

Another class of methods—two-dimensional Fourier transform, spin warp, echo planar—uses gradients of varying magnitude in time and space to impart spatial information by effectively shifting the phases of spins through momentary frequency changes. Thus, spatial information is encoded and this is extracted by Fourier transformations of the detected signals. The signal-to-noise ratio per unit acquisition time and the performance time are close to those of projection imaging. Extensions of this method for spectroscopic imaging are straightforward, though the computational and data acquisition problems are large.

Echo planar imaging offers faster imaging acquisition than other techniques and, after first being used to produce images of the beating heart, is now being employed to give information on organ perfusion and to reduce motion artifacts. This technique employs multiple spin echoes produced in oscillating gradients superimposed on the static magnetic field by a gradient coil during the single transient signal following one RF pulse.

A comprehensive conceptualization of the MR imaging process has been used to describe and analyze both new and old imaging strategies. The method is known as the k-trajectory formulation. The time-varying field gradients impart spatially dependent phase information to the spin density distribution of the object. The observed signal is the superposition of the oscillations corresponding to the concentration of spins at each spatial position but with a phase shift imposed by the gradients used to probe the image space. The strategy or path of sampling in the spatial-frequency domain is known as the k-trajectory. The observed signal in this type of imaging experiment is composed of a time

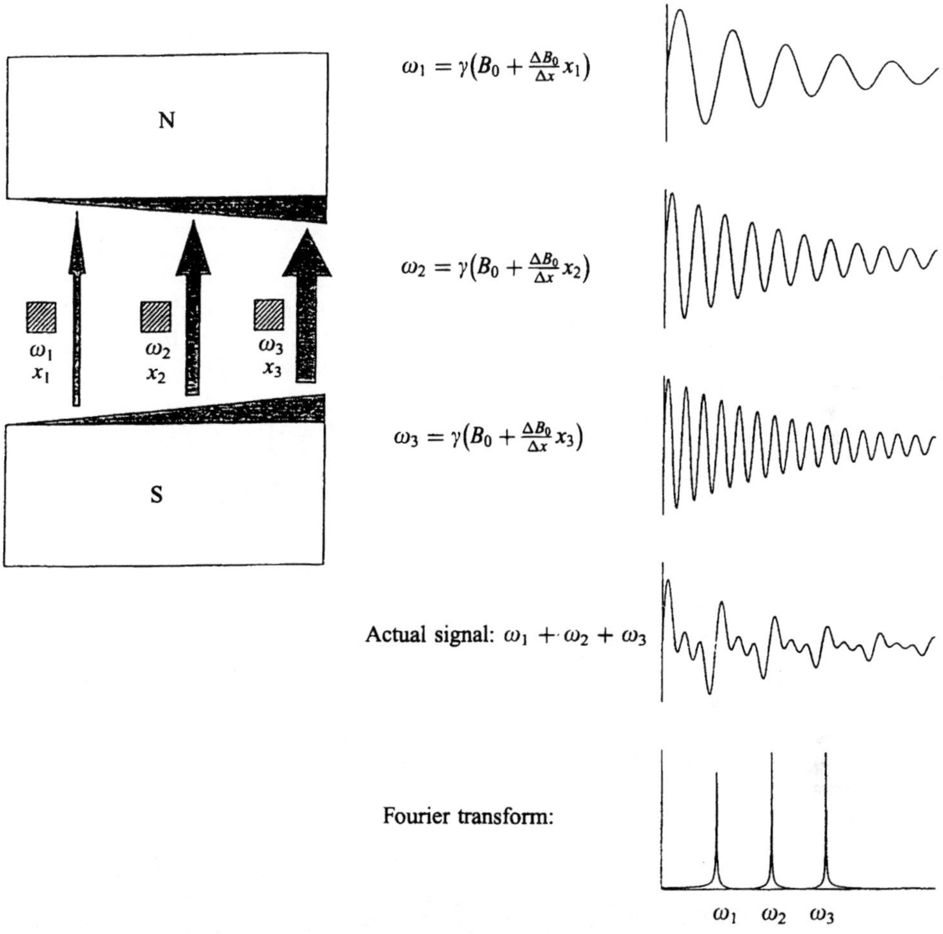

Fig. 12.16 The concept of acquiring spatial information in MR imaging (MRI).

varying signal given as

$$S(t) = \eta \int \mathrm{d}r\, \rho(r) \mathrm{e}^{-\mathrm{i}2\pi r k(t)}, \qquad (12.11)$$

where η represents the efficiency of the receiver coil detection circuit, $\rho(r)$ the spin density, and $k(t)$ the time integral of the gradient field:

$$k(t) = \gamma \int_0^t G(t')\, \mathrm{d}t'. \qquad (12.12)$$

Here G is the spatial gradient of the magnetic field. We neglect the T_2 relaxation term e^{-t/T_2}, and the off-resonance frequencies are neglected as well.

The most commonly used strategies of MR image reconstruction involve acquisition of 256 signals per transverse section from 16 sections or more. Each signal is the response to an input pulse in which the magnetic field gradients are varied along two or three orthogonal axes. If the intensity of the static magnetic field component in say the x-direction is varied momentarily during the initial RF pulses by application of a magnetic field gradient in the x-direction, the frequency of the nuclear spins will change for the period of the field change (i.e. eqn. 12.9). Application of a spatial gradient in the y-direction during signal acquisition will acquire different frequencies corresponding to the position of the spins in the y-direction. The resulting set of the 256 time varying signals is then Fourier-transformed in the x- and y-directions—2-D spatial transform—to give the MR image (MRI).

12.7 MR spectroscopy and chemical shift imaging

The chemical environment of a nucleus causes a small change in the local magnetic field due to the shielding effects of the electrons and the influence of nearby magnetic spins of other nuclei (Fig. 12.17). The phenomenon of chemical shift is given by

$$\omega = 2\pi\nu = \gamma B_0(1 - \sigma), \qquad (12.13)$$

where σ is the shielding constant (a dimensionless quantity).

Thus when a sample is irradiated by a band of RF frequencies, the Fourier transform of the signal will give a spectrum characteristic of the chemical compounds in the sample or in the body (Fig. 12.18). Spatial localization of the chemical shift information involves the use of a surface coil or more complex techniques wherein gradient sequencies used for imaging are combined with spectroscopic data acquisition to give coarse resolution images depicting the intensity of the specific spectral peaks. The resolution is poor because the concentration of the important compounds is low for the MR method. This might come as a surprise, since we can achieve resolutions less than 1 mm in MR imaging of protons. The difference here is that the concentration of protons in tissue is 10^4 times greater than that of the important phosphorus compounds such as creatine phosphate and adenosine triphosphate whose concentrations are indicators of the energy state of the tissues. Additionally, if we take into acccount the 1 per cent abundance of ^{13}C, the proton concentration is 10^7 times

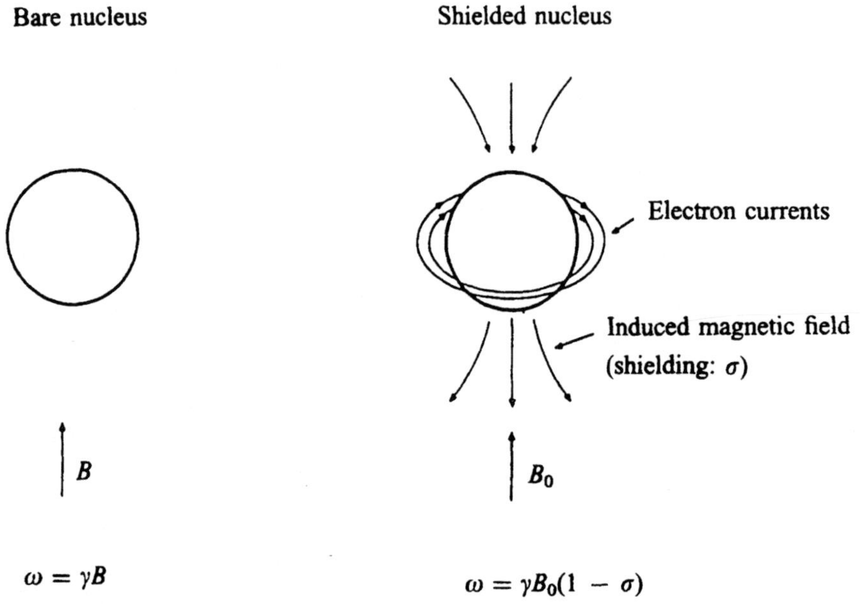

Fig. 12.17 Electron shielding results in a small change in the magnetic field with a corresponding change in frequency: the chemical shift.

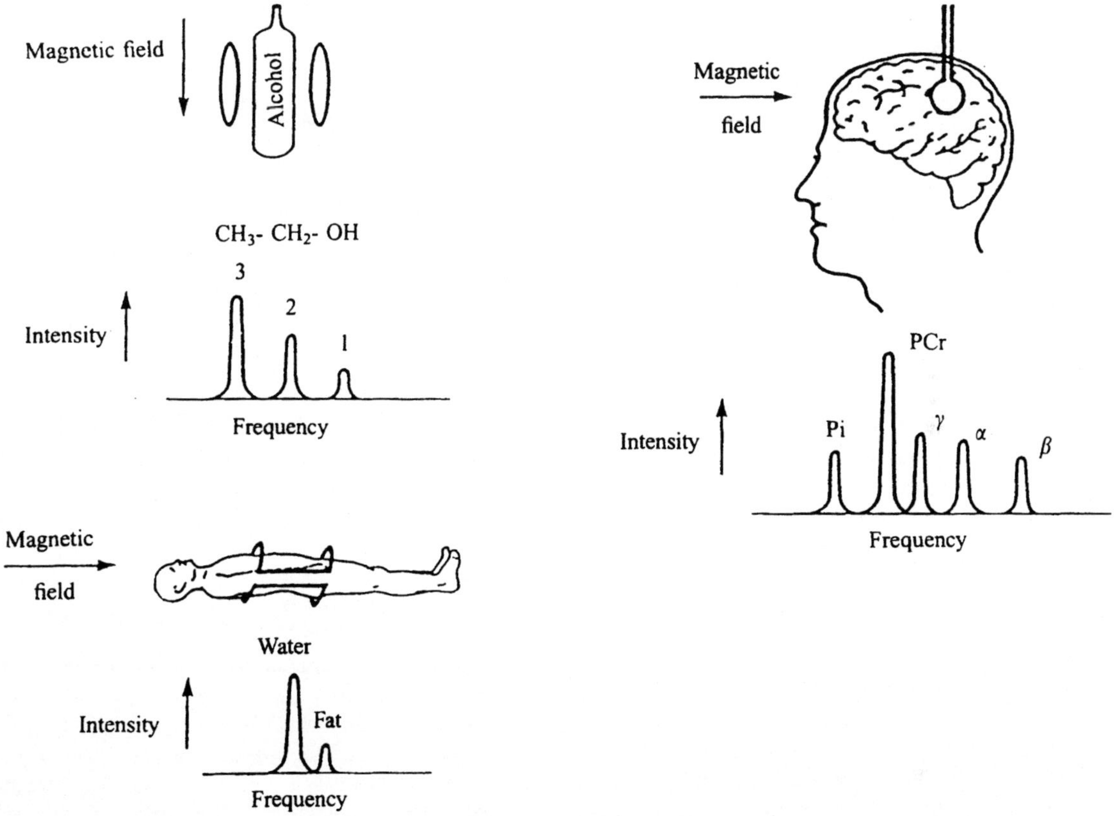

Fig. 12.18 Three types of NMR chemical shift spectroscopy. Note the three peaks from the protons located on three different carbon atoms of ethyl alcohol. The chemical shift of protons associated with fat is a few parts per million or a frequency shift of about 150 Hz. The phosphorous spectrum typically shows inorganic phosphorous, Pi; phosphocreatine, PCr; and the three phosphorous nuclei (γ, α, β) of adenosine triphosphate

greater than that of the important carbohydrate metabolites. In tissue, the concentration of biological compounds is in the millimolar range, whereas the concentration of water protons is nearly 100M.

12.8 Ultrasound

Ultrasound denotes the use of acoustical waves at frequencies greater than 20 kHz. Generally, medical ultrasound is performed at frequencies in the range of 1 MHz. The technique is used to determine the location of surfaces within tissues by measuring the time interval between the production of an ultrasonic pulse and the detection of its echo resulting from reflections from those surfaces. By measuring the time interval between the transmitted and detected pulse, we can calculate the distance between the transmitter and the object using the following equation:

$$d = \frac{ct}{2}, \tag{12.14}$$

where d is the distance, c is the velocity of the sound in the medium, and t is time. The velocity of sound is not equal in all soft tissues; however, generally a constant velocity is assumed at 1540 m s^{-1} for operation of commercial equipment. The ultrasound is both produced and detected by a piezoelectric crystal, which has the property of changing its physical dimensions in response to an electric field and can produce an emf if its physical dimensions are changed mechanically. Thus, ultrasonic vibrations or compression waves are produced by applying an oscillating potential across the crystal. The reflected ultrasound imposes a distortion on the crystal, which in turn produces an oscillating voltage in the crystal. The same crystal is used for both transmission and reception. The expected axial and lateral resolutions are 1 to 2 mm.

The differences in tissue properties which lead to ultrasound reflections are the density and compressibility of tissue (see Section 7.5.7). When an ultrasonic beam is incident upon a boundary between two tissue interfaces, part of the beam is reflected and part is transmitted into the second medium. If within the second medium another interface is encountered, the same processes will also occur. The fraction of the incident beam intensity that is reflected depends upon the acoustic impedances of the two tissues. Acoustic impedance is a fundamental property of matter (Section 7.5.7). For longitudinal ultrasonic waves (vibrational motion in the same direction as the wave propagates), acoustic impedance is equal to the product of the material density and the velocity of sound in the material,

$$Z = \rho c, \tag{12.15}$$

where Z is the acoustic impedance, ρ is the density, and c is the velocity of sound. (In the cgs system, the acoustic impedance is given in rayls, density in g cm^{-3}, and velocity in cm s^{-1}.) For an acoustic wave at normal incidence on a tissue interface between, for example, blood and heart muscle, the reflected power is given by:

$$R = \left(\frac{Z_1 - Z_2}{Z_2 + Z_2}\right)^2, \tag{12.16}$$

where Z_1 and Z_2 represent the two impedances of the different tissues (Section 7.5.7).

Another aspect of acoustic imaging is tissue attenuation. The general expression for attenuation is the same form as eqn (12.1):

$$I(d) = I_o e^{-\mu d}, \tag{12.17}$$

where the attenuation coefficient is μ and the tissue thickness is d. The attenuation coefficient increases with frequency. Bone has a high attenuation at the frequencies needed for imaging. This is one of the major reasons why high-resolution brain imaging is not possible with ultrasound. The method of ultrasound imaging is very similar to that of radar ranging. The distance between the CRT trace origin and the reflected signals on the CRT allow one to map the shape of the tissue interfaces; and if the trajectory of the sound beam is emulated by the CRT trace, a two-dimensional image of reflecting surfaces is easily obtained (Fig. 12.19).

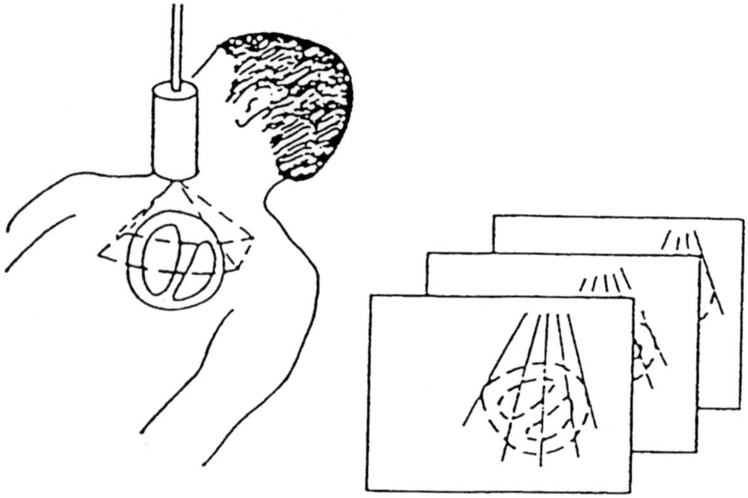

Fig. 12.19 Coordination of the position of a CRT trace with the position of a transducer allows one to display the spatial distribution of surfaces.

12.9 Velocity measurements by ultrasound

If a structure is stationary, the frequency of the reflected wave will be identical to the impinging wave. A moving structure will cause the back-scattered signal frequency to be shifted up or down by an amount approximately proportional to the component of the interface's velocity aligned with the sound beam axis. This shift is given by the following equation:

$$\Delta f = \frac{-2vf_o}{c+v}\cos\theta \approx \frac{-2vf_o}{c}\cos\theta, \qquad (12.18)$$

where Δf is the Doppler shift (Hz), v is the vector velocity at the interface, f_o is the frequency of the impinging sound beam (Hz), c is the velocity of sound in the tissue medium, and θ is the angle between the sound beam axis and the velocity vector. Thus Δf is negative if the interface is moving away from the sound source and positive if it is moving toward the source. The actual received frequency from a moving structure will be:

$$f_r = f_o + \Delta f. \qquad (2.19)$$

When the impinging sound beam passes through a blood vessel, scattering occurs. In this process, small amounts of sound energy are absorbed by each red cell and reradiated in all directions. If the red cell is moving with respect to the sound source, the back-scattered energy returning to the receiving transducer will be shifted in frequency; the magnitude and direction are proportional to the velocity of the respective cell. If we use ultrasound to image the cross-sectional area of the blood vessel, then volume flow (F) can be calculated from

$$F = vA, \qquad (12.20)$$

where v calculated from eqn (12.18) is the mean velocity orthogonal to the cross-sectional area A.

12.10 Generalized theory of image formation

It is convenient to consider the lens of an imaging system mathematically, as a resolution function of an imaging system. The resolution function of an imaging system can be characterized as the system's spatial impulse response (see Chapters 6 and 7 for time-domain examples) or the system's point spread function (Fig. 12.20). By definition, the Fourier transform of the point spread function is the system's spatial transfer function. The modulus of the Fourier components is the modulation transfer function (MTF). This function tells us how well each spatial frequency inherent in an object is passed through the imaging system to the image.

For x-ray CT, emission CT, and ultrasound, the system components and parameters that provide the resolution function are the collimators and the size of the detectors or the positioning circuits associated with the detector. For MRI, the uniformity of the magnetic field and the magnitude of the switched magnetic field gradients determine the lens or resolution function.

As shown in Fig. 12.20, the Fourier transform of the point spread function exists in the back focal plane of a thin optical lens. In Chapter 6 and Appendix A, the Fourier transform is used to decompose a time domain signal to the frequencies, amplitudes, and phases of the sinusoids, which when added together give the time domain signal. Similarly the spatial domain Fourier transform gives the one- two- or three-dimensional distribution of frequencies, amplitudes and phases of the sinusoids comprising a spatial signal or an image. In image processing we work with the two-dimensional Fourier transform, which in its discrete form is defined as

$$F(k) = \sum_{0}^{N-1} A(x)\exp(-i2\pi k\cdot x)/(N/2). \qquad (12.21)$$

Because the Fourier transform is separable, it is convenient to perform the 2-D Fourier transform by first transforming all of the rows in an array and then transforming the resulting array column by column.

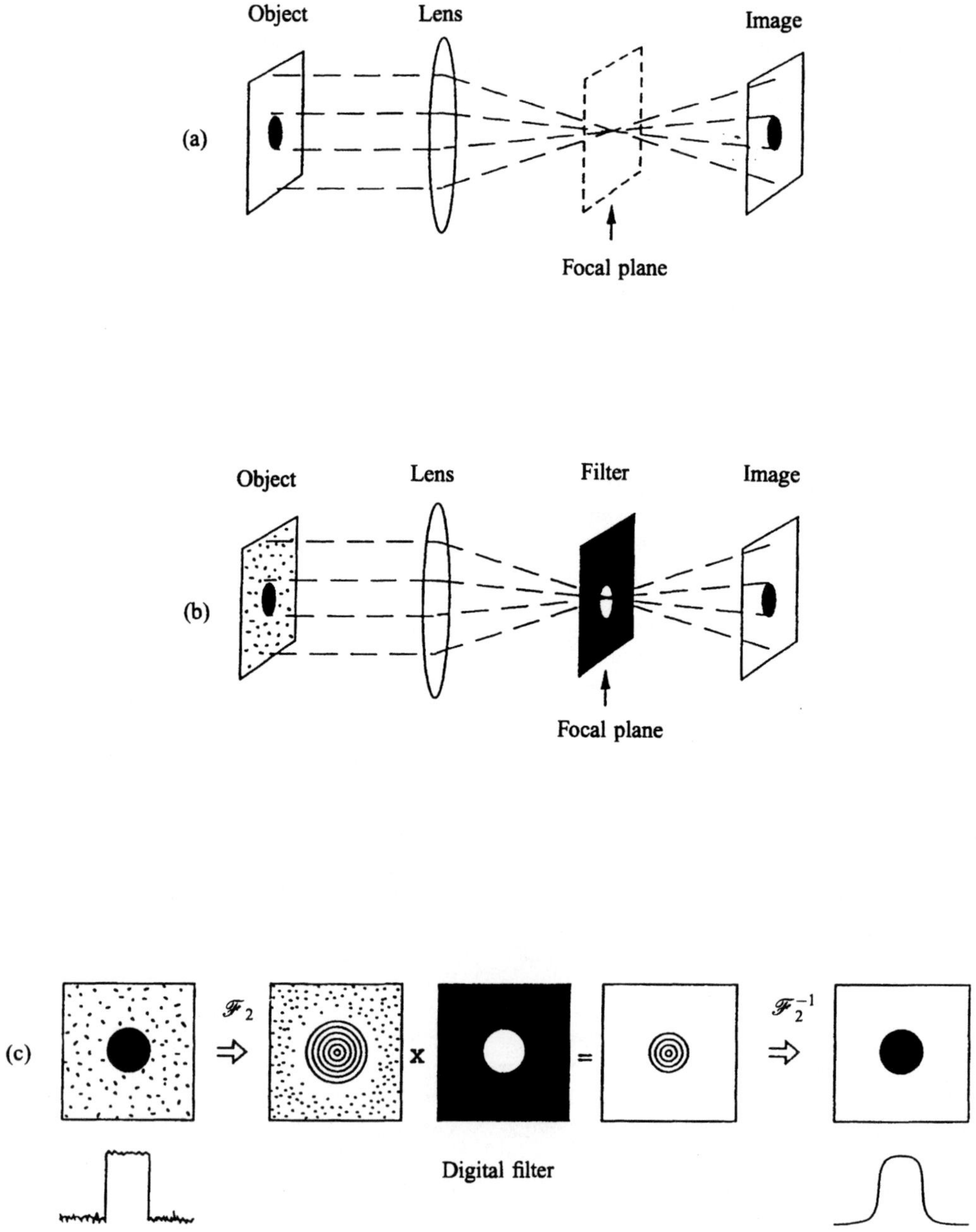

Fig. 12.20 The spatial frequencies of an object appear in the back focal plane of a thin refractive lens where filtering can lead to image sharpening or blurring. Digital Fourier transformation, digital filtering, and inverse Fourier transformation can be seen as an analogous process. (a) Generalized theory of image formation; (b) analog filtering; (c) digital filtering.

Thus, although the medical imaging lenses discussed here do not have a physical refracting lens as shown in Fig. 12.20, we can evaluate the behavior of the resolution function by observing the result of imaging a source (e.g. attenuation object, radionuclide source, spin concentration, impendance mismatch area) that is sufficiently small to approximate a point. In the spatial domain the result is a point spread function, $h(x, y)$. At the back focal plane of the lens, in response to a point source, is found the Fourier transform of $h(x, y)$, which is the transfer function of the system (Fig. 12.20). According to the principle of superposition (see Sections 6.3.1 and Appendix A.2.1) we expect the relation between the spatial distribution of intensity over the observed image, $g(x, y)$, and that, $f(x, y)$, over the actual object being observed (the true object or image) to be given by the convolution of $f(x, y)$ and $h(x, y)$,

$$g(x, y) = \int_{-\infty}^{\pm\infty}\int_{-\infty}^{\pm\infty} f(x', y')h(x - x', y - y')\,\mathrm{d}x'\mathrm{d}y', \tag{12.22}$$

which can be abbreviated as

$$g(x, y) = f(x, y) * h(x, y), \tag{12.23}$$

where $*$ denotes the convolution operation shown above (see Sections 6.3.1, Appendix A.2.1). Just as convolution in the time domain translates to multiplication in the frequency domain of the Laplace transform (Appendix A.3), convolution in the spatial domain translates to multiplication in the frequency domain of the Fourier transform,

$$\mathscr{F}\{g(x, y)\} = \mathscr{F}\{h(x, y)\} \cdot \mathscr{F}\{f(x, y)\}, \tag{12.24}$$

where $\mathscr{F}$ is the 2-D Fourier transform operation, g is the observed image, h is the point spread function, and f is the true object or true image. Note that $\mathscr{F}\{g(x, y)\}$ can be considered to exist in the back focal plane of the lens. This can be demonstrated easily, using a laser source of coherent light and an appropriate lens system.

This generalized theory of image formation is useful in three areas. First, we use this theory to understand the properties of imaging systems because through its use, we can analyze system performance, the origin of image distortions, and strategies to improve resolution. Secondly, we now have a basis for image processing by techniques such as low-pass filtering (for example, noise suppression), high-pass filtering (for example, edge sharpening) and notch filtering (for example, suppression of an artifact with periodicity). Thirdly, through the generalized theory of image formation we can understand the general method of image reconstruction from projections.

12.11 Image processing

With the advent of disk operating systems and lower-priced array processors 20 years ago, theoretically well known optical processing methods were adopted by the digital signal processing scientists and used for image restoration and enhancement. Image *restoration* is the process whereby the true two-dimensional data (source intensity distribution) are approximated by manipulation of the observed image data. Image *enhancement* is the result of an operation on the observed image whose purpose is to emphasize some particular feature (for example, edge enhancement, feature extraction). Image *segmentation* is the process whereby features of particular interest are extracted from the image. Image *segmentation* is now generally used when seeking to determine the volume of three-dimensional images of tumors or multiple sclerosis lesions from a sequence of NMR, PET, or x-ray CT images. Additional processes such as thresholding, high- and low-pass filtering and color coding of gray levels (i.e. pseudo-color) are commonly used methods of manipulating 2-D images.

We can perform effective image processing using what we learned about the generalized theory of image formation. Suppose we wish to suppress noise spikes which appear as 'grass' on an image. An effective method is to suppress the higher frequencies of the spatial frequency transform of the image. This will result in blurring the high-frequency noise. Unfortunately, the suppression of high frequencies will also remove high frequencies inherent in the object; none the less, a judicious choice of frequency filter can substantially improve the visual impact of an image without loss of content. The method involves removing the higher frequencies from the two-dimensional spatial transform of the image. After this filter operation, known as low-pass filtering, the inverse Fourier transform gives the noise-suppressed image. This is illustrated in Fig. 12.20.

12.12 Reconstruction tomography

12.12.1 General theory of 3-D reconstruction by convolution or filtering

The main task, indeed the essence of computed tomography, is to remove or filter the blurring from the back projection image. In x-ray transmission, emission imaging, and some forms of MRI, the data are projections of a spatial distribution onto a line for 2-D distribution or a plane for 3-D distribution. The 1-D projection profile from a 2-D distribution (transverse section) can be back-projected into the 2-D plane and this operation gives a back-projection image. The simple back-projection of projection profiles gives only a blurred or smeared representation of the true object illustrated in Fig. 12.6. The reason is that each point in the reconstructed transverse section will consist of the superposition of a set of straight lines corresponding to each projection value from that point in the true object (Fig. 12.21). The resulting back-projection image is a convolution of the true distribution with a function, called a kernel, that smoothes each point by a weighting value which is inversely proportional to the distance from that point. This results in an image which appears similar to the true transverse section but with syrup poured over it. Mathematically we represent this as

$$B(x, y) = \iint \frac{A(x', y')}{\{(x - x')^2 + (y - y')^2\}^{1/2}} \mathrm{d}x' \mathrm{d}y' = A * \frac{1}{r}, \qquad (12.25)$$

where B is a back-projected image and A is the true image. This smearing results in the middle image of Fig. 12.21, which in this case is a simulation of the heart and lungs in the thorax.

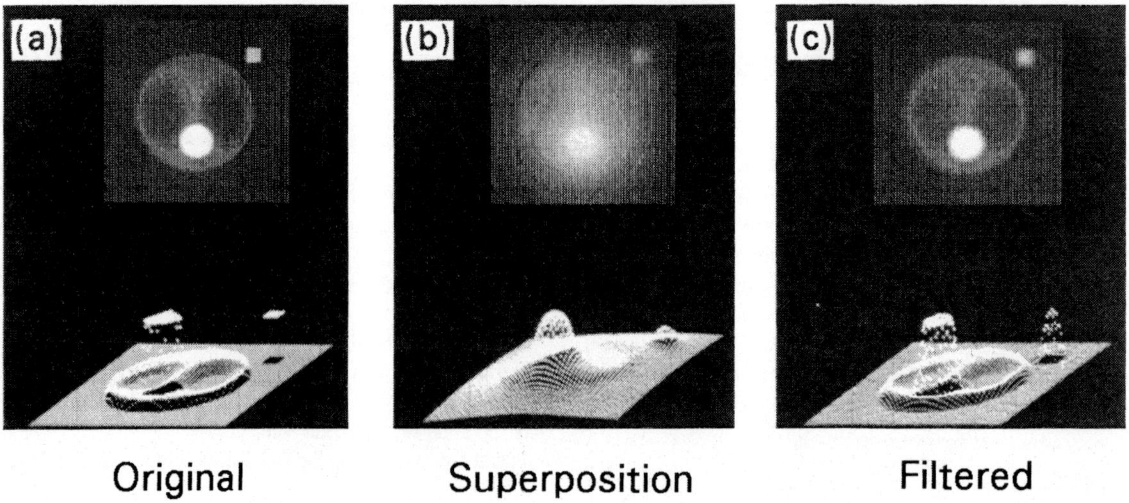

Fig. 12.21 The process of back-projection results in a bluured representation of the true object. This blurring is removed by a high-pass filter operation, which is the operational essence of image reconstruction from projections. (a) Original. (b) Superposition. (c) Filtered.

Since we know the image smoothing kernel, we can find a kernel that reverses the process, unsmoothing the back-projection image. This deblurring function (or kernel) is applied to the blurred image to give the result shown at the right in Fig. 12.21. To find the deblurring function, we seek a technique whereby $1/r$ can be removed (by a process known as deconvolution) from $B(x, y) = B(r, \phi)$, where $B(r, \phi)$ is the back-projected image expressed in polar coordinates (r, ϕ).

From the convolution theorem, we have

$$\mathscr{F}\{B(r, \phi)\} = \mathscr{F}\{A(r, \phi)\} \cdot \mathscr{F}\left\{\frac{1}{r}\right\}. \tag{12.26}$$

Given that

$$\mathscr{F}\left\{\frac{1}{r}\right\} = 2\pi \int \frac{1}{r} J_0(2\pi Rr) r \, \mathrm{d}r = R^{-1}, \tag{12.27}$$

where R is the frequency space radius and J_o is a Bessel function of zeroth order, the true image is related in the back-projection image as

$$A(r, \phi) = \mathscr{F}^{-1}\{\mathscr{F}\{A(r, \phi)\}\} = \mathscr{F}^{-1}\{|R|\mathscr{F}\{B(r, \phi)\}\}. \tag{12.28}$$

The operations of eqn (12.28) involve the following steps:

(1) obtain a series of projections—P;
(2) derive the back-projected image by simple linear superposition—Σ;
(3) Fourier-transform the two-dimensional image—$\mathscr{F}_2$;
(4) multiply the Fourier coefficients by the spatial frequency radius filter—R_2;
(5) Fourier-transform (invert) the result of step (4) to obtain the true image—$\mathscr{F}_2^{-1}$.

The entire sequence can be written in operator construction as

$$A = \mathscr{F}_2^{-1} R_2 \mathscr{F}_2(\Sigma P), \tag{12.29}$$

where A is the true image. Note that ΣP is the back-projection image.

12.12.2 The convolution or back-projection of filtered projection technique

Because the 2-D Fourier transform was not easily implemented in the early 1970s when computed tomography was introduced, an equivalent method was then and is still being employed. This method involves filtering the projections before back-projection. As the back-projection and filtering operations are linear, the sequence of operations in eqn (12.29) can be altered without affecting the final result. Thus,

(1) obtain the projections—P;
(2) Fourier-transform the one-dimensional projections—$\mathscr{F}_1$;
(3) multiply the Fourier coefficients by the frequency radius—R_1;
(4) perform the inverse Fourier transform of the results of step (3)—$\mathscr{F}_1^{-1}$;
(5) derive the reconstructed image of filtered projections by linear superposition—Σ over each angle (e.g. Fig. 12.21).

In symbols we have

$$A = \Sigma(\mathscr{F}_1^{-1} R_1 \mathscr{F}_1 P). \tag{12.30}$$

Of course the operations of eqns (12.29) and (12.30) can be accomplished using convolution operations, but these are not as fast, due to the availability of the fast Fourier transform.

12.12.3 The Fourier projection theorem

Another method of image reconstruction from projections lies at the basis of the mathematical derivations leading to the previously discussed techniques. The Fourier projection theorem equates the Fourier transform components of the object along a section (or a line) in the spatial frequency domain (i.e. Fourier space) through the origin and normal to the projection, to the Fourier transform of the projection (Fig. 12.22). Consider a proof for a three-dimensional distribution from which two-dimensional projections in plane integrals are obtained. $A(x, y, z)$ is a three-dimensional distribution, and the two-dimensional projection is defined as

$$P(x, y) = \int A(x, y, z) \, \mathrm{d}z. \tag{12.31}$$

The three-dimensional Fourier transform is

$$\mathscr{F}\{A(x, y, z)\} = \iiint A(x, y, z) \mathrm{e}^{[-2\pi \mathrm{i}(x \cdot X + y \cdot Y + z \cdot Z)]} \mathrm{d}x \, \mathrm{d}y \, \mathrm{d}z, \tag{12.32}$$

where X, Y, and Z are the spatial frequencies of the

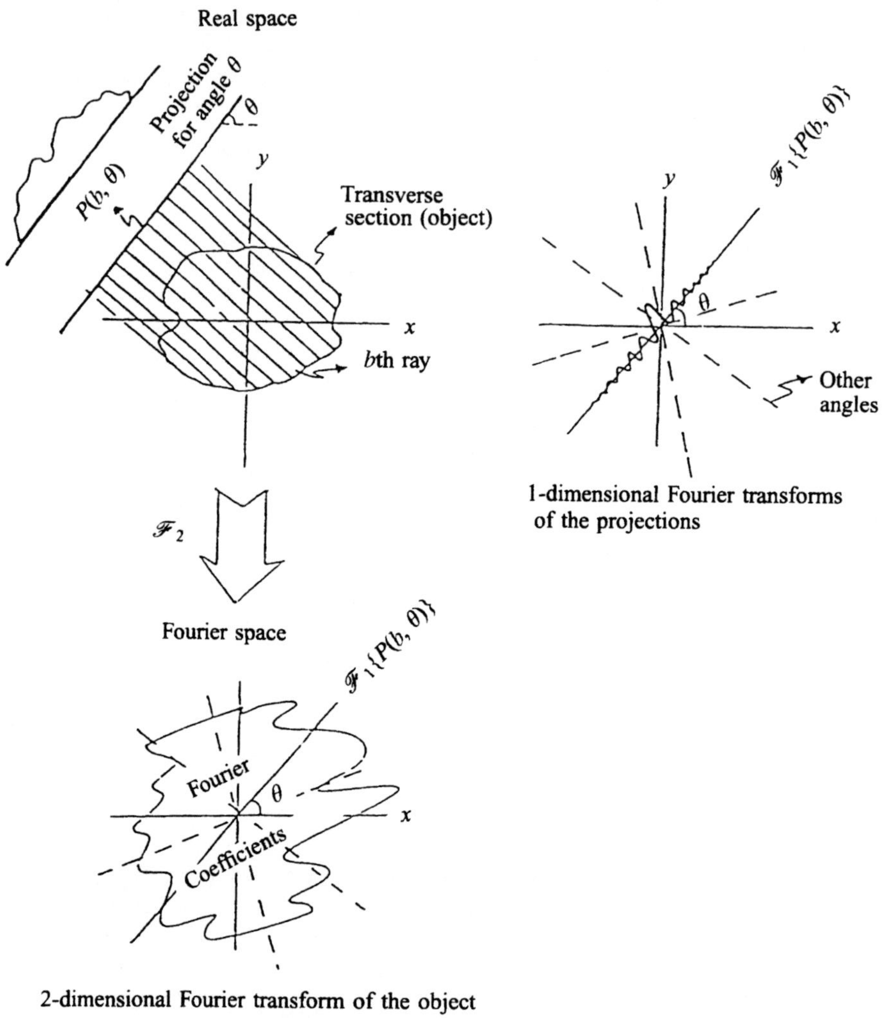

Fig. 12.22 An illustration of the Fourier projection theorem in two dimensions, which relates the Fourier coefficients of one-dimensional projections to the Fourier coefficients of the two-dimensional object.

Cartesian coordinates. For $z = 0$, we have

$$\mathscr{F}\{A(x,y,z)\}_{z=0} = \iint\left\{\int A(x,y,z)\,\mathrm{d}z\right\}\mathrm{e}^{[-2\pi\mathrm{i}(x\cdot X+y\cdot Y)]}\,\mathrm{d}x\,\mathrm{d}y$$
$$= \iint P(x,y)\mathrm{e}^{[-2\pi\mathrm{i}(x\cdot X+y\cdot Y)]}\,\mathrm{d}x\,\mathrm{d}y. \tag{12.33}$$

Q.E.D.

Direct use of the Fourier projection theorem in digital image reconstruction involves correct interpolation between Fourier coefficients to properly discretize the data from the radial distribution of coefficients before performing the 2-D or 3-D discrete Fourier transforms.

12.12.4 Arithmetic reconstruction methods

The first use of the Fourier transform methods in reconstruction was by R. Bracewell for astrophysical studies prior to the introduction of x-ray CT. The first x-ray CT systems used algebraic techniques which are

A_{11}	A_{12}	$P_1 = A_{11} + A_{12}$
A_{21}	A_{22}	$P_2 = A_{21} + A_{22}$

$P_3 = A_{11} + A_{21}, P_4 = A_{12} + A_{22}$

5	5	$P_1 = 3,\ R_1 = 10$
5	5	$P_2 = 7,\ R_2 = 10$

$P_3 = 4, \quad P_4 = 6$

$R_3 = 10, \quad R_4 = 10$

$\frac{3}{10} \times 5$	$\frac{3}{10} \times 5$	$P_1 = 3,\ R_1 = 3$
$\frac{7}{10} \times 5$	$\frac{7}{10} \times 5$	$P_2 = 7,\ R_2 = 7$

$P_3 = 4, \quad P_4 = 6$

$R_3 = 5, \quad R_4 = 5$

$\frac{3}{2} \times \frac{4}{5}$	$\frac{3}{2} \times \frac{6}{5}$
$\frac{7}{2} \times \frac{4}{5}$	$\frac{7}{2} \times \frac{6}{5}$

=

1.2	1.8	$P_1 = 3,\ R_1 = 3$
2.8	4.2	$P_2 = 7,\ R_2 = 7$

$P_3 = 4, \quad P_4 = 6$

$R_3 = 4, \quad R_4 = 6$

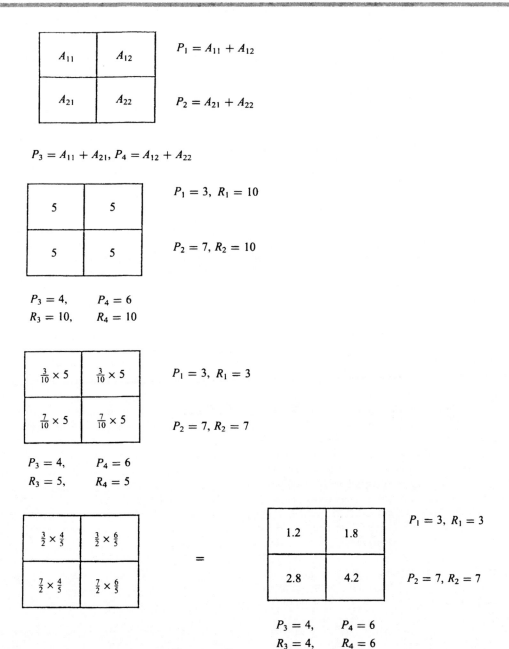

Fig. 12.23 The multiplicative arithmetic reconstruction technique is demonstrated for a 2 × 2 array. This method was used from 64 × 64 arrays in the early 1970s when x-ray CT was first introduced. Note: simple algebraic inversion of this problem with the geometry shown cannot be accomplished, because the determinant is zero.

simple to understand but require many more computational steps than the convolution or Fourier transform methods. Thus, these algebraic methods have not been used for the past 20 years, but now the capability of using these methods with prior information has led to a resurgence in interest.

The simplest form of algebraic reconstruction, known as ART, is shown in Fig. 12.23. The essence of the technique is to estimate (guess) the value for each element of the image array and modify these estimates by how much the projections from the esimates deviate from the measured projections. Thus, in Fig. 12.23 our first guess at the values in each array box is based on the fact that we know the total density is 10 (i.e. each projection sums to 10, as it should for perfect transmission data). For convenience our initial estimate is 5 for each pixel. But note that the projection R_1 for estimate deviates from the measured P_1 by a factor

$$\frac{R_1}{P_1} = \frac{10}{3}. \qquad (12.34)$$

Thus to correct the first estimate, we modify the corresponding pixels by a multiplicative factor of 3/10. By multiple iterations we find that an image distribution can be estimated, and indeed this distribution has been shown to closely correspond to that found by the techniques of the previous section using the same projection data.

A general formulation of the iterative algebraic reconstruction techniques includes the weighting factors f_{ij} which give the fractional contribution of each pixel to a particular projection bin. In an array of 100 × 100 pixels, there are 10^4 pixels, but if we include the weighting factors for each (say) 100 projections the reconstruction problem requires 100 arrays of 10^4 elements. From this argument we see that the reconstruction method requires rapid access to 10^6 words—a requirement not easily met on practical commercial machines in the 1970s and early 1980s. Iterative methods are now being employed, particularly when the geometry of projection rays is divergent and when attenuation factors are needed, such as in the single-photon tomography (SPECT) problem.

At the nth iteration, the algebraic reconstruction technique gives estimated projections $R_b^n(\theta)$ for bin b and projection angle θ:

$$R_b^n(\theta) = \sum_{i,j \in \text{ray } b(\theta)} f_{ij}^{\theta} A^n(i,j). \qquad (12.35)$$

Here θ designates a particular projection. The simple algebraic technique discussed above modifies $A^n(i,j)$ by the ratio $P_b(\theta)/R_b(\theta)$.

A more general algebraic technique improves the estimate of each pixel value $A(i,j)$ from one iteration to the next in a least-squares sense. That is, the residual is minimized between the estimated projection values of all ray sums which pass through a particular pixel $A(i_o, j_o)$ and the actual measured projections,

$$\mathcal{R}\{A^{n+1}(i_o,j_o)\} = \sum \frac{\{P_b(\theta) - R_b^n(\theta)\}^2}{\sigma_b^2(\theta)}. \qquad (12.36)$$

Here $\mathcal{R}$ is the residual minimized and $\sigma_b^2\theta$ is the variance for each projection. The operational equation for the estimate of $A(i_o, j_o)$ at the $(n+1)$th iteration is

$$A^{n+1}(i_o,j_o) = A^n(i_o,j_o) + \delta\Delta^n A(i_o,j_o), \qquad (12.37)$$

where

$$\Delta^n A(i_o,j_o) = \sum_\theta f_{i_o j_o}^{\theta} \frac{(P_b(\theta) - R_b^n(\theta))^2}{\sigma_b^2(\theta)} \Bigg/ \sum_\theta \left(\frac{f_{i_o j_o}^{\theta}}{\sigma_b(\theta)}\right)^2. \qquad (12.38)$$

This method allows incorporation of weighting factors which take into account the pixel versus bin geometry, resolution variation and attenuation factors of particular importance in emission tomography. An example of the use of transmission tomography to determine attenuation coefficients and the iterative least-squares technique are shown in Fig. 12.24. An alternative iterative approach is the maximum-likelihood approach, which gives similar results to those of the iterative least-squares method and is applicable to the situation of Poisson statistics generally assumed in SPECT imaging.

These algebraic reconstruction algorithms involve iterative solutions to the classic inverse problem,

$$P = FA, \qquad (12.39)$$

where P is the projection matrix, A is the matrix of true data being sought, and F is the projection operation. The inverse is

$$A = F^{-1}P,$$

which is computed by iteratively estimating the data A' and modifying the estimate by comparison of the calculated projection set P' to the true observed projections P. The expectation-maximization algorithm solves the inverse problem by updating each pixel value a_i in accord with

$$a_i^{n+1} = \sum_j p_j \frac{a_i^n f_{ij}}{\sum_i a_i^n f_{ij}}, \qquad (12.40)$$

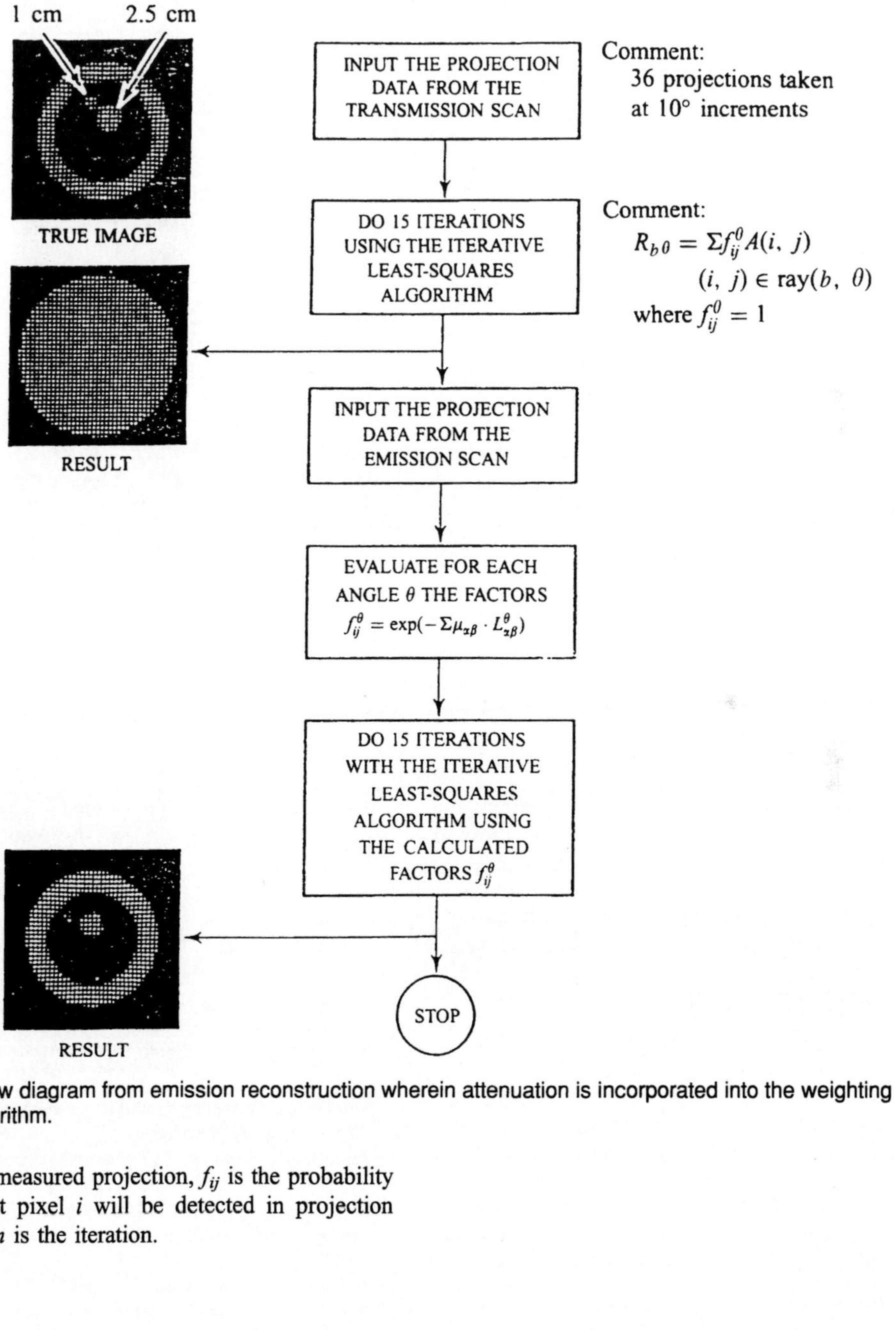

Fig. 12.24 A flow diagram from emission reconstruction wherein attenuation is incorporated into the weighting factors for an iterative algorithm.

where p is the measured projection, f_{ij} is the probability that a source at pixel i will be detected in projection detector j, and n is the iteration.

12.13 Synopsis

The four major modalities of medical imaging have different applications in accord with their individual capabilities for detecting physiological and anatomical processes. X-ray imaging detects density and, to some extent, elemental composition. Emission imaging (SPECT and PET) detects concentration of tracers specific for different metabolic functions. MR imaging (MRI) detects distribution and relaxation of proton spins, which is reflective of anatomical changes associated with water composition changes. MR spectroscopy (MRS) is unique in its ability to measure chemical composition, but with a spatial resolution far less than that of other modalities which do not have this unique capability. Ultrasound measures anatomical surfaces, blood movement, and the movement of surfaces with practical applications in many but not all parts of the body.

The medical imaging techniques rely on data manipulation methods currently available in modern computing systems. Both image reconstruction and image processing can be understood in their major applications by a study of the generalized theory of image formation. This theory relates the behavior of an optical lens to the transfer function of an imaging system and then to the digital spatial Fourier transform. Image reconstruction is seen as a type of high-pass filtering, and image processing is cast in the context of high- and low-pass filtering of spatial frequencies.

Acknowledgments

This chapter was written with the assistance of Miguel M. Colina De Vivero and Ismail S. Khalil, former students, who helped cast it into a level appropriate for upper division students interested in bioengineering. This work was supported in part by National Institutes of Health Training grant no. HL07367.

Selected reading

Bracewell, R. N. (1995). *Two-dimensional imaging*. 689 pages. Prentice Hall, Englewood Cliffs, NJ.

Bronzino, J. D. (ed.) (1995). *The biomedical engineering handbook*. 2862 pages. CRC Press and IEEE Press, Boca Raton, FL.

Budinger, T. F. and Lauterbur, P. C. (1984). Nuclear magnetic resonance technology for medical studies. *Science*, **226**, 288–98.

Budinger, T. F., Gullberg, G. T. and Huesman, R. H. (1979). Emission computed tomography. In *Image reconstruction from projections*, (ed. G. T. Herman), pp. 147–246. Springer–Verlag, Berlin, New York.

Castleman, K. R. (1996). *Digital image processing*. 667 pages. Prentice Hall, Englewood Cliffs, NJ.

Krestel, E. (ed.) (1990). *Imaging systems for medical diagnostics*. 636 pages. Siemens Aktiengesellschaft, Berlin.

Wehrli, F. W., Shaw, D. and Kneeland, J. B. (1988). *Biomedical magnetic resonance imaging principles, methodology and applications*. VCH, New York.

BIOLOGICAL APPLICATIONS OF IONIZING RADIATION

Selig N. Kaplan and Howard Maccabee

13

Contents

13.1 Radiation physics

13.1.1 The nature of ionizing radiation

To understand and deal with what is commonly called *ionizing radiation* it is first necessary to think of radiation in the quantum sense, that is, as individual packets or *quanta* of energy. Even to the uninitiated, thinking this way is straightforward if one is talking about electrons. We can picture a single electrically charged particle traveling through an electrical circuit, or being accelerated to a high speed and directed to the specially coated screen in a video monitor where it induces the emission of light. We must now, however, think of this light also as having a particulate nature. Quanta of light are called *photons*. Just as a moving electron has kinetic energy, so does a photon. Its energy is related to its frequency, ν, through Planck's constant, h. The photon energy $E = h\nu$.

Ordinary energy units are too large to describe the energy of individual, submicroscopic quanta, so a convenient unit has been devised for that purpose, the *electronvolt*. When an electron falls through an electrical potential, as in the video monitor described above, it acquires a kinetic energy, eV joules, where e is the electron charge measured in coulombs, and V the magnitude of the potential measured in volts. We say that this electron has a kinetic energy of V *electronvolts* (eV): 1 eV $= 1.602 \times 10^{-19}$ joules.

The average wavelength of visible light is 500 nm. This corresponds to a frequency of 6×10^{14} Hz and to a photon energy of 2.48 eV. X-ray photons used for medical x-ray imaging have an average energy of the order of 50 keV (5×10^4 eV). Nuclear γ-rays, identical in character to light and x-rays but with shorter wavelengths, have typical energies of the order of 1 MeV (10^6 eV). Much-higher-energy photons can be produced with high-energy particle accelerators.

When quanta have sufficient kinetic energy and collide with stationary atoms, they can knock bound electrons out of the atom, thus ionizing the atom—hence the name ionizing radiation. To produce ionization the particle must be able to transfer to the atom sufficient energy to overcome the electronic binding. In a single atom of atomic number Z, the electronic binding energies range from a few eV for the valence electrons to approximately $13.6Z^2$ eV for a K-shell electron. As we shall see, the ionization-producing reactions by radiation passing through matter are the principal energy-transfer mechanisms between the moving particle and the stationary medium. These interactions are responsible both for the attenuation, or stopping, of the radiation and for the radiation damage produced in the matter.

The process by which a photon causes ionization is called the photoelectric effect. In this process the photon is absorbed by the atom, and the atom emits one of its electrons. The kinetic energy of the emitted electron is equal to the energy of the incident photon minus the energy required to release the electron from the atom (the electron binding energy). For energetic x-rays the kinetic energy of the emitted electron will be much greater than the binding energy, and this electron will, in turn, cause more ionization. As it passes through matter it will undergo Coulomb scattering interactions with the constituent nuclei and electrons. Energy transfer from the electron to the massive nucleus is very inefficient, much like trying to transfer energy from a ping-pong ball to a bowling ball. However, the transfer of energy between electrons is quite efficient, electron–electron collisons result in further ionization, and the photoelectron gradually loses its energy and comes to rest. A 50 keV photon, for example, will ionize of the order of 2000 atoms in air, one of these being from the initial photoelectric process and the remainder from the Coulomb interactions of the photoelectron. This succession of Coulomb scatterings, either directly or by interaction products of high-energy quanta (as illustrated in the x-ray example above), is the mechanism by which virtually all ionization is produced.

Sources of ionizing radiation

Ionizing radiation can be either natural or man-made. Natural radiation originates from both terrestrial and extraterrestrial sources. Terrestrial sources are called simply *natural radioactivity*. Extraterrestrial radiation is called *cosmic rays*. The earliest man-made ionizing radiation sources were x-rays, which have been used in medical diagnosis and treatment almost since their discovery in 1895. The other principal sources of man-made radiation are particle accelerators and nuclear reactors. These latter two sources produce ionizing radiation directly, and also produce radioactive isotopes of natural elements and new radioactive elements—*artificial radioactivity*. Some of this radioactivity production is deliberate, for research and for medical and industrial purposes. Much, however, is an unavoidable by-product of accelerator and reactor operation.

Radioactivity. While nearly all of the familiar chemical elements and their natural isotopes are stable, there are some whose nuclei undergo spontaneous transmutations, or decay. These isotopes are called radioactive. All isotopes of elements with atomic number, Z, greater than 83 (bismuth) are radioactive. Radioactive decay is a stochastic process, which means that, while we cannot predict precisely when a particular atom will decay, the decay characteristics of a large ensemble of identical nuclides can be predicted quite accurately. The probability per unit time of decay is a constant, called the decay rate, λ. Each radionuclide is characterized by a unique λ. The decay rate of a sample, N, of identical radionuclides is

$$dN/dt = N\lambda. \qquad (13.1)$$

It follows that if there are initially N_o nuclides, the number remaining after time t is

$$N = N_o e^{-\lambda t}. \qquad (13.2)$$

The *average* or *mean life* of a radionuclide, $\tau = 1/\lambda$. The time required for half of a sample of radionuclides to decay, the *half-life*, is $t_{1/2} = \ln 2 \times \tau = 0.693\tau$. Most commonly, references will only list $t_{1/2}$.

The decay rate, or *activity*, of a radioactive sample is

$$A = N\lambda = N_o\lambda e^{-\lambda t} = A_o e^{-\lambda t}. \qquad (13.3)$$

The SI unit for radioactivity is the becquerel (Bq) named for Henri Becquerel, who was the first person to report the observation of natural radioactivity: 1 Bq = 1 decay per second. An earlier, and still more commonly used, unit of measurement for radioactivity is the curie, Ci, named for Marie Curie, probably the most famous of the early pioneers in the study of radioactivity, and the co-discoverer, with her husband Pierre, of radium. One curie was originally defined to be the activity of one gram of pure ^{226}Ra, and later rounded off by an international committee to be $1\ \text{Ci} = 3.7 \times 10^{10}$ decays per second $= 3.7 \times 10^{10}$ Bq.

Radionuclides emit one or more of three types of radiation, called respectively α, β, and γ-radiation. Alpha particles are ^{4}He nuclei, β-rays are electrons, and γ-rays are photons identical in character to x-rays. As a result of α- or β-decay the radionuclide *transmutes*, changes chemically, to a different element. In α-decay the atomic number, Z, decreases by 2 units. In addition, the atomic mass number, A, decreases by 4 units. In β-decay, Z changes by 1 unit and A remains the same. For the β-decay that occurs in natural radioactivity an ordinary electron is emitted and Z increases. However, some artificially produced radionuclides emit positively-charged electrons, *positrons*, thereby decreasing Z of the product nuclide by 1 unit. The emission of γ-rays accompanies transitions between energy states of the same nucleus, in analogy with the characteristic light and x-ray radiation that accompany atomic transitions. Gamma emission ordinarily merely accompanies α- and β-decay, although α- and β-decay also occur without accompanying γ-emission. There are rare cases of radionuclides that emit only γ-rays. These are called isomers. An example of such an isomer that has important applications in nuclear medicine is $^{99}Tc^{m}$.

Some examples of radioactive decay that also illustrate how one may write decay equations are:

$$^{226}\mathrm{Ra} \rightarrow {}^{222}\mathrm{Rn} + \alpha$$

$$^{90}\mathrm{Sr} \rightarrow {}^{90}\mathrm{Y} + \beta^{-} + \bar{\nu}$$

$$^{18}\mathrm{F} \rightarrow {}^{18}\mathrm{O} + \beta^{+} + \nu$$

$$^{99}\mathrm{Tc}^{\mathrm{m}} \rightarrow {}^{99}\mathrm{Tc} + \gamma$$

The principal conservation laws that govern these decay equations are conservation of energy, A, and Z. Conservation of A means that the sum of the atomic mass numbers on both sides of the equation must agree. Conservation of Z means that the sums of atomic numbers on both sides of the equation must agree. (An electron is considered to have $Z = -1$, and a positron $Z = +1$).

Invoking conservation of energy requires converting all masses into their equivalent energies, according to the Einstein relationship, $E = Mc^2$. In fact, when dealing with masses and energies on this submicroscopic level it is more convenient to express the masses of particles as energies directly, that is Mc^2 instead of M. Strictly speaking, we would call this the rest energy of a particle, but colloquially we usually just call it the mass.

$$1\ \mathrm{amu} = 931.494\ \mathrm{MeV}/c^2$$

$$M_{\mathrm{proton}} = 938.272\ \mathrm{MeV}/c^2$$

$$M_{\mathrm{electron}} = 0.511\ \mathrm{MeV}/c^2$$

$$M(^{1}\mathrm{H}) = 938.783\ \mathrm{MeV}/c^2$$

Mass tabulations always give atomic masses, not nuclear masses. The atomic masses are the appropriate quantities to use in an energy balance calculation, but in beta-decay the atomic electron count must also be balanced.

The difference between the rest energy of the radionuclide and the rest energy of the decay products appears as kinetic energy of the decay products. The kinematics of the decay are governed by the law of conservation of momentum. In α-decay, for example, the α-particle and the product nucleus are emitted at 180° with respect to each other and have kinetic energies in inverse proportion to the ratios of their respective masses. Beta-decay is more complicated, because there are three particles in the final state. As a result the energy and momentum carried off by any one of the particles is not unique, but can vary in a statistically described way within constraints imposed by conservation of energy and momentum. The third particle in β-decay is the neutrino, or ν. The ν is a massless, chargeless particle that always accompanies β-decay. It was first inferred to explain the observed final-state kinematics in β-decay, and was later observed directly. On the average, more than half of the kinetic energy expected in β-decay is carried off by the ν.

The positron is a stable particle in the sense that is not radioactive. However, on coming to rest in matter it interacts with an electron by the reaction

$$e^{+} + e^{-} \rightarrow \gamma + \gamma.$$

The gammas are emitted in diametrically opposite directions, and each has an energy exactly equal to the electron rest energy, 0.511 MeV. This interaction is called positron annihilation. The apparent conversion of matter into *pure* energy is explained by calling the positron antimatter—hence the mutual annihilation of matter (electron) and antimatter (positron). The positron camera is based on detection of the two 0.511 MeV collinear gamma rays that accompany the annihilation.

Particle accelerators. The earliest high-energy particle accelerators were high-voltage machines that imparted to a particle of electrical charge q a kinetic energy equal to $|q|V$, where V is the magnitude of the high voltage. A particle with a charge e (one electron charge) attains a kinetic energy of V electronvolts, V eV. Electrostatic high-voltage machines can produce electrons and protons with kinetic energies up to several MeV. The cyclotron, invented by E. O. Lawrence, achieved high particle kinetic energies through a succession of cleverly applied lower-voltage accelerations, eliminating the need for very high voltages, but requiring instead a radiofrequency

(RF) voltage oscillator and a magnet to bend the particle in a path that would expose it to repeated accelerations.

To achieve very high kinetic energies, sophisticated electromagnetic focusing techniques have been developed to confine the particles to their path over very long transit distances. The highest-energy particle accelerator today produces protons with energies of 1 TeV (10^{12} eV). In accelerating to this energy the proton circumscribes repeatedly an almost circular path, traveling a total distance of approximately 20 million kilometers before reaching full energy. Such ultra-high-energy accelerators are used only for basic research, However, electrostatic accelerators and cyclotrons in the 10 MeV range are routinely used for radiation therapy, and for medical radioisotope production. In addition, at a few major research laboratories radiotherapy is performed with α-particles and other heavier ions accelerated to kinetic energies in excess of 1 GeV (10^9 eV).

Cosmic rays. Galactic and extragalactic natural particle accelerators produce high-energy nuclides, principally protons, that continuously bombard our upper atmosphere. These *primary* cosmic-ray nuclides collide with the nuclei of the air molecules, producing *secondary* radiation that reaches the surface of the Earth. This secondary radiation includes electromagnetic *showers* (electrons and γ-rays in a burst, as opposed to individually), neutrons, and muons. Muons are the principal constituent of secondary cosmic radiation at the surface of the Earth. The muon (μ) is a *heavy electron* ($m_\mu = 206.8\ m_e$) that decays into an ordinary electron with a half-life, $t_{1/2} = 1.52$ μs. Muons exist only as the transient products of cosmic-ray or high-energy particle accelerator interactions. The surface of the Earth is bombarded by approximately one muon; per square centimeter per minute. These muons have an average kinetic energy greater than 1 GeV. Very few muons with energy less than 500 MeV reach the Earth's surface because lower-energy muons decay in the atmosphere or are bent away by the Earth's magnetic field.

The production of radioactivity

Most naturally occurring radioactivity results from the decay of ^{232}Th, ^{235}U, and ^{238}U. (Table 13.1). The half-lives of these isotopes are of the order of the life of the Earth, so some still remain from primordial times. Each of these isotopes undergoes a succession of α- and β-decays to final stable products: ^{208}Pb, ^{209}Bi, ^{206}Pb respectively. Other heavy, shorter-lived radioactive elements such as radium and radon are intermediate decay products of these three *decay chains*. The most significant other radionuclide that still exists from primordial times is ^{40}K. There are a few other, even longer-lived nuclides in nature, but because of their very long half-lives and consequent very low activities, they are of little practical interest.

All other radionuclides are produced by nuclear reactions. Two nuclides can interact, producing as a product any rearrangement of the constituent protons and neutrons, subject to the conservation of energy and momentum. The likelihood of a particular rearrangement is described by a reaction *cross section*, σ. If we think of one of the two reacting nuclides as a point projectile, then

Table 13.1 Sources of natural radioactivity

Radioisotope	Radiation; max. energy (MeV)	Half-life (years)	Isotopic abundance (%)	Comments
^{232}Th	α; [4.01]	1.41×10^{10}	100	Primordial. Note that the half-lives are of the order of, or greater than, the age of the Earth.
^{235}U	α; [4.60]	7.04×10^8	0.72	
^{238}U	α; [4.47]	4.47×10^9	99.275	
^{40}K	β; [1.33](89%) EC + γ; 1.46(11%)	1.28×10^9	0.0117	
^{87}Rb	β^-; 0.27	4.72×10^{10}	27.83	
^{226}Ra	α; 4.78	1600		Byproducts of ^{238}U decay
^{222}Rn	α; 5.49	1.05×10^{-2}		
^{14}C	β; 0.16	5730		Produced in the air by cosmic-ray neutrons

σ is the area of the target presented by the other nuclide. The common unit for measurement of reaction cross sections is the *barn*, b, which is of the order of the geometrical cross-sectional area of a nucleus: $1\ \mathrm{b} = 10^{-28}\ \mathrm{m}^2$. Only a small fraction of the possible reactions are of any practical interest. The remainder have unobservably small cross sections. Reaction cross sections, under certain conditions, called *resonances*, can be much larger than geometric cross sections.

The equation for a reaction is written

$$\mathrm{a} + \mathrm{X} \rightarrow \mathrm{Y} + \mathrm{b},$$

where X and a are the reactants and Y and b are the products. Ordinarily the lower-case letters are used to designate the lighter, or less complex nuclides, or, in the case of a, the projectile, as opposed to the target. A more abbreviated, and frequently preferred, reaction notation is X(a, b)Y. A very important naturally occurring radionuclide is produced by the interaction of cosmic-ray neutrons in the atmosphere, $^{14}\mathrm{N(n, p)}^{14}\mathrm{C}$.

At low relative kinetic energies the reaction cross section for two nuclides becomes vanishingly small because the repulsive Coulomb force between the nuclides keeps them apart. On the other hand, neutrons, being uncharged, are not repelled. It can be shown that, at low kinetic energies, neutron reaction cross sections are velocity-dependent, and vary as $1/v$, where v is the relative velocity between the neutron and the target nucleus.

A neutron in the atmosphere, or in condensed matter, behaves not unlike a fast atom in a gas. It undergoes a succession of collisons, and ultimately comes to thermal equilibrium with the other material. We call this a *thermal neutron*. Although the neutron is radioactive, with a half-life of 10.25 minutes, this half-life is so long that in matter a neutron will never decay. It will always *thermalize* and react, unless it reacts even before the thermalization. The most common neutron reaction is simply neutron *capture*, accompanied by γ-emission, written

$$^{A}\mathrm{X}(\mathrm{n}, \gamma)^{A+1}\mathrm{X},$$

where $^{A}\mathrm{X}$ is the isotope of the element X, with atomic mass number A. The product nuclide $^{A+1}\mathrm{X}$ is frequently, but not necessarily, radioactive. In the case of certain heavy nuclides, notably $^{235}\mathrm{U}$ and artificially produced $^{233}\mathrm{U}$ and $^{239}\mathrm{Pu}$, the product nuclide breaks in two: it *fissions*. The fission products will promptly emit both neutrons and γs, and are invariably radioactive.

Common reactions for producing radionuclides in particle accelerators are (d, n), (d, p), (d, α), (p, n), and (p, α). Particle accelerators allow production of a greater range of radionuclides. In addition, with a particle accelerator the target nuclide can be chemically different from the radioactive product, allowing chemical separation. For example, consider the reaction

$$\mathrm{d} + {}^{34}\mathrm{S} \rightarrow {}^{32}\mathrm{P} + \alpha$$

as an alternative to

$$\mathrm{n} + {}^{31}\mathrm{P} \rightarrow {}^{32}\mathrm{P} + \gamma.$$

13.1.2 The interaction of radiation in matter

Ionization

The primary mechanism for energy transfer by radiation to matter (called the *absorber*) is through ionization of the absorber's atoms by an electrically charged component of the radiation. Charged particles (α, p, e, etc.) transfer their energy directly. Neutral particles (n, γ, etc.) make large energy transfers to charged particles (nuclides in the case of neutrons, and electrons in the case of gammas), which then transfer their energy by ionization.

Ionization energy transfer by a heavily charged particle passing through an absorber is described by the Bethe–Bloch, or *stopping-power*, equation,

$$\frac{\mathrm{d}E}{\mathrm{d}x} = \left(\frac{e^2}{4\pi\epsilon_0}\right)^2 \frac{4\pi z^2}{mv^2}\left(\ln\frac{2mv^2}{I(1 - v^2/c^2)} - v^2/c^2\right)N_e, \tag{13.4}$$

where z is the electronic charge of the particle, v is its velocity, m is the electron mass, e is the electron charge, c is the velocity of light, N_e is the number of electrons per unit volume in the absorber, and I is the average energy required to ionize these electrons.

This formula overestimates the energy transfer when the particle velocity falls below the velocity of its own bound K-shell electrons, or of the K-shell electrons of the absorber. In the former case the particle will begin to capture electrons into atomic orbits, thereby reducing its effective z. In the latter case, the particle will no longer be able to ionize some of the electrons on the absorber atom, thereby reducing the effective number of electrons available for energy transfer. In most applications, corrections for these effects are not important, because they occur only near the very end of the particle's *range*.

The range, the distance the particle travels before coming to rest, may be calculated as

$$R = \int_{E_0}^{0} -\frac{\mathrm{d}E}{\mathrm{d}E/\mathrm{d}x}, \qquad (13.5)$$

where E_0 is the initial kinetic energy of the radiation. In practice the integral is not taken to zero kinetic energy. The stopping-power equation is not well behaved, as v becomes comparable to the average velocity of bound electrons in the absorber atom, so the integral must be truncated in some fashion, for example by setting the lower limit of the integral to the kinetic energy corresponding to $2mv^2 = I$. The result for the calculated range is not very sensitive to the exact choice of truncation velocity.

Because the energy loss of charged particles occurs as a result of many interactions, each of which involves a very small energy transfer, energy loss is uniform along the particle track and, subject only to small statistical fluctuations, all particles of the same z, A, and kinetic energy will have exactly the same range, that is, travel exactly the same distance before coming to rest. This effect is clearly observable for heavy charged particles. For electrons, however, because these light particles scatter easily, the total path length and the crow's-flight distance do not correspond, so the uniformity of energy loss is not as easily evident.

The stopping power and range of an α-particle in water are shown in Figure 13.1. The stopping power, in biological contexts, is more commonly called *unrestricted linear energy transfer*, which is abbreviated or symbolized as LET, L_∞[1] or simply L. Water is a very good analog for soft tissue.

Stopping power is a function only of the velocity and the charge of the fast ion.

$$L = z^2 f(v), \qquad (13.6)$$

so Fig. 13.1 can be used to determine L and R for other ions using the formulae

$$L(z, A, T) = \frac{z^2}{4} L\left(2, 4, \frac{4}{A}T\right), \qquad (13.7)$$

$$R(z, A, T) = \frac{A}{Z^2} R\left(2, 4, \frac{4}{A}T\right), \qquad (13.8)$$

where z is the ion's charge (atomic number), A its atomic mass number, and T its kinetic energy. (Note that an α-particle of kinetic energy $\frac{4}{A}T$ has the same velocity as an ion of atomic number A and kinetic energy T.)

The kinetic energies of most β-rays from radioactive sources are comparable to or greater than their rest energy. As a result their velocities approach a constant, the speed of light, and their stopping power is also constant, about 2 MeV cm^{-1} in water.

Uncharged particle interactions

Uncharged radiation, principally neutrons and gamma rays, interact more sporadically, passing through matter unimpeded for some significant distance, and then undergoing a single, sudden interaction involving a large energy transfer. Microscopically, such interactions are also characterized by a cross section, σ. Macroscopically, however, it is usually more convenient to characterize them by an *attenuation coefficient*, μ, which is the probability of an interaction per unit length of path. The attenuation coefficient and the cross section are related by

$$\mu = N\sigma, \qquad (13.9)$$

where N is the number of *targets* per unit volume, each with cross section σ. For a single-element absorber (more precisely a single-isotope absorber, because neutron cross sections are different for different isotopes of the same element) the number of targets per unit volume is just the number of atoms per unit volume, or

$$N = N_A \frac{\rho}{A}, \qquad (13.10)$$

where N_A is the Avagadro constant, A is the atomic mass number, and ρ is the mass density. The linear attenuation coefficient is then

$$\mu = 0.602 \frac{\rho\sigma}{A} \mathrm{cm}^{-1} \qquad (13.11)$$

if ρ is in g cm^{-3} and σ in barns. Frequently attenuation is expressed as μ/ρ. This is called the *mass attenuation*

[1]For biological effects, the radial distribution of the ionization density is of interest, so a distinction is made between L_∞ and L_Δ, the *restricted linear energy transfer*, which includes only the energy transferred into a cylindrical volume of unit length and radius Δ about the track of the particle.

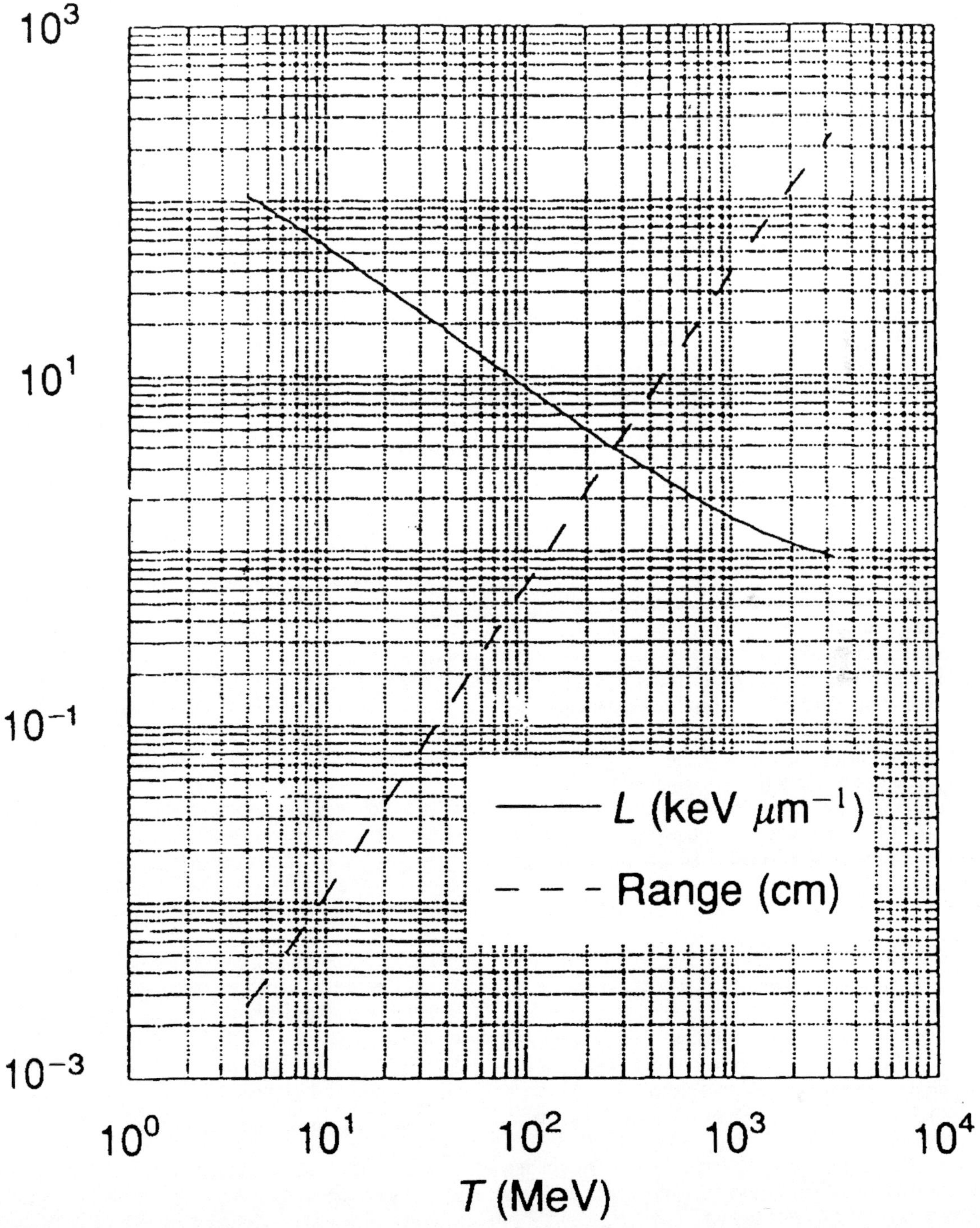

Fig. 13.1 Linear energy transfer (L) and range of an α-particle in water, calculated using the formula and the method described in the text.

coefficient and has the dimensions $cm^2\ g^{-1}$. The reciprocal of μ is the average distance a particle travels before a collision, and is called the *mean free path.*

For example, in Pb ($A = 207.2$, $\rho = 11.35$, and taking the total cross section from Fig. 13.2) the mean free path of 1 MeV photon is

$$1/\mu = 207.2/[11.35 \times 0.602 \times (17 + 6)] = 1.3\ \text{cm}.$$

This means that a 1 MeV photon will, on the average, travel 1.3 cm in Pb before it has a collision. (Photon cross sections are described in more detail in the next section.)

Neutral (uncharged) radiation of intensity ϕ will be attenuated according to the relationship,

$$\frac{d\phi}{\phi} = -\mu\ dx,$$

or

$$\phi = \phi_o e^{-\mu x}, \tag{13.12}$$

where ϕ_o is the incident intensity, and ϕ is the intensity after passing through a thickness x of absorber. This attenuation in intensity is the basis for radiation shielding. The interactions themselves do not transfer significant energy directly to the bulk matter. They merely transfer energy to charged particles, electrons or nuclei, which then produce the ionization.

Photons have three principal interactions in matter: the photoelectric effect, Compton scattering, and pair production. Each is characterized by an energy-dependent cross section. Neutrons interact through billiard-ball-type scattering, collisions with the atomic nuclei of the absorbing material. The attenuation coefficient for neutrons is sometimes called the *macroscopic cross section*, Σ.

Photon interactions

The most important photon interaction at low energies is the photoelectric effect, in which an atom absorbs the photon and emits an orbital electron. This electron will have a kinetic energy equal to the energy of the incident photon less the initial binding energy (or ionization potential) of the emitted electron. The atom then returns to its ground state by the emission of characteristic X-rays or Auger electrons. At energies above the binding energy of the K-shell electron the photoelectric cross section varies approximately as $E^{-3.5}Z^5$.

At intermediate energies (approximately 0.2 to 5 MeV) the photon interaction process is dominated by direct collisions with individual atomic electrons, Compton scattering. The electron binding energy is small compared to the energy transfer, and to a very good approximation the process can be described as an elastic collision between a photon and a stationary free electron (a relativistic billiard-ball collision). The energy of the scattered photon can be described as a function of its scattering angle,

$$E' = \frac{E}{E/m(1 - \cos\theta) + 1}. \tag{13.13}$$

At high photon energies ($E > 2mc^2$), the photon in a nuclear collision will *break up* into an e^+–e^- pair, *pair production* (the inverse process to annihilation, mentioned earlier). Because total energy is conserved, the sum kinetic energies of the two electrons is equal to the photon kinetic energy less the rest energy of the two electrons, $2mc^2$.

Figure 13.2 shows the cross sections for the three different types of photon interactions as a function of energy for a water molecule ($2\sigma_H + \sigma_O$), and for a lead atom. The break in the Pb photoelectric cross section is at 82 keV, the binding energy of a K-shell electron in Pb. This is called the K edge. Such a break occurs in the cross section for all elements. It also occurs at the binding energies of the higher electronic shells and subshells. A break does not appear in the water cross section, because the K-shell binding energy for O is less than 1 keV, off-scale on the graph. The electronic Compton cross section, σ_e, in the free-electron approximation, is independent of the atomic absorber. The atomic Compton cross section, shown in the figure, is equal to $Z\sigma_e$. A comparison of the Pb and H_2O Compton cross sections at 1 MeV, for example, should show that they are in the ratio

$$Z(\text{Pb})/[Z(\text{O}) + 2Z(\text{H})] = 82/10.$$

The pair-production cross section has an approximate Z^2-dependence. At very high energies the pair-production cross section of an electron is approximately equal to that of a proton, so we would expect, for example, that at 100 MeV the ratio of the pair cross sections for Pb and H_2O would be of the order of

$$[Z(\text{Pb})^2 + Z(\text{Pb})]/\{[Z(\text{O})^2 + Z(\text{O}) + 2Z(\text{H})]\} \approx 90.$$

Lead is the most commonly used shielding material around gamma sources and x-ray machines because of its high atomic number and relatively low cost. Water can be used to approximate soft tissue, but, because of the strong Z-dependence of the photoelectric cross section in

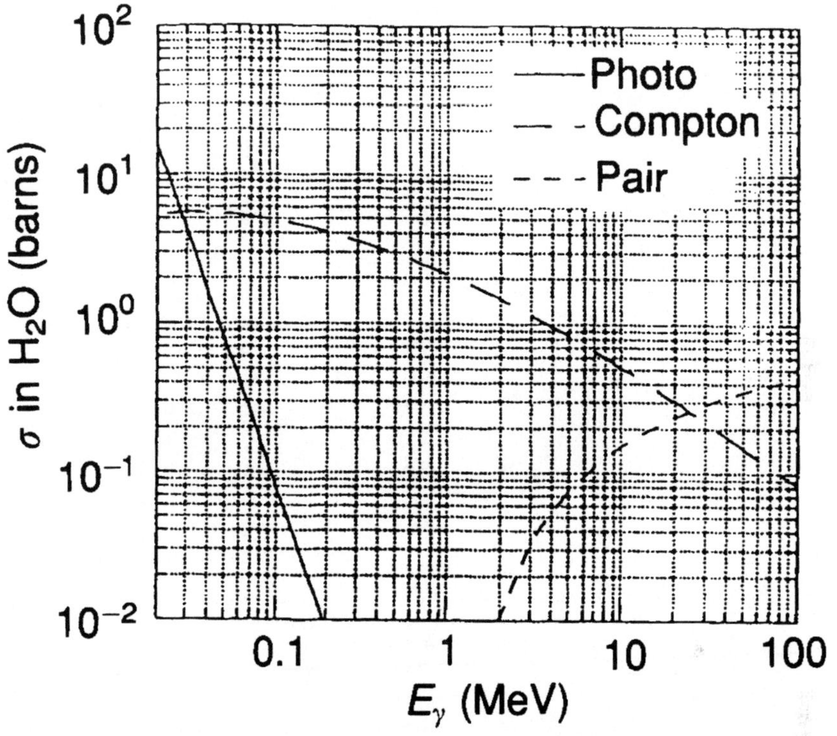

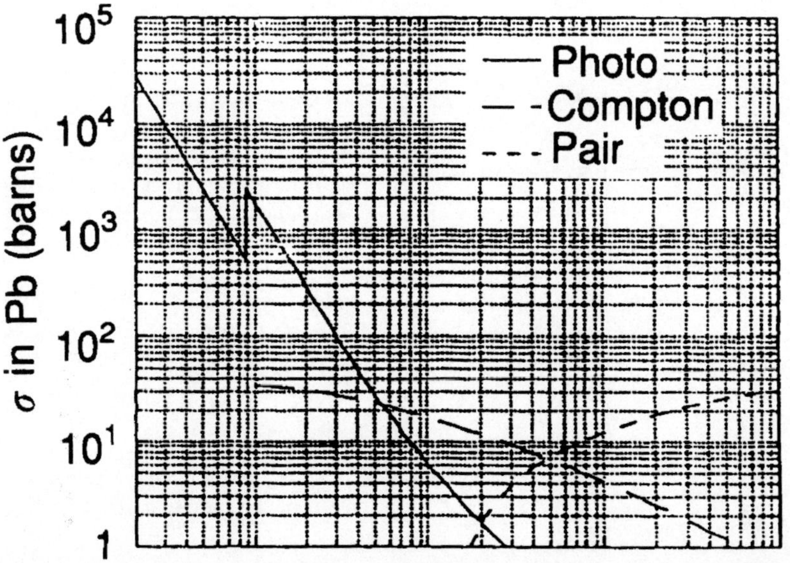

Fig. 13.2 Gamma-ray cross sections for lead and water. The total cross section at any energy is obtained by summing the individual contributions (photo, Compton, and pair). At most energies, however, only one of the three is important. The data for these curves as well as for other materials can be found in Storm and Israel (1970).

particular, so-called *tissue-equivalent* mixtures of elements are frequently used for certain calculations and measurements. The calcium ($Z = 20$) content of bone gives it a much higher attenuation coefficient than soft tissue.

13.1.3 Radiation dosimetry

The *specific energy* (the energy per unit mass) transferred by radiation to matter is called *dose*, and the procedure of calculating or measuring dose is called *dosimetry*. The deposition of energy in a unit mass is best pictured by imagining a rectangular volume of dimension 1 cm × 1 cm × $1/\rho$ cm. This volume will have a mass of 1 g (if the units of ρ are g cm^{-3}). If $\mu_a E$ is the energy deposited by a quantum of radiation per cm of track length, then the rate at which energy is deposited per g, or the *dose rate*, is

$$\frac{dD}{dt} = \frac{\phi \mu_a E}{\rho}, \qquad (13.14)$$

that is, the product of the number of quanta incident on 1 cm^2 times the energy each quantum deposits in a distance $1/\rho$. While this description of dose rate applies to charged particles too, the quantity μ_a is most frequently used in connection with photons, and is called the *linear absorption coefficient*. It is related to the linear attenuation coefficient by

$$\mu_a = \frac{\overline{\Delta E}}{E}\mu, \qquad (13.15)$$

where $\frac{\overline{\Delta E}}{E}$ is the average energy transfered to ionizing particles in a collision.

Dose is generally inferred from direct measurements of ionization with radiation detectors, which are then converted to dose using conversion factors that relate average ionization density to energy deposited. The ionization density is called *exposure*. The historical unit of exposure is the roentgen, R (named for Wilhelm Roentgen, the discoverer of x-rays):

$$1\ \text{R} = 1\ \text{esu of separated charge per cm}^3\ \text{of dry air at stp}$$
$$= \frac{1}{3} \times 10^{-3}\,\text{C m}^{-3}.$$

The SI unit of absorbed dose is the gray (Gy), 1 Gy = 1 J^{-1}. In most literature on dosimetry, however, the old unit the rad is still used:

$$1\ \text{rad} = 10^{-2}\,\text{J kg}^{-1}$$
$$= 10^{-2}\,\text{Gy} \equiv 1\ \text{cGy}.$$

The symbol rad is an acronym for 'radiation absorbed dose'. In air, 1 R of exposure produces a dose approximately equal to 1 rad.

Two other quantities closely related to dose are also used in dosimetry: the kerma and dose equivalent.

Kerma. When the primary radiation is electrically neutral, it is not always easy to calculate, or to measure precisely how much energy is transferred to bulk matter directly. For example, when a γ-ray has a Compton interaction in a volume element, it is a comparatively easy matter to calculate or to measure the energy transferred to an electron. The scattered electron then loses energy according to the stopping-power equation, and has a finite range. So it will carry energy away from the initial point of interaction. Accounting for all of the interaction details is very complicated. An easier quantity to determine is the amount of energy that is converted to ionizing radiation. The specific energy transfer to ionizing radiation is called kerma (an acronym for 'kinetic energy released in matter'). The concept of kerma simplifies and standardizes the bookkeeping for certain energy-transfer calculations. Kerma is usually close to, but not the same magnitude as, dose. As a clearly defined quantity it can be calculated and compared for a variety of circumstances.

Dose equivalent. Radiation doses are found not to be equivalent in producing certain biological effects. Many biological effects depend, for example, on the energy transferred to, or the ionization produced in, a specific sensitive target, rather than that averaged over a unit volume. For biological effects, therefore, energy transfer along the track of a single ionizing particle is very important. This characteristic of the radiation is called its relative biological effectiveness, RBE, and is quantified by a quality factor, $Q(L)$. The dose equivalent.

$$H = \bar{Q} \times D. \qquad (13.16)$$

where $\bar{Q}$ is the quality, suitably averaged over the range of L of the radiation producing the dose D, that is,

$$\bar{Q} = \int_{L_{min}}^{L_{max}} \frac{Q(L)\,D(L)\,dL}{D}. \qquad (13.17)$$

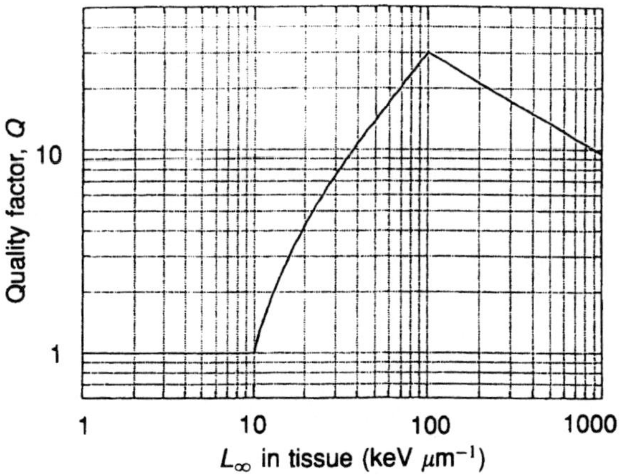

Fig. 13.3 Quality factor as a function of L (ICRP 1991).

Table 13.2 Radiation weighting factors ($\bar{Q}$) recommended for use as alternatives to detailed integration over L

Radiation	$\bar{Q}$
X-rays, γ-rays	1
Electrons, muons	1
Neutrons	
$E \leq 10$ keV	5
10 keV $< E <$ 100 keV	10
100 keV $< E <$ 2 MeV	20
2 MeV $< E$	10
Protons	
$E <$ 30 MeV	20
30 MeV $< E$	1
α-particles and heavy ions	20

Source: ICRP (1991)

Table 13.3 Radiation exposure to the US population from natural and man-made sources

	μ rem h^{-1}	mrem a^{-1}	mSv a^{-1}
NATURAL			
Cosmic radiation			
San Francisco, CA	4.7	41	0.41
Denver, CO	8.6	75	0.75
Airplane (12 km)	500		
Terrestrial radioactivity			
External			
San Francisco, CA	3.2	28	0.28
Denver, CO	10.3	90	0.90
^{14}C	0.16	1.4	0.014
^{87}Rb	0.10	0.98	0.01
U, Th, products	3.2	28	0.28
^{40}K	2.16	19	0.19
Inhaled (from radon)	23	200	2.0
TECHNOLOGICALLY ENHANCED			
Building materials and water supply	0.8	7.5	0.075
Other consumer products (TV, smoke alarms, etc.)	$<10^{-2}$	<0.1	$<10^{-3}$
Nuclear power	$<10^{-4}$	$<10^{-3}$	$<10^{-5}$
MEDICAL PROCEDURES (population av.)			
Diagnostic x-rays	4.5	39	0.39
Nuclear medicine	1.6	14	0.14

Sources: NCRP (1987b); Oakley (1972).

The Sievert (Sv) is the SI unit of dose equivalent. A dose of 1 Gy gives a dose equivalent of $\bar{Q}$ Sv. In much of the radiation literature, dose equivalent is given in rem (an acronym for 'roentgen equivalent man'). A dose of 1 rad gives a dose equivalent of Q rem.

The weighting, correlation, and evaluation of radiation injury data are quite complicated, and subject to varying interpretations. Therefore, the assignment of Q to various types and energies of radiation continues to be an evolving process. However, for consistency in radiation safely evaluations it is important to have consensus values. Periodically, such concensus values are set and recommended for use by national or international advisory bodies, the NCRP (National Council on Radiation Protection and Measurement) and ICRP (International Commission on Radiological Protection) respectively. Figure 13.3 shows the recommended relationship between Q and L (ICRP 1991).

Evaluation of $\bar{Q}$, however, can be tedious, even when the exact nature and energy of the radiation are known. Therefore an alternative, simplified but conservative, procedure for assigning $\bar{Q}$ is also recommended (Table 13.2.)

Radiation exposure

People are continuously exposed to a number of sources of ionizing radiation. A composite of recent estimates is given in Table 13.3.

Individuals may be exposed to considerably more radiation than the population average, either through normal occupational or medical procedure, or in connection with some accident. In addition, there are places on the Earth where radon and other terrestrial radiation exposure can be as much as an order of magnitude greater than the values in the table.

Examples of typical doses from x-ray and nuclear medicine procedures are given in Tables 13.4 and 13.5.

13.1.4 Biological effects of radiation

The ionization produced by the absorption of radiation energy in human tissue can cause biological effects and medically observable changes. Effects that are relevant to radiation protection are discussed in this section; processes relevant to radiation therapy are covered in Section 13.2.2.

Effects can be produced anywhere or everywhere in the body, depending on the nature of the irradiation. Examples include total body exposure due to highly penetrating radiation, exposure limited to the skin or an internal membrane due to minimally penetrating particles or *soft* x-rays, localized exposure of a small volume of tissue by a collimated external beam, or internal exposure by absorption of a radioactive isotope. The nature of the effect clearly will be dependent on the type and volume of tissue involved and may not be a simple function of total absorbed energy. Remember that dose is an energy density (erg g^{-1} or J kg^{-1} etc.). Exposure can be acute or chronic. Acute exposures cover time periods from instantaneous to minutes or hours, i.e. short compared to the time involved in biological repair processes. Chronic exposures occur over weeks, months, or years. Because of *dose-rate effects*, doses which could cause serious medical problems if given acutely may not even cause observable effects if given chronically.

Massive tissue damage from acute high-dose exposure of a large portion of the body can cause severe illness or death due to organ failure. By contrast, small doses of radiation given chronically to specific sensitive tissues can cause subtle damage at the sub-cellular level, which may not be observable for a generation. Examples of this include birth defects and effects in offspring due to genetic mutations in sperm and egg cells, and *transformations* by mutation of normal somatic cells into malignant cells which can develop into a tumor, with a *latency* time of 10–30 years. These have been called *late effects* in the past, but are now called *stochastic effects*.

Stochastic effects

Stochastic effects are dominated by randomness in the processes which cause the effects. They are governed by statistical laws, but because of the wide variability in the biological and physical processes involved, one cannot discover the specific cosmic-ray particle hit that caused a birth defect, and one cannot trace a cancer case back to a specific x-ray dose. For each increment of radiation exposure there is a very small but nonzero probability of inducing such a late effect. Confounding estimates of these relationships is the fact that mutagenesis and carcinogenesis are not unique to man-made radiation. Environmental background radiation exposure to all life on Earth causes a number of mutations and cancers, and a much larger *background* level of cancers and mutations exists due to other processes as universal as thermal vibration of DNA molecules, and mutagens contained in natural foods. Since stochastic effects of doses of

Table 13.4 Collective effective dose equivalents for the USA in 1980 from diagnostic medical x-ray examinations

Examination	Annual number of examinations (thousands)	Average effective dose equivalent per examination (μSv)[b]	Annual collective effective dose equivalent[a] (person-Sv)[c]
Computed tomography (head and body)	3300	1100	3700
Chest	64 000	60	4100
Skull and other head and neck	8200	200	1800
Cervical spine	5100	200	1000
Biliary tract	3400	1900	6400
Lumbar spine	12 900	1300	16 400
Upper gastrointestinal tract	7600	2450	18 500
Kidney, ureter, bladder	7900	550	4400
Barium enema	4900	4050	19 900
Intravenous pyelogram	4200	1600	6600
Pelvis and hip	4700	650	3000
Extremities	45 000	10	450
Other[d]	8400	500	4200
Rounded total	180 000	500	~91 000

Source: NCRP (1987b).
[a] Numbers obtained from product of two previous columns but using unrounded figures.
[b] 1 μSv = 0.1 mrem.
[c] 1 person-Sv = 100 person-rem.
[d] Estimated from the average of all examinations.

Table 13.5 Collective effective dose equivalent for the USA in 1982 from diagnostic nuclear medicine tests

Examination	Annual number of examinations (thousands)	Average effective dose equivalent per examination (μSv)[b]	Annual collective effective dose equivalent[a] (person-Sv)[c]
Brain	810	6500	5300
Hepatobiliary	180	3700	700
Liver	1400	2400	3400
Bone	1800	4400	8000
Lung	1200	1500	1800
Thyroid	680	5900	4000
Kidney	240	3100	700
Tumor	120	12 000	500
Cardiovascular	950	7100	6700
Rounded total	7400	4300	32 000

Source: NCRP (1987b).
[a] Numbers obtained from product of two previous columns but using unrounded figures.
[b] 1 μSv = 0.1 mrem.
[c] 1 person-Sv = 100 person-rem.

radiation which are small (compared to environmental background) are not generally observable, it is necessary to extrapolate downward from effects observed at high doses. If one assumes that no threshold dose is necessary for a minimum biological effect, and accept a linear dose–effect relationship (i.e. the probability of effect is exactly proportional to the dose) it is possible to extrapolate down to very small doses for the purposes of radiation protection calculations. This is called the *linear hypothesis* and is the most convenient and the most commonly used model. Another common assumption is that of interchangeability of person-rads. That is, we posit that the effect (e.g. total number of cancers) from exposing a population of 1000 people to 10 rad of acute whole-body radiation is the same as exposing 10 000 people to 1 rad, or 1 million to 0.01 rad (cGy).

Much of the data on stochastic effects come from the continuing studies of survivors of the bombings of Hiroshima and Nagasaki in 1945. This is by far the largest population that has received radiation doses high enough to give statistically significant results. Even in these studies the doses have been calculated by inference (not measured) and there is some uncertainty in the estimate, resulting in occasional reassessments of risk, based on refinements of dosimetric data analysis.

It should be noted that there are pitfalls in extending the linear hypothesis too far. For example, in Denver, Colorado, the population receives twice the natural cosmic-ray and terrestrial dose compared to San Francisco, California. Despite the large population samples involved and good epidemiological data, the expected differential carcinogenesis is not observed. In fact, some cancers known to be potentially radiogenic are more prevalent in San Francisco, presumably due to other environmental factors.

A simplified dose–response model that accounts for the time delay before cancer appears (latency) and for the uncertainty in the time of occurrence (plateau) is illustrated in Figure 13.4. Some model values for these parameters are given in Table 13.6. During the plateau period the risk to an individual is described by a risk factor expressed in deaths per year per rem per 10^6

Table 13.6 Model values for risk estimation

Age of irradiation (years)	Type of cancer	Duration of latent period (years)	Duration of plateau region[a] (years)	Risk estimate	
				Absolute risk[b]	Relative risk[c]
In utero	Leukemia	0	10	25	50
	All others	0	10	25	50
0–9	Leukemia	2	25	2.0	5.0
	All others	15	(a) 30 (b) Life	1.0	2.0
10+	Leukemia	2	25	1.0	2.0
	All others	15	(a) 30 (b) Life	5.0[d]	0.2

Source: BEIR (1972, Table 3.2). Both numerical and modeling details have been superseded in later work, but the present authors felt that the added sophistication of the later modelings would not change significant conclusions sufficiently to warrant the added complication to the pedagogy. The latest consensus conclusions about risk can be found in BEIR (1990).

[a] Plateau region: interval following latent period during which risk remains elevated; (a) and (b) are alternative model assumptions.

[b] Deaths per year per rem per 10^6 population.

[c] Per cent increase in deaths per rem.

[d] The absolute risk for those aged 10 or more at the time of irradiation for all cancers excluding leukemia can be broken down into the respective sites as follows:

Breast	1.5
Lung	1.3
GI incl. stomach	1.0
Bone	0.2
All others	1.0

The figure for breast cancer is derived from a reference value of 6.0 incidences corrected for a 50% cure rate and the inclusion of males as well as females in the population.

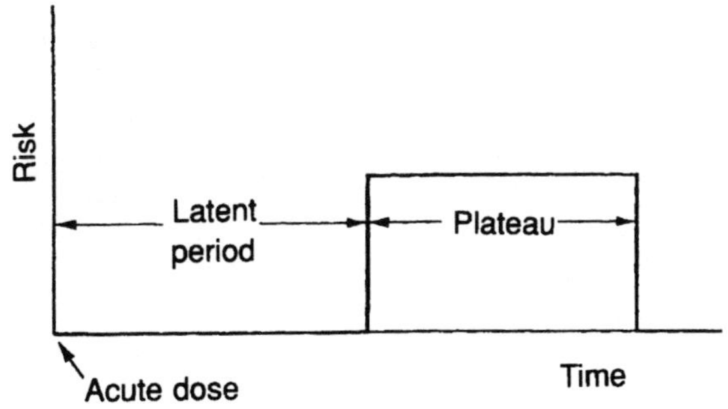

Fig. 13.4 Risk vs. time relationship for simplified dose–response model.

exposed population. For some cancers the length of the plateau is demonstrably finite; for others the risk may continue thoughout life. The relationship of the radiation risk to the risk from the other cancer causes can be assumed to be additive (absolute risk) or multiplicative (relative risk). The difference in consequences of the two assumptions is only observable in later years of life when the incidence of *spontaneous* cancer increases (approximately doubling with every eight years of life). The present recommendations of the BEIR committee[2] is to assume that radiogenic cancers also increase with age in the same proportion.

Nonstochastic effects

Nonstochastic radiation effects can be likened to sunburn, not just at the skin, but wherever the radiation penetrates in the body. Since the effect is nearly always observed for a given high dose, the probability of an effect is essentially unity, by contrast with the small probabilities for stochastic effects. The severity of nonstochastic effects increases with dose, but not necessarily in a linear relation. Erythema (skin reddening) is an easily observed example, and was used historically for measuring clinical x-ray treatment doses, before modern dosimetric techniques were developed. Other nonstochastic effects include cataract formation and suppression of bone marrow production of blood cells.

Acute whole-body exposure can be fatal at high doses. At doses from 10 000 to 100 000 rad, fatality occurs in minutes, due to damage to the central nervous system (CNS death). Acute abdominal doses of about 500 rad can cause death in a matter of days to weeks by destroying a critical fraction of the radiosensitive cells of the lining of the intestines (gut death). Doses to the whole body as low as 300 rad can cause death in a period of several weeks by destroying the stem cells in the bone marrow which produce white cells (resulting in overwhelming infection), red cells (resulting in severe anemia), and platelets (resulting in bleeding).

The common expression for lethal dose is LD 50/30, meaning the acute whole body dose that is expected to be fatal for 50 per cent of the exposed population within 30 days. For humans, the LD 50/30 is 300–500 rad, depending on age, health, and medical care available. We emphasize that this applies only to whole-body exposures received in a short time. The same dose applied over weeks or months might not produce any symptoms, but changes in blood counts and chromosome abnormalities would be measurable.

13.1.5 Recommended limit on radiation exposure

For the purpose of recommending limits on radiation exposure in the workplace (occupational exposure) the NCRP (1987a) assumes that a whole-body dose equivalent of 10 mSv (1 rem) results in a blanket lifetime risk of 10^{-4} for fatal stochastic effects to an adult worker, and concludes that an annual *effective dose equivalent* limit of 50 mSv (5 rem) would constitute a workplace risk comparable to that of a worker from other work-related causes in a *safe* industry. For the purpose of calculating an effective dose equivalent, H_E, localized doses are assigned weighting factors according to Table 13.7, and H_E is defined as

$$H_E = \sum_T H_T w_T, \qquad (13.18)$$

and represents the whole-body dose that would present the same risk as the local tissue dose H_T. For example, a thyroid dose equivalent of 10 rem is assumed to represent the same risk as a uniform dose to the whole body of 0.3 rem.

The NCRP sets additional limits for nonstochastic effects of 150 mSv (15 rem) for the crystalline lens of the eye, and 500 mSv (50 rem) for all other tissues or organs including the skin and extremities'.

[2]The National Research Council's committee on the Biological Effects of Ionizing Radiations (BEIR)

Table 13.7 Weighting factors (w_T) for calculating effective dose equivalent.

Tissue (T)	w_T[d]	
	NCRP (1987)	ICRP (1991)
Gonads	0.25	0.20
Breast	0.15	0.05
Red bone marrow	0.12	0.12
Lung	0.12	0.12
Thyroid	0.03	0.05
Bone surfaces	0.03	0.01
Colon	[a,b]	0.12
Stomach	[a]	0.12
Bladder	[a]	0.05
Liver	[a]	0.05
Oesophagus	[a]	0.05
Skin	[a]	0.01
Remainder	0.30[a]	0.05[c]
Total	1.00	1.00

[a] At 0.06 per tissue, excluding extremities, skin (0.01), and lens of the eye, to a maximum of 5 additional tissues.
[b] The upper large intestine is considered an additional tissue.
[c] Includes brain, upper large intestine, small intestine, kidney, muscle, pancreas, spleen, thymus, and uterus.
[d] The differences between ICRP and NCRP are judgmental, not based on different information, and are within the ranges of uncertainty of such assignments.

The recommended nonoccupational limits (that is, for the general public) for whole-body-equivalent exposure from man-made sources is an order of magnitude lower, 5 mSv (0.5 rem), for infrequent annual exposure, and 1 mSv (0.1 rem) for continuous annual exposure. The rationale here also is that the risk presented by such exposure is less than that from natural radiation, and is in the same range as, or less than, a 'manifold of mortability risks faced by members of the general public [that] seem ... to be accepted'. A detailed tabulation of the NCRP recommendations is given in Table 13.8. The NCRP, as well as regulatory bodies like the Nuclear Regulatory Commission, have the position that even if radiation exposure within the recommended limits is justifiable, that such exposure should be '*as low as reasonably achievable*' (ALARA principle).

Finally, in the interest of allaying sometimes irrational concerns about the risks from radiation exposure it cannot be too strongly emphasized that eons of natural radiation exposure provide a benchmark that rules definitively against any surprise hazard that could be associated with these exposure levels.

13.1.6 Radiation shielding and control—a few guidelines

Radiation dose is controlled by limiting either the intensity of the radiation or the time of exposure, or both. A comprehensive development of the principles of radiation shielding is beyond the intended scope of this chapter, but the following guidelines should prove useful for working with radioactivity, in particular with gamma-ray sources.

Distance. Most isotropic sources are effectively point sources. Intensity will therefore obey a $1/r^2$ law. Twice the distance will give $\frac{1}{4}$ the dose. A simple, approximate relationship between source activity and exposure is

$$\dot{R} = 6CE/r^2, \qquad (13.19)$$

where $\dot{R}$ is the exposure rate (in R h^{-1}), C is the source activity (in Ci), E is the γ energy (in MeV), and r is the distance from the source (in feet). Not SI, but a very useful thing to know.

Shielding. To a first order of approximation, photon radiation will be attenuated by shielding according to the relationship,

$$\frac{\phi}{\phi_0} = e^{-\mu x}, \qquad (13.20)$$

where x is the shielding thickness. For $\mu x = \ln 2$, ϕ falls by a factor of two. This quantity, $\ln 2/\mu$, is called the half-thickness of the shield or, symbolically, $t_{1/2}$. Each additional half-thickness reduces intensity by another factor of two. We would therefore, for example, expect $20t_{1/2}$ to reduce intensity by approximately 10^6. This is not quite true, because the photon doesn't necessarily lose all of its energy in a collision. The cumulative error due to this incomplete energy transfer becomes relatively larger for larger attenuation factors. This effect is called dose *buildup*, and one must take it into account in designing shields with large attenuation. To minimize shield mass and cost, the shield should be as close as possible to the source. One might say that an r^2 law applies, that is, if r is the distance of a (spherical) shield from a source, the required shield mass will increase as r^2.

Time. Exposure to a source of radiation should only be for as long as it takes to get the job done. Dose is directly proportional to time of exposure.

Table 13.8 Summary of recommendations by the NCRP on radiation exposure limits[a]

A. Occupational exposure (annual)[b]		
1. Effective dose equivalent limit (stochastic effects)[c]	50 mSv	(5 rem)
2. Dose equivalent limits for tissues and organs (nonstochastic effects)		
(a) Lens of eye	150 mSv	(15 rem)
(b) All others (e.g. red bone marrow, breast, lung, gonads, skin, and extremities)	500 mSv	(50 rem)
3. Guidance: cumulative exposure	10 mSv × age (in years)	(1 rem × age)
B. Planned special occupational exposure, effective dose equivalent limit[b]	(NCRP 1987, Section 15)	
C. Guidance for emergency occupational exposure[b]	(NCRP 1987, Section 16)	
D. Public exposures (annual)		
1. Effective dose equivalent limit continuous or frequent exposure[b]	1 mSv	(0.1 rem)
2. Effective dose equivalent limit, infrequent exposure[b]	5 mSv	(0.5rem)
3. Remedial action recommended when:		
(a) Effective dose equivalent[d]	>5 mSv	(>0.5 rem)
(b) Exposure to radon and its decay products	>0.007 j h m^{-3}	(>2WLM)
4. Dose equivalent limits for lens of eye, skin and extremities[b]	50 mSv	(5 rem)
E. Education and training exposure (annual)[b]		
1. Effective dose equivalent limit	1 mSv	(0.1 rem)
2. Dose equivalent limit for lens of eye, skin and extremities	50 mSv	(5 rem)
F. Embryo-fetus exposures[b]		
1. Total dose equivalent limit Effective dose equivalent[d]	5 mSv	(0.5 rem)
2. Dose equivalent limit in a month	0.5 mSv	(0.05 rem)
G. Negligible individual risk level (annual)[b]		
Effective dose equivalent per source or practice	0.01 mSv	(0.001 rem)

Source: NCRP (1987a, Table 22.1).
[a] Excluding medical exposures, and using values in Table 13.2 for $\bar{Q}$.
[b] Sum of external and internal exposures.
[c] ICRP (1991) recommends that this limit, based on a 5-year average, be reduced to 20 mSv.
[d] Including background but excluding internal exposures.

13.2 Radiation biology

13.2.1 Radiation interaction with biological structures

Ionizing radiation affects living cells and organisms in multiple ways. We limit our discussion to processes that are biologically or medically significant, especially those that are quantifiable. The ionization process itself is the key initial mechanism. If an atom in a vital molecule is ionized by a passing photon or charged particle, the molecule itself will be disrupted along the path of the radiation quantum (direct action) or a *free radical* can be formed which will be transported some distance away, disrupting a different molecule (indirect action). Although there has been some attention to cellular membranes and vital proteins etc. as sensitive targets, the most important structure in the cell for radiation damage is the double-stranded DNA molecule which carries the information to direct protein and enzyme production within the cell, and must be replicated properly for cell division to be successful. If there is ionization damage to the DNA bases which form the 'rungs of the helical ladder', the coded information (for protein assembly from amino acids) will be erroneous and point mutations, for example, will result. If the sugar–phosphate chains that make the long members of the 'twisted ladder' are broken, cutting of one or both strands (single- or double-stranded scission) can occur. If both strands are broken, the two ends of the DNA molecule may separate. If they are reassembled improperly, a chromosome aberration can occur, which is usually fatal to a dividing cell. If they are not reassembled, it is impossible to copy (replicate) the DNA, and cell division stops (*reproductive death*). These processes are complex and are affected by the physical dose rate and the chemical environment within the cell and outside the cell, by the density of energy deposition (LET) and ionization events along the track, and by active biological mechanisms. Perhaps the most interesting of these is the *repair* process.

Life on earth evolved in a continuous but unsteady 'bath' of background radiation, both ultraviolet from the sun and ionizing radiation from cosmic rays, the sun, and the radioactive elements in the earth itself. In order to preserve a more faithful reproduction of offspring from the continuing induced mutations that would result, special mechanisms evolved, which repair radiation damage to DNA, including enzymes that find a faulty sequence, cut out the error-containing portion, and fill in and reconnect the remaining ends properly. The effectiveness of the repair process varies from cell to cell, with differences between tissues, different radiation dose rates, time between radiation exposures, and *quality* of the radiation (different local ionization densities). Double-strand scission caused by high-LET particle tracks is more often unrepairable.

Because of repair and other biological and chemical processes which modify the response to a given physical dose of radiation, it is impossible to predict exactly the response to a given amount of energy deposited in a cell or an organism. In addition to biological variability, the stochastic nature of the radiation absorption process itself (secondary electron tracks from photons, *delta-rays* from particle tracks, etc.) causes variation in the way that the energy is deposited in DNA or any sensitive structure.

Nevertheless, it is desirable and useful to try to describe the biological effects of radiation (especially cell-killing) in quantitative ways. The basic concept is the *survival curve*. This can be measured by exposing a group of similar organisms in a controlled environment to an identical dose of radiation and measuring the fraction surviving by observation of how many of them are able to reproduce and form colonies when transplanted to a new growth medium. Multiple measurements are performed for different dose exposures, resulting in data of a type that was first presented by Puck and Marcus (1956) for mammalian cells, as shown in Figure 13.5.

The resulting data points fall on a straight line on a semi-log plot, for x-ray doses over 200 R. If the type of radiation is changed to neutrons or another with very high local energy transfer, the data approaches a straight line sloping downward from a 100 per cent survival at zero dose, suggesting a relationship

$$S = e^{-\alpha D}, \tag{13.21}$$

where S is the surviving fraction, D is the absorbed radiation dose, and α is a measured parameter.

This is also the shape of the data measured from inactivation of *simple* targets such as enzymes, viruses, etc. There have been multiple attempts, with varying success, to explain this type of data from theory, by mathematically describing the basic process of radiation interacting with biological structures. Theoretical models can be useful to analyze and unify the large amount of biological data, to improve the protection of people from low doses of radiation, to increase the effectiveness of radiation therapy, and perhaps also to understand the very nature of the structures and processes involved.

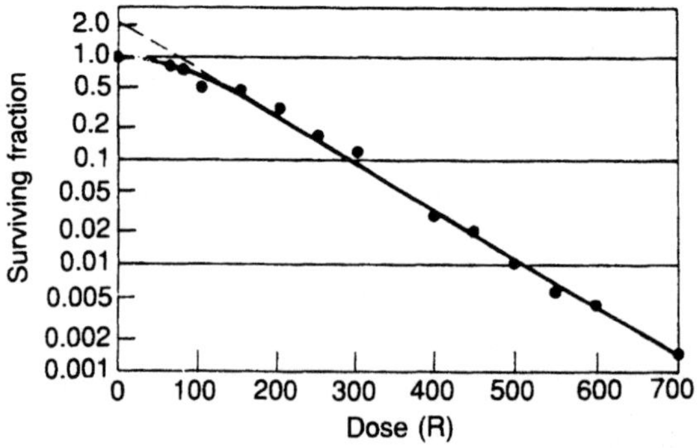

Fig. 13.5 Survival curve of HeLa cells (a human cancer cell line) in culture exposed to x-rays. The small initial shoulder may be interpreted as a result of repair process occurring at doses less than 200 R (Puck and Marcus 1956).

Hit theory

Consider N_o as the initial number of cells, and N the number that survive a dose D. Then the proportion not yet hit, which will be hit by an increment dD in the dose, will be given by

$$\frac{dN}{N} = -\frac{dD}{D_0}, \tag{13.22}$$

where D_0 is the average dose required to score one fatal hit per organism. Integration yields

$$\ln\left(\frac{N}{N_0}\right) = -\frac{D}{D_0}. \tag{13.23}$$

If S, the surviving fraction, equals N/N_0,

$$S = e^{-D/D_0}. \tag{13.24}$$

Thus D_0, the mean lethal dose, is the dose which leaves 1/e or 37 per cent surviving fraction on a purely exponential (one-hit) survival curve.

Generalizing for multiple hits, if the sensitive target has volume v, and the product vD represents the mean number of hits in the sensitive volume per dose D, the probability of exactly n hits is given by the Poisson distribution,

$$P(n) = \frac{(vD)^n e^{-vD}}{n!}. \tag{12.25}$$

If n hits are required for inactivation, any organism receiving $n-1$ or fewer hits will survive. The surviving fraction is the sum over 0, 1, 2, ... , $n-1$ hits,

$$\frac{N}{N_0} = S = e^{-vD}\sum_{k=0}^{n-1}\frac{(vD)^k}{k!}. \tag{13.26}$$

For a single hit, this becomes

$$S = e^{-vD}. \tag{13.27}$$

We thus can identify the sensitive volume with the reciprocal of the mean lethal dose for a single-hit process ($v = D_0^{-1}$). For a two-hit process,

$$S = e^{-D/D_0}(1 + D/D_0); \tag{13.28}$$

for a three-hit process,

$$S = e^{-D_0}[1 + D/D_0 + (D/D_0)^2/2!]; \tag{13.29}$$

etc.

At doses large compared to D_0, the curve approaches *exponential* survival, but at small doses it *flattens out* to form a *shoulder* such as that shown in Fig. 13.5. The size of the shoulder is a function of the number of hits needed to inactivate the target structure. The shoulder can also be interpreted as a result of a repair process; if there is recovery from the first hit, it may take two or three hits to inactivate the organism.

Target theory

The theory can be generalized even further to account for the possibility that there is more than one target structure in the organism. If each of m targets requires n hits for inactivation, the survival characteristic will be

$$\frac{N^+}{N_0} = \left(1 - e^{-vD}\sum_{k=0}^{n-1}\frac{(vD)^k}{k!}\right)^m, \tag{13.30}$$

where N^+ is the number of nonsurvivors, and $S = 1 - N^+/N_0$.

Expressions can also be written to describe variation of hit-number and targets of varying size, but the inherent difficulties in measuring exact dose–response curves mean that in practice the errors involved prevent unambiguous determination of the parameters v, n, and m. The most instructive result from the theory is the simplified case of only *one* hit required for each of m targets. Then,

$$\begin{aligned} N/N_0 = S &= 1 - (1 - e^{-vD})^m \\ &= 1 - (1 - me^{-vD} + \ldots e^{-mvD}). \end{aligned} \tag{13.31}$$

At high doses the terms beyond me^{-vD} are negligible, leading to

$$\ln(N/N_0) = -vD + \ln m. \tag{13.32}$$

Thus at high doses the curve becomes linear on a semi-log plot, with a slope of $-v$. At low doses the shoulder becomes horizontal, but if the exponential is extrapolated back to $D = 0$ in the above equation, the vertical intercept is m, the number of targets, also called the *extrapolation number*. This value was about 2 for the data of Puck and Marcus (1956). Although most of the data from *in vitro* studies of mammalian cell killing have been fitted satisfactorily with multi-target curves down to survival levels of the order of 10^{-5}, the more important portion of the data for radiation protection is from low doses. In this region the multitarget theory is not useful, as it predicts no cell killing, or less than is observed in practice. Radiation therapy is also *fractionated* into multiple small doses, not described well by target theory.

Linear quadratic model

In order to interpret the data from low-dose exposures, where there is a finite initial downward slope of the survival curve (that is, no threshold), the concept of *dual radiation action* was developed by Rossi and Hall (Hall 1988). If there are two components to radiation effects on cells, one of which is proportional to dose and the other to the square of the dose, the survival characteristic is

$$S = e^{-\alpha D - \beta D^2}, \qquad (13.33)$$

where α and β are parameters measuring the relative importance of the two components. When their effects are equal; that is,

$$\alpha D = \beta D^2,$$

then

$$D = \alpha/\beta. \qquad (13.34)$$

This characteristic bends continuously with no final straight portion, so it is not useful for survival data below 10^{-4}, but it does represent data adequately in the region of daily doses for radiation therapy and the much lower doses received environmentally. It has the advantage of only two parameters, and a nonzero slope approaching $-\alpha$ at near zero dose (no threshold).

Effects on tissues and organs

The response of tissues and organs to radiation is more complex and harder to quantify than that of cultured cells in a controlled environment. In addition to the repair process described briefly above, there are three other processes at the tissue level that affect response: reoxygenation, repopulation, and reassortment. These four processes are collectively known as the 4 R's of radiobiology.

Reoxygenation refers to the fact that well-oxygenated cells are up to three times more sensitive to x-rays (and other sparsely ionizing radiation) than hypoxic cells. At the center of a tumor which has outgrown its blood supply, there may be a significant portion of cells which are quiescent due to hypoxia, and relatively resistant to a single dose of radiation. After one or several radiation treatments, however, which kill the more sensitive (competing) tumor cells which are closer to functioning blood capillaries, some of the previously hypoxic cells may become reoxygenated and enter growth phase again, causing failure to cure. If a much larger number of treatments is given, however, there is a greater probability of cure, as there is a greater chance that all the hypoxic cells will be eventually reoxygenated and killed by succeeding radiation fractions.

Repopulation refers to the concept that both tumor cells and many normal cells are in the process of continual *turnover* through cell division, and the number of each type of cells will increase between the radiation doses. Normal cells in normal tissues will repopulate under growth regulation mechanisms that are not present in tumors, which by definition are growing out of control. This does not mean that all tumor cells divide more rapidly than normal cells. There are some tumors, such as sarcomas, where cell division times are much longer, but tumor growth exceeds normal tissue growth because of other factors such as decreased tumor cell loss, or normal cell populations remaining in a nondividing stage (interphase).

Reassortment refers to the fact that all cells have different stages in their life cycle between and during the process of cell division, and that cells in interphase (not cycling) may be induced to start dividing again after a sublethal radiation dose. Furthermore, cells that are cycling have different radiation sensitivities when they are in DNA synthesis or the actual mitosis (division) phase, as compared to the gaps between them. After a sublethal radiation dose the remaining cells undergo reassortment into different phases of the cell cycle, possibly under active stimulation by growth factors present in tissues and induced by cell depletion.

13.2.2 Radiation in medicine

Radiation therapy

A year after Roentgen discovered in 1895 that it was possible to make an image of a part of the human body with x-rays, Grubbe in Chicago found that a prolonged exposure to these rays would cause regression of the surface of a breast cancer. The first cure of a tumor with radiation was reported by Sjögren in Sweden in 1899. In subsequent years, special x-ray machines were developed specifically for radiation treatment, with maximum photon energies of 100–500 kilo-electronvolts (keV). There was success in using these *ortho-voltage* techniques for treatment of superficial cancers, but because of the rapid absorption of these beams with depth in tissue, it was not possible to give curative doses of radiation to tumors deep in the body, except with quasi-surgical techniques using radium sources implanted in the tumor or adjacent to it.

It was recognized that photon beam energies in excess of 1 million electronvolts (*megavoltage*) were necessary, but with several exceptions, this technology became widely available only after several unexpected developments from World War II era research. Enrico Fermi built an atomic pile (nuclear reactor) as part of the development of the nuclear fission weapon, and it was discovered that cobalt-59, when irradiated with neutrons in a reactor, was activated to cobalt-60, which then decayed by emission of a gamma ray of average energy about 1 MeV, with a half-life of about 5 years. Treatment machines with Co-60 sources were developed in the USA and Canada, and were introduced around the world in the decade after 1950.

Wartime development of radar resulted in the refinement of microwave technology, which was then applied to the development of the linear accelerator, which could reliably generate electron beams of energies greater than 4 MeV. Henry Kaplan, a physican at Stanford University Medical School, recognized the therapeutic potential of this, and with engineering and physics colleagues, built linear accelerators specifically for tumor treatment purposes. Accelerators with external photon beams from 4 up to 25 MeV became widely available in the decades after 1960.

The application of these technologies resulted in breakthroughs in the treatment of several types of cancer. For example, the 5-year survival rate for early stage Hodgkin's lymphoma improved from approximately 10% in 1950 to approximately 90 per cent in 1970. The complexity of the new techniques and the required understanding of oncology (study and treatment of cancer) increased rapidly, resulting in the formation of a new subspeciality, *radiation oncology*, which grew out of the therapeutic component of general radiology.

The impact of radiation therapy in the treatment of cancer is not widely appreciated. For example, of the approximately eight hundred thousand people who develop solid tumors in the USA per year, approximately one-half (400 000) will be treated with radiation, either alone or along with surgery and/or chemotherapy. Roughly a third of these will have regionally advanced or widely disseminated disease and are treated *palliatively* for pain control, tumor shrinkage, life extension, etc., when cure is not possible. Approximately one-half will have localized or limited regional disease, with possibility of cure, and will be treated *definitively*, with curative intent. The actual cure rate varies widely (for example, from less than 10 per cent to greater than 90 per cent) with the specific stage and type of cancer, but the net result is that roughly 100 000 lives per year are saved by the use of radiation therapy either alone or with other modalities. When cure is not possible, there are also large populations benefited. We estimate that roughly 100 000 lives are extended and an additional 100 000 are benefited in the form of palliation of symptoms, on an annual basis in the United States alone. World-wide estimates could be up to an order of magnitude greater.

Technical aspects of this powerfully beneficial technology are discussed in the following section.

Tumor kinetics

The essence of curative treatment of a tumor with radiation (or surgery or chemotherapy) is the killing of all of the malignant cells present, or the reduction in number sufficient that a few remaining cells can be destroyed by the patient's immune system. Even for palliation, it is often necessary to kill or impede the further cell division of 80 to 90 per cent of the tumor cells; remember that reduction of the diameter of a spherical tumor by half requires the disappearance of seven-eighths of the original tumor volume. The numerical magnitude of this task is appreciated when we estimate that the total number of cells in a tumor volume of 1 cm^3 is about one billion.

Since the characteristic diameter of a cell (normal or malignant) is about 10 μm (10^{-3} cm), the characteristic cell volume is about 10^{-9} cm^3. Thus there may be 10^9 cells in a tumor of diameter 1 cm, which is relatively small and often hard to find in a clinical setting. A 1 kg tumor of 10 cm diameter, which is very large by clinical standards and would usually have deadly physiological effects on its host, would have roughly 10^{12} cells.

If, for example, each radiation treatment kills or inactivates half of the living (remaining) cancer cells (that is, the surviving fraction per dose is $\frac{1}{2}$), 30 treatments are needed to reduce the malignant cell population of a 1 cm tumor from 10^9 to 1, since

$$(\tfrac{1}{2})^{30} = 2^{-30} \approx 10^{-9}.$$

This concept can be generalized, if we make the simplifying assumptions (as above) that there is no proliferation of tumor cells *between* treatments, and that the radiation sensitivity of the tumor cells remains constant over the duration of the course of treatment. Thus:

$$N = N_0(S)^n, \tag{13.35}$$

where N is the number of cells remaining, N_0 the initial number of cells, S the surviving fraction per treatment, and n the number of treatments.

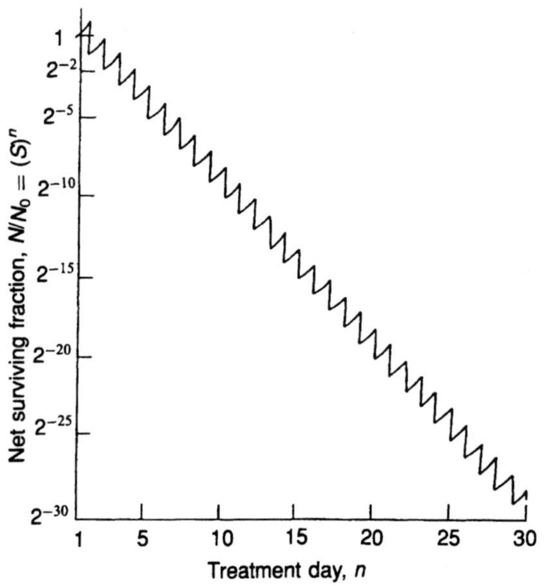

Fig. 13.6 Idealized tumor treatment kinetics, with simplifying assumption of one daily treatment resulting in half of the previous day's cells surviving (i.e. $S = 2^{-1}$).

Figure 13.6 is an idealized representation of this type of tumor treatment kinetics. The sawtooth nature of the plot is a representation of the fact that if treatments are given on a daily basis, there is some growth of the remaining tumor cell number in the time between treatments. Clearly if there is a longer gap between treatments (for example, one week) there could be enough tumor cell replication to require several more treatments for cure, or even to make cure impossible within tolerable dose. The uniform downward slope of the sawtooth is related to the assumption that the surviving fraction per treatment remains approximately constant. If there is a selection process such that more-radioresistant cells remain after the initial treatments, the slope would decrease and require additional treatments to achieve cure.

Note that the above examples are illustrative of actual clinical practice. Prescriptions of 30 to 35 treatments of 200 rads (2 Gy) each are often given to attempt cure of tumors with diameters of 1 to 5 cm. Clearly it is desirable to reduce the expected number of surviving tumor cells to *less* than one, but the random stochastic nature of the process implies that it is difficult to reach zero. In practice, if 30 treatments of 200 cGy each give an occasional cure, one would actually use three or four more treatments, i.e. 6600 to 6800 cGy, to try to achieve cure probabilities over 80 per cent. Because of biological variabilities, it is rare to achieve local control rates in excess of 95 per cent.

Radiation is often given in a prophylactic or *adjuvant* situation, where all gross tumor has been removed, and the purpose of treatment is to prevent regrowth of *microscopic* residual disease which is too small to be detected by usual methods. As an example, consider a 1 mm tumor, which would contain 10^6 cells but would not be palpable. In this case, if surviving fraction is $\frac{1}{2}$ per 200 cGy treatment, 20 treatments for a total of 4000 cGy would reduce the expected tumor cell number to 1.

In practice, 4400 to 5000 cGy are given in 22 to 28 treatments, with a high expectation of long-term success.

In concept, one could increase the treatment number and dose indefinitely to control very large masses, but in fact, for multiple reasons including maximum tolerance of surrounding normal tissue, doses of more than 7000 cGy are rarely used; it is very difficult to control most tumors greater than 5 cm in diameter with radiation alone.

13.3 Conclusion

Radiation applied to biology and medicine is now a 'mature technology' with myriad areas of successful use, resulting in major economic value as well as service to humanity. Just as x-ray imaging (conventional radiograpy) revolutionized medical diagnosis in the first half of the 20th century, computerized x-ray tomography and its 'cousin', magnetic resonance imaging, are revolutionizing current medicine, with positron emission tomography and nuclear magnetic resonance techniques becoming clinically useful in this decade.

X-ray crystallography revealed the structure of the double helix of DNA, revolutionizing molecular biology, and radioactive tracer techniques modernized physiology studies of how body systems work. Radioactive labelling is still a keystone of current research in fundamental biology.

Although the continuing use of radiation physics to explore the structure of matter is outside our scope, the associated exposure of personnel to radiation brings biomedical engineering into the picture, with the related discipline of *health physics*. Design of facilities for radiation safety is paramount in these areas, as well as in research and in commercial and military applications of nuclear power. In view of the continuing worldwide need for additional sources of reliable, economic non-fossil-fuel energy, there may be a new growth of nuclear power technology in the twenty-first century. Nuclear fusion may also come 'online' as a power source, with many new problems of radiation protection associated with the high neutron fluxes involved. Even if there is no major growth in these areas, there will be continuing need for research and development in measuring low doses, disaster prevention, and management.

Preservation of food by irradiation is also a mature technology, waiting for public acceptance to allow major commercial growth. Similarly, high-dose irradiation for waste sterilization and water-recycling is on the threshold of widespread application.

In the field of radiation oncology (the study and treatment of tumors) there are several areas of potential rapid expansion. The use of x-ray and electron beams from linear accelerators is a fully developed technology, but several new areas of technique are emerging. The direct use of digital information from CT or MRI tumor localization leads to computerized treatment planning, with the possibility of digital control of beams in three dimensions. This is leading to 'dynamic' therapy, in which the size and shape of the beam may be changed during the treatment, as a function of rotation and translation about the tumor target area.

Stereotactic placement of very narrow collimated beams allows the physician to 'operate with an atomic knife' inside the brain without opening the skull or drawing blood. The use of intra-operative radiation therapy (IORT) allows the treatment of abdominal masses, after exposure by the surgeon, to doses not safely attainable with an external beam. The development of beams of heavy charged particles (protons, helium ions, neon ions, etc.) enables tumoricidal doses to be delivered to targets adjacent to very sensitive normal organs (the eye, spinal cord, etc.) by stopping the beam at a specific range specified by the particle energy. Beams of neutrons, pi-mesons, and charged particles with high local LET within a tumor volume may offer biological advantages by discouraging repair of tumor cells relative to normal tissue.

Aggressive approaches with implantation of radioactive isotopes in tumor regions are now in rapid evolution, with development of high-dose-rate applications and combination with hyperthermia, generating locally higher temperatures using RF antenna, microwave, and ultrasound technologies. It has been found that raising the temperature to 40°C (104°F) not only kills some tumor cells, but makes the remainder more sensitive to ionizing radiation.

Finally the new 'biotechnology', combining radiopharmacology, immunology, and genetic engineering, has yielded isotopically labelled monoclonal and polyclonal antibodies which are capable of seeking out tumor cells or clumps within the body, attaching to them long enough for radioactive decay, resulting in destruction of certain malignancies with acceptable side effects.

Problems

P13.1 The kinetic energy of a heavy ion is frequently expressed as the ratio T/A (called the *kinetic energy per nucleon*). For example, a $^{20}_{10}$Ne nucleus with a total kinetic energy $T = 2000$ MeV would have a kinetic energy per nucleon $T/A = 100$ MeV/A. Using eqn (13.7) and Fig. 13.1, find L for a 100 MeV/A $^{20}_{10}$Ne nucleus and for a 100 MeV/A proton.

P13.2 Compare the $\bar{Q}$'s for the two particles in the previous problem using Fig. 13.3. How do these values compare with those in Table 13.2?

P13.3 To be useful for radiation therapy, beams of protons or $^{20}_{10}$Ne should have a range of at least 10 cm of water. Use eqn (13.8) and Fig. 13.1 to find the kinetic energies required. Express your results in MeV/A.

P13.4 Show that the relationships, eqn (13.7) and eqn (13.8), follow directly from eqn (13.5) and eqn (13.6).

P13.5 A beam of α-particles used for radiation therapy in soft tissue is to deposit a dose of 1 Gy averaged over the last 1 cm of its range. Find the beam fluence (number of particles per cm^2) required to produce this average dose. Assume the stopping power of soft tissue is equivalent to that of water.

P13.6 From Fig. 13.2 find the thickness of lead shielding required to reduce the intensity of 50 keV x-rays by a factor of 1000. Estimate an uncertainty in your result due to reading the data from the graph.

P13.7 From eqn (13.13) find the maximum energy transfered to an electron in tissue as a result of a Compton-scattering collision of a 50 keV x-ray and of a 500 keV γ-ray.

For problems P13.8 through P13.12, on risk from radiation exposure, unless otherwise instructed, use the data in Table 13.6 assuming simple flat plateaus (the 'Absolute risk' column). For problems P13.10 through P13.12 assume the risk to persist for a lifetime (alternative (b) in the fourth column of Table 13.6). Where an average loss of life expectancy is to be calculated, assume that, if not for exposure to the risk, a person would have lived exactly 75 years, and that the loss may be treated as small compared to the 75 years.

P13.8 Estimate the loss of life expectancy, in days of a person who was exposed, *in utero*, to 10 mSv (1 rem) of x-radiation from a pelvic radiograph of the mother.

P13.9 Children in Denver, Colorado, receive, from conception, a whole-body radiation dose equivalent that is approximately 1 mSv (100 mrem) per year greater than that received by children in the San Francisco area. Calculate the expected excess number of deaths per year, due to leukemia, for 5-year-olds in the Denver area compared to San Francisco. Express your result as excess deaths/10^6 children.

P13.10 In response to an interviewer's question about latent injury from the Chernobyl accident, a scientist observed that, even to a person receiving a near-fatal acute exposure, the added risk of delayed cancer death was relatively small compared to other life risks. Estimate the loss of life expectancy of a 20-year-old worker who survived an acute whole-body dose of 5 Sv (500 rem) at the time of the accident.

P13.11 It can be inferred from the 1979 Report by the Surgeon on Smoking and Health that smoking one cigarette per day for a year produces a mortality risk of $1.3 \times 10^{-5}\ a^{-1}$ with no latency and a plateau of 15 a. Ascribing equal values to risks that produce the same loss in life expectancy, find, for a 30-year-old person, the radiation dose received over a year that will produce a risk equivalent to smoking one cigarette a day for that year.

P13.12 The most recently updated recommendation by the ICRP (ICRP 1991) sets a limit of 20 mSv per year for the average equivalent whole-body occupational dose. The rationale for this new and lower limit is a model calculation that predicts a loss of life expectancy of 0.5 a for a worker receiving this annual dose over a working life of 47 years (from 18 to 65), and the assignment of this as a maximum acceptable risk value. Find the loss of life expectancy obtained from these assumptions using the absolute-risk-model data in Table 13.6 and compare with the ICRP result.

P13.13 After a course of conventional x-ray treatment a radiation oncology physican estimates that 100 cancer cells are left alive. She decides to give a boost with neutrons, which are known to have purely exponential cell-killing with a mean lethal dose (D_0) of 1 Gy (100 rad) for this type of cancer cell. What dose of boost is required to reduce the estimated surviving cancer cell number to one? (*Hint*: try eqn (13.24).)

P13.14 High-intensity γ-rays are used to irradiate food to keep it from spoiling. It is found that the mean lethal dose D_0 to kill bacteria by a single hit process is 3 Gy. If it is desired to achieve 99.999% sterilization, what dose of γ-rays is required? If it is desired to achieve the same level of sterilization of viruses, which obey single-hit kinetics but have a volume 200 times smaller, what dose is required? (*Hint*: see discussion following eqn (13.27).)

P13.15. If the mean lethal dose D_0 for neutrons is 1 Gy for a certain biological effect, and the dose for x-rays is 3 Gy for the same effect, what is the RBE (relative biological effectiveness) for neutrons compared to x-rays in this situation? (*Hint*: see discussion before eqn (13.16).)

P13.16 Radiation biologists have observed a maximum of certain biological effects (per unit dose) at LET (L_∞) of 100 keV μm^{-1}. What quality factor has been assigned to this quality of radiation? (*Hint*: see Fig. 13.3.)

P13.17 A health physicist is asked to calculate the fraction of bone-marrow stem cells killed in an air crew that received a dose of 10 mrad (10^{-4} Gy) from flying a transatlantic round trip. Because of the small dose involved, he uses the linear quadratic equation (13.33), with parameters. $\alpha = 1$ Gy^{-1} and $\beta = 0.1$ Gy^{-2}. What is the result? (*Hint*: fraction killed = 1-surviving fraction.)

Discussion: given that the average North American receives about 100 mrad per year from natural background (excluding inhaled radon) and that flight crews receive more radiation dose from multiple flights than the vast majority of radiation workers, who are required to wear film badges, do you believe that air crews should be required to wear film badges to measure their occupational radiation exposure? Note that bone-marrow stem cells are continuously replaced by natural body processes.

P13.18 In the 'breast-conserving' treatment of breast cancer, radiation therapy is given after surgical excision of the gross tumor mass (lumpectomy), in order to kill any microscopic disease in the remaining breast tissue. One common practice is to give 28 treatments of 180 rad each (5040 rad total) to the breast. This has a success rate of 95 per cent; that is, only 5 per cent of the patients get a recurrence in the breast over ten years of follow-up. If each treatment kills 1/3 of the remaining cancer cells, what is the expected remaining number of cancer cells after such a course of treatment if the surgeon leaves behind a pinhead-sized mass of 1 mm diameter containing 10^6 cells? (*Hint*: see discussion preceding eqn (13.35).)

References

BEIR (1972). *The effects on populations of exposure to low levels of ionizing radiation*, Report of the Advisory Committee on the Biological Effects of Ionizing Radiation (BEIR Report), NAS–NRC. National Academy Press.

BEIR (1990). *Health effects of exposure to low levels of ionizing radiation* (BEIR V), Report of the Committee on the Biological Effects of Ionizing Radiation, NAS–NRC. National Academy Press.

Hall, E. J. (1988). *Radiobiology for the radiologist*, (3rd edn), pp. 23 and 65. Lippincott, Philadelphia.

ICRP (1991). *1990 Recommendations of the international commission on radiological protection*, ICRP Publication 60, Annals of the ICRP, 21. Pergamon Press, Oxford.

NCRP (1987a). *Recommendations on limits for exposure to ionizing radiation*, NCRP Report No. 91. National Council on Radiation Protection and Measurement, Bethesda, MD.

NCRP (1987b). *Ionizing radiation exposure of the population of the United States*, NCRP Report No. 93. National Council on Radiation Protection and Measurement, Bethesda, MD.

Oakley, D. T. (1972). *Natural radiation exposure in the United States*, ORP/SID 72-1. US Environmental Protection Agency, Washington, DC.

Puck, T. T. and Marcus, T. I. (1956). Action of x-rays on mammalian cells. *Journal of Experimental Medicine*, **103**, 653.

Storm, E. and Israel, H. I. (1970). 'Photon cross sections from 1 keV to 100 MeV for elements $Z = 1$ to $Z = 100$'. *Nuclear data tables*, A7, pp. 565–681.

Further reading

Krane, K. S. (1988). *Introductory nuclear physics*. Wiley, New York. (This is a good first source for more detailed information on radioactivity, radiation, and the interaction of radiation with matter, as well as for additional references.)

Hacker, B. J. (1987). *The dragon's tail—radiation safety in the Manhattan Project, 1942–1946*. University of California Press. (This is an excellent history of the evolution of health physics, the science of radiation safety.)

Hall, E. J. (1988). *Radiobiology for the radiologist*. Lippincott, Philadelphia. (Already noted above, this is an excellent text and a comprehensive reference on the biological and medical effects of radiation.)

Nias, A. H. W. (1990). *An introduction to radiobiology*. Wiley, Chichester, UK (A comprehensive and quantitative text on the subject, with chapters on radiation therapy, diagnostic radiology, environmental radiation, and radiation protection.)

BIOEFFECTS OF NONIONIZING ELECTROMAGNETIC FIELDS

14

Charles Süsskind

Interactions between electromagnetic fields and biological tissue are classified as *ionizing* if the energy $E = h\nu$ (where h is Planck's constant and ν is the frequency) is high enough to cause ionization to occur, and as *nonionizing* if E is too low for that. Ionizing effects are associated, for example, with γ- and x-rays, and nonionizing effects with microwave, radio, and on down to powerline frequencies and below.

The most obvious effect, familiar from microwave ovens, is heating resulting from the energy absorbed when electromagnetic energy passes through a 'lossy' dielectric such as water, a major component of most biological tissues. In dielectrics that are not very lossy, such as glass, the dielectric constant κ_e that characterizes them is a dimensionless number; when multiplied by the permittivity ε_0 of free space, it yields the permittivity of the dielectric, $\varepsilon = \kappa_e\varepsilon_0$. (The permittivity is related to the refractive index n familiar from geometrical optics by $n = \sqrt{\kappa_e}$.)

In lossy dielectrics, the situation is somewhat more complex. The permittivity (and hence the dielectric constant) becomes a complex number: $\varepsilon = \varepsilon' - j\varepsilon''$ and $\kappa_e = \kappa_e' - j\kappa_e''$, where $j = \sqrt{-1}$; κ_e' accounts for the dielectric properties and κ_e'' for the losses that occur when an electromagnetic wave passes through the dielectric. For low-loss dielectrics, $\kappa_e'' \to 0$ and $\kappa_e = \kappa_e'$; but for lossy dielectrics, κ_e'' must be taken into consideration. Moreover, both κ_e' and κ_e'' depend on frequency. This dependence, which is the basis for microwave spectroscopy, can be quite complicated. In the simplest case, it means that the absolute value of κ_e' remains constant up to a certain frequency and then falls off to unity (that is, the material ceases to behave as a dielectric, which is why glass lenses cannot focus x-rays), exhibiting absorption bands along the way; whereas κ_e'' remains constant except for peaks in the same absorption bands (which account, for example, for the color of glass to which certain substances have been added).

Pure (distilled) water is not very lossy, but salt water, tap water, and the water contained in biological tissues contain other elements and compounds that make it quite lossy, so that heat is dissipated in it. We may then ask: how much heat can be dissipated in a living organism before it is irreversibly damaged? The earliest approach to this question was by a calculation based on human metabolic processes. It was estimated that a human body dissipated a certain number of calories under normal conditions, and about double that under thermal stress, as long as its normal defenses (circulation, perspiration, reradiation) were unimpaired, but that it could not deal with larger stresses. Translated into power density of nonionizing radiation impinging on a unit area of body surface, that is approximately 100 mW cm^{-2}, so we arrive at a permissible power density (after reduction by a safety factor of 10) of 10 mW cm^{-2}.

Another way to answer the question is to measure the increase in temperature exhibited by a body under nonionizing radiation. An increase of 6–7°C can be fatal; an increase of 1°C can be tolerated for a long time. The power density that produces an increase of 1°C is also approximately 100 mW cm^{-2}, so that by applying the same safety factor as before, we again arrive at 10 mW cm^{-2} as the maximum value permissible from the purely 'temperature' viewpoint.

We have arrived at this result by a consideration of the effect on a body placed in a previously undisturbed electromagnetic field. To describe the field, we measure its power density, a process called *densitometry.* But not all the energy impinging on the body is absorbed by it: some is absorbed by clothing, some is reflected from the clothing and from the body itself, and the absorption is in any case not uniform because of the body's irregular surface and constitution. Moreover, resonance effects affecting the value of absorption may set in when the size of the body stands in certain proportions to the wavelength λ, which is related to the frequency by $\lambda = c/v$, where c is the velocity of light in free space.

It would thus be better if we could relate the power density to some measure that would express the energy actually absorbed by the body, a process known as *dosimetry*. One way to do that is to make certain assumptions that lead to a specific absorption rate (SAR), stated in terms of power per unit body weight, in units of watts per kilogram of body weight. For an adult man, 10 mW cm^{-2} corresponds to 4 W kg^{-1}, 2.5 mW cm^{-2} to 1 W kg^{-1}, and so on in proportion.

The difficulty with this sort of averaging is that the absorption is actually not uniform over the body but can vary up or down from the average by a factor of as much as 4. For example, it is higher than the average at the extremities, and lower over the flat portions. There are other considerations as well. Does the field originate from a rotating or scanning antenna, so that the irradiation is intermittent? Is the radiation steady, in a sinusoidal continuous wave (CW), or is it pulsed (that is, does it come in periodic short bursts followed by zero radiation), as in most radars? If so, can we assume that the total energy is the average of the bursts and the silent periods between them? In that case, a 1 kW burst lasting 1 μs, followed by 999 μs of no radiation, would be the equivalent of a CW power of 1 W, which is the conventional way of computing average power; but has the above example of a 1 μs pulse at 1000 pps (pulses per second) the same biological effect as a 10 W pulse at 100 pps? And what happens when the wavelength of the radiation is of the same order of magnitude as the size of the irradiated obstacle, such as a human body: do the resulting resonance effects set in and increase the absorption?

Such uncertainties have caused the US National Institute of Standards and Technology to revise its original recommendation as to what is safe from a flat 10 mW cm^{-2} at any frequency to a different form that takes frequency into consideration. In particular, the power density drops in the 'well' of the resonant range; and the permissible value rises at very low and very high frequencies, since radiation at these frequencies is known not to penetrate as deep as in the intermediate range (Table 14.1 and Fig. 14.1).

As new discoveries are made, it is likely that the maximum permissible levels will be lowered still further. In some countries they are already lower, although they may not have been set to the same criteria: for example, not according to the lowest field intensity at which a potentially detrimental biological effect has been observed, as in the USA, but to the highest at which no effect, detrimental or not, has been reported. One consideration must be how far the restrictions can go if they are not to prohibit RF radiation altogether, with devastating consequences on its many uses in communications, industry, medicine, and countless other fields.

Table 14.1 IEEE Standard C95.1.1991 for continuous exposure in uncontrolled environments (general public) between 3 kHz and 300 GHz

Frequency, f (MHz)	Electric field, E (V m^{-1})	Magnetic field, H (A m^{-1})	Power density, P (W m^{-2})
0.003–0.1	614	163	–
0.1–1.5	614	$16.3/f$	–
1.5–30	$823.8/f$	$16.3/f$	–
30–100	27.5	$158.3/f^{1.688}$	–
100–300	27.5	0.073	2
300–10 000	–	–	$f/150$
10 000–300 000	–	–	100

The heating effects of an RF electromagnetic field result from interactions of its electric component with biological tissue; since its magnetic permeability is almost the same as that of free space, the magnetic component plays virtually no part in these 'thermal' effects. However, we cannot regard a living organism as an inert sack of tissue: it contains organs, cells, and still smaller particles that exist in and contribute to an environment replete with electric and magnetic phenomena and that can certainly interact with outside fields without necessarily causing heating. As one example of the resulting 'nonthermal' effects, even a weak magnetic field (of the order of the Earth's magnetic field) could conceivably affect the motion of charged particles in a manner detrimental to the organism's functioning.

As a result, attention has shifted from the thermal to the nonthermal effects of RF radiation and has led to investigations of a number of observations in which systemic heating should not be a problem. (A corollary is that neither incident power density nor energy absorption can then serve as a criterion for the hazard.) Among these studies are effects on cellular and even subcellular systems; on the blood and immunological systems; on

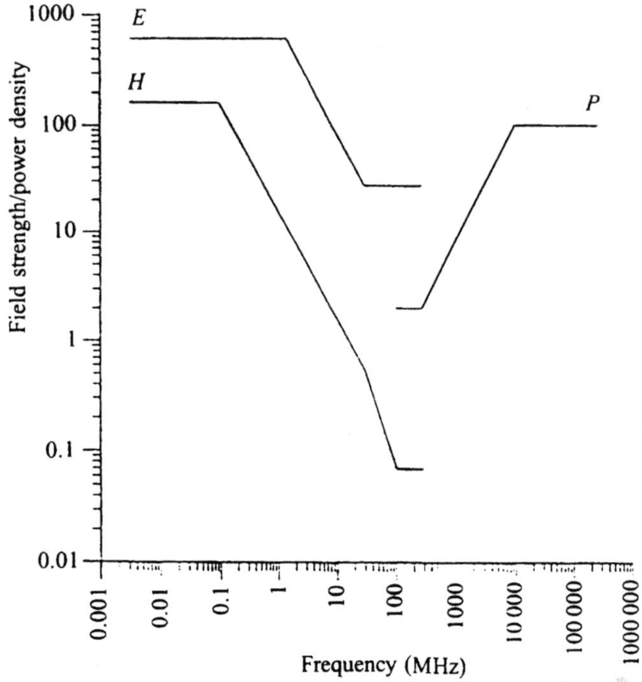

Fig. 14.1 Graph of exposure limits listed in Table 14.1, expressed in terms of field strengths (below 300 MHz) or power density (above 100 MHz).

the reproductive system and the embryo; on the nervous system; on behavior; on endocrine, physiological, and biochemical effects; on genetics and mutagenesis; on vision, hearing, and cutaneous perception; on life span; and on carcinogenesis. Since deliberate exposure of human subjects is not feasible, most of these studies have been conducted on animals and animal preparations, but there have also been some retrospective studies on specific human populations such as groups known to have been occupationally exposed to RF radiation. These studies have tended to center on carcinogenesis, the uncontrolled division of cells that can cause a tumor and that can be carried elsewhere in the body and cause new growths there. Other endpoints investigated have been cardiovascular disease and eye cataracts. None of these investigations has yielded convincing evidence that prolonged exposure to RF radiation at energies below those causing heating has led to adverse biological effects in humans.

One finding related to cellular and subcellular effects has attracted particular attention: that nonthermal RF radiation modulated by extremely low frequencies (ELF), of the order of 10 Hz, affects the flow of chemicals in and out of the cell. ('Modulated' in this context means that the amplitude of the RF wave is varied sinusoidally at a much lower frequency.) The interest in this phenomenon is fourfold. First, many natural functions (for example, heartbeat, brain waves) occur at comparable frequencies, so that interference with them becomes plausible. Second, some applications, such as communications with submerged submarines, are carried out at ELF. Third, it has long been known that wounds and broken bones heal quicker if electrical currents are applied to them. Fourth, powerline frequencies (50 to 60 Hz) also fall in this range, so that proximity to appliances such as electric blankets and video terminals, and immersion in the strong fields associated with powerlines outside and inside dwellings, may be hazardous.

These concerns have led to numerous investigations in the USA and other countries, as well as the formation of at least two professional societies devoted to these and allied questions: the Bioelectromagnetics Society and the Society for Physical Regulation in Biology and Medicine (formerly the Biological Repair and Growth Society). In the main, the studies fall into two categories: epidemiology and mechanisms.

In epidemiological studies, the approach is to select a group believed to be at risk and compare it with a like group not subject to the same hazard (say, maintenance workers at electric utilities with office personnel); or to take a subset of a group known to have suffered a disability and establish whether the subset's circumstances differ from those of the group at large (say, whether children that have contracted leukemia in a certain locality tend to differ from a peer group in that they live near high-voltage powerlines). Some early studies have suggested correlations between exposure to ELF fields and adverse effects, but many scientists remain unconvinced. Other, more carefully designed studies are currently under way.

In studies of mechanisms, the approach is to establish, in the laboratory, whether subjecting living tissues and organisms to an ELF field produces an observable effect on them, and whether the effect is deleterious. That is a much more difficult task than conducting epidemiological studies, but one that is likely to lead to more convincing results. Such studies are also under way.

These matters have created considerable public interest, due in part to the publicity given to the subject in popular media, to a point at which some government action is expected. On the other hand, groups whose operations would be impeded by restrictions on the use of nonionizing electromagnetic fields (such as the military, broadcasters, and public utilities) can be expected to urge caution in the promulgation of unnecessarily severe restrictions. The subject has thus acquired a distinctly political aspect, with environmentalists, users, and regulatory agencies all endeavoring to arrive at a generally acceptable outcome. For a more extensive (and impartial) account of this problem, readers are referred to a series of three articles by Robert Pool in *Science* (7 September, 21 September, and 5 October 1990); for a broader account of world-wide research, see the compendium of review and research papers presented at the First World Congress for Electricity and Magnetism in Biology and Medicine: Martin Blank (ed.), *Electricity and magnetism in biology and medicine*, San Francisco Press, 1992 (936 pp).

APPENDIX: LINEAR TRANSFORMS

Contents

Linear transforms provide powerful languages with which to deduce and describe key properties of linear network and linear system models. In this appendix, three linear transforms are introduced. Although there are many computational tools available with which the modern engineer can insert linear network and linear system models directly into computers and allow the analytical manipulations to proceed automatically, knowledge of a small repertoire of analytical algorithms continues to be valuable. Among other things, it allows the engineer to understand what the computer is accomplishing and to generalize the results in a logical fashion.

A.1 Definitions

Linear function. A linear function exhibits the properties of additivity and homogeneity (i.e., the graph of such a function has constant slope and passes through the origin).

Linear network. A linear network model or linear system model is one in which the constitutive relationships all are linear functions.

State variable. The state variables of a network or system model are the dependent variables of the differential equations or difference equations describing the dynamics of that model. For compartmental models, for example, they are the accumulated quantities $\{Q_i\}$ in the various compartments and the flows $\{J_{ij}\}$ between compartments; and for a Markov chain they are the probabilities of occupancy $\{P_i\}$ of the various states.

Source or input variable. The source or input variables of a network or system model are the dependent variables of the driving functions that make the model's differential or difference equations nonhomogeneous. The time course of a source or input variable is independent of the dynamics of the model. The source variables of a compartmental model, for example, are the flows $\{U_i\}$ into the various compartments from sources outside the model.

Response variable. A response variable is any state variable whose value as a function of time one wishes to deduce in response to initial conditions or in response to one or more specified source variables.

Zero-state network or system model. A zero-state network or system model is one in which all state variables initially (that is, at $t < 0$) equal zero.

Zero-state response. A zero-state response is the response of a zero-state network model or system model to one or more specified source variables. The response may be described in terms of one or more response variables.

Single-input–single-output analysis. This is deduction of the relationship between a single, designated source variable and a single, designated response variable.

Impulse. For continuous-time models, an impulse is a pulse whose duration is very short compared to the shortest interval of time one wishes to resolve in the deductions from the model. For a discrete-time model, an impulse is a pulse that takes place entirely within one time increment.

Unit impulse. The unit impulse is defined to be an impulse at $t = 0$ for which the integral of amplitude over time equals 1.0. Thus, for continuous-time compartmental models it is an impulse delivering one unit of conserved stuff to a particular compartment at $t = 0$. For discrete-time models, the unit impulse delivers one unit of stuff or one unit (1.0) of probability to a particular state during the 0th time increment.

Unit-impulse response. This is the time function or functions describing one or more response variables of a zero-state network model or system model to which a unit impulse has been applied. Every continuous-time function $h(t)$ or discrete-time function $h(n)$ describing a unit-impulse response is taken to be zero when $t < 0$ or $n < 0$.

A.2 Superposition

A.2.1 Continuous-time models

Homogeneity and additivity are properties of entire, linear, zero-state network models and system models, applicable to the relationship between any set of zero-state responses (e.g. $\{Q_i(t), J_{ij}(t)\}$, or $\{P_i(n)\}$, or $\{N_i(n)\}$ and the corresponding set of source variables ($\{U_i(t)\}$, or $\{U_i(n)\}$). Let $Q_i(t) = f_1(t)$ be the zero-state response of a particular continuous-time model to the single source $U_j(t) = x_1(t)$; let $Q_i(t) = f_2(t)$ be the zero-state response of the same model when $U_j(t) = x_2(t)$; and let a and b be scalar constants. Then linearity implies all of the relationships of Table A.1.

Homogeneity and additivity are applicable to all single-input–single-output analyses of linear, zero-state network models and system models, and provide the foundation for a very convenient analytical procedure for such models. That procedure consists of decomposition of the time function of the single, designated source variable into a series of impulses, followed by summation of the corresponding impulse responses to yield the time function of the single, designated response variable.

Let the source variable for a zero-state compartmental model be the flow $U_j(t)$ injected into compartment j; let

Table A.1 Linearity

If a model is linear and its responses to stimuli $x_1(t)$ and $x_2(t)$ are $f_1(t)$ and $f_2(t)$, respectively,

then its response to...	is...	
$ax_1(t)$	$af_1(t)$	homogeneity
$bx_2(t)$	$bf_2(t)$	homogeneity
$x_1(t) + x_2(t)$	$f_1(t) + f_2(t)$	additivity
$ax_1(t) + bx_2(t)$	$af_1(t) + bf_2(t)$	linearity

the response variable be the resulting accumulation, $Q_i(t)$, of conserved stuff in compartment i. Consider the function $U_j(t)$ to comprise the sum of the pulses in the set $\{p_n(t)\}$,

$$U_j(t) = \sum_{n=0}^{\infty} p_n(t), \tag{A.1}$$

where

$$\begin{aligned} p_n(t) &= 0 \quad \text{when} \quad -\infty < t \leqslant n\Delta t, \\ p_n(t) &= U_j\left(n\Delta t + \frac{\Delta t}{2}\right) \quad \text{when} \quad n\Delta t < t \leqslant (n+1)\Delta t, \\ p_n(t) &= 0 \quad \text{when} \quad (n+1)\Delta t < t, \end{aligned} \tag{A.2}$$

and Δt is very much less than the minimum time interval that must be resolved in the deductions from the model. Thus $p_n(t)$ can be considered to be an impulse of flow, delivering an amount, Q_n, of conserved stuff equal to

$$Q_n = U_j\left(n\Delta t + \frac{\Delta t}{2}\right)\Delta t. \tag{A.3}$$

If the response at compartment i to a unit impulse of flow into compartment j is

$$Q_i(t) = h_{ij}(t), \tag{A.4}$$

then homogeneity implies that the response to $p_n(t)$ is

$$Q_i(t) = \left[U_j\left(n\Delta t + \frac{\Delta t}{2}\right)\Delta t\right][h_{ij}(t - n\Delta t)], \tag{A.5}$$

and additivity implies that the response to the complete source function, $U_j(t)$, is

$$Q_i(t) = \sum_{n=0}^{\infty}\left[U_j\left(n\Delta t + \frac{\Delta t}{2}\right)\Delta t\right][h_{ij}(t - n\Delta t)]. \tag{A.6}$$

Recall that the function $h(\cdot)$ is taken to be zero whenever its argument is less than zero. Here, the average value of U_j during each pulse was chosen to be its value at the *center* of the pulse; the starting time of the corresponding impulse response was chosen to be at the *beginning* of the pulse. Because Δt is presumed to be beyond the limit of temporal resolution required of the model, this difference is not important. Taking the limit of the summation as Δt approaches zero, one finds

$$Q_i(t) = \int_{\tau=0}^{t} U_j(\tau) h_{ij}(t - \tau)\, d\tau. \tag{A.7}$$

The operation represented by the integral in eqn (A.7) is *convolution* and often is abbreviated as follows:

$$Q_i(t) = U_j(t) * h_{ij}(t). \tag{A.8}$$

Convolution is a commutative operation:

$$U_j(t) * h_{ij}(t) = h_{ij}(t) * U_j(t), \tag{A.9}$$

where the functions $U_j(\cdot)$ and $h_{ij}(\cdot)$ are defined to be zero for all values of their arguments less than zero. The continuous-time convolution operation often is more easily carried after the time functions have been transformed. When $U_j(\cdot)$ is an arbitrary input function and $h_{ij}(\cdot)$ is a function determined by the system being modeled, $h_{ij}(\cdot)$ sometimes is called the *kernel* of the convolution integral.

A.2.2 Discrete-time models

Let $N_i(n) = f_1(n)$ be the zero-state response of a particular discrete-time, linear network model to the single source $U_j(n) = x_1(n)$; let $N_i(n) = f_2(n)$ be the zero-state response of the same model when $U_j(n) = x_2(n)$; and let a and b be scalar constants. Then all of the statements of Table A.1 are true when the continuous, independent variable t is replaced by the discrete, independent variable n.

Let the source variable to a zero-state, discrete-time network model be the flow $U_j(n)$ injected into state j; let the response variable be the resulting expected number of particles, $N_i(n)$, in state i. The function $U_j(n)$ comprises a sum of impulses, each occurring during a different time increment. If the response to the unit impulse is

$$N_i(n) = h_{ij}(n), \tag{A.10}$$

then homogeneity implies that the response to $U_j(k)$ (k being a particular value of the integral variable n) is

$$N_i(n) = U_j(k) h_{ij}(n - k) \tag{A.11}$$

and additivity implies that the response to the complete source function, $U_j(n)$, is

$$N_i(n) = \sum_{k=0}^{n} U_j(k) h_{ij}(n - k). \tag{A.12}$$

The operation represented on the right-hand side of eqn (A.12) is discrete convolution. As in the case of continuous convolution, the operation often is abbreviated as follows:

$$N_i(n) = U_j(n) * h_{ij}(n), \tag{A.13}$$

where the functions $U_j(\cdot)$ and $h_{ij}(\cdot)$ are defined to be zero for all values of their arguments less than zero. As is true of continuous-time convolution, discrete-time convolution is commutative and it often is more easily carried after the time functions have been transformed.

A.3 Laplace transform

The Laplace transform $L[f(t)]$ of the function $f(t)$ is defined to be

$$L[f(t)] = \int_{t=0}^{\infty} f(t)\mathrm{e}^{-st}\,\mathrm{d}t = F(s), \tag{A.14}$$

where $f(\cdot)$ is defined to be zero whenever the argument is less than zero. From these definitions, one can deduce the properties displayed in Table A.2.

The entries in the left-hand column of Table A.2 are the Laplace transforms of the corresponding functions on the right. The process of transformation has mapped the functions of time into functions of another variable, s, which sometimes is called *complex frequency*. The functions in the right-hand column are the *inverse Laplace transforms* of the corresponding functions on the left. The process of inverse transformation maps functions of complex frequency into functions of time. In both directions, the mapping is unique. In this text, the inverse transform is denoted L^{-1}. Thus if

$$L[f(t)] = F(s), \tag{A.15}$$

then

$$L^{-1}[F(s)] = f(t) \tag{A.16}$$

Table A.2 Properties of Laplace transforms for zero-state conditions

Let $F(s)$ and $G(s)$ be the Laplace transforms of $f(t)$ and $g(t)$, respectively, and let a and b be scalar constants;

then...	is the Laplace transform of...	
$aF(s)$	$af(t)$	homogeneity
$F(s) + G(s)$	$f(t) + g(t)$	additivity
$aF(s) + bG(s)$	$af(t) + bg(t)$	linearity
$e^{-as}F(s)$	$f(t - a)$	delay[a]
$F(s)G(s)$	$f(t) * g(t)$	convolution
$sF(s)$	$df(t)/dt$	time derivative
$F(s)/s$	$\int f(t)dt$	time integral
$-dF(s)/ds$	$tf(t)$	multiply by t
$aF(as)$	$f(t/a)$	scale t
$F(s + a)$	$e^{-at}f(t)$	translate s

[a] $f(t - a) = 0$ for $t < a$.

A.4 Ordinary differential equations with constant coefficients

A.4.1 Homogeneous equations

The solution, $f(t)$, of a homogeneous, linear, ordinary differential equation with constant (real) coefficients and time (t) as the independent variable typically is

$$f(t) = \sum_{i=1}^{N} K_i \mathrm{e}^{s_i t}, \tag{A.17}$$

where, in general, the coefficients K_i and s_i may be real, imaginary, or complex. Recall that the members of $\{s_i\}$ are the natural frequencies of the network model or system model. In the case of solutions to differential equations describing compartmental models, the members of $\{K_i\}$ and $\{s_i\}$ are real; and the exponential coefficients $\{s_i\}$ are less than or equal to zero.

Whether the natural frequency s_i is real or complex,

$$L[K_i e^{s_i t}] = \frac{K_i}{s - s_i} \tag{A.18}$$

and

$$L^{-1}\left[\frac{K_i}{s - s_i}\right] = K_i e^{s_i t}. \tag{A.19}$$

When a member of the set $\{s_i\}$ of natural frequencies is complex, it always is accompanied by its complex conjugate; and the corresponding members of $\{K_i\}$ also are complex conjugates. Thus, if the set $\{s_i\}$ contains

$$s_k = -\alpha + j\beta, \tag{A.20}$$

where

$$j = \sqrt{-1}, \tag{A.21}$$

then it also contains

$$\bar{s}_k = -\alpha - j\beta \tag{A.22}$$

and the corresponding terms in $L[f(t)]$ are

$$\frac{K_k}{s - s_k}, \quad \frac{\bar{K}_k}{s - \bar{s}_k}, \tag{A.23}$$

where

$$K_k = A + jB, \quad \bar{K}_k = A - jB, \tag{A.24}$$

and A, B, α, and β are positive or negative real numbers.

The corresponding terms in $f(t)$ sum to yield

$$Ae^{-\alpha t}[e^{j\beta t} + e^{-j\beta t}] + jBe^{-\alpha t}[e^{j\beta t} - e^{-j\beta t}], \tag{A.25}$$

which is interpreted to be

$$2Ae^{-\alpha t}\cos[\beta t] - 2Be^{-\alpha t}\sin[\beta t]. \tag{A.26}$$

The corresponding pair of terms in $L[f(t)]$ sum to yield

$$\frac{2[A(s+\alpha) - B\beta]}{(s+\alpha)^2 + \beta^2}. \tag{A.27}$$

Although it is possible to have other types of terms in the solutions to homogeneous, linear, ordinary differential equations with constant coefficients (for example, terms corresponding to multiple occurrences of the same natural frequency), such terms are extremely unlikely to occur in practical modeling studies. Furthermore, they always can be approximated as precisely as one wishes with terms involving arbitrarily close, but not identical, natural frequencies. Therefore, the engineer usually needs to be familiar with just two types of terms; negative real exponential functions of time ($e^{-\alpha t}$) and variously damped sinusoids (for example, $e^{-\alpha t}\cos[\beta t]$). The appropriate Laplace transforms and inverse transforms are displayed in Table A.3.

Table A.3 Laplace transform pairs for homogeneous equations

$f(t)$	$F(s)$
e^{-at}	$1/(s+a)$
$\cos[bt]$	$s/(s^2+b^2)$
$e^{-at}\cos[bt]$	$(s+a)/[(s+a)^2+b^2]$
$\sin[bt]$	$b/(s^2+b^2)$
$e^{-at}\sin[bt]$	$b/[(s+a)^2+b^2]$

Notice that the last three transform pairs in Table A.3 can be derived easily from the second pair by application of properties in Table A.2. For linear, lumped-element network models and for system models one usually can generate any required pair by this method—starting just with the transform pairs of Table A.3.

A.4.2 Nonhomogeneous equations

The source variables that a bioengineer includes in network models or system models often are representations of conditions that have been or can be imposed during an experimental study. For example, an experimenter might suddenly initiate a constant flow of a tracer into some locale in a system being represented by a compartmental model. Comparing dynamic behavior deduced from the model with the behavior actually observed in the system, the engineer then can test the validity of the model and modify it if necessary. The sudden initiation of a constant flow, usually called a *step flow*, is one of several commonly applied source functions. Others are the pulse flow, the impulse of flow, and sinusoidally modulated flow. The Laplace-transform

Table A.4 Laplace transform pairs for common source functions

$f(t)$	$F(s)$
unit impulse	1
unit step $f(t) = 0,\ t<0$ $f(t) = 1,\ t \geqslant 0$	$1/s$
unit ramp $f(t) = 0,\ t<0$ $f(t) = t,\ t \geqslant 0$	$1/s^2$
unit-amplitude pulse $f(t) = 0,\ t<0$ $f(t) = 1,\ 0 \leqslant t < T$ $f(t) = 0,\ t \geqslant T$	$(1 - e^{-sT})/s$
cosine $f(t) = 0,\ t<0$ $f(t) = \cos[bt],\ t \geqslant 0$	$s/(s^2 + b^2)$
sine $f(t) = 0,\ t<0$ $f(t) = \sin[bt],\ t \geqslant 0$	$b/(s^2 + b^2)$

pairs corresponding to each of these source functions are displayed in Table A.4.

In zero-state, single-input–single-output analysis, the engineer algebraically reduces the differential equations for a network model or system model to a single equation with one source variable and one response variable. The general form of the equation is

$$\frac{d^m f_i}{dt^m} + a_{m-1}\frac{d^{m-1} f_i}{dt^{m-1}} + \ldots + a_1\frac{df_i}{dt} + a_0 f_i = b_h\frac{d^h U_j}{dt^h} + \ldots + b_1\frac{dU_j}{dt} + b_0 U_j. \tag{A.28}$$

Taking the Laplace transform of each side, assuming zero-state conditions and applying the time-derivative pair in Table A.2, one obtains

$$P_{ij}(s)F_i(s) = Q_{ij}(s)U_j(s), \tag{A.29}$$

where

$$P_{ij}(s) = s^m + a_{m-1}s^{m-1} + \ldots + a_1 s + a_0,$$
$$Q_{ij}(s) = b_h s^h + \ldots + b_1 s + b_0, \tag{A.30}$$
$$U_j(s) = L[U_j(t)], \quad F_i(s) = L[f_i(t)].$$

Thus

$$F_i(s) = \frac{Q_{ij}(s)}{P_{ij}(s)}U_j(s) \tag{A.31}$$

and the solution, $f_i(t)$, is

$$f_i(t) = L^{-1}[F_i(s)] = L^{-1}\left[\frac{Q_{ij}(s)}{P_{ij}(s)}U_j(s)\right]. \tag{A.32}$$

The corresponding unit impulse response and its Laplace transform are found by setting $U_j(s) = 1$ (see Table A.4):

$$h_{ij}(t) = L^{-1}\left[\frac{Q_{ij}(s)}{P_{ij}(s)}\right], \quad H_{ij}(s) = \frac{Q_{ij}(s)}{P_{ij}(s)}. \tag{A.33}$$

If the source function is something other than an impulse, the corresponding response function can be found from the following relationship:

$$f_i(t) = L^{-1}[H_{ij}(s)U_j(s)], \tag{A.34}$$

$$F_i(s) = H_{ij}(s)U_j(s). \tag{A.35}$$

A.4.3 Partial-fraction expansion

The step remaining in the solution represented by eqn (A.35) is the process of inverse Laplace transformation. For most problems with lumped-element linear network models or linear system models, inverse Laplace transformation can be accomplished by a very simple procedure known as partial-fraction expansion. In general,

$$U_j(s) = \frac{N_j(s)}{D_j(s)}, \tag{A.36}$$

where $N_j(s)$ and $D_j(s)$ are polynomials in s with real coefficients:

$$D_j(s) = s^w + \ldots + c_1 s + c_0. \tag{A.37}$$

Combining this expression for $U_j(s)$ with eqns (A.33) and (A.35), one obtains

$$F_i(s) = \frac{Q_{ij}(s)N_j(s)}{P_{ij}(s)D_j(s)}. \tag{A.38}$$

The zeros of $D_j(s)$ (that is, the values of s for which $D_j(s) = 0$) and those of $P_{ij}(s)$ may be real numbers or conjugate pairs of complex numbers. Thus,

$$F_i(s) = \frac{Q_{ij}(s)N_{ij}(s)}{(s - s_1)(s - s_2)(s - s_3)\dots(s - s_{w+m})}, \quad (A.39)$$

where $s_1, s_2, \dots, s_{w+m}$ are the zeros of the polynomials $D_j(s)$ and $P_{ij}(s)$ and are either real numbers or members of pairs of complex conjugate numbers. The *partial-fraction expansion* of this expression is

$$F_i(s) = \sum_{k=1}^{w+m} \frac{K_k}{s - s_k}, \quad (A.40)$$

where

$$K_k = [(s - s_k)F_i(s)]_{s=s_k}. \quad (A.41)$$

The corresponding inverse Laplace transform is found by applying the additivity property:

$$f_i(t) = \sum_{k=1}^{w+m} L^{-1}\left[\frac{K_k}{s - s_k}\right]. \quad (A.42)$$

For terms in which s_k is a real number, the inverse Laplace transform is a real exponential function (see Table A.3). For terms in which s_k is a complex number, inverse transforms normally are not given in tables. Conventionally, each pair of terms with complex conjugate values of s_k is combined into a single term with a quadratic polynomial in s as its denominator (eqn (A.27)). The corresponding inverse transform (eqn (A.26)) can be obtained by applying the additivity property and using the appropriate transform pairs in Table A.3.

The zeros of the polynomial $P_{ij}(s)$ are the natural frequencies of the model; and the corresponding time functions (with different scaling factors) generally are present regardless of the form of the source function. An exception would occur, for example, if $N_j(s)$ happened to contain a binomial factor $(s - s_k)$ identical to one in $P_{ij}(s)$.

Table A.5 shows some transform pairs derivable by partial-fraction expansion.

Example. Solve the following differential equation under zero-state conditions when $U(t)$ is a step of flow with amplitude K:

$$\frac{d^3f}{dt^3} + 4\frac{d^2f}{dt^2} + 6\frac{df}{dt} + 4f = U(t). \quad (A.43)$$

From the Laplace transform for a unit step and the homogeneity property in Table A.2,

Table A.5 A few transform pairs easily derived by partial-fraction expansion

$F(s)$	$f(t)$
$1/s(s + a)$	$(1 - e^{-at})/a$
$1/(s + a)(s + b)$	$(e^{-at} - e^{-bt})/(b-a)$
$1/s(s + a)(s + b)$	$(1/ab) + (be^{-at} - ae^{-bt})/ab(a - b)$
$1/s^2(s + a)$	$(e^{-at} + at - 1)/a^2$
$1/s^2(s + a)(s + b)$	$-(a + b)/a^2b^2 + t/ab + (-b^2e^{-at} + a^2e^{-bt})/a^2b^2(a - b)$
$1/(s + a)^n$	$(t^{n-1}e^{-at})/(n - 1)!$

$$U(s) = \frac{K}{s}. \quad (A.44)$$

From the linearity and zero-state time-derivative properties in Table A.2,

$$(s^3 + 4s^2 + 6s + 4)F(s) = U(s). \quad (A.45)$$

Combining these two expressions, one obtains

$$F(s) = \frac{K}{s^4 + 4s^3 + 6s^2 + 4s}. \quad (A.46)$$

Factoring the denominator (with the help of a digital computer if necessary), one obtains

$$F(s) = \frac{K}{s(s+2)(s+1+\mathrm{j}1)(s+1-\mathrm{j}1)}. \quad (A.47)$$

Partial-fraction expansion leads to

$$F(s) = \frac{K/4}{s} - \frac{K/4}{s+2} + \frac{\mathrm{j}K/4}{s+1-\mathrm{j}1} - \frac{\mathrm{j}K/4}{s+1+\mathrm{j}1}. \quad (A.48)$$

The two terms with conjugate complex natural frequencies correspond to a partial-fraction expansion of eqn (A.27) with $\alpha = -1$, $\beta = 1$, $A = 0$, and $B = K/4$. Applying the linearity property from Table A.2 and the transform pairs of Table A.3 and eqn (A.27), one obtains

$$f(t) = \frac{K}{4}\,[1 - e^{-2t} + 2e^{-1}\sin(t)]. \quad (A.49)$$

Recall that according to the convention established at the outset of this section, this solution applies only to values of time greater than or equal to zero, and $f(t)$ is taken to be zero for all other values of time. Thus, the solution should be written

$$f(t) = \frac{K}{4}[1 - e^{-2t} + 2e^{-t}\sin(t)], \quad t \geqslant 0, \qquad (A.50)$$
$$f(t) = 0, \quad t < 0.$$

A.5 Z transform

The Z transform $Z[f(n)]$ of the discrete-time function $f(n)$ is defined to be

$$Z[f(n)] = \sum_{n=0}^{\infty} f(n)z^{-n} = F(z), \qquad (A.51)$$

where $f(\cdot)$ is defined to be zero whenever the argument is less than zero. For zero-state analysis, the value of each response variable is taken to be zero for all values of the argument that are less than zero. From these definitions, the properties of the Z transform displayed in Table A.6 can be deduced. In this chapter, the inverse Z transform is denoted Z^{-1}. Thus if

$$Z[f(n)] = F(z), \qquad (A.52)$$

then

$$Z^{-1}[F(z)] = f(n). \qquad (A.53)$$

Table A.6 Properties of Z transforms for zero-state conditions

Let $F(z)$ and $G(z)$ be the Z transforms of $f(n)$ and $g(n)$, respectively, and let a and b be scalar constants;

then...	is the Z transform of...	
$aF(z)$	$af(n)$	homogeneity
$F(z) + g(z)$	$f(n) + g(n)$	additivity
$aF(z) + bG(z)$	$af(n) + bg(n)$	linearity
$z^{-a}F(z)$	$f(n - a)$	delay[a]
$F(z)G(z)$	$f(n) * g(n)$	convolution
$F(z)/(1 - z^{-1})$	$\sum f(n)$	sum over time
$-z\, dF(z)/dz$	$nf(n)$	multiply by n
$F(az)$	$a^{-n}f(n)$	scale z

[a] $f(n - a) = 0$ for $n < a$.

A.6 Ordinary difference equations with constant coefficients

A.6.1 Homogeneous equations

The solution, $f(n)$, of a homogeneous, linear, ordinary difference equation with constant (real) coefficients and discrete time (n) as the independent variable typically is

$$f(n) = \sum_{i=1}^{N} K_i r_i^n, \qquad (A.54)$$

where, in general, the coefficients $\{K_i\}$ and the common ratios $\{r_i\}$ may be real, imaginary, or complex. Whether the common ratio is real or complex,

$$Z[K_i r_i^n] = \frac{K_i}{1 - r_i z^{-1}} \tag{A.55}$$

and

$$Z^{-1}\left[\frac{K_i}{1 - r_i z^{-1}}\right] = K_i r_i^n. \tag{A.56}$$

As is true of natural frequencies in continuous-time models, when a member $\{r_i\}$ is complex, it always is accompanied by its complex conjugate, and the corresponding coefficients also are complex conjugates:

$$K_k r_k^n + \bar{K}_k \bar{r}_k^n. \tag{A.57}$$

The sum of these two terms is oscillatory and grows, declines, or remains constant in amplitude, depending on the magnitude of r.

As in the case of continuous-time models, terms corresponding to multiple occurrences of the same common ratio are extremely unlikely to occur in practical modeling studies. Furthermore, they always can be approximated with terms involving arbitrarily close, but not identical, common ratios. Therefore, for solutions of homogeneous, linear difference equations with constant coefficients, one needs to be familiar with just two types of discrete-time functions: simple geometric sequences of real numbers (with common ratios that are positive real numbers), and simple growing, decaying, or conservative oscillations (with common ratios that are complex numbers or negative real numbers). The appropriate Z transforms and inverse transforms are displayed in Table A.7.

Table A.7 Z transform pairs for homogeneous equations

$f(n)$	$F(z)$
ar^n	$1/(1 + az^{-1})$
$\cos[bn]$	$(1 - \cos[b]\, z^{-1})/(1 - 2\cos[b]\, z^{-1} + z^{-2})$
$a^n \cos[bn]$	$(1 - a\cos[b]\, z^{-1}))/(1 - 2a\cos[b]\, z^{-1} + a^2 z^{-2})$
$\sin[bn]$	$\sin[b]\, z^{-1}/(1 - 2\cos[b]\, z^{-1} + z^{-2})$
$a^n \sin[bn]$	$a\sin[b]\, z^{-1}/(1 - 2a\cos[b] z^{-1} + a^2 z^{-2})$

A.6.2 Nonhomogeneous equations

Source variables commonly used in discrete-time models are the flow impulse, the flow step, and the flow pulse. The corresponding Z-transform pairs are displayed in Table A.8. As with continuous-time models, in zero-state, single-input–single-output analysis, the engineer algebraically reduces the equations for a network model or a system model to a single equation with one source variable and one response variable. In this case, however, the equation is a linear difference equation, the form of which generally is

$$\begin{aligned} &f_i(n-m) + a_{m-1} f_i(n-m+1) + \ldots \\ &+ a_1 f_i(n-1) + a_0 f_i(n) \\ &= b_h U_j(n-h) + \ldots + b_1 U_j(n-1) \\ &+ b_0 U_j(n). \end{aligned} \tag{A.58}$$

Taking the Z transform of each side of eqn (A.58), assuming zero-state conditions and applying the delay pair in Table A.6, one obtains

$$P_{ij}(z)F_i(z) = Q_{ij}(z)U_j(z), \tag{A.59}$$

where

$$\begin{aligned} P_{ij}(z) &= z^{-m} + a_{m-1}z^{-m+1} + \ldots + a_1 z^{-1} + a_0, \\ Q_{ij}(z) &= b_h z^{-h} + \ldots + b_1 z^{-1} + b_0, \\ U_j(z) &= Z[U_j(n)], \quad F_i(z) = Z[f_i(n)]. \end{aligned} \tag{A.60}$$

Thus

$$F_i(z) = \frac{Q_{ij}(z)}{P_{ij}(z)} U_j(z) \tag{A.61}$$

and the solution, $f_i(n)$, is

$$f_i(n) = Z^{-1}[F_i(z)] = Z^{-1}\left[\frac{Q_{ij}(z)}{P_{ij}(z)} U_j(z)\right]. \tag{A.62}$$

The corresponding unit impulse response and its Z transform are found by setting $U_j(z) = 1$ (see Table A.8):

$$\begin{aligned} h_{ij}(n) &= Z^{-1}\left[\frac{Q_{ij}(z)}{P_{ij}(z)}\right], \\ H_{ij}(z) &= \frac{Q_{ij}(z)}{P_{ij}(z)}. \end{aligned} \tag{A.63}$$

If the source function is something other than an impulse, the corresponding response function can be found from

Table A.8 Z transform pairs for common source functions

	U(n)		U(z)
unit impulse			
$U(n)=0,$	$n<0$		
$U(n)=1,$	$n=0$		1
$U(n)=0,$	$n>0$		
unit step			
$U(n)=0,$	$n<0$		$1/(1-z^{-1})$
$U(n)=1,$	$n\geqslant 0$		
unit-amplitude pulse			
$U(n)=0,$	$n<0$		
$U(n)=1,$	$0\leqslant n<T$		$(1-z^{-T})/(1-z^{-1})$
$U(n)=0,$	$n\geqslant T$		

the following relationship:

$$f_i(n) = Z^{-1}[H_{ij}(z)U_j(z)], \tag{A.64}$$

$$F_i(z) = H_{ij}(z)U_j(z). \tag{A.65}$$

A.6.3 Partial-fraction expansion

Once again, partial-fraction expansion can be used to accomplish the remaining step in the solution represented by eqn (A.64), namely inverse transformation. In general,

$$U_j(z) = \frac{N_j(z)}{D_j(z)}, \tag{A.66}$$

where $N_j(z)$ and $D_j(z)$ are polynomials in z^{-1} with real coefficients:

$$D_j(z) = z^{-w} + \ldots + c_1 z^{-1} + c_0. \tag{A.67}$$

Combining this expression for $U_j(z)$ with eqns (A.63) and (A.65), one obtains

$$F_i(z) = \frac{Q_{ij}(z)N_j(z)}{P_{ij}(z)D_j(z)}. \tag{A.68}$$

The zeros of $D_j(z)$ and $P_{ij}(z)$ may be real numbers or conjugate pairs of complex numbers. Thus,

$$F_i(z) = \frac{Q_{ij}(z)N_j(z)}{(1-r_1z^{-1})(1-r_2z^{-1})\ldots(1-r_{m+w}z^{-1})}, \tag{A.69}$$

where $r_1, r_2, \ldots, r_{m+w}$ are real numbers or pairs of complex conjugate numbers. The partial fraction expansion of this expression is

$$F_i(z) = \sum_{k=1}^{m+w} \frac{K_k}{1-r_kz^{-1}}, \tag{A.70}$$

where

$$K_k = [(1-r_kz^{-1})F_i(z)]_{z=r_k}. \tag{A.71}$$

The corresponding inverse Z transform is found by applying the additivity property and applying the transform pair of eqn (A.55):

$$f_i(n) = \sum_{k=1}^{m+w} Z^{-1}\left[\frac{K_k}{1-r_k^{-1}}\right], \tag{A.72}$$

$$f_i(n) = \sum_{k=1}^{m+w} K_k r_k^n. \tag{A.73}$$

For each term in which r_k is a positive, real number, the inverse Z transform is a positive, real geometric sequence. For each term in which r_k is a negative, real number, the inverse Z transform is a real geometric sequence with alternating sign (representing oscillation between positive and negative values—at the highest frequency that can be represented with the selected interval of discrete time). For each pair of terms in which the r_k's are conjugate complex numbers, the coefficients (K_k's) also are complex conjugates:

$$\frac{K_m}{1-r_mz^{-1}} + \frac{\bar{K}_m}{1-\bar{r}_mz^{-1}} = 2\frac{C(1-\alpha\cos(\beta)z^{-1}) - D\sin(\beta)z^{-1}}{(1-2\alpha\cos(\beta)z^{-1}+\alpha^2z^{-2})}, \tag{A.74}$$

where

$$r_m = \gamma + \mathrm{j}\delta, \quad K_m = C + \mathrm{j}D, \quad \alpha^2 = \gamma^2 + \delta^2, \quad \cos(\beta) = \frac{\gamma}{\alpha}. \tag{A.75}$$

The corresponding inverse transform is obtained by applying the additivity property and using the appropriate transform pairs in Table A.7:

$$Z^{-1}\left[2\frac{C(1-\alpha\cos(\beta)z^{-1}) - D\sin(\beta)z^{-1}}{(1-2\alpha\cos(\beta)z^{-1}+\alpha^2z^{-2})}\right] = 2\alpha^n[C\cos(\beta n) - D\sin(\beta n)]. \tag{A.76}$$

The function $f(n)$ in this case describes oscillation at a frequency less than the maximum that can be represented with the discrete time interval of the model.

The zeros of the polynomial $P_{ij}(z)$ are the common ratios of the model itself; and the corresponding time functions (with different scaling factors) generally are present regardless of the form of the source function. As is true for natural frequencies of continuous-time models, an exception occurs if $N_j(z)$ happens to contain a binomial factor identical to one in $P_{ij}(z)$.

Example. Solve the following difference equation when $U(n)$ is a step of flow with amplitude K:

$$-\frac{1}{4}f(n-3)+f(n-2)-\frac{3}{2}f(n-1)+f(n)=U(n). \tag{A.77}$$

From Table A.8,

$$U(z)=\frac{K}{1-z^{-1}}. \tag{A.78}$$

From the linearity and delay properties in Table A.6,

$$\left[-\frac{1}{4}z^{-3}+z^{-2}-\frac{3}{2}z^{-1}+1\right]F(z)=U(z). \tag{A.79}$$

Combining these two expressions and factoring the denominator, one obtains

$$F(z)=\frac{K}{(1-z^{-1})(1-\frac{1}{2}z^{-1})(1-\frac{1}{2}[1+\mathrm{j}]z^{-1})(1-\frac{1}{2}[1-\mathrm{j}]z^{-1})}. \tag{A.80}$$

Partial-fraction expansion leads to

$$F(z)=\frac{4K}{1-z^{-1}}-\frac{K}{1-\frac{1}{2}z^{-1}}-\frac{K}{1-\frac{1}{2}[1+\mathrm{j}]z^{-1}}-\frac{K}{1-\frac{1}{2}[1-\mathrm{j}]z^{-1}}. \tag{A.81}$$

Applying the property of additivity and the transform pairs of Table A.7, one obtains the inverse transform

$$f(n)=4K-K\left[\frac{1}{2}\right]^n-2K\left[\frac{1}{\sqrt{2}}\right]^n\cos\frac{\pi n}{4}. \tag{A.82}$$

In this case, $f(n)$ grows in an oscillatory fashion toward $4K$. The first term in $f(n)$ is constant; the second term decays geometrically with increasing n; the third oscillates and decays geometrically with increasing n.

A.7 Phasor transforms

A bioengineer occasionally has reason to model a linearly operating system that is only partially accessible. Physiological systems, for example, often fall in this category. A powerful method for establishing constraints on the structure of the model involves studying the Fourier transform of any impulse response of the system to be modeled. Under favorable circumstances, the amplitude and phase characteristics of that transform bear strong clues to the number and distribution of natural frequencies in the system. In the inevitable presence of noise, however, recording the impulse response and subsequently transforming it is not always an effective approach. Often it is much more effective to collect the amplitude and phase data directly in the frequency domain, one frequency at a time. To do this, the observer applies a stimulus (to be equated with a source variable in the model) that is a sinusoidal function of time, then waits for two things; (1) the transient responses of the system (corresponding to its natural frequencies) to become vanishingly small, and (2) the purely sinusoidal response to the stimulus to emerge from the noise. Step (2) often requires averaging over many cycles of the stimulus sinusoid. Special-purpose electronic apparatus and special-purpose computer programs are available for this averaging process. To understand the use of sinusoids in this way, and to relate the results to prospective models, the bioengineer will find phasor transforms especially useful. They also are useful for designing and analyzing the network models associated with transducers.

Consider a linear, time-invariant, continuous-time model driven by a cosinusoidal flow of unit amplitude. The differential equation describing a single-input–single-output response of the model will have the following form:

$$\frac{\mathrm{d}^n f_i}{\mathrm{d}t^n}+a_{n-1}\frac{\mathrm{d}^{n-1}f_i}{\mathrm{d}t^{n-1}}+\ldots+a_1\frac{\mathrm{d}f_i}{\mathrm{d}t}+a_0 f_i =b_m\frac{\mathrm{d}^m U_j}{\mathrm{d}t^m}+\ldots+b_1\frac{\mathrm{d}U_j}{\mathrm{d}t}+b_0 U_j, \tag{A.83}$$

where

$$U_j(t) = \cos[\omega t] \quad (A.84)$$

and the frequency (ω) is given in radians per second. If all of the natural frequencies of the model either are negative real numbers or else complex numbers with negative real parts (which is to be expected of a large proportion of models of biological processes), then the transient responses of $f_i(t)$ to the onset of the cosine function eventually will vanish, and what remains will be a pure sinusoid:

$$f_i(t) = A\cos[\omega t + \beta], \quad (A.85)$$

where the amplitude (A) and phase (β) both depend on the frequency (ω) chosen for the sinusoid. The functions $A(\omega)$ and $\beta(\omega)$, when plotted (on log–log and semi-log graphs, respectively) over all values of ω for which response is considered significant, are called Bode diagrams. They compose truncated versions of the Fourier transform of the solution to eqn (A.83) when $U_j(t)$ is a unit impulse. They are, in fact, the functions of interest in this section.

If one postulates that e^{jx} (j being $\sqrt{-1}$) obeys the same rules of exponentiation as e^x where x is a real number, specifically that e^{jx} can be evaluated by substituting jx for x in the Maclaurin series for e^x, then

$$\cos(\omega t) = \mathrm{Re}\{e^{j\omega t}\} \quad (A.86)$$

and

$$A\cos(\omega t + \beta) = \mathrm{Re}\{Ae^{j(\omega t+\beta)}\} = \mathrm{Re}\{(Ae^{j\beta})e^{j\omega t}\}, \quad (A.87)$$

where Re$\{z\}$ is the real part of the complex number z.

The phasor transform applies only to a purely sinusoidal time function with fixed frequency (ω). It is defined by its inverse:

$$f(t) = \mathrm{Re}\{F(\omega)e^{j\omega t}\}$$

and has the following form:

$$F(\omega) = Ae^{j\beta} = A[\cos(\beta) + j\sin(\beta)]. \quad (A.88)$$

Based on this definition, the phasor-transform properties in Table A.9 can be deduced.

Table A.9 Properties of phasor transforms

Let $F(\omega)$ and $G(\omega)$ be the phasor transforms of $f(t)$ and $g(t)$, respectively, and let a and b be scalar constants;

then...	is the phasor transform of...	
$aF(\omega)$	$af(t)$	homogeneity
$F(\omega) + G(\omega)$	$f(t) + g(t)$	additivity
$aF(\omega) + bG(\omega)$	$af(t) + bg(t)$	linearity
$e^{-ja\omega}F(\omega)$	$f(t - a)$	delay
$F(\omega)/j\omega$	$\int f(t)\,dt$	integration
$j\omega F(\omega)$	$df(t)/dt$	differentiation

A.8 Ordinary differential equations with constant coefficients

A.8.1 Sinusoidal steady state

Once the transient responses corresponding to the model's natural frequencies have become vanishingly small, and only the pure sinusoidal response remains, the model is said to be in the condition of *sinusoidal steady state*. In principle, establishment of sinusoidal steady state requires infinite time. Thus, in contrast to zero-state analysis, where the model is taken to be unexcited at times less than zero, sinusoidal steady-state analysis presumes that the model has been subjected to a pure sinusoidal excitation for an indefinitely long period. In practice, sinusoidal steady state is established as soon as the transient responses have diminished to amplitudes

Table A.10 Phasor transform pairs for common source functions

$U(t)$	$U(\omega)$
$\cos[\omega t]$	1
$\sin[\omega t]$	$-\mathrm{j}$
$A\cos[\omega t+\gamma]$	$A\mathrm{e}^{\mathrm{j}\gamma}$

that are very small in comparison to the pure sinusoidal response. Source variables used for sinusoidal steady-state analysis and their phasor transforms are displayed in Table A.10.

Using the properties of the phasor transform given Table A.9, one can transform both sides of eqn (A.28), which describes the generalized single-input–single-output response:

$$P_{ij}(\omega)F_i(\omega)=Q_{ij}(\omega)U_j(\omega), \tag{A.89}$$

where

$$P_{ij}(\omega)=(\mathrm{j}\omega)^n+a_{n-1}(\mathrm{j}\omega)^{n-1}+\ldots a_1(\mathrm{j}\omega)+a_0,$$
$$Q_{ij}(\mathrm{j}\omega)=b_m(\mathrm{j}\omega)^m+\ldots+b_1(\mathrm{j}\omega)+b_0. \tag{A.90}$$

$U_j(\omega)$ is the phasor transform of $U_j(t)$ and $F_i(\omega)$ is the phasor transform of $f_i(t)$. Thus

$$F_i(\omega)=\frac{Q_{ij}(\omega)}{P_{ij}(\omega)}U_j(\omega) \tag{A.91}$$

and the solution, $f_i(t)$, is

$$f_i(t)=\mathrm{Re}\{F_i(\omega)\mathrm{e}^{\mathrm{j}\omega t}\}. \tag{A.92}$$

The sinusoidal steady-state response to $\cos[\omega t]$ is found by setting $U_j(\omega)=1$ (see Table A.10):

$$h_{ij}(t)=\mathrm{Re}\left\{\frac{Q_{ij}(\omega)}{P_{ij}(\omega)}\mathrm{e}^{\mathrm{j}\omega t}\right\},$$
$$H_{ij}(\omega)=\frac{Q_{ij}(\omega)}{P_{ij}(\omega)}. \tag{A.93}$$

$H_{ij}(\omega)$ is the Fourier transform of $f_i(t)$ when $U_j(t)$ is a unit impulse.

If the source function is a sinusoid other than $\cos[\omega t]$, then the corresponding response function can be found from the following relationship:

$$f_i(t)=\mathrm{Re}\{H_{ij}(\omega)U_j(\omega)\mathrm{e}^{\mathrm{j}\omega t}\}, \tag{A.94}$$

$$F_i(\omega)=H_{ij}(\omega)U_j(\omega). \tag{A.95}$$

Example. Solve the following differential equation when $U(t)=A\cos(\omega t)$:

$$\frac{\mathrm{d}^3f}{\mathrm{d}t^3}+4\frac{\mathrm{d}^2f}{\mathrm{d}t^2}+6\frac{\mathrm{d}f}{\mathrm{d}t}+4f=U(t). \tag{A.96}$$

From Tables A.9 and A.10,

$$[-\mathrm{j}\omega^3-4\omega^2+\mathrm{j}6\omega+4]F(\omega)=A,$$

$$F(\omega)=A/(a+\mathrm{j}b), \tag{A.97}$$

$$a=-4\omega^2+4,\quad b=-\omega^3+6\omega;$$

$$F(\omega)=\frac{(a-\mathrm{j}b)A}{(a+\mathrm{j}b)(a-\mathrm{j}b)}$$
$$=\frac{(a-\mathrm{j}b)A}{a^2+b^2}; \tag{A.98}$$

$$(t)=\frac{aA}{a^2+b^2}\cos(\omega t)+\frac{bA}{a^2+b^2}\sin(\omega t),$$
$$=\sqrt{\frac{(a^2A^2+b^2A^2)}{(a^2+b^2)^2}}\cos(\omega t+\theta), \tag{A.99}$$
$$=\frac{A}{\sqrt{a^2+b^2}}\cos(\omega t+\theta);$$

$$\theta=\tan^{-1}[-b/a]. \tag{A.100}$$

Bibliography

Bateman, H. (1954). *Tables of integral transforms*, Vol. I. McGraw-Hill, New York.

Cadzow, J. A. (1973). *Discrete-time systems*. Prentice Hall, Englewood Cliffs, NJ.

Chua, L. O., Desoer, C. A., and Kuh, E. S. (1987). *Linear and nonlinear circuits*. McGraw-Hill, New York.

Desoer, C. A. and Kuh, E. S. (1969). *Basic circuit theory*. McGraw-Hill, New York.

Gardner, M. F. and Barnes, J. L. (1942). *Transients in linear systems*. Wiley, New York.

Jury, E. I. (1964). *The theory and application of the z transform methods*. Wiley, New York.

Kuo, T. (1967). *Linear networks and systems*. McGraw-Hill, New York.

Roberts, G. E. and Kaufman, H. (1966). *Table of Laplace transforms*. Saunders, Philadelphia.

Skilling, H. H. (1957). *Electrical engineering circuits*. Wiley, New York.

ANSWERS TO SELECTED PROBLEMS

Chapter 1

P1.1 $F_{\text{seat}_{\text{vert}}} = 508$ N (up); $F_{\text{seat}_{\text{horiz}}} = 140$ N (to left); $F_{\text{board}} = 228$ N (up).

P1.2 $R = 110.6$ N; $P = 73.2$ N

P1.3 **(a)** $F_c = 389$ N; $F_s = 338$ N
(b) (i) $F_c = 4347$ N; $F_s = (-)627$ N
(ii) $F_c = 3266$ N; $F_s = (-)214$ N

P1.4 **(a)** $F_{\text{biceps}} = 1493$ N; $F_{\text{humerus}} = 1425$ N at 42.3° to the right and down.
(b) $p = 1.19$ MPa

P1.5 **(a)** Spec. grav. = 1.07; density = 1.07×10^3 kg/m^3.
(b) $W_{\text{arm}} = 36.7$ N; scale reading = 71.5 kg.

P1.6 **(a)** 0.252 m
(b) 1.109 m
(c) Using (a), $m = 3.357$ kg, $W = 32.9$ N; using alternative, $m = 3.217$ kg, $W = 31.56$ N.
(d) Using (a), spec. grav. = 1.108; using alternative, spec. grav. = 1.062

P1.7 $x_2 - x_2' = L(Q - Q_1)/(W - W_1)$.

P1.8 **(a)** $m_{\text{trunk}} = 8653$ kg
(b) $\sigma(\text{top}) = 1.561$ MPa; $\sigma(\text{bottom}) = 0.266$ MPa.
(c) $y = 10.71$ m
(d) $\sigma\langle y\rangle = [5.115 \times 10^6 - 17.65 \times 10^3 (19.2y - y^2/2)]/[19.2 - y]$.

P1.9 **(a)** $M_k = m[R \times (\ddot{r}_{O'} - g) + k^2\dot{\Omega}]$, Ω angular velocity of O′xyz, positive ccw.
(b) $M_k = m[(\frac{k^2}{h} - R_1)a_1 - R\Omega^2\delta_1 - \frac{k^2}{h}a_2$
R is distance from O′ to center of gravity of leg;
$r_{O'}$ = displacement OO′.
$\underline{k}$ = radius of gyration of leg
$\bar{\Omega}$ = angular velocity of leg

P1.10 **(a)** $W = 1478$ N; power = 520 W = 0.697 hp.
(b) $v_{\max} = 0.2117$ m s^{-1}

P1.11 $F = 73.79$ N

P1.12 **(a)** $M = 156.5$ N m;
(b) This is attainable.
(c) $F_{\max} = 6770$ N

P1.13 **(a)** $F_{\text{forward}} = 205.3$ N
(b) Torque = 205.3 N m
(c) $F_{\text{pedal}} = 1173$ N
(d) Power = 680 W (= 0.912 hp)

P1.14 **(a)** $v_B = 31.3$ m s^{-1}
(b) $d = 4.2$ m above the water; person remains dry.
(c) Amplitude = 15.5 m
(d) Half period is 1.535 s
(e) $\epsilon = 0.36$; $\sigma_{\max} = 26$ MPa
(f) $v_{\text{term}} = 31.3$ m s^{-1} (by chance the same as *a*)

P1.15 **(a)** $v_0 = 13.51$ m s^{-1}
(b) $F = 982$ N
(c) $M = 500$ N m

P1.16 **(a)** 0.0221 m
(b) 0.00107 m

P1.17 **(a)** $m\ddot{y} + c\dot{y} + ky = m(2c_1 b\Omega \cos \Omega t + n^2 b \sin \Omega t)$

(b) $Q = \sqrt{\dfrac{\phi^2 + n^4}{\phi^2 + \mu^2}}$

$\phi \equiv 2c_1\Omega;\ \mu \equiv n^2 - \Omega^2.$

P1.18 $k_2\ddot{\theta}(L+r)\hat{e}_\theta - kr\hat{e}_r - mg\boldsymbol{j} = m\boldsymbol{a}_1 = m_1[a_2\boldsymbol{i} + \ddot{\theta}(L+r)\hat{e}_\theta - \dot{\theta}^2(L+r)\hat{e}_r]$
where $\hat{e}_\theta$ and $\hat{e}_r$ are the orthogonal unit vectors in the θ and r directions, respectively; $\boldsymbol{i}$ and $\boldsymbol{j}$ are the horizontal left and downward vertical unit vector. $\boldsymbol{i}$ and $\boldsymbol{j}$ are related to $\hat{e}_\theta$ and $\hat{e}_r$ through angle θ. r is the increase in L due to spring extension in the direction θ from the vertical.

P1.19 **(b)** Energy/volume = 17.82 kPa
(c) Energy/volume = 17.63 kPa
(d) $\alpha = 425$ rad/s^2
(e) $\sigma = 140$ kPa < 186 kPa. Will not fail. Other muscles also resist the blow.

P1.20 **(a)**

	σ, MPa	ε	
Point A	1.24	1.55×10^{-4}	
Point B	3.68	4.45×10^{-4}	Point B is the point
Point C	2.12	2.56×10^{-4}	of maximum stress.
Point D	0.692	0.837×10^{-4}	

(b) 3.68 MPa $\ll$ 172 MPa Well below failure stress; hence not dangerous.

P1.21 **(a)** $G = 3.10$ MPa; $\lambda = 6.02$ GPa
$K = 8.09$ MPa

P1.22 $J\langle t\rangle = 2.5 \times 10^{-6} + 2.0 \times 10^{-6} \times \sum_{n=1}^{\infty}[(-0.04082t)^n - (-1.2844t)^n]/n \cdot [n!]$

P1.23 **(a)** 115 J
(b) (i) 9.89 KN
(ii) 7.77 kN
(iii) 4.944 kN

(c)

	For 400 g	For 200 g	For 150 g
(i)	Pass	Pass	Pass
(ii)	Pass	Pass	Pass
(iii)	Pass	Pass	Pass

(d)

(i)	19.78 kN	Pass	Pass	Pass
(ii)	15.54 kN	Pass	At limit	Pass
(iii)	9.894 kN	Pass	Pass	Fail

P1.24 **(a)** $h = 4.809$ m
(b) $v = 9.160$ m s^{-1}
(c) $F = 2831$ N @ 107.8° ccw from positive x-axis
(d) $k = 2472$ N/m

P1.25 **(a)** (i) $v = 141.1$ m s^{-1} (ii) $v = 46.16$ m s^{-1}
(b) Free fall distance is 228.6 m

P1.26 Pick two points, Λ_1 at F_1 and Λ_2 at F_2. Then $(1-\Lambda_1)/(1-\Lambda_2) = [\ln(F_1/A_1b)+1]/\ln[(F_2/A_2b)+1]$ Solve for b by trial and error. Then solve for a from $a = \{\ln[F_1/a_1b)+1]\}/\{\Lambda_1 - 1\}$

P1.27 **(a)** $V = 44.59 + 27.29x$ N; x in meters from free end
$M = -44.59x - 13.64x^2$ N m
(b) $y = -0.0582 \times 10^{-3}\,[710 - 334x + 7.433x^3 + 1.1375x^4]$ N m, x in meters from free end.
(c) $y_{max} = -0.0413$ m $= -41.3$ mm
(d) $\sigma_{max} = 6.052$ MPa

P1.28 **(a)** $\sigma = 4.167$ MPa
(b) $\sigma_B = 62.5$ MPa
(c) (i) 1.49 (ii) 1.16

P1.29 **(a)** $R_{min} = 7.49$ mm
(b) 4.427 rad s^{-1}

P1.30 **(a)** $\epsilon = 0.0436$
(b) $\Delta L = 1.308$ mm
Viscous effects are negligible.

Chapter 3

P3.3 **(c)** $\alpha = 2a_0^2/Eh$

P3.11 $u/V = [\{(1+3\alpha)/(1+\alpha)\}\{1 - (r/a)^{(1+\alpha)/\alpha}\}]$

P3.15 **(b)** -0.2×10^3 N m^{-2} at the top, 1.3×10^3 N m^{-2} half-way up, and 2.8×10^3 N m^{-2} at the bottom.
(c) 200 N m^{-2}
(d) -1.2×10^3 N m^{-2} at the top, 0.3×10^3 N m^{-2} half-way up, and 1.8×10^3 N m^{-2} at the bottom.
(e) Totally closed at the very top, totally open at the very bottom, capillaries begin to close

partially at some distance up the lung until they close completely near the top.

(f) 3 cm above the right ventricle.

P3.16 **(a)** 1.1 W or 1.5×10^{-3} hp

(b) 10 per cent

Chapter 4

P4.1 See Table 4.1, eqns (4.1D)–(4.1F) and (4.1J)–(4.1L).

P4.2 —

P4.3 —

P4.4 (c) 19.1 g/g-moles.

P4.5 Experimental 6.9×10^{-7} cm^2 s^{-1} in water [C. Tanford, *Physical chemistry of macromolecules*, (1967), p. 358. Wiley, New York.]

P4.6 $P_{XM} = (\mathscr{D}_{XM}\phi)/\delta$

P4.7 (a) $\delta_D^2 + \left(\dfrac{2\delta_H\mathscr{D}_{dm}}{K\mathscr{D}_{db}}\right)\delta_D = 2\dfrac{c_m}{c_o}\mathscr{D}_{dm}\left(1 - \dfrac{c_b}{Kc_m}\right)t.$

(b) $W_d = AJ_d \approx Ac_o\left|\dfrac{d\delta_D}{dt}\right|$

$$= \frac{A\mathscr{D}_{dm}\mathscr{D}_{db}c_mK\left(1 - \dfrac{c_b}{Kc_m}\right)}{\delta_H\mathscr{D}_{dm} + K\delta_D(t)\mathscr{D}_{db}}.$$

(c) For $K = 10^{-1}$, 10^4 s (2.8 hours).

For $K = 10^{-2}$, 10^6 s (11.6 days).

For $K = 10^{-3}$, entire period (10^8 s > 3 years, in principle).

(d) $t_{max} \approx \dfrac{H\delta_H}{2K\mathscr{D}_{db}}\left(\dfrac{c_o}{c_m}\right)$ IF constant rate prevails until wafer is exhausted.

(e) For $K = 10^{-1}$, $t_{max} = 25\,000$ s (not valid since rate is not constant throughout).
For $K = 10^{-2}$, $t_{max} = 2.5 \times 10^5$ s (2.89 days).
For $K = 10^{-3}$, $t_{max} = 2.5 \times 10^6$ s (28.9 days).

P4.8 Pseudo-steady-state analysis gives $W_d \approx 2\pi\mathscr{D}_{dm}c_ba_i$ when $R(t) \gg a_i$ (after the initial transient).

P4.9 (c) $c^* = 1 - y^* - (-1)^n \sum_{n=1}^{\infty} \dfrac{2(-1)^n}{\pi n} e^{-n^2\pi^2 t^*} \sin n\pi y^*$

(d) When $t^* = 0.5$, c^* is within about 99.3 per cent of its final distribution, so the transient lasts about 5 seconds.

P4.10 (a) 10.8 mg%; 3.32 meq/L.

(b) {12.38 g = 391 meq}PO_4 and {85.1 g} urea removed.

(c) *urea:* $c_{B2} = 3050 - 15.7\, c_{D1} \ldots$ mg%.

PO_4: $c_{B2} = 105 - 15.7\, c_{D1} \ldots$ meq/L.

P4.11 Blood resistance = 2.66 per cent (R_B/R).
Dialysate resistance = 1.50 per cent (R_D/R).
Membrane resistance = 95.84 per cent (R_M/R).

P4.12 $\dfrac{c - c_B^i}{c_D^i - c_B^i} = \dfrac{\ln(r/r_B)}{\ln(r_D/r_B)}.$

$W = (A_rN_r) = 2\pi L\phi\mathscr{D}_M \dfrac{(c_D^i - c_B^i)}{\ln(r_D/r_B)}.$

P4.13 —

P4.14 (a) $c = c_c - \dfrac{(-C)}{4\mathscr{D}_{Ot}}\left[2r_t^2 \ln\dfrac{r}{r_c} - (r^2 - r_c^2)\right]$ where $C < 0$ (oxygen consumption).

(b) $c = c_c \left[\dfrac{I_o(x) + \dfrac{I_1(x_t)}{K_1(x_t)} \cdot K_o(x)}{I_o(x_e) + \dfrac{I_1(x_t)}{K_1(x_t)} \cdot K_o(x_c)}\right]$

where: $x \equiv r\sqrt{\dfrac{k_f}{\mathscr{D}_{Ot}}}$

P4.15 (a) $c_{B2} = c_{B1}(\exp[(-AK_X)/Q_B]$

P4.16 (c) $(1 - x_A) = (1 - x_{AL})^{z/L}$

(d) $N_{Az} = c\mathscr{D}_{AB}\dfrac{\ln(1 - x_{AL})}{L}$,

and

$W_A = N_{Az} \cdot \dfrac{\pi}{4}D^2$

(e) $W_A = c_A^{(l)} \cdot \dfrac{\pi}{4}D^2\left(-\dfrac{dL}{dt}\right)$; $c_A^{(l)}$ = *liquid* concentration

(f) $L = \left[-2\dfrac{c}{c_A^{(l)}}\mathscr{D}_{AB}\ln(1 - x_{AL})\right]^{\frac{1}{2}} \cdot t^{\frac{1}{2}}.$

P4.17 —

Chapter 5

P5.2 $U = [1/h_A + \Delta x_a/k_a + \Delta x_b/k_b + 1/h_B]^{-1}$.

P5.3 $\Delta x_h = 0.21$ cm

P5.4 $q_{seated} = 6.64\ W$; $q_{walking} = 102.5$ W

P5.5 12 W

P5.6 $\dot{m} = 5.03 \times 10^{-2}$ g s^{-1}

P5.7 $R_{eq} = 0.019/A$

P5.8 $t = 1956$ s $= 0.54$ h

P5.9 $t = 2329$ s $= 0.65$ h

P5.12 90 W m^{-2}

P5.13 $K_{eff} = 379$ W m^{-2}°C^{-1}

Chapter 6

P6.1 **(a)** $n(t)$ would be given in spikes or discharges s^{-1} (frequency).

(b) $$\frac{X(s)}{N(s)} = \frac{20}{s+2} = \frac{10}{1+(1/2)s},$$

time constant = 1/2 s

(c) With feedback we have

$$\frac{X}{N} = \frac{\dfrac{20}{s+2}}{1+\dfrac{20}{s+2}} = \frac{20}{2+22} = \frac{20/22}{1+(1/22)s}.$$

The time constant is now = 0.045 s (over 10 times faster), while gain has decreased to 0.91.

(d) The feedback could come from muscle length receptors (spindles) feeding back to motoneurons.

P6.2 **(a)** To open the loop, measure eye position and use this signal to control an electronic feedback of +1 to control the target position. Then total feedback = −1 + 1 = 0 (open loop).

(b) Loop again $s \cdot A \cdot \dfrac{1}{s} \cdot \dfrac{1}{1+Ts} = \dfrac{A}{1+Ts}$.

(c) Yes—maximum phase lag is only 90°.

(d) $$\text{Output} = \text{input} \cdot \frac{\dfrac{A}{1+Ts}}{1+\dfrac{A}{1+Ts}}$$

$$= \frac{100}{s} \cdot \frac{\dfrac{A}{A+1}}{1+\dfrac{Ts}{A+1}}.$$

Applying the final value theorem to find the steady state:

$$\text{output}\left(\lim_{x\to\infty}\right) = \lim_{s\to 0} \frac{100s}{s} \cdot \frac{\dfrac{A}{A+1}}{1+\dfrac{Ts}{A+1}}$$

$$100 \cdot \frac{A}{A+1} = 100 \cdot \frac{100}{101} = 99° \text{ s}^{-1}$$

(e) The open-loop transfer function is now $100e^{-0.2}s/1 + 0.1s$. When we plot this open-loop transfer function as a Bode plot, we find the gain curve crosses −180° at a frequency of about 12 rad s^{-1} and the gain at this frequency is about 35 dB. Therefore, the system is unstable in the closed loop and will oscillate at about 2 Hz.

(f) We have $\dot{E}_{measured} \times 0.95\dot{E}_{actual}$.

Let $\dfrac{\dot{E}_{\text{measured}}}{T} = x$ and $x < 1$; $\dfrac{\dot{E}_{\text{actual}}}{T} = y$,

$x = 0.95y$.

Then $\dfrac{A}{A+1} = x$ and $\dfrac{B}{B+1} = y$,

where B = actual A

= 100 (from part d)

$$Ax + x = A \text{ (estimated } A)$$
$$Ax - A = -x$$
$$A(x - 1) = -x$$

$$A = +\frac{x}{1-x} = \frac{0.95y}{1-0.95y}, \quad B = \frac{y}{1-y},$$

$y = 0.99$

$$\% \text{ error} = \frac{B-A}{B} \times 100$$
$$= \frac{\dfrac{y}{1-y} - \dfrac{0.095y}{1-0.95y}}{\dfrac{y}{1-y}}$$
$$= \frac{y(1-0.95y) - 0.95y(1-y)}{y(1-0.95y)}$$
$$= \frac{y - 0.95y^2 - 0.95y + 0.95y^2}{y - 0.95y^2}$$
$$= \frac{y - 0.95y}{y - 0.95y^2} = \frac{1-0.95}{1-0.95y}$$
$$= \frac{0.05}{1-0.95y} = \frac{0.05}{1-(0.95)(0.99)}$$
$$\approx 83\% \text{ error}.$$

P6.3

	Low frequency	High frequency
(a)	gain = $1/a$	falling off, 20 dB/dec
	phase = 0	constant at $-90°$
(b)	gain = b/a	constant at 0 dB.
	phase = 0	constant at 0°
(c)	gain = increasing at 20 dB/dec	constant at $1/T$
	phase ≈ 90° lead	constant at 0°
(d)	gain = 1	falling off, 40 dB/dec
	phase = 0	increasing lag linearly with ω.
(e)	gain = $1/a$	falling off at 20 dB/dec
	phase = 0	constant at +90°

P6.4 $G(s) = \dfrac{1 - \dfrac{s}{a}}{1 + \dfrac{s}{a}} = \dfrac{1 - \mathrm{j}\dfrac{\omega}{a}}{1 + \mathrm{j}\dfrac{\omega}{a}}$,

$$\text{gain} = \frac{1^2 + \left(-\dfrac{\omega}{a}\right)^2}{1^2 + \left(\dfrac{\omega}{a}\right)^2} = 1.$$

The gain is 1 or 0 dB for all frequencies; so this looks like a pure delay.
The phase due to the denominator goes to $-90°$; call this ϕ_d.
The phase due to the numerator is

$$\phi_n = \tan^{-1}\left(-\frac{\omega}{a}\right) = -\tan^{-1}\left(\frac{\omega}{a}\right).$$

This merely adds another phase curve that looks exactly like the denominator. So, as shown in Fig. S.1, up to $\omega = a$ the transfer function (G) phase curve looks like a pure delay. It will simulate a delay of τ, where $\phi = 90° = \omega\tau$:

$$\tau = 90/57.3a = 1.57/a$$

P6.5 (a) $Y(s) = \dfrac{1}{s} \cdot \dfrac{10}{1+Ts} = \dfrac{1}{s} \cdot \dfrac{10/T}{s + 1/T}$.

From the table, $y(t) = T \cdot \dfrac{10}{T}(1 - \mathrm{e}^{-(1/T)t})$

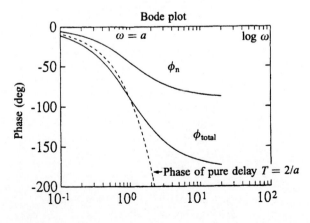

$= 10(1 - \mathrm{e}^{-(1/T)t})$.

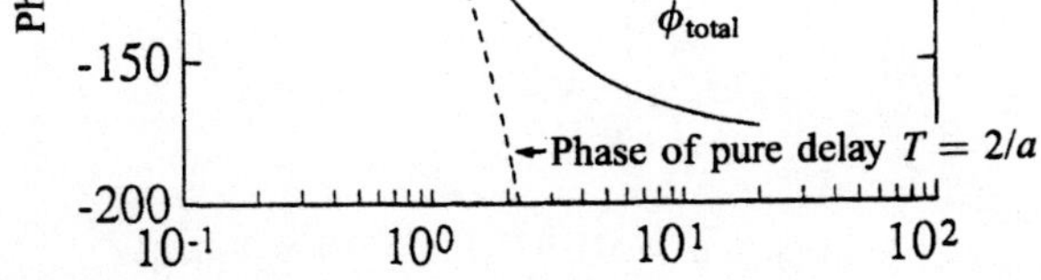

Fig. S.1

(b) New output $= Y(s) \cdot s = N(s)$

$$= \frac{10}{1+Ts} = \frac{10/T}{s+\frac{1}{T}}$$

$$n(t) = 10/T e^{-(1/T)t}$$

By direct differentiation in the time domain,

$$\frac{d}{dt}(10 - 10e^{-(1/T)t}) = \frac{10}{T}e^{-(1/T)t}.$$

Thus, we obtain the same answer with either method.

(c) Plot the frequency response, then use the frequency response curves to estimate gain and phase of output for an input of 10 sin ωt: system gain at $\omega = 1$ is 40 dB or a factor of 100 (and gain multiplies the input magnitude);
system phase at $\omega = 1$ is $+90°$ (and phase adds to input phase);
output $= 1000 \sin(\omega t + 90°)$.

P6.6 (a) $BW = \frac{1}{2\pi T}$ where T is the lowest frequency pole

$$\therefore BW = \frac{1}{2\pi T_1} = \frac{1}{2\pi 10} = 0.1\frac{1}{2\pi} \approx 0.016 \text{ Hz}.$$

(b) The high-frequency pole is at a frequency of

$$\frac{1}{2\pi T_2} = \frac{1}{2\pi(0.005)} \approx 32 \text{ Hz}.$$

Therefore, we can ignore it for natural head movements.

$$\text{approx. trans. function} = \frac{k}{1+T_1 s}$$

(c) For an input of 0.5 Hz = 3.14 rad s^{-1},

$$\frac{1}{T_1} = 0.1,$$

$\therefore$ input is well above break frequency $\Rightarrow$ system looks like an integrator $\Rightarrow$ phase lag $\approx -90°$,
$\therefore$ output in phase with head velocity.
Input is at 0.001 Hz = 0.00628 rad s^{-1}. This is well below break frequency $\Rightarrow$ phase lag with respect to head acceleration input $\Rightarrow$ output is in phase with head acceleration.

(d) Shown in Fig. S.2 is a positive impulse of head acceleration at $t = 0$ (starts the head rotating at a constant velocity). The head receives a negative impulse acceleration at $t = 100$ (stops the head rotation). The time response of the canal sensor to the positive impulse is a positive decaying exponential (with a time constant of 10 s). Therefore, the response almost decays away by 25 s. Second impulse at $t = 100$ generates a canal response in the opposite direction.

(e) $$\frac{1}{1+10s} \cdot \frac{a}{1 - \frac{a}{1+10s}} = \frac{a}{1+10s-a}$$

$$= \frac{a/1-a}{1+\frac{10}{1-a}s};$$

$$\text{new } TC = \frac{10}{1-a} > 10,$$

$\therefore$ low frequency pole is moved to lower frequency.

$\Rightarrow$ Output remains in phase with head velocity to a lower frequency.

P6.7 (a) Block diagram of stretch reflex is shown in Fig. S.3. Notice that I = constant and let $x = \Delta x$.

(b) $f = kx^2$; energy stored $= \int_0^x kx^2 \, dx$

$$= \frac{kx^3}{3}.$$

(c) $f = kx$; energy stored $= \int_0^x kx \, dx$

$$= \frac{kx^2}{2}.$$

P6.8 $$G_{CL} = \frac{G}{1+GH},$$

$$G_{OL} = G = \frac{1/2}{1+2s},$$

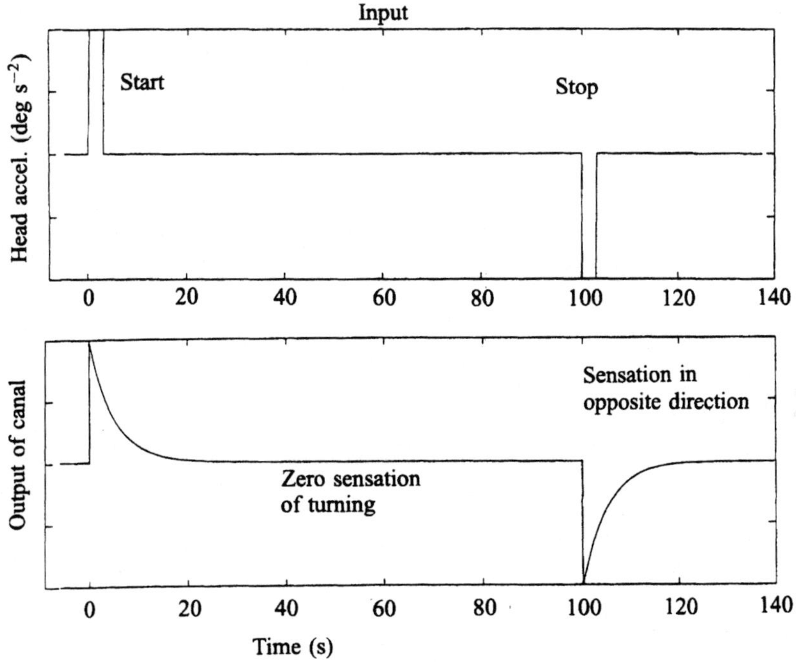

Fig. S.2

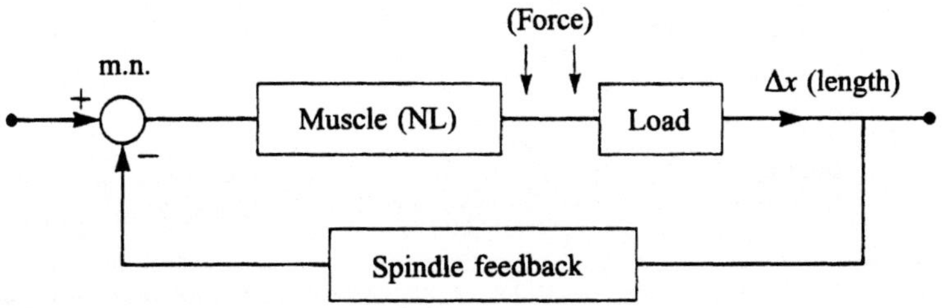

Fig. S.3

$$G_{CL} \text{ (from Bode plot)} = \frac{1/3}{1+3s}.$$

Thus, from (1) above,

$$\frac{1/3}{1+3s} = \frac{\dfrac{1/2}{1+2s}}{1+\dfrac{1/2}{1+2s}\cdot H}.$$

Solving for H,

$$\frac{1/3}{1+3s} = \frac{1/2}{1+2s+1/2H}$$

$$\frac{1}{3}+\frac{2}{3}s+\frac{1}{6}H = \frac{1}{2}+\frac{3}{2}s$$

$$2+4s+H = 3+9s$$

$$H = 1+5s$$

P6.9 Let $R(s) = \dfrac{1}{s}\cdot\dfrac{1}{s^2+b^2} = \dfrac{1}{s}\cdot H(s)$

Then by the final value theorem:

$$\lim_{t\to\infty} r(t) = \lim_{s\to 0} s\left[\frac{1}{s}\cdot\frac{1}{s^2+b^2}\right] = \frac{1}{b^2} \ \textbf{wrong;}$$

$$\lim_{t\to 0} r(t) = \lim_{s\to\infty} s\left[\frac{1}{s}\cdot\frac{1}{s^2+b^2}\right] = 0 \quad \textbf{okay}.$$

Note that $h(t) = 1/b \sin bt$ is not defined as $t \to \infty$. Restrict H to be a function for which $|h(t)| \to 0$ as $t \to \infty$.

P6.10 The gain curve suggests that the spindle response has a zero and pole.

For the zero: $1 + T_1 s$ where

$$T_1 = \frac{1}{0.7 \times 2\pi} = 0.23 \text{ s}$$

For the pole:

$$\frac{1}{1 + T_2 s}$$

where

$$T_2 = \frac{1}{7 \times 2\pi} = 0.023 \text{ s}$$

The phase curve shows nonminimum characteristics. The zero–pole combination would provide a phase curve such that $\phi \approx 0$ at $f = 100$ Hz. But the actual phase curve shows about 360° of phase lag at 100 Hz. Therefore a non-minimum phase element must be present. Try e^{-sT_3}, $\omega T_3 = 360° \,|_{\omega = 628}$:

$$628 \times 57.3 \times T_3 = 360,$$
$$T_3 = 10 \text{ ms}.$$

P6.11 **(a)** $k + \dfrac{1}{s} = \dfrac{sk+1}{s} = \dfrac{1+ks}{s}$

(b) $\dfrac{E(s)}{N(s)} = \dfrac{1+ks}{s}\cdot\dfrac{1}{1+ks} = \dfrac{1}{s}.$

Overall transfer function $= 1/s$. This type of transfer function is called an integrator.

(c) Note that phase $= -90°$ for all frequencies. Therefore expected phase angle between head velocity and response E is 90° phase lag over all frequencies.

P6.12 **(a)** Delay of 200 ms; gain = 100. Integrator due to linear increase. Overall model (in open loop) is shown in Fig. S.4.

(b) Closed-loop system is shown in Fig. S.5.

Input $= 1/s$ = unit step function.

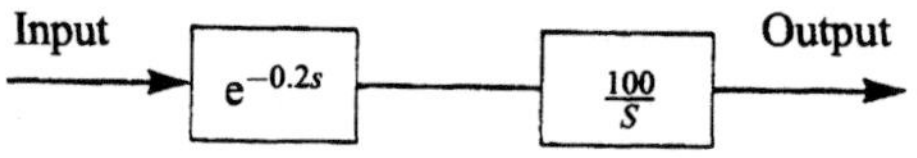

Fig. S.4

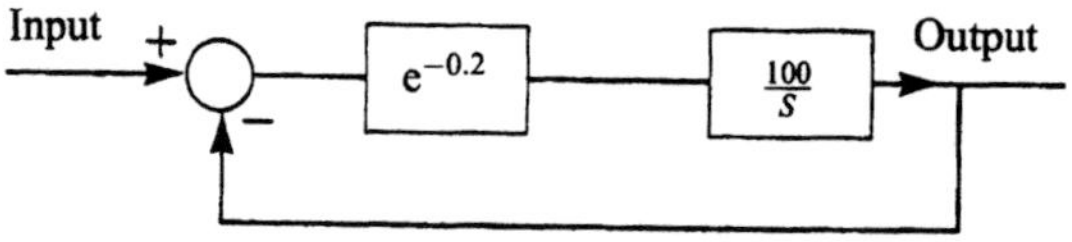

Fig. S.5

$$H_{CL} = \frac{100e^{-0.2s}/s}{1 + 100e^{0.2s}/s} = \frac{100e^{0.2s}}{s + e^{0.2s}(100)}$$

$$\text{Response}_{CL} = \frac{100}{s}\cdot\frac{e^{-0.2s}}{s + e^{0.2s}(100)}$$
$$= \frac{1}{s}\cdot\frac{1}{\frac{1}{100}s + e^{-0.2s}}\cdot e^{-0.2s}$$

(c) Output

$$= \lim_{t\to 0} s\cdot\frac{100}{s}\cdot\frac{e^{-0.2s}}{s + e^{-0.2s}(100)}$$
$$= \frac{100}{100} = 1$$

Difference between input and output → 0.

(d) No, it is unstable. The phase curve crosses −180° at about 9 rad^{-1}. The gain value at this frequency is about 40 dB ⇒ unstable system.

(e) $TC = 0.01$ s = 10 ms

(f) Open-loop transfer function

$$= H_{OL} = \frac{100e^{-0.2s}}{1 + Ts}, \quad T = 0.01.$$

Closed-loop transfer function

$$= H_{CL} = \frac{100e^{-0.2s}/(1+Ts)}{1 + 100e^{-0.2s}/(1+Ts)}$$
$$= \frac{100e^{-0.2s}}{1 + Ts + 100e^{-0.2s}}$$

P6.13 Let $s = j\omega$

$$F(j\omega) = (j\omega)^k = (j\omega)^{1/2} = |M|e^{j\phi}$$

$$
\begin{aligned}
|M| &= 20 \log |j\omega|^{1/2} \\
&= 1/2 \cdot 20 \log \omega \\
&= 10 \log \omega.
\end{aligned}
$$

(This has a slope of +10 dB/dec for all frequencies.)

$$
\begin{aligned}
\text{Phase angle} &= \tan^{-1}\frac{\operatorname{Im} F(j\omega)}{\operatorname{Re} F(j\omega)} \\
&= \tan^{-1}\frac{\omega^k}{0} = k\tan^{-1}\frac{\omega}{0} \\
&= k(90°) \\
&= +45°.
\end{aligned}
$$

(Constant phase lead of 45° for all frequencies.)

P6.14 **(a)** Bode plots show −60 dB/dec at high frequency with a break frequency at $\omega = 10$ rad s^{-1} and 180° of phase lag at about 10 rad s^{-1}.

(b) No! OL gain $\ll 1$ for all frequencies.

(c) With a gain of 20 in the feedback path, the OL DC gain $\approx (0.2)(20) = 4 \Rightarrow$ unstable.

The system will oscillate at the 180° crossover frequency of about 2 Hz.

(d) $$\lim_{s\to 0} s\left[\frac{1}{s}\left(\frac{\dfrac{0.2e^{0.18s}}{(1+0.1s)^3}}{1+\dfrac{0.2e^{0.18s}}{(1+0.1s)^3}}\right)\right] = \frac{0.2}{1+0.2} = 0.167$$

(e) $G_B = 1$

$$
\begin{aligned}
G_A &= \frac{\dfrac{e^{-0.18s}}{(1+0.1s)^2}}{1+\dfrac{0.2e^{-0.18s}}{(1+0.1s)^3}} \\
&= \frac{e^{-0.18s}(1+0.1s)}{(1+0.1s)^3 + 0.2e^{-0.18s}}
\end{aligned}
$$

If we let $s \equiv j\omega$ and take the absolute value at $\omega = 2\pi(10) \approx 62$, we find that noise injected at point A is attenuated by a factor of about 1/50 at 10 Hz, measured by pupil size. Thus, point B is the more likely site of the noise.

Chapter 7

P7.1 **(a)** $\boldsymbol{RT}\ln[a_i(0)/a_j(0)] = \mu_i - \mu_j$.

(b) $F_{ij} = [\boldsymbol{RT}/a_i(0)][\Delta a_i - \Delta a_j]$.

(c) $\Delta F_i = [\boldsymbol{RT}/Q_i(0)]\Delta Q_i, \quad a_i = \gamma_i Q_i$.

(d) $$\frac{d\Delta F_j}{dt} = \frac{\Delta F_i}{R_{ij}C_j} - \frac{\Delta F_j}{R_{ij}C_j},$$
$$\frac{d\Delta F_j}{dt} = -\frac{\Delta F_i}{R_{ij}C_i} + \frac{\Delta F_j}{R_{ij}C_i}.$$

(e) $$\frac{dQ_i}{dt} = -k_{ji}Q_i + k_{ij}Q_j, \quad \frac{dQ_j}{dt} = k_{ji}Q_i - k_{ij}Q_j.$$

(f) $C_iR_{ij} = 1/k_{ji}, \quad C_jR_{ij} = 1/k_{ij}$,
$k_{ij}Q_j(0) = k_{ji}Q_i(0)$,
$C_i = Q_i(0)/\boldsymbol{RT}, \quad C_j = Q_j(0)/\boldsymbol{RT}$,
$R_{ij} = 1/k_{ji}C_i = 1/k_{ij}C_j$.

(g) $dW = F_{ij}\, dQ$

$$W = \int_{\delta Q=0}^{\Delta Q} F_{ij}(\delta Q)\, d(\delta Q) = \left[\frac{C_i + C_j}{2}\right]\Delta Q^2$$

(h) $J_s(s) = \Delta Q$,

$$\Delta F_j(s) = \frac{\Delta Q}{C_i + C_j}\left[\frac{1}{s} - \frac{1}{s + \dfrac{1}{C_iR_{ij}} + \dfrac{1}{C_jR_{ij}}}\right]$$

$$\Delta F_j(t) = \Delta Q\,\frac{1 - \exp\left(-\dfrac{C_i + C_j}{C_iC_jR_{ij}}t\right)}{C_i + C_j}.$$

(i) $L_1 = \dfrac{-k_{ji}}{s}, \quad L_2 = \dfrac{k_{ij}k_{ji}}{s^2}, \quad L_3 = \dfrac{-k_{ij}}{s},$

$$P_1 = \frac{k_{ji}}{s^2},$$

$$\frac{\Delta Q_j(s)}{J_s(s)} = \frac{P_1}{1 - L_1 - L_2 - L_3 + L_1 L_3},$$

$$\Delta Q_j(t) = \frac{k_{ji}\Delta Q}{k_{ji} + k_{ij}}\{1 - \exp[-(k_{ij} + k_{ji})t]\}.$$

(j) In the signal flow graph of part (i), remove all directed lines that begin at state j; remove the factor $1/s$ from each directed line that ends on state j (that node now represents the flow, J_j, into state j); find $[J_j(s)/J_s(s)]$ and set $J_s(s) = 1$.

$$H_{sr}(s) = \frac{J_j(s)}{J_s(s)} = \frac{k_{ji}}{s + k_{ji}},$$

$$h_{rs}(\tau) = k_{ji}\mathrm{e}^{-k_{ji}\tau}, \quad M_1(\tau) = \frac{1}{k_{ji}}, \quad cv = 1.0.$$

(k) Substitute $j\omega$ for s, then determine final and intermediate asymptotic relationships. The amplitude approaches 1.0 at low values of ω, and k_{ji}/ω at high values of ω. On a log–log plot (amplitude Bode diagram), these asymptotes intersect at $\omega = k_{ji}$, which therefore is a *corner frequency* in the diagram. The total range of phase shift is $\pi/2$ rad (90°).

P7.2 (a) Begin at the end farthest from the source and work toward the source, one branch at a time. This produces an implicit relationship between F_n or J_n and every other branch variable (for example, F_i/F_n and U_S/F_n). The explicit relationship between any variable and the source variable then is found by taking the appropriate ratio (for example, $F_i/U_S = (F_i/F_n)/(U_S/F_n)$).

$$J_n = Y_n F_n \quad J_{n-1} = J_n = Y_n F_n$$
$$F_{n-1} = Z_{n-1}J_{n-1} = Z_{n-1}Y_n F_n$$
$$F_{n-2} = F_n + F_{n-1} = (1 + Z_{n-1}Y_n)F_n \ldots$$

(b) $Y_j = sC_j, \quad C_j = \dfrac{AL}{E_B};$

for fully-developed flow,

$$Z_i = R_i = \frac{8\eta\pi L}{A^2};$$

for undeveloped flow,

$$Z_i = sI_i, \quad I_i = \rho\frac{L}{A};$$

where η = dynamic viscosity; ρ = density.

(c) The boundary occurs approximately where $\omega I = R$.

(d) $Y_n = 0$; J_n = vibration velocity of top layer; U_S = vibration velocity of floor; $Z_i = j\omega m_i$, m_i being the mass of the ith layer; $Y_j = 1/(R_j + K_j/j\omega)$, R_j being the viscous resistance and K_j the stiffness of the jth viscoelastic layer.

(f) $Z_i = R_i = \dfrac{RTL_i}{c_0 D_k A}, \quad Y_j = sC_j,$

$$C_j = \frac{AL_j c_0}{RT},$$

where L_i and L_j are the lengths of the diffusion paths represented by Z_i and Y_j, D_k is the diffusivity of species k, and c_0 is its resting concentration.

(h) $G_j = \sigma_m \dfrac{\pi DL}{\delta}, \quad C_j = \epsilon_m \dfrac{\pi DL}{\delta},$

$$R_i = \rho_c \frac{4L}{\pi D^2},$$

where δ is the thickness of the membrane, σ_m is its electric conductivity, and ϵ_m is its dielectric permittivity; ρ_c is the electric resistivity of the material in the core of the dendrite.

(i) $y = g + sc = \sigma_m \dfrac{\pi D}{\delta} + s\epsilon_m \dfrac{\pi D}{\delta},$

$$z = r = \rho_c \frac{4}{\pi D^2}, \quad Z_0 = \sqrt{\frac{z}{y}}.$$

For maximum transfer of signal power,

$$Y_{op} = \sum_{n=1}^{N} Y_{on} \quad D_p^{\frac{3}{2}} = \sum_{n=1}^{N} D_n^{\frac{3}{2}},$$

where Y_{op} and D_p are the characteristic admittance and diameter of the parent dendrite, and Y_{on} and D_n are the characteristic admittance and diameter of the nth branch.

P7.3 (a) The motion of the mass (J_1) is the same motion that elicits viscous drag (in the resistance); and it is the same motion that excites the transducer and is absorbed in the compliance.

(b) $Z_{DP}(1) = z_{11} - \dfrac{z_{12}z_{21}}{z_{22} + Z_L} = 0,$

$$F_s = \left[R + j\omega m + \frac{1}{j\omega C}\right] J_1,$$

$$\frac{J_1}{F_s} = \frac{j\omega C}{1 - \omega^2 mC + j\omega RC}.$$

(c) Critical damping corresponds to

$$RC = 2\sqrt{mC}, \quad R^2 = 4\frac{m}{C}.$$

Under these circumstances, the natural frequencies of the translational mechanical circuit are two identical, real numbers ($-R/2m$). When the circuit is overdamped, ($R^2 > 4m/C$), the natural frequencies are two different, real numbers, whose mean is $-R/2m$. When the circuit is underdamped, the natural frequencies are two complex conjugate numbers.

(d) Connect an appropriate electric resistor (R_L) across the electrical port.

$$r_{DP}(1) = z_{11} - \frac{z_{12}z_{21}}{z_{22} + R_L} = \frac{k^2}{R_L},$$

$$\left[R + \frac{k^2}{R_L}\right]^2 = 4\frac{m}{C}.$$

(e) $F_2 = kJ_1 = \dfrac{kF_s}{j\omega m + R + \dfrac{1}{j\omega C}},$

$$R = 2\sqrt{\frac{m}{C}}, \quad \frac{F_2}{F_s} = \frac{kj\omega C}{[1 + j\omega\sqrt{mC}]^2}.$$

(f) $F_2 = \dfrac{j\omega Cj\omega mv}{[1 + j\omega\sqrt{mC}]^2}.$

When $\omega\sqrt{mC} \gg 1$, then $F_2 \approx -kv$, which is the desired relationship between F_2 and v.

P7.4 (a) For the piston

$$h_{11}(a) = \frac{j\omega m}{A^2}, \quad h_{12}(a) = \frac{1}{A},$$

$$h_{21}(a) = -\frac{1}{A}, \quad h_{22}(a) = 0.$$

For the strain gauge

$$z_{11}(b) = \frac{1}{j\omega C}, \quad z_{12}(b) = \frac{Q_0}{j\omega A\epsilon},$$

$$z_{21}(b) = \frac{Q_0}{j\omega A\epsilon}, \quad z_{22}(b) = \frac{D_0}{j\omega A\epsilon}.$$

For the system

$$z_{11} = \frac{1}{j\omega A^2 C} + \frac{j\omega m}{a^2} \quad z_{12} = \frac{Q_0}{j\omega A^2\epsilon}$$

$$z_{21} = \frac{Q_0}{j\omega A^2\epsilon} \quad z_{22} = \frac{D_0}{j\omega A\epsilon}$$

P7.5 (a) Rate constant = (100 mL min^{-1})/(500 mL) = 0.2 min^{-1}
time to reach $1/e^2$ = 10 min

(c) 46 000 min

(e) RBC state 1 to RBC state 2 = 230 h^{-1}
plasma to RBC state 1 = 0.084 h^{-1}
RBC state 2 to plasma = 0.020 h^{-1}

(h) 0.263 gal/min

Chapter 8

P8.1 $M = 3000\ \rho RT/E$

Substituting $\rho = 500$ kg m^{-3}
$T = 20$ degrees C = 293 degrees K
$R = 8.314$ J/degree/mole

one has $M = 2030$ kg mol^{-1}
$= 2.03$ g mol^{-1}

The calculated value is 2030/3200 = 63 per cent of the measured value.

P8.2 (a) The total strain energy density for the steel column loaded to the yield point is

$$U = \tfrac{1}{2}\,\sigma_y^2/E.$$

Substituting the yield stress and modulus values gives

$$U = 4.5 \times 10^5 \text{ J/m}^3.$$

This is less than one-quarter of the stated value for resilin: 2×10^6 J m^{-3}.

(b) If the spring is in the form of a cantilever beam, one must integrate through the beam to obtain its total strain energy. Let y be the distance from the neutral axis and x the distance from the fixed end. The cross-sectional moment of inertia of the beam is $I = h^4/6$. One has:

$$U = \frac{1}{2}\sigma(x, y)^2/E,$$

$$\sigma(x, y) = Fxy/I = 6Fxy/h^4.$$

Integrating U over the beam's thickness (from $-h/2$ to $+h/2$) and along its length (from 0 to $16h$) yields

$$U_T = 16\,384\ F^2/Eh.$$

The mean energy/unit volume for the beam is, since the volume is $32h^3$,

$$U_m = 512\ F^2/Eh^4.$$

The load at a yield will be determined by the yield stress at the top or bottom surface at the fixed end of the beam:

$$\sigma_Y = FLh/2I = 48\ F/h^2,\ F = \sigma_Y h^2/48.$$

Substituting this into the previous equation gives

$$U_m = 128(\sigma_Y h^2/48)^2/Eh^4 = 0.222\sigma_Y^2/E$$

Substituting for σ_Y and E gives the result

$$U_m = 2.01 \times 10^5(425 \times 10^6)^2/200 \times 10^9$$
$$= 5.02 \times 10^4 \text{ J m}^{-3}$$

This is only 10 per cent of the energy density that can be stored in resilin.

(c) 13.6 cm

P8.3 (a) Let the subscripts 1 and 2 denote bone and implant, respectively. Since the strain in the bone and implant are equal, the stresses in the two materials are

$$\sigma_1/E_1 = \sigma_2/E_2 = \varepsilon.$$

From this, one has for the stress in the bone

$$\sigma_1 = \sigma_2(E_1/E_2),$$

and the greater the implant modulus, the less stress the bone will carry.

(b) The load on the bone-implant system is distributed as follows:

$$\text{applied load} = F = \sigma_1 A_1 + \sigma_2 A_2.$$

Solving this for the stress in the bone gives

$$\sigma_1 = F/A_1 - \sigma_2 A_2/A_1.$$

P8.4 For the Voigt model, materials 1 and 2 have the same strain, ε. Let the stresses and cross-sectional areas of the two materials be σ_i and A_i, respectively, where $i = 1{,}2$. One has two relationships:

$$\sigma_1/E_1 = \sigma_2/E_2 = \varepsilon = \sigma/E,$$

and

$$\text{applied load} = \sigma_1 A_1 + \sigma_2 A_2 = \sigma(A_1 + A_2),$$

where σ and E are defined as the composite stress and modulus, respectively. By solving the first of these for σ_1 and σ_2, respectively, in terms of σ and substituting the results into the second equation, one gets

$$\sigma E_1 A_1/E + \sigma E_2 A_2/E = \sigma(A_1 + A_2),$$

which reduces to

$$E = V_1 E_1 + V_2 E_2,$$

where it has been noted that the volume fractions of the two materials are $V_i = A_i/(A_1 + A_2)$.

For the Reuss model, the two materials have the same stress, σ, and strains ε_i The equivalent two relationships to those for the Voigt model are

$$\varepsilon_1 E_1 = \varepsilon_2 E_2 = \sigma = \varepsilon E$$

and

$$\text{total strain} = (\varepsilon_1 L_1 + \varepsilon_2 L_2)$$
$$(L_1 + L_2) = \sigma/E.$$

Here, the Ls represent the heights of the two materials in their stacked configuration. Again, one substitutes from the first of these into the second; the result is

$$(\sigma L_1/E_1 + \sigma L_2/E_2)/(L_1 + L_2) = \sigma/E,$$

which reduces to

$$E = E_1 E_2/(V_1 E_2 + V_2 E_1)$$

when one considers that the volume fractions are $V_i = L_i/(L_1 + L_2)$.

P8.5 Letting the mineral be material 1 and the collagen material 2, one has:
Voigt model, upper bound, parallel components

$$E = V_1E_1 + V_2E_2 = (0.45)(114) + (0.55)(1.3)$$
$$= 52.0 \text{ GPa}.$$

Reuss model, lower bound, stacked or series components

$$E = E_1E_2/(V_1E_2 + V_2E_1)$$
$$= (114)(1.3)/[(0.45)(1.3) + (0.55)(114)]$$
$$= 2.34 \text{ GPa}.$$

Note the large difference between these bounds and the small contribution of collagen to the results. The modulus for compact bone (15–20 GPa) is somewhat closer to the lower bound.

P8.6 The graph is shown as a semi-log plot in the figure below. At a strain rate of 1 s^{-1}, the equation predicts moduli of 30.3 GPa for cortical bone and 102 MPa for cancellous bone. The former value is considerably higher than measured values. A 30 per cent reduction in trabecular bone apparent density would change the modulus from 102 to 35 MPa, a very significant decrease.

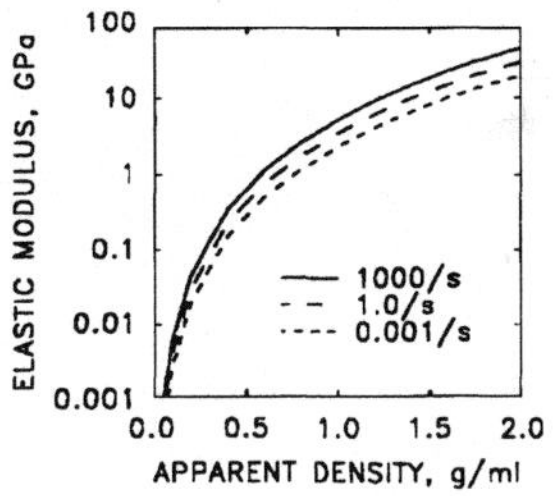

Figure P8.6.

Chapter 9

P9.1 Compressive stress −60 MPa
Tensile stress 60 MPa

P9.2 **(a)** 1.033 MPa
(b) Maximum stress 270 MPa
(c) Maximum stress with stress concentration site 864 MPa

P9.3 **(a)** Deflection (at load point) 1.3×10^{-4} m
Deflection (at end) 2.4×10^{-4} m

Chapter 10

P10.1 $(gl)^{0.5}$

P10.4 1.67 meters

P10.5 1.28 meters/second

P10.11 **(a)** dorsiflexors; about 2 times
(b) plantarflexors; about 8 times
(c) about 1000 N
(d) about 0.3 m (use problem 10.4)
(e) (i) $F_{sol}/F_{gas} = 6610/4160$
(ii) all soleus
(iii) $(6610/4160)^2$
(iv) all soleus

P10.12 **(a)** soleus: 1322 N; gastrocnemius: 832 N
(b) 215.2 N m

P10.13 **(a)** asymptotes of hyperbola
(b) $(F + a)(V + b) = (V_{max} + b)a$

or $(FV_{max} + F_0b)(V + b) = (V_{max} + b)F_0b$
or $(F + a)(F_0V + aV_{max}) = (F_0 + a)aV_{max}$
(c) $\delta V_{max}/\delta F_0 = b/a$
$\delta V_{max}/\delta b = F_0/a \approx 4\ \delta V_{max}/\delta F_0$
$\delta V_{max}/\delta a = -F_0b/a^2 \approx 4\ \delta V_{max}/\delta F_0$

P10.14 **(a)** 0.09 F_0V_{max}, attained at 0.31 V_{max}
(b) maximum power $\propto F_0V_{max}$; maximum velocity is independent of F_0

P10.18 **(a)** many
(b) slower
(c) no difference

P10.19 **(a)** (0.7) (9.5) (225); about 1
(b) no

Chapter 11

P11.2 **(a)** $T_0 - T_1 = 0.43°C$
(b) $Ra = 419 \gg 1$, so would very much expect convection.
(d) 56 μm

P11.3 $R = 2.49$

Chapter 13

P13.1 75 keV μm^{-1}, 0.75 keV μm^{-1}

P13.2 23, 1

P13.3 325 MeV/A, 125 MeV/A

P13.5 4.5×10^7 cm^{-2}

P13.6 0.084 cm

P13.7 8.2 keV, 331 keV

P13.8 13 days

P13.9 $2.5 \frac{\text{fatalities}}{10^6}$. *Note*: the observed difference is significantly larger in magnitude, and in the opposite direction

P13.10 2.5 years

P13.11 25 mSv (2.5 rem)

P13.12 approximately 3 months

P13.13 4.6 Gy (or 460 rad)

P13.14 $D_{bacteria} = 34.5$ Gy, and $D_{virus} = 6900$ Gy

P13.15 3

P13.16 30

P13.17 10^{-4}

P13.18 $(2/3)^{28} \times 10^6 \approx 12$

INDEX